AMERICAN BEGINNINGS

Crazy Notch, Maclaren River Valley, Alaska. (*Photo courtesy of Frederick H. West.*)

AMERICAN BEGINNINGS

The Prehistory and Palaeoecology of Beringia

Edited by

Frederick Hadleigh West

With the assistance of

Constance F. West
Brian S. Robinson
John F. Hoffecker
Mary Lou Curran
Robert E. Ackerman

The University of Chicago Press
Chicago and London

FREDERICK HADLEIGH WEST is Director of Archaeology at the Peabody Essex Museum, Editor of *The Review of Archaeology*, and author of *The Archaeology of Beringia* (1981).

The University of Chicago Press, Chicago 60637
The University of Chicago Press, Ltd., London
© 1996 by The University of Chicago
All rights reserved. Published 1996
Printed in the United States of America

05 04 03 02 01 00 99 98 97 96   1 2 3 4 5

ISBN: 0–226–89399–5   (cloth)

Library of Congress Cataloging-in-Publication Data

American beginnings: the prehistory and palaeoecology of Beringia /
   edited by Frederick Hadleigh West; with the assistance of Constance
   F. West . . . [et al.].
      p.   cm.
      Includes bibliographical references and index.
      1. Man, Prehistoric—Bering Land Bridge.   2. Paleo-Indians—Bering
   Land Bridge.   3. Excavations (Archaeology)—Russia (Federation)
   4. Excavations (Archaeology)—Alaska.   5. Paleontology—Bering Land
   Bridge.   6. Palynology—Bering Land Bridge.   7. Bering Land Bridge—
   Antiquities.   I. West, Frederick Hadleigh.   II. West, Constance F.
   GN885.A44   1996
   909'.0964'51—dc20   96–11719
                                                            CIP

CONTENTS

PREFACE

Every book begins with a conviction. In the present case it is the conviction that far too much of the debate on the earliest peopling of the New World proceeds with only perfunctory reference to the very region where it all began and even less to the significance of that fact. If we wish to know the origins and affinities of the earliest Americans, that knowledge will be found in the results of research carried out in those distant regions. But, that knowledge simply cannot be fully appreciated without constant reference to an environmental context for which there is no modern counterpart. The circumstances that attended the population movements considered here were extraordinary—the time, the place, would be utterly foreign to people of today. While the region of former Beringia may impose severe stresses on its inhabitants today, in the past these environmental stresses were many times greater. It was a different world—one which, only now, it is becoming possible to portray.

New findings and new interpretations in Beringian research have been a constant accompaniment over the past thirty years. It has been difficult even for researchers to keep up with these changes that have often required the complete reworking of previously held ideas. Within the bounds of present knowledge, our goal here is to present an objective and accurate depiction of late Quaternary Beringia. Something of the nature of this undertaking—its challenge *and* its excitement—is metaphorically captured in the fact that a major portion of the Beringian landscape lies under 60 meters of water, as it has lain for the past 10,000 years.

The authors of this monograph are, or have been, active field investigators in the areas about which they write. They know thoroughly well the complexity of each individual research task they have undertaken. They know, as well, that this complexity results from the inherent and unavoidable ambiguity of these tasks. Every single record, whether archaeological, palynological, or geological, provides evidence that is more or less equivocal. The field investigator interprets his record to the best of his ability, reconstructing its bits and pieces into a comprehensible structure. The judgment of its integrity is provided by its durability. But, if scholars can quarrel over the interpretation of the *written* record, what then when the record consists of fragments, traces, scars, or piles of debris in the landscape—and where, moreover, there may be no immediate indication of placement in time?

Although each is abbreviated, these articles may be read as primary sources. Each is unique and has its own particular scientific value. The composite of them allows a sweeping reconstruction of Beringia as it was at the time of its discovery and occupation by Siberian hunters. Owing to these many individual labors we are now allowed a vista—clouded to be sure—of that world at a moment of crucial importance in world prehistory.

The fifty-six contributors to this volume have devoted an aggregate of approximately 1,130 years to Northern research, with individual values ranging from five to fifty years. The average works out to some twenty years, well sufficient for this to be characterized as a group of experts.

American Beginnings is, by design, a compendium. It is not pretended that every study that could or should be included is to be found here. It is, however, a very full representation of the basic scientific data relating to Beringia. Especially in Siberia, where distances are vast and the number of field scientists restricted, there are some areas where data are relatively sparse. Offsetting those deficiencies may be accomplished by giving particular attention to the character of the evidence that is available. It is the coherence of all these data that confers their particular strength.

Beringia began as a theory, a logical construct. Field investigations have given it form, so that now it is possible to portray it almost as fully as the ancient past of a land still existing. It is hoped that the presentation of these evidences in the form of articles written by their investigators will convey a feeling for the nature of these research sciences and allow the reader better to judge the validity of the grand ideas that flow from them. They are ideas fundamental to understanding the making of a New World.

———◆◆———

It is a pleasant duty to acknowledge here the gratitude owed the many people whose efforts and cooperation have made possible the completion of this volume. First, and foremost, to the contributors, all working and productive scientists, whose faith and diligence in this enterprise, and willingness to take an active part in it, brought it to reality. Particular thanks, of course, go to those colleagues who actively assisted the program in a variety of fundamental ways. These are Robert E. Ackerman, Mary Lou Curran, John F. Hoffecker, Brian S. Robinson, and Constance F. West. Each contributed mightily of his and her professional knowledge and judgement. Without their help this book would assuredly be at least another year in the making and would be immeasurably poorer. An early and enthusiastic adherent to this work, Valery Alexeev was precluded from participating by his passing in 1991. He is missed. We regret as well the passing of N. N. Dikov just months before the appearance of this volume.

To those who labored in the trenches at the Peabody Essex Museum and elsewhere: to Louise M. Sullivan who brought her rich professional experience to the formidable job of copy editing and production; to Margaret A. Dorsey who, in addition to her responsibilities as managing editor of *The Review of Archaeology*, undertook to type, and retype, the articles in their successive formulations while aiding with the copy editing of the whole; to Robert C. Forget, whose skills in cartography are well known to many scholars, and who, ably assisted by his son, Peter, drafted most of the original line drawings and put all into final form. Here special mention must be made of Brian Robinson who, aided by Jeffrey A. Williams, recast many of the artifact drawings and formatted all of them. Many of the artifact drawings were done by Dianne Gardner whose skill and dexterity are evident. Most of the translation of the Russian manuscripts was carried out by Alex Gitlin, with significant aid rendered by Natasha Selinvanova. Finally, the photographs reproduced here have all benefited from Rolf Hansen's wizardry—a fact that knowledgeable readers will quickly recognize. It has been a pleasure, and an education, to be able to work with all these people.

American Beginnings has been a very large undertaking. The number of people hours involved in its production would be, literally, beyond calculation. For me, it has provided the opportunity to be in contact with colleagues whom I have known for years and to become acquainted with others whom I knew only by reputation but had not previously met. In all cases, the great pleasure has been to be in constant contact with scholars whose research interests are similar to my own, to be able to exchange views, to be made aware of research results and problems as these were just being learned, and to be able, thus, to refine ideas long held and to change those now shown as wanting. It has been an enriching experience on a subject that has been a consuming concern of mine for many years—for all of which, I thank all my colleagues. I only hope that some of them, at least, feel as I do about this experience.

Finally, to my wife, Constance: beyond her purely archaeological contributions, over the years that this work wore on, preempting so many normal activities, she cheerfully and willingly provided absolutely fundamental moral and intellectual support. Without that support *American Beginnings* would not have begun, would not have ground—and sometimes spun—along, and surely would never have come to an ending.

Frederick Hadleigh West

ACKNOWLEDGMENTS

The editors wish to acknowledge the cooperation and generosity of the following individuals and publishers in granting permission for the reproduction of illustrations and tabular material. Where a contributor has supplied his or her own photographs and tables, or where they are part of an article for which credit is given in a footnote, no credit is given in either this list or in the text.

Figure 3 of the Preamble: R. Dale Guthrie (1990), *Frozen Fauna of the Mammoth Steppe*, p. 254. Published by the University of Chicago Press. © 1990 by The University of Chicago.

Table 1–1, Figure 1–1: T. D. Hamilton, G. M. Ashley, K. M. Reed, and C. E. Schweger (1993), Late Pleistocene Vertebrates and Other Fossils from Epiguruk, Northwestern Alaska, *Quaternary Research* 39:381–87. Published by Academic Press, Inc., Orlando, Florida.

Table 1–2, Figure 1–2: T. D. Hamilton and G. M. Ashley (1993), Epiguruk: A Late Quaternary Environmental Record from Northwestern Alaska, *Geological Society of America Bulletin* 105:583–602.

Figures 2–1, 2–2, 2–8, 2–9: Reprinted from *Quaternary Science Reviews*, Vol. 31, P. M. Anderson and L. B. Brubaker, Vegetation History of Northcentral Alaska, pp. 71–92, Copyright 1994, with kind permission from Elsevier Science Ltd, The Boulevard, Langford Lane, Kidlington 0X5 1GB, UK.

Figure 2–3: Reprinted from *Review of Palaeobotany and Palynology*, Volume 46, P. M. Anderson and L. B. Brubaker, Modern Pollen Assemblages from Northern Alaska, pp. 273–91, Copyright 1986, with kind permission from Elsevier Science BV, P. O. Box 211, 1000 AE Amsterdam, The Netherlands.

Figure 2–4: P. M. Anderson, P. J. Bartlein, L. B. Brubaker, K. Gajewski, and J. C. Ritchie (1988), Modern Analogues of Late-Quaternary Pollen Spectra from the Western Interior of North America, *Journal of Biogeography* 16:573–96. Published by Blackwell Science Ltd, Oxford, UK.

Figure 2–5: P. M. Anderson (1988), Late Quaternary Pollen Records from the Kobuk and Noatak River Drainages, Northwestern Alaska, *Quaternary Research* 29:263–76. Published by Academic Press, Inc., Orlando, Florida.

Figure 2–6: L. B. Brubaker, H. L. Garfinkel, and M. E. Edwards (1983), A Late-Wisconsin and Holocene Vegetation History from the Central Brooks Range, *Quaternary Research* 20:194–214. Published by Academic Press, Inc., Orlando, Florida.

Figure 2–7: P. M. Anderson, R. E. Reanier, and L. B. Brubaker (1988), Late Quaternary Vegetational History of the Black River Region, Northeastern Alaska, *Canadian Journal of Earth Sciences* 25:84–94.

Table 2–1: Reprinted with permission from J. Brown and P. V. Sellman, Radiocarbon Dates on Frozen Peats from Barrow, Alaska, *Science* 153:299–300. Copyright 1966 American Association for the Advancement of Science.

Figure 2–13: P. A. Colinvaux (1964), The Environment of the Bering Land Bridge, *Ecological Monographs* 34:297–325.

Photograph on page 124: R. Dale Guthrie (1990), *Frozen Fauna of the Mammoth Steppe*, p. 73. Published by the University of Chicago Press. © 1990 by The University of Chicago.

Figures 4–16 and 4–20: Adapted from N. A. Shilo, N. N. Dikov, and A. V. Lozhkin, The First Data on the Stratigraphy of the Palaeolithic of Kamchatka. *Arctic Anthropology*, Volume 5, Number 1, pp. 196, 210. Copyright 1968. Reprinted by permission of The University of Wisconsin Press.

Photograph on p. 297: Reprinted with permission of Bradford Washburn. Boston Museum of Science photo #2403.

Figure 6–1: Compiled from illustrations found in *The Campus Site: A Prehistoric Camp at Fairbanks, Alaska*, by Charles M. Mobley. University of Alaska Press, 1991.

Table 6–1: Compiled from material found in *The Campus Site: A Prehistoric Camp at Fairbanks, Alaska*, by Charles M. Mobley. University of Alaska Press, 1991.

Table 6–2: Adapted from J. Erlandson et al. (1991), Two Early Sites of Eastern Beringia. *Radiocarbon* 33(1):35–50. Reprinted by permission of *Radiocarbon*.

Figure 7–6: N. H. Bigelow and W. R. Powers (1994) New AMS Dates from the Dry Creek Paleoindian Site, Central Alaska, *Current Research in the Pleistocene* 11:114–16.

Figure 7–7: After W. R. Powers and J. F. Hoffecker, Late Pleistocene Settlement in the Nenana Valley, Central Alaska. Reproduced by permission of the Society for American Archaeology from *American Antiquity* Vol. 54, No. 2, 1989.

Figures 7–13, 7–17: Reprinted with permission from J. F. Hoffecker, W. R. Powers, and T. Goebel, The Colonization of Beringia and the Peopling of the New World, *Science* 259:46–53. Copyright 1993 American Association for the Advancement of Science.

Figure 7–15, 7–16: Adapted from J. F. Hoffecker (1985), The Moose Creek Site, *National Geographical Society Research Reports* 19:33–48. Reprinted with permission of the National Geographic Society, Washington, D.C.

Figure 9–1: Adapted from D. N. Swanston (1989), Glacial Stratigraphic Correlations and Late Quaternary Chronology, in *The Hidden Falls Site, Baranof Island, Alaska*, edited by S. D. Davis, Aurora, No. 5, pp. 47–60. Published by the Alaska Anthropological Association.

Figure 9–2: Adapted from S. D. Davis (1989), Research Objectives and Field Methods, in *The Hidden Falls Site, Baranof Island, Alaska*, edited by S. D. Davis, Aurora, No. 5, pp. 31–39. Published by the Alaska Anthropological Association.

Table 9–2: S. D. Davis (1989), Cultural Component I, in *The Hidden Falls Site, Baranof Island, Alaska*, edited by S. D. Davis, Aurora, No. 5, pp. 159–198. Published by the Alaska Anthropological Association.

Figure 9–5: R. E. Ackerman, T. D. Hamilton, and R. Stuckenrath (1979), Early Cultural Complexes on the Northern Northwest Coast, *Canadian Journal of Archaeology* 3:195–209.

Figure 9–15: Adapted from A. P. McCartney and C. G. Turner II, Stratigraphy of the Anangula Unifacial Core and Blade Site, *Arctic Anthropology*, Volume 3, Number 2, pp. 28–40. Copyright 1966. Reprinted by permission of The University of Wisconsin Press.

Figure 9–16(*b*): From W. S. Laughlin and J. S. Aigner, Preliminary Analysis of the Anangula Unifacial Core and Blade Industry, *Arctic Anthropology*, Volume 3, Number 2, pp. 41–56. Copyright 1966. Reprinted by permission of The University of Wisconsin Press.

Figure 9–16(*c–e, h–j*): J. S. Aigner and T. Del Bene (1982), Early Holocene Maritime Adaptations in the Aleutian Islands, in *Peopling of the New World*, J. E. Ericson, R. E. Taylor, and R. Berger, eds, pp. 35–67. Los Altos, California: Ballena Press.

Figure 9–16(*g, k, l*): J. S. Aigner, Early Holocene Evidence for the Aleut Maritime Adaptation, *Arctic Anthropology*, Volume 13, Number 2, pp. 32–45. Copyright 1976. Reprinted by permission of The University of Wisconsin Press.

Photograph on page 483 and Figure 11-1: From H. Larsen (1968), Trail Creek: Final Report on the Excavation of Two Caves on Seward Peninsula, Alaska, *Acta Arctica*, Fasc. 15, pp. 7–79. Published by the Danish Arctic Institute, Copenhagen.

Photograph on page 485: From D. D. Anderson (1988), Onion Portage: The Archaeology of a Stratified Site from the Kobuk River, *Anthropological Papers of the University of Alaska*, Vol. 22, No. 1–2.

Figures 11–2, 11–3: From D. D. Anderson (1970), Akmak: An Early Archaeological Assemblage from Onion Portage, Northwest Alaska, *Acta Arctica*, Fasc. 16. Published by the Danish Arctic Institute, Copenhagen.

Figures 11–4, 11–5, 11–7: S. C. Gerlach and E. S. Hall, Jr. (1986), The View from the Hinterland. Report prepared for Cominco Alaska's Red Dog Operations, Kotzebue, Alaska.

Figure 11–11: Adapted from H. L. Alexander (1987), *Putu: A Fluted Point Site in Alaska*. Publication No. 17. Courtesy: Archaeology Department, Simon Fraser University, B. C., Canada.

Figure 11–12: Adapted from J. Cinq-Mars (1990), *La Place des Grottes du Poisson-Bleu dans la Prehistoire Beringienne, Revista de Arqueología Americana*, No. 1, p. 21. Published by the Pan American Institute of Geography and History, Mexico.

Figure 12–2: Drawings in third column reproduced with the permission of Richard Michael Gramly, Curator, Great Lakes Artifact Repository, Buffalo, N.Y.

THE CONTRIBUTORS

Robert E. Ackerman Department of Anthropology, Washington State University, Pullman

Larry D. Agenbroad Department of Geology, Northern Arizona University, Flagstaff

Patricia M. Anderson Quaternary Research Center, University of Washington, Seattle

Glenn H. Bacon Alaska Heritage Research Group Inc., Fairbanks

Nancy H. Bigelow Department of Anthropology, University of Alaska, Fairbanks

Linda B. Brubaker College of Forest Resources, University of Washington, Seattle

Paul A. Colinvaux Smithsonian Tropical Research Institute, Panama

John P. Cook Bureau of Land Management, Fairbanks, Alaska

Mary Lou Curran Peabody Essex Museum, Salem, Massachusetts

Stanley D. Davis U.S. Forest Service, La Junta, Colorado

Anatoly P. Derevianko Institute of Archaeology and Ethnography, Siberian Branch, Russian Academy of Sciences, Novosibirsk

Nikolai N. Dikov Northeastern Interdisciplinary Scientific Research Institute, Russian Academy of Sciences, Magadan

Thomas E. Dilley Department of Geosciences, University of Arizona, Tucson

R. Greg Dixon Office of History and Archaeology, Alaska Division of Parks, Anchorage

Scott A. Elias Institute of Arctic and Alpine Research, University of Colorado, Boulder

Svetlana A. Fedoseeva Department of Archaeology and Human Palaeoecology, Academy of Sciences of the Sakha Republic, Yakutsk

S. Craig Gerlach Department of Anthropology, University of Alaska, Fairbanks

Ted Goebel Department of Sociology/Anthropology, Southern Oregon State College, Ashland

Joseph H. Greenberg Department of Anthropology, Stanford University, Stanford, California

R. Dale Guthrie Institute of Arctic Biology, University of Alaska, Fairbanks

Edwin S. Hall, Jr. Department of Anthropology, State University of New York, Brockport

Thomas D. Hamilton U.S. Geological Survey, Anchorage, Alaska

Andrew S. Higgs Fairbanks, Alaska

John F. Hoffecker Argonne National Laboratory, Argonne, Illinois

Charles E. Holmes Office of History and Archaeology, Alaska Division of Parks, Anchorage

David M. Hopkins Department of Geology and Geophysics, University of Alaska, Fairbanks

Maureen L. King Desert Research Institute, Las Vegas, Nevada

Margarita A. Kiryak Northeastern Interdisciplinary Scientific Research Institute, Russian Academy of Sciences, Magadan

Michael L. Kunz Bureau of Land Management, Fairbanks, Alaska

Yaroslav V. Kuzmin Pacific Institute of Geography, Russian Academy of Sciences, Vladivostok

Anatoly M. Kuznetsov Department of History, Ussuriisk State Teachers Training College, Ussuriisk, Russia

Ralph A. Lively U. S. Forest Service, Ketchikan, Alaska

Allen P. McCartney Department of Anthropology, University of Arkansas, Fayetteville

Jim I. Mead Department of Geology, Northern Arizona University, Flagstaff

Charles M. Mobley Charles M. Mobley Associates, Anchorage, Alaska

Yuri A. Mochanov Department of Archaeology and Human Palaeoecology, Academy of Sciences of the Sakha Republic, Yakutsk

T. D. Morozova Institute of Geography, Russian Academy of Sciences, Moscow

Peter G. Phippen New London, New Hampshire

Mark E. Pipkin Office of History and Archaeology, Alaska Division of Parks, Anchorage

W. Roger Powers Department of Anthropology, University of Alaska, Fairbanks

Richard E. Reanier Reanier and Associates, Seattle, Washington

Douglas R. Reger Office of History and Archaeology, Alaska Division of Parks, Anchorage

Brian S. Robinson Archaeology Research Center, University of Maine, Farmington

G. M. Savvinova Department of Archaeology and Human Palaeoecology, Academy of Sciences of the Sakha Republic, Yakutsk

Sergei B. Slobodin Northeastern Interdisciplinary Scientific Research Institute, Russian Academy of Sciences, Magadan

Ralph S. Solecki Department of Anthropology, Texas A & M University, College Station

Savelli V. Tomirdiaro Russian Academy of Sciences, St. Petersburg

Valentina V. Ukraintseva Botanical Institute, Russian Academy of Sciences, St. Petersburg

Richard VanderHoek Department of Anthropology, University of Illinois at Urbana-Champaign

Ruslan S. Vasilievsky Institute of Archaeology and Ethnography, Siberian Branch, Russian Academy of Sciences, Novosibirsk

Andrej A. Velichko Institute of Geography, Russian Academy of Sciences, Moscow

Douglas W. Veltre Department of Anthropology, University of Alaska, Anchorage

Christopher F. Waythomas U.S. Geological Survey, Anchorage, Alaska

Thompson Webb III Department of Geological Sciences, Brown University, Providence, Rhode Island

Constance F. West Peabody Essex Museum, Salem, Massachusetts

Frederick H. West Peabody Essex Museum, Salem, Massachusetts

INTRODUCTION:
THE CONCEPT OF BERINGIA

David M. Hopkins

In 1920, a young Eric Hultén embarked with his new bride on an exciting, in fact, hair-raising, honeymoon on the Kamchatka Peninsula—a wedding trip punctuated by a shipwreck, skirmishes between remnants of the old tsarist government and irregulars of the new Communist regime, and encounters with the Kamchatkan brown bears that padded about in the undergrowth beneath the scrubby birch forest (Hultén n.d.).

Over the next three decades, Hultén's wedding trip grew into a series of surveys and inventories of the floras of Kamchatka, northeast Siberia, Alaska, and the Yukon Territory. New volumes of the Lunds University Årsskrift containing chapters of Hultén's *Flora of Alaska and the Yukon* (1941–1950) were still being eagerly awaited when Bob Sigafoos and I were beginning to acquaint ourselves with the vegetation and the landscape of Alaska's Seward Peninsula in 1948. Later, Hultén's grand compilation, *Flora of Alaska and Neighboring Territories* (1968), was published by Stanford University Press almost simultaneously with my Bering Land Bridge book (1967b).

Among the most valuable as well as most interesting features of Hultén's papers (Hultén n.d., 1937, 1941–1950, 1968) were their many range maps. As Hultén toiled through his enormous plant collections, and as he compared his materials with plants collected elsewhere in the Arctic and Subarctic, he began to notice that "a great number of plants were proved to have ranges symmetrical to a line through Bering Strait. If they spread a little bit to the east of this line, they also spread a short bit to the west of it. Should they spread far east, they also spread far west" (Hultén n.d.:14). He noted that regions on both sides of Bering Strait were

largely unglaciated and that the growth of Pleistocene ice caps elsewhere should have resulted in sea level lowered enough to create a land bridge in place of the present shallow Bering Sea (Hultén 1937). Thus was the concept of Beringia born.[1]

Thanks largely to Hultén (but thanks also to Bob Sigafoos, to Louis Giddings, to molluscan palaeontologist Stearns MacNeil, and to Inupiat historian and naturalist William Oquilluk), this drowned land bridge became one of my lifelong obsessions.

Hultén regarded the Bering Land Bridge as only a hypothesis, though certainly a rather strong one. But by the time my book *The Bering Land Bridge* (1967b) went to press, the reality of the former Bering Land Bridge was firmly established. Knowledge of the bathymetry of the Bering Sea and Bering Strait had improved, and the amplitude of Pleistocene glacio-eustatic sea level fluctuations was better established. It had also become clear that the Bering Land Bridge could have existed only during the cold intervals marked by glaciers in other parts of the world—during the last few hundred thousand years, at least. The clincher was Joe Creager's and Dean McManus's recognition of remnants of an ancient drainage system on the Chukchi Sea floor (Creager and McManus 1965).

The interesting question now became: to what extent could we reconstruct ancient environments on or near the land bridge? And, particularly, to what extent could we reconstruct the environment of the most *recent* land bridge, the land connection

---

[1] I define Beringia as that territory that lies west of the Mackenzie River in northwestern Canada and east of the Kolyma River in northeastern Siberia.

that existed during some, perhaps all, of the Wisconsinan cold period? Here, Hultén's plant lists seemed of only limited usefulness, since many Beringian endemics are calciphiles, plants found mainly on limestone substrates. Still, it was informative that there are no living tree species common to both eastern and western Beringia[2]—this showed that the land bridge was *not* forested. Still more informative was the observation that certain steppe plants that are widespread in Yakutia are also found isolated in very small, disjunct populations on south-facing river bluffs in Alaska and the Yukon (Murray 1981). This suggested that the land-bridge landscape must have been quite dry, perhaps even steppe-like (Young 1982; Yurtsev 1982).

Vertebrate palaeontologists had long been convinced of the reality of Cenozoic dispersals of mammals across the Bering Land Bridge (Davies 1934). In 1967, Charles Repenning showed that late Pliocene and Pleistocene mammalian dispersals were, in fact, episodic. Later, Repenning elaborated this idea in a series of publications (Repenning 1987 and references cited therein) showing that episodic intercontinental dispersals in various microtine lineages could be used to correlate and date palaeogeographic events in Beringia. Repenning also showed that as time went on, directions of mammalian dispersals became progressively more biased toward movement of Asian elements into North America. Perhaps more importantly, the migrating faunas became more and more indicative of both steppe-like and tundra-like conditions on the land bridge. By 1968, Guthrie (1968a, 1968b) could show that late Pleistocene land-bridge vertebrate faunas were indicative of widespread tundra-steppe conditions. In 1990, Guthrie and Stoker confirmed Vereshchagin's and Baryshnikov's 1982 suggestion that foot and hoof morphologies show that Pleistocene ungulates lived on a landscape with a firm and rather abrasive substrate, a substrate evidently quite unlike the soggy, mushy tundra that covers much of present-day Beringia.

Guthrie (1968a) pointed out that the late Pleistocene Beringian mammal fauna was far more diverse than modern subarctic and arctic faunas, and that it was dominated by large, gregarious ungulates which were preyed upon by diverse predators, some of them also large and gregarious. Although it seemed quite impossible to estimate late Pleistocene ungulate population densities, the character of the fauna appeared to call for a vegetation much more productive than present-day tundra vegetation, despite a climate that was certainly very much colder (Brigham-Grette and Miller 1982). Russian palaeoecologists postulated that the large ungulates subsisted on a *tundra-steppe* vegetation, but some North Americans now began to speak of *arctic steppe*, invoking as an analogue Africa's Serengeti Plain with its huge herds of antelope preyed upon by lions, hyenas, and wild dogs (Matthews 1976). Enthusiasm for the arctic steppe concept reached its high-water mark at a symposium entitled "Hot and Cold Deserts during the Last Glaciation" held at the 1976 biennial meeting of the American Quaternary Association in Tempe, Arizona.

North American palynologists—specialists in interpreting ancient vegetation (and, hopefully, climate) from pollen preserved in lake sediments—would have none of this. Paul Colinvaux (1964) had already shown that late Pleistocene pollen floras were very depauperate, devoid of large trees, and with few if any large shrubs. Les Cwynar and Jim Ritchie confirmed the depauperate character of late Pleistocene pollen floras in northern Yukon Territory and showed that pollen influx rates were very low, indicating a dearth of wind-pollinated plants and suggesting that the vegetation cover was discontinuous and unproductive (Cwynar and Ritchie 1980; Cwynar 1982). They argued that Beringian vegetation, in fact, was best described as polar desert, rather than highly productive arctic steppe.

At a Burg Wartenstein Conference on the palaeoecology of Beringia, held in 1979 at the beautiful Austrian conference center of the Wenner Gren Foundation, it was pointed out that most of the Pleistocene bones now in museum collections were found washed out and redeposited on river bars, beaches, and the floors of placer pits. Therefore, polar desert advocates said, few if any could be reliably dated, and they certainly did not establish that

[2]Except for *Populus balsamifera,* a clone of which grows in an isolated mountain valley north of Providenniya and near the Bering Sea coast of the Chukchi Peninsula (B. A. Yurtsev, personal communication, 1973).

the large mammal fauna was present during the severe climate indicated by late Wisconsinan pollen spectra. But compilations of the few well-dated faunas known in 1982 (Matthews 1982: Fig. 2; Schweger, et al. 1982:Table 1) and Dale Guthrie's later program to directly date Pleistocene bones (Guthrie 1990) firmly established that mammoth, horse, bison, musk-ox, caribou, mountain sheep, and ground squirrels were all present in Beringia throughout the last cold stage of the Pleistocene. (The polar positions in this debate were articulated most clearly in a series of exchanges published in *The Quarterly Review of Archaeology* [Colinvaux and West 1984; Guthrie 1985; Colinvaux 1986]). Thus was born the concept of the *productivity paradox* (Schweger et al. 1982).

This argument could best be settled, of course, if we could rent a time machine and pay a personal visit to the ice-age land bridge. Unfortunately, time machines have not yet been invented, but we *have* found a 1,000-square-kilometer fragment of intact land-bridge landscape and its vegetation cover preserved intact and frozen beneath volcanic tephra erupted 17,500 years ago from the Devil Mountain Lake maar on northern Seward Peninsula. The buried surface bears a continuous veneer of plant remains a few millimeters thick, but there is no accumulation of peat (Goetcheus et al. 1994). Clumps of sedge or grass are interspersed amongst a network of prostrate willow stems and a more or less continuous carpet of mosses. Seeds and a few stems and leaves indicate the presence of a broad variety of forbs. The mosses and the higher plants both indicate a mosaic of moist, meadowy and dry, windswept micro-environments. The plants were growing on a shallow loessial soil charged with fine, herbaceous roots (Hoefle et al. 1994; Höfle 1995). A now-frozen active layer (the layer which once froze and thawed annually) was 30–40 cm deep. The soil displays evidence of seasonal waterlogging, perhaps during spring and autumn, and seasonal dessication, perhaps during mid- to late summer. Small-scale hummocky microrelief on the buried surface is also suggestive of seasonal dessication. The buried vegetation can be matched in Sylvia Edlund's zone of prostrate shrubs (Edlund 1989). Both vegetation and soils are suggestive of

dry, cloudless summers and heavy rains just before freeze-up. I believe that we will ultimately be able to show that both vegetation and soils can be matched on Banks Island at the southwest corner of the Canadian Arctic Archipelago and on Wrangel Island, home of the Holocene pygmy mammoth (Vartanyan et al. 1993).

Ideas about the antiquity of humans in Beringia have evolved significantly through the years. In the mid-1960s, many of us (including me [Hopkins 1967a]) thought it likely that the earliest humans had passed through Beringia more than 20,000 years ago. Most of the North and South American sites that in the 1960s had seemed to provide credible evidence for a very early presence of humans were discredited by 1980, but numerous new candidates had appeared. In the synthesis chapter of *Palaeoecology of Beringia* (Schweger et al. 1982) we offered a much more tentative statement that still favored the presence of humans in Beringia more than 20,000 years ago. Now, most if not all of the evidence cited there in 1982 has also been discredited. To me, it now seems crystal clear that human prehistory in Beringia began not 20,000 but closer to 12,000 years ago, as in fact, argued by West in 1981.

How, then, did the first humans disperse from northeastern Asia into northwestern North America? Did they walk on dry ground? Did they arrive crossing over smooth ice? Or did they arrive in boats? Until recently, study of sediment cores lifted from the Bering and Chukchi Seas seemed to indicate that the land bridge was drowned by rising sea level some 14,000 years ago (McManus et al. 1983), more than two millennia before the oldest recognized east-Beringian archaeological sites were occupied. But dating of these cores was based on radiocarbon analyses of dispersed carbon extracted from bulk sediments—carbon that included uncertain but significant quantities of redeposited detrital coal. New cores recently obtained by Larry Phillips of the U.S. Geological Survey from waters off Icy Cape in the Chukchi Sea penetrated alluvial sediments containing layers of detrital plant remains. The alluvium is as deep or deeper than the Bering Strait. The radiocarbon-dated plant remains in the alluvium show that sea level did not rise high enough to drown Bering Strait until after 10,000 radiocarbon years ago (Elias, et al. 1992)! It is now clear that the earliest human invaders could easily have passed over dry land, quite unconscious that they were crossing to a new continent. The prehistory of Beringia seems to have begun with hunters and foragers who ranged over a narrowing Beringian isthmus at a time when the summer climate was rapidly warming and environmental change was progressing with great rapidity.

References Cited

Brigham-Grette, J., and G. F. Miller. 1982. Paleotemperature Estimates of the Alaskan Arctic Coastal Plain during the Last 125,000 Years. In *Permafrost*, pp. 80–85. Washington, D.C.: National Academy of Science Press.

Colinvaux, P. A. 1964. The Environment of the Bering Land Bridge. *Ecological Monographs* 34:297–329.

———. 1986. Plain Thinking on Bering Land Bridge Vegetation and Mammoth Populations. *Quarterly Review of Archaeology* 7(1):8–9.

Colinvaux, P. A., and F. H. West. 1984. The Beringian Ecosystem. *Quarterly Review of Archaeology* 5(3):10–16.

Creager, J. S., and D. A. McManus. 1965. Pleistocene Drainage Patterns on the Floor of the Chukchi Sea. *Marine Geology* 3:279–90.

Cwynar, L. C. 1982. A Late-Quaternary Vegetation History from Hanging Lake, Northern Yukon. *Ecological Monographs* 52:1–24.

Cwynar, L. C., and J. C. Ritchie. 1980. Arctic Steppe-Tundra: A Yukon Perspective. *Science* 208:375–77.

Davies, J. L. 1934. *Tertiary Faunas*, Vol. 2: *The Sequence of Tertiary Faunas*. London: Murby and Co.

Edlund, S. A. 1989. Regional Congruence of Vegetation and Summer Climate Patterns in the Queen Elizabeth Islands, Northwest Territories, Canada. *Arctic* 42(1):3–23.

Elias, S. A., S. K. Short, and R. L. Phillips. 1992. Paleoecology of Late-Glacial Peats from the Bering Land Bridge, Chukchi Sea Shelf Region, Northwestern Alaska. *Quaternary Research* 38:371–78.

Goetcheus, V. G., D. M. Hopkins, M. E. Edwards, and D. H. Mann. 1994. Window on the Bering Land Bridge: A 17,000-Year-Old Paleosurface on the Seward Peninsula, Alaska. *Current Research in the Pleistocene* 11:131–32.

Guthrie, R. D. 1968a. Paleoecology of the Large-Mammal Community in Interior Alaska during the Late Pleistocene. *American Midland Naturalist* 79(2):236–68.

———. 1968b. Paleoecology of a Late Pleistocene Small Mammal Community from Interior Alaska. *Arctic* 21:223–44.

———. 1985. Woolly Arguments against the Mammoth Steppe—A New Look at the Palynological Data. *Quarterly Review of Archaeology* 6(3):9–16.

———. 1990. *Frozen Fauna of the Mammoth Steppe*. Chicago: University of Chicago Press.

Guthrie, R. D., and S. F. Stoker. 1990. Paleoecological Significance of Mummified Remains of Horses from the North Slope of the Brooks Range. *Arctic* 43(3):159–73.

Hoefle, C., C. L. Ping, D. Mann, and M. Edwards. 1994. Buried Soils on Seward Peninsula: A Window on the Paleoenvironment of the Bering Land Bridge. *Current Research in the Pleistocene* 11:134–36.

Höfle, C. 1995. Buried Soils on Seward Peninsula, Northwest Alaska: A Window into the Late Pleistocene Environment of the Bering Land Bridge. M.S. thesis, University of Alaska.

Hopkins, D. M. 1967a. The Cenozoic History of Beringia: A Synthesis. In *The Bering Land Bridge*, ed. D. M. Hopkins, pp. 451–84. Stanford, Calif.: Stanford University Press.

Hopkins, D. M., ed. 1967b. *The Bering Land Bridge*. Stanford, Calif.: Stanford University Press.

Hultén, E. n.d. Typescript autobiography in possession of author.

———. 1937. *Outline of the History of Arctic and Boreal Biota during the Quaternary Period*. Stockholm: Borförlag Aktiebolaget Thule.

———. 1941–1950. *Flora of Alaska and the Yukon*, 1–10. Lunds Universitet Årrskrift B, F, m avd. 2, v. 37:1–46–1.

———. 1968. *Flora of Alaska and Neighboring Territories*. Stanford, Calif.: Stanford University Press.

Matthews, J. V., Jr. 1976. Arctic Steppe: An Extinct Biome. AMQUA *Abstracts* 4:73–77.

———. 1982. East Beringia during Late Wisconsin Time: A Review of the Biotic Evidence. In *Paleoecology of Beringia*, ed. D. M. Hopkins, J. V. Matthews, Jr., C. E. Schweger, and S. B. Young, pp. 127–50. New York: Academic Press.

McManus, D. A., J. S. Creager, R. J. Echols, and M. L. Holmes. 1983. The Holocene Transgression on the Arctic Flank of Beringia: Chukchi Valley to Chukchi Estuary to Chukchi Sea. In *Quaternary Coastlines and Marine Archeology*, ed. P. M. Masters and M. C. Flemming, pp. 365–88. New York: Academic Press.

Murray, D. F. 1981. The Role of Arctic Refugia in the Evolution of the Arctic Vascular Flora: A Beringian Perspective. In *Evolution Today, Proceedings of the 2d International Congress of Systematic and Evolutionary Biology*, ed. G. G. Scudder and J. V. Reveal, pp. 11–20. Pittsburgh: Hunt Institute for Botanical Documentation, Carnegie-Mellon University.

Repenning, C. A. 1967. Palearctic-Nearctic Mammalian

Dispersals in the Late Pleistocene. In *The Bering Land Bridge,* ed. D. M. Hopkins, pp. 288–311. Stanford, Calif.: Stanford University Press.

———. 1987. Biochronology of the Microtine Rodents in the United States. In *Cenozoic Mammals of North America: Geochronology and Biostratigraphy,* ed. M. O. Woodburne, pp. 236–68. Berkeley: University of California Press.

Schweger, C. E., J. V. Matthews, Jr., D. M. Hopkins, and S. B. Young. 1982. Paleoecology of Beringia: A Synthesis. In *Paleoecology of Beringia,* ed. D. M. Hopkins, J. V. Matthews, Jr., C. E. Schweger, and S. B. Young, pp. 425–44. New York: Academic Press.

Vartanyan, S. L., V. E. Garutt, and A. V. Sher. 1993. Holocene Dwarf Mammoths from Wrangel Island in the Siberian Arctic. *Nature* 362:337–40.

Vereshchagin, N. K., and G. F. Baryshnikov. 1982. Paleoecology of the Mammoth Fauna in the Eurasian Arctic. In *Paleoecology of Beringia,* ed. D. M. Hopkins, J. V. Matthews, Jr., C. E. Schweger, and S. B. Young, pp. 267–89. New York: Academic Press.

West, F. H. 1981. *The Archaeology of Beringia.* New York: Columbia University Press.

Young, S. B. 1982. The Vegetation of Land-Bridge Beringia. In *Paleoecology of Beringia,* ed. D. M. Hopkins, J. V. Matthews, Jr., C. E. Schweger, and S. B. Young, pp. 179–91. New York: Academic Press.

Yurtsev, B. A. 1982. Relics of the Xerophyte Vegetation of Beringia in Northeastern Asia. In *Paleoecology of Beringia,* ed. D. M. Hopkins, J. V. Matthews, Jr., C. E. Schweger, and S. B. Young, pp. 157–58. New York: Academic Press.

AMERICAN BEGINNINGS

THE STUDY OF BERINGIA

Frederick H. West

Possibly the most far-reaching changes ever witnessed by man unfolded in the time and in the place that form the setting for this study of American beginnings. The evidence of these environmental changes and the record of the human witness are set out in the articles that follow. Those changes took place in a region then, as now, remote from the centers of human population, but a region which in this time played a determining role in the future of that population; in its numbers, distribution, and composition, and, ultimately, thus, in the formation of the complex mosaic of New World cultures.

In his original usage of the term *Beringia*, Eric Hultén restricted its application to the then (1937) hypothetical intercontinental land connection at Bering Strait. Subsequently, this came to be known as the "Bering Land Bridge." As research in various fields burgeoned, augmenting the concept, the name Beringia came to be loosely extended to *include* these adjacent regions of Alaska and Siberia. In a strict sense then, it could be said that at that point the subject became *Greater Beringia*. But as the concept of the province was, is, and must be, plastic, the qualifier seems unnecessary.

As used here, eastern Beringia includes most of modern mainland Alaska and much more restricted portions of Yukon Territory and Northwest Territories. It is bounded on the east by the Mackenzie River valley beyond which lay continental ice, on

the south by the mountains of southern Alaska whose ice joined the great cordilleran ice sheet of western Canada. Note, however, that certain portions of the coastal region south of this glaciation were open, accessible, contiguous, and therefore a piece of eastern Beringia. The western boundary of Beringia is placed in the valley of the Lena River. From there a line drawn generally southeast encompasses the Aldan plateau and the maritime region of Primorye.

The time period of concern is the latter portion of the last glacial episode. The fourth glacial was broadly marked by two major cold oscillations with glacial advances, separated by a rather deep interstadial. This interstadial, the "Eemian," had its onset about 70,000 years ago and gave way to the final glaciation perhaps 30,000 years ago. That final episode is estimated to have begun its climax about 11,500 years ago. The end of the Pleistocene is by common consent set at 10,000 years ago—somewhat arbitrary, but a time around which a number of important Pleistocene–Holocene transition phenomena tend to cluster (Fig. 1).

The fourth glacial cycle left sequences of records that are generally quite clear and which may be correlated coherently from region to region in the northern hemisphere. The familiar sequence names, Warthe–Weichsel (in Fennoscandia), Wisconsin I and II (in America), Würm I and II (in the

SELECTED RADIOCARBON-DATED GLACIAL AND CLIMATIC RECORDS
Compiled by Thomas D. Hamilton

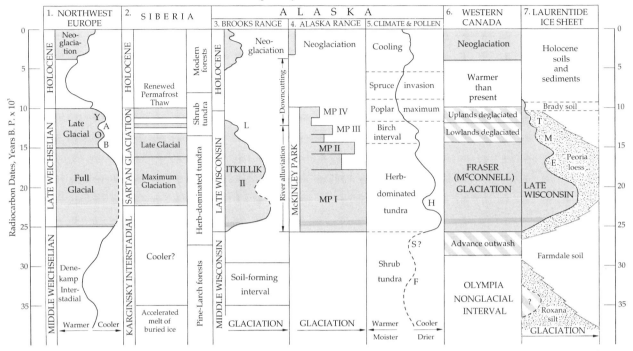

Figure 1 Glacial chronology chart.

Notes: (1) Glacial sequence from Mangerud (1991). Weichselian paleotemperature curve based on Woillard and Mook (1982); Holocene curve follows Nesje and Kvamme (1991). Abbreviations: A, Allerod; B. Bolling; O, Older Dryas; Y, Younger Dryas.

(2) See Arkhipov (1984) and Glushkova (1994) for glacial record. Pollen based on northeast Siberian record of Anderson et al. (1994); thermokarst record from Astakhov (1992).

(3) Itkillik II glacial record from Hamilton (1986; 1994); Neoglacial record from Ellis and Calkin (1984). Soil and river data from Hamilton and Ashley (1993). L, Late Itkillik II readvance.

(4) Late Wisconsin glacial advances from Ten Brink and Waythomas (1985); Neoglacial records summarized in Calkin (1988).

(5) Data generalized for northern and central Alaska. Climatic curve from Hamilton and Fulton (in prep.); pollen from Anderson and Brubaker (1993, 1994). Thermal events abbreviated: F, Fox; H, Hanging Lake; S?, Sixtymile.

(6) Data compiled by Clague (1989).

(7) Glacial record compiled from Richmond and Fullerton (1986); loess and paleosols from Leigh (1994) and May and Holen (1993). Interstades abbreviated: E, Erie; M, Mackinaw; T, Two Creeks.

Alpine system of southern Europe), have many local counterparts, e.g., Delta–Donnelly in the north central Alaska Range and Itkillik I–Itkillik II of the central Brooks Range in Alaska (Péwé 1975; Hopkins 1982). In the Verkhoyansk Mountains of Sakha (Yakutia) the local sequence is Zyrian–Sartan. In all cases, an important interstadial of long duration is recorded separating these two glacial events. For this interstadial, called Karginsky in Siberia, David Hopkins has proposed extending the name "Boutellier interval" to apply to all of Beringia (Hopkins 1982:7–9). For the "extremely dry, probably cold, periglacial interval in the unglaciated parts of Be-

ringia," i.e., the temporal equivalent of Wisconsin II or the Weichsel, the same author has suggested the designation "Duvannyy Yar" (Hopkins 1982:9–10). It is important to emphasize that Beringia remained subaerial and unbreached throughout the last glacial period. Moreover, recent sub-marine research in Bering Strait provides proof that its late Pleistocene inundation was significantly later than previously estimated (Elias, this volume).

Late Pleistocene glaciation in Beringia is shown in Figure 2. The most striking feature shown is the contrast between the very restricted, localized glaciation and the vast, uninterrupted expanse of ungla-

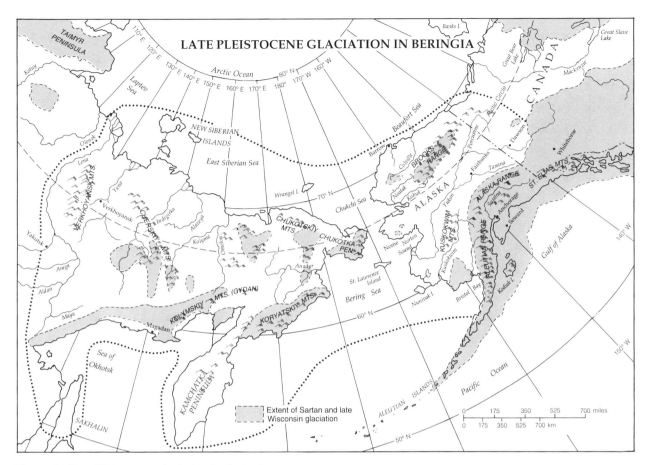

Figure 2 Late Pleistocene glaciation in Beringia.

ciated country extending across, indeed defining, the entire east–west dimension of Beringia. These, essentially geological, distributional relationships are generally acknowledged. The great question that then arises is, what was the nature of this unglaciated environment? At this point, as will be seen in the various discussions bearing on it, there is some controversy. This is in no small part owing to the fact that the types of evidence which contribute to the requisite reconstructions derive from several disciplines, meaning, of course, that the disparate evidences they adduce are not themselves immediately comparable but require rectification.

There are variously conflicting interpretations of the Beringian environment of late Würmian time. The argument turns primarily on the interpretation of two ubiquitous observations in the biotic record: the presence of high proportions of *Artemisia* in the

pollen profiles and the high numbers of ungulate fauna. Paul Colinvaux (1967) adumbrated the nature of the discussion just then on the horizon when he noted that Russian investigators, citing as examples Giterman and Golubeva in the same volume, interpreted *Artemisia* in Siberia as one of the important indicators of steppe, albeit cold. Colinvaux, in contrast, was inclined to read that occurrence at Imuruk in Alaska as local and more likely indicative of bare ground, meaning, presumably, as Ritchie and Cwynar (1982) would later maintain, that there was no climatic significance to be read into such presence. Armed with new data on Alaskan fossil identifications and the ruminations of other scholars, R. D. Guthrie became the most vigorous supporter of the *Artemisia* steppe-tundra position, based upon his interpretation of an abundant, varied fossil fauna that demanded grasses to account for their

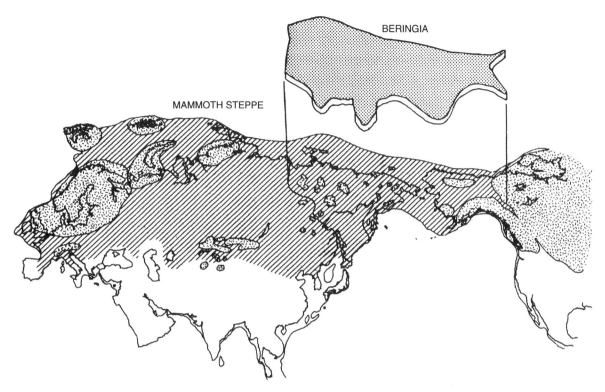

Figure 3 The "mammoth steppe" of late Pleistocene time, showing the distribution of mammoth across Eurasia and Beringia. This distribution was to reduce in the late Würm to extreme northeastern Asia, which is to say, Beringia. (*Source: University of Chicago Press. See Acknowledgments.*)

presence. This interpretation was enunciated clearly in several articles, culminating in 1990 with a full exposition of what he came to call the "mammoth steppe" (Guthrie 1990) (Fig. 3).

Still advocated by some palynologists is the position that sees Würmian Beringia as being composed of a mosaic of vegetational associations rather than continuous steppe. Such an interpretation would only permit of restricted, localized reconstructions. As has been pointed out, however, the potency of this counterargument is in large measure dependent upon its exponents being able to suggest convincingly—if not demonstrate—that the unusual array of large mammalian fauna described by the palaeontologists did not, in fact, actually coexist. But if coexistence can be proven, then an explanation other than the presence of extensive grasslands becomes difficult to sustain.

The most important constituents of the late Beringian fauna were ungulates, and where these late Pleistocene faunal remains are found, they are commonly dominated by the triad of grazers—mammoth, bison, and horse. These are normally accompanied, in much lower frequencies, by such other forms as musk-ox, caribou, wapiti (red deer), saiga antelope, and Dall sheep. Table 1 summarizes large mammal occurrences in both parts of Beringia. Table 2 presents some representative radiocarbon assays on these most important grazers, indicating their age placement in the late Würm. The weight of evidence appears to favor the steppe argument but, as its proponents insist, certainly it was unlike any modern grassland, but perhaps therein approximating the mosaic. The appellation "mammoth steppe" seems quite reasonable. It addresses both the fauna and the flora, but simultaneously emphasizes its non-modern character.

The depiction of the holarctic flora and the explanation of its distribution by means of such a land bridge was Eric Hultén's life work. That bridge was

Table 1 Summary of Large Mammal Occurrences in Both Parts of Beringia

Species	Siberia	Alaska–Yukon Territory
Smilodon (saber-toothed cat)	x	x
Mammuthus primigenius (mammoth)	x	x
Equus lambei (small horse)		x
E. hemionus (horse)	x	
Coelodonta antiquitatus (woolly rhinoceros)	x	
Cervus elaphus (red deer/wapiti)	x	x
Bison priscus (steppe bison/ large-horned bison)	x	x
Bootherium sp./*nivicolens* (musk-ox)	x	x
Symbos sp./*cavifrons* (musk-ox)	x	x

equally important in the movement of animals. By far the greater part of the migration of mammals between continents was from Eurasia to America. For the entire Pleistocene, Bjorn Kurtén found only six forms moving from America to Eurasia and but two of these in Würm times. By contrast, twenty-two mammal species migrated to America, nine of those in the Würm. These included mammoth, musk-ox, caribou, moose, grizzly bear, polar bear, and saiga antelope (Kurtén 1968:268–69).

———— ◆◆◆ ————

The question of fauna is, indeed, of fundamental importance. The ultimate concern here is human prehistory. It is a safe assumption that it was not the character of the vegetation that brought people into the most difficult environment the genus *Homo* ever came to colonize. But it needs to be emphasized that this is, after all, the period of the Upper Palaeolithic, one characterized across Eurasia by the application of highly evolved, highly efficient techniques for the hunting of this very fauna. It is beginning to appear, furthermore, that those fundamental attributes of the Upper Palaeolithic may be at least as early in central Asia as in Europe, and that observation may turn out to hold also for the northernmost and easternmost Upper Palaeolithic folk who are the subjects of this volume.

After the existence of Bering Strait became generally known and, with it, the near proximity to each other of the two continents, older, often fanciful, notions about the origins of the native Americans were gradually supplanted by ones now seen as more scientifically plausible. In time, Bering Strait came to be accepted as the only feasible route of entry. But the step from logical deduction to demonstration has been a large one. In fact, "step" is misleading: there have been many steps over a long period of time to achieve the present condition of knowledge. There were fundamental matters of time and resemblances to be dealt with.

Time was not in the purview of the archaeologist; data were needed from other disciplines, notably geology. It may be said, however, that, early on, it was perhaps not recognized by archaeologists just how critically important that time scale and its text actually were. Thus, simplistically, it was thought that the passage across the water of Bering Strait was accomplished either by watercraft in summer months or by foot over the ice in winter. The plausibility of this theory could be readily illustrated by reference to the Eskimos of the region who had been making just such passages for (as is now known) several thousand years. Yet, prior to the acquisition of knowledge of the nature of late Pleistocene Bering Strait, there was required a kind of disjunct logic in taking such necessarily maritime people and transforming them into the plains hunters who were the recognized "Early Man in America." As strained as some of these constructs may now appear, there is in them the essential point that—in the absence of explanations by way of lost continents or rafting across the Atlantic

Table 2   Representative Radiocarbon Assays on Large Mammal Occurrences*

Species	Locality	Age (yrs BP)	Sample Identification Number
Mammuthus sp.	Central Alaska	32,700 ± 980	ST-1632
——	Yukon Territory	32,350 ± 1750	I-4226
——	Yukon Territory	15,500 ± 130	GSC-3053
——	Central Alaska	15,380 ± 300	SI-453
——	Yukon Territory	13,820 ± 840	Beta-13867
Equus sp.	Yukon Territory	31,450 ± 1300	I-10935
——	Central Alaska	26,760 ± 300	SI-355
——	Yukon Territory	12,900 ± 100	GSC-2881
Bison sp.	Yukon Territory	26,460 ± 280	TO-393
——	Yukon Territory	22,200 ± 1400	I-3570
——	Central Alaska	21,063 ± 1363	SI-839
——	Central Alaska	18,000 ± 200	SI-841
——	Central Alaska	17,170 ± 840	SI-838

*Alaskan data from Péwé 1975: Table 13; Yukon Territory data from Harington 1989: Table 2.

Ocean—the earliest ancestors of the earliest Americans must have entered the continent by way of Bering Strait. But, again, between the logical assumption and the demonstration is a gulf to be bridged only by a span constructed of small, solid, independent blocks of evidence that will bear the weight of the closest critical examination. Ideally, to satisfy the requirements of scientific proof, several independent variables must be resolved, then set right, which is to say, the evidences found must be shown to be of appropriate antiquity and they must then be shown to array themselves in an appropriate order geographically.

Fortunately, when American scholars began earnestly pursuing these questions in the North, there was already established a standard against which to assess their finds. Whether Clovis was interpreted as the earliest human presence in America or simply as one of the earliest, by the 1960s it was already abundantly documented, coherent, extraordinarily widespread, *and* well-dated. Against this standard, then, the task of the Northern researcher was to make a reasonable case that there were, in the North, sites in which the typology *and* dating might allow an assumption of antecedency. It was not necessary, on this level, to show any sort of direct genetic relationship between such Northern sites and Clovis. In 1960, Alaska and the Yukon Territory were archaeologically still largely *terra incognita*. The area is immense, in total almost one-fourth that of the forty-eight contiguous United States. To that time, a bare handful of systematic surveys had been undertaken, most of them in coastal regions historically home to the Eskimos, Aleut, and Tlingit. The work that had been accomplished was of first order importance, but still the knowledge of Alaskan prehistory was meagre; that of its earliest phases practically nil. The decades of the 1960s through the 1980s saw a great deal of field research carried out, much of it in the interior—that region characterized by Froelich Rainey as "undoubtedly one of the most discouraging areas for archaeological research in North America" (Rainey 1953:43). Rainey knew whereof he spoke: along with his other pioneering field research in Alaska, between 1937 and 1940 he carried out the first successful systematic surveys in the interior. (Aleš Hrdlička had surveyed portions of the Yukon and Kuskokwim rivers in the early 1930s but without result.) In any event, there was little more known of interior archaeology in 1960 than in 1940. The growth of field research after that time paralleled that of American archaeology itself. With more people now willing to take up the challenge, new findings were bound to result.

Clearly, in order to address the question of Old World relationships, the most fundamental requirement was to demonstrate that resemblances between archaeological complexes on both sides of Bering Strait do, in fact, exist. Dating and, with that, "time slope," would come later. It is therefore of genuine interest that the first assertion of intercontinental resemblance came in what must be described as the first truly scientific article on interior Alaskan archaeology. This, of course, was N. C. Nelson's very judicious assessment of the Campus site in 1937, entitled "Notes on Cultural Relations Between Asia and America," in which he compared the Campus site microblade cores with those he had excavated in the Gobi desert in 1926 and 1928. Even though his typological assertions require a bit of correction, his overall conclusions bear up remarkably well. In an earlier brief note, Nelson wrote, "The specimens furnish the first sure evidence we have of early migrations to the American continent, apparently during the final or Azilian–Tardenoisian stage of the Palaeolithic culture horizon, possibly 7,000–10,000 B.C." (Nelson 1935). Nelson, then curator of prehistoric archaeology at the American Museum, was probably the only American who might have been able to make an assessment of such prescience. It was only owing to an unusual arrangement involving the collection for that museum of Pleistocene faunal remains from placer mines—this by Otto Geist of the University of Alaska—that the Campus site collections of 1934 and 1935 were sent to Nelson for examination. In 1936 Froelich Rainey, then a professor at the University of Alaska, excavated at the Campus site as part of his research in the interior. In his 1939 report of his 1936–1937 surveys, he repeated Nelson's suggestion of relations with Mongolia and noted his discovery at Dixthada—far upstream on the Tanana River—of simi-

lar microblade cores amongst the midden debris of this recent Athapaskan site. Rainey made no suggestion then of possible mixture, although this was voiced in 1951.

The puzzling "small polyhedral cores and their micro-blades" continued to niggle at Rainey, eventuating in his 1951 paper, "The Significance of Recent Archaeological Discoveries in Inland Alaska," presented at the symposium "Asia and North America: Transpacific Contacts" (Rainey 1953). New findings prompted a quasi-synthesis: J. L. Giddings's Denbigh site, W. N. Irving's on the Anaktuvuk, H. Larsen's Trail Creek Caves, R. S. Solecki's finds on the Kukpowruk, F. Johnson and H. M. Raup's findings in the Shakwak Valley of Yukon, and W. S. Laughlin's at Ananiuliak (Anangula) were all related. While on the one hand suggesting a relationship among these small collections, Rainey at the same time recognized that Irving's Anaktuvuk material most closely resembled Denbigh. It is quite possible that the then current estimation of the age of Denbigh at 10,000 years or more led to the proposition that all were probably Mesolithic; this was construed as a more conservative interpretation ("instead of stretching a single kind of industry over a period of several thousand years" [Rainey 1953: 44]). Finally, and to the present point, while urging their ultimate relationship, Rainey noted resignedly, "The best one can do with [this grouping] is to point out that we have no evidence for those migrations across the Strait which we have all assumed took place during the Neolithic period" (Rainey 1953:45).

The interior remained largely neglected from the turn of the century until the 1960s. Much was learned in those years of Eskimo and Aleut prehistory, but the interior remained effectively unknown. The major exceptions to that generalization were the investigations of Ralph Solecki in the Brooks Range and those of R. S. MacNeish in Yukon Territory. In both these cases, the thread discovered by Nelson and followed by Rainey was again encountered.

In the southwestern Yukon Territory, MacNeish described a number of sites containing microblade cores, but generally in forms and combinations rather unlike the Alaskan occurrences—a predominance of conical and tabular forms with relatively few wedge-shaped specimens and all often co-occurring. These formed the basis of the "Northwest Microblade Tradition" which MacNeish dated to approximately 5,000 years BP (MacNeish 1964). Both the typology and the dating suggested attenuation from the more western forms.

Solecki reported two sites discovered in 1949 while working in the Kukpowruk region in which cores similar to those of the Campus site were recovered. Site 65 produced both conoidal and wedge forms, the latter quite resemblant to that from Campus but all well beyond the "micro" range in size. The second site, 121, yielded microblade cores more nearly like those of the Campus site. Solecki suggested both sites might be referred to a "University of Alaska Campus" phase and discussed at some length the Bering Land Bridge—but without, of necessity, any site or sites proximate in Eurasia with which to make comparison (Solecki 1950, 1951, this volume). Solecki's later expedition (1961) recovered no core and blade material (Solecki, this volume).

The 1960s witnessed the awakening of activity in interior Alaska which has continued to the present. The results of some of this activity are detailed in the following pages. In summary, it could be said that by 1960 there had been accumulated a small body of evidence of a sort that substantially reinforced Nelson's evaluation and that, typologically, clearly pointed to the Old World, but which allowed little discussion beyond a reiteration of Nelson's original statement. There was nothing comparable on record east of the Lena River; the only sites in Siberia that bore any resemblance whatever were Verkholenskaya Gora, Afontova Gora, and perhaps one or two others west of Lake Baikal. Verkholenskaya Gora was mentioned by Rainey (1939:389) and had been known for some years, but it was distant and far from identical in artifact forms (and well beyond the western boundary of Beringia). That situation had not altered appreciably when, in 1967, the core and blade Denali complex was proposed (West 1967)—with the important exception that N. N. Dikov had now issued preliminary reports on the Ushki locality on the Kamchatka Peninsula. Of the few Siberian sites available for comparison, Afontova Gora 2, Krasnyy Yar, and

Verkholenskaya Gora were proposed as showing certain resemblances but obviously attenuated in character.

It is of great interest that in this same period the Mochanovs were beginning their intensive surveys in the Aldansk and eastward—as detailed in this volume—that would provide just the sort of comparative materials that were required, and these squarely in western Beringia. That was demonstrated most graphically in 1973 when, at the Bering Land Bridge Conference in Khabarovsk, it was possible for Russian and American field archaeologists to compare directly collections from Alaska and northeastern Siberia. In the minds of at least some of those fortunate few conferees, all doubts were laid to rest. Moreover, there came the suspicion that these might be American progenitors. Compiling further and sufficient proof and convincing others of its soundness was to require time and effort—and ultimately was to make necessary this volume.

In his judgment of the resemblances between the Campus site microblade forms and those of his "Dune Dwellers of the Gobi," Nelson was as near to being correct as anyone could have been at the time. His brief note on the matter was presented so authoritatively and so unequivocally that it has never been seriously questioned.

Nelson never completed his final report on the Gobi. But Walter Fairservis had promised him many years ago that he would produce the report that Nelson's ill health forced him to abandon. That monograph, *The Archaeology of the Southern Gobi of Mongolia,* appeared in late 1993, a year before Fairservis's passing. It was Fairservis's final publication. Froelich Rainey died in 1993.

———————————•◆•———————————

Human evolution has been primarily a phenomenon of the Quaternary. The pace of hominid evolution has surely exceeded that of any other large mammal. But, while the history of the genus *Homo* was largely played out in the two million or so years of the Pleistocene, the modern species only made its appearance in the final phases of that epoch. Granting the European bias in this record, for the northern hemisphere the appearance is that

Neanderthalian *Homo sapiens*—probably cold-adapted—only gave way to *Homo sapiens sapiens* towards the middle of the last glacial episode. The association of Middle Palaeolithic–Mousterian industry with Neanderthalers is not 1:1, but it is near to it. The disappearance of that industry corresponds closely to the disappearance of the Neanderthalers. The industries that followed—characterizing the Upper Palaeolithic—are conceded generally to represent not merely change, but advance, in lithic technology, indicating certainly a more efficient articulation with the environment. The result was that these were superior hunters possessing a superior weaponry. The question of whether the Upper Palaeolithic folk were partly, or wholly, descendants of the Neanderthalers appears still to be moot. Clearly, though, by whatever means, Neanderthal cold-adaptation survived in their successors—and thereby hangs the tale.

BIBLIOGRAPHY

Anderson, P. M., and L. B. Brubaker. 1993. Holocene Vegetation and Climatic Histories of Alaska. In *Global Climates Since the Last Glacial Maximum*, ed. H. E. Wright, Jr., et al., pp. 386–400. Minneapolis: University of Minnesota Press.

———. 1994. Vegetation History of Northcentral Alaska. *Quaternary Science Reviews* 13:71–92.

Anderson, P. M., A. V. Lozhkin, and L. B. Brubaker. 1994. A Late Quaternary Vegetation History from Elikchan Lake, Northeast Siberia. In *Program and Abstracts*, 13th Biennial Meeting of the American Quaternary Association, p. 196. Minneapolis: University of Minnesota Limnological Research Center.

Arkhipov, S. A. 1984. Late Pleistocene Glaciation of Western Siberia. In *Late Quaternary Environments of the Soviet Union*, ed. A. A. Velichko, pp. 13–19. Minneapolis: University of Minnesota Press.

Astakhov, V. 1992. The Last Glaciation in West Siberia. *Sveriges Geologiska Undersökning*, Ser. Ca 81, pp. 21–30.

Calkin, P. E. 1988. Holocene Glaciation of Alaska (and Adjoining Yukon Territory, Canada). *Quaternary Science Reviews* 7:159–84.

Clague, J. J., comp. 1989. Quaternary Geology of the Canadian Cordillera. In *Quaternary Geology of Canada and Greenland*, ed. R. J. Fulton, pp. 15–96; Vol. K-1, *The Geology of North America*. Boulder, Colorado: Geological Society of America.

Colinvaux, P. A. 1967. Quaternary Vegetational History of Arctic Alaska. In *The Bering Land Bridge*, ed. D. M. Hopkins, pp. 207–31. Stanford: Stanford University Press.

Ellis, J. M., and P. E. Calkin. 1984. Chronology of Holocene Glaciation, Central Brooks Range, Alaska. *Geological Society of America Bulletin* 95:897–912.

Fairservis, W. A., Jr. 1993. *The Archaeology of the Southern Gobi of Mongolia*. Durham: Carolina Academic Press.

Flerow, C. C. 1967. On the Origin of Mammalian Fauna of Canada. In *The Bering Land Bridge*, ed. D. M. Hopkins, pp. 271–80. Stanford: Stanford University Press.

Glushkova, O. Yu. 1994. Paleogeography of Late Pleistocene Glaciation of Northeastern Asia. In *Proceedings, International Conference on Arctic Margins*, Anchorage, ed. D. K. Thurston and K. Fujita, pp. 339–44. Anchorage: Minerals Management Service OCS Study MMS 0040.

Guthrie, R. D. 1968. Paleoecology of the Large Mammal Community in Interior Alaska during the Late Pleistocene. *The American Midland Naturalist* 79(2):346–63.

———. 1972. Re-creating a Vanished World. *National Geographic Magazine* 141(3):294–301.

———. 1990. *Frozen Fauna of the Mammoth Steppe*. Chicago: University of Chicago Press.

Hamilton, T. D. 1986. Late Cenozoic Glaciation of the Central Brooks Range. In *Glaciation in Alaska: The Geologic Record*, ed. T. D. Hamilton, K. M. Reed, and R. M. Thorson, pp. 9–49. Anchorage: Alaska Geological Society.

———. 1994. Late Cenozoic Glaciation of Alaska. In *The Geology of Alaska*, ed. G. Plafker and H. C. Berg, pp. 813–44; Vol. G-1, *The Geology of North America*. Boulder, Colorado: Geological Society of America.

Hamilton, T. D., and G. M. Ashley. 1993. Epiguruk: A Late Quaternary Environmental Record from Northwestern Alaska. *Geological Society of America Bulletin* 105:583–602.

Hamilton, T. D., and R. J. Fulton. In preparation. Middle and Late Wisconsin Environments of Eastern Beringia. U.S. Geological Survey.

Harington, C. R. 1971a. Ice Age Mammals in Canada. *The Arctic Circular* 22(2):66–89.

———. 1971b. Ice Age Mammal Research in the Yukon Territory and Alaska. In *Early Man and Environments in Northwest North America*, ed. R. A. Smith and J. W. Smith, pp. 35–51. Calgary: Archaeological Association of the University of Calgary.

———. 1989. Pleistocene Vertebrate Localities in the Yukon. In *Late Cenozoic History of the Interior Basins of Alaska and the Yukon*, ed. C. L. David, T. D. Hamilton, and J. P. Galloway, pp. 93–98. U.S. Geological Survey Circular 1026.

Hopkins, D. M. 1982. Aspects of Paleogeography of Beringia during the Late Pleistocene. In *Paleoecology of*

Beringia, ed. D. M. Hopkins, J. V. Matthews, Jr., C. E. Schweger, and S. B. Young, pp. 3–28. New York: Academic Press.

Klein, R. G. 1971. The Pleistocene Prehistory of Siberia. *Quaternary Research* 1(2):133–61.

Kurtén, B. 1968. *Pleistocene Mammals of Europe.* Chicago: Aldine.

Leigh, D. S. 1994. Roxana Silt of the Upper Mississippi Valley: Lithology, Source, and Paleoenvironment. *Geological Society of America Bulletin* 106:430–42.

MacNeish, R. S. 1964. Investigations in Southwest Yukon: Archaeological Excavations, Comparisons, and Speculations. *Papers of the R. S. Peabody Foundation for Archaeology* 6(2):201–488. Andover, Massachusetts: Phillips Academy.

Mangerud, J. 1991. The Last Interglacial/Glacial Cycle in Northern Europe. In *Quaternary Landscapes,* ed. L. C. K. Shane and E. J. Cushing, pp. 38–75. Minneapolis: University of Minnesota Press.

Martin, P. S. 1967. Prehistoric Overkill. In *Pleistocene Extinctions: The Search for a Cause,* ed. P. S. Martin and H. E. Wright, Jr., pp. 75–120. New Haven: Yale University Press.

Martin, P. S., and J. E. Guilday. 1967. A Bestiary for Pleistocene Biologists. In *Pleistocene Extinctions: The Search for a Cause,* ed. P. S. Martin and H. E. Wright, Jr., pp. 1–62. New Haven: Yale University Press.

May, D. W., and S. R. Holen. 1993. Radiocarbon Ages of Soils and Charcoal in Late Wisconsinan Loess, South-Central Nebraska. *Quaternary Research* 39:55–58.

Nelson, N. C. 1935. Early Migration of Man to America. *Natural History* 35(4):356.

———. 1937. Notes on Cultural Relations Between Asia and America. *American Antiquity* 2:4:267–72.

Nesje, A., and M. Kvamme. 1991. Holocene Glacier and Climate Variations in Western Norway: Evidence for Early Holocene Glacier Demise and Multiple Neoglacial Events. *Geology* 19:610–12.

Péwé, T. L. 1975. *Quaternary Geology of Alaska.* U.S. Geological Survey Professional Paper 835.

Rainey, F. 1939. Archaeology in Central Alaska. *Anthropological Papers of the American Museum of Natural History* 36:4:351–405.

———. 1953. The Significance of Recent Archaeological Discoveries in Inland Alaska. In *Asia and North America: Transpacific Contacts,* ed. M. W. Smith, pp. 43–61. Memoirs of the Society for American Archaeology No. 9, Vol. 18, No. 3, Pt. 2. Salt Lake City.

Richmond, G. M., and D. S. Fullerton, eds. 1986. Quaternary Glaciations in the United States of America. *Quaternary Science Reviews* 5:3–196.

Ritchie, J. C., and L. Cwynar. 1982.The Late Quaternary Vegetation of the North Yukon. In *The Paleoecology of Beringia,* D. M. Hopkins, J. V. Matthews, Jr., C. E. Schweger, and S. B. Young, eds., pp. 113–26. New York: Academic Press.

Sher, A. V. 1973. Mammals and the Beringian Land during the Late Cenozoic: Puzzling Problems and Some Ways to Solve Them. In *The Bering Land Bridge and Its Role for the History of Holarctic Floras and Faunas in the Late Cenozoic (Abstracts).* Academy of Sciences of USSR, Far-Eastern Scientific Centre, Khabarovsk.

Solecki, R. S. 1950. A Preliminary Report of an Archaeological Reconnaissance of the Kukpowruk and Koklik Rivers in Northwest Alaska. *American Antiquity* 16:1:66–69.

———. 1951. Archaeology and Ecology of the Arctic Slope of Alaska. *Annual Report Smithsonian Institution 1950,* pp. 69–95. Washington, D.C.

Ten Brink, N. W., and C. F. Waythomas. 1985. Late Wisconsin Glacial Chronology of the North-Central Alaska Range: A Regional Synthesis and Its Implications for Early Human Settlements. *National Geographic Society Research Reports* 19:15–32.

Vangengeim, E. A. 1967. The Effect of the Bering Land Bridge on the Quaternary Mammalian Faunas of Siberia and North America. In *The Bering Land Bridge,* ed. D. M. Hopkins, pp. 281–87. Stanford: Stanford University Press.

West, F. H. 1967. The Donnelly Ridge Site and the Definition of an Early Core and Blade Complex in Central Alaska. *American Antiquity* 32:3:360–82.

———. 1981. *The Archaeology of Beringia.* New York: Columbia University Press.

Woillard, G. M., and W. G. Mook. 1982. Carbon-14 Dates at Grande Pile: Correlation of Land and Sea Chronologies. *Science* 215:159–61.

Part One

PALAEOENVIRONMENTAL RESEARCH

RECONSTRUCTING THE ENVIRONMENT

Paul A. Colinvaux

AMERICAN BEGINNINGS were in the arctic of a glacial age. It was at an arctic place that world-wide lowering of sea levels during the Pleistocene exposed the Bering and Chukchi shelves during times of world glaciation (Hopkins 1959). Global Circulation Models (GCMs) and the contemporary astronomic theory of ice ages describe the mean climate at high latitudes as significantly colder than that of the present day. Beringia was a cold arctic place.

Geological evidence reinforces this elementary observation of a cold climate by showing that eastern Siberia and western Alaska were then deep within a continental-scale land mass, whereas now they are maritime. It is likely that continental extremes of climate produced warmer summers than those of the present day. Also, the strong warming of late-glacial times about 13,000 years ago came before the flooding of the land bridge 10,000 years ago, giving centuries of relative warmth that possibly correlate with the first arrival of humans in the Americas.

Before this late-glacial window of opportunity, the glaciation had lasted a long time (70,000 years) and its climates changed frequently around its mean of glacial cold. This history has left its mark on the longer geological sections, particularly in the comparatively warm episode called the Karginsky interstadial in Siberia (Morozova and Velichko, this volume). This episode is usually simply referred to as the last interstadial or, in Alaska, the mid-Wisconsin interstadial. In a typical stream bluff exposure the effects of this milder climate are evident in soil profiles or as traces of increased thermokarst activity as described by both Waythomas and Hamilton (this volume).

The interstadial began a few millennia before 30,000 BP—thus at the farthest reach of radiocarbon dating—making it difficult to date the interval pre-

cisely. Nevertheless, dates for the interstadial in both east and west Beringia are converging, with a span of 33,000 to 28,000 BP being offered for Siberia and 35,000 to 24,000 for Alaska.

For the peopling of the Americas, the environment following the interstadial, ca. 25,000–10,000 years BP, is of greatest interest. But, the 5,000 to 10,000 years of warmer climate represented by the Karginsky may hold the key to understanding some of the more perplexing evidence of ancient Beringian climates and biogeography, from larger ice-wedges to lion bones (Ukraintseva et al., this volume).

The land surface of Beringian lowlands that remains above the sea surface in both Siberia and Alaska is covered with a mantle of silt, more or less mixed with ice. Freeze-thaw cycles rework the surface of the silt, producing solifluction terraces, patterned ground, and thaw lakes. Except when reworked in these manifestations of freeze-thaw cycles, silt is not accumulating in modern environments. Both radiocarbon dating and stratigraphic evidence indicate that the last period of silt accumulation was in glacial times: the Sartan of Siberia and the Itkillik of Alaska, both broadly synchronous with the Wisconsin of North America and the Würm of Europe.

Near Fairbanks in central Alaska, masses of frozen silt up to 70 m thick collected in valley bottoms—burying some of the world's richest gold placer deposits. Gold miners called the frozen silt "muck" as they attacked it with steam hoses. Mixed with the gold and gravel on the drainage gates of the mines were numerous bones of extinct megafauna (Guthrie, this volume). Eventually, Troy Péwé demonstrated that the silt was, in fact, loess, thus providing the first evidence of a dry, windswept environment in the Beringia of the last glaciation (Péwé 1955).

The extensive silt deposits in Siberia earned a

special name, the "yedoma." Among the characteristics of yedoma are massive inclusions of ice, frequent preservation of mammoth bones and tusks, and a pattern of shallow basins and ponds called "alasy" (in Alaska called "thaw lakes"). Early Russian geologists interpreted this vast landscape of yedoma and alasy as representing lacustrine deposits—relict of ponded land surfaces. N. A. Shilo and his students later demonstrated that the yedoma silts were windblown loess (Shilo 1964; Tomirdiaro 1980). The yedoma was the Siberian equivalent of the "muck" of the Alaskan gold miners, and was, likewise, an aeolian deposit.

This thick mantle of windblown silt stretched all across Beringia, from its far west to its far east. It is the heritage of more than one glaciation. Successive glacial periods added their quota of loess to the land surface as is shown by the placement, above many meters of loess in central and southwestern Alaska, of the Old Crow tephra, determined by isothermal plateau fission-track dating to be about 140,000 years old (Westgate 1982). In both Siberia and Alaska the silt mantle included remnants of ice wedges (Morozova and Velichko, this volume).

The bones of large mammals are preserved in the frozen loess and have perhaps accumulated from the beginning of silt deposition. This huge loessic plain is evidence of a former dry, windswept landscape. The work of Péwé in Alaska and the Shilo team in Siberia, therefore, shows that the palaeoecology of Beringia before the late-glacial warming was set as colder and drier than any environment now prevailing on the modern remnants of the old land bridge.

Dry glacial times would have provided ample sources for the loess. Outwash of the large northward-flowing rivers of Siberia provided silt for wind to carry to the interfluves. The Yukon, Tanana, and Kuskokwim river systems were the source of loess for large parts of Alaska; glacial outwash was available on the southern edge of Beringia, while the northern rim could have been supplied from the exposed continental shelves of Siberia and Alaska. Yet, there are some oddities about the yedoma deposits of the Far North that open the way to some other ideas.

In its extreme northerly parts—on the Alaskan

coastal plain, coastal Siberia, and particularly in such Siberian islands as Novaya Zemlya—the frozen loess is actually mostly ice. Tomirdiaro (this volume) observes that in parts of this coastal region the ice content is in the order of 90 percent of total soil volume and residual columns of more-or-less mineral soil may be separated by bodies of clear ice up to 8 m wide. It is as if the land were actually the surface of a subterranean glacier on which dusty loess had fallen. Some argue that these icy soils of the Far North are merely the extreme manifestation of ice-wedge polygons in which the wedges grew to widths of 8 m. Tomirdiaro prefers the hypothesis that in glacial times dust deposition over preexisting and underground glacial formations left a narrow band of soil over ice along the Beringian north coast. A now-vanished portion of this formation may have spread over part of the Arctic Ocean, where dust is postulated to have covered permanent sea ice as well as the underground glacier. The extreme cold of the arctic coast of Beringia in glacial times remains apparent—whatever the interpretation of the origins of the ice.

The basic theory of Beringian origins comes from the physical data of geology: the land connection was established by eustatic drop of sea level; the chronology of Beringia is the chronology of ice ages, and the data of geomorphology describe the land surface as a dry, windswept plain of patterned ground.

A time lag in the melting of glaciers let the land connection continue to exist into late-glacial time, even after drastic environmental changes subsequent to the spectacular global warming that brought the Sartan–Itkillik glaciation to an end. For further details we must rely on reconstructions from biological data, from biogeography, from fossil assemblages, and from pollen.

Erik Hultén (1937) noted that Beringia was a significant barrier to the distribution of trees and showed that the forests of Alaska and Siberia had not merged at any time in the Pleistocene. Any naturalist with the good fortune to visit both Alaska and Siberia can see how right he was. Modern subarctic forests on opposite sides of the Bering Strait are distinctively different: on the Siberian side are open woodlands of larches (*Larix*) mixed with spreading, shrub-like pines (stone pines, *Pinus pum-*

illa); on the Alaskan side are spruce forests, either the black spruce (*Picea mariana*) in boggy bottom lands or the white spruce (*Picea glauca*) on well-drained slopes or at the treeline. As Hultén said, none of the trees had crossed the land bridge to colonize each other's side of Beringia. It follows that the central windswept plains defined by geology were nowhere forested. Nearly sixty years after Hultén wrote, his observation still stands.

Pollen analysis in Beringia has been of two traditions: direct examination of the fine matrix of stratigraphic sections chosen by geological field parties, and primary biological or palaeoclimatological research using cores of lake sediments. Pollen analysis of stratigraphic sections has the great advantage that it allows direct correlation of the results with glacial stratigraphy and chronology deduced by other means. On the other hand, pollen analysis of lake sediments is more likely to yield an accurate description of regional vegetation, is more amenable to statistical manipulation, and allows more objective calibration of the pollen data.

Pollen analysis of stratigraphic sections has provided much of what we know about the vegetation history of western Beringia in Siberia. The maps of southwestern Beringia provided by Kuzmin (this volume) are derived from such analyses, showing hardwood forest near modern Vladivostok in the Karginsky interstadial and birch forest on the peninsula that was to become Hokkaido Island changing to treeless tundra in the Sartan glaciation. South of 60° N latitude, the interior of Siberia had a larch–stone pine–tundra vegetation in the last glacial maximum comparable to the modern tundra ecotone of 5°–10° further north (Savvinova et al., this volume; Webb, this volume).

But the matrix of sections in stream bluffs and road cuts is seldom ideal for pollen collection and preservation. Kuzmin notes that in one important stratum of his sections only 34–39 pollen grains and spores were recovered. It is generally accepted that counts of 200–300 pollen grains are necessary to yield reliable results at the level usually allowed by pollen analysts. Even these counts are subsamples of populations of pollen grains numbering in the thousands.

Some of the uncertainties inherent in this geological approach to pollen analysis are eliminated when pollen evidence from numerous sections is combined. In this way the pollen sections from Siberia confirm that the Siberian approaches to the Bering Land Bridge were treeless, just as Hultén had deduced from biogeographical data. What was left was the pollen signature of typical tundra ingredients, such as sedges, grasses, willows, dwarf birches, and heaths. But the treeless vegetation of glacial times also had peculiarities. Notable is the apparently significant presence of the arctic sage *Artemisia*, which suggests a steppe environment such as that which now borders the Siberian tundras to the southwest in the dry heartland of the continent. *Artemisia* and the loess deposits of the yedoma seem to have such a natural affinity that discussion of a vanished "arctic steppe" or "tundra steppe" began to appear in the literature (Tomirdiaro, this volume).

In the eastern parts of Beringia that are now Alaska, pollen analysis started later. Some sections in bluffs and stream banks were examined, as in Siberia—for instance, C. E. Schweger's analyses of the bluffs near Onion Portage described here by Hamilton. But the different tradition of pollen analysis of lake sediments to test biological hypotheses became dominant in Alaska, probably because of the pioneering studies of Daniel Livingstone, who designed specifically for Alaskan pollen analysis the piston sampling equipment that carries his name (Livingstone 1955a).

With only minimal radiocarbon control and Alaskan glacial geology still in its infancy, Livingstone (1955b) treated his pollen diagrams from the Brooks Range and Arctic coastal plain as being restricted to late-glacial and Holocene time, although modern glacial studies in the Brooks Range suggest that his Chandler Lake sequence begins in full glacial times. The broad vegetation changes he identified have since been shown to be typical of the pollen history in lake sediments throughout the Alaskan parts of Beringia (Anderson and Brubaker, this volume; Colinvaux, this volume). An herb zone, with willow (*Salix*) and birch (*Betula*) pollen, attributable to dwarf plants, characterized glacial times; a steep rise in percentage of *Betula* records late-glacial warming; and a zone with *Alnus* or *Picea* records the eventual arrival of forest trees, or, in tundra sites, the approach of distant treelines.

An herb pollen zone, roughly assignable to Livingstone's Zone I, has since been found to characterize Beringian tundras of glacial age at all latitudes in Alaska, from the south land-bridge coast at the Pribilof Islands to the Arctic coast at Point Barrow. Yet most examples have the peculiarity of including prominent percentages of *Artemisia* pollen. *Artemisia* is the genus of sage brush. At least nine species of *Artemisia* are on the Alaskan species list, although the sage brush itself is not. Very likely the *Artemisia* pollen came from an assemblage similar to those from Siberian sections that suggested "steppe" to Russian workers. However, the pollen might as well have come from the tiny *Artemisia globularia* found on bare ground in modern Alaskan montane tundra as from the species of grassland or steppe.

In the long pollen record from Imuruk Lake on Seward Peninsula, *Artemisia* approaches 20 percent of total pollen in an early glacial interval, when it is associated with maximal reduction of pollen of dwarf shrubs and with increased percentages of grass pollen (Colinvaux, this volume). Thirty years after the Imuruk Lake record was published, pollen spectra made up largely of sedge, grass, and *Artemisia* pollen have turned out to be characteristic of glacial age lake sediments from all the Alaskan parts of Beringia (Anderson and Brubaker, this volume). The incidence of *Artemisia* certainly can be associated with the dry conditions suggested by the loess of muck and yedoma. But the actual vegetation that left the pollen signature of these *Artemisia*-rich spectra remains a puzzle. Nothing quite like it exists in the contemporary Arctic.

Pollen analysts still dream of more detailed reconstructions of ancient vegetation from pollen data. Lake sediments provide more or less standard samples of the ancient pollen rain. Theoretically, pollen productivity and dispersal to lake sediments can be measured. Moreover, lake sediments allow good enough radiocarbon control to calculate sedimentation rates for the last 20,000 years or so, which in turn allows calculation of pollen influx rates, taxon by taxon.

However, in the Alaskan Arctic calculations of influx of sedge, grass, and *Artemisia* in the glacial sections are still few. The most dramatic calculations are from sites near Old Crow in the Yukon where L. Cwynar (1982) calculated total pollen influx of full glacial times as being equivalent to that of the stark barren-land tundras of the modern Canadian Arctic archipelago. In Cwynar's view, the high *Artemisia* percentage in these samples was simply a result of sparse sedge and grass pollen production, which left the pollen of *Artemisia* from the few barren-land sage plants like *A. globularia* and *A. frigida* as important ingredients in a much-depleted pollen sum. In short, the vegetation of ice-age Beringia was a polar desert.

The polar desert hypothesis, however, will not explain the *Artemisia*-rich pollen spectra at Imuruk Lake, at the many lakes of the southern Brooks Range studied by Anderson and Brubaker (this volume), nor the Tanglefoot Lakes east of Fairbanks studied by T. Ager (1975). In these sites, *Artemisia* influx was high and could not be explained away by reduction of the productivity of all other taxa. Thus, the task of botanists is to explain the nature of treeless, tundra-like vegetation that grew on the loess and ice surface of Beringia in times of extreme cold and which had populations of *Artemisia* denser than any known in contemporary arctic Alaska.

Anderson and Brubaker (this volume) summarize the effort to match pollen spectra from the principal pollen zones of the southern Brooks Range to surface samples taken as pollen in the mud-water interfaces of a large suite of Alaskan lakes. This work includes the most determined effort yet to identify the vanished Beringian vegetation. The closest matches are on the Alaskan coastal plain—the bleakest, most unproductive tundras in mainland Alaska. These are, of course, places with moist, cold maritime tundras, and the modern spectra do not have the high *Artemisia* percentages of the glacial period. The old Beringian vegetation, therefore, was not quite like that of the coastal plain. The *Artemisia* pollen is consistent with dryness, and the enhanced grass-to-sedge ratios suggest a less tussocky sward. The close fit of these spectra with those of the arctic coast gives a strong hint of a barren unproductive land.

Such is the evidence of pollen analysis, now based on perhaps several dozen lake sections from Alaska. I know of no pollen history that offers a different interpretation. Yet a very different account of the old Beringian enviroment has been developed by the palaeoecological record of large Beringian mammals (Guthrie, Ukraintseva et al., this volume).

Beginning in the 1930s, the Alaskan naturalist Otto Geist collected eight tons of bones from the drainage gates of gold mines in the vicinity of Fairbanks. The collection eventually came into the hands of Dale Guthrie, who presents here his analysis of the species composition of the bones in Table 2–5. Many of the animals appear to have been—at least partially—grazers that required grass in their diets, and their numbers must have been significant because bones of their predators, lions and sabretoothed cats, were included. Guthrie uses this analysis as a basis for his hypothesis that an immense grassland covered Beringia, finding confirmation in the fact that large mammal bones are found in yedoma and frozen silt throughout both western and eastern Beringia. The size of the animals, and their presumed numbers, implies that this grassland must have been highly productive.

A different view of the palaeoecological significance of the Beringian mammal fauna is offered by Ukraintseva, Agenbroad, and Mead (this volume). Their analysis is based on stomach contents of frozen carcasses, as well as on plant fossils in associated sediments. They find that the megafauna of the "mammoth epoch" in Siberia was at its richest in the interstadial period and that stomach contents of animals, particularly mammoths, from the last glacial maximum, show that they were feeding on tundra plants.

Despite this suggestion of a tundra habitat (rather than a grassland) for the mammal megafauna, the difficulty remains of accounting for the nutrient requirements of large animals if Beringia was an unproductive place (Bliss and Richards 1982). Thus the vegetation and environment suggested by mammal remains are quite different from the vegetation and environment reconstructed by all the botanists who have looked at the Alaskan pollen record. This is not the place to try to resolve the argument, though some of the uncertainties on both sides can be pointed out.

Against the botanists it can be said that the interpretation of Beringian herb pollen is most tenuous, for the truth remains that Beringian spectra still have no modern analog. Pollen analysts have not yet measured those ancient Beringian pollen rains in a way that will let them estimate the productivity of the treeless vegetation that the pollen rains represent.

Against the mammalogists' theory it can be said that coexistence of dense herds of fauna is not established, since the frozen bones discovered represent an accumulation of remains from the 70,000 years or so of the last glacial period. In central Alaska, the most provocative fossils, like the frozen bison carcass with lion claw marks, may antedate the local pollen records. The frozen bison, for instance, has a radiocarbon age at the extreme limit of radiocarbon dating (at 36,000 BP), which suggests the Karginsky interstadial, if not before.

Neither side in the debate should wish away the other's data. Increasingly, the pollen records appear to demonstrate that in late Sartan and Itkillik II time the Beringian vegetation was a dry facies version of the highly unproductive tundra of the modern arctic coastal plain. The mammalian fossils also demonstrate that a more productive vegetation should have been available at some time, at least in central Alaska. Final resolution of this dilemma probably requires a detailed chronology of the mammal fossils, coupled with much longer pollen records than we now have.

From the archaeological point of view the issue is surely moot. The more exciting mammal finds are of Karginsky age or older, although mammoths existed at the beginning of the Holocene. If the American beginnings were late glacial, the Beringian people lived long after the episode of lions attacking bison, and in an environment well within the reach of contemporary pollen analysis.

Of more interest to archaeology is the striking event in the pollen diagrams of the late glacial in arctic Alaska: the birch (*Betula*) rise. This is seen in Colinvaux's Figure 2–13 for a lake in a modern tundra setting and in Anderson and Brubaker's Figure 2–9 for a forested region of the southern Brooks Range. In both settings it is best explained as expansion of local tundra dwarf birch populations that had persisted in the tundras of glacial times. The birch rise happened everywhere in Alaska between

12,000 and 13,000 BP, almost certainly in response to the sudden warming that ended the last glaciation and that has been recorded all over the northern hemisphere.

Independent evidence for warming in the 12,000–13,000 year interval comes from Elias's evidence of beetle remains. The modern distributions of carnivorous beetles, particularly ground beetles of the family Carabidae, are narrowly constrained by air temperature. In a classic study in England, G. R. Coope (1977) demonstrated the northward advance of ground beetles in response to warming in the late glacial, well ahead of the signal of climate change in pollen diagrams.

Scott Elias, John Matthews, R .E. Nelson, and others have applied Coope's methods in Alaska. Use of these methods is restricted to deposits that will yield a kilogram of matrix rich in beetle fragments. This means that the methods cannot be applied to lake sediments used in Alaskan pollen analysis, but are applicable only to peat deposits or organic-rich strata in exposures. However, where likely deposits are found, beetles provide one of the most sensitive environmental measures in palaeoecology.

Carabid beetle remains show not only that the late-glacial birch rise was indeed a response to warming but that this warming continued into the early Holocene, when temperatures in arctic Alaska of 10,000 years ago were 3–5° C warmer than present (Elias, this volume). The period 13,000 to 9,000 years ago in Beringia should have included colder intervals associated with the Younger Dryas event but otherwise was not only warmer than any that had gone before in the time span plausible for American beginnings, but was also warmer than any time since.

In this warm window of opportunity, the south land-bridge coast may actually have been partly forested. In Sartan–Itkillik II time of the last glacial, the south coast at the Pribilof Islands was unremittingly cold and bleak, despite its maritime position. But by 10,000 BP the pollen spectra were of a forest ecotone equivalent to Livingstone's Zone III (Colinvaux, this volume). This forest ecotone community had no future, of course, because the rising sea drowned it. Subsequently, the modern maritime tundra of the Pribilofs was established.

The origin of the Pribilof spruce remains one of the puzzles of Beringian reconstruction. Pollen diagrams of central Alaska suggest that spruce arrived in the early- to mid-Holocene, after spreading progressively from populations that may have persisted in glacial times in the delta of the Mackenzie River (Anderson and Brubaker, this volume). Presumably a different source is needed for the south land-bridge spruce population of the late glacial.

In reviewing the palaeoecology of Beringia for any particular purpose it is important to keep time constraints firmly in mind. Beringia existed as a land connection between the continents for most of the two million or so years of the Pleistocene, with only short-lived interruptions of flooding during brief interglacials such as the present one. Geomorphological evidence leaves little doubt that it was a bitterly cold, windswept place, where the wind-borne silt of loess accumulated. But climate changed constantly, even within glaciations. The most prominent changes were perhaps interstadials like the Karginsky. Evidence of shorter intervals of aberrant warmth a few centuries long may have been missed in spite of the present density of sampling in this huge area. Once there were significant numbers of big mammals in parts of Beringia; we do not yet know for what length of time or how local were their populations. The big mammals are probably of interest to anthropology only if people can be shown to have lived in eastern Beringia in Karginsky time or earlier. The Beringia of 14,000 to 9,000 years ago, though often warmer than its contemporary remnants, was a place of shrub tundras, providing more bushes than contemporary tundras but not significantly more productive.

REFERENCES CITED

Ager, T. 1975. Late Quaternary Environmental History of the Tanana Valley, Alaska. Ph.D. dissertation, Ohio State University.

Bliss, L. C., and J. H. Richards. 1982. Present Day Arctic

Vegetation and Ecosystems as a Predictive Tool for the Arctic-Steppe Mammoth Biome. In *Palaeoecology of Beringia*, ed. D. M. Hopkins, J. V. Matthews, Jr., C. E. Schweger, and S. B. Young, pp. 241–57. New York: Academic Press.

Coope, G. R. 1977. Fossil Coleopteran Assemblages as Sensitive Indicators of Climatic Changes during the Devensian (Last) Cold Stage. *Philosophical Transactions of the Royal Society of London*, Ser. B, 280:313–40.

Cwynar, L. C. 1982. A Late Quaternary Vegetation History from Hanging Lake, Northern Yukon. *Ecological Monographs* 52:1–24.

Hopkins, D. M. 1959. Cenozoic History of the Bering Land Bridge. *Science* 129:1519–28.

Hultén, E. 1937. *Outline of the Arctic and Boreal Biota during the Quaternary Period.* Stockholm: Borförlag Aktiebolaget Thule.

Livingstone, D. A. 1955a. A Lightweight Piston Sampler for Lake Deposits. *Ecology* 36:137–39.

———. 1955b. Pollen Profiles from Arctic Alaska. *Ecology* 36:587–600.

Péwé, T. L. 1955. Origin of the Upland Silt near Fairbanks, Alaska. *Geological Society of America Bulletin* 66:699–724.

Shilo, N. A. 1964. About the Development History of Northeastern Asian Subarctic Belt Downs. Transactions of the USSR Academy of Sciences, Magadan, Vol. 11, pp. 154–69. (In Russian)

Tomirdiaro, S. V. 1980. *Loess-Glacial Formation of Eastern Siberia in the Late Pleistocene and Holocene.* Moscow: Nauka. (In Russian)

Westgate, J. A. 1982. Discovery of a Large-Magnitude, Late Pleistocene Volcanic Eruption in Alaska. *Science* 328:789–90.

Chapter 1

GEOLOGICAL RECORDS

LATE PLEISTOCENE STRATIGRAPHIC SECTIONS FROM NORTHERN ALASKA

Thomas D. Hamilton

EPIGURUK

A conspicuous river bluff locally known as *Epiguruk* (an Inupiat term meaning "big bank") extends for about 3.5 km along a meander bend of the Kobuk River about 170 km inland from Kotzebue Sound (Fig. 1–1). Epiguruk is flanked to the south and west by the active Kobuk sand dunes and to the north by the Brooks Range. The bluff also is near the Onion Portage archaeological site (D. D. Anderson 1970, 1988; Schweger 1985; see also this volume), which contains one of the oldest known artifact assemblages in Alaska. Epiguruk varies in height from 12 m at its downstream end to more than 30 m in central and upstream portions (Fig. 1–2). It contains an unusually rich record of interacting fluvial, aeolian, and soil-forming environments that fluctuated in response to climatic changes during the late Quaternary (Hamilton and Ashley 1993; Ashley and Hamilton 1993). It also contains an abundant fossil record (Hamilton et al. 1993).

History of Research

Epiguruk was first studied in 1952 by A. T. Fernald (1964), who measured a section near its upstream end. C. E. Schweger subsequently mapped the northern 900 m of the bluff, dated its major units by radiocarbon, and analyzed pollen from four measured sections (Schweger 1976, 1982). In 1981 and 1982, Hamilton and others (1984) mapped the principal sedimentary units that occur in the northern 2,600 m of the bluff and collected radiocarbon (Hamilton and Ashley 1993), fossil (Hamilton et al. 1993), and sediment samples (Ashley and Hamilton 1993).

Stratigraphy and Dating

The soils and sediments at Epiguruk can be grouped into seven major stratigraphic units (Fig. 1–2 and Table 1–1).

Basal Sand and Silt (Unit 1) This unit is locally exposed at the base of the bluff. Fine sand and organic-rich silt predominate, and deformation structures suggest severe frost activity above a former permafrost table. The sediments are interpreted as probably floodplain deposits of the Kobuk River.

The Lower Palaeosol (Unit 2) The lower palaeosol is a compact, dark-colored, peat and humic sand

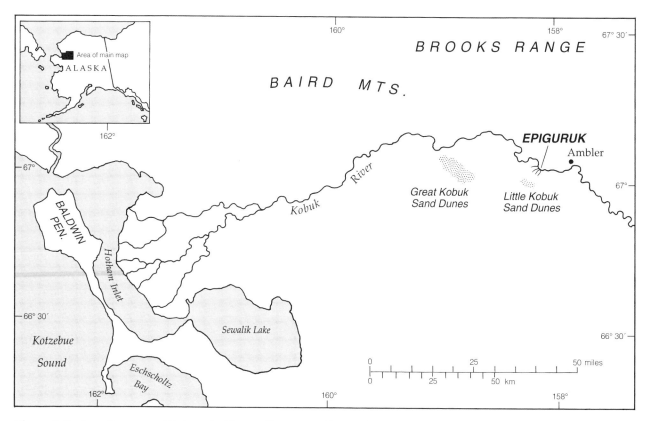

Figure 1–1   Location map of Epiguruk. (*Source:* Quaternary Research. *See Acknowledgments.*)

1.5–2.0 m thick. Between 1,960 and 1,500 m (see base-line scale on Fig. 1–2) the palaeosol surface forms a regular series of mounds with intervening trenches that are identical to the thermokarst mounds in Alaska and Siberia that result from the thaw of polygonal ice-wedge systems in permafrost (Péwé 1954; Czudek and Demek 1970). Radiocarbon age determinations suggest that the palaeosol probably is older than about 44 ka (thousand years BP).

The Lower Alluvial-Aeolian Complex (Unit 3)
This unit was formed subsequently by aggradation of fluvial and aeolian sediments. The base of this unit locally is a loess-like, structureless, very fine aeolian sand. Overlying cross-bedded, medium-to-fine sand represents channel and overbank deposits that formed during alluviation of the Kobuk River. The fluvial section commonly is overlain by thick, windblown sand and locally by lens-shaped pond deposits. Two radiocarbon ages suggest that alluviation began before 39 ka (Table 1–2).

Table 1–1   Major Stratigraphic Units at Epiguruk (From Hamilton et al. 1993. *See Acknowledgments.*)

Unit Designation (Number and Name)	*Changes in River Level*	*Ages (thousand ^{14}C yr BP)*
7. Surface soil	Stable	3.6–0
6. Holocene alluvialaeolian complex	Downcutting, then local alluviation	10.5–4
5. Late Wisconsinan	Alluviation	24–10.5
4. Upper palaeosol	Downcutting, then stability	33–24
3. Lower alluvialaeolian complex	Alluviation	>44–35
2. Lower palaeosol	Downcutting, then stability	>44
1. Basal sand and silt	Alluviation	>44

The Upper Palaeosol (Unit 4)   The upper palaeosol is a compact, resistant, organic soil 1.5–6 m thick that was formed primarily by accretion of organic and aeolian deposits. Thick peaty or silty facies formed under relatively wet conditions in low-lying sites; sandy soils developed on higher and drier sites. The palaeosol is locally deformed by solifluction structures (e.g., asymmetric and recumbent folds) and by collapse into ice-wedge voids. Eleven radiocarbon ages (documented in Hamilton and Ashley 1993) provide a generally concordant record of soil development. Fluvial interbeds at the base of a palaeosol section near river level contain wood and bones dated at about 34.6 to 32.8 ka; other ages on upland facies of the palaeosol are 34.9 to 33.7 ka. Soil formation near river level ceased by 24 ka when the alluviating Kobuk River flooded the site.

The Late Wisconsinan Alluvial-Aeolian Complex (Unit 5)   This unit consists dominantly of fluvial and aeolian sand about 5 to 10 m thick that formed during renewed alluviation of the Kobuk River. Floodplain deposits typically form laminated beds that contain abundant windblown sand (Ashley

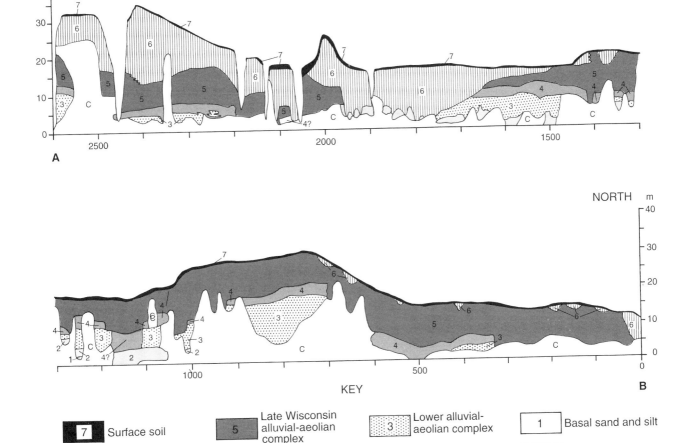

Figure 1–2   Stratigraphic section at Epiguruk. Numbers at base of section are distances in meters along base line. Vertical exaggeration 8x. (*Source:* Geological Society of America Bulletin. *See Acknowledgments.*)

Table 1–2 Selected Radiocarbon Ages from Epiguruk. Ages marked with Asterisk (*) Are on Vertebrate Remains or Associated Organic Matter. (From Hamilton and Ashley 1993. *See Acknowledgments.*)

Age	Laboratory and No.	Material Dated	Stratigraphic and Base-line Positions	Comments
3650 ± 90	I-13,044	Wood (*Salix*)	Basal part of unit 7; 2100 m	
8635 ± 210	GX-1442	Wood fragments	Near base of Holocene channel; 000 m	From Schweger (1976, 1982)
8800 ± 210	GX-1445	Twigs, detrital	Near base of Holocene pond deposit; 650 m	From Schweger (1976, 1982)
10,370 ± 150	I-12,438	Wood (*Salix*)	Base of Holocene channel; 1900 m	
*14,130 ± 60	USGS-1658	Wood (*Salix*)	Unit 5; local channel margin (+17 m); 955 m	Rooted shrubs. Associated with *Mammuthus*
14,980 ± 200	I-12,437	Wood fragments (*Salix*)	Unit 5 (+10 m); 095 m	Rooted shrubs on point bar
17,440 ± 280	I-13,091	Wood (*Salix*)	Unit 5 (+10 m); 100 m	Rooted shrubs on point bar
18,100 ± 550	I-4776	Twigs, detrital	Unit 5 (solifluction facies); 650 m	From Schweger (1976, 1982)
18,250 ± 350	I-13,093	Wood (*Salix*?)	Unit 5 (+6 m); 100 m	Rooted shrubs on point bar
18,500 ± 320	I-13,209	Wood (*Salix*)	Unit 5 (+14.5 m); 1475 m	Rooted shrubs
*18,520 ± 120	USGS-1451	Wood fragments	Unit 5 (+13 m); 1142 m	Associated with *Mammuthus*
*18,560 ± 70	USGS-1485	Bone (*Mammuthus*)	Unit 5 (+13 m); 1142 m	Fibula
*18,970 ± 170	USGS-1441	Peat	Unit 5 (+13 m); 1170 m	Associated with *Mammuthus*
*19,060 ± 90	USGS-1439	Bone (*Mammuthus*)	Unit 5 (+13 m); 1170 m	Portion of rib
19,900 ± 170	USGS-1446	Wood (*Salix*)	Unit 5 (+12 m); 1960 m	Rooted shrubs
20,290 ± 90	USGS-1449	Wood (*Salix*?)	Unit 5 (+12 m); 1200 m	Rooted shrubs
20,870 ± 190	USGS-1655	Wood (*Salix*)	Unit 5 (+9 m); 1960 m	Rooted shrubs
21,020 ± 460	I-13,208	Wood (*Salix*)	Unit 5 (+13.2 m); 1320 m	Rooted shrubs
21,450 ± 230	USGS-1653	Peat	Basal part of unit 5 (+8 m); 2404 m	
22,770 ± 100	USGS-1445	Peat	Contact between units 4 and 5 (+6 m); 450 m	Interfingering relationship
22,970 ± 230	USGS-1663	Wood (*Salix*)	Lower unit 5 (+8 m); 450 m	Detrital fragments
*23,560 ± 160	USGS-1442	Wood fragments (*Salix*)	Channel fill at top of unit 4 (+9 m); 1540 m	Mostly *in situ* roots. Associated with *Mammuthus* (USGS-1438)
*23,620 ± 110	USGS-1438	Bone (*Mammuthus*)	Channel fill at top of unit 4; 1540 m	Portion of jaw
24,050 ± 620	I-12,471	Wood (*Salix*)	Basal part of unit 5 (+8 m); 600 m	Rooted shrubs
24,290 ± 720	GX-1446	Wood fragments	Basal part of unit 5 (+6 m); 600 m	From Schweger (1976, 1982)
30,650 ± 1200	I-13,043-C	Peat and wood fragments	Unit 4 (clast in unit 5a); 2150 m	Reworked from palaeosol
32,830 ± 730	USGS-1661	Wood (*Salix*) and peat	Alluvium near base of unit 4; 410 m	Detrital
*33,670 ± 280	USGS-1443	Twigs (*Salix*)	Alluvium near base of unit 4; 2250 m	Detrital. Associated with *Mammuthus*
33,690 ± 960	USGS-1659	Peat and wood fragments	Within unit 4; 725 m	
*34,620 ± 560	USGS-1440	Bone (*Mammuthus*)	Alluvium near base of unit 4; 2250 m	Portion of rib
34,900 ± 320	USGS-1444	Wood (*Salix*)	Within unit 4; 2585 m	
>39,000	USGS-1726	Woody peat	Unit 3; 2585 m	
39,500 ± 1200	USGS-1727	Woody peat	Unit 3; 2585 m	
43,900 ± 1600	USGS-1730	Woody peat	Near top of unit 2; 1750 m	Very dense, compact peat

and Hamilton 1993). Much of the bluff is mantled by a layer of faintly laminated loess 3 to 9 m thick. Between about 685 and 540 m, where the loess was redeposited by solifluction on a palaeoslope, it thickens to a maximum of 11.5 m, is darker and more humic than normal loess, and contains fold and shear structures resulting from downslope flowage.

Thirty-five radiocarbon ages, more than half of them on rooted willow shrubs (Hamilton and Ashley 1993), demonstrate that this phase of alluviation of the Kobuk River took place during late Wisconsinan time. The beginning of the alluvial episode may be indicated indirectly by loess that began to cover the upper palaeosol about 24 ka. Fluvial deposits of the aggrading Kobuk River built up to about 6 m above present river level by 22.8 ka and to 12 m height about 20 ka. The river evidently stood at a maximum height of about 13 m above its present level for the next 1,500 years, then it downcut abruptly around 18.4 ka. A second episode of alluviation, during which the river evidently meandered across the north end of the bluff while aggrading to about 10 m in height, lasted from about 18 to 15 ka.

The Holocene Alluvial-Aeolian Complex (Unit 6)   The Holocene complex consists of several deep channel fillings, two dune-like deposits of aeolian sand, and local pond sediments. A channel complex between about 2,100 and 1,500 m was incised nearly to present river level about 10.4 ka. The channels have sharp, erosional, lower contacts and are filled with rippled, cross-bedded sand that contains abundant fragments of detrital wood. A separate channel in the Holocene floodplain at the north end of the bluff had incised to near modern river level by 8.1 ka.

The two dune-like deposits are of distinctive yellowish aeolian sand and are cut by periglacial sand wedges. The larger sand body, near the south end of the bluff, is up to 17 m thick and contains abundant accretionary mounds that probably formed around clumps of shrubby vegetation. The smaller sand deposit, between 2,040 and 1,960 m, is up to 12 m thick and has distinctive dunal surface relief. Pond deposits 1 to 4 m thick occur within the loess along the top of the bluff north of 710 m. One pond has a basal radiocarbon age of 8.8 ka.

The Surface Soil (Unit 7)   The surface soil unit of sandy peat and sod interlaced with spruce roots forms a cohesive mat 0.3 to 1.0 m thick along the entire length of Epiguruk. This unit has a basal radiocarbon age of 3.6 ka.

Flora and Fauna

Vertebrate remains from Epiguruk include mammoth, bison, caribou, an equid, a canid, arctic ground squirrel, and voles (Hamilton et al. 1993). Radiocarbon ages on mammoth bones, confirmed by concordant ages on associated peat and wood, show that mammoths inhabited the Epiguruk area around 34 ka, 23.5 ka, 19 ka, and 14.1 ka. Radiocarbon ages on rooted *Salix* (willow) shrubs and grasses confirm that the Epiguruk area supported a rich riparian vegetation through the full-glacial phase of the late Wisconsin.

The pollen record from the upper palaeosol shows Cyperaceae (sedges) dominant, with *Salix*, *Betula* (birch), *Artemisia* (sage), and Gramineae (grasses) also present. A closely similar pollen record was obtained from late Wisconsinan floodplain sediments in the bluff that have an estimated age of 18 ka. Two pollen profiles from Holocene deposits show basal assemblages like those of the late Pleistocene followed by successive increases in *Betula*, *Alnus* (alder), and *Picea* (spruce).

Discussion

The two major palaeosols at Epiguruk probably represent cool, moist, peat-forming interstadial intervals during the last (Wisconsin) glacial stage. Absence of wood from trees or large shrubs suggests that neither palaeosol is as old as the last interglacial maximum (marine oxygen–isotope substage 5e) because trees and large shrubs extended to or beyond their present limits in northwestern Alaska at that time (Hamilton and Brigham-Grette 1992). During both soil-forming intervals, bogs and marshes were common, aeolian activity diminished, vegetation cover was continuous, and permafrost was widespread. The lower palaeosol probably is older than about 44 ka. Subsequent alluviation must have begun sometime prior to 39 ka and terminated about 34 ka when the upper palaeosol began to form.

Thick peat accumulation and the growth of large woody shrubs in the upper palaeosol began between 34 and 33 ka, a time of general interstadial conditions in northern and central Alaska. For example, peat and large ice wedges formed during a hiatus in loess deposition about 35 to 31 ka near Fairbanks (Hamilton et al. 1988a). Alluvium, directly beneath till and outwash of the late Wisconsinan glacial advance in the Noatak River valley, has similar radiocarbon ages of about 35 to 30 ka (see Noatak section following).

The last major alluviation of the Kobuk River began about 24 ka, synchronous with reactivation of at least part of the Kobuk sand dunes (Hamilton et al. 1988b). These events coincide with the independently dated beginning of the late Wisconsinan (Itkillik II) glaciation of the Brooks Range (Hamilton 1982, 1986) and with alluviation in the upper Kobuk and Koyukuk river systems (Hamilton 1982). Pollen data of P. M. Anderson (1988) show decrease in *Populus* about 27 ka and in both *Betula* and total pollen after 26 ka, suggesting onset of conditions colder and (or) drier than those of middle Wisconsinan time.

Downcutting by the Kobuk River about 18.5 ka has no clear analogs in the known geologic record from eastern Beringia, but pollen data from northeastern Alaska and the northern Yukon show a change to more continuous and more diverse tundra cover about 18.5 to 18.0 ka (Cwynar 1981; P. M. Anderson et al. 1988). This change probably represents greater winter precipitation, possibly combined with warmer summer temperatures.

The Kobuk River meandered at a level 5 to 10 m above present until some time after 15 ka and had cut down close to its present surface by 10.3 ka. This record corresponds generally to alluvial history of the upper Koyukuk River valley and to deglaciation records of the Brooks Range (Hamilton 1982) and the north-central Alaska Range (Ten Brink and Waythomas 1985), where glacier retreat and ice disintegration were nearly complete by about 11.8 ka. An abrupt transition from herbaceous tundra to birch-dominated shrub tundra began about 13.5 ka in northwestern Alaska, triggered by a major change to warmer and perhaps moister summers (Barnosky et al. 1987; P. M. Anderson 1988). The Kobuk dunes may have become largely stabilized by vegetation in response to increasing moisture at this time and consequently would have furnished less sediment to the Kobuk River.

Holocene peat accumulation began about 3.6 ka at Epiguruk, synchronous with peat formation at Onion Portage and at other sites in northern and central Alaska (Hamilton and Robinson 1977) and with heightened storm activity on the Seward Peninsula (Mason 1989). Radiocarbon ages from the central Brooks Range show that alluviation of upper mountain valleys by outwash from expanding cirque glaciers also began about 3.6 ka (Hamilton 1981a, 1986).

The vertebrate and pollen records from Epiguruk, together with the abundant willow shrubs, suggest that the central Kobuk River valley supported a rich riparian flora and fauna during the height of the last glaciation as well as during the interstadial and late-glacial intervals that preceded and followed it. These records contrast with the sparse mammal remains and other evidence for severe aridity during late Wisconsinan time on the Alaskan North Slope (Carter 1981, 1983; Guthrie and Stoker 1990) and in the northern Yukon (Cwynar and Ritchie 1980; Cwynar 1982). The central Kobuk River valley, which was situated close to the eastern end of the Bering Land Bridge, may have provided an unusually favorable environment for the early influx of humans into North America.

NOATAK MORAINE BELT SITE

A late Pleistocene glacier that filled the upper Noatak River valley of northwestern Alaska terminated on the valley floor near Douglas Creek (about 67°50′N, 156°45′W), forming a prominent end moraine (shown as Unit id$_2$ on Fig. 1–3). The Noatak River cuts through the end moraine, associated outwash, and moraine-dammed lake deposits, which are exposed in a series of bluffs along the north side of the river. Radiocarbon dates from this set of bluffs provide the best bracketing ages presently available on the last major glaciation of northern Alaska, which has been termed the Itkillik II ice advance (Hamilton and Porter 1975) and is broadly equivalent to late Wisconsinan glaciation (Hamilton 1982, 1986).

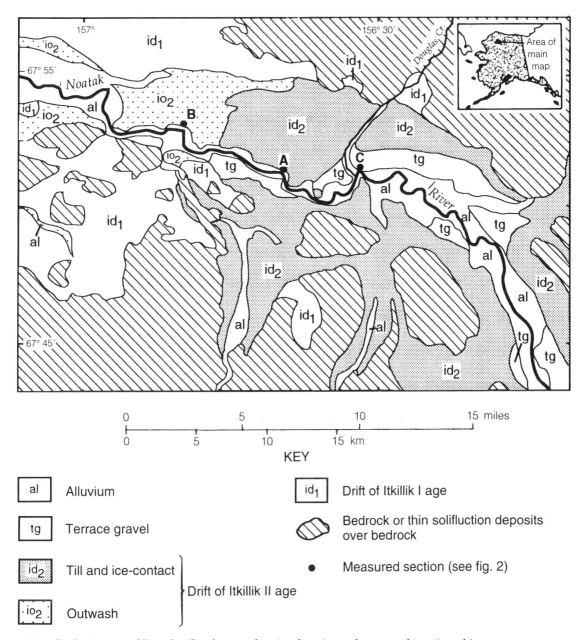

Figure 1–3 Geologic map of Douglas Creek area, showing locations of measured stratigraphic sections. (*Modified from Hamilton et al. 1987.*)

History of Research

The bluff exposures were measured and sampled in 1982 and 1983 during helicopter-supported field mapping of the upper Noatak valley by the U.S. Geological Survey (Hamilton 1981b, 1984). The following summary is a condensed version of a final report on the bluffs and their glacial record by Hamilton et al. (1987).

Stratigraphy and Dating

Three prominent river bluffs, designated A, B, and C, respectively, intersect the end-moraine belt, the outwash apron that formed in front of it, and a lake basin that subsequently developed as the glacier retreated upvalley. The bluffs range in height from 18 to 70 m; they expose glacial and interstadial diamicton, alluvium, outwash, and lacustrine sediments

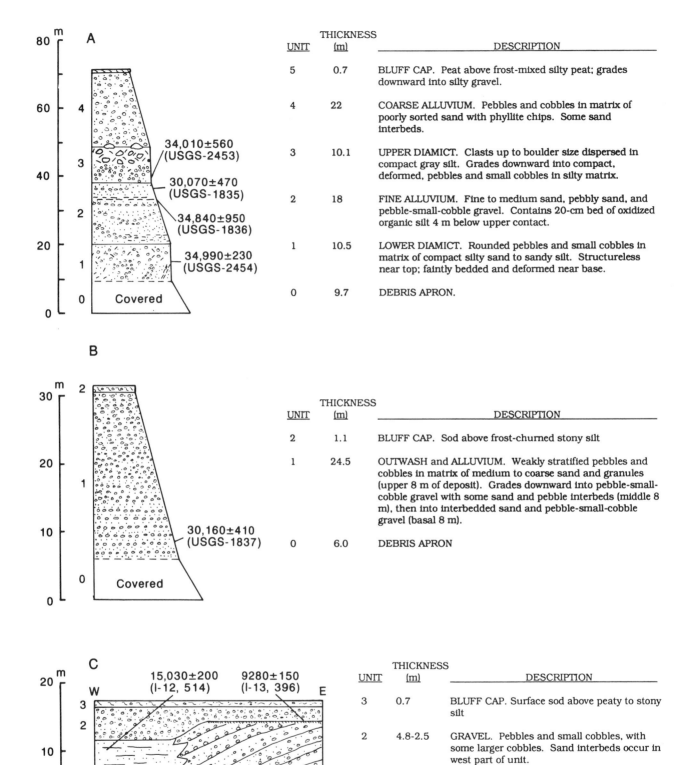

THICKNESS

UNIT	(m)	DESCRIPTION
5	0.7	BLUFF CAP. Peat above frost-mixed silty peat; grades downward into silty gravel.
4	22	COARSE ALLUVIUM. Pebbles and cobbles in matrix of poorly sorted sand with phyllite chips. Some sand interbeds.
3	10.1	UPPER DIAMICT. Clasts up to boulder size dispersed in compact gray silt. Grades downward into compact, deformed, pebbles and small cobbles in silty matrix.
2	18	FINE ALLUVIUM. Fine to medium sand, pebbly sand, and pebble-small-cobble gravel. Contains 20-cm bed of oxidized organic silt 4 m below upper contact.
1	10.5	LOWER DIAMICT. Rounded pebbles and small cobbles in matrix of compact silty sand to sandy silt. Structureless near top; faintly bedded and deformed near base.
0	9.7	DEBRIS APRON.

Section A dates: 34,010±560 (USGS-2453); 30,070±470 (USGS-1835); 34,840±950 (USGS-1836); 34,990±230 (USGS-2454)

THICKNESS

UNIT	(m)	DESCRIPTION
2	1.1	BLUFF CAP. Sod above frost-churned stony silt
1	24.5	OUTWASH and ALLUVIUM. Weakly stratified pebbles and cobbles in matrix of medium to coarse sand and granules (upper 8 m of deposit). Grades downward into pebble-small-cobble gravel with some sand and pebble interbeds (middle 8 m), then into interbedded sand and pebble-small-cobble gravel (basal 8 m).
0	6.0	DEBRIS APRON

Section B date: 30,160±410 (USGS-1837)

THICKNESS

UNIT	(m)	DESCRIPTION
3	0.7	BLUFF CAP. Surface sod above peaty to stony silt
2	4.8-2.5	GRAVEL. Pebbles and small cobbles, with some larger cobbles. Sand interbeds occur in west part of unit.
1	12.0-14.3	LACUSTRINE BEDS. Steeply dipping deltaic sand and fine gravel; grades westward (downvalley) into horizontally bedded fine sand, silty fine sand, and clayey silt.

Section C dates: 15,030±200 (I-12, 514); 9280±150 (I-13, 396)

Figure 1–4 Measured stratigraphic sections with radiocarbon dates, Noatak River near Douglas Creek. Radiocarbon laboratories: I = Teledyne Isotopes, Westwood, N. J.; USGS = U.S. Geological Survey, Menlo Park, Calif. (*Modified from Hamilton et al. 1987.*)

and are capped by about a meter of Holocene sod, peat, and frost-churned stony silt.

Exposure A   Exposure A (Fig. 1–4a), a south-facing, 70-m bluff, intersects the end moraine about 5.5 km below Douglas Creek (Fig. 1–3). The lowest stratigraphic unit (unit 1) is a compact diamicton which forms vertical buttresses near the base of the section. Fine (pebble to small cobble) gravel occurs in a matrix of silty sand. The upper part of the deposit is structureless, but overturned folds and a near-vertical, 30-cm bed of deformed peaty silt are present at greater depth. The peaty silt bed has an apparent radiocarbon age of 34,990 ± 230 yr BP.

Overlying alluvium (unit 2) consists of parallel-bedded to cross-bedded sand, pebbly sand, and fine gravel. Beds of oxidized organic silt with *in situ* rootlets 4 m and 2 m below the top of the alluvium are dated at 34,840 ± 950 BP and 30,070 ± 470 BP, respectively. Gravel beds at greater depth dip upvalley at angles of 12–34°, then level out and fine laterally to form a palaeobasin floor composed of muddy sand. This closed depression within the gravel could have resulted from continued melt of buried glacier ice during deposition of unit 2.

The upper diamicton (unit 3) is interpreted as massive till above redeposited river gravel. Unsorted, round to subrounded, striated clasts up to boulder size dispersed in compact, grey silt grade downward at about 3.5 m depth into compact fine gravel with deformed bedding that resembles the diamicton of unit 1. Eroded fragments of peaty silt incorporated in the diamicton 70 cm above its base are dated at 34,010 ± 350 BP, and evidently were redeposited from unit 2.

Unit 4 consists of rounded to subrounded alluvial pebbles and cobbles in poorly sorted matrix of sand with phyllite chips. The alluvium contains several near-horizontal sand beds 10–30 cm thick that are continuous laterally for at least 20 m. A lag concentration of boulders up to 60 cm diameter occurs at the base of the alluvium and on the upper surface of the underlying diamicton.

Exposure B   Exposure B (Fig. 1–4b), 13 km down-valley from Douglas Creek, intersects the outwash terrace that originated at the end moraine. The terrace surface is planar near the exposure, but closer to the end moraine it degenerates into kame-and-kettle topography where the outwash evidently overlapped buried glacier ice. Exposure B consists of alluvium which coarsens progressively upward into glacial outwash. In the basal part of the exposure, beds of sand alternate with beds of sandy fine gravel. The central part of the section consists of faintly bedded rounded pebbles, small cobbles, and sparse medium cobbles, with some interspersed beds of coarse sand and muddy gravel. Pebbles, small to medium cobbles, and sparse larger cobbles predominate in the upper part of the deposit. Progressive changes in cobble lithology show increasing dominance of upvalley sediment sources upward in the section: locally derived carbonate rocks decrease, granitic rocks from glacial source areas near the valley head increase. A bed of silty alluvium in the fine preglacial gravel 8.5 m above the river level contains sparse rootlets and organic detritus that are dated at 30,160 ± 410 BP.

Exposure C   Exposure C (Fig. 1–4c), near the mouth of Douglas Creek, intersects an 18-m alluvial terrace that is inset within the end-moraine belt. The eastern (upvalley) part of the exposure is dominated by deltaic foreset beds of sand and fine gravel that dip downvalley at angles as steep as 26°. Near the center of the exposure, the deltaic sediments grade into grey, horizontally bedded, well-sorted fine sand with some interspersed beds of clayey silt to silty fine sand. Near the west end of the bluff, dark grey organic silt grades upward into sandy fine gravel. Small wood fragments and rootlets from along bedding planes 1.4 m below the top of this unit are dated at 15,030 ± 200 BP. Unit 2 is fluvial gravel composed of pebbles, small cobbles, and some larger cobbles in a sparse sandy matrix. The gravel thickens downvalley, where an inactive fan of Douglas Creek is graded to the gravel terrace. A bed of black *in situ* peat 2–8 cm thick at the contact between the gravel and the underlying deltaic beds is dated at 9280 ± 150 BP. The peat extends through the eastern part of the bluff, indicating that deltaic sediments were exposed subaerially for some time before the Noatak River overran the site.

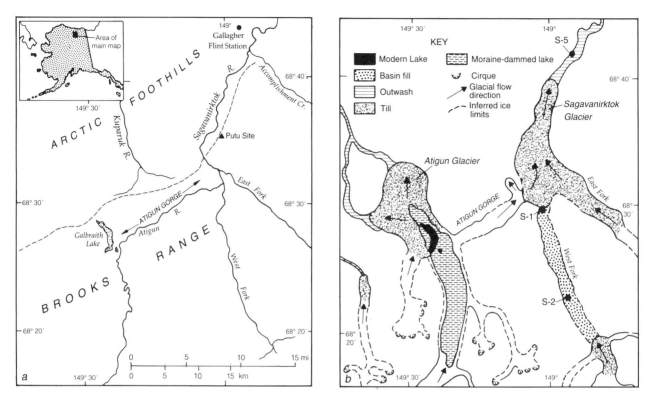

Figure 1–5   Atigun–Sagavanirktok rivers area. (*a*) Modern lakes and major drainages. Dashed line separates Brooks Range from arctic foothills. (*b*) Reconstruction of late Itkillik II glacier and related deposits.

Discussion

Radiocarbon dates and stratigraphic relations in the three bluff exposures demonstrate that the end moraine near Douglas Creek is of late Wisconsinan age and is assignable to the Itkillik II glacial phase as defined by Hamilton and Porter (1975) and dated by Hamilton (1982, 1986). Deposition of fine alluvium prior to the ice advance began some time before about 34.8 ka and continued until some time after 30 ka. Subsequent expansion of the Itkillik II glacier is recorded by coarsening-upward sediments with increasing upvalley lithologies at site B and by the upper diamicton at site A. The glacier initially deformed and redeposited preexisting alluvium from the valley center, then later deposited massive, bouldery silt-rich till at the site. A minimum age for the beginning of deglaciation is provided by the date of about 15 ka on lake sediments that formed as the glacier retreated from its end moraine.

The older (lower) diamicton at Section A ap-

pears to have been deformed by glacier overriding. If this interpretation is correct, then the radiocarbon age of $34{,}990 \pm 230$ BP on this unit very likely is too young. The remaining four radiocarbon ages of 34.8 to 30.1 ka from Sections A and B were all obtained from interstadial sediments, and similar age ranges have been reported from Epiguruk and the Fox permafrost tunnel near Fairbanks (Hamilton et al. 1988a). These dated deposits indicate that a mild interstadial interval occurred throughout northern and central Alaska prior to late Wisconsinan glacial expansion.

UPPER SAGAVANIRKTOK VALLEY SITE

The Sagavanirktok River originates in the Brooks Range and flows north to enter the Arctic Ocean at Prudhoe Bay. Near the north flank of the range, the Sagavanirktok is joined by the Atigun River (Fig. 1–5a), which was dammed by a massive moraine complex near Galbraith Lake (shown as "Till" on Fig. 1–5*b*) and deflected eastward through Atigun Gorge. Farther north, the Sagavanirktok River crosses a complex of moraines and related glacial deposits of the late Pleistocene Itkillik glaciation (Hamilton 1978, 1986). The youngest (southernmost) moraine of the Itkillik complex was formed by a glacial readvance of late Itkillik II age (Hamilton 1986). Glaciers advanced to the range front in many of the larger drainages at that time, but elsewhere they terminated deep within mountain valleys (Hamilton 1986).

During the late Itkillik II readvance, a large ice stream flowed north for about 50 km down the Atigun River valley and terminated just north of the present Galbraith Lake (Fig. 1–5*b*). A distributary ice stream flowed eastward through Atigun Gorge, then spread out to block what is known as the "west fork" of the Sagavanirktok River valley. Glaciers near the head of the west fork were surprisingly restricted at this time; they extended only 15–25 km from their cirques and formed an end moraine about 30 km south of Atigun Gorge.

History of Research

Stratigraphic sections within the Sagavanirktok Valley were studied initially in 1972 in collaboration with E. James Dixon (Anthropology Department, University of Alaska). A broader investigation of local Quaternary geology later was carried out as part of a regional mapping project of the U.S. Geological Survey (Hamilton 1978, 1979).

Stratigraphy and Dating

Three exposures along the Sagavanirktok River provide radiocarbon ages that closely date the Itkillik II readvance (Fig. 1–6). These exposures intersect outwash of that readvance and deposits dammed by the Atigun ice lobe.

Exposure S-1   Exposure S-1 is located where the ice stream that flowed eastward through Atigun Gorge blocked the Sagavanirktok's west fork (Fig. 1–5*b*). Most of the 25-m-thick sediment sequence at this locality was formed by accumulation of sandy alluvium in a basin that formed behind the dam of glacial ice and debris. Units 1, 2, and lower unit 3

were formed during rapid alluviation of channel deposits along the west fork. Fluvial features include layers and lenses of rounded and well-sorted pebbles, detrital wood deposited along bedding planes, and lateral facies changes from relatively coarse bar deposits to relatively fine-grained channel fillings. Absence of weathering and scarcity of fine-grained overbank deposits indicate that alluviation probably proceeded without major pauses or reversals. Upper unit 3 and unit 4 reflect decreasing rates of sedimentation, probably as glacial activity in Atigun Gorge began to wane. Rooted plant remains become increasingly abundant upward, and the uppermost 3 m of unit 4 consists of massive peat. Four radiocarbon ages show that rapid alluviation took place about 12.8 ka and continued at a decreasing rate until about 11.4 ka. By about 9 ka, the Sagavanirktok River no longer was alluviating, and sediments above the 20-m level in the bluff consist almost entirely of highly organic aeolian sand and silt.

Exposure S-2   Located 13 km above the mouth of Atigun Canyon, Exposure S-2 contains fluvial sand at its base (unit 1) that probably correlates with the basin-filling sand of upper unit 3 or lower unit 4 at exposure S-1. The radiocarbon age of $11,760 \pm 200$ BP is close to the oldest date from unit 4 of exposure S-1, which seems to confirm that slow alluviation of the upper Sagavanirktok River behind the moraine dam continued until about 11.4 ka. Absence of gravel suggests that outwash deposition from the upper Sagavanirktok Valley glacier either had ceased or was unable to extend this far north in the sand-filled depositional basin. A sharp unconformity at the top of unit 2 probably marks downcutting of the Sagavanirktok River after breaching of the moraine dam near the mouth of Atigun Gorge. The date of this event is uncertain. Later accretion of aeolian sand began sometime before 2.3 ka and continues to the present. An undated archaeological site at 1.3 m depth in the aeolian sand consists of a compacted occupation surface littered with fire-cracked rocks, river cobbles, broken caribou bones, charcoal fragments, and sparse chert chips.

Exposure S-5   Exposure S-5 intersects an outwash terrace, which is traceable for 6 km upvalley into

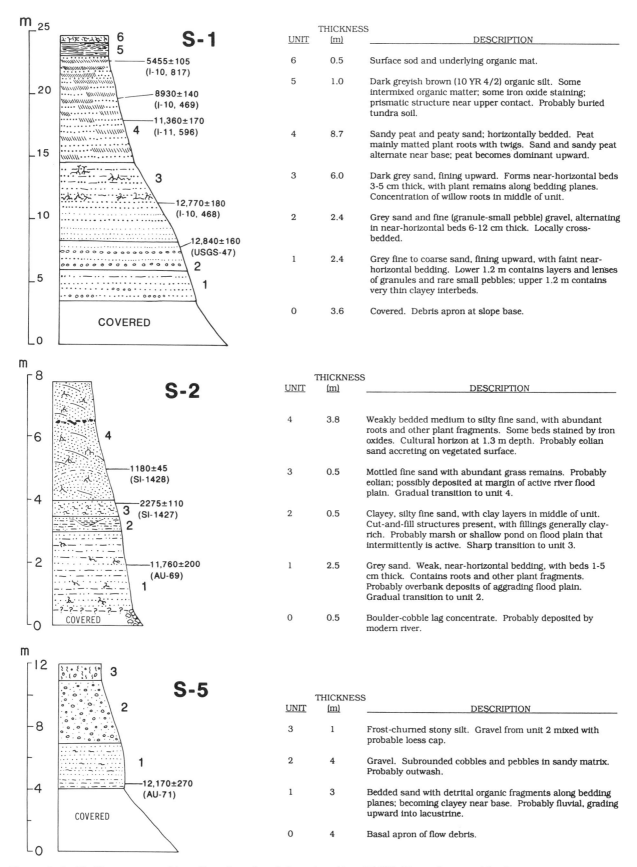

S-1

UNIT	THICKNESS (m)	DESCRIPTION
6	0.5	Surface sod and underlying organic mat.
5	1.0	Dark greyish brown (10 YR 4/2) organic silt. Some intermixed organic matter; some iron oxide staining; prismatic structure near upper contact. Probably buried tundra soil.
4	8.7	Sandy peat and peaty sand; horizontally bedded. Peat mainly matted plant roots with twigs. Sand and sandy peat alternate near base; peat becomes dominant upward.
3	6.0	Dark grey sand, fining upward. Forms near-horizontal beds 3-5 cm thick, with plant remains along bedding planes. Concentration of willow roots in middle of unit.
2	2.4	Grey sand and fine (granule-small pebble) gravel, alternating in near-horizontal beds 6-12 cm thick. Locally cross-bedded.
1	2.4	Grey fine to coarse sand, fining upward, with faint near-horizontal bedding. Lower 1.2 m contains layers and lenses of granules and rare small pebbles; upper 1.2 m contains very thin clayey interbeds.
0	3.6	Covered. Debris apron at slope base.

Dates in S-1: 5455±105 (I-10, 817); 8930±140 (I-10, 469); 11,360±170 (I-11, 596); 12,770±180 (I-10, 468); 12,840±160 (USGS-47)

S-2

UNIT	THICKNESS (m)	DESCRIPTION
4	3.8	Weakly bedded medium to silty fine sand, with abundant roots and other plant fragments. Some beds stained by iron oxides. Cultural horizon at 1.3 m depth. Probably eolian sand accreting on vegetated surface.
3	0.5	Mottled fine sand with abundant grass remains. Probably eolian; possibly deposited at margin of active river flood plain. Gradual transition to unit 4.
2	0.5	Clayey, silty fine sand, with clay layers in middle of unit. Cut-and-fill structures present, with fillings generally clay-rich. Probably marsh or shallow pond on flood plain that intermittently is active. Sharp transition to unit 3.
1	2.5	Grey sand. Weak, near-horizontal bedding, with beds 1-5 cm thick. Contains roots and other plant fragments. Probably overbank deposits of aggrading flood plain. Gradual transition to unit 2.
0	0.5	Boulder-cobble lag concentrate. Probably deposited by modern river.

Dates in S-2: 1180±45 (SI-1428); 2275±110 (SI-1427); 11,760±200 (AU-69)

S-5

UNIT	THICKNESS (m)	DESCRIPTION
3	1	Frost-churned stony silt. Gravel from unit 2 mixed with probable loess cap.
2	4	Gravel. Subrounded cobbles and pebbles in sandy matrix. Probably outwash.
1	3	Bedded sand with detrital organic fragments along bedding planes; becoming clayey near base. Probably fluvial, grading upward into lacustrine.
0	4	Basal apron of flow debris.

Date in S-5: 12,170±270 (AU-71)

Figure 1–6 Bluff exposures with radiocarbon-dated deposits of late Itkillik II age. Sagavanirktok River valley. S-1, North bank of Sagavanirktok River, 3 km above mouth of Atigun River. S-2, West side of Sagavanirktok River, 13 km above mouth of Atigun River. S-5, West bank of Sagavanirktok River, 0.5 km above mouth of Accomplishment Creek.

the moraine of late Itkillik II age (Hamilton 1978). The lowest exposed sediments are sandy, basin-filling deposits, dated 12.2 ka, that accumulated after the Sagavanirktok Valley glacier retreated from the outermost Itkillik II moraine complex. The overlying outwash (unit 2) and frost-churned loess (unit 3) are assigned respectively to the maximum stand and subsequent recession of the late Itkillik II glacier. The radiocarbon age is incompatible with dates from exposures S-1 and S-2, which seem to demonstrate that the late Itkillik ice tongue was at its maximum extent about 13–12 ka.

Discussion

The radiocarbon ages from exposures S-1, S-2, and S-5 are part of a concentration of eight radiocarbon ages between 12.8 and 11.5 ka from six localities in the upper Sagavanirktok River valley (Hamilton 1979). Most of these ages are on rooted willow shrubs or on peat, and they represent an episode particularly favorable for plant growth and preservation in the region. Large segments of the upper Sagavanirktok drainage were deglaciated and revegetated by about 12 ka.

Two concordant ages from S-1 and two other concordant ages from the Anaktuvuk River valley (Hamilton 1986) appear to demonstrate that glaciers of late Itkillik II age began to readvance about 13 ka and were at their maximum positions about 12.8 ka; they began to stagnate or retreat shortly after that time. Radiocarbon ages of about 11.3 to 11.8 ka from fluvial deposits exposed in S-1 and S-2 may reflect possible late fluctuations of the Atigun Gorge ice tongue or some other factor that delayed downcutting by the Sagavanirktok River following culmination of the late Itkillik II ice advance. The age of 12.2 ka from S-5, on the other hand, seems to be too young. This site should be resampled and redated because of its importance in providing a maximum age limit for the late Itkillik II readvance.[1]

---

[1]The radiocarbon ages with AU laboratory designations were early experimental analyses by the newly established radiocarbon laboratory at the University of Alaska's Institute of Marine Science. Some of these ages have later proved to be inaccurate (see Hamilton et al. 1983).

The episode of abundant shrub growth and peat accumulation in the upper Sagavanirktok River valley coincides with a widespread transition from herbaceous tundra to shrub tundra beginning about 13.5 ka, which is shown by pollen data elsewhere in northern Alaska. This vegetation shift was triggered by a major change to warmer and perhaps moister summers (Barnosky et al. 1987; Anderson 1988), perhaps associated with deeper winter snows (Ager 1983). Increased snowfall in upper valleys of the Brooks Range would have increased glacier nourishment, perhaps leading to the late Itkillik II glacial advance.

REFERENCES CITED

Ager, T. A. 1983. Holocene Vegetational History of Alaska. In *Late Quaternary Environments of the United States*. Vol. 2, *The Holocene*, ed. H. E. Wright, Jr., pp. 128–41. Minneapolis: University of Minnesota Press.

Anderson, D. D. 1970. Akmak: An Early Archaeological Assemblage from Onion Portage, Northwest Alaska. Copenhagen: *Acta Arctica*, Fasc. 16.

———. 1988. Onion Portage: The Archaeology of a Stratified Site from the Kobuk River, Northwest Alaska. *Anthropological Papers of the University of Alaska*, 22 (1–2).

Anderson, P. M. 1988. Late Quaternary Pollen Records from the Kobuk and Noatak River Drainages, Northwestern Alaska. *Quaternary Research* 29:263–76.

Anderson, P. M., R. E. Reanier, and L. B. Brubaker. 1988. Late Quaternary Vegetational History of the Black River Region in Northeastern Alaska. *Canadian Journal of Earth Sciences* 25:84–94.

Ashley, G. M., and T. D. Hamilton. 1993. Fluvial Response to Late Quaternary Climatic Fluctuations, Central Kobuk Valley, Northwestern Alaska. *Journal of Sedimentary Petrology* 63:814–27.

Barnosky, C. W., P. M. Anderson, and P. J. Bartlein. 1987. The Northwestern U.S. during Deglaciation: Vegetational History and Paleoclimatic Implications. *In* North America and Adjacent Oceans during the Last Deglaciation, ed. W. F. Ruddiman and H. E. Wright, Jr. *The Geology of North America* K–3:289–321. Boulder, Colorado: Geological Society of America.

Carter, L. D. 1981. A Pleistocene Sand Sea on the Alaskan Arctic Coastal Plain. *Science* 211:381–83.

———. 1983. Fossil Sand Wedges on the Alaskan Arctic Coastal Plain and Their Paleoenvironmental Significance. *4th Permafrost International Conference Proceedings, Fairbanks, Alaska, July, 1983*, pp. 109–14. Washington, D.C.: National Academy Press.

Cwynar, L. C. 1982. A Late-Quaternary Vegetation History from Hanging Lake, Northern Yukon. *Ecological Monographs* 52:1–24.

Cwynar, L. C., and J. C. Ritchie. 1980. Arctic Steppe-Tundra: A Yukon Perspective. *Science* 208:1375–77.

Czudek, T., and J. Demek. 1970. Thermokarst in Siberia and Its Influence on the Development of Lowland Relief. *Quaternary Research* 1:103–20.

Fernald, A. T. 1964. Surficial Geology of the Central Kobuk River Valley, Northwestern Alaska. *U.S. Geological Survey Bulletin* 1181-K:K1–K31.

Guthrie, R. D., and S. Stoker. 1990. Paleoecological Significance of Mummified Remains of Pleistocene Horses from the North Slope of the Brooks Range, Alaska. *Arctic* 43:267–74.

Hamilton, T. D. 1978. *Surficial Geologic Map of the Philip Smith Mountains Quadrangle, Alaska.* U.S. Geological Survey Miscellaneous Field Studies Map MF-879-A, Scale 1:250,000.

———. 1979. Radiocarbon Dates and Quaternary Stratigraphic Sections, Philip Smith Mountains Quadrangle, Alaska. U.S. Geological Survey Open-File Report 79-866.

———. 1981a. Episodic Holocene Alluviation in the Central Brooks Range: Chronology, Correlation, and Climatic Implications. In *The United States Geological Survey in Alaska: Accomplishments during 1979,* ed. N. R. D. Albert and T. Hudson, pp. 21–24. U.S. Geological Survey Circular 823-B.

———. 1981b. *Surficial Geologic Map of the Survey Pass Quadrangles, Alaska.* U.S. Geological Survey Miscellaneous Field Studies Map MF-1320, scale 1:250,000.

———. 1982. A Late Pleistocene Glacial Chronology for the Southern Brooks Range: Stratigraphic Record and Regional Significance. *Geological Society of America Bulletin* 93:700–16.

———. 1984. *Surficial Geologic Map of the Ambler River Quadrangle, Alaska.* U.S. Geological Survey Miscellaneous Field Studies Map MF-1678, scale 1:250,000.

———. 1986. Late Cenozoic Glaciation of the Central Books Range. In *Glaciation in Alaska: The Geologic Record,* ed. T. D. Hamilton, K. M. Reed, and R. M. Thorson, pp. 9–49. Anchorage: Alaska Geological Society.

Hamilton, T. D., and G. M. Ashley. 1993. Epiguruk: A Late Quaternary Environmental Record from Northwestern Alaska. *Geological Society of America Bulletin* 105:583–602.

Hamilton, T. D., and J. Brigham-Grette. 1992. The Last Interglaciation in Alaska—Stratigraphy and Paleoecology of Potential Sites. *Quaternary International* 10–12:49–71.

Hamilton, T. D., and S. C. Porter. 1975. Itkillik Glaciation in the Brooks Range, Northern Alaska. *Quaternary Research* 5:471–97.

Hamilton, T. D., and S. W. Robinson. 1977. Late Holocene (Neoglacial) Environmental Changes in Central Alaska. *Geological Society of America Abstracts with Programs* 9:1003.

Hamilton, T. D., J. L. Craig, and P. V. Sellmann 1988a. The Fox Permafrost Tunnel: A Late Quaternary Geologic Record in Central Alaska. *Geological Society of America Bulletin* 100:948–69.

Hamilton, T. D., J. P. Galloway, and E. A. Koster. 1988b. Late Wisconsin Eolian Activity and Related Alluviation, Central Kobuk River Valley. In *Geologic Studies in Alaska by the U.S. Geological Survey during 1987,* ed. J. P. Galloway and T. D. Hamilton, pp. 39–43. U.S. Geological Survey Circular 1016.

Hamilton, T. D., G. A. Lancaster, and D. A. Trimble. 1987. Glacial Advance of Late Wisconsin (Itkillik II) Age in the Upper Noatak River Valley: A Radiocarbon-Dated Stratigraphic Record. In *Geologic Studies in Alaska by the U.S. Geological Survey during 1986,* ed. T. D. Hamilton and J. P. Galloway, pp. 35–39. U.S. Geological Survey Circular 998.

Hamilton, T. D., G. M. Ashley, K. M. Reed, and C. E. Schweger. 1993. Late Pleistocene Vertebrates and Other Fossils from Epiguruk, Northwestern Alaska. *Quaternary Research* 39:381–87.

Hamilton, T. D., G. M. Ashley, K. M. Reed, and D. P. Van Etten. 1984. Stratigraphy and Sedimentology of Epiguruk Bluff: A Preliminary Account. In *The United States Geological Survey in Alaska: Accomplishments during 1982,* ed. K. M. Reed and S. Bartsch-Winkler, pp. 12–15. U.S. Geological Survey Circular 939.

Mason, O. K. 1989. Heightened Storm Activity in the Late Holocene as Reflected by Northwest Alaska Beach Ridges. *Geological Society of America Abstracts with Programs* 21 (6):A344.

Péwé, T. L. 1954. Effect of Permafrost on Cultivated Fields, Fairbanks Area, Alaska. *U.S. Geological Survey Bulletin* 989-F:315–51.

Schweger, C. E. 1976. Late Quaternary Paleoecology of the Onion Portage Region, Northwestern Alaska. Ph.D. dissertation, University of Alberta.

———. 1982. Late Pleistocene Vegetation of Eastern Beringia: Pollen Analysis of Dated Alluvium. In *Paleoecology of Beringia,* ed. D. M. Hopkins, J. V. Matthews, Jr., C. E. Schweger, and S. B. Young, pp. 95–112. New York: Academic Press.

———. 1985. Geoarchaeology of Northern Regions: Lessons from Cryoturbation at Onion Portage, Alaska. In *Archaeological Sediments in Context,* ed. J. K. Stein and W. R. Farrand, pp. 127–41. Peopling of the Americas Series, Vol. 1. Orono: University of Maine, Center for the Study of Early Man.

Ten Brink, N. W., and C. F. Waythomas. 1985. Late Wisconsin Glacial Chronology of the North-Central Alaska Range: A Regional Synthesis and Its Implications for Early Human Settlements. In *North Alaska Range Early Man Project,* ed. W. R. Powers et al. *National Geographic Society Research Reports* 19:15–3

LATE QUATERNARY AEOLIAN DEPOSITS OF THE HOLITNA LOWLAND, INTERIOR SOUTHWESTERN ALASKA

Christopher F. Waythomas

Recent investigations in the lowlands of southwestern Alaska (Fig. 1–7) have documented thick stratigraphic sequences of unconsolidated sediment of Quaternary age exposed in river cutbanks in the Holitna and Nushagak lowlands and in sea bluffs along the Bristol Bay coast (Fig. 1–7; Lea 1989, 1990a, 1990b; Waythomas 1990; Lea et al. 1991; Waythomas et al. 1993). Upper Pleistocene deposits in these areas are dominated by well-sorted, subhorizontally stratified, silty-to-sandy sediments mostly of aeolian origin that form a blanketing surficial mantle over low bedrock hills, thickening to primary basin-fill deposits in the lowlands. Primary aeolian deposits of late Pleistocene and Holocene ages are particularly well represented in the Holitna lowland. Analysis of these sediments provides new information about the chronology of late glacial and postglacial environmental changes. It may, as well, provide contextual evidence by which to assess archaeological remains found within them.

Prior knowledge of the Quaternary deposits in the Holitna lowland was limited to a brief summary given by Cady et al. (1955) as part of their geologic reconnaissance of the region. However, no descriptions of aeolian deposits were presented. In the early 1980s, an archaeological survey was done along the Holitna River (Ackerman 1982, this volume), but was focused primarily on the loess deposits capping low bedrock hills and promontories in the western portion of the lowland. Artifacts were discovered in loess deposits of Holocene age (Ackerman 1982, this volume). At that time the stratigraphic relations and depositional framework of aeolian deposits in the Holitna lowland were unknown, although some preliminary stratigraphic data were collected (Lea and Elias 1983). Data for this article were collected during the summers of 1983, 1984, and 1985.

PHYSICAL SETTING

The Holitna lowland is an elongate, northeast-trending, riparian basin partially enclosed by formerly glaciated mountains and unglaciated rolling uplands (Fig. 1–8). The southern sector of the lowland abuts the low foothills and foreland mountains of the northwestern Alaska Range and lies within several kilometers of late Pleistocene glacier systems (Fig. 1–8). Unconsolidated Quaternary deposits that blanket the lowland are well exposed along river cutbanks in bluffs up to 20 m high.

This region has a modern climate that is transitional between continental and maritime. The mean annual temperature is about -2°C, and mean annual precipitation is about 40 cm (Arctic Environmental Information and Data Center 1986). Most of the lowland is poorly drained and is characterized by extensive tracts of muskeg, bog, and numerous lakes. The lowland is in the zone of discontinuous permafrost (Péwé 1975b); however, neither seasonal ground ice nor relict permafrost was observed in river-bluff exposures during summer fieldwork. The modern vegetation in the lowland consists of an open-canopy, spruce-hardwood forest dominated by paper birch (*Betula paperifera*) and white spruce (*Picea glauca*). Dense thickets of willow (*Salix*) and alder (*Alnus*) and groves of cottonwood (*Populus balsamifera*) border riparian corridors.

During Pleistocene glacial maxima, the Holitna lowland was a few tens of kilometers beyond the large piedmont-glacier complexes of the northwestern Alaska Range and Ahklun Mountains (Fig. 1–8). While not invaded by Pleistocene ice, the lowland functioned as a depositional basin in which many meters of aeolian, fluvial, and colluvial deposits and tephra were deposited. Upland massifs along the western margin of the lowland, such as the Chuilnuk and Kiokluk mountains (Fig. 1–8), bore small alpine valley glaciers. At least one late Pleistocene and one middle or late Pleistocene ice advance are recognized in this area (Waythomas 1984, 1990) and elsewhere in the region (Fernald 1960; Kline and Bundtzen 1986). Local moraine records remain undated, however, and age assignments primarily are based on relative-age criteria (Waythomas 1990).

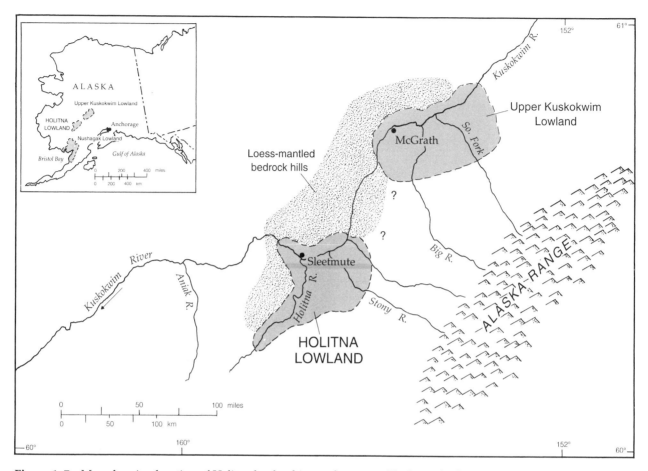

Figure 1–7   Map showing location of Holitna lowland in southwestern Alaska and other place names mentioned in text.

STRATIGRAPHY

Kulukbuk Bluffs

The Kulukbuk Bluffs section (61°N, 157°W; Fig. 1–8) is one of the most important stratigraphic exposures in the Holitna lowland because it contains the widespread Old Crow tephra as well as deposits of middle Pleistocene through Holocene age (Waythomas 1990; Waythomas et al. 1993). At this locality, approximately 20 meters of vertical section are exposed along an 800-m-long cutbank of the Holitna River (Fig. 1–9). Five lithostratigraphic units are present in the bluff and extend laterally for several tens of meters along the bluff face (Fig. 1–10).

Unit A   The basal sediments in the sequence (unit A, Fig. 1–10) consist of fine-to-medium, horizontally laminated sand (Miall 1977) with small-scale cross-laminae and medium-scale planar cross-strata. This facies association and palaeocurrent indicators (dip of cross-strata) indicate a shallow, sandy braided river (cf. Miall 1977) flowing parallel to the course of the presently meandering Holitna River. Unit A likely accumulated as distal outwash derived from glaciers in the northwestern Alaska Range, northern Ahklun Mountains, and from local alpine glacier systems (Fig. 1–8).

Unit B   Unit B is 7 to 9 meters thick and comprises most of the basal part of the Kulukbuk Bluffs section (Fig. 1–10). The lowermost beds are gradational with unit A, and mark a change from planar cross-bedded sands (unit A) to horizontally and subhorizontally bedded, fine-to-medium, well-sorted sand containing thin, laterally discontinuous, medium-to-coarse sand stringers and lenses. The basal part

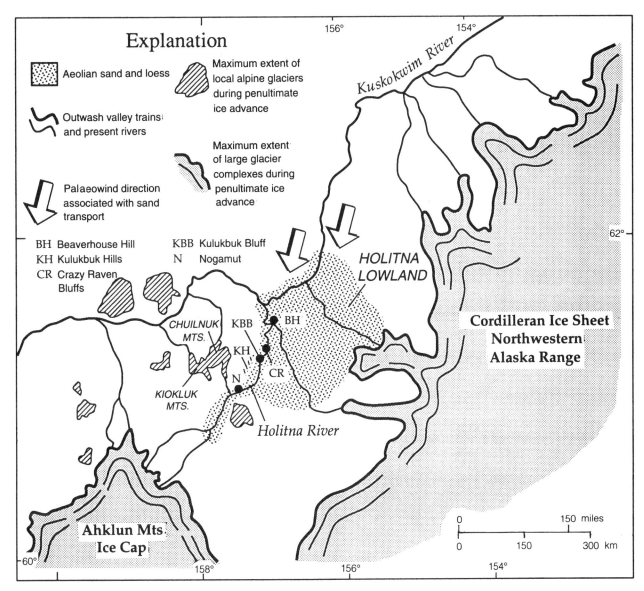

Figure 1–8 Palaeogeographic map of interior southwest Alaska during the penultimate glaciation. Extent of glaciers during last glaciation only slightly less than indicated for penultimate glaciation.

of unit B is well stratified and is characterized by thin parallel laminae in 5-to-20-cm-thick packages or sets. Bounding surfaces are commonly denoted by lenses of medium-to-coarse sand and a dispersed aeolian lag that truncates the strata at low angles (<15°). Sediments in the basal part of unit B are interpreted as aeolian sand-sheet deposits produced by wind-ripple migration (Hunter 1977; Kocurek and Dott 1981; Lea 1990a).

Sand-sheet deposits in the lower part of unit B grade upward into parallel-bedded, fine-to-me-

dium sand, sand-silt intergrades, and finally into massive silt over a vertical distance of approximately 5 meters. Alternating beds of medium-to-fine sand and very fine sand-to-silt overlie the basal sand-sheet deposits and form the middle portion of the sequence (subunit B2, Fig. 1–10). Individual beds in subunit B2 range in thickness from several millimeters to several centimeters and are either massive or show faint internal lamination. Bounding surfaces between beds are wavy to irregular. The sand-dominated beds are laterally discontinu-

Figure 1–9   Kulukbuk Bluffs exposure on Holitna River. Prominent white band in middle of exposure is Old Crow tephra. Bluff is about 20 meters high, downstream is from left to right.

ous and tend to pinch and swell along strike. The proportion of sandy beds diminishes up-section, and the rhythmic bedding pattern becomes less distinct.

The upper 3 to 4 meters of unit B consist of massive silt and faintly laminated sandy silt that contains Old Crow tephra (Waythomas et al. 1993). The sediments beneath the tephra consist of light brown to grey-brown sandy silt in sharp contact with the base of the tephra. Overlying Old Crow tephra is about 2 meters of massive blue-grey silt. The contact between Old Crow tephra and the overlying silt is gradational. Pods and lenses of blue-grey silt are locally present within the tephra and indicate that the tephra has been partially reworked.

Unit B Cryogenic Features

ICE-WEDGE PSEUDOMORPHS   Downward tapering, wedge-shaped structures associated with com-

pressively deformed upwarped and downwarped host strata are interpreted as ice-wedge casts or pseudomorphs (Figs. 1–10, 1–11). These features are common near the base and top of unit B. Ice-wedge pseudomorphs are typically about one meter wide at the top of the feature and 1 to 6 meters long. Material that forms the wedge cast consists of medium-to-fine sand and silt, and, occasionally, Old Crow tephra that has slumped into the void produced by the melting ice wedge. Sediment comprising the ice-wedge pseudomorphs was likely derived from the overlying and adjacent host strata. The fabric of the wedge-filling sediment consists of slumped infills of massive, structureless sand and silt. However, only rarely do the fills consist of vertically oriented layers of fine-to-medium sand.

Ice-wedge pseudomorphs in present or former permafrost terrain form by (1) progressive deposition of sediment in open thermal contraction cracks (primary infilling) as the wedge structure develops

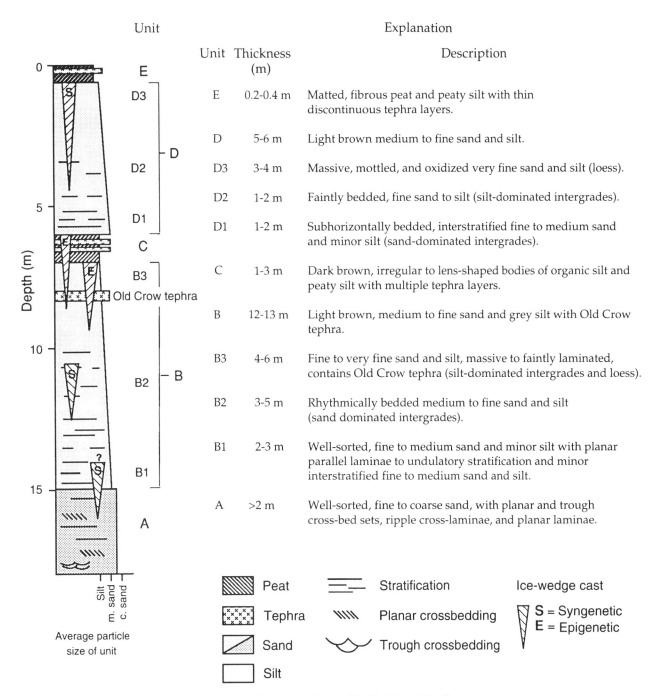

Figure 1–10 Stratigraphic profile. Kulukbuk Bluffs.

Unit

		Explanation	
Unit	Thickness (m)		Description
E	0.2-0.4 m		Matted, fibrous peat and peaty silt with thin discontinuous tephra layers.
D	5-6 m		Light brown medium to fine sand and silt.
D3	3-4 m		Massive, mottled, and oxidized very fine sand and silt (loess).
D2	1-2 m		Faintly bedded, fine sand to silt (silt-dominated intergrades).
D1	1-2 m		Subhorizontally bedded, interstratified fine to medium sand and minor silt (sand-dominated intergrades).
C	1-3 m		Dark brown, irregular to lens-shaped bodies of organic silt and peaty silt with multiple tephra layers.
B	12-13 m		Light brown, medium to fine sand and grey silt with Old Crow tephra.
B3	4-6 m		Fine to very fine sand and silt, massive to faintly laminated, contains Old Crow tephra (silt-dominated intergrades and loess).
B2	3-5 m		Rhythmically bedded medium to fine sand and silt (sand dominated intergrades).
B1	2-3 m		Well-sorted, fine to medium sand and minor silt with planar parallel laminae to undulatory stratification and minor interstratified fine to medium sand and silt.
A	>2 m		Well-sorted, fine to coarse sand, with planar and trough cross-bed sets, ripple cross-laminae, and planar laminae.

(Black 1976; Carter 1983; Harry and Gozdzik 1988), or (2) deposition of sediment adjacent to the ice wedge (secondary infilling) as thawing proceeds from the ice margin inward (Harry and Gozdzik 1988). Because the host and infill sediments are similar, and vertical layering is rare, ice-wedge pseudo-

morphs in the lower portion of unit B probably formed mainly by secondary infilling of the wedge cavity.

Growth of ice wedges in the lower portion of unit B was accompanied by the accumulation of aeolian sediment. Ice wedges that form in this man-

Figure 1–11   Casts of syngenetic ice wedges in aeolian sand-sheet deposits, unit B, Kulukbuk Bluffs. Trowel for scale about 20 cm long.

Figure 1–12   Cast of epigenetic ice wedge cross-cutting Old Crow tephra, unit B, Kulukbuk Bluffs. Scale is 15 cm.

ner have been termed *syngenetic* (Gallwitz 1949; Dostovalov and Popov 1966; French and Gozdzik 1988) and imply cyclical sedimentation and ice wedge development associated with a periglacial climate (Fig. 1–11). Conversely, *epigenetic* ice wedges form after deposition of the host strata and provide evidence for a periglacial climate some time after sediment accumulation. Syngenetic ice wedges are almost always associated with sediments that are genetically related to periglacial conditions, whereas epigenetic ice wedges can form in any sediment type. Discriminating between epigenetic and syngenetic ice wedges based on characteristics of ice-wedge pseudomorphs is sometimes difficult.

Syngenetic ice-wedge pseudomorphs in the

lower part of unit B have several important palaeo-environmental implications. First, the association between aeolian sand-sheet deposits and syngenetic ice-wedge pseudomorphs would be expected if the deposits are periglacial in origin. Second, the presence of syngenetic ice-wedge pseudomorphs rather than sand wedges indicates that sand and silt were not available for transport during middle and late winter, when thermal contraction cracks are most likely to be open (Foscolos et al. 1977; Carter 1983). Also, sufficient active-layer moisture must have been available to form ice wedges while contraction cracks were still open, probably during late spring.

Ice-wedge pseudomorphs in the upper part of unit B are characteristically different from those found in the lower part of the unit. Wedge casts in the upper part of the unit tend to be wider and shorter, infilling sediment is massive blue-grey silt, and—where preserved—deformation of host sediments indicates only downwarped strata. Ice-wedge pseudomorphs in the upper portion of unit B are epigenetic features that developed in the silty beds at the top of the unit and penetrated downward through underlying Old Crow tephra (Fig. 1–12).

FROST CRACKS   Narrow (< 2 cm), 1-to-2-m long, vertically oriented linear fissures—similar to syngenetic ice-wedge pseudomorphs, but smaller—also are common in unit B. These features are interpreted as frost cracks (Washburn 1980:102) and are additional evidence for periglacial conditions during the formation of unit B. The formation of frost cracks rather than ice wedges is probably related to the inability of ground cracking to occur in the same place over time.

CONDUIT STRUCTURES   The term *conduit structure* was introduced by Lea (1986) to describe near-circular, tube-like features in periglacial aeolian deposits that are discordant with primary stratification. These features are common in unit B and are composed of massive, slumped, or crudely stratified sediment that is coarser and less well-sorted than the surrounding host strata. The diameter of conduit structures in unit B ranges from 2 to 20 cm, but is typically 6 to 10 cm. Such features resemble rodent burrows (*krotovina*), but differ in that they commonly exhibit multiple infills (i.e., crude stratification), lack recognizable organic remains, and tend to be associated with sagging or downwarped strata that overlie the structure.

The origin of conduit structures in unit B is unclear. Lea (1986, 1989:196) has suggested that conduit structures may develop by subsurface piping during thaw of permafrost. Piping structures associated with thermokarst development in loess and reworked loess of Holocene age are common near Fairbanks, Alaska (Péwé 1982). The conditions that result in subsurface soil piping include temporarily high hydraulic gradients, easily erodible, dispersible sediment, and an impermeable zone below a piping-susceptible horizon (e.g., unfrozen sandsheet deposits above the permafrost table; Jones 1981; Lea 1989:200). Piping in aeolian sand-sheet

deposits could result from snowmelt and rapidly thawing ice-bonded sediment becoming readily dispersed within a thickening active layer (Lea 1986, 1989:200). However, it is unclear how the required hydraulic gradient can be generated in essentially flat-lying deposits, unless the deposits are incompletely thawed, allowing conduit systems to become established only where sediment is not ice-cemented.

Conduit structures may be initiated by differential thawing and are likely to form in sandy sediment because of the differences in thaw rate between sand and fines (Washburn 1980:168). The upper portions of conduit structures in unit B are almost always associated with layers of medium-to-coarse sand, indicating that these layers may be the sites of conduit structure development.

An additional constraint on the development of conduit structures by piping is related to the availability of water having the erosive capacity to form a subcircular "pipe," and to transport fine-to-coarse sand which eventually fills the conduit structure. A rapidly melting, thick snow cover could provide the necessary amount of water; however, a thick winter snow cover is inconsistent with the evidence for ice-wedge development and accumulation of aeolian sand-sheet deposits. Furthermore, unless the conduit structures are initiated at the surface, a mechanism is required for the movement of appropriate quantities of groundwater through texturally stratified aeolian deposits to the sites of conduit structure formation.

Unit C   Unit C is composed of organic-bearing-to-peaty silt with discontinuous layers, lenses, and thin pods of fine-grained volcanic ash, and subordinate thin interbeds of slightly organic grey-brown silt (Figs. 1–10, 1–12). The thickness of unit C varies from 10 cm to 3.5 m, reflecting its preservation as discontinuous, pod-shaped masses that rest on an irregular low-angle surface of unconformity that dips gently (<10°) to the north.

Downward-pointing, wedge-shaped structures with irregular margins in unit C are pseudomorphs of epigenetic ice wedges (Fig. 1–12). The pseudomorphs consist of massive-to-slumped infills of organic-rich silt and tephra that range from 1 to 1.5 m

Unit   Explanation

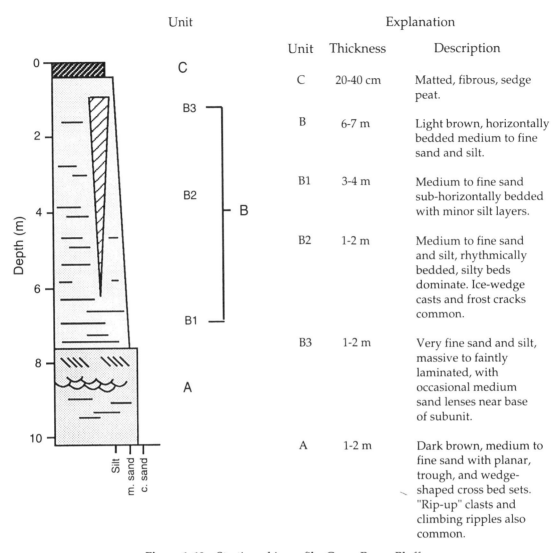

Unit	Thickness	Description
C	20-40 cm	Matted, fibrous, sedge peat.
B	6-7 m	Light brown, horizontally bedded medium to fine sand and silt.
B1	3-4 m	Medium to fine sand sub-horizontally bedded with minor silt layers.
B2	1-2 m	Medium to fine sand and silt, rhythmically bedded, silty beds dominate. Ice-wedge casts and frost cracks common.
B3	1-2 m	Very fine sand and silt, massive to faintly laminated, with occasional medium sand lenses near base of subunit.
A	1-2 m	Dark brown, medium to fine sand with planar, trough, and wedge-shaped cross bed sets. "Rip-up" clasts and climbing ripples also common.

Figure 1–13 Stratigraphic profile. Crazy Raven Bluffs.

in length and 30 to 80 cm in width. These features are interpreted as the casts of epigenetic ice wedges because (1) the upper portions of the ice-wedge pseudomorphs coincide with the top of unit C, indicating that ice-wedge development occurred sometime after deposition of unit C; (2) unit C probably formed during a late Pleistocene warm interval (oxygen isotope stages 3, 5, or both) when ice wedges likely were not forming; and (3) only down-warped host strata have been observed adjacent to the ice-wedge pseudomorphs.

Unit D Unit D is a 4-to-6-m-thick sequence of fine-to-medium sand, loamy sand, and silt that is

similar to unit B. The unit shows an overall fining-upward change in particle size from medium sand to silt and possesses many of the same sedimentary structures and facies types found in unit B (Fig. 1–10). The contact between units D and C is sharp and probably represents an unconformity.

The basal 1 to 2 m of unit D consists of weakly oxidized, evenly laminated, fine-to-medium sand with occasional lenses of medium-to-coarse sand. Individual sedimentation units range in thickness from 10 to 50 cm and are truncated by low angle (>5°) bounding surfaces that are discordant with the primary stratification. Above this basal zone, the unit grades upward into subhorizontally bed-

Figure 1–14 Cast of syngenetic ice wedge in fining-upward aeolian deposits, unit B, Crazy Raven Bluffs. Shovel for scale is about 60 cm long.

ded, sand-dominated, sand-silt intergrades—or "loamy coversand"—and finally into massive, oxidized and cryoturbated, very fine sand and silt.

Evidence of cryogenic activity is less common in unit D than in other units. However, irregular and contorted bedding indicative of cryoturbation and sediment reworking by frost-heaving are common features in unit D. Oxidized laminae that define micro-scale undulations and vertically offset bedding are other common features of unit D. These structures may represent local adjustments caused by settling, possibly associated with melt-out of buried snow layers (Koster and Dijkmans 1988), or they could represent the effects of repeated freezing and thawing (Coutard and Mücher 1985).

Unit E   Unit E consists of 20 to 50 cm of matted,

fibrous sedge peat and silty peat with multiple tephra layers (Fig. 1–10). The tephras are whitish in color and form laterally discontinuous layers ranging in thickness from 0.5 to 1 cm.

Crazy Raven Bluffs

The informally designated "Crazy Raven Bluffs" (61°21'N, 157°4'W) are located along a prominent cutbank of the Holitna River approximately 12 km downstream from the Kulukbuk Bluffs (Fig. 1–8). At this locality, an exposure several hundred meters in length reveals 10 to 11 vertical meters of subhorizontally bedded sand-sheet deposits, massive silt, and minor fluvial deposits (Fig. 1–13).

Three stratigraphic units are recognized here. Unit A, like unit A of Kulukbuk, is a fluvial deposit formed in a sandy braided environment. There are no organics present and no cryogenic features. Unit B is generally resemblant to the corresponding unit of Kulukbuk. As with the latter, cryogenic features are common, consisting of numerous ice-wedge casts, frost cracks, and several micro-to-meso–scale structures of probable cryogenic origin. The ice-wedge pseudomorphs appear to be entirely epigenetic in origin—unlike unit B at Kulukbuk. Smaller ice-wedge pseudomorphs are also present in unit B, their size suggesting some physical repression of development, perhaps by the accumulation of sediment over active ice wedges which would have inhibited ground cracking. Conduit structures are present here too. Unit C is the youngest exposed at Crazy Raven Bluffs. It consists generally of a sedge peat mat, 20–40 cm thick, that is a continuous cover over the exposure (see Figs. 1–13, 1–14).

Beaverhouse Hill

Beaverhouse Hill is an isolated erosional remnant that forms a low rounded bluff within the floodplain of the lower Holitna River (157°2'W, 61°31'N; Fig. 1–8). The bluff is approximately 20 meters high and about 100 m long, and contains a sequence of horizontally bedded sand, silt, and minor organic detritus (Fig. 1–15).

Unit A, the base of the sequence here, is at least 1.7 m thick and consists of interbedded blue-grey organic-bearing clayey silt, silt, and medium-to-fine

Figure 1–15   Beaverhouse Hill section. Maximum height of bluff is about 15 m. Radiocarbon-dated bone was collected from the interval just above swallow nests in the middle of the photograph. Downstream is to left.

sand. The unit is weakly laminated to massive, and is locally iron-stained. Organic detritus (disseminated organic matter, wood fragments, and twigs) is dispersed throughout the upper portion of unit A. Ice-wedge pseudomorphs of epigenetic origin penetrate to a depth of at least 1.7 m. They are truncated by the upper contact of unit A and indicate an interval of cold climate that postdates the deposition of unit A.

Unit B is the thickest sequence of aeolian sediment exposed in the Holitna lowland. Three subunits are recognized on the basis of textural and sedimentological variations. The sands and minor silt fractions of subunits B1 and B2 show evidence of wind deposition while the upper portion of unit B (subunit B3) mainly consists of silt and very fine

sand. The silt-dominated deposits contrast markedly with the underlying sand-dominated deposits of subunit B2 (Fig. 1–16). Features indicative of former periglacial conditions are less common in unit B than in equivalent deposits elsewhere in the Holitna lowland. Well-developed ice-wedge pseudomorphs are present within subunits B2 and B3, but are uncommon. These features exhibit all of the attributes of the casts of syngenetic ice wedges that were described for the Kulukbuk Bluffs and Crazy Raven localities.

Unit C consists of about 20 cm of brown-black, fibrous, matted sedge peat capping the bluff (Fig. 1–16). The peat forms a continuous mantle over the exposure and is in sharp contrast with underlying massive loess.

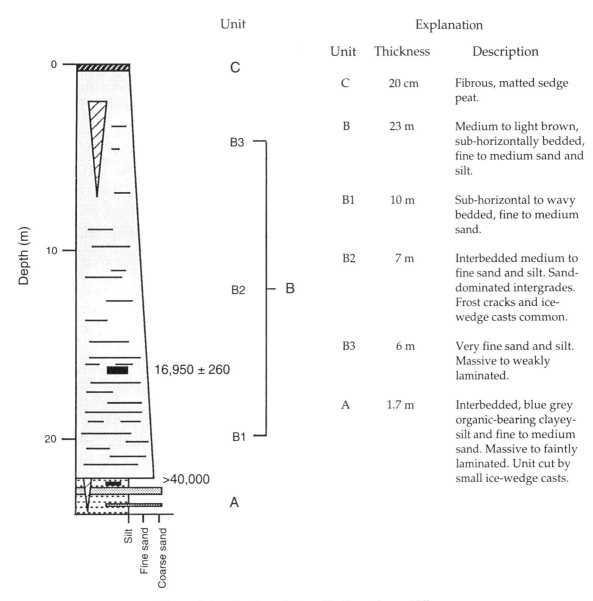

Figure 1–16   Stratigraphic profile. Beaverhouse Hill.

Explanation

Unit	Thickness	Description
C	20 cm	Fibrous, matted sedge peat.
B	23 m	Medium to light brown, sub-horizontally bedded, fine to medium sand and silt.
B1	10 m	Sub-horizontal to wavy bedded, fine to medium sand.
B2	7 m	Interbedded medium to fine sand and silt. Sand-dominated intergrades. Frost cracks and ice-wedge casts common.
B3	6 m	Very fine sand and silt. Massive to weakly laminated.
A	1.7 m	Interbedded, blue grey organic-bearing clayey-silt and fine to medium sand. Massive to faintly laminated. Unit cut by small ice-wedge casts.

Nogamut

The Nogamut section is named for a cutbank ex-poure within an abandoned meander loop of the upper Holitna River near the settlement of Noga-mut (61°3′N, 157°41′W; Fig. 1–8).

Three units are recognized at this locality (Fig. 1–17). Unit A is a fluvial deposit, probably outwash, and it records an aggradational phase of the Holitna River that was probably associated with local alpine glaciation in the nearby Chuilnuk, Taylor, and northeastern Ahklun mountains (Fig. 1–8).

Ice-wedge pseudomorphs are common in unit A (Fig. 1–17). These features end abruptly at the top of the unit. They are the casts of epigenetic ice wedges that developed on an inactive segment of a gravel outwash plain along the Holitna River. Unit B is a loess deposit and is probably correlative with thicker, possibly last-glacial, aeolian deposits at Ku-

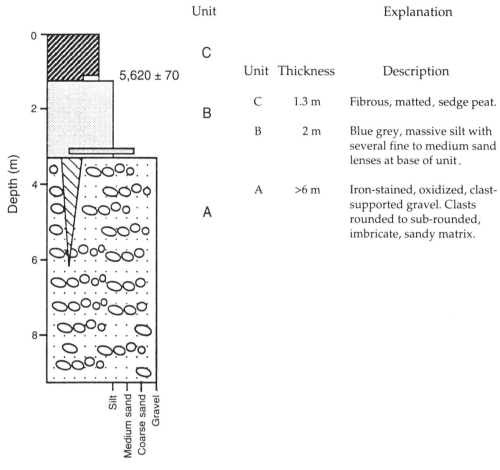

Figure 1–17   Stratigraphic profile. Nogamut.

lukbuk Bluffs, Crazy Raven Bluffs, and Beaverhouse Hill. A weakly developed soil has formed in the upper 40 cm of unit B. This soil consists of an incipient Bw horizon (10 cm thick) over a Cox horizon (30 cm thick). Unit C consists of about 1 to 1.5 meters of fibrous, matted sedge peat (Fig. 1–17).

STRATIGRAPHIC CORRELATION AND GEOCHRONOLOGY

Correlation of stratigraphic sequences within the Holitna lowland is shown in Figure 1–18. These correlations are based on radiocarbon-dated organic matter (Table 1–3) and Old Crow tephra. Three names are proposed for lithostratigraphic units in the Holitna lowland. The names are used infor-

mally and are intended to facilitate correlation with stratigraphic sequences elsewhere in Alaska.

Kulukbuk Formation

The Kulukbuk formation is defined for aeolian sand-sheet deposits, sand-loess intergrades, minor fluvial sand, and loess that contains Old Crow tephra. Deposits of the formation make up the basal portion of the stratigraphic sequence at the Kulukbuk Bluffs (units A and B, Fig. 1–10). The lower portion of the formation contains casts of syngenetic ice wedges, whereas casts of epigenetic ice wedges are present at the top of the formation.

Deposits correlative with the Kulukbuk formation are not recognized elsewhere in the Holitna lowland. Thus, the term *formation* is used in an in-

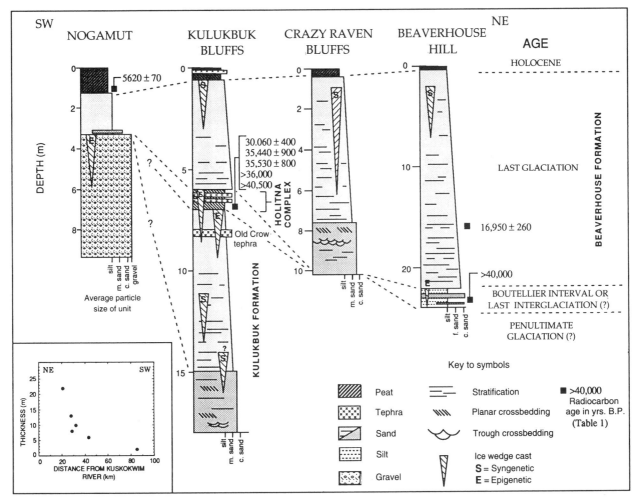

Figure 1–18 Correlation of stratigraphic sequences within the Holitna lowland.

Table 1–3 Radiocarbon Dates on Organic Matter from Deposits in the Holitna Lowland

Radiocarbon Age (BP)	Laboratory and No.	Material Dated	Location/Unit (Fig. 1–8)
5620 ± 70	Beta-5232	Basal peat	Nogamut
15,740 ± 120	GX-13092-A	Bone, apatite fraction	Beaverhouse Hill
16,950 ± 260	GX-13092-C	Bone, collagen fraction	Beaverhouse Hill
30,060 ± 400	Beta-5234	Peaty silt	Kulukbuk Bluffs, Unit C
35,440 ± 900	Beta-5233	Wood fragments of *Salix*	Kulukbuk Bluffs, Unit C
35,530 ± 800	Beta-5236	Disseminated organics in silt	Kulukbuk Bluffs, Unit C
>36,000	GX-13409	Organic silt	Kulukbuk Bluffs, Unit C
>40,000	GX-13408	Wood fragments	Beaverhouse Hill
>40,500	GX-10688	Peaty silt	Kulukbuk Bluffs, Unit C
>44,000	GX-13549	Peat, organic silt	Kulukbuk Bluffs, Unit A

formal sense to indicate the regional importance of these deposits because they contain the widespread Old Crow tephra (Westgate et al. 1985; Hamilton and Brigham-Grette 1991; Hamilton, this volume).

The Old Crow tephra is a prominent marker bed within the upper portion of the Kulukbuk formation, providing geochronological control and a basis for regional correlation. Isothermal plateau fission-track ages on hydrated glass shards of Old Crow tephra (including samples from the Holitna River locality) yield a weighted mean age of 140,000 ± 10,000 yrs (Westgate 1988; Westgate et al. 1990). A thermoluminescence (TL) age of 170,000 ± 27,000 yrs on the glass fraction of the tephra was recently reported (Berger 1991). TL ages of 110,000 ± 32,000 yrs for loess 10 cm above the tephra and 140,000 ± 30,000 yrs for loess 60 cm below the bottom of the tephra support these results (Berger et al. 1992).

Stratigraphic profiles from many Old Crow tephra localities indicate that deposition of the tephra occurred prior to the onset of warm interglacial conditions recorded by forest beds and peaty deposits that formed during marine oxygen isotope stage 5 (Hamilton et al. 1991; Hamilton and Brigham-Grette 1991; Hamilton this volume). The association of Old Crow tephra with cold-climate aeolian deposits of the Kulukbuk formation and its position beneath organic sediments that accumulated during the last interglacial or younger warm periods is evidence that the Kulukbuk formation was most likely deposited during marine oxygen isotope stage 6.

Holitna Complex

Holitna complex designates that sequence of organic silts and tephra-bearing silt that overlies the Kulukbuk formation at the Kulukbuk Bluffs. The term *complex* is used to indicate that the deposit is reworked and may include sediment of multiple ages. Deposits of the Holitna complex crop out only at the Kulukbuk Bluffs (unit C, Fig. 1–10).

Radiocarbon dates on plant macrofossils, peat, woody debris, and organic silt from the Holitna complex range from ca. 28 ka to >44 ka. Similar deposits downstream from Beaverhouse Hill

yielded a radiocarbon date of 28,090 ± 540 BP. The Holitna complex thus was probably deposited during the ca. 30–70 ka Boutellier nonglacial interval (Hopkins 1982) or last interglacial (Hamilton and Brigham-Grette 1991; Hamilton, this volume).

Beaverhouse Formation

The Beaverhouse formation is named for a 20-meter-thick sequence of aeolian sediments, consisting of sand-sheet deposits, sand-loess intergrades, and loess exposed at Beaverhouse Hill. Equivalent deposits are found at Crazy Raven Bluffs, the Kulukbuk Bluffs, and at Nogamut (Fig. 1–18), as well as at numerous other locations along the Holitna River.

A variety of cryogenic and ground-ice features are common in the Beaverhouse formation, their expression in part a function of sediment grain size. The Beaverhouse formation was deposited during a lengthy interval of aeolian deposition that was mostly coeval with the last glaciation and early Holocene in southwestern Alaska. Radiocarbon dates on bone fragments from the basal portion of the formation indicate that aeolian deposition was in progress by about 17,000 BP (Table 1–3). A date of 5620 ± 70 BP (Beta-5232) was obtained on peat directly overlying the Beaverhouse formation, indicating that aeolian deposition had ceased by that time.

DEPOSITIONAL FRAMEWORK AND PALAEOECOLOGY

Kulukbuk Formation

The Kulukbuk formation records minor glacifluvial aggradation and a lengthy interval of aeolian sediment accumulation that culminated with the deposition of the Old Crow tephra. Although the formation has been recognized only at this one locality, it has regional significance because it contains the widespread Old Crow tephra.

Limited palaeoenvironmental data from the Kulukbuk formation indicates relatively cold, dry-to-mesic conditions, with sparse graminoid tundra and scattered grasses and herbs (Waythomas 1990). Although it appears that environmental conditions associated with the Kulukbuk formation were generally dry, xeric pollen taxa (e.g., *Artemisia*) are only minor components of the pollen spectra. Furthermore, the predominance of sand-sheet deposits, rather than dunes, implies that loose, dry sand was not available for aeolian transport. The casts of syngenetic ice wedges, common in the Kulukbuk formation, also indicate that moisture must have been seasonally available to allow ice-wedge development. Thus, although conditions were much drier than present, the area must have remained somewhat mesic to account for the aeolian facies types, cryogenic features, and pollen remains.

Old Crow tephra in the upper part of the formation indicates that most of the Kulukbuk formation was deposited by about 149,000 BP (Westgate 1988; Waythomas et al. 1993). Thus, the Kulukbuk formation is probably correlative with marine oxygen isotope stage 6, and aeolian deposits in the formation are likely coeval with regional glaciation.

Holitna Complex

Organic-bearing silt and peat of the Holitna complex record an interval of climate that was intermediate between full-glacial and modern. Variable amounts of grass, sedge, and birch in the pollen spectra indicate transient fluctuations in local climate, moisture regime, or both. However, since the Holitna complex is reworked, stratigraphic interpretations of the pollen sequence are limited. Deposits of the Holitna complex indicate a period of stability when soils, peat, and tephra could accumulate on the landscape.

Beaverhouse Formation

Sediments of the Beaverhouse formation are the most widespread in the Holitna lowland. They are similar to deposits of the Kulukbuk formation, containing similar facies types, cryogenic features, and pollen remains.

Age control on the Beaverhouse formation is provided by a radiocarbon date of 16,950 ± 260 BP obtained on the collagen fraction of bone fragments from the lower portion of the formation and an infinite radiocarbon age on deposits of the Holitna complex underlying the formation at Beaverhouse

Hill (Fig. 1–16; Table 1–2). At Kulukbuk Bluffs, the Beaverhouse formation overlies deposits of the Holitna complex that date to the Boutellier interval or last interglacial.

Accumulation of last-glacial aeolian sediment in the western portion of the Holitna lowland was probably related to the development of a sandy-braided outwash train along the glacial Kuskokwim River, which provided an ample source for aeolian sand and silt. Northeasterly palaeowinds stripped sand and silt from unvegetated, outwash surfaces of the glacial Kuskokwim and deposited this sediment as sand sheets and loess over most of the Holitna lowland. As glaciers of the northwestern Alaska Range began to retreat from their full-glacial positions, incision of the distal outwash reach began, and migration of the coarse-proximal and fine-distal segments of the outwash train followed ice retreat. During the ice-retreat phase, the source-belts for aeolian sediment migrated upvalley in association with retreating ice margins, and portions of the lowland that formerly experienced sand-dominated aeolian deposition received mainly loess. Aeolian deposition in the lowland ended when the Kuskokwim River lost its braided character as a result of ice retreat, incision of the outwash train, and decreased sediment loads.

REGIONAL CORRELATIONS

Late Quaternary deposits in the Holitna lowland can be correlated with other sequences from Quaternary basins in Alaska, using the radiocarbon ages and Old Crow tephra. Aeolian deposits that are broadly similar to the Beaverhouse and Kulukbuk formations are documented in the Nushagak Bay region of coastal southwestern Alaska (Lea 1989; Lea 1990a; Lea et al. 1991). Last-glacial aeolian deposits correlative with the Beaverhouse formation (Igushik formation) are particularly well represented in the Nushagak lowland, where major periods of aeolian accumulation occurred between about 17,000 and 12,500 BP (Lea 1989:218).

The southern Nushagak lowland also contains organic silt and peaty deposits (Etolin complex of Lea 1989:106) that date to >40,000 BP and are probably correlative with the Holitna complex. Both deposits are at least partially equivalent to the ca. 30–70 ka Boutellier nonglacial interval and record mesic-to-wet, tundra environments, and climatic conditions intermediate between full glacial and modern. Similar deposits dating to this time period are recognized throughout Alaska and generally record forested interstadial-like conditions as warm as, or warmer than, present (Hopkins 1982; Schweger and Matthews 1985; Hamilton et al. 1988; Hamilton, this volume).

Aeolian deposits containing Old Crow tephra are lacking in coastal southwestern Alaska (Lea et al. 1986). Correlation of the Kulukbuk formation with deposits in this area is further inhibited by inadequate geochronological control. In the southern Nushagak lowland, deposits of the Flounder Flat complex (Lea 1989:100) are beneath organic silt and peat equivalent to the Holitna complex, and are thus in the same stratigraphic position as the Kulukbuk formation. However, the Flounder Flat complex records depositional environments that were characterized by the accumulation of tundra lake sediments and minor aeolian deposition during an interval of colder-than-modern but not full-glacial climate (Lea 1989:136). The Flounder Flat complex remains undated; however, Lea (1989:134) suggests that it probably postdates the last interglacial.

One of several drift units that are stratigraphically below the Flounder Flat complex in the southern Nushagak lowland may correlate with the Kulukbuk formation. If so, this correlation would imply extensive ice cover in the Nushagak lowland at a time when aeolian sand sheets and loess were accumulating in the Holitna lowland.

Correlation of Quaternary deposits in the Holitna lowland with late Quaternary stratigraphic sequences elsewhere in Alaska is enhanced by the presence of the Old Crow tephra. Regional correlation of Quaternary stratigraphic sequences using Old Crow tephra are discussed in Hamilton and Brigham-Grette (1991). Old Crow tephra is documented in the upper portion of the Gold Hill loess near Fairbanks (Péwé 1975a, 1989), where it crops out beneath interglacial deposits of the Eva formation dated to >56.9 ka (Péwé 1989). The Gold Hill loess is thought to be of Illinoian age (Péwé 1989) and correlates at least in part with the Kulukbuk

formation. Both Gold Hill loess and the Kulukbuk formation record a lengthy interval of periglacial aeolian sedimentation and are believed to have formed during a major episode of regional glaciation (Péwé 1989; Waythomas et al. 1993).

Deposits of early Wisconsinan age (ca. 73,000 to 59,000 BP; Martinson et al. 1987) are lacking in the Holitna lowland. The apparent absence of a depositional record that dates to this time interval suggests that early Wisconsinan deposits were either not preserved or constitute a minor and obscure component of the stratigraphic record. Aeolian deposits that date to this time period are not recognized. This may indicate restricted glaciation in the northwestern Alaska Range.

References Cited

Ackerman, R. E. 1982. The Archaeology of the Central Kuskokwim Region. Unpublished report to the National Geographic Society.

Arctic Environmental Information and Data Center. 1986. *Climatological Data for Alaska.* Anchorage.

Berger, G. W. 1991. The Use of Glass for Dating Volcanic Ash by Thermoluminescence. *Journal of Geophysical Research* 96:19705–20.

Berger, G. W., B. J. Pillans, and A. S. Palmer. 1992. Dating Loess up to 800 ka by Thermoluminescence. *Geology* 20:403–406.

Black, R. F. 1976. Periglacial Features Indicative of Permafrost: Ice and Soil Wedges. *Quaternary Research* 6:3–26.

Cady, W. M., R. E. Wallace, J. M. Hoare, and E. J. Webber. 1955. *The Central Kuskokwim Region.* U.S. Geological Survey Professional Paper 268.

Carter, L. D. 1983. Fossil Sand Wedges on the Alaskan Arctic Coastal Plain and Their Paleoenvironmental Significance. In *Proceedings of the Fourth International Conference on Permafrost,* pp. 109–14. Washington, D.C.: National Academy Press.

Coutard, J. P., and H. J. Mücher. 1985. Deformation of Laminated Silt Loam Due to Repeated Freezing and Thawing Cycles. *Earth Surface Processes and Landforms* 10:309–19.

Dostovalov, B. N., and A. I. Popov. 1966. Polygonal Systems of Ice Wedges and Conditions for their Development. In *Proceedings of the First International Conference on Permafrost,* pp. 102–105. National Research Council of Canada Publication 1287. Ottawa: National Academy of Science.

Fernald, A. T. 1960. Geomorphology of the Upper

Kuskokwim Region. *U.S. Geological Survey Bulletin* 1071-G:191–279.

Foscolos, A. E., N. W. Rutter, and O. L. Hughes. 1977. *The Use of Pedological Studies in Interpreting the Quaternary History of Central Yukon Territory.* Geological Survey of Canada Bulletin 271.

French, H. M., and J. S. Gozdzik. 1988. Pleistocene Epigenetic and Syngenetic Frost Fissures, Belchatón, Poland. *Canadian Journal of Earth Sciences* 25:2017–27.

Gallwitz, H. 1949. Eiskeile und glaziale sedimentation. *Geologica* Bd. 2.

Hamilton, T. D., and J. Brigham-Grette. 1991. The Last Interglaciation in Alaska: Stratigraphy and Paleoecology of Potential Sites. *Quaternary International* 10–12:49–71.

Hamilton, T. D., J. L. Craig, and P. V. Sellmann. 1988. The Fox Permafrost Tunnel: A Late Quaternary Geologic Record in Central Alaska. *Geological Society of America Bulletin* 100:948–69.

Hamilton, T. D., J. A. Westgate, and J. E. Begét. 1991. The Old Crow Tephra: A Stratigraphic Marker for the Last Interglaciation in Alaska and the Yukon Territory? *Geological Society of America Abstracts with Programs* 23:62.

Harry, D. G., and J. S. Gozdzik. 1988. Ice Wedges: Growth, Thaw Transformation and Paleoenvironmental Significance. *Journal of Quaternary Science* 3:39–55.

Hopkins, D. M. 1982. Aspects of the Paleogeography of Beringia during the Late Pleistocene. In *Paleoecology of Beringia*, ed. D. M. Hopkins, J. V. Matthews, Jr., C. E. Schweger, and S. B. Young, pp. 3–28. New York: Academic Press.

Hunter, R. E. 1977. Basic Types of Stratification in Small Eolian Dunes. *Sedimentology* 24:361–87.

Jones, J. A. A. 1981. *The Nature of Soil Piping: A Review of the Research.* British Geomorphological Research Group Monograph 3. Norwich, England: Geobooks.

Kline, J. T., and T. K. Bundtzen. 1986. Two Glacial Records from West-Central Alaska. In *Glaciation in Alaska: The Geologic Record*, ed. T. D. Hamilton, K. M. Reed, and R. M. Thorson, pp. 123–50. Anchorage: Alaska Geological Society.

Kocurek, G., and R. H. Dott. 1981. Distinctions and Uses of Stratification Types in the Interpretation of Eolian Sand. *Journal of Sedimentary Petrology* 51:579–95.

Koster, E. A., and J. W. A. Dijkmans. 1988. Niveo-Eolian Deposits and Denivation Forms, with Special Reference to the Great Kobuk Sand Dunes, Northwestern Alaska. *Earth Surface Processes and Landforms* 13:153–70.

Lea, P. D. 1986. Subsurface Piping Structures in Periglacial Sand-Sheet Deposits (Coversands). *Geological Society of America Abstracts with Programs* 18:669.

———. 1989. Quaternary Environments and Depositional Systems of the Nushagak Lowland, Southwestern Alaska. Ph.D. dissertation, University of Colorado.

———. 1990a. Pleistocene Periglacial Eolian Deposits in Southwestern Alaska: Sedimentary Facies and Depositional Processes. *Journal of Sedimentary Petrology* 60:582–91.

———. 1990b. Pleistocene Glacial Tectonism and Sedimentation on a Macrotidal Piedmont Coast, Ekuk Bluffs, Southwestern Alaska. *Geological Society of America Bulletin* 102:1230–45.

Lea, P. D., and S. A. Elias. 1983. Late-Quaternary Environments of the Holitna Lowland, Interior Southwestern Alaska. *Geological Society of America Abstracts with Programs* 15:624–25.

Lea, P. D., S. A. Elias, and S. K. Short. 1991. Stratigraphy and Paleoenvironments of Pleistocene Nonglacial Deposits in the Southern Nushagak Lowland, Southwestern Alaska, U.S.A. *Arctic and Alpine Research* 23:375–91.

Lea, P. D., C. F. Waythomas, R. C. Walter, and J. A. Westgate. 1986. Distribution of Old Crow Tephra in Southwestern Alaska: Implications for the Old Crow Eruption and Late-Pleistocene Glacial History of Eastern Beringia. *Geological Society of America Abstracts with Programs* 18:668.

Martinson, D. G., N. G. Pisias, J. D. Hays, J. Imbrie, T. C. Moore, Jr., and N. J. Shackleton. 1987. Age Dating and the Orbital Theory of the Ice Ages: Development of a High Resolution 0 to 300,000-Year Chronostratigraphy. *Quaternary Research* 27:1–29.

Miall, A. D. 1977. A Review of the Braided-River Depositional Environment. *Earth Science Reviews* 13:1–62.

Péwé, T. L. 1975a. *Quaternary Stratigraphic Nomenclature in Unglaciated Central Alaska.* U.S. Geological Survey Professional Paper 862.

———. 1975b. *Quaternary Geology of Alaska.* U.S. Geological Survey Professional Paper 835.

———. 1982. *Geological Hazards of the Fairbanks Area, Alaska.* Alaska Geological and Geophysical Surveys Special Report 15.

———. 1989. Quaternary Stratigraphy of the Fairbanks Area. In *Late Cenozoic History of the Interior Basins of Alaska and the Yukon*, ed. L. D. Carter, T. D. Hamilton, and J. P. Galloway, pp. 72–77. U.S. Geological Survey Circular 1026.

Schweger, C. E., and J. V. Matthews. 1985. Early and Middle Wisconsin Environments of Eastern Beringia: Stratigraphic and Paleoecological Implications of the Old Crow Tephra. *Geographie physique et Quaternaire* 39:275–90.

Washburn, A. L. 1980. *Geochronology.* New York: John Wiley and Sons.

Waythomas, C. F. 1984. Quaternary Glacial Sequence in the Chuilnuk and Kiokluk Mountains, Southwest Alaska. *Geological Society of America Abstracts with Programs* 6:339.

———. 1990. Quaternary Geology and Late-Quaternary Environments of the Holitna Lowland and Chuilnuk–Kiokluk Mountains Region, Interior South-western Alaska. Ph.D. dissertation, University of Colorado.

Waythomas, C. F., P. D. Lea, and R. C. Walter. 1993. Stratigraphic Context of Old Crow Tephra, Holitna Lowland, Interior Southwest Alaska. *Quaternary Research* 39:20–29.

Westgate, J. A. 1988. Isothermal Plateau Fission-Track Age of the Late Pleistocene Old Crow Tephra, Alaska. *Geophysical Research Letters* 15:376–79.

Westgate, J. A., B. A. Stemper, and T. L. Péwé. 1990. A 3 m. y. Record of Pliocene–Pleistocene Loess in Interior Alaska. *Geology* 18:858–61.

Westgate, J. A., R. C. Walter, G. W. Pearce, and M. P. Gorton. 1985. Distribution, Stratigraphy, Petrochemistry, and Paleomagnetism of the Late Pleistocene Old Crow Tephra in Alaska and the Yukon. *Canadian Journal of Earth Sciences* 22:893–906.

LOESS-ICE FORMATION IN NORTHEASTERN ASIA

T. D. Morozova and Andrej A. Velichko

The extraordinary formations found uniquely in northeasternmost Asia—the *yedoma*—represent a late Pleistocene periglacial sedimentary complex that consisted originally of frozen loess-like deposits, containing well-preserved syngenetic vein and ground ice. The northern boundary of these sediments coincides with the present-day coastline (Fig. 1–19), but during the late Pleistocene the coastline would have lain farther northward.

The yedoma of Siberia may be regarded as part of a formerly immense periglacial hyperzone that existed in the Northern Hemisphere. Permafrost has never melted entirely from this region since its first appearance (Velichko 1973). These huge areas were once one palaeogeographic landmass, Beringia, with the same type of periglacial ice-loessic lithogenesis throughout all late Pleistocene glacial phases. It is thus reasonable to assume that one may be able to correlate similarly derived northeast Siberian and Alaskan loessic deposits.

There are two major hypotheses that have been proposed by Russian scientists to explain the origin of these loessic soil formations in northeastern Asia. A. I. Popov (1953) and some other scientists have proposed a lacustrine-alluvial source that emphasizes the accumulation processes in floodplain-lacustrine tundra wetlands subjected to periodic flooding. S. V. Tomirdiaro (1980, this volume) suggests a subaerial source in which sediments are assumed to have formed in arid periglacial arctic steppe tundra or steppe environments with active aeolian processes.

The formation of these deposits was influenced by the following sequence of events. Worldwide sea level lowered by 130 to 100 m, as water contributed to glacier growth. Shelves became exposed and fine loessic silt began accumulating on their surfaces in eastern regions. The barren mountain ridges of Yakutia and Chukotka, which lacked even lichens under the severe glacial conditions, may have been the sources of loessic silt.

A unique "Siberian sphynx" thus appeared in Siberia that was quite different from typical loesses. Tomirdiaro argues for an immense subterranean complex—or loessic-ice body—that was composed of 70–80 percent ice. The initial loess accumulation may have been enhanced by constant winds blowing from the same direction in both summer and winter. In that way wind-borne material accumulated in the high-latitude, nonglaciated plains of eastern Siberia and Alaska. Ice-saturated loamy sands and loams located between continuous walls of fossil vein ice still exist today.

Climatic warming and an increase in moisture in the late glacial period promoted the development of thermokarst, leading to the destruction of the original loessic-ice plains. Thawing created alasy-like lacustrine depressions, which were separated by remnants of elevated loessic-ice plains. These remnants resemble plateaus, tablelands, or hills with gentle slopes. High residual uplands composed of ice-saturated sediments and interspersed by thermokarst lakes or erosional gullies received the local name "yedoma." This term is now widely used in scientific publications for sediments (or suites of sediments) of the provenance and properties described here.

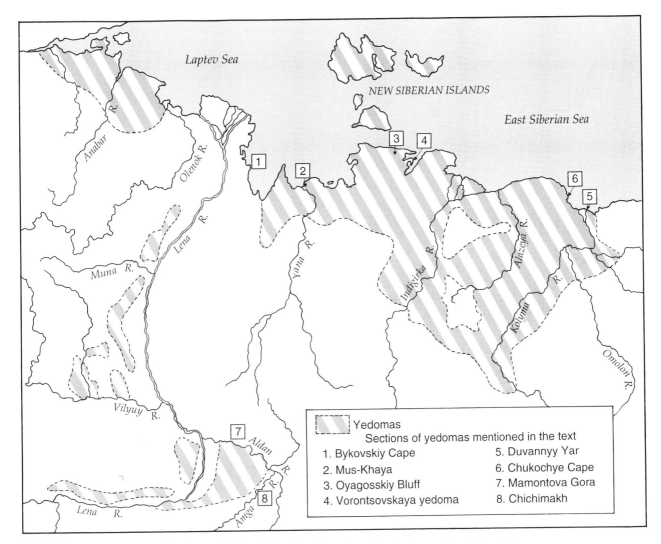

Figure 1–19   Occurrence of yedomas in the plains of northeastern Siberia (according to Tomirdiaro 1993, with authors' supplements).

CHRONOSTRATIGRAPHY OF THE LOESS-ICE FORMATION OF NORTHEASTERN SIBERIA

The yedoma formation consists of three horizons—identified throughout almost the entire area. These correspond to the Zyrian, Sartan, and Karginsky Wisconsinan glacial period of the Upper Pleistocene (Table 1–4).

1. The oldest horizon, the Zyrian (local name, "Oyagosskiy") is represented by loess-like loams or sandy loams with ice veins.
2. The Karginsky (local name, "Molotkovsky") cor-

responds to the mid-Wisconsinan warm interval. It is represented by peatbogs, gleyed lacustrine thermokarst facies, and palaeosols.

3. The Sartan (local name, "Muskhainsky") horizon (late Wisconsinan) is represented by loessic loams.

MORPHOLOGICAL FEATURES OF LOESS-ICE FORMATIONS

Two types of morphological features have been recognized within the yedoma by Tomirdiaro (1993).

Table 1–4   Correlation of Stratigraphic Schemes for the Late Pleistocene

Section	Age	Northeastern Siberia		Eastern Europe		North America	
Holocene	10,000						
Upper Pleistocene		Yedoma	Sartan (Muskhainsky)	Valday	Upper	Wisconsin	Upper
			Karginsky (Molotkovsky)		Briansk		Middle
			Zyrian (Oyagossky)		Lower		Lower
	80,000						
	120,000	Krest-Yuriakhsky		Mikulino		Sangamon	

The Arctic-Shelf (Loessic-Ice) Type   The arctic-shelf type of yedoma is characterized by the following set of properties: ice content in the upper 30–35 m reaches 85–93 percent of the whole volume; ice vein widths are 8–9 m; while the intervein earthen blocks are only 2–3 m wide, creating, in effect, earthen columns within the ice (Fig. 1–20).

With the beginning of the Holocene the formation began to thaw 0.5–0.6 m yearly, so that the entire thickness of ice became subjected to thermic planation. Thus, the surface of such yedoma acquires a peculiar microrelief represented by heavy loamy hummocks with continuous vegetative cover. This type of sediment is disappearing now because its thermal regime is incompatible with the present-day climate. These sediments were encountered on the islands of the Novosibirsk archipelago and on the northern Yakutia coast between the Lena and Indigirka deltas. Their southern border, according to S. V. Tomirdiaro, coincides with 72°N latitude.

The oldest horizon, Zyrian, was observed in the Bykovskiy cape and Oyagosskiy bluff yedomas of the northeast Siberian plains. The many radiocarbon dates available suggest the horizon is quite old. Palynological reconstructions indicate the presence of a periglacial steppe lacking arboreal species. Mammoth remains were numerous. Cryolithological data testify to a rather humid climate, relative to the more recent Sartan period. Thermokarst lakes,

Table 1–5   Results of ^{14}C Dating of Plant Residues (from Tomirdiaro 1993)

Yedoma Section	Geological Age (Horizon)	Depth in m	^{14}C Date (years ago)
Bykovskiy	Sartan		22,070 ± 410 (LC-1263)
	Karginsky		28,500 ± 1690 (LC-1329)
			33,040 ± 810 (LC-1330)
Mus-Khaya	Sartan	2	11,500 ± 210 (MAG-137)
		5	15,500 ± 50 (GIN-541)
		15.5	23,360 ± 720 (MAG-175)
Yakutsk	Sartan	3	13,700 ± 580 (IM-433)
		14	17,800 ± 850 (IM-431)
Duvannyy Yar	Sartan	3	15,850 ± 150 (MAG-592)

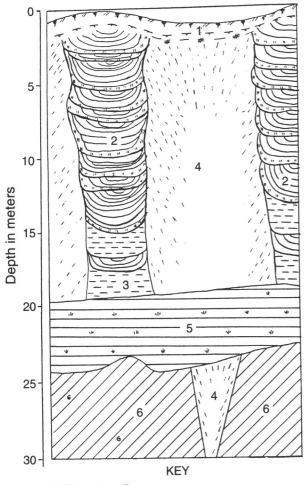

KEY

1 Recent soil

2 Packets of ice and ice-earthy layers in
   earthy columns

3 Packets of fine ice lamellae

4 Ice in ice veins

5 Talus

6 Marine sediments

Figure 1–20 Section of a yedoma complex of the arctic type. (*From Tomirdiaro 1993.*)

rapidly drying out, are presumed to have been common during that period (Tomirdiaro 1980).

In the Bykovskiy section, the Karginsky horizon, composed of moss peats, dates to approximately 33,000–28,000 years. The radiocarbon age of the Sartan horizon (determined on the basis of fossil grass root residues) is approximately 22,000–21,000 years (Table 1–5).

Investigations of arctic yedomas demonstrated a continuous but extremely weak accumulation of very fine and light silt, with a negligible accumulation rate, probably several microns per year.

The Subarctic (Continental or Ice Loess) Type This type of yedoma is characterized by lower ice content, particularly low in the Sartan horizon. Ice veins are no wider than 2–3 m. The cryogenic material is represented by loamy sand or silty sand. There are fragments of shale, sandstone, aleurites, various quartzites, as well as ferruginated calcareous rocks in the silt fraction.

The subarctic yedomas usually occur more to the south of the arctic ones. They may be found, however, within the areas of the latter. The deposition of loessic silt is far more advanced in subarctic yedomas. In the Sartan period the rate of loess deposition drastically increased, reaching 1 mm per year.

The upper Sartan horizon is the thickest in the yedoma sequence. It is composed of a loess which contains no wood macroresidues. Pollen and spores distinctive of this horizon indicate the absolute dominance of typical steppe or tundra-steppe, cryoxerophytic vegetation (Tomirdiaro 1980). Radiocarbon-dated root remains show these sediments to be 17,000–11,000 years old (Table 1–5). The loessic Sartan horizon is underlain by Karginsky deposits: palaeosols, alas peatbogs, or lacustrine-thermokarst sediments. The lower loess-ice unit displays higher ice content, and ice veins up to 5–6 m wide are common. Spore-pollen spectra are similar to those of the Sartan period. They testify to the dominance of open steppe landscapes, with a lower proportion of xerophytes. Cryolithological studies reveal periodic excessive wetness caused by thawing, creating small, ephemeral lakes (Tomirdiaro 1993). All of the subarctic yedomas may be regarded as cryolithological bisequa, with continuous microstreaks, in incompletely stratified cryotexture in the upper layer, combined with coarse, streaky, completely stratified cryotexture in the lower bisequa member. Yedomas of this type occur from Mamontova Gora in southern Yakutia to Mus-Khaya in the lower reaches of the Yana and to Duvannyy Yar and Chu-kochye Cape in the lower Kolyma River.

COMPOSITION OF YEDOMA SEDIMENT

The sediments in question are mainly loams in granulometric composition, but occasionally sandy loams. Coarse silt (0.05–0.01 mm) (the loessic fraction) dominates. According to data collected by S. V. Gubin (1987), this fraction makes up 40 to 65% of the arctic yedomas. Our observations of subarctic yedoma (loess-like sediments) in the Mamontova Gora section indicate the proportion of this fraction ranged within 30 to 35%. Loess-like sandy loams display higher amounts of fine sand, up to 7% and more. Rather high clay content (particles <0.001 mm) has been reported as frequent in subarctic yedomas. Therefore, granulometric composition of the northeast Siberian sediments under consideration indicates they belong to the loess group of sediments.

Mineralogical analyses of some subarctic yedoma sections reveal quartz and feldspar as dominant minerals; the main components of the heavy fraction are as follows: mica, ilmenite, sillimanite, limonite, sphene, and amphiboles. T. A. Khalcheva, in quantifying the primary mineral of subarctic yedomas in central Yakutia, found the same groups of rock-forming minerals, although some neoformed minerals were also present: quartz, secondary calcite, and secondary rutile (Morozova 1971). At the same time, concentrations of other minerals—iron oxides and hydroxides—are very high; this feature distinguishes the periglacial loess yedoma sediments from the loess-like Russian plain deposits.

The arctic shelf–type yedoma (observed at Oyagosskiy Bluff) chiefly has montmorillonite-hydromica-chlorite associations in the clay fraction, whereas subarctic continental yedoma (as at Vorontsovskiy Bluff) has a chlorite-hydromica association (Chernenky 1982).

Bulk chemical composition is also of a loessic type: SiO_2, 60–70%; Al_2O_3, 3–17%; Fe_2O_3, 5–8%; CaO, 1–1.8%; MgO, 0.5–2.0%. The pH (water extraction) is weakly alkaline to neutral; CO_2 content is not high and does not usually surpass 1%. The humus content is rather elevated: 1–2%; in the sub-arctic yedoma palaeosol it may reach 4%.

MICROMORPHOLOGICAL DATA

Micromorphological investigations, carried out with a mineralogical microscope and thin sections of samples with undisturbed structure, provide additional details of the characteristics of these unique ice-loess formations. Micromorphological features of loess-ice arctic yedomas have been discussed by S. V. Gubin and others (1987). They found that yedoma material in the Kolyma lowland (the Duvannyy Yar and Great Chukochye Cape sections) is peculiar in the complete absence of soil plasma and two forms of organic matter.

The organic matter is represented by evenly comminuted and weakly mineralized plant residues. No organic remains of unequivocal fluviatile origin were identified. The organic residues were represented by fragments of the above-ground parts of grasses and weakly decomposed roots of herbaceous plants. Humic compounds barely migrated from their places of origin, some precipitated on mineral grain surfaces as fine compact zones of dark flakes. Their coagulation and precipitation were enhanced by freezing. The other specific organic matter form is very fine brownish scaly segregates of sesquioxide hydrates. The yedoma layers with the latter form of organic matter (coagulated humus) seem to have higher grades of pedality than the other layers. However, microaggregates resembling the "loessic" ones, which might form coarse silt-size aggregates with humus as a cement, were not found.

Micromorphological studies of *subarctic* loessic yedoma deposits have been performed by T. D. Morozova for some profiles located in the Lena–Amga interfluve (Fig. 1–21). The loessic sediments of this area are pale grey, rather homogeneous, fine silty, weakly calcareous, and porous, with a fine platy structure produced by cryogenic processes. A cinnamon-grey palaeosol of Karginsky age (?) is regarded as the boundary between the Sartan and Zyrian loesses. Both sediments and palaeosols contain numerous well-preserved organic residues, fre-

(a)

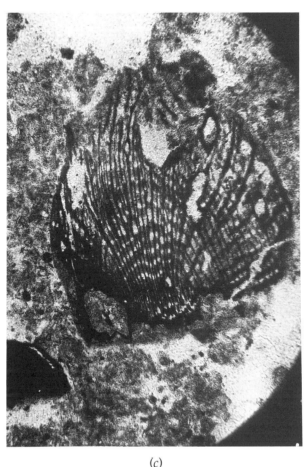

(c)

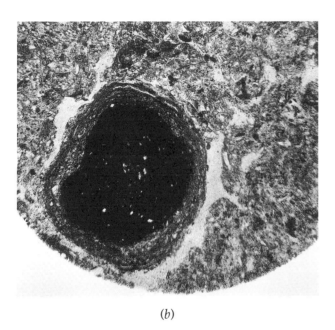

(b)

Figure 1–21 Micromorphology of subarctic yedoma in Chichimakh section: (*a*) micromorphology of a loess-like loam at the depth of 2.6 m; 45×, PPL; (*b*) rounded subangular aggregates and an iron nodule in the palaeosol; 45×, PPL; (*c*) plant residue in the palaeosol; 45×, PPL.

quently found in the course of terrain observations. There were no carbonate nodules, which are expected in typical loesses.

Micromorphologically, the loess-like sediments in the region discussed are conspicuous by good sorting of rock-forming minerals according to their size, the coarse silt fraction being dominant (Morozova 1971). Sediments lack compact, intergranular porosity. The characteristic of loesses usually prevails, although there are few biogenic voids with plant residues in them. Plant residues are common in the groundmass; some of them have their cell tissue fabric discernible, clay plasma performing the role of cement in aggregates of various sizes, while micrite is entirely absent. Clay has no optical orientation.

SUMMARY

The fabric, composition, and micromorphology of the northeast Siberian loess formation sediments allow us to conclude that these unique late Pleistocene periglacial phenomena are undoubtedly subaerial accumulations that may be regarded as loesses in the initial stages of loessic reworking ("loessification," as defined by I. P. Gerasimov 1962). They differ from the typical loesses occurring within the Russian plain or Central Asia because of the minimal degree of reworking of the initial material. This is attributed to the conserving effect of both ancient and present-day cryoaridic climatic conditions and of permafrost upon the mineral and organic material. Primitive soil-forming processes have been operating here under extremely cold and dry conditions; hence, the secondary carbonates content is low. Plant residues remain well-preserved and, as in sediments, so in palaeosols. Rock-forming minerals are weakly weathered. Good granulometric sorting of loessic rock-forming minerals, strong pedality, intrapedal and biogenic porosity are the common features of the yedoma sediments and loesses. Loessic reworking is more pronounced in the subarctic yedoma sediments.

The loess-like sediments of northeastern Siberia are thought to belong to the gigantic periglacial hyperzone of the northern hemisphere that existed in the late Pleistocene in the northern half of Europe, where typical loesses originated.

REFERENCES CITED

Chernenky, B. I. 1982. Comparative Mineralogical and Geochemical Analysis of Sediments Composing Yedomas of Shelf and Continental Types. In *Cryogenic-Geological Processes and Palaeogeography of Lowlands of North-Eastern Asia*, pp. 69–75. Magadan. (In Russian)

Gerasimov, I. P. 1962. Loess and Soil Formation. Izv. Academy of Sciences USSR, *Ser. Geogr.* 2:3–9. (In Russian)

Gubin, S. V. 1987. *Palaeogeographic Aspects of Soil Formation in the Coastal Lowland of Northern Yakutia.* Pushchino. (In Russian)

Morozova, T. D. 1971. Characteristics of Loess-like Sediments of Central Yakutia. Izv. Academy of Sciences USSR, *Ser. Geogr.* 7:95–104. (In Russian)

Popov, A. I. 1953. Lithogenesis Features in Fluvial Plains under Severe Climatic Conditions. Izv. Academy of Sciences USSR, *Ser. Geogr.* 2:50–72. (In Russian)

Tomirdiaro, S. V. 1980. *Loess-Ice Formation of Eastern Siberia in the Late Pleistocene and Holocene.* Moscow: Nauka. (In Russian)

———. 1993. Palaeogeography of Northern Plains (According to Data of Loess-Ice Complex Research). Development of Landscapes and Climate of Northern Eurasia. Late Pleistocene–Holocene. Elements of Forecast. *Regional Palaeogeography* 1:66–70. (In Russian)

Velichko, A. A. 1973. *Natural Process in the Pleistocene.* Moscow: Nauka. (In Russian)

PALAEOGEOGRAPHY OF BERINGIA AND ARCTIDA

Savelii V. Tomirdiaro

On the basis of extensive field and laboratory research, the author postulates the former existence of a continuous loess-covered land surface underlain by permanently frozen sea ice. This vast land area connected Asia and America in the late Pleistocene, in addition to the now more conventionally accepted Bering Land Bridge. The unique geological deposits that represent the remnants of this loess-ice bridge—Arctida—are found scattered over the landscape of a number of islands of the Novosibirsk Archipelago and along the shoreline of the eastern Siberian lowlands. The relative scarcity of these remnants called *yedoma*, seen as high relict hills and plateaus surrounded by boggy Holocene lowlands, is the result of melting in the Holocene of the extensive ground-ice formations by lake and land thermokarst processes (Tomirdiaro 1978, 1980).

The structure of these remnants of Arctida is unique. The hills and plateaus investigated consist of 90 percent underground ice and only 10 percent mineral loam. The loam is found as earthen columns two to three meters in diameter and forty me-

Figure 1–22   General view of a remnant of Arctida: Dmitri Laptev Strait (Region A on Fig. 1–26). The dark sections of earthen columns are visible in the wall of underground ice. The exposure is 25 m in height. The longitudinal stripes revealed in the exposure as the sun melts the ice indicate its veined structure. At the very top of the exposure a thin layer of tundra soil protects the ice from melting. In the foreground are small mounds of soil—the remains of earthen columns—left by the receding of the ice wall.

ters or more in height as observed in shoreline exposures (Figs. 1–22, 1–23, 1–24). In plan, these columns are positioned in a chess-like arrangement surrounded by massive ice formations up to eight meters wide. Thus are seen here polygonally veined ice formations developed to incredible size but in form rather typical ice-wedge polygons.

Patterned ground is still developing today in the plains of the far north due to the yearly cracking of perennially frozen soils into polygonal blocks as a result of expansion and contraction. These deep cracks were formed by infilling with sublimated and congealed ice, forming ice wedges. The size of the polygonal pattern is a function not only of the annual average temperature, but also of the amplitude of seasonal temperature fluctuation of the permafrost. In the maritime climate of the Arctic today there is found patterned ground with polygonal nets as large as 15×20 meters or more (Fig. 1–25). The Pleistocene patterned ground, as preserved in the Arctida remnants, displays nets twice as large (Fig. 1–23).

The nature of the Pleistocene patterned ground provides unambiguous evidence of the dramatic climatic changes in the arctic and sub-arctic that occurred at the Pleistocene–Holocene boundary, from a strongly continental climate to a maritime one. The formation of such massive ground ice as found

Figure 1–23   Aerial view of a remnant of Arctida: shore of Dmitri Laptev Strait. The same exposure as shown in Figure 1–22 is seen as the ice bank in the foreground with the dark sections of earthen columns showing. Mounds on the tundra surface in the distance are the tops of earthen columns. These are visible as a result of contemporary melting of the underground ice under a very thin soil cover. The location of the mounds shows the net of the Pleistocene patterned ground.

in the remnants of Arctida would be virtually impossible under contemporary conditions.

It is very fortunate for the study of the palaeogeography of the Pleistocene that these exceptional deposits have been preserved with their unique palaeoecological and climatic record. However, they have survived only in a very specific, limited region of the coastal plains of far northern Yakutia between the Taimyr Peninsula and the mouth of the Indigirka River, north of 72°N latitude (Fig. 1–26). The historical descriptions and photographs of "ice walls" found along riverbanks and eroded sea cliffs of Chukotka, Alaska, and the more southerly regions of Yakutia are of much thinner ice veins de-

veloped in the sandy loess. The "solid" ice visible in those exposures is illusory, caused by the collapse of the bank along the pre-existing frost cracks. As these veins melt under the summer sun, it is seen that they are quite narrow, no more than 1.5 to 2.5 meters wide (Figs. 1–27, 1–28).

Based on the various characteristics of the Yakutian deposits, late Pleistocene permanently frozen loess deposits of northeastern Asia can be divided into two categories. The first, predominantly ice—loessic ice—is designated "arctic" (Figs. 1–22, 1–24); the second, predominantly loess—icy loess—is designated "subarctic" (Figs. 1–27, 1–28). Detailed data on the mineralogical, palynological,

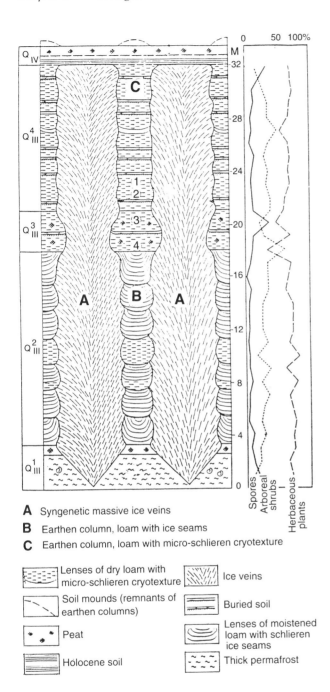

Figure 1–24 Section of an Arctida remnant—"arctic" loessic ice deposit—with pollen and spore diagram. The lower portion of the earthen column (B) consists of loam separated by ice seams. The curvature of these seams indicates the compression of the soil in the center of the polygonally patterned ground formations as a result of the growth of surrounding ice (frost) wedges. The disconformity of the bedding of the loam and the presence of schlieren ice seams within it indicates periodic melting and moistening of the loam during the Zyrian period (Tomirdiaro 1980, 1982, 1987). The upper 10-meter portion of the earthen column section (C) shows no evidence of periodic moisture in its loam with micro-schlieren cryotexture. This indicates the extreme dryness of the climate in Sartan times. Several interbedded buried soils are found in the upper deposits. (*From Tomirdiaro and Chernenky 1987.*)

A Syngenetic massive ice veins

B Earthen column, loam with ice seams

C Earthen column, loam with micro-schlieren cryotexture

Lenses of dry loam with micro-schlieren cryotexture

Ice veins

Soil mounds (remnants of earthen columns)

Buried soil

Peat

Lenses of moistened loam with schlieren ice seams

Holocene soil

Thick permafrost

palaeontological, granulometrical, glaciological, and other analyses of the deposits of both categories have been published previously (Tomirdiaro 1978, 1980, 1982a, 1984, 1987; Tomirdiaro and Chernenky 1987).

Of special interest are the most recent data concerning the petrography of the mineral content of the coarse-grained dust fraction which dominates both the "arctic" and the "subarctic" deposits. In the latter case, this fraction appears to contain tiny fragments of metamorphic sedimentary rocks of terrigenous origin, which are found in the mountain masses that surround the north Yakutian plains. In the former case, the earthen columns of the arctic zone, this fraction contains fragments of volcanic origin: tuffs, lavas, and abundant accumulations of volcanic glass in the form of undisturbed layers of volcanic ash (Tomirdiaro and Chernenky 1987). To explain this in the absence of volcanoes in the Asian arctic and subarctic requires resorting to an entirely different system of atmospheric circulation in the late Pleistocene arctic in order to transport products of volcanic activity from remote regions.

The very small mineral content and the very strong development of the ice veins in the loessic ice deposits of Arctida (Figs. 1–22, 1–24) indicates a very slow rate of accumulation of mineral dust. The average thickness of soil on thermo-terraces is only 1.5–2 meters, having developed since the melting of the ice accumulated during the Zyrian Q^2III and the Sartan Q^4III cold stages of the late Pleistocene. Such a slow rate of deposition is commensurate with the global rain of volcanic and meteoric dust. Other indications of the exotic origins of the dust fraction of the earth columns in Arctida are the lack of arboreal pollen and abundance of herbaceous pollen in the deepest cold periods Q^2III and Q^4III (Figs. 1–24, 1–28) in the high arctic with the simultaneous exis-

Figure 1–25   Aerial view of modern polygonal ground on tundra lowland located between Arctida remnants (Region B on Fig. 1–26). Earthen columns forced above ground by modern ice veins comprise the micro-relief of the boggy Holocene lowlands. Beneath each mound is an epigenetic ice vein, 3–5 m wide and 5–7 m deep. The polygonal net measures approximately 15 × 20 m.

tence of ''refugia'' of forest species in the mountain valleys to the south of the mountains (Tomirdiaro 1978, 1980, 1984; Kaplina 1979; Tomirdiaro and Chernenky 1987) and the almost complete lack of the sand fractions in their granulometric composition. The direction of the transport of the ''exotic dust'' appears to have been mostly from the north, from the direction of the Arctic Ocean which was frozen at that time period (Tomirdiaro 1984; Tomirdiaro and Chernenky 1987).

There is extensive development of loessic ice deposits of the type shown in Figures 1–22, 1–23, and 1–24 on present-day islands in the Laptev and East Siberian seas (Fig. 1–26). Some evidence has been lost due to coastal erosion during this century, as is the case for the neighboring Vasilievsky and

Semenovsky islands (Tomirdiaro 1974). The sea islands of Mercurius and Diomida, which were historically mapped, and Sannikov Land, which was known locally, have disappeared (Sukhotsky 1972). Their place now is occupied by dangerous shoals known among sailors as the infamous ''icy bottoms'' because of their slippery banks, which will not hold anchors.

This foregoing evidence points to the conclusion that a loessic-ice formation of uniform structure covered the entire continental shelf of northern Yakutia and reached over the Arctic Ocean on the thick, perpetually frozen pack ice as recently as before the beginning of the Holocene. In the southern regions, the ice overlying the continental shelf is conventionally included as a part of the Bering

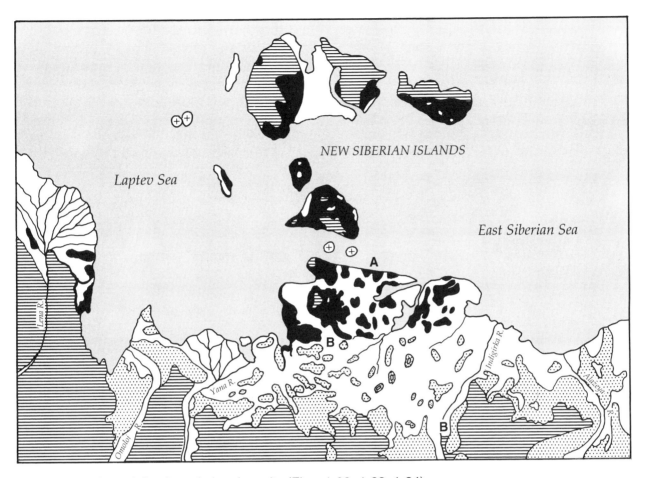

A = location of "arctic" or loessic-ice deposits (Figs. 1-22, 1-23, 1-24)

B = location of "subarctic" or icy loess deposits (Figs. 1-27, 1-28)

Plateaus and relicts of plateaus with "subarctic" icy loess deposits (relatively low ice content)

Relicts of plateaus with "arctic" loessic ice deposits (high ice content) – remnants of Arctida

Boggy Holocene lowlands created by thermokarst processes (see Fig. 1-25)

Mountain masses

Approximate locations of Arctida remnants recently destroyed by the sea: Vasilievsky, Semenovsky, Mercurius , and Diomida islands

Figure 1–26   Schematic map of the location of surviving remnants of "arctic" loessic-ice deposits of Arctida.

Land Bridge; in the northern region the ice covered with aeolian dust constituted a floating land, "Arctida"—"floating" in the sense that it was not immediately underlain by terrestrial deposits but by permanent pack ice and, beneath that, water. The

remnants of this vast cover found on the arctic coast and islands attest to its uniqueness and the lack of contemporary analogs.

Within the relicts of Arctida (loessic-ice deposits) are found abundant mammoth remains. This

Figure 1–27   General view of an exposure of a "subarctic" icy loess deposit, 20 m high (Region B on Fig. 1–26). Narrow vertical ice veins cut perpendicularly to their axes (cut horizontally) are visible in the central portion of the exposure. To the right and left extremes of the exposure, ice walls formed from the same thin veins but cut vertically are visible. In the foreground is the snow-covered base of the exposure.

contrasts sharply with the significantly more rare occurrence of mammoth finds in the more southern subarctic icy loess deposits. As the deposits are of the same age, it indicates not only better preservation in the north but also a greater abundance of fauna.

The study of macro- and micro-remains of flora in the deposits indicates that within the limits of Arctida during the cold phases of the late Pleistocene there existed here extremely rich grassy meadowlands. In contrast, to the south, dry *Artemisia* steppes were widespread, giving way in southern Yakutia to sandy, stony deserts (Kolpakov 1970, 1973, 1982; Kaplina 1979).

Thick permafrost underlay the thin soil layer during the accumulation of the Arctidan loessic-ice deposits (Figs. 1–22, 1–23), and the thickness of the ice wedges did not diminish even during the coldest and driest times of the Siberian Sartan—late

Würm in Europe (Fig. 1–24). South of approximately 72° north latitude (Fig. 1–26) in sub-arctic deposits of the same time period—Q⁴III—the thickness of ice wedges decreased rapidly, accompanied by a rapid increase in the loess accumulation rate (Tomirdiaro 1984; Tomirdiaro and Chernenky 1987). This acceleration in the accumulation rates of periglacial loess was a global phenomenon seen also in Quaternary records of Europe and North America (Kriger et al. 1972; Velichko 1973; Péwé 1975) and caused by the sudden cooling and drying; in other words, the continentalization of the climate. In contrast, however, in the arctic regions of eastern Siberia, loess continued to accumulate at the same slow rate as during the more moderate Zyrian phase (early Würm Q²III), which allowed the continued growth of the immense ice wedges (Fig. 1–24).

The giant ice wedges of the relict Arctida—and even occasionally some of those found further to

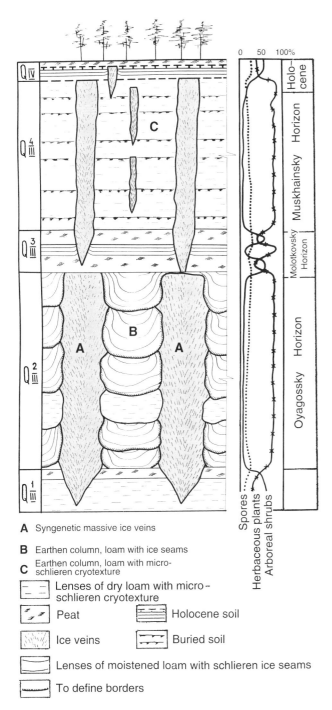

A Syngenetic massive ice veins

B Earthen column, loam with ice seams

C Earthen column, loam with micro-schlieren cryotexture

Lenses of dry loam with micro-schlieren cryotexture

Peat

Ice veins

Holocene soil

Buried soil

Lenses of moistened loam with schlieren ice seams

To define borders

Figure 1–28 Section of "subarctic" icy loess deposit with pollen and spore diagram. (*From Tomirdiaro and Chernenky 1987.*)

the south—are composed of dry, loose ice crystals of a sublimational genesis rather than solid ice formed from water entering into the cracks during the summer (Davidenko 1982; Tomirdiaro and Chernenky 1987). The ice crystals formed on the walls of the cracks in the early spring as a result of condensation directly from the air, this because of the temperature differential between the sun-warmed air and the winter-cooled permafrost—a process analogous to the familiar frost build-up in a freezer.

Despite their formation under similar palaeoclimatic conditions of extreme continentality and aridity, the remarkable difference in the geological structure of the aeolian-cryogenic deposits of the arctic and subarctic types produced vastly differing landscapes. In the Arctic, moist highly productive tall-grass meadows abounded with herds of mammoths, bison, and large horses. Evidence from the sizes of mammoth tusks found in the relicts of Arctida, particularly those in the Novosibirsk Archipelago, indicates that these animals were significantly larger than the "mountain mammoths" of the southern regions. It is possible that these highly productive meadows extended across marine Arctida as a continuous land surface connecting eastern Siberia to the arctic territories of Canada. The abundance of ice in the Arctidan deposits supplying the necessary moisture might explain the rich plant growth.

In the southern steppes of the sub-Arctic the source of the aeolian dust was the nearby mountain ranges. Even with the prevailing northeastern winds of that time, the rate of deposition of loess was ten times that of the arctic regions. Under these conditions ice wedges did not grow to the breadth they achieved in the arctic deposits (Fig. 1–28). Thus, with a significantly smaller proportion of ice in the subsoil providing significantly less moisture to plant life, dry *Artemisia* plains and even semi-arid deserts developed. These supported saiga, muskoxen, bison, and smaller horses.

Recent research on the composition and structure of the ice wedges has contributed new data to the interpretation of the palaeogeography of the region. As the porosity of the ancient ice veins is often as high as 9 to 10 percent (which is possible only in ice formed by sublimation), a detailed analytical study of the ancient air contained in the numerous bubbles trapped in the ice crystals was carried out. Large quantities of samples were collected from the various arctic relics which were then analyzed in the gas analytic laboratory. It was found that the oxygen content of the ancient air was two times lower than that of modern air (Tomirdiaro and Chernenky 1987:157). As it is postulated that the ice formed by sublimation in the polygonal frost cracks in early spring, it is supposed that this ancient air was trapped before the beginning of the growing season and thus represents wintertime conditions of low oxygen content. During the summer's 24-hour daylight, the rapid growth of grasses undoubtedly saturated the air with oxygen.

The ice-wedge polygons appear to be excellent "palaeo-thermometers." The structure of the individual ice wedges as well as the overall size of the polygonal grid allows reconstruction of winter palaeotemperature conditions in the arctic that could reach as low as −100°C or lower. The summer temperatures could reach 20°C or higher, indicating an extreme continental climate that has no modern analogs (Tomirdiaro 1984; Tomirdiaro and Chernenky 1987). The cause for such extreme climatic conditions is found in the existence of dry land in the place of the modern Arctic Ocean. A solid armor of thick marine ice covered the ocean for millennia. In Europe and western Siberia the continental glaciation extended onto this ice shield. In eastern Siberia, however, due to the extreme aridity, glaciers formed only in the mountains (Kolpakov 1973; *Geological Map* 1978). As a result, aeolian loess deposits with a high content of underground ice accumulated in the lowlands (Figs. 1–22, 1–27). In the deposits of Arctida the ice content is so high (Figs. 1–22, 1–23, 1–24) that the first visitors to the region took them to be buried glaciers (Toll 1897). Soviet geologists soon replaced this explanation of the origin of these deposits with a theory of alluvial origins that persisted for some time. However, evidence against alluvial origins of the frozen loess deposits of Yakutia is found in (1) the abundance of typical steppe faunal remains, (2) the absence of remains of fish or other aquatic fauna, (3) the impossibility of depositing alluvium of the same age

on numerous terraces of different ages and elevations, (4) the absence of riverbed faciae in the sections of the loess deposits and the typical loess micro-aggregation of these deposits, and, finally, (5) in the uniformity of granulometric and mineralogic composition in these deposits. A discussion of the aeolian origins of the loess deposits of Yakutia and Chukotka is found in Tomirdiaro 1980 and 1982b. The aeolian origins of the subarctic type of cryogenic loess were recognized earlier in Alaska (Péwé 1968, 1975).

In the late Pleistocene, Yakutia, Chukotka, and Alaska were united into one palaeogeographic landscape (Péwé et al. 1977). The discussion of the cryogenic-aeolian genesis of the so-called underground glaciation of Yakutia allows the postulation of a unique loessic-ice land expanse—Arctida—that occupied the Yakutian continental shelf as well as huge ocean areas to the north (Tomirdiaro and Chernenky 1987).

Evidence indicates that although Arctida of the late Pleistocene could have been a trans-arctic bridge for mammoth fauna and meadow and steppe flora, for humans it would have been an insuperable barrier. Extremely low winter temperatures combined with air containing half the modern norm of oxygen would have proved fatal for humans. The mammoth, however, was well adapted to these conditions, having flourished in the area throughout the late Pleistocene. Mammoth bones were found *in situ* in a frozen loess relict of Arctida in the region of the delta of the Lena River (Tomirdiaro et al. 1984). These were dated to 21,600 years ago by radiocarbon. Mammoth tusks and a tooth found by V. M. Makeev on the now glacier-covered Severnaya Zemlya are approximately 19,300 years old (Makeev et al. 1979). Notably, this was the coldest and harshest period of the entire Pleistocene and these remains were found in the deepest Arctic. Thus a migration across this land of peoples who would become the Palaeo-Indians of America could not have occurred during Sartan times—Q⁴III—but only at the end of the Pleistocene or the beginning of the Holocene when Arctida was destroyed, the Arctic Ocean uncovered, and the climate changed drastically.

In the late Pleistocene, a vast land developed

over an extensive portion of the Arctic Ocean connecting Eurasia and North America, helping to create an immense supercontinent. In that process the pole of cold shifted to the region of the geographic north pole (where it is supposed to be according to the radiational balance of the planet). The continentality of climate of this supercontinent became truly unearthly. The global circulation resembled an hourglass in appearance, with a perpetual arctic anticyclone in the lower layers of the atmosphere representing the bottom of the hourglass. In this lower layer the air flowed from the zone of barometric maximum near the geographic pole towards the south. The Coriolis force caused the winds to blow from the east towards the west. The evidence of these predominantly east and northeast winds is seen in the structure of the continental dunes that developed in Yakutia, the depositional and mineralogical characteristics of the arctic and subarctic aeolian loess deposits, and the pollen diagrams derived from these deposits, as described above (Figs. 1–24, 1–28).

To compensate for the outflow of air to the south, air was drawn down from the upper layers of the atmosphere and troposphere. Thus, above the permanent arctic anticyclone formed a permanent arctic cyclone, resembling the upper portion of the hourglass. It was this cyclone that drew the cosmic and volcanic dust from the upper layers of the atmosphere of the entire northern hemisphere into the arctic. This might explain the large clay fraction found by granulometric analysis of the earthen columns of the arctic deposits. Weathering conditions in the north do not normally permit the breakdown of materials to the clay fraction. Thus, little clay is found in the loess deposits of the subarctic type. (Relevant plots of the granulometric composition of the arctic and subarctic deposits have been published previously [Tomirdiaro 1980, 1984].) The abundant clay fraction of the arctic deposits suggests an exotic origin—from remote southern regions. The terrigenous material, containing fine-grained as well as coarse-grained dust which is also found in the earthen columns of arctic deposits, was drawn up into the upper layers of the atmosphere by means of storms, tornadoes, and hurricanes.

The global circulation here postulated would

have supplied the Arctic with oxygen-poor air drawn down from the upper atmosphere. In summer the photosynthesis of plants would have eased the problem. But in winter in Arctida and northern Beringia, horrific ecological conditions were created: temperatures of $-100°C$ or more, the darkness of the three-month-long polar night, and half the normal content of oxygen in the air.

These severe conditions in the depths of the Pleistocene cold period have raised the question of hibernation of mammoths. It is possible that these animals experienced a kind of torpor similar to that observed among modern penguins in Antarctica. A recent publication by a group of biologists from Moscow and Yakutia bears directly on this question. They have discovered a substance in the brains of modern horses of the Yakutian breed which, during severely cold weather, decreases locomotor activity, significantly decreases the consumption of oxygen, and, as well, suppresses the synthesis of protein when a reduction in the mass of fat occurs. Further, injecting this substance into non-hibernating mice reduced their rectal temperature by 7°C and their oxygen consumption by 60 percent (Sukhova et al. 1990). It is important to note that the authors of this study were not familiar with the palaeogeographical hypothesis set forth in this paper, nor were they aware of the recently established fact that the Yakutian breed of horse is a direct descendant of the horses of the mammoth world of Yakutia.

The remains of Arctida are not only being rapidly destroyed by the sea, lakes, and rivers, but also by simple settling as a consequence of the melting of the tips of the ice veins during the summer. This process is very visible in the aerial photographs of tundra showing the tops of earthen columns as raised mounds on the surface of the tundra (Fig. 1–23). The contemporary climate of the Arctic has become too warm for the loessic-ice deposits of the late Pleistocene.

The massive destruction of the whole of Arctida, however, did not result from the increase in solar radiation—that is, warming from above—rather it resulted from the breakthrough of warm Atlantic waters into the Arctic Ocean—that is, warming from below—that occurred at the Pleistocene–Holocene boundary. The warming of the cold waters of the Arctic Ocean below the loessic-ice armor of Arctida led to its thawing and disintegration. Evidence for this is seen in the sudden appearance of mollusks (malacofauna) and Atlantic Foraminifera in the early Holocene sediments of the Arctic Ocean.

After the destruction of Arctida, the warming of the Arctic Ocean was so intense that an ocean lacking or nearly lacking ice might have taken its place in the Holocene optimum. Thus, the forest growth of the Holocene optimum spread not only across the plains of arctic Yakutia, but also as far north as the northernmost islands of the Novosibirsk Archipelago, as shown by large quantities of remains of arboreal willow and birch found in peat bog deposits of that period (Khotinsky 1977). The transformation of the ice-free Arctic Ocean of the Holocene optimum to the contemporary Arctic Ocean occurred during the second half of the Holocene, in consequence of the reduction of the inflow of Atlantic waters. Contemporary landscapes of the arctic tundra formed during this period likewise (Khotinsky 1977).

CONCLUSION

In conclusion, the destruction of such an immense land area in the Arctic—that formed by Arctida in the eastern sector and by the glacial shields extending onto the ocean in the western sector—resulted in the complete reconstruction of atmospheric circulation in the northern hemisphere. The uncovering of a huge area of arctic ocean which had previously been buried under an ice or loessic-ice armor resulted in a winter barometric minimum in place of the previous barometric maximum. Powerful cyclones replaced the anticyclonic regime; a pluvial maritime climate replaced the polar arid climate. Dry steppe landscapes of the sub-Arctic were replaced by poorly drained landscapes of continuous bogs and lakes. Mammoth and Siberian musk-oxen became extinct; horses and bison were transformed into the early Holocene forest species. Even the insect fauna changed (Kiselev 1981).

Such a major catastrophe in the Arctic and sub-Arctic could not have taken place without dramatically altering the natural environment of the mid-latitudes of the northern hemisphere. Perhaps it is not coincidental that in the Biblical and Sumerian legends the Flood may be ascribed to a period of the early Holocene. Echoes of these phenomena can be heard in the legends of American Indians. Such significant global events were explained by many explorers as the destruction of the mythical continent Atlantis. However, in such relatively recent times, it would be impossible for a continent to disappear into the Atlantic Ocean without leaving a trace. But in the Arctic a vast, super-marine landmass composed mostly of ice and comparable in size to some continents did, indeed, vanish simply by melting. In that event may lie the roots of some legends of catastrophe.

REFERENCES CITED

Davidenko, V. P. 1982. Signs of Sublimational Genesis of Veined Ice of the Yano-Omoloyan Complex. In *Cryogeological Processes and the Palaeogeography of Lowlands in Northeastern Asia*, pp. 66–69. Works of SVKNII DVNC AN USSR. Magadan. (In Russian)

Geological Map of the Northeast USSR. 1:1,500,000. 1978. Leningrad. (In Russian)

Kaplina, T. M. 1979. Pollen Spore Spectra of Ice Complex Deposits in the Maritime Lowlands of Yakutia. Izvestiya, AN USSR Geographic Series #2, pp. 85–93. (In Russian)

Khotinsky N. A. 1977. *The Holocene of Northern Eurasia*. Moscow: Nauka. (In Russian)

Kiselev, S. V. 1981. *Late Cenozoic Coleoptera of Northeastern Siberia*. Moscow: Nauka. (In Russian)

Kolpakov, V. V. 1970. Concerning Fossil Deserts of the Lower Lena River. *Bulletin of the Commission on Quaternary Research* 37:75–84. (In Russian)

———. 1973. Palaeogeographic Significance of Quaternary Eolian Deposits of Northern Eastern Siberia. In *Some Questions of Regional Geology*, pp. 38–42. Moscow: MGU. (In Russian)

———. 1982. Conditions of the Distribution and Position of the Yedoma Formation. In *Cryogeological Processes and Palaeogeography of the Lowlands of Northeastern Asia*, pp. 22–30. Works of SVKNII DVNC AN USSR, Magadan. (In Russian)

Kriger, N. I., N. E. Kotelnikova, and V. V. Sevastyanova. 1972. Regarding the Question of the Nature of Yedoma Deposits. In *Abstracts*, 14th Pacific Scientific Congress, Cenozoic Section, Vol. 2: 158.

Makeev, V. M., K. A. Arslanov, and V. E. Garutt. 1979. The Age of Mammoth from the Northern Land and Some Questions about Pleistocene Palaeogeography. Papers of the Academy of Sciences, USSR 245(2):421–24. (In Russian)

Péwé, T. L. 1968. Loess Deposits of Alaska. 23d International Geological Congress, Prague. Vol. 8, pp. 297–309.

———. 1975. *Quaternary Stratigraphic Nomenclature in Unglaciated Central Alaska.* U. S. Geological Survey Professional Paper 862.

Péwé, T. L., A. Journaure, and R. Stuckenrath. 1977. Radiocarbon Dates and Late Quaternary Stratigraphy from Mamontova Gora, Unglaciated Central Yakutia, Siberia. USSR. *Quaternary Research*, No. 7.

Sisko, R. K. 1970. Thermoabrasional Shores of Arctic Seas as Exhibited by the Novosibirsk Archipelago Examples. *Works of the Arctic and Antarctic Scientific Research Institute* 294:183–94. Leningrad. (In Russian)

Sukhotsky, V. I. 1972. New Traces of Diomida Island in the Laptev Sea. In *Chronicle of the North*, pp. 126–128. Moscow: Misl. (In Russian)

Sukhova, G. S., D. A. Ignatiev, A. K. Akremenko, et al. 1990. Cardiotropic, Hypometabolic and Hypothermal Properties of Peptide Components in the Tissues of Cold-Adapted Animals. *Magazine of Evolutionary Biochemistry and Physiology* 26:623–29. (In Russian)

Toll, E. D. 1897. Fossil Glaciers of the Novosibirsk Archipelago: Their Relationship to Mammoth Remains and to the Ice Age. *Notes of the Russian Geographical Society* 32(1):130–52. St. Petersburg. (In Russian)

Tomirdiaro, S. V. 1974. Holocene Thermoabrasion Formation of the Shelf, East Arctic Seas of the USSR. *Papers of the Academy of Sciences USSR* 219(1):179–82. (In Russian)

———. 1975. Loess-Ice Formation of the Late Pleistocene Hyperzone in the Northern Hemisphere. In *Geological Research in the Northeastern USSR.* Works of SVKNII DVNC AN USSR, Vol. 68. Magadan. (In Russian)

———. 1978. *Natural Processes and the Opening of Permafrost Territories.* Moscow: Nedra. (In Russian)

———. 1980. *Loess-Ice Formation in Eastern Siberia during the Late Pleistocene–Holocene.* Moscow: Nauka. (In Russian)

———. 1982a. Arctic and Subarctic Types of Cryoloess and Distinguishing the Shelf and Continental Types of Yedoma Formation. In *Natural Development of the Territory of the USSR during the Late Pleistocene and Holocene*, pp. 134–42. Moscow: Nauka. (In Russian)

———. 1982b. Evolution of Lowland Landscapes in Northeastern Asia during Late Quaternary Time. In *Paleoecology of Beringia*, ed. D. M. Hopkins, J. V. Matthews, Jr., C. E. Schweger, and S. B. Young, pp. 29–37. New York: Academic Press.

———. 1983. Loess-like Yedoma Complex Deposits in Northeastern USSR: Stages and Interruptions in Their Accumulation and Their Cryotextures. *Proceedings,* Fourth International Permafrost Conference. Washington, D.C.: National Academy Press.

———. 1984. Researching the New Cryogenic-Eolian Geological Formation in the Northeastern USSR. In *Geology and Ore Minerals of Northeastern Asia*, pp. 43–56. DVNC AN USSR. Vladivostok. (In Russian)

———. 1987. Periglacial Landscapes and Loess Accumulation in the Late Pleistocene Arctic and Sub-Arctic. In *Late Quaternary Environments of the Soviet Union*, pp. 141–45. Minneapolis: University of Minnesota Press.

Tomirdiaro, S. V., and B. I. Chernenky. 1987. *Cryogenic-Eolian Deposits in the Eastern Arctic and Sub-Arctic.* Moscow: Nauka. (In Russian)

Tomirdiaro, S. V., K. A. Arslanov, and B. I. Chernenky. 1984. New Data on the Formation of Loess-Ice Sequences in North Yakutia and the Condition of Mammoth Habitat in the Arctic during the Late Pleistocene. *Papers of the Academy of Sciences, USSR* 278(6):1446–49. (In Russian)

Velichko, A. A. 1973. *Natural Process in the Pleistocene.* Moscow: Nauka. (In Russian)

Chapter 2

BIOTIC RECORDS

LATE PLEISTOCENE AND EARLY HOLOCENE POLLEN RECORDS FROM THE SOUTHERN BROOKS RANGE

Patricia M. Anderson and Linda B. Brubaker

During the transition between the Pleistocene and Holocene, the Beringian landscape changed dramatically. The complexity of vegetation response to these climatic changes is now apparent for some areas of eastern Beringia (see reviews by Ritchie 1984, 1987; Ager and Brubaker 1985; Barnosky et al. 1987; Anderson and Brubaker 1994). For example, palynological studies from northern Alaska indicate: (1) that major plant taxa responded individualistically to postglacial climatic amelioration (Anderson 1985, 1988; Anderson and Brubaker 1994), and (2) that geographic variations in late Quaternary vegetation have had strong longitudinal (i.e., east-west) and latitudinal (i.e., north-south) components (Anderson and Brubaker 1994). Furthermore, assemblages from this transitional period lack clear analogs in modern tundra or boreal forest regions of North America (Anderson et al. 1989). Such changes surely challenged human adaptive strategies as subsistence resources fluctuated in response to the shifting environments.

Palynological research has focused on regions within or near the Brooks Range of northern Alaska for over three decades, beginning with the pioneering work of Livingstone (Livingstone 1955; Ager and Brubaker 1985; Anderson and Brubaker 1994). From this suite of sites (Fig. 2–1), we selected three representative pollen records to illustrate regional vegetation patterns across the southern Brooks Range between ca. 14,000 and 8,000 BP. Readers are referred to Anderson and Brubaker (1994) for presentation of the entire data set in mapped form and to Barnosky et al. (1987) for comparison of computer simulations of palaeoclimates to the late Pleistocene/early Holocene pollen records.

PRESENT-DAY LANDSCAPE

This report presents selected pollen data from sites in the southern foothills of the Brooks Range and adjacent lowlands and plateaus. We refer to this area as the southern Brooks Range, although technically it encompasses parts of the Intermontane Province (Wahrhaftig 1965). The region is cross-cut by numerous rivers, which flow from the mountains to the Bering or Chukchi sea. The largest drainages, such as those associated with the Kobuk, Noatak, Koyukuk, Chandalar, Porcupine, and Yukon rivers (Fig. 2–1), probably were important migratory routes for plant species during postgla-

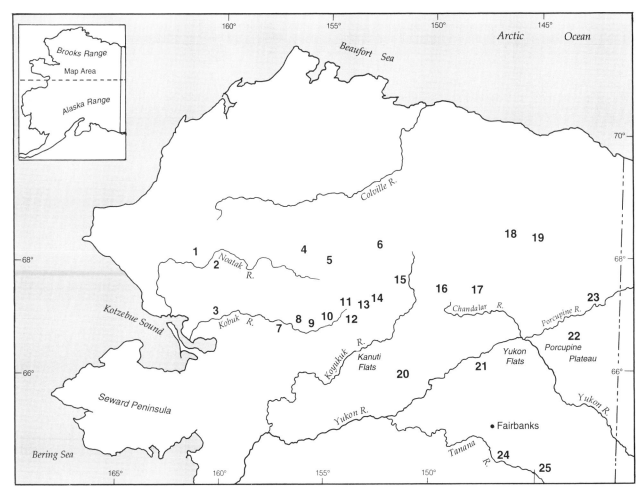

Figure 2–1 Map of the southern Brooks Range with site information. (*Source:* Quaternary Science Reviews. *See Acknowledgments.*)

cial times. Numerous periglacial features, such as frost heaves, thaw lakes, and patterned ground, indicate the importance of the annual freeze-thaw cycle in shaping this northern landscape (Washburn 1980). Permafrost, although discontinuous, underlies extensive portions of the study area (Ferrians 1965) and strongly affects temperature, moisture, and nutrient availability in soils. These soil conditions, in turn, act as important controls over the composition and distribution of the local vegetation.

Boreal forest characterizes much of the study area. The forest-tundra transition is abrupt in the south-central and southeastern Brooks Range, where continuous white spruce forests extend almost to the absolute limit of tree growth on mountain slopes. In the southwestern Brooks Range, open white spruce woodland or mixed forest-tun-

dra forms a broad transitional zone limited to the lower elevations of the foothills. In both regions, continuous forests extend up river valleys, and tree densities decrease markedly with altitude or distance from rivers.

Throughout the study area, hardwood species, such as balsam poplar, aspen, and tree birch, are locally abundant, especially in riparian settings and on sites recently disturbed by fire. Without disturbance, these species are replaced by white spruce and, after significant build-up of organic layers, black spruce.

Tundra dominates the higher elevations and poorly drained soils of the study area. As observed by Murray (1978; Fig. 2–2), local conditions and regional climatic variations, however, create a mosaic of tundra types. Low-shrub heath tundra is the

Key to Figure 2–1

Site #	Site Name	Location	Elev.(m)	Age[1]	Reference
1	Kaiyak	68°09′N, 161°25′W	190	18*	Anderson 1985
2	Niliq	67°52′N, 160°26′W	274	14	Anderson 1988
3	Squirrel	67°06′N, 160°23′W	91	18*	Anderson 1985
4	Etivlik	68°08′N, 156°02′W	631	14	Anderson unpublished data
5	Headwaters	67°56′N, 155°02′W	820	11	Brubaker et al. 1983
6	Redondo	67°43′N, 154°33′W	460	5	Brubaker et al. 1983
7	Joe	66°46′N, 157°13′W	183	18*	Anderson 1988
8	Kollioksak	66°58′N, 156°27′W	213	14	Anderson unpublished data
9	Selby	66°51′N, 155°43′W	145	18*	Anderson unpublished data
10	Minakokosa	66°55′N, 155°02′W	122	18*	Anderson unpublished data
11	Ruppert	67°04′N, 154°15′W	210	12	Brubaker et al. 1983
12	Angal	67°08′N, 153°54′W	820	14	Brubaker et al. 1983
13	Ranger	67°09′N, 153°39′W	820	18*	Brubaker et al. 1983
14	Redstone	67°15′N, 152°36′W	914	14	Edwards et al. 1985
15	Screaming Yellowlegs	67°35′N, 151°25′W	650	14	Edwards et al. 1985
16	Rebel	67°25′N, 149°48′W	914	18*	Edwards et al. 1985
17	Sakana	67°26′N, 147°51′W	640	13	Brubaker unpublished data
18	Crowsnest	68°20′N, 146°29′W	881	10	Anderson unpublished data
19	Seagull	68°16′N, 145°13′W	637	14	Brubaker unpublished data
20	Sithylemenkat	66°07′N, 151°26′W	213	13	Anderson et al. 1990
21	Sands of Time	66°02′N, 147°31′W	250	18*	Lamb and Edwards 1988
22	Tiinkdhul	66°35′N, 143°09′W	189	18*	Anderson et al. 1988
23	Ped	67°12′N, 142°04′W	211	12	Edwards and Brubaker 1986
24	Birch	64°19′N, 146°50′W	274	15	Ager 1975
25	George	63°47′N, 144°30′W	389	17	Ager 1975

[1]Maximum age (years BP × 1,000) used in mapping based on linear interpolation of radiocarbon dates.
*Maximum core age exceeds 18,000 BP.

most common type, occurring on relatively dry sites in the rolling uplands. Low shrubs, primarily dwarf/resin birch and willows, form an open canopy with a ground cover of sedge, grass, forbs, and heaths. Upland sites with intermediate drainage support a low-shrub tussock tundra. Tussocks are formed by sedge species, with shrubs such as heaths and birch often growing on the sides of the tussocks. This tundra type is also found in drained thermokarst basins. Tall shrub thickets composed of resin birch, alder, and willows typify riparian settings. Dry, windswept hilltops support sparse plant communities dominated by *Dryas octopetala, D. integrifolia*, grass, sedge, and dwarf willow.

DESCRIPTION AND INTERPRETATION OF PAST VEGETATION FROM POLLEN DATA

The earliest palynologists working in Alaska lamented the "blunt" nature of pollen data (e.g., Colinvaux 1964), referring to the coarse taxonomic classification and low pollen productivity of several important tundra and boreal forest taxa. While

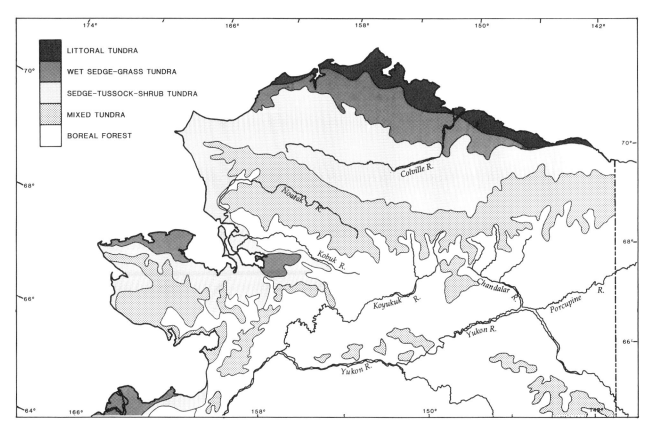

Figure 2–2　Vegetation map. (*Source:* Quaternary Science Reviews. *See Acknowledgments.*)

many of the reservations about the limitations of pollen data for interpreting high-latitude vegetation are well founded, recent research has significantly improved the interpretation of arctic and subarctic pollen records. Three types of analysis have proved especially helpful: pollen taxonomy, modern pollen collections, and application of new quantitative techniques.

Better taxonomic resolution of northern pollen types has increased the detail of vegetational reconstructions. For example, careful attention to pollen morphology has improved the definition of heaths and minor herb taxa that are so important for differentiating tundra types (e.g., Cwynar 1982). Relative abundances of black and white spruce can now be determined morphologically (Hansen and Engstrom 1985) or statistically (Brubaker et al. 1987). Such separation of spruce species is vital for distinguishing between well-drained, warm sites occupied by white spruce forests and cool, poorly drained black spruce muskegs.

Collections of modern pollen samples from surficial lake muds have proved an invaluable tool for describing regional vegetation patterns (Anderson and Brubaker 1986; Anderson et al. 1991). Visual comparisons of mapped, modern pollen percentages to vegetation maps show that certain isopolls (contours of pollen percentages) consistently delimit species range limits. For example, the 10% spruce isopoll generally represents treeline in northern Alaska (Figs. 2–3: *a, b*). Once modern pollen-vegetation relationships are established, key isopolls can be examined through time. Thus, mapping the location of the 10 percent spruce isopoll throughout the Holocene can give an approximate idea of treeline movement (Fig. 2–3: *c*).

A second, more quantitative, means of comparing modern and fossil data is with dissimilarity measures (Overpeck et al. 1985). Unlike the qualitative evaluation of individually mapped taxa, this method compares pollen assemblages. Results of this analysis can be presented as maps or time se-

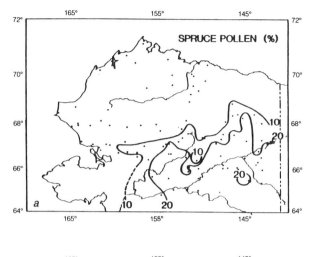

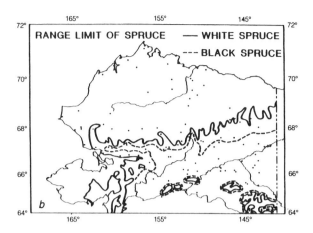

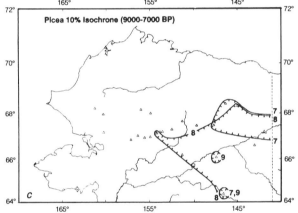

Figure 2–3   Maps showing (a) spruce range limits, (b) modern isopoll map, (c) 10% isochrone map of spruce (9,000 to 7,000 bp). Sites used in mapping are shown in Figure 2–1. (*Source:* Review of Palaeobotany and Palynology. *See Acknowledgments.*)

ries (Fig. 2–4). Particular times are represented in the maps, which compare a single fossil assemblage to each of the modern pollen samples. A sequence of maps can illustrate the presence or absence of good analogs and the movement of the most analogous vegetation type through time. Plotting squared chord distances as time series facilitates intersite comparisons by emphasizing mutual periods of good or poor analogs. This method, however, does not provide information on location of analogous vegetation types. Of note, analogs for the period discussed in this volume are generally the weakest ones of any over the last 18,000 years.

The Data

Late Quaternary pollen diagrams from arctic and subarctic Alaska are typically divided into three zones based on changes in percentages of major pollen taxa (Livingstone 1955). Our discussion focuses on Zone 2 and early Zone 3. The earliest zone (Zone 1 or the herb zone) is dominated by sedge, grass, and *Artemisia* pollen. Vegetational interpretations for this period are particularly difficult, because the predominant taxa have wide ecological tolerances and the pollen assemblages lack strong analogs on the modern landscape. Many palynologists believe that the full-glacial vegetation of Zone 1 was a mosaic of tundra types, with at least some areas supporting little or no vegetation (e.g., Cwynar 1982; Ritchie and Cwynar 1982; Schweger 1982; Anderson 1985, 1988; Barnosky et al. 1987; Anderson and Brubaker 1994). However, this view conflicts with palaeontological data that suggest that herds of large mammals roamed eastern Beringia. If significant numbers of Pleistocene megafauna were

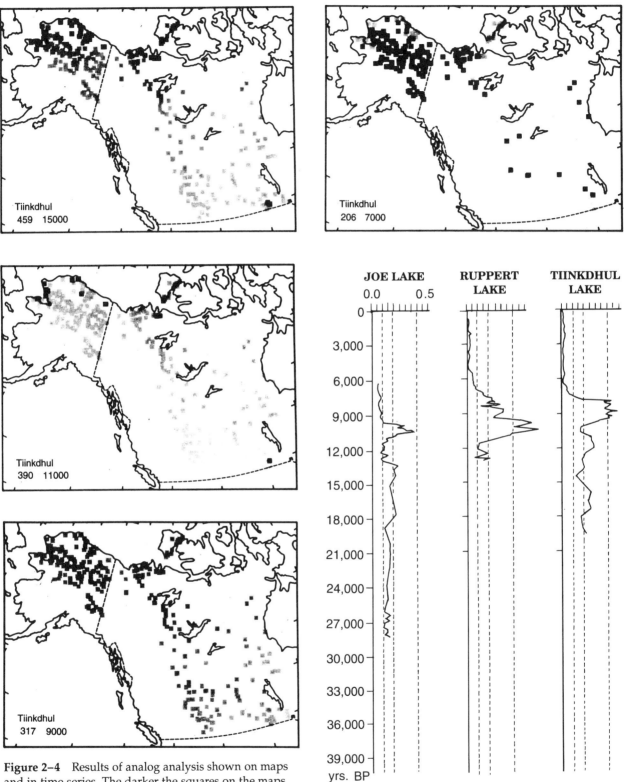

Figure 2–4 Results of analog analysis shown on maps and in time series. The darker the squares on the maps the stronger the analog; the larger the squared chord distance in the time series, the weaker the analog. (*Source:* Journal of Biogeography. *See Acknowledgments*.)

JOE LAKE

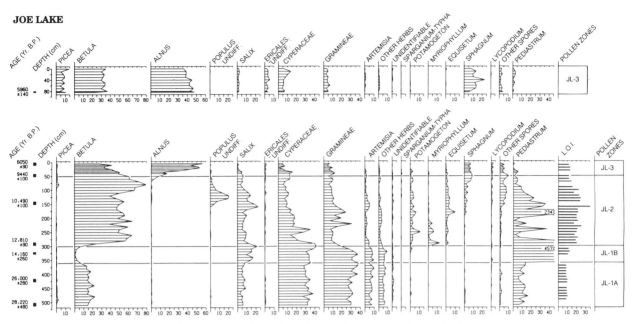

Figure 2–5 Percentage pollen diagram from Joe Lake. (*Source:* Quaternary Research. *See Acknowledgments.*)

present, the vegetation must have been more productive than generally proposed by palynologists (see Guthrie 1982, 1990, this volume).

Zone 2 (birch zone) marks the initial response of vegetation to postglacial climatic warming, ice retreat, and rising sea levels. Assemblages with high percentages of birch pollen replace the herb-dominated spectra of Zone 1. The abundance of birch pollen suggests the widespread presence of a birch-shrub tundra across northern Alaska. Many Alaskan pollen diagrams indicate a time of poplar pollen increase during the latter part of this zone, suggesting that poplar woodlands were at least locally common (Anderson et al. 1988).

The most recent zone (Zone 3, the spruce/alder zone) represents the first widespread appearance of modern pollen assemblages and, by inference, modern vegetation communities. The arrival and spread of spruce and alder in the southern Brooks Range is time-transgressive. That is, spruce typically preceded alder in eastern sites and alder arrived before spruce in the west.

The Sites

Three lakes have been chosen to illustrate the vegetation histories of the southwestern, south-central,

and southeastern portions of the study area: Joe Lake; Ruppert Lake; and Tiinkdhul Lake (see Fig. 2–1), respectively. Cores were obtained with a modified Livingstone piston corer described by Wright (1967), and samples were processed according to standard procedures established for silt-rich sediments from arctic and subarctic lakes (see Ritchie 1984). Pollen sums, which include all identified and unidentified grains, generally exceed 300. Spores and aquatics are expressed as percent of pollen sum.

Joe Lake (66°46′N, 157°13′W; elevation 183 m) is located in mixed forest-tundra of the Selawik uplands between the Kobuk and Selawik drainages. A small outlet flows from this 2-ha lake, and seasonal seepages enter the lake from nearby slopes. The lake consists of a gently sloping basin with an elongated, narrow, steep-sided depression. The record comes from two overlapping cores: one spanning the period 6,000 to 28,000 BP and the second the last 6,000 years (Fig. 2–5). The basal 447 cm of the core are silts, whereas the uppermost sediments are organic-rich gyttja.

Ruppert Lake (67°04′N, 154°14′W; elevation 210 m) is a closed basin lying within the terminal moraine complex that confines the southern end of Walker Lake. It is approximately 5 ha in area. Today

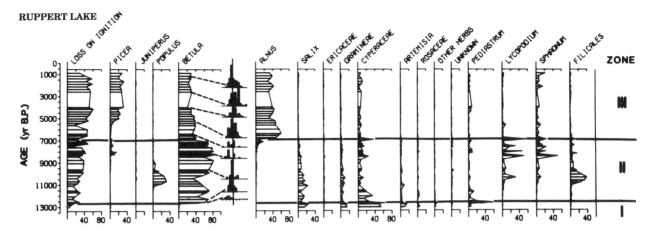

Figure 2–6   Percentage pollen diagram from Ruppert Lake. (*Source:* Quaternary Research. *See Acknowledgments.*)

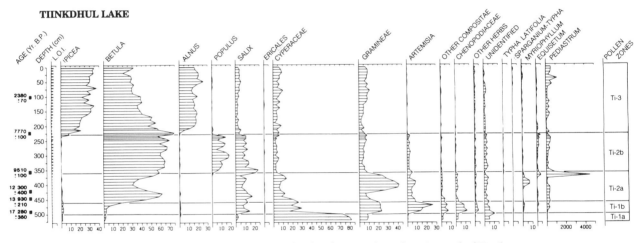

Figure 2–7   Percentage pollen diagram from Tiinkdhul Lake. (*Source:* Canadian Journal of Earth Sciences. *See Acknowledgments.*)

the site is surrounded by black spruce woodland. The pollen record is ca. 13,000 years old. No sediment description is available for this core.

Tiinkdhul Lake (66°34′N, 143°09′W; elevation 189 m), at ca. 13 ha, is the largest of the three sites. It is volcanic in origin and has single inlet and outlet streams. The lake is located on the western edge of the Porcupine Plateau bordering the Yukon Flats. The regional vegetation is mixed white and black spruce forest. The Tiinkdhul record dates to ca. 18,000 BP. The uppermost 60 cm of sediment are gyttja, with organic-rich silts between 60 and 385 cm, and silt between 385 and 540 cm. Small rounded clasts, informally referred to as "peds," occurred

between 160 and 250 cm (see also Edwards and Brubaker 1986).

The Pollen Diagrams

Although complete percentage diagrams are presented in Figures 2–5 through 2–7, we discuss only Zone 2 and lower Zone 3 (i.e., the birch and earliest spruce/alder zones).

Zone 2—Birch Zone   Birch pollen at Joe Lake has a maximum value of 80% with a minimum of 24%. The record from Ruppert Lake displays a similar range of 82% to 35%. The Tiinkdhul diagram contains slightly lower maximum and minimum percentages of birch pollen (74% and 18%, respectively). Willow, a typically under-represented pollen taxon, has relatively high percentages at all three sites (Joe Lake, 23% to 4%; Ruppert Lake, 23% to 2%; Tiinkdhul Lake, 23% to 4%). Heaths, also low-pollen producers, appear consistently in all records, although in frequencies less than 5%.

Grass and sedge pollen, dominant taxa of Zone 1 assemblages, continue to be important components in Zone 2. Grass pollen reaches a maximum of 31% at Joe Lake, but values fall more typically between 10% to 20%. Sedge percentages, like grass, usually are between 10% to 20%, but with a single spike of 35%. Compared to the other two sites, Ruppert Lake has the lowest percentages of grass pollen (10% to 2%). Sedge values (34% to 2%), however, are similar to the other sites. At Tiinkdhul Lake, sedge and grass pollen have maximum frequencies of 26% and 41% and minimum values of 2% and 3%, respectively. At each site, percentages of graminoid pollen tend to be greatest in lower Zone 2 but gradually decrease in more recent sediments. *Artemisia* pollen is the most common minor herb taxon, but it is usually less than 10%. Other minor herb taxa are relatively uncommon.

In all three cores, poplar pollen becomes a codominant type with birch within Zone 2. Maximum frequencies for poplar are 21%, 31%, and 18% at Joe, Ruppert, and Tiinkdhul lakes, respectively. At Joe and Ruppert lakes, poplar pollen declines in the uppermost portions of Zone 2. At Joe Lake, the latest Zone 2 pollen spectra are similar to those in lower Zone 2. At Ruppert Lake, however, upper Zone 2 contains a slight increase in spruce percentages (10% to 15%).

Zone 3—Lower Spruce/Alder Zone Increased percentages of alder pollen at Joe and Ruppert lakes and alder and spruce pollen at Tiinkdhul Lake mark the beginning of Zone 3. Percentages of alder pollen are ca. 10% to 20% at all three sites, increasing to maximum values of ca. 50% to 60% in the upper parts of this zone. At Joe Lake, willow, sedge, and grass pollen remain similar to their upper Zone 2 values, although *Artemisia* pollen decreases. Willow, *Artemisia*, and grass percentages decrease be-

tween Zones 2 and 3 at both the Ruppert and Tiinkdhul sites, but sedge percentages are similar between zones.

Chronology

Hopkins (1982) has referred to the late Pleistocene/early Holocene transition, the period of concern in this volume, as the Birch Interval (14,000 to 9,000 BP), drawing the name from the ubiquitous presence and dominance of birch in all eastern Beringian pollen records. Based on the southern Brooks Range pollen data, it is proposed here to further subdivide the Birch Interval into 2 parts: Early Birch Interval (14,000 to 11,000 BP) and Late Birch Interval (11,000 to 9,000 BP). We have referred to Zone 3 times as the Boreal Interval (9,000 BP to present) with 2 subdivisions: an Early (9,000 to 6,000 BP) and a Late Boreal Interval (6,000 BP to present), corresponding to the subzones of Zone 3.

Although this general chronology is useful, the exact timing of zone boundaries varies from site to site, as is evident from the three pollen diagrams (Figs. 2–5 through 2–7). Birch pollen percentages increase at ca. 13,000 BP at Joe Lake, 12,500 BP at Ruppert Lake, and 14,000 BP at Tiinkdhul Lake. Pollen accumulation rates, however, indicate that significant numbers of birch were not present in the Tiinkdhul region until ca. 10,000 BP. The poplar period varies between 10,500 to 8,700 BP (Joe Lake), 10,500 to 9,400 BP (Ruppert Lake), and 9,300 to 8,000 BP (Tiinkdhul Lake). Alder arrived near Joe Lake at ca. 9,500 BP, Ruppert Lake ca. 6,600 BP, and Tiinkdhul Lake ca. 7,000 BP.

INTERPRETATION

A suite of twenty-four lake records comprise the total southern Brooks Range data set, with a total of 114 radiocarbon dates used for chronological control (see Fig. 2–8 for histogram of dates). Mapped pollen data from the combined records provide a stronger basis from which to interpret late Quaternary vegetation history than is possible from an individual site. The following vegetation history, schematically summarized in Figure 2–9, relies on

DISTRIBUTION OF RADIOCARBON DATES

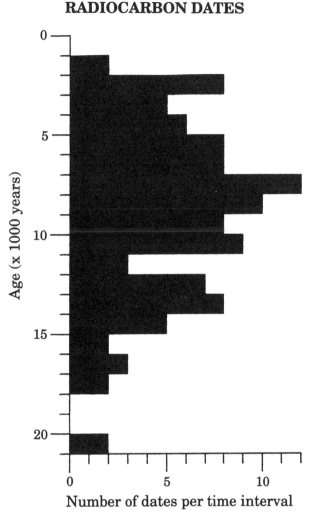

Figure 2–8   Histogram of ¹⁴C dates. (*Source:* Quaternary Science Reviews. *See Acknowledgments.*)

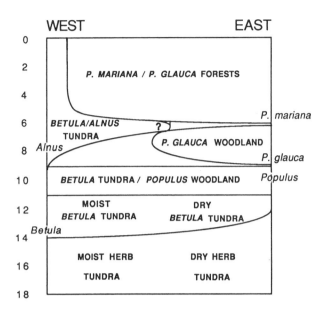

Figure 2–9   Schematic representation of the late Quaternary vegetation history of the southern Brooks range. (*Source:* Quaternary Science Reviews. *See Acknowledgments.*)

the combined data set, but the importance of the three sites described previously to the broader context is clear.

Early Birch Interval (14,000 to 11,000 BP)

Birch-shrub tussock tundra replaced the earlier, full-glacial herb tundra at ca. 14,000 BP in the southwestern Brooks Range. The presence of tussocks is indicated by increases in heath pollen and *Sphagnum* spores. These communities probably were most common in areas of intermediate drainage,

such as uplands and locally drained thermokarst basins. Tall birch and willow shrubs likely formed dense thickets along streams. Balsam poplar possibly was present in riparian settings.

In the southeastern Brooks Range, between 14,000 and 12,000 BP, xeric tundra communities, characteristic of the earlier full-glacial period, perhaps expanded into areas previously too harsh for plant growth, and mesic tundra may have become more common locally. As in the southwestern Brooks Range, birch pollen percentages increase at ca. 14,000 BP, but pollen accumulation data indicate that in the southeast these shrubs were relatively rare on the landscape until 12,000 BP. After 12,000 BP, dry, birch-shrub tundra dominated the southeastern landscape and was perhaps similar to the modern high-shrub tundra found in areas of the northern Alaska Range.

Late Birch Interval (11,000 to 9,000 BP)

Birch-shrub tundra continued to dominate the landscape, with moist tussock communities in the west and drier shrub assemblages in the east. Local habitats, however, were dramatically changed with the expansion of poplar and heaths. Differences between eastern and western regions indicate a less pronounced vegetation gradient than during any other time during the late Quaternary. Balsam poplars (*P. balsamifera*) probably expanded as gallery forests along rivers, while quaking aspen (*P. tremuloides*) perhaps formed parklands on south-facing, well-drained hillslopes. Expansion of heath communities from the southwestern to southeastern Brooks Range suggests the development of moist shrub tundra in the thermokarst terrain of the eastern lowlands and/or the presence of local heath tundra communities on uplands. On an even more local scale, range extensions of *Typha* and *Myriophyllum* indicate local changes in aquatic environments, lowering of lake levels, and/or a rise in water temperature.

Early Boreal Interval (9,000 to 6,000 BP)

White spruce invaded the Tanana and upper Yukon River drainages between 10,000 and 9,000 BP, reaching the south-central Brooks Range by 8,000 BP. White spruce forests in northwestern Canada were probably the seed source for colonization of eastern Alaska (Ritchie 1977, 1982, 1984, 1985a, 1985b; Spear 1983; Cwynar and Spear 1991). River valleys provided extensive migration corridors and excellent sites for seedling establishment and tree growth. Dense white spruce forests probably dominated floodplains and river courses in the southeastern Brooks Range. The newly established gallery forests would have provided seed sources for further spruce expansion into birch-shrub communities on south-facing slopes and on well-drained soils away from the rivers. Hardwood stands of poplar and tree birch may have been an important component of the southeastern forests or woodlands. Poplar probably remained common along valley bottomlands, although its abundance declined after 8,000 BP.

In the southwestern Brooks Range, birch tussock tundra and poplar woodlands continued to dominate the landscape. By 8,000 BP, alder appeared throughout much of the region. These shrubs were probably restricted to floodplain thickets and mid-

elevations of mountain slopes. Tree birch may also have formed scattered stands on south-facing slopes or well-drained alluvial soils with poplars.

REFERENCES CITED

Ager, T. A. 1975. *Late Quaternary Environmental History of the Tanana Valley, Alaska.* Institute of Polar Studies, Report No. 54. Columbus: Ohio State University.

Ager, T. A., and L. B. Brubaker. 1985. Quaternary Palynology and Vegetation History of Alaska. In *Pollen Records of Late-Quaternary North American Sediments,* ed. V. M. Bryant, Jr., and R. G. Holloway, pp. 353–84. Dallas: American Association of Stratigraphic Palynologists Foundation.

Anderson, P. M. 1985. Late Quaternary Vegetational Change in the Kotzebue Sound Area, Northwestern Alaska. *Quaternary Research* 24:307–21.

———. 1988. Late Quaternary Pollen Records from the Kobuk and Noatak River Drainages, Northwestern Alaska. *Quaternary Research* 29:263–76.

Anderson, P. M., and L. B. Brubaker. 1986. Modern Pollen Assemblages from Northern Alaska. *Review of Palaeobotany and Palynology* 46:273–91.

———. 1994. Vegetation History of Northcentral Alaska: A Mapped Summary of Late-Quaternary Pollen Data. *Quaternary Science Reviews* 13:71–92.

Anderson, P. M., R. E. Reanier, and L. B. Brubaker. 1988. Late Quaternary Vegetational History of the Black River Region in Northeastern Alaska. *Canadian Journal of Earth Sciences* 25:84–94.

———. 1990. A 14,000-Year Pollen Record from Sithylemenkat Lake, North-Central Alaska. *Quaternary Research* 33:400–404.

Anderson, P. M., P. J. Bartlein, L. B. Brubaker, K. Gajewski, and J. C. Ritchie. 1989. Modern Analogs of Late Quaternary Pollen Spectra from the Western Interior of North America. *Journal of Biogeography* 16:573–96.

———. 1991. Vegetation-Pollen-Climate Relationships for the Arcto-Boreal Region of North America and Greenland. *Journal of Biogeography* 18:565–82.

Barnosky, C. W., P. M. Anderson, and P. J. Bartlein. 1987. The Northwestern U.S. during Deglaciation: Vegetational History and Paleoclimatic Implications. In *North America and Adjacent Oceans during the Last Deglaciation,* ed. W. F. Ruddiman and H. E. Wright, Jr., pp. 289–321. Boulder, Colorado: Geological Society of America.

Brubaker, L. B., L. J. Braumlich, and P. M. Anderson. 1987. An Evaluation of Statistical Techniques for Discriminating *Picea glauca* from *Picea mariana* Pollen in Northern Alaska. *Canadian Journal of Botany* 63:616–26.

Brubaker, L. B., H. L. Garfinkel, and M. E. Edwards. 1983. A Late Wisconsin and Holocene Vegetation History from the Central Brooks Range: Implications for Alaskan Paleoecology. *Quaternary Research* 20:194–214.

Colinvaux, P. A. 1964. The Environment of the Bering Land Bridge. *Ecological Monographs* 34:297–329.

Cwynar, L. C. 1982. A Late-Quaternary Vegetation History from Hanging Lake, Northern Yukon. *Ecological Monographs* 52:1–24.

Cwynar, L. C., and R. W. Spear. 1991. Reversion of Forest to Tundra in the Central Yukon. *Ecology* 72:202–12.

Edwards, M. E., and L. B. Brubaker. 1986. Late Quaternary Environmental History of the Fishhook Bend Area, Porcupine River, Alaska. *Canadian Journal of Earth Sciences* 23:1765–73.

Edwards, M. E., P. M. Anderson, H. L. Garfinkel, and L. B. Brubaker. 1985. Late Wisconsin and Holocene Vegetational History of the Fishhook Bend Area, Porcupine River, Alaska. *Canadian Journal of Earth Sciences* 23:1765–73.

Ferrians, O. J., (compiler) 1965. *Permafrost Map of Alaska.* U.S. Geological Survey Miscellaneous Investigations Series, Map I-44.

Guthrie, R. D. 1982. Mammals of the Mammoth Steppe as Paleoenvironmental Indicators. In *Paleoecology of Beringia,* ed. D. M. Hopkins, J. V. Matthews, Jr., C. E. Schweger, and S. B. Young, pp. 307–26. New York: Academic Press.

———. 1990. *Frozen Fauna of the Mammoth Steppe: The Story of Blue Babe.* Chicago: University of Chicago Press.

Hansen, B. C. S., and D. R. Engstrom. 1985. A Comparison of Numerical and Qualitative Methods of Separating Pollen of Black and White Spruce. *Canadian Journal of Botany* 63:2159–63.

Hopkins, D. M. 1982. Aspects of the Paleogeography of Beringia during the Late Pleistocene. In *Paleoecology of Beringia,* ed. D. M. Hopkins, J. V. Matthews, Jr., C. E. Schweger, and S. B. Young, pp. 3–28. New York: Academic Press.

Lamb, H. F., and M. E. Edwards. 1988. The Arctic. In *Vegetation History,* ed. B. Huntley and T. Webb III, pp. 519–55. Boston: Kluwer.

Livingstone, D. A. 1955. Some Pollen Profiles from Arctic Alaska. *Ecology* 36:587–600.

Murray, D. F. 1978. Vegetation, Floristics and Phytogeography of Northern Alaska. In *Vegetation and Production Ecology of an Alaskan Arctic Tundra,* ed. L. T. Tieszen, pp. 19–36. Ecological Studies 29. New York: Springer-Verlag.

Overpeck, J. T., I. C. Prentice, and T. Webb III. 1985. Quan-titative Interpretations of Fossil Pollen Spectra: Dissimilarity Coefficients and the Method of Modern Analogs. *Quaternary Research* 23:87–108.

Ritchie, J. C. 1977. The Modern and Late-Quaternary Vegetation of the Campbell–Dolomite Uplands, near Inuvik, N. W. T., Canada. *Ecological Monographs* 47:401–23.

———. 1982. The Modern and Late-Quaternary Vegetation of the Doll Creek Area, North Yukon, Canada. *New Phytologist* 90:563–603.

———. 1984. *Past and Present Vegetation of the Far Northwest of Canada.* Toronto: University of Toronto Press.

———. 1985a. Late-Quaternary Climatic and Vegetational Change in the Lower Mackenzie Basin, Northwest Canada. *Ecology* 66:612–21.

———. 1985b. Quaternary Pollen Records from the Western Interior and the Arctic of Canada. In *Pollen Records of Late-Quaternary North American Sediments,* ed. V. M. Bryant, Jr. and R. G. Holloway, pp. 327–52. Dallas: American Association of Stratigraphic Palynologists Foundation.

———. 1987. *Postglacial Vegetation of Canada.* Cambridge: Cambridge University Press.

Ritchie, J. C., and L. C. Cwynar. 1982. The Late Quaternary Vegetation of the North Yukon. In *Paleoecology of Beringia,* ed. D. M. Hopkins, J. V. Matthews, Jr., C. E. Schweger, and S. B. Young, pp. 113–26. New York: Academic Press.

Schweger, C. E. 1982. Late Pleistocene Vegetation of Eastern Beringia: Pollen Analysis of Dated Alluvium. In *Paleoecology of Beringia,* ed. D. M. Hopkins, J. V. Matthews, Jr., C. E. Schweger, and S. B. Young, pp. 95–112. New York: Academic Press.

Spear, R. W. 1983. Paleoecological Approaches to a Study of Treeline Fluctuation in the Mackenzie Delta Region, Northwest Territories: Preliminary Results. In *Treeline Ecology Nordicana* 46, ed. P. Morisson and S. Payette, pp. 61–72. Quebec: Centre d'études nordiques, Université Laval.

Viereck, L. A., and E. L. Little, Jr. 1975. *Atlas of the United States Trees.* Vol. 1. *Alaskan Trees and Common Shrubs.* Miscellaneous Publication No. 1293, U.S. Forest Service.

Wahrhaftig, C. 1965. *Physiographic Divisions of Alaska.* U.S. Geological Survey Professional Paper 482.

Washburn, A. L. 1980. *Geocryology: A Survey of Periglacial Processes and Environments.* New York: John Wiley and Sons.

Wright, H. E., Jr. 1967. A Square-Rod Piston Sampler for Lake Sediments. *Journal of Sedimentary Petrology* 27:975–76.

POLLEN RECORDS: POINT BARROW, PRIBILOF ARCHIPELAGO, AND IMURUK LAKE

Paul A. Colinvaux

POINT BARROW

Footprint Lake is on the north Alaskan coastal plain in the vicinity of Point Barrow, about 156°40′W, 71°18′N. Surface pollen was collected here and from an unnamed lake by Livingstone (1955) during the research that established the first pollen profiles for arctic Alaska. Pollen profiles from these two spectra gave the first definition of pollen rains from herb tundra that was essentially free of dwarf birch (*Betula nana*). They were compared by Livingstone with his glacial or late glacial pollen zones from the Brooks Range to suggest that comparable herb tundras free of *Betula nana* and associated plants occupied the interior of Alaska in glacial times. Quite probably the low importance of *Artemisia* in the modern vegetation of the arctic coastal plain, with its proportionate lack of visibility in the Barrow surface samples, served to obscure the significance of 5 to 10 percent *Artemisia* pollen in his Chandler Lake diagram. Although Livingstone did raise short cores from Barrow thaw lakes, the main thrust of the early limnological work was to explain the remarkable parallel orientation of lakes on the Alaskan coastal plain, a problem of palaeoecological significance because likely explanations invoked the force and direction of prevailing winds (Leahy et al. 1958).

A long pollen profile from the region developed only when the U.S. Army Cold Regions Research and Engineering Laboratory (CRREL) drilled tunnels for studies of the origin and stability of permafrost on the coastal plain, discovering in the process lenses of peat that were polleniferous and that could be individually radiocarbon dated (Brown and Sellmann 1966). Pollen from these samples was used to falsify the Ewing and Donn theory of ice ages, which required an open Arctic Ocean as a condition for glacial advance by providing the first direct evidence that the Arctic Ocean was per-

manently ice-covered in glacial times (Ewing and Donn 1956; Livingstone 1959; Colinvaux 1964, 1965). The Barrow pollen sequence remains as the only long pollen record from an area close to the Arctic Ocean in Alaska.

Stratigraphy and Dating

Peat samples in the CRREL tunnels and drill cores were assumed not to be in sequence because of the possibilities of movement within the permafrost. Thus each peat sample was radiocarbon-dated, both to establish the ages of the peats and to use the resulting chronology to reconstruct the stratigraphic history of the permafrost (Brown and Sellmann 1966). This yielded a total of seven samples spanning from 25,000 to 1,700 BP, together with another sample of apparently glacial age that was radiocarbon infinite (Table 2–1).

Table 2–1 Radiocarbon Dates on Frozen Peats from Barrow, Alaska (From Brown and Sellmann, 1966. *See Acknowledgments.*)

1775 ± 120	(I-699)
8715 ± 250	(I-1182)
9155 ± 240	(I-1183)
9550 ± 240	(I-700)
10,525 ± 280	(I-701)
14,000 ± 500	(I-1171)
25,300 ± 2300	(I-1384)
>36,300	(I-1394)

Pollen Stratigraphy

The six samples dated from 14,000 to 1,700 BP all yielded pollen of an herb tundra comparable with the surface sample from Footprint Lake, though with variations in the Ericaceae and *Salix* spectra that are as easily attributed to very local site differences as they are to environmental change. None had notable quantities of *Artemisia* pollen. The only provocative change in this last 14,000 years of Point Barrow tundra history—at least at this crude resolution—was a rise in *Betula* percentage from a background of 5% or less to over 10% at 8715 ± 250 BP,

Reindeer at Imuruk Lake, Seward Peninsula.

giving some suggestion of a slightly drier or warmer interval in the early Holocene. More interesting are the two samples of glacial age (25,300 and >36,000) because they define the vegetation of the northern coast of Beringia. They reveal sedge tundra with grass and a very little *Artemisia*. Cyperaceae pollen is at 71% and 84% in the two sites, with Gramineae down to 14% and 9% (Colinvaux 1965). *Artemisia* is at 5% in the 25,000 BP sample and only a trace in the older sample. All other pollen combined is less than 10% in each sample, and most of this is herb pollen, with *Betula*, *Alnus*, *Picea*, *Salix*, and even Ericaceae reduced to trace representation. These data seem to make it inescapable that the arctic coast in Beringian times supported a moist herb tundra largely made up of sedges.

PRIBILOF ARCHIPELAGO

The Pribilof Islands of St. Paul and St. George are the last land remnants of the south land-bridge coast outside the Aleutian chain. Unlike the Aleutians, however, they were never glaciated, so that a land bridge surface is preserved. Water between the most inland of the two islands, St. Paul, and the southern edge of the shelf is now between 70 and 90 m deep. This indicates that land between the islands and the coast would have been very low-lying in land-bridge times and subject to flooding with the first significant rise of sea level. A submarine canyon shows that the Yukon River had its land-bridge delta just to the south and east of St. George Island. The Pribilofs thus are sited on what should

Cagaloq Lake, St. Paul Island, in the Pribilof Archipelago.

have been a key region in the economy of Beringia: low-lying, coastal, and close to the great river system draining the Alaskan interior. The principal palaeoecology record from the archipelago comes from Cagaloq Lake, which occupies an explosion crater (maar) at 220-m elevation on Lake Hill, near the center of St. Paul Island.

History of Research

Early references to Pribilof lakes are associated with the extensive literature about the fur seal industry (Elliot 1881), which interest also led to early geological surveys (Stanley-Brown 1982; Dawson 1984). More recently, Hopkins and Einarsson (1966) found evidence for Pleistocene glaciation on St. George Island but none on St. Paul Island, contrary to the findings of Barth (1956). The modern synthesis appears to be that St. Paul Island was never glaciated, with evidence from remanent magnetism and K/Ar dating to show that all rocks on the island are of late Pleistocene age (Cox et al. 1966). A program of core drilling of sediments in all the lakes and bogs of both islands was undertaken in 1963 with the object of discovering an environmental history of the south land-bridge coast. No ancient sites were discovered on St. George Island (Parrish 1979). On St. Paul Island only Cagaloq Lake held a long history, but this was sufficient to describe the land-bridge vegetation of the south coast of Beringia in full-glacial times. It also revealed a warm interval in the late-glacial period when the spruce ecotone

reached the islands several thousand years before the spruce rise in the interior of Alaska (Colinvaux 1967, 1981).

Stratigraphy and Dating

The sediments of Cagaloq Lake are 14 m thick: 5.5 m of algal gyttja, followed by 2.5 m of clay-rich silty gyttja over 6 m of silt, clay, and sand. Basal sediment is black sand, fully compatible with black sand beaches of the modern lake. Eight bulk radiocarbon dates are available (Table 2–2). Five dates in the upper unit of algal gyttja show that this spans roughly the 10,000 years of the Holocene. Radiocarbon dates on the lowermost sediments are age-inverted as a result of enrichment with modern car-

Table 2–2 Radiocarbon Dates from Cagaloq Lake Sediments

Lab. No.	Depth (m)	Core A (yr BP)	Core B (yr BP)	$\delta^{13}C$
Y-1388	1.4 – 1.6	2620 ± 160		
Y-1389	3.2 – 3.4	3510 ± 100		
Y-1990	4.6 – 4.9		9250 ± 150	-24.4
Y-1390	5.4 – 5.6	9570 ± 160		
I-3031	7.4 – 7.8	$10,600 \pm 470$		
Y-1391	10.5 –10.8	$17,800 \pm 700$		
I-1848	13.0 –13.2	7630 ± 270		-23.4
I-1846	13.65–13.95	5760 ± 180		

Table 2–3   Tritium Assay of Interstitial Water of Cagaloq Lake Sediments

Lab. No.	Depth (m)	Lake	Core A (Tritium units)	Core B
	0	920 ± 25		
TC2525	4.15– 4.3			188 ± 10
TC2524	5.0 – 5.2		331 ± 13	
TC2490	6.0 – 6.25			235 ± 13
TC2526	8.0 – 8.2			222 ± 12
TC2567	10.15–10.4		1130 ± 40	
	12.72–12.92		1220 ± 30	
TC2568	13.6 –14.0		940 ± 30	

bon through groundwater flow in the basal sands, a process that also enriched the bottom sediments with tritium (Table 2–3; Colinvaux 1981). Despite this enrichment, an apparent age of $17,800 \pm 700$ years was obtained for a sample in the tenth meter, providing a minimum age for the sediments. It seems safe to conclude that the organic sediment of the top 5.5 meters spans the Holocene, that the clay and silt unit below that encompasses at least the late-glacial period between 14,000 BP and 10,000 BP, and that the thick deposit of sand and clay of the bottom half of the core represents a significant part of the full-glacial period.

Pollen Stratigraphy

Holocene sediments of the top 5.5 m yield pollen spectra with 10–30 percent Umbelliferae pollen, evidently representative of wet tundra characteristic of oceanic islands of the South Bering Sea. Traces of *Alnus*, *Betula*, and *Picea* pollen throughout the Holocene deposits are attributed to long-distance transport, since none of these taxa are known from the islands in modern times (Fig. 2–10; Colinvaux 1981).

The sediments of the silt-clay unit spanning the late-glacial interval include significant percentages of *Alnus*, *Betula*, and *Picea* pollen. Calculations of a range of plausible spruce pollen accumulation rates demonstrated that *Picea* influx was high, showing that high pollen percentages cannot be the result of percentage statistics. The parsimonious explanation

of high influx is that the parent plants are growing close to the site, suggesting that a spruce forest ecotone occupied the south land-bridge coast in late glacial time (Colinvaux 1967). This observation was taken by Hopkins (1972) to imply the former existence of a "forest refugium" near the south land-bridge estuary of the Yukon–Kuskokwim river system, an idea later abandoned in favor of a late arrival of spruce that left the Pribilof spruce maximum unexplained except by long-distance transport in exceptional circumstances of high winds (Hopkins et al. 1981).

Pollen concentrations were extremely low in the sand and silt deposits of full-glacial time, but suggest sparse herb tundra with *Artemisia* of the kind found to be ubiquitous in pollen spectra of land-bridge age. This vegetation type is thus shown to have existed close to the south coast of the Bering land bridge, thus denying the region to the coastal mesic strip postulated by some authors (e.g., Young 1982).

Imuruk Lake

Imuruk Lake lies on central Seward Peninsula, roughly on the latitude of the narrows of Bering Strait, at 65°21'32" N, 163°12'W, at 309 m elevation. The site is separated from Norton Sound to the south by the Bendeleben Mountains (ca. 1,500 m) and from Kotzebue Sound to the north by the Asses Ears range (ca. 600 m). Both of these mountain ranges run east and west. In Bering Land Bridge times, the lake was thus close to the center of the land mass joining the continents. It was far from maritime influences, on a ridge of land tapering from about 400 m elevation above contemporary sea level at Imuruk Lake westwards some 250 km to the 50 m lowlands of Bering Strait itself. With mountains to both north and south, the lake lay in the path of a migration route alternative to the bottomlands that then occupied what are now Kotzebue and Norton sounds.

Imuruk Lake is also critically placed with reference to major vegetation boundaries. The closest known spruce tree is 25 km away and the closest alder bush 40 km. The modern pollen rain detects the distant presence of spruce and alder popula-

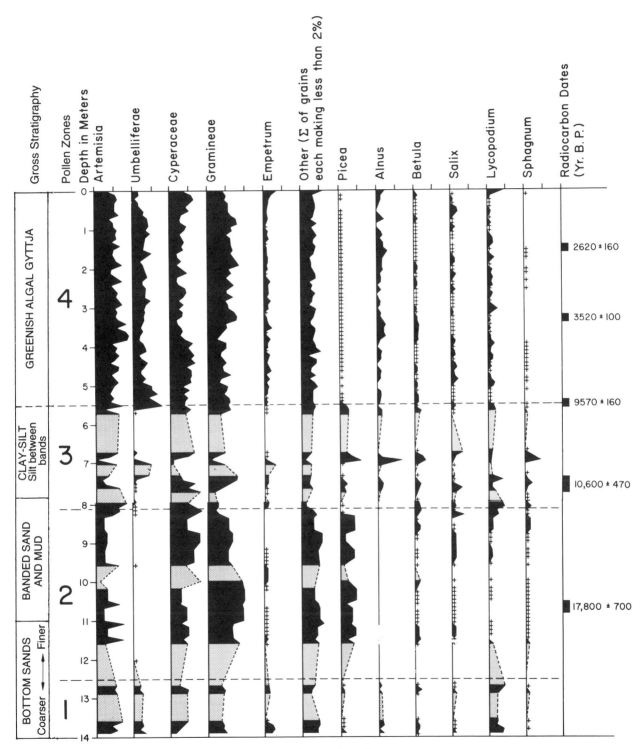

Figure 2–10 Cagaloq Lake pollen diagram.

tions beyond these outliers as low percentages of total pollen. Richly diverse tundra covers the catchment itself with fields of *Eriophorum vaginatum* tussocks interlaced with *Betula nana* alongside the lake, various rock field communities on the younger lava flows, and alpine tundra at higher elevations. Thus the site is excellently placed for pollen studies to detect shifts in the main arctic vegetation types in response to climatic change as tree, bush, and shrub tundra boundaries close or retreat from the lake and are recorded in the rain of spruce, alder, and birch pollen.

History of Research

Imuruk Lake first appears in the scientific literature following the Alaskan gold rush, when the large and elevated lake was seen as a supplemental source of water for placer operations on the lower Inmachuk and Pinnel rivers (Henshaw and Parker 1913). A small dam was constructed where Imuruk Lake feeds into the gorge of the Kugruk River, with the Fairhaven Ditch carrying water northwards to the Pinnel River. Traces of both dam and ditch remain. The importance of Imuruk Lake to Beringian research, however, began with a geological survey by Hopkins (1959, 1963), who established the probable great antiquity of the lake. Hopkins demonstrated that the lake and its immediate basin had never been glaciated, whereas various of the lava flows forming the basin were of early- or mid-Pleistocene age. Furthermore, a series of warped terraces and an underfit stream valley, traceable from the northwest corner of the lake to the Goodhope River, showed that the lake had once had an outlet at its northwest end. The basin had subsequently been tilted until the old outlet was abandoned and the modern outlet to the Kugruk River cut at the southeast end. Both a constructional terrace (the intermediate terrace) containing mammoth bones and a wave-cut scarp many meters above it had been inclined by these movements. The apparent antiquity of these terraces confirmed that the lake had held water continuously since at least some time early in the last glacial cycle, making plausible an even earlier antiquity, as suggested by the ages of lava flows.

Following Hopkins's (1959) declaration of the

then unprecedented antiquity of the Imuruk Lake sediments, coring operations were mounted in the summer of 1960, and again in November to December of that year. The winter operations represented the first attempt to core an Alaskan arctic lake through the winter ice (Colinvaux 1964a). Pollen analysis of the 8-m core taken in the winter operation yielded the first pollen history of land-bridge environments (Colinvaux 1963, 1964b). These results were subsequently elaborated by pollen studies of the suite of cores taken in the summer coring operations (Colbaugh 1968; Shackleton 1979) and by studies of remanent magnetism (Noltimier and Colinvaux 1976; Marino 1977). The presence of the Old Crow tephra as a distinct stratum in these cores was demonstrated by Westgate (1982), confirming the considerable antiquity of the deposits.

Morphometry and the Sedimentary Regime

Imuruk Lake, some 15 km long and 10 km wide, occupies about 70 square km of a primary depression and multiple graben in lava flows, the total area of the drainage being about 285 square km. Lake bathymetry is simple, with a flat bottom under 3 m of water extending throughout the main basin and sloping gently to the shorelines on all sides. Well-developed beaches of quartz-monzonite sands extend for 5 km or so on the northwest side, the remaining shore being mostly lined by beaches of rounded lava cobbles. On the south shore, lava cliffs some 10 m high mark where the Lost Jim lava flow entered the lake late in its history. Inlets are minimal, consisting of a slow-moving stream 1 to 2 m across at the remains of the ancient outlet on the northwest end and a few minor or ephemeral water courses lined with willow bushes. Water lost to the outlet into the Kugruk River at the southeast end is probably balanced by snowmelt and surface run-off along the extensive shorelines. Water level has fluctuated by about .5 m over the last thirty years.

Sediments are stiff, clay-rich, silty gyttja, best explained as the product of reworking the thick loess mantle that covers the older lava flows that make up most of the catchment. Because the loess mantle is both permanently frozen and covered with tussock tundra, the release of silt to the lake should be extremely slow. The lake is turbid throughout the open water season, as the force of waves resuspends bottom sediments. Hopkins (1959) postulated that suspended sediment might be lost in the outlet so that sediment accumulation in modern times has been minimal. This hypothesis was tested by Colbaugh (1968), who used radiocarbon dating of core tops to show that sediment of the last few thousand years is missing from the record. Apparently a steady state has been established between sediment inputs from melting out of the loess mantle and loss of resuspended sediment in the outlet. This steady state was first established in mid-Holocene time when a combination of infilling and stream downcutting at the outlet reduced the depth of the lake to the critical depth when wave energy resuspended sediment in the main basins.

The silty-clay sediment is nearly 5 m thick at the site of Core 1 within little more than a kilometer of the sand beaches of the eastern shore (Fig. 2–11), where they overlie a sequence of alternating old beach sands interspersed with older silty-clay deposits (Colinvaux 1964b). At the site of Core 5 the surficial silty-clay is at least 7 m thick and no old beach sands were encountered. The complete sediment body has not yet been penetrated by any of the cores.

Stratigraphy and Dating

Twenty-three radiocarbon determinations on bulk sediment demonstrate that deposits below 1.5–2.0 m are more than 30,000 years old (Table 2–4). Three dates in the seventh meter of Core 1, from underneath the sequence of beach sands, yielded slightly positive ages with large errors. These apparent ages are parsimoniously explained as the result of contamination by younger carbon introduced in groundwater flow through the sand layers. A similar circumstance of an inverted youthful age under buried beach sands in another arctic lake was investigated by tritium analysis, when the postulated groundwater flow was demonstrated (Colinvaux 1981).

An alternative time-stratigraphic datum is provided by the Old Crow tephra, which has been identified as a layer 10 cm thick at 5.3 m in Core 5 (Westgate 1982). A similar ash layer at 3.5 m in Core 1 can be expected to represent the same ash fall,

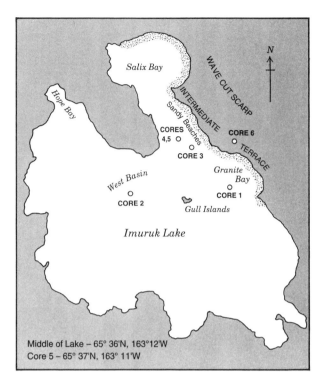

Figure 2–11   Sketch map of the Imuruk Lake region.

although the material was not formally identified in this core. That the ash appears under 3 m of silty clay at one site and under 5 m at the other is reasonable, because the two core sites are more than 2 km apart in a lake of complex morphology and tectonic history. The Old Crow tephra has now been positively dated by isothermal plateau fission track dating at 140,000 ± 10,000 BP (Westgate 1988).

Accepting the Old Crow tephra age as a reliable time-stratigraphic marker yields sediment accumulation rates for the silty clay gyttja of 26 years per mm in Core 5 and 40 years per mm in Core 1. A check on the validity of these sedimentation rates comes from calculating the ages of the top four radiocarbon-dated intervals of Core 1. Assuming 40 yrs/mm and a core top age of 6,000 BP assigns ages to the three positive dates at 12 cm, 35 cm, and 55 cm that are close to the stated errors of the radiocarbon dates. The calculation yields an age of 40,000 BP for the 115-cm level that was radiocarbon dated to more than 34,500 BP. Thus sedimentation rates calculated from the time-stratigraphic datum of the Old Crow tephra are consistent with rates allowed by radiocarbon dating.

Table 2–4   Radiocarbon Dates from Imuruk Lake Sediments

Depth (cm)	Lab. No.	Source of Sample	Age in Years BP
4–20	I-3629	Core 1, Granite Bay	8620 ± 300
30–40	Y-1144	Core 1, Granite Bay	12,355 ± 160
50–60	I-588	Core 1, Granite Bay	13,250 ± 700
110–120	Y-1142	Core 1, Granite Bay	>34,500
255–263	Y-1143	Core 1, Granite Bay	>37,000
719–730	I-801	Core 1, Granite Bay	24,060 ± 3,000*
730–750	Y-1417	Core 1, Granite Bay	29,300 ± 1,000*
745–755	I-415	Core 1, Granite Bay	21,700 ± 2,000*
5–19	I-3630	Core 2, West Basin	12,160 ± 300
100–114	I-3661	Core 2, West Basin	>28,700
95–115	I-7545	Core 3, Salix Bay	10,610 ± 150
5–12	I-3631	Core 4, Salix Bay	10,700 ± 150
96–115	I-3632	Core 4, Salix Bay	11,910 ± 180
172–184	I-3662	Core 4, Salix Bay	25,850 ± 2,400
17–26.5	DIC-906	Core 5, Salix Bay	10,370 ± 175
71–81	DIC-907	Core 5, Salix Bay	10,780 ± 580
120–133	DIC-908	Core 5, Salix Bay	11,330 ± 1,210
160–173	DIC-909	Core 5, Salix Bay	24,190 ± 1,070
199–210	DIC-910	Core 5, Salix Bay	26,400 ± 4,030
277.2–315.6	I-11,550	Core 5, Salix Bay	>24,400
630.6–678.1	I-11,551	Core 5, Salix Bay	>25,900
	W-1235	Intermediate terrace	7400 ± 300
	W-1213	Intermediate terrace	9900 ± 400

Extrapolating these sedimentation rates below the level of the Old Crow tephra gives a basal age of 190,000 BP for Core 5. A similar extrapolation is not possible for Core 1 because of the presence of strata of old beach sands. However, if the unrealistic assumption is made that the deposition of each sand horizon was instantaneous, the sedimentation rate can be applied to the intervening silty clay deposits, resulting in a minimum age for the base of Core 1 of 270,000 BP. If the more rapid sedimentation rate of Core 5 (25 yrs/mm) is applied to the deep deposits of Core 1 in this way, a basal age of 220,000 BP is obtained. We have no reason for thinking that either core penetrated to the base of the sediments. Hopkins's original (1959) prediction that the Imuruk Lake sediments should reveal a history of a significant part of the Pleistocene is rather robustly upheld by these calculations.

This new chronology of the Imuruk Lake cores shows that the entire history of the Wisconsinan Bering Land Bridge is contained within the top 3 m of Imuruk Lake sediments.

Magnetic Stratigraphy

The Imuruk Lake sediments have been shown to hold a record of magnetic palaeosecular variations (PSV) comparable to the record from Lake Biwa (Verosub and Bannerjee 1977). Noltimier and Colinvaux (1976) demonstrated that an excursion occurred in Core 4 at an extrapolated radiocarbon age of about 18,000 BP, thus strongly suggesting the Biwa event itself. Following this demonstration, Marino (1977) measured remanent magnetism on 177 samples from Core 5 and demonstrated that a number of PSV anomalies were present, including a major excursion at 6.2 m (Fig. 2–12). Marino discussed the possibility that this excursion could represent the Blake event. We cannot assert that this PSV anomaly is indeed the Blake event, nor is the anomaly bracketed by satisfactory time-stratigraphic markers. However, extrapolating from the time-stratigraphic datum of the Old Crow tephra yields boundary ages for the anomaly of 160,000–170,000 BP.

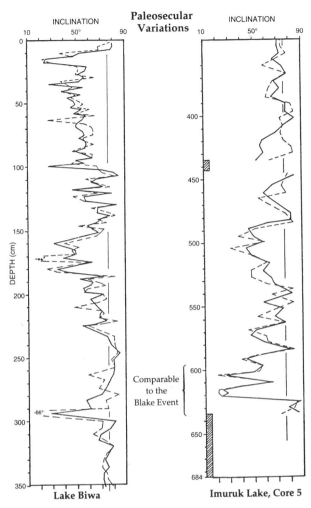

Figure 2–12 Magnetic palaeosecular variations (PSV) from Lake Biwa and Imuruk Lake Core 5. (*Adapted from Marino 1977.*)

Pollen Stratigraphy

Pollen percentage diagrams are available for four of the Imuruk Lake cores. The longest succession of pollen zones comes from Core 1, in which 12 pollen zones (A–L) are recognized (Fig. 2–13; Colinvaux 1964b). Following the Colbaugh (1968) demonstration that sediment ceased to accumulate in mid-Holocene times, a surface zone, M, in which *Picea* pollen has declined to 1 or 2 percent can be added to the sequence, making 13 pollen zones in all. Pollen diagrams from the other cores represent portions of this sequence, the longest being from Core 5, which spans zones G through L (Fig. 2–14; Shackleton 1979).

The pollen zones are usefully compared to the three-zone sequence established for late-glacial and Holocene sections in tundra regions of arctic Alaska by Livingstone (1955). Livingstone's system recognized that tundra pollen rains were always characterized by herb pollen produced locally in the tundra and by birch (*Betula*) and willow (*Salix*) pollen from dwarf shrubs of the tundra. Alder (*Alnus*) and spruce (*Picea*) pollen could make up significant percentages of tundra pollen rains even when the nearest outlier was many kilometers from the sample site. The late-glacial and Holocene history of tundras north of the Brooks Range is thus a three-zone sequence of Zone I (mostly herb pollen), Zone II (herbs with significant birch from dwarf tundra shrubs), and Zone III (*Alnus* or *Picea* or both added to herb and birch tundra pollen). This three-zone record represented a history of warming that resulted in first a dwarf-birch expansion and then the arrival of spruce and alder populations to near their present positions.

The sequence of Zones J, K, L, and M at Imuruk Lake is closely comparable to the Zones I, II, and III sequence north of the Brooks Range. Zones A to i can also all be referred to one of Livingstone's three zonal types, even if in extreme form, so that the whole long pollen history of the last 200,000 years at Imuruk Lake is largely written in the relative prominence of the tundra dwarf birch community or as the advance or retreats of distant alder and spruce populations. The local landscape was occupied continuously by tundra communities throughout these 200,000 years, though spruce and alder treelines would have been closer to the site than they are now in three intervals represented by Zones A, i, and L.

Apart from its length, the Imuruk Lake pollen history differed from any previously published Alaskan diagram in that herb zones had large percentages of *Artemisia* pollen. Similar *Artemisia*-rich herb zones are now known from virtually all glacial-age pollen spectra from arctic Alaska, but they are still without convincing modern analogs. Zone J, dating from radiocarbon infinity to about 13,000 BP, is of this type. Colinvaux (1964b) argued that Zone J represented the vegetation of the old Bering Land Bridge, a form of tundra with abundant *Artemisia* not present on the contemporary

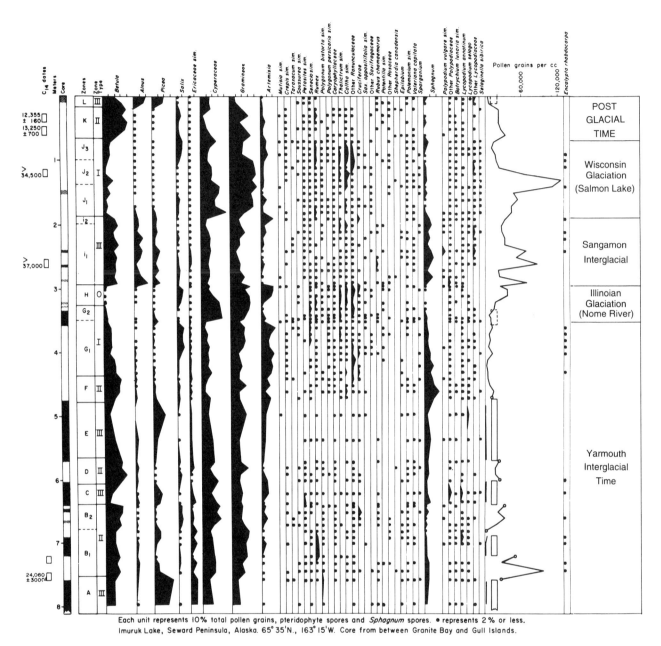

Each unit represents 10% total pollen grains, pteridophyte spores and *Sphagnum* spores. • represents 2% or less.
Imuruk Lake, Seward Peninsula, Alaska. 65°35'N., 163°15'W. Core from between Granite Bay and Gull Islands.

Figure 2–13   Partial pollen diagram and sediment stratigraphy from Imuruk Lake Core 1. (*Source: Ecological Monographs. See Acknowledgments.*)

earth. An earlier manifestation of this zonal type in Zones G and H represented an earlier land bridge episode.

Dating and Vegetation of Land-Bridge Episodes

That the pollen Zone J represents at least the later history of the Wisconsinan-age Bering Land Bridge is demonstrated unequivocally by radiocarbon dates that show it to have been present from at least 30,000 BP until the onset of the birch rise at 13,000 BP (Figs. 2–13, 2–14). The age of the onset of this *Artemisia*-rich herb Zone J of land bridge times can also now be reliably dated.

The onset of Zone J environmental conditions is presented as the Zone i/Zone J boundary, the age of which can be explored separately in Cores 1 and 5. In Core 1 interpolation between the dated Old Crow tephra and the dated birch rise at 13,250 BP yields an age of 74,000 BP. Simple application of the Core 1 sedimentation rate of 40 yrs/mm to the boundary depth less an assumed core top age of 6,000 BP yields essentially the same age: 75,400 BP for the Zone i/Zone J boundary. A radiocarbon determination from the middle of Zone J of greater than 34,500 is consistent with this (Fig. 2–13).

Radiocarbon dating of Core 5, however, appears at first sight to conflict with the 74,000 BP age for the base of Zone J because direct radiocarbon dating at the Zone i/Zone J boundary yields an age of only 24,600 (Fig. 2–14). But this radiocarbon determination has a huge estimated error of 8,000 years—one of three dates with very large error spreads for which the laboratory reported technical problems requiring recounting (DIC-908–910; Table 2–4). Accumulated experience with radiocarbon dating suggests that the results should properly have been reported as radiocarbon infinite. Dating the base of Zone J in Core 5 by interpolating between the core top and the Old Crow tephra yields an age of 64,000 BP, more in keeping with the dating of Core 1.

Interpolated ages of 74,000 BP and 64,000 BP for the Zone i/Zone J boundary in Cores 1 and 5 respectively are in rough agreement, granted the unsatisfactory assumption of uniform sedimentation rates that had to be made for each core. Thus it is reasonable to take as a working hypothesis that vegetation described by the *Artemisia*-rich herb pollen of Zone J occupied central Seward Peninsula from about 70,000 BP to 13,000 BP, which is to say through the whole period in which pre-Clovis populations might have been thought to have crossed the land bridge.

The most recent warm interval before the onset of late-glacial and Holocene times is Zone i, a Livingstone type-III zone in which *Alnus* and *Picea* approached most closely to Imuruk Lake, perhaps even growing widely on Seward Peninsula. The climate of Zone i was certainly the warmest in the last 140,000 years since the deposition of the Old Crow tephra, except for a short interval of similar warmth in the early Holocene (Zone L). Schweger and Mat-

thews (1985) argued that Zone i was representative of a warm interval that they called the Koy-Yukon thermal event and which they associated with the Boutellier interval of Hopkins (1982). They assigned this interval to isotope stage 2 and dated it from 60,000 to 25,000 BP. This interpretation is now shown to be untenable by the new chronology of Imuruk Lake made possible by the firm dating of the Old Crow tephra. Zone i ended before 60,000 BP, not at 25,000 BP as Schweger and Matthews supposed. Their error was compounded of three parts: an incorrect age estimate for the Old Crow tephra, reliance on the accuracy of the suspect date DIC-910 despite the contrary evidence of the radiocarbon dating of Core 1, and reliance on a unpublished speculation that a PSV anomaly in Core 5 was the Blake event with an age of 110,000 BP. The result was an error in dating this critical pollen boundary by roughly 40,000 years.

Zone i began after the deposition of the Old Crow tephra at 140,000 years BP. Interpolating between the Old Crow tephra and the birch rise in Core 1 yields ages of 125,000 BP and 74,000 BP for its beginning and end respectively, strongly suggesting that Zone I represents the whole of isotope stage 5. Zone i thus represents the last interglacial as first proposed nearly thirty years ago (Colinvaux 1964b).

This new chronology shows that the whole of the last Bering Land Bridge period is represented in the Imuruk Lake pollen diagrams by the *Artemisia*-rich Zone J. There were no warm intervals with spruce and alder approaching from the interior. The last warming of the kind assigned by Schweger and Matthews (1985) to the proposed Koy-Yukon thermal event, and by Hopkins (1982) to the Boutellier Interval was experienced at Imuruk Lake only in the course of the interglacial periods of isotope stage 5.

(*Figure 2–14 follows on pp. 94–95.*)

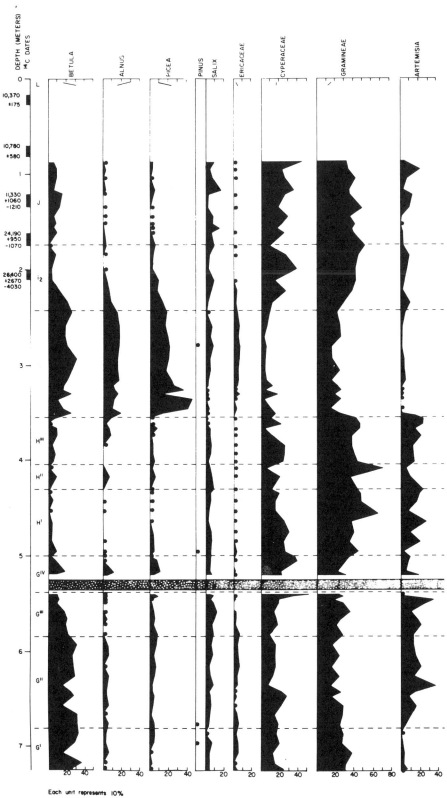

Each unit represents 10%
● Represents 2% or less
▨ = Ash layer

POLLEN PERCENTAGE DIAGRAM,

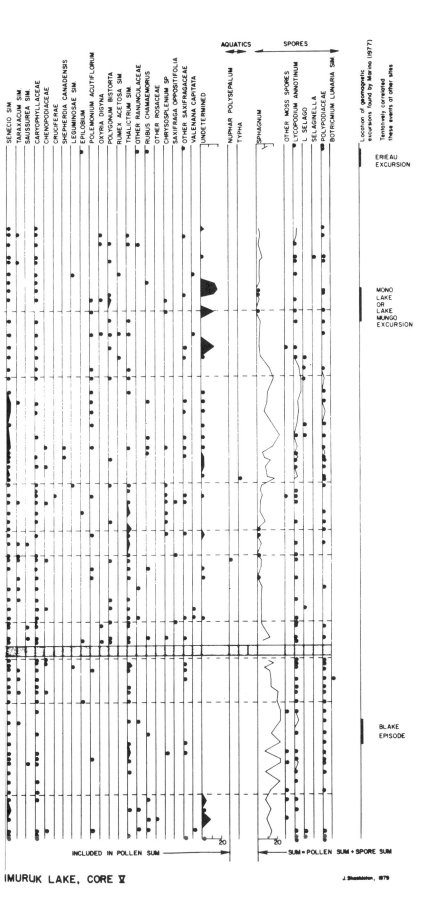

Figure 2–14 Partial pollen diagram and sediment stratigraphy from Imuruk Lake Core 5. (*From Shackleton 1979.*)

IMURUK LAKE, CORE V

J. Shackleton, 1979

REFERENCES CITED

Barth, T. F. W. 1956. Geology and Petrology of the Pribilof Islands, Alaska. *U.S. Geological Survey Bulletin 1028-F*:101–60.

Brown, J., and P. V. Sellmann. 1966. Radiocarbon Dating of Coastal Peat, Barrow, Alaska. *Science* 153:299–300.

Colbaugh, P. R. 1968. The Environment of the Imuruk Lake Area, Seward Peninsula, Alaska, during Wisconsin Time. M.S. thesis, Department of Zoology, Ohio State University.

Colinvaux, P. A. 1963. A Long Pollen Record from Arctic Alaska Reaching Glacial and Bering Land Bridge Times. *Nature* 198:609–10.

———. 1964a. Sampling the Stiff Sediments of an Ice-Covered Lake. *Limnology and Oceanography* 9:262–64.

———. 1964b. The Environment of the Bering Land Bridge. *Ecological Monographs* 34: 297–325.

———. 1965. Pollen from Alaska and the Origin of Ice Ages. *Science* 147:632–33.

———. 1967. Bering Land Bridge: Pollen Evidence for Spruce in Late-Wisconsin Times. *Science* 156:380–83.

———. 1981. Historical Ecology in Beringia: The South Land Bridge Coast at St. Paul Island. *Quaternary Research* 16:18–36.

Cox, A., D. M. Hopkins, and G. B. Dalrymple. 1966. Geomagnetic Polarity Epochs, Pribilof Islands. *Geological Society of America Bulletin* 77:883–909.

Dawson, G. M. 1884. Geological Notes on Some of the Coasts and Islands of the Bering Sea and Vicinity. *Bulletin of the Geological Society of America* 5:117–46.

Elliot, H. W. 1881. *The History and Present Condition of the Fishery Industries: The Seal Islands of Alaska*. Washington, D.C.: Government Printing Office.

Ewing, M., and W. L. Donn. 1956. Theory of Ice Ages III. *Science* 123:1061–66.

Henshaw, F. F., and G. L. Parker. 1913. *Surface Water Supply of Seward Peninsula, Alaska*. Water-Supply Paper 314. Washington, D.C.: U.S. Geological Survey.

Hopkins, D. M. 1959. History of Imuruk Lake, Seward Peninsula, Alaska. *Bulletin of the Geological Society of America* 70:1033–46.

———. 1963. *Geology of the Imuruk Lake Area, Seward Peninsula, Alaska*. U.S. Geological Survey Bulletin 1141-C.

———. 1972. The Paleogeography and Climatic History of Beringia during Late Cenozoic Time. *Inter-Nord* 12:121–50.

———. 1982. Aspects of the Paleogeography of Beringia during the Late Pleistocene. In *Paleoecology of Be-*

ringia, ed. D. M. Hopkins, J. V. Matthews, Jr., C. E. Schweger, and S. B. Young, pp. 3–28. New York: Academic Press.

Hopkins, D. M., and T. Einarsson. 1966. Pleistocene Glaciation on St. George, Pribilof Islands. *Science* 152:343–45.

Hopkins, D. M., P. A. Smith, and J. V. Matthews. 1981. Dated Wood from Alaska and the Yukon: Implications for Forest Refugia in Beringia. *Quaternary Research* 15:217–49.

Leahy, R. G., D. A. Livingstone, and K. Bryan. 1958. Effects of an Arctic Environment on the Origin and Development of Freshwater Lakes. *Limnology and Oceanography* 3:193–214.

Livingstone, D. A. 1955. Pollen Profiles from Arctic Alaska. *Ecology* 36:587–600.

———. 1959. Theory of Ice Ages. *Science* 129:436–65.

Marino, R. J. 1977. Paleomagnetism of Two Lake Sediment Cores from Seward Peninsula, Alaska. M.S. thesis, Department of Geology and Mineralogy, Ohio State University.

Noltimier, H. C., and P. A. Colinvaux. 1976. Geomagnetic Excursion from Imuruk Lake, Alaska. *Nature* 259:197–200.

Parrish, L. L. 1979. A Record of Holocene Climatic Changes from St. George Island, Pribilof Islands, Alaska. M.S. thesis, Ohio State University.

Schweger, C. E., and J. V. Matthews. 1985. Early and Middle Wisconsinan Environments of Eastern Beringia: Stratigraphic and Paleoecological Implications of the Old Crow Tephra. *Geographie Physique et Quaternaire* 39:275–90.

Shackleton, J. 1982. Paleoenvironmental Histories from Whitefish and Imuruk Lakes, Seward Peninsula, Alaska. Institute of Polar Studies Report 76, Ohio State University.

Stanley-Brown, J. 1982. Geology of the Pribilof Islands. *Geological Society of America Bulletin* 3:496–500.

Westgate J. A. 1982. Discovery of a Large-Magnitude, Late Pleistocene Volcanic Eruption in Alaska. *Science* 328:789–90.

———. 1988. Isothermal Plateau Fission-Track Age of the Late Pleistocene Old Crow Tephra, Alaska. *Geophysical Research Letters* 15:376–79.

Verosub, K. L., and S. K. Bannerjee. 1977. Geomagnetic Excursions and Their Paleomagnetic Records. *Review of Geophysics and Space Physics* 15:145–55.

Young, Stephen B. 1982. The Vegetation of Land Bridge Beringia. In *Paleoecology of Beringia*, ed. D. M. Hopkins, J. V. Matthews, Jr., C. E. Schweger, and S. B. Young, pp. 179–94. New York: Academic Press.

POLLEN RECORDS FROM ARCHAEOLOGICAL SITES IN THE ALDANSKIY REGION, SAKHA REPUBLIC

G. M. Savvinova, Svetlana A. Fedoseeva, and Yuri A. Mochanov

The pollen records presented here were derived from major Dyuktai archaeological sites described in this volume. In addition to their value as sensitive environmental indicators, a property common to all pollen profiles, these diagrams correlate directly with particular periods of the Siberian Upper Palaeolithic, in sites that sometimes contain important elements of extinct fauna, and in which the carefully worked out stratigraphy is often buttressed with radiocarbon dates. The stratigraphic profiles accompanying these diagrams are repeated in the archaeological reports.

Dyuktai Cave

The cave is located on the right bank of the Dyuktai River 112 meters from its confluence with the Aldan River at km 1031 in the Ust-Mayaskiy region of the Sakha Republic (formerly Yakutia) (59°18′N, 132°36′E).

Forty-nine samples were collected for spore-pollen analysis from the most representative stratigraphic column just in front of the cave (Fig. 2–15). All evidence found here—stratigraphy, faunal remains, cultural remains, and radiocarbon dates—indicate these to be eroded Sartan deposits overlain by Holocene accumulations. Eight phases of vegetation development were recognized in this section.

Phase I (samples 49–46) is characterized by the spread of small forest, consisting of pine, fir, larch, cedar, and flatleafed birch. These small forests were interspersed by thickets of shrubby birch, alder, and willow. Phase I appears to coincide with one of the early Sartan interstadials.

Phase II (samples 45–42) indicates the reduction of forest vegetation. Arboreal pollen is repre-

sented by isolated grains from pine, cedar, arborescent and shrubby birch, and alder. Meadow of low herbs occurred. Green mosses and sphagnum were widely distributed, as were Lycopodium.

Phase III (samples 41–34) is indicative of the presence of pine-larch forests with admixture of fir and cedar creeper; shrubby birch, alder, and willow are extensive. *Carex* is predominant among herbs. Phase III is, most probably, related to the interstadial period.

Phase IV (samples 33–24) points to a reduction of arboreal and herbaceous vegetation. More frequent are cryptogams, represented by mosses and ferns. The spectra of this phase are indicative of severe climatic conditions.

Phase V (samples 23–22) is characterized by the spread of larch forests with admixtures of fir, cedar, shrubby birch, alder, and willow. The herbage has meadow-steppe characteristics. This is probably an interstadial phase.

Phase VI (samples 21–16) is indicative of a decrease in arboreal vegetation, while the meadow-steppe associations are still preserved.

Phase VII (samples 15–9) is indication of further climatic deterioration. Green mosses, horsetail, and Lycopodium (club moss) predominate in the vegetation.

Phase VIII (samples 8–1) is related to the Holocene and represented by vegetation of contemporary appearance. The Holocene spectra of Phase VIII have a species composition very similar to that of the spectra of interstadial Sartan Phases I, III, and V.

Ust-Mil 2

The site is located on the left bank of the Aldan River two kilometers below the confluence of the Aldan and Mil rivers, 971 km from the mouth of the Aldan, in the Ust-Mayaskiy region of the Sakha Republic (59°38′N, 133°07′E).

One hundred samples taken from various strata at the Ust-Mil 2 site provide the basis for the spore-pollen diagram (Fig. 2–16). Spores predominate over pollen (88%–93% of the spectra) in the lower part of the geological strata 4–5 (1.75–5.0 meters

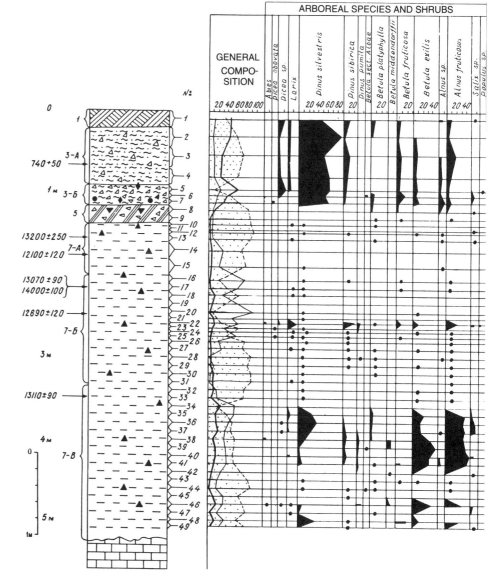

Figure 2–15 Dyuktai Cave. Stratigraphic profile and pollen spore diagram.

below surface). Arboreal plant and shrub pollen fluctuate between 11% and 39%, dominating at 2.5–5.0 meters.

Judging by stratigraphy and radiocarbon dates, deposits of units 4 and 5 should have been accumulating during the Karginsky era. For the entire period, pine and larch forests persisted. In the lower part of unit 4 and unit 5, the admixture of fir (up to 45%) and cedar is most notable.

According to stratigraphy and radiocarbon dates, geological unit 3 deposits were accumulating at the end of the Sartan era on the eroded surface of the Karginsky horizon. This period is characterized

by a closed pine forest with a small admixture of larch, cedar, and arboreal birch. Shrub birch were widespread. The uppermost loam (geological unit 2) was deposited during the Holocene. The spread of almost modern arboreal vegetation (birch, pine, and larch) occurred during this period.

Ust-Timpton

The site is located on a promontory on the left bank of the Timpton River, at its confluence with the Aldan River at km 1547 (58°42′N, 127°07′E) in the

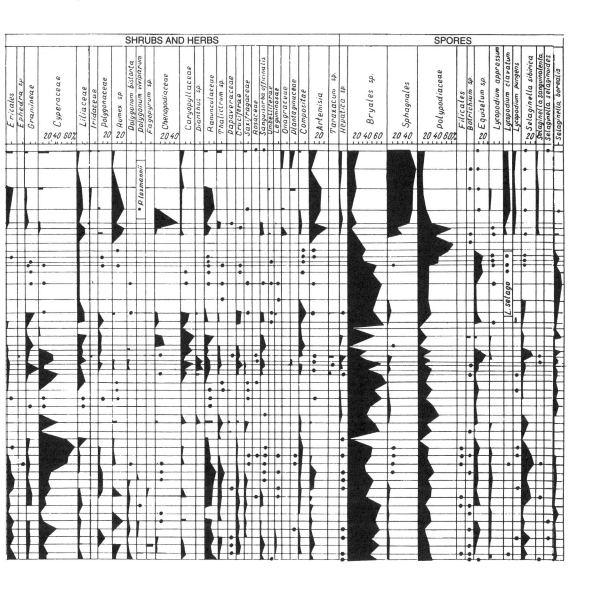

Aldanskiy region of the Sakha Republic. Ust-Timpton is located on the second (12 m) terrace of the Aldan, which may be traced for 350–400 m along the Aldan and for 120–150 m along the Timpton.

Eighty-five samples were collected for spore-pollen analysis (75 from one stratigraphic column) (Fig. 2–17). The samples selected from geological unit 3 indicate that ferns and mosses dominated the assemblages in the lower and middle sections of the unit. Arboreal, shrub, and herbaceous plant pollen are represented by isolated specimens. The available evidence (stratigraphy, radiocarbon dates, and vegetation characteristics) suggests that the deposits of unit 3 were forming at the end of the Sartan glaciation, under cold climatic conditions. Additional probes for samples equivalent to samples 53–61 may yet reveal pollen spectra corresponding with the Taimyr (Allerød) interstadial.

Pollen-spore samples from the upper section of geological unit 3, which corresponds with Cultural Strata VI and VII, indicate the presence of arboreal vegetation (pine, larch, and birch); shrubs; dwarf shrubs; and herbaceous plants. In Cultural Stratum VI, pine and larch tree trunks (15–17 cm diameter), which still retain their charred bark, had been used in construction of a dwelling. This phase of vegeta-

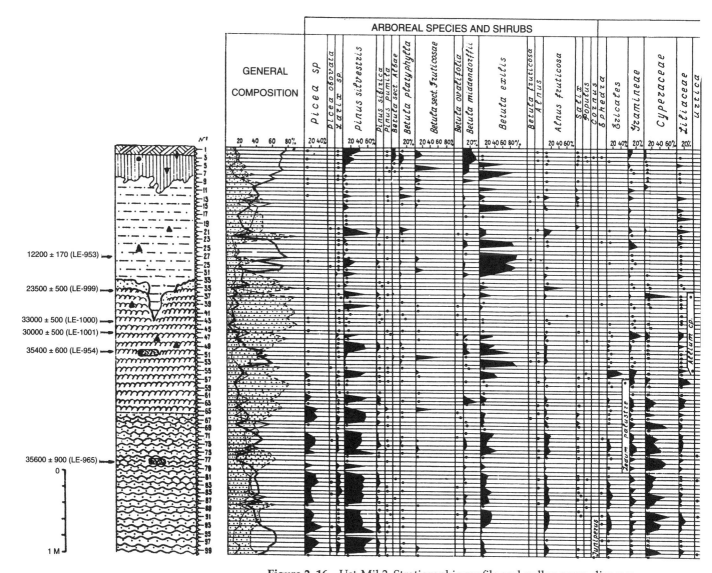

Figure 2–16   Ust-Mil 2. Stratigraphic profile and pollen spore diagram.

tional development obviously delineates the Pleistocene–Holocene boundary; it may represent the first Holocene warming (the beginning of the Pre-Boreal period).

The pollen-spore spectra for Cultural Strata Va–v (buried soil in the upper part of geological unit 3), indicate forest degradation as a consequence of rapid cooling. Clearly delineated frost-related cracks provide good supporting evidence. These spectra may thus represent the Pitsko-

Igarkinsky cold snap phase (the second half of the Pre-Boreal period).

IKHINE 1

Ikhine 1 is located high on the right bank of the Aldan River, just downstream from the point where the Ikhine Stream enters it, 283 km from the mouth of the Aldan, in the Tattinskiy region of the Sakha Republic (63°07′N, 133°36′E).

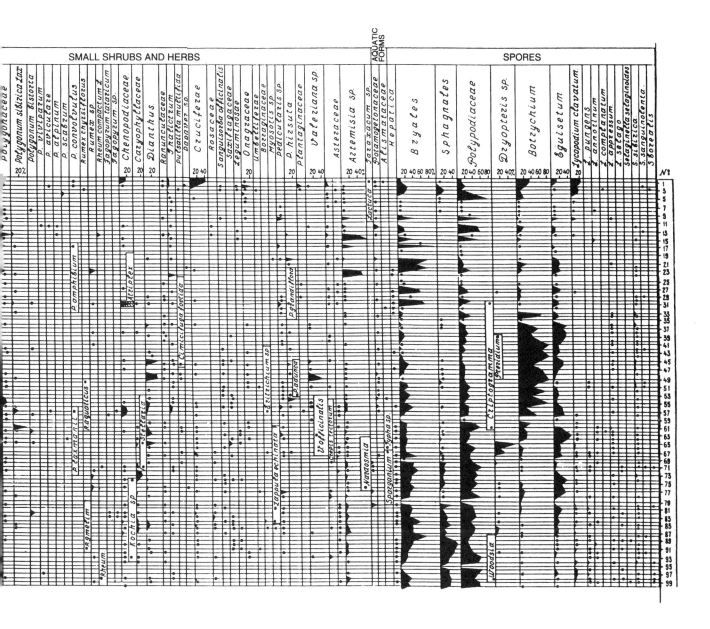

A pollen-spore analysis was made of stratigraphic section deposits to a depth of 160 cm (Fig. 2–18). The deposit spectra for Cultural Strata I, II, and III are dominated by spores (90%); the pollen count for arboreal and herbaceous plants is low. Among spores, green mosses predominate (up to 82% of the sample); ferns constitute 13%–35% of the sample. In the lower half of the stratigraphic section, arboreal and shrubby plant pollen range from 3% to 19% of the samples. Pine, fir, Siberian cedar, cedar creeper, larch, birch, and alder are present. These palynological characteristics indicate that the third terrace alluvium of the Aldan was forming during a cold, humid climate.

Ikhine 2

This site, across the stream from Ikhine 1, was discovered two years after Ikhine 1.

Analysis of fifteen palynological specimens col-

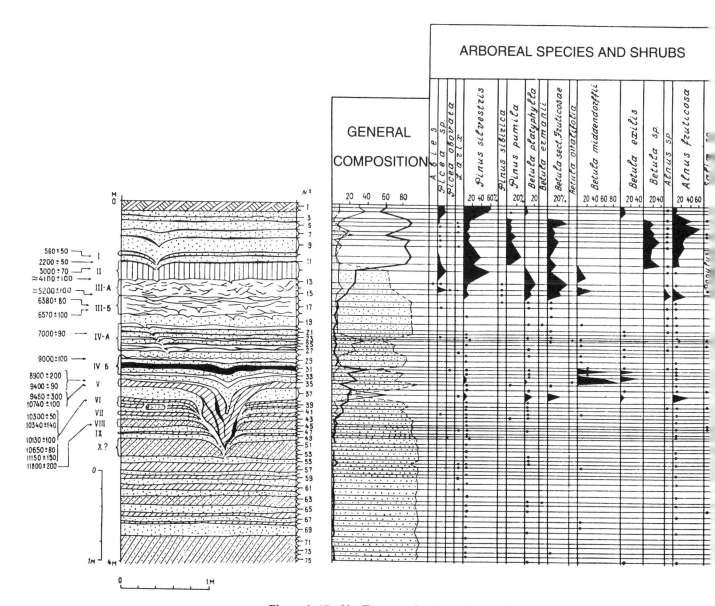

Figure 2–17   Ust-Timpton. Stratigraphic profile and pollen spore diagram.

lected from Ikhine 2 permits identification of several phases of vegetational development in the lower part of the stratigraphic section which contains Palaeolithic Cultural Stratum II (Fig. 2–19). Samples from the bottom of the third terrace alluvium indicate a spread of larch–pine woodlands, with a significant fir component. The spectra, however, contain up to 80% spores. This phase should thus correspond to the Malokhetsky warming of

the Karginsky mid-Ice Age. Pollen-spore spectra from the middle stratigraphic section consist of Compositae pollen, with decreasing amounts of arboreal pollen. This is indicative of forest degradation and the development of steppe vegetation. The spectra for the upper section of Stratum II indicate the development of spore plants dominated by ferns. There are only isolated pollen grains of arboreal and herbaceous plants. The deposits of Palaeo-

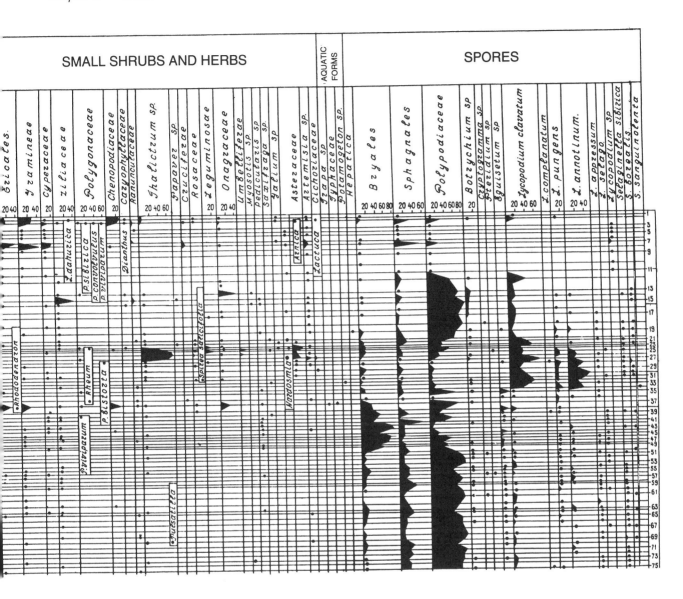

lithic Stratum I indicate the beginning of a spread of arboreal and shrubby plants; ferns are well preserved. This represents a phase related to the end of the Pleistocene.

EzHANTSY

The Ezhantsy site is located on the right bank of the Aldan River 794 km from its mouth in the settle-

ment of Ezhantsy in the Ust-Mayaskiy region of the Sakha Republic (60°28′N, 135°08′E).

Thirteen pollen-spore samples were obtained from various strata at the site (Fig. 2–20). Spores dominated the assemblage (71%–94% of the sample), especially those of green mosses and horsetails in the lower section and ferns in the upper section. Pollen of arboreal and shrubby plants constitute 2%–16% of the samples, represented by isolated

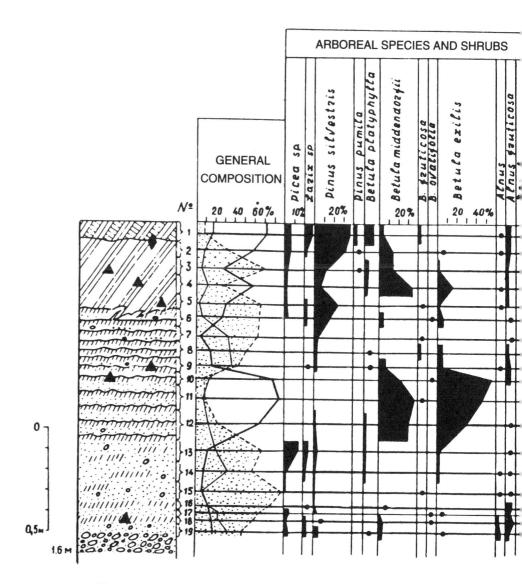

Figure 2–18   Ikhine 1. Stratigraphic profile and pollen spore diagram.

grains of pine, alder, fir, and shrub birch. According to Tomskaya and Savvinova, the pollen diagram is indicative of a cold climate.

Tumulur

The Tumulur site is on the left bank of the Aldan River at km 1456 in the Ust-Mayaskiy region (58°47′N, 128°11′E).

Fifty-four samples were collected from various strata for pollen-spore analysis. Five phases of vegetation development were recognized in the stratigraphic section (Fig. 2–21). Phase I (the lower horizon of geological stratum 4) is characterized by the presence of larch-pine forest, with an admixture of fir, cedar, and cedar creeper. Birch, alder, and willow likely grew along the banks of ponds, and open areas were occupied by meadow-steppe associations, with a significant spread of ferns, club moss (Siberian *Lycopodium*), and liverwort. This

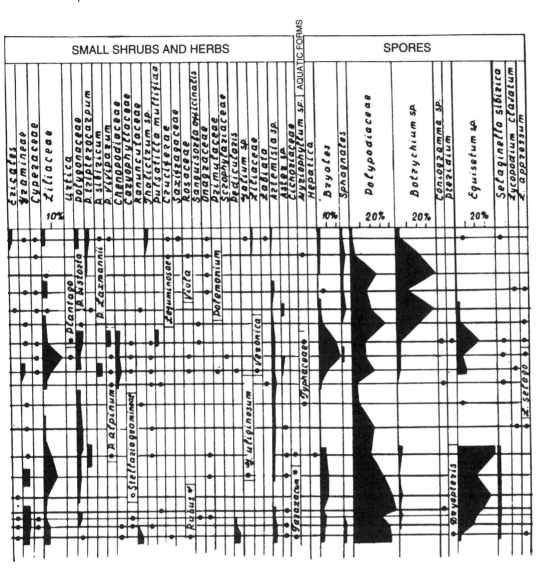

phase is probably related to the early Sartan inter-glacial.

Phase II (the middle and upper section of the lower horizon of geological stratum 4, the overbank facies) is characterized by an almost complete disappearance of arboreal vegetation; it was preserved only in isolated pockets. This phase is probably associated with the Sartan glaciation. Green mosses, ferns, and, in lesser quantity, club moss continued to spread.

Phase III (the upper horizon of geological

stratum 4; Upper Palaeolithic Cultural Stratum) exhibits arboreal/shrub vegetation and steppe associations. This phase is obviously associated with the Sartan interstadial, dated to 13,000–12,000 years BP, based on archaeological data.

Phase IV represents the deterioration of arboreal and herbaceous vegetation at the end of the Sartan. Phase V is related to the development of Holocene forest vegetation.

(Figures 2–19 to 2–21 follow on pp. 106–107.)

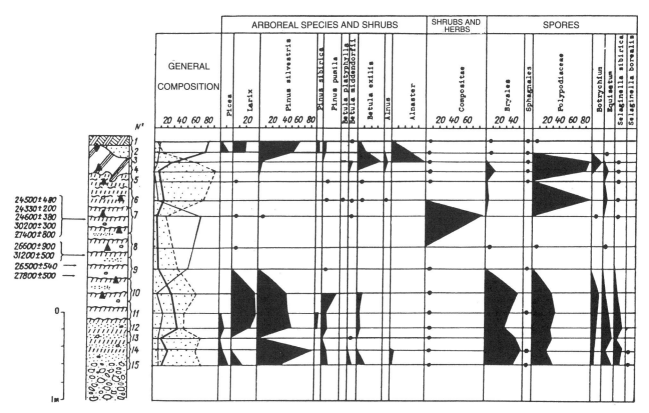

Figure 2–19   Ikhine 2. Stratigraphic profile and pollen spore diagram.

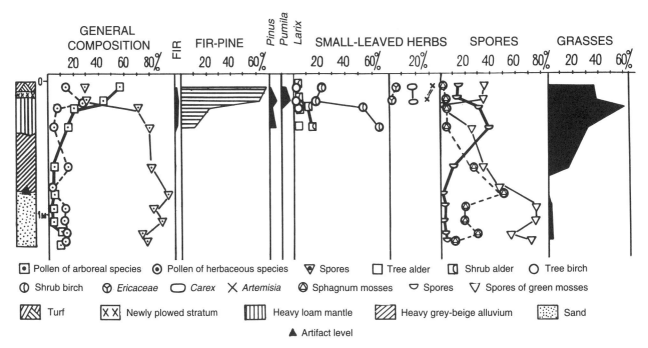

Figure 2–20   Ezhantsy. Stratigraphic profile and pollen spore diagram.

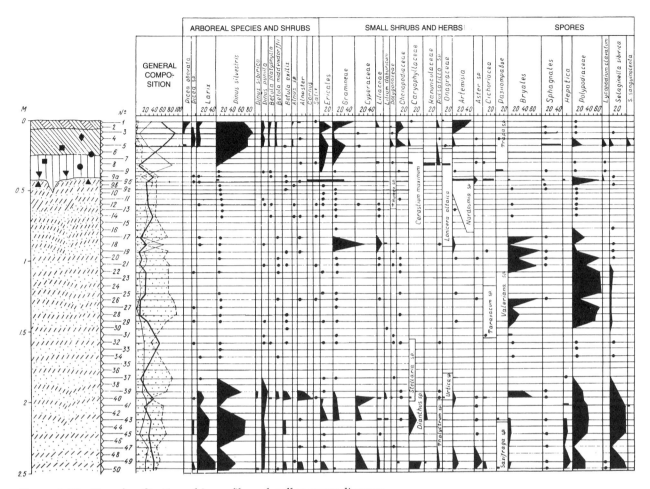

Figure 2–21   Tumulur. Stratigraphic profile and pollen spore diagram.

INTERPRETATION OF THE POLLEN RECORDS FROM LATE QUATERNARY ARCHAEOLOGICAL SITES OF THE ALDAN*

Thompson Webb III

SITES AND DATING

Four of the archaeological sites with pollen data are located along the Aldan River between 58°42'N and 59°38'N and between 127°7'E and 133°7'E (Savvinova, Fedoseeva, and Mochanov, this volume). Two sites (Ust-Timpton, the youngest, and Tumulur) are in the second (13 m) terrace above the river. Ust-Mil 2, the oldest site, is in the third (16–18 m) terrace, and the sediments from Dyuktai are from a cave in karst topography along the Dyuktai River 112 m from its confluence with the Aldan River. Three other more northern sites include the Ezhantsy site (60°28'N, 135°8'E) and the Ikhine sites 1 and 2 (63°7'N, 133°36'E). These are also located along the Aldan River and, like the Ust-Mil 2 site, are in the third (16–18 m) terrace. All sites were investigated by Y. A. Mochanov and S. A. Fedoseeva between 1964 and 1982, and my description and interpretation are based entirely on the pollen diagrams and site descriptions that they have included in this volume.

The pollen records for the three radiocarbon-dated sections south of 60°N cover intermittently the period from 36,000 years ago to present. The record at Ust-Mil 2 is the oldest, with dates back to 35,600 ± 900 BP. Its record extends up to 30,000 to 33,000 years ago, and then dates to 23,500 ± 500 and finally 12,200, where it fills in a gap between the records at Dyuktai Cave (13,000 to 14,000 years ago) and Ust-Timpton with its oldest date at 11,800 ± 200

BP. The 26 radiocarbon dates from Ust-Timpton indicate the longest sequence of almost continuous sedimentation at this site from 11,800 to 560 BP, but these dates and those from the other sites show that sedimentation in the terrace and cave sections was far from continuous. Dating reversals at all three sites indicate potential overturning in the sediments. At the one undated site south of 60°N, morphostratigraphic and archaeological interpretations allow dating of the Tumulur sediments to 13,000 to 12,000 years ago. Its pollen record therefore overlaps with that from Dyuktai Cave.

Of the three northern sites, only Ikhine 2 has radiocarbon dates. These are from 40 cm in the middle of the 3-m section, and the nine dates range from 24,330 ± 200 to 31,200 ± 500 BP, with a date of 27,800 ± 500 BP at the bottom and one of 24,500 ± 480 BP at the top. These dates are consistent with other dates for the third terrace and with the archaeological dating of this site and the other two northern sites. Mochanov and Fedoseeva conclude that the archaeological remains at Ikhine 1 and 2 date to 35,000 to 25,000 years ago, and that Ezhantsy has an age of 35,000 BP and is "one of the most ancient Dyuktai sites currently known."

POLLEN RECORDS

The pollen sequences are plotted in the Russian fashion with a column showing the total composition with categories of tree and shrub pollen, other shrub and herb pollen, and spores, all in the pollen sum. The individual taxa are then plotted as percentages for each of these categories, where tree and shrub pollen total to 100%, other shrub and herb pollen total to 100%, and spores total to 100%. Trace amounts are shown by black dots, and for some samples only trace amounts are recorded for tree, shrub, and herb pollen. For these samples, only the spores add up to 100%. The alluvial and cave sediments seem to alternate between periods of good and/or poor accumulation and preservation of pollen.

The record at Ust-Mil 2 starts out with abundant pine, sedge, spruce, birch, alder, and larch pollen at ca. 36,000 years ago. By 35,400 ± 600 BP, birch

*An NSF Climate Dynamics Grant supported the writing of this paper. P. Anderson and A. V. Lozhkin helped with the interpretations. My contribution is based entirely on the published work of G. M. Savvinova, S. A. Fedoseeva, and Y. A. Mochanov and has been made in the spirit of making their work more widely available to English-reading scientists.

pollen has risen to dominate the tree and shrub sum, and spruce and sedge pollen are no longer abundant. By 33,000 to 30,000 BP, spores dominate the record. At 23,500 BP, the one sample dated to just before the last glacial maximum contains abundant pine, birch, alder, and sedge pollen with a trace of larch pollen, but no spruce or willow pollen. Spores continue to dominate the total sum. At 12,200 ± 170 BP, tree pollen again dominates the record with almost all of it being birch (*Betula exilis*) pollen. This pollen spectrum, however, differs from those at Dyuktai Cave, where a mixture of pine, birch, alder, heath, grass, sedge, Ranunculaceae, *Thalictrum*, and Compositae pollen dominate before 13,100 BP and little or no tree, shrub, and herb pollen are found at 12,100 BP. Dating reversals, however, make the pollen record from the cave site difficult to interpret. The pollen record from Tumulur, the stratigraphically dated site, is similar to that at Dyuktai Cave before 13,100 BP but contains more larch and pine pollen. The larch values at 20% to 40% are unusually large.

After 12,000 BP, the record from Ust-Timpton begins. It contains a dozen cultural units in 2.5 m of alluvial sediments with 75 samples collected for pollen from one 4-m stratigraphic column. Cultural units VI to X dating from 9,000 to 11,800 ± 200 BP are in permanently frozen overbank sand deposits. The 30 lowermost pollen samples contain almost no pollen and are dominated by spores. Birch, alder, heath, grass, and other herb pollen increases in abundance about 10,000 ± 500 BP, and birch pollen dominates alone about 9,000 ± 500 BP. Little pollen except *Thalictrum* and Asteraceae is found between 9,000 and 6,000 ± 500 BP. After this time, a mixture of tree and shrub pollen types dominate the record, including spruce, pine, birch, alder, and heath pollen. Spores are relatively rare in these upper sediments. Only traces of larch pollen are found even though this is a major tree type today in the region, but larch pollen is generally underrepresented in the pollen record.

At the sites north of 60°N, spores dominate most of the record at the undated but purportedly 30,000-plus-years section at the Ezhantsy site. A mixture of tree and shrub pollen appears in the upper 20 cm of the section. Farther north, tree pollen of *Pinus sylvestris*, *Larix*, and *Picea* was moderately abundant before and after the dated interval of 24,000 to 30,000 years ago at Ikhine 2. Moss and fern spores are abundant throughout the record except at the top. At Ikhine 1, spores including ferns and *Equisetum* (horsetails) are abundant throughout, shrub birches dominate in the middle of the undated section, and pine, larch, and spruce trees with some birch shrubs are dominant in the upper meter.

DISCUSSION

Maps in two chapters of Velichko's (1984) *Late Quaternary Environments of the Soviet Union* provide interpretations of the past vegetation for this region. Grichuk (1984) maps open larch and birch forest with tundra elements as growing in this region at the maximum of the last glacial period, which dates from 21,000 to 14,000 BP, and Khotinsky (1984) maps light coniferous forest of larch–pine middle taiga with spruce and steppe as growing around the sites during the Boreal (9,000 to 8,000 BP) and Atlantic (6,000 to 4,600 BP) periods. The pollen records are generally consistent with these interpretations. Only the record from Ust-Timpton for 8,000 to 9,000 BP seems sparse in its pollen content during the Boreal period; Khotinsky (1984) maps light forest in the region, but poor accumulation and preservation of pollen in these sediments may be the cause of this discrepancy. The presence of sedge pollen with tree types at 23,500 BP at Ust-Mil 2 and at before 13,000 BP at Dyuktai Cave is consistent with tundra/open-forest interpretation, and the mixture of tree and shrub pollen with some grass, pigweed, and sage pollen after 6,000 BP at Ust-Timpton is consistent with light coniferous forest with spruce and steppe growing near the site. Steppe pollen types are somewhat rarer than this interpreted vegetation might imply, however.

The decrease in the rich tree, shrub, and herb pollen flora about 35,000 BP at Ust-Mil 2 indicates major vegetation and, presumably, climate changes in this region then. The decrease in spruce and sedge pollen abundance is consistent with a cooling and perhaps drying of the climate. The pollen data from Ikhine 1 and 2 and from Ezhantsy that are dated to 35,000 to 25,000 years ago are indicative of cold climates and are therefore consistent with this interpretation. Only the dated record after 6,000 BP at Ust-Timpton shows as rich a pollen flora as that at Ust-Mil 2 for 35,000 years ago.

REFERENCES CITED

Grichuk, V. P. 1984. Late Pleistocene Vegetation History. In *Late Quaternary Environments of the Soviet Union,* ed. A. A. Velichko, pp. 155–78. Minneapolis: University of Minnesota Press.
Khotinsky, N. A. 1984. Holocene Vegetation History. In *Late Quaternary Environments of the Soviet Union,* ed. A. A. Velichko, pp. 179–200. Minneapolis: University of Minnesota Press.
Velichko, A. A., ed. 1984. *Late Quaternary Environments of the Soviet Union.* Minneapolis: University of Minnesota Press.

INSECT FOSSIL EVIDENCE ON LATE WISCONSINAN ENVIRONMENTS OF THE BERING LAND BRIDGE*

Scott A. Elias

The study of Quaternary insect fossils has grown rapidly during the last thirty years. This research has revealed that insect species have remained remarkably constant for more than a million years (Matthews 1977; Coope 1978). Hence, our studies deal with the fossil remains of organisms that still have living descendants, which means that fossil identification does not involve the naming of extinct forms, but rather the matching of fossils to modern specimens. The modern examples of the insects found in Quaternary assemblages provide vital, exceptionally accurate information on climatic toler-

*This study was funded in part by the Minerals Management Service through interagency agreement with the National Oceanic and Atmospheric Administration, as part of the Outer Continental Shelf Assessment Program. Support for manuscript preparation was provided by a grant to S. A. Elias and S. K. Short by the National Science Foundation, DPP-8921807.

ances, habitat requirements, and distribution patterns. This information is applied directly to the fossil assemblages. The link between fossil and modern insects is important in the development of accurate reconstructions of the often-baffling environments of the Quaternary (Elias 1993).

Most Quaternary insect studies have focused on the remains of hard-bodied groups. Chief among these are the beetles, or Coleoptera, although other insects are playing an increasingly important part in palaeoenvironmental reconstructions. The main exoskeletal parts that preserve in sediments are the head capsule, the pronotum (the dorsal thoracic shield), and the wing covers, or elytra. These plates are often ornamented with a number of features that are preserved for as long as the sclerites themselves are preserved. These surface features, as well as the overall shape and size of body parts, are the principal characteristics used to identify the fossils.

Insects are arguably the most successful group of organisms ever to have lived on this planet. The current group of insect orders has existed since before the dinosaurs; it comprises about three-quarters of all animal species. Beetles alone account for 20 percent of all known species of organisms, which is more than all flowering plants combined. The diversity of beetles is staggering. There are probably more than a million species of beetles in 168 families (Crowson 1981), and many tropical forms remain undescribed. This taxonomic diversity corresponds to an equally astounding ecological complexity. Beetles occupy almost every conceivable ecological niche in terrestrial and freshwater habitats, from polar deserts to the tropics.

Many beetle species are quite sensitive to environmental change, as are some caddisflies, midges, and other insect groups. Predators and scavengers receive the most attention in palaeoenvironmental reconstructions because they are not tied to the vegetation. Many species are stenotherms, adapted to only a narrow range of temperatures. These insects may colonize a region as long as the climate is suitable. When climatic conditions change, stenotherms depart with equal rapidity. For instance, studies of late-glacial faunas in Britain (Coope 1977) and North America (Elias 1990) have shown that the replacement of cold-adapted species with warm-adapted species took place within a few decades at the end of the last glaciation.

Over the last few decades, the ubiquity of well-preserved insect fossils in many archaeological settings has been demonstrated. The principal types of deposits containing insect fossils are water-logged sediments. Insect exoskeletons in water-logged sediments show excellent preservation and are often very abundant. Insect fossils have also been found in domestic debris, stored products, human coprolites in dry caves, stored products in tombs and dwellings, and on human mummies (Elias 1993). Until recently, insect fossil studies were applied to archaeological sites mainly in Europe (chiefly Great Britain). In the 1980s, a number of North American projects adopted insect fossil analyses. These include the author's involvement in such research at Lamb Spring, Colorado; Lubbock Lake, Texas; and False Cougar Cave, Montana (Elias 1986, 1990; Elias and Johnson 1988; Elias and Toolin 1989). Insect fossils from natural deposits adjacent to archaeological sites (pond and lake sediments, peat bogs, etc.) have helped delineate palaeoenvironmental conditions for the time of human occupation.

About thirty Quaternary insect fossil studies from eastern Beringian sites have been published. These and other studies have shown that eastern Beringia was not a monolithic steppe-tundra ecosystem during the Pleistocene but rather a mosaic of different biological communities, many of which no longer exist in the same form. For instance, recent work in southwestern Alaska has shown that mesic and moist tundra habitats persisted there, even during the height of the last glaciation (Lea et al. 1991; Elias 1992).

INSECT FOSSILS FROM THE LAND BRIDGE

The writer has recently had the privilege of studying insect fossils extracted from peats that grew on the exposed shelf of the Chukchi Sea. That shelf formed the northern sector of the Bering Land Bridge during many intervals of lowered sea level in the Pleistocene.

The cores sampled for insect fossils came from the Chukchi Shelf, west of Icy Cape and north of

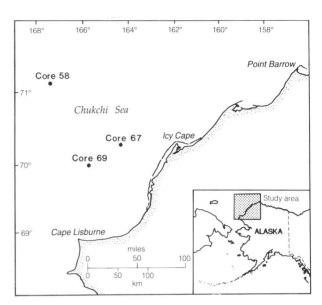

Figure 2–22   Map of northwestern Alaska, showing location of study cores on the Chukchi Shelf.

Cape Lisburne, northwestern Alaska (Fig. 2–22). The late Pleistocene peat horizon in the cores is about 50 m below modern sea level.

HISTORY OF LAND BRIDGE RESEARCH

During 1985, R. L. Phillips of the U.S. Geological Survey, Menlo Park, California, carried out geologic and geophysical surveys of the Chukchi Sea from the National Oceanic and Atmospheric Administration research ship, *Discoverer*. Four major stratigraphic units were identified from a transect of sites: a basal Cretaceous terrestrial sequence, a (possibly) Pliocene marine unit, the Pleistocene terrestrial sequence representing the land bridge, and an overlying Holocene marine deposit. During the investigations, twenty-two vibracore samples of shelf sediments were obtained. The cores were 9 cm in diameter and up to 6 m long. The cores were stored and subsampled at the U.S. Geological Survey in Menlo Park. Phillips submitted peat samples for radiocarbon age determinations and allowed the writer and his colleague, Susan Short, to study fossil remains from three cores. Fossil insects, pollen, and plant macrofossils were extracted from peat samples in cores 58, 67, and 69. These studies were performed at the Institute of Arctic and Alpine Research, University of Colorado, Boulder. The initial results have recently been published (Elias et al. 1992).

STRATIGRAPHY AND DATING

Peaty silts and peat laminations were found in a total of ten cores. The thickness of this sequence is variable, and the unit thins to the west. The sequence rests on an erosional contact above Cretaceous tephras and nonmarine strata or (possibly) Pliocene marine strata (Phillips and Colgan 1987; Phillips et al. 1987). The upper part of this unit was eroded by the Holocene marine transgression.

The stratigraphic sequence of the land-bridge organic deposits is composed of interbedded gravel, sand, muddy silt and sand, silty mud, peat beds, and organic-rich laminations. The peat beds were sampled for this insect fossil study. The peat is composed of partially decomposed sedges and wood fragments. In addition to insects and plant remains, freshwater ostracods (and in a few samples, brackish-water ostracods) were also found in the peat samples (Phillips and Brouwers 1990).

The peat layers in the cores are of latest Pleistocene age. Radiocarbon ages were obtained on peat macrofossils from two cores. The base of Pleistocene peat in core 54 yielded an age of $11,330 \pm 70$ BP (Beta-43952), and peat laminations from near the top of the nonmarine sequence in core 69 yielded an age of $11,000 \pm 60$ BP (Beta-43953).

FAUNA AND FLORA IDENTIFIED IN THE CHUKCHI CORES

The pollen spectra from the peats are dominated by grasses and sedges, with moderate percentages of birch. These spectra are consistent with the age assignment to the Birch Zone (14,000–9,000 BP). The data suggest a meadow-like graminoid tundra, with birch shrubs and some willow shrubs growing in sheltered areas (Elias et al. 1992).

The three cores yielded five samples containing fossil insect remains, mostly beetles. No single peat sample yielded sufficient specimens to allow a detailed palaeoenvironmental reconstruction. However, the samples are nearly contemporaneous; the dates on peat from two cores suggest that this peat horizon falls within the interval 11,330–11,000 BP. As such, I will discuss all identified specimens from the peat samples as an assemblage from one 300-year chronozone.

Several beetle species were identified from the assemblages, providing valuable palaeoenvironmental data. Among these are predators in the families Carabidae (ground beetles), Staphylinidae (rove beetles), and Dytiscidae (predaceous diving beetles). The carabid, *Bembidion grapii*, lives on dry ground with sparse vegetation. It ranges today across the boreo-arctic and alpine regions, as far north as the Mackenzie Delta region (Fig. 2–23a). According to Lindroth (1963), *B. grapii* does not occur in true tundra habitats in North America. However, in Greenland, populations of this species are found as far north as 76°53′N in the high arctic zone (Böcher 1988).

The dytiscid beetle, *Hydroporus polaris*, is found today on the arctic coast of Canada, as far west as the Alaskan border, and north through the Canadian arctic archipelago (Danks 1981) (Fig. 2–23b).

Three species of rove beetles were identified from the Chukchi Sea cores. *Eucnecosum brachypterum* is a northern, Holarctic species. Its modern range includes the Canadian Northwest Territories west through Alaska, in arctic, subarctic, and boreal forest regions (Fig. 2–23c). Modern specimens of *E. brachypterum* have been collected from a variety of

Figure 2–23 Known modern North American distributions of species found in the Chukchi Shelf samples: *a, Bembidion grapii* (after Lindroth 1963); *b, Hydroporus polaris*, (from data in Danks 1981); *c, Eucnecosum brachypterum*, (after Campbell 1984); *d, Holoboreaphilus nordenskioeldi*, (after Campbell 1978); *e, Tachinus brevipennis*, (after Ullrich and Campell 1974 and Campbell 1988); *f, Helophorus splendidus* (after Smetana 1985).

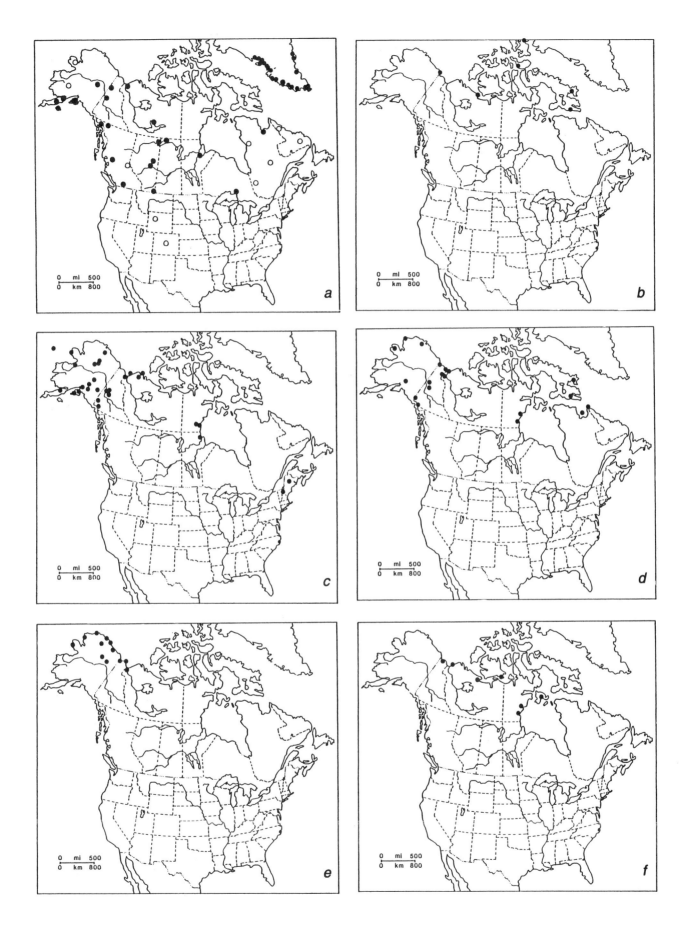

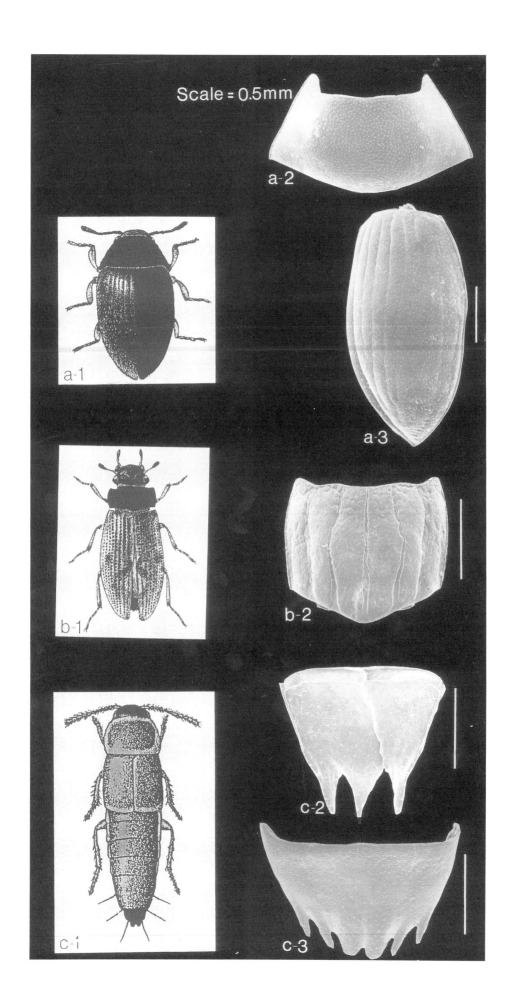

Scale = 0.5mm

a-1

a-2

a-3

b-1

b-2

c-1

c-2

c-3

moist habitats, especially alder and willow leaf litter and submerged clumps of sedges (Campbell 1984).

Holoboreaphilus nordenskioeldi is another arctic staphylinid beetle. Its modern range is similar to that of *E. brachypterum*, but it also ranges east to northern Labrador and southern Baffin Island (Fig. 2–23d). Its ecological requirements are likewise similar to *E. brachypterum's* (Campbell 1978). The third rove beetle (Fig. 2–24c-1) identified from the fossil assemblages is *Tachinus brevipennis* (Fig. 2–24c-2, c-3). This is an amphi-Beringian species, occurring today on both sides of the Bering Strait. In North America, it lives only in arctic Alaska and east as far as the Mackenzie Delta region (Fig. 2–23e). This species is commonly found in Wisconsin-age fossil assemblages from Alaska (Elias 1992).

Scavenging beetles are also reliable indicators of climate change, because they are able to invade new regions as soon as the ''bare bones'' ecosystem, including pioneer vegetation species, becomes established. For instance, water scavenger beetles (Hydrophilidae) in the genus *Helophorus* (Fig. 2–24b-1) often invade newly formed pools of water in such disturbed sites as gravel pits and roadside ditches. *Helophorus splendidus* (Fig. 2–24b-2) was found in the Chukchi peat samples. It lives in the arctic tundra regions along the coast of northern Canada, from the Mackenzie Delta region east to Hudson Bay (Fig. 2–23f). It may also live in arctic Alaska, but has yet to be found there (Smetana 1985).

The pill beetle (Byrrhidae), *Simplocaria tessellata* (Fig. 2–24a-1, a-2, a-3), is a northern species that feeds on mosses. This species is probably the same species as the European *S. metallica*. If so, then its range includes southern Greenland and the boreo-alpine regions of Scandinavia, as well as boreo-arctic and alpine regions of North America (Böcher 1988).

DISCUSSION

The insect fossils from the Chukchi cores are indicative of mesic to moist habitats with mosses, sedges, and deciduous shrubs surrounding the ponds in which the sediments accumulated. The upland beetle species are indicative of drier conditions, however, with only sparse vegetation.

The modern distributions of the identified species from the assemblages overlap in a narrow region of the arctic coast, on the Mackenzie Delta in Canada. Mean July temperatures in the overlap region range from 10.6° C at Tuktoyaktuk to 14° C at Aklavik (Environment Canada 1982). This reconstruction suggests that at 11,000 BP the climate of the exposed Chukchi Sea Shelf region was substantially warmer than is the modern North Slope of Alaska. Mean July temperatures there range from 3.8° C (38.9° F.) at Barrow to 4.4° C (39.9° F.) at Barter Island (National Oceanic and Atmospheric Administration 1982). So summer temperatures on the Chukchi Shelf may have averaged 6–10° C warmer than modern temperatures on the North Slope.

Fossil insect records show that the northern regions of eastern Beringia were subject to rapid warming at the end of the last glacial interval. The North Slope of Alaska today is an arctic tundra underlain by permafrost. The most easterly of the late Wisconsin–early Holocene sites was described by Matthews (1975), from Clarence Lagoon, just east of the U.S.–Canadian border on the Yukon coast (Fig. 2–25: 5). Wood in the sample yielded an age of 10,900 BP. The insect assemblage is essentially a tundra fauna with some species found today in the northern boreal forest. This species composition is very similar to the modern regional fauna, which suggests that the climate at 10,900 BP was, likewise, similar to the modern one.

Peat from a thaw lake in the Niguanak Uplands region of the northeastern Alaskan North Slope

---

Figure 2–24 Line drawings of modern specimens and scanning electron micrographs of fossil specimens of species found in the Chukchi peats: *a-1,* modern specimen of *Simplocaria tessellata* (blackened regions denote body parts preserved as fossils); *a-2,* fossil pronotum of *S. tessellata*; *a-3,* fossil right elytron of *S. tessellata*; *b-1,* modern specimen of *Helophorus* (blackened region denotes body part preserved as fossil); *b-2,* fossil pronotum of *H. splendidus*; *c-1,* modern specimen of *Tachinus* (blackened regions denote body parts preserved as fossil); *c-2,* fossil eighth abdominal tergite (female) of *T. brevipennis*; *c-3,* fossil eighth abdominal sternite (female) of *T. brevipennis*.

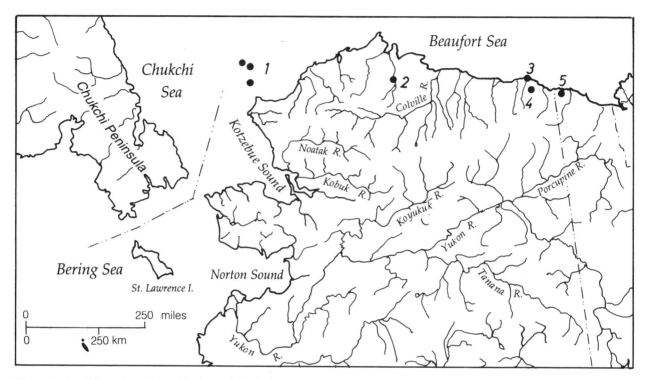

Figure 2–25   Map of northern Alaska and the Yukon Territory, showing locations of fossil insect sites discussed in text.

(Fig. 2–25) was studied by Wilson and Elias (1986) for insect fossils. The age of the samples ranged from 10,400 to 1,320 BP. Even the basal sample yielded an insect assemblage indicative of modern conditions.

Farther west on the coastal plain (Fig. 2–25), Nelson and Carter (1987) discussed an insect fauna from the Ikpikpuk River, dating to 9,400 BP. The deposit also contained *Populus* macrofossils and pollen. Since *Populus* does not grow this far north today, the botanical evidence is suggestive of climatic conditions warmer than present ones. The insect fauna contained large numbers of carabid beetles, which are found today on arctic tundra. It also contained more warm-adapted species. These beetles are all found today in the boreal zone of north-central Alaska to adjacent Canada. The *Populus* data might only be taken as an indication of increased moisture at the site, but the beetle data clearly indicate the North Slope climatic conditions in the early Holocene were 3–5° C warmer than present.

The depositional environment of the Chukchi peats was a coastal plain. The presence of both fresh- and brackish-water-adapted ostracods in the peats suggests that sea level was approaching the core sites at about 11,000 BP. The late glacial peats lie at depths of about 50 m below modern sea level (Fig. 2–22). If the − 50 m contour on the Chukchi Shelf was still above water until after 11,000 BP, then substantial parts of the Bering Land Bridge may have remained exposed at that time (Fig. 2–26).

Recent global sea-level reconstructions place sea level at about − 70 m at 11,000 BP (Fairbanks 1989). However, global sea-level curves are of little use on a regional basis, especially in a region as tectonically active as the Bering and Chukchi shelves. There have been two other regional studies on the timing of late Wisconsinan inundation of the Bering Land Bridge. Nelson (1982) described terrestrial peats and peaty silts from the Chirikov Basin, which he interpreted as late Pleistocene tundra and freshwater silt with occasional alluvial deposits. Based on radiocarbon assays of peat macrofossils, the uppermost land-bridge deposits in these cores range in ^{14}C age from 13,200 BP at about 46 m water depth to 12,700 BP at about 31 m water depth.

McManus et al. (1983) and McManus and Creager (1984) also estimated past sea levels for the Bering and Chukchi shelves. Their reconstructions suggested that sea level had risen to the −55 m level by 16,000 BP, although a stillstand at about −50 m was interpreted for the Chukchi Sea. However, their sea-level chronology is suspect, because it is based on radiocarbon ages of bulk sediments. McManus et al. (1983) observed microscopic particles of "coal-like material" in the sediments before they were dated and attempted to correct for inactive carbon contamination by following Broecker and Kulp's (1956) procedure. McManus et al. (1983) cautioned that their dates must be viewed as questionable, because of the need for such corrections and because large thicknesses of core section were required to provide sufficient carbon for dating.

The radiocarbon ages on peat macrofossils

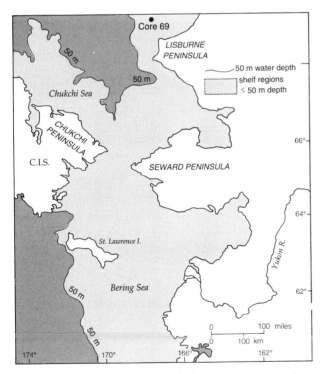

Figure 2–26 Regions of the Chukchi and Bering shelves above the −50 m contour that are postulated to have remained above sea level until after 11,000 bp.

from the Chukchi Shelf suggest that the previous dates on bulk sediments are as much as 3,800 years too old. The dates on the Chukchi Shelf peats are more reliable than the bulk sediment dates because the peat macrofossils were not subject to contamination by inert, microscopic carbon. It appears likely that additional radiocarbon ages based on macrofossils from other locations and other stratigraphic horizons on the Chukchi and Bering shelves will necessitate revisions in regional sea-level chronologies for the Pleistocene/Holocene transition. The Chukchi Shelf data represent only one short segment in the regional sea-level curve, and additional radiocarbon assays of macrofossils from land-bridge terrestrial deposits are needed to clarify the timing of the regional Pleistocene–Holocene marine transgression.

Apparently, the inundation of the land bridge coincided closely with the warmer-than-modern climatic episode on the North Slope, rather than predating it by several thousand years, as McManus and Creager's (1984) sea-level curve would suggest. The land bridge itself probably acted to increase the continentality of climates in eastern Beringia during the late Pleistocene. It blocked the oceanic circulation between Pacific and Arctic waters, and maritime influences on climate were removed—by more than 1,000 km—from the center of the land bridge and adjacent continental regions. As a corridor of cold, dry environments, the land bridge probably served as a conduit for steppe biota (Matthews 1982).

As the land bridge was inundated, the reestablishment of circulation between the Pacific and Arctic oceans through the Bering Strait probably played a major role in climatic amelioration in eastern Beringia around 11,000 BP. The Chukchi Shelf evidence suggests that this warming was both rapid and intense, not unlike the amelioration seen in British insect faunas upon the return of Gulf Stream waters to northwest Europe, following the Younger Dryas oscillation (Coope 1977).

There are some important implications for Beringian archaeology in the Chukchi Shelf fossil data. The previous sea-level chronologies for the Bering Strait region suggested that the land bridge was inundated as early as 14,500 BP. Even though

McManus and Creager cautioned that their chronology was a preliminary estimate, based on suspect radiocarbon ages, theirs was the only published sea-level chronology for the Bering Strait region for almost a decade. It has been widely cited in the archaeological literature, to the point where the original cautions have lost their potency. Under the McManus and Creager model, the humans crossing the land bridge would have had to make their journey sometime before 14,500 BP, despite the lack of hard evidence that humans were in Alaska at that time.

Additional dates need to be obtained from land-bridge terrestrial peats. If these dates confirm the Chukchi radiocarbon chronology, then it will be apparent that people could have crossed the land bridge as late as 11,000 BP. Several unequivocal occupation horizons in eastern Beringia predate 11,000 BP by up to several centuries.

The other salient point brought forth by the Chukchi data is that environmental conditions just prior to the inundation of the land bridge were probably quite conducive to migration of both game animals and humans. Insect fossil data from elsewhere in Alaska (Elias 1992) suggest that climatic amelioration actually began by 12,500 BP.

REFERENCES CITED

Böcher, J. 1988. The Coleoptera of Greenland. *Meddelelser om Grønland. Bioscience* 26.

Broecker, W. S., and J. L. Kulp. 1956. The Radiocarbon Method of Age Determination. *American Antiquity* 22:1–11.

Campbell, J. M. 1978. A Revision of the North American Omaliinae (Coleoptera: Staphylinidae). 1. The Genera *Haida* Keen, *Pseudohaida* Hatch, and *Eudectoides* New Genus. 2. The Tribe Coryphiini. *Memoirs of the Entomological Society of Canada* 106.

———. 1984. A Revision of the North American Omaliinae (Coleoptera: Staphylinidae), the Genera *Arpedium* Erichson and *Eucnecosum* Reitter. *Canadian Entomologist* 116:487–527.

———. 1988. New species and Records of North American Tachinus Gravenhorst (Coleoptera: Staphylinidae). *Canadian Entomologist* 120:231–95.

Coope, G. R. 1977. Fossil Coleopteran Assemblages as Sensitive Indicators of Climatic Changes during the Devensian (Last) Cold Stage. *Philosophical Transactions of the Royal Society of London,* Ser. B, 280:313–40.

———. 1978. Constancy of Insect Species Versus Inconstancy of Quaternary Environments. In *Diversity of Insect Faunas,* ed. L. A. Mound and N. Waloff, pp. 176–87. Royal Entomological Society of London Symposium.

Crowson, R. A. 1981. *Biology of the Coleoptera.* New York: Academic Press.

Danks, H. V. 1981. *Bibliography of the Arctic Arthropods of the Nearctic Region.* Ottawa: Entomological Society of Canada.

Elias, S. A. 1986. Fossil Insect Evidence for Late Pleistocene Paleoenvironments of the Lamb Spring Site, Colorado. *Geoarchaeology* 1:381–86.

———. 1990. The Timing and Intensity of Environmental Changes during the Paleoindian Period in Western North America: Evidence from the Fossil Insect Record. In *Megafauna and Man,* ed. L. D. Agenbroad, J. I. Mead, and L. W. Nelson, pp. 11–14. Mammoth Site, Hot Springs, South Dakota.

———. 1992. Late Quaternary Beetle Fauna of Southwestern Alaska: Evidence of a Refugium for Mesic and Hygrophilous Species. *Arctic and Alpine Research* 24:133–44.

———. 1993. *Quaternary Insects and Their Environments.* Washington, D. C.: Smithsonian Institution Press.

Elias, S. A., and E. Johnson. 1988. Pilot Study of Fossil Beetles at the Lubbock Lake Landmark. *Current Research in the Pleistocene* 5:57–59.

Elias, S. A., and L. J. Toolin. 1989. Accelerator Dating of a Mixed Assemblage of Late Pleistocene Insect Fossils from the Lamb Spring Site, Colorado. *Quaternary Research* 33:122–26.

Elias, S. A., S. K. Short, and R. L. Phillips. 1992. Paleoecology of Late Glacial Peats from the Bering Land Bridge, Chukchi Sea Shelf Region, Northwestern Alaska. *Quaternary Research* 38:371–78.

Environment Canada. 1982. *Canadian Climate Normals. Vol. 2. Temperature 1951–1980.* Downsview, Ontario: Environment Canada, Atmospheric Environment Service.

Fairbanks, R. G. 1989. A 17,000-Year Glacio-Eustatic Sea Level Record: Influence of Glacial Melting Rates on the Younger Dryas Event and Deep-Ocean Circulation. *Nature* 342:637–42.

Lea, P. D., S. A. Elias, and S. K. Short. 1991. Stratigraphy and Paleoenvironments of Pleistocene Nonglacial Units in the Nushagak Lowland, Southwestern Alaska. *Arctic and Alpine Research* 23:375–91.

Lindroth, C. H. 1963. The Ground Beetles of Canada and Alaska, Part 3. *Opuscula Entomologica* Supplement 24:201–408.

Matthews, J. V., Jr. 1975. Incongruence of Macrofossil and Pollen Evidence: A Case from the Late Pleistocene of the Northern Yukon Coast. *Geological Survey of Canada Paper* 75-1, B:139–46.

————. 1977. Tertiary Coleoptera Fossils from the North American Arctic. *Coleopterists Bulletin* 31:297–308.

————. 1982. East Beringia during Late Wisconsin Time: A Review of the Biotic Evidence. In *Paleoecology of Beringia,* ed. D. M. Hopkins, J. V. Matthews, Jr., C. E. Schweger, and S. B. Young, pp. 127–50. New York: Academic Press.

McManus, D. A., and J. S. Creager. 1984. Sea-Level Data for Parts of the Bering–Chukchi Shelves of Beringia from 19,000 to 10,000 ^{14}C yr B. P. *Quaternary Research* 21:317–25.

McManus, D. A., J. S. Creager, R. J. Echols, and M. L. Holmes. 1983. The Holocene Transgression of the Arctic Flank of Beringia: Chukchi Valley to Chukchi Estuary to Chukchi Sea. In *Quaternary Coastlines and Marine Archeology,* ed. P. M. Masters and M. C. Flemming, pp. 365–88. New York: Academic Press.

National Oceanic and Atmospheric Administration. 1982. Climatography of the United States. Monthly Normals of Temperature, Precipitation, and Heat and Cooling Degree Days by State, 1951–1980. Washington, D. C.

Nelson, C. H. 1982. Late Pleistocene–Holocene Transgressive Sedimentation in Deltaic and Non-Deltaic Areas of the Northeastern Bering Epicontinental Shelf. *Geologie en Mijnbouw* 61:5–18.

Nelson, R. E., and L. D. Carter. 1987. Paleoenvironmental Analysis of Insects and Extralimital *Populus* from an Early Holocene Site on the Arctic Slope of Alaska. *Arctic and Alpine Research* 19:230–41.

Phillips, R. L., and E. M. Brouwers. 1990. Vibracore Stratigraphy of the Northeastern Chukchi Sea, Alaska. Nineteenth Arctic Workshop, Boulder, Colorado. *Program and Abstracts*: 63–64.

Phillips, R. L. and M. W. Colgan. 1987. Vibracore Stratigraphy of the Northeastern Chukchi Sea. In *Geologic Studies in Alaska by the U.S. Geological Survey during 1986,* ed. T. D. Hamilton and J. P. Galloway, pp. 157–60. U.S. Geological Survey Circular 998.

Phillips, R. L., L. G. Pickthorn, and D. M. Rearic. 1987. Late Cretaceous Sediments from the Northeast Chukchi Sea. In *Geologic Studies in Alaska by the U.S. Geological Survey during 1987,* ed. J. P. Galloway and T. D. Hamilton, pp. 187–89. U.S. Geological Survey Circular 1016.

Smetana, A. 1985. Revision of the Subfamily Helophorinae of the Nearctic Region (Coleoptera: Hydrophilidae). *Memoirs of the Entomological Society of Canada* 131.

Ullrich, W. G., and J. M. Campbell. 1974. Revision of the *Apterus*-Group of the Genus *Tachinus* Gravenhorst (Coleoptera: Staphylinidae). *Canadian Entomologist* 106:627–44.

Wilson, M. J., and S. A. Elias. 1986. Paleoecological Significance of Holocene Insect Fossil Assemblages from the North Coast of Alaska. *Arctic* 39:150–57.

FOUR LATE PLEISTOCENE LARGE-MAMMAL LOCALITIES IN INTERIOR ALASKA

R. Dale Guthrie*

The four populations reported here were collected as a by-product of the extensive placer gold mining carried out in Alaska's interior in the 1930s through the 1950s. It must be noted that these collections were made under conditions less than ideal scientifically. The frozen, fossil-rich silt overburden (up to 50 m in depth) overlying the gold-bearing gravels was thawed and washed away with high-pressure hoses. Seldom was it ever possible to record the stratigraphy of these Pleistocene deposits, let alone record fossil occurrences in terms thereof. For the most part these specimens—in vast quantities—ended up on the drainage grates of the placer operations. There they were collected, notably by O. W. Geist, for the Frick Laboratory of the American Museum of Natural History which today houses thousands of these specimens.

Despite the constraints so imposed, a great deal of invaluable scientific evidence does inhere in these collections, and subsequent work has resulted, in some cases, in a delineation of the stratigraphy. In most cases it was determined that the Pleistocene sediments so removed were deposits representing a relatively short duration. At many Pleistocene exposures in interior Alaska, only deposits of Wisconsinan age and later are present; at others the Illinoian is also represented (Péwé 1952). The greatest concentrations of fossil large mammals occur in the upper part of the sections—those beds dated as Wisconsinan. So, although detailed stratigraphic information is often not available, the narrow unit of time involved permits the palaeosynecologist to consider these fossil assemblages as remnants of a community, which occu-

*Abridged by the editors from "Palaeoecology of the Large-Mammal Community in Interior Alaska during the Late Pleistocene." Originally published in *The American Midland Naturalist,* Vol. 79, No. 2, April 1968.

pied the immediate area where they were recovered, and to be essentially contemporaneous.

Even though the Alaskan assemblages are late Pleistocene in age and are well preserved (dried blood and other tissues are present on some bones), biases still exist that must be considered in any attempt at a detailed faunal analysis. Retransport is always a problem in faunal reconstruction, but in this case the bias is not so important, as the skeletal elements of large mammals are resistant to retransportation over long distances. The drainage basins from which the deposits were derived are all rather small.

Differential preservation of species, individuals, and elements due to size differences is surely an important consideration. My hypothesis of a grassland environment increases in likelihood, because most of the nongrazers are larger than or as large as the grazers. Organisms with longer generation lengths contribute correspondingly fewer individuals to the death assemblage (Van Valen 1964). Again, there are not great differences between the various species. The grazers have a longer or as long a life expectancy in the wild as the nongrazers.

The preservation differential based on size would tend to skew the observed proportion of proboscidians in one direction and the bias of species longevity would skew it in the opposite direction. These counterbiases are no doubt unequal, but I am unable to weigh their relative importance, and the relative proportion of proboscidians was therefore left unweighted. However, the exact number of proboscidians does not materially affect the major conclusion as to the nature of the habitat. Collecting biases are also difficult to analyze. The collector and his assistants attempted at each locality to collect all specimens of the elements used for my computations. Also, different individual collectors worked at different localities over a number of years. These facts tend to moderate any collecting bias. No screen-washing was done, so the series are composed primarily of large-mammal elements.

Although mammals are not as environmentally specific as many groups of organisms, some do show strong environmental correlations. For example, ungulates fall into either of two main dietary categories: browsers or grazers, with only a few in-

termediates. A fossil assemblage consisting chiefly of members of one of these categories can, as in this study, be reasonably assumed to have occupied a habitat corresponding to the dietary demands of its component members.

Only the limb bones (humerus, femur, radius, tibia, metapodials, and phalanges) were used to make the calculations. Although the individual limb elements differ in their preservation and probability of recovery, they do not vary much between species of similar size. Corrections were made for the differences in element number between species (e.g., wolf: 4 metapodials per foot; bison: 1 per foot).

Four localities were chosen from among the many where specimens have been collected. These localities have the four largest series of specimens and all are near Fairbanks, Alaska (Fig. 2–27). Multiple samples taken from roughly the same area provide a check on the likelihood of only one sample deviating drastically from the actual percentages in the fossil assemblage by chance alone. Detailed comparisons of the four assemblages show differences between any two. However, more general comparisons point up their strong similarities. Some of the differences between assemblages may be due to the previously mentioned biases acting differently at each locality. There are several physiographic differences between the four localities. The localities are:

1. Fairbanks Creek Mine: 65°04′–147°10′. Valley runs NW–SE. Drains SW. 30 km from Tanana River floodplain. Altitude approx. 300 m above sea level. Drainage basin approx. 40 square km. Mainly Wisconsin-age silt deposits, but pre-Wisconsin sediments present (Péwé 1965).

2. Engineer Creek Mine: 64°57′–147°38′. Valley runs NW–SE. Drains NW. 7 km from Tanana River floodplain. Altitude approx. 250 m above sea level. Drainage basin approx. 12 square km. Age of deposits similar to those at Fairbanks Creek (Péwé 1952).

3. Cripple Creek Mine: 64°49′–148°01′. Valley runs NE–SW. Drains NE. 5 km from Tanana River floodplain. Drainage basin approx. 6 square km. Altitude approx. 200 m above sea level. Age of deposits similar to those at Fairbanks Creek (Péwé 1965).

4. Gold Hill Mine: 64°51′–147°59′. Valley runs NE–
SW. Drains NE. 3 km from Tanana River floodplain. Drainage basin difficult to determine
precisely, because gold-bearing gravels represented somewhat different channel than today's,
but approx. 40 square km. Altitude approx. 200
m above sea level. This cut is on a bench of Cripple Creek and, although Wisconsinan sediments
are present, the major portions are pre-
Wisconsin (Péwé 1965).

Fairbanks Creek is somewhat higher than the
other three, and farther removed from the river valley. This may partially account for its being the locality with the least typical species composition.

The Faunal Community

The fossil mammals comprise an impressive array
in comparison with the poverty of the modern
fauna of this region. Whether or not all species were
exact contemporaries, at least they all inhabited the
area during part of, or throughout, a very short geological time span.

Before discussing the community let us look
briefly at some of the individual members. *Elephas
primigenius*, the woolly mammoth, is widely
known; in all probability it is a circumpolar form.
Frozen carcasses have been recovered from both Siberia and Alaska. Literally thousands of mammoth
elements have been found in Alaskan Pleistocene
deposits.

The large extinct bison fossils in Alaska probably represent one species, *Bison priscus* (= *B. crassicornis*). At least this species is far in the majority.
Again we are probably dealing with a circumpolar
species (Skinner and Kaisen 1947). The male had
very large horns, although not as large as *B. latifrons*
of the Great Plains. Bison are the most abundant
vertebrate fossils found in Alaska. Although these
animals were large-horned, they were not much
larger than *B. bison* in body size.

As yet no review has dealt directly with the
Alaskan horse. There are at least two size groups,
which may represent contemporaneous species or,
as Vangengeim (1961) suggests, two forms in one

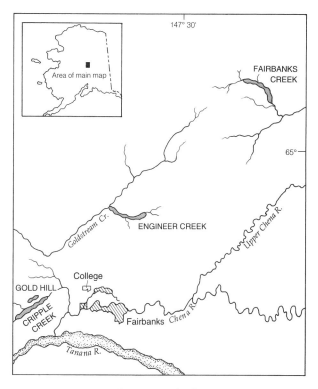

Figure 2–27 Map illustrating the four mining cuts
(striped areas) from which the fossil samples were
obtained. Inset shows the area of Alaska covered by the
map.

phylogenetic line. They appear to be closely related
to the quite variable Eurasian *Equus caballus*. The
bear, *Ursus arctos*; moose, *Alces alces* (= *A. americanus*); caribou, *Rangifer tarandus*; musk-ox, *Ovibos
moschatus*; wapiti, *Cervus elaphus* (= *C. canadensis*);
and sheep, *Ovis nivicola* (= *O. dalli*), are thought to
be the same as their modern counterparts, and all
were probably present in both Eurasia and North
America.

These mammals, which comprise the majority
of the assemblage, as well as the minority ("trace")
species, are listed in Table 2–5 with an estimate of
their weights and diets. The weights represent approximate annual averages of all age classes. The
presence of the saiga antelope, *Saiga tatarica*, is noteworthy. The nearest distributional perimeter of the
living Eurasian saiga is over 6,000 km from the fossil localities in interior Alaska, where its bones were
recovered from sediments of Wisconsinan age.

Table 2–5   Related Percentages and Biomass of Fauna at Four Fossil Localities

Species	Common Name	Diet	Est. Wt. (kg)	Fairbanks Creek		Engineer Creek		Cripple Creek		Gold Hill	
				I	B	I	B	I	B	I	B
Elephas (mammuthus) primigenius	wooly mammoth	G	3,000	4.7	25.9	6.7	36.0	10.9	50.8	3.4	20.9
Bison priscus (= B. crassicornis)	giant bison	G	500	64.6	59.4	44.9	40.2	35.9	27.9	54.9	56.4
Equus caballus	horse	G	290	17.3	9.2	36.8	19.1	41.0	18.5	35.4	21.1
Rangifer tarandus	caribou	N	100	4.7	.9	2.9	.5	6.6	1.0	3.0	.1
Ovibos moschatus	musk-ox	N	180	.9	.5	6.5	3.4	1.7	.8	1.2	.7
Alces alces (= A. americanus)	moose	N	370	4.7	3.3	.3	.2	.7	.4	.3	.2
Cervus elaphus (= C. canadensis)	wapiti or elk	G	220	.3	.1	.7	.3	.7	.2	.4	.2
Ovis nivicola (= O. dalli)	sheep	G	80	.1	.0*	.0*	.0*	.0*	.0*	.0*	.0*
Cervalces sp.	moose stag	N	350	.1	.1	.2	.1	.3	.2	.0	.0
Canis lupus	wolf	C	30	.9	.1	.5	.0	1.2	.1	.7	.0*
Canis latrans	coyote	C	15	.0	.0	.0	.0	.1	.0*	.0	.0
Felis sp.	lion	C	110	.8	.2	.2	.1	.5	.1	.0	.0
Symbos sp.; *Bootherium* sp.	woodland musk-ox	N	180	.6	.2	.2	.1	.0*	.0*	.2	.1
Saiga tatarica	saiga antelope	G	50	.0	.0	.0	.0	.0*	.0*	.0*	.0*
Camelops sp.	camel	G	180	.0*	.0*	.0*	.0*	.1	.1	.3	.2
Bos sp.	yak	G	300	.0*	.0*	.3	.2	.0*	.0*	.0*	.0*
Mastodon americanus	mastodon	N	2,500	.0*	.0*	.0	.0	.0	.0	.0	.0
Ursus arctos	brown bear	C, G	150	.0*	.0*	.0*	.0*	.2	.1	.2	.1
Smilodon sp.	saber-toothed cat	C	100	.0	.0	.0	.0	.0*	.0*	.0	.0
Percent grazer biomass				94.5		95.3		97.2		98.4	
Sample size (number of individual elements)				2,073		595		2,293		1,004	

Abbreviations:
   I—Percent of individuals in sample.   N—Predominantly nongrass-eaters.
   B—Percent of biomass.   C—Carnivores.
   G—Predominantly grass-eaters.

*Indicates species was present, but less than 0.05% of sample.

Saiga and *Bos* are the only Palaearctic ungulate genera that entered North America in the Pleistocene but are not found south of Alaska.

The relative compositions of the four faunas are presented in Table 2–5 and are illustrated graphically in Figure 2–28. In every case the bison and horse are the two most abundant species, with mammoth ranking—or tied for—third. Bison outnumber the horse in all of the faunas except one—Cripple Creek. At Fairbanks Creek there were

almost four times as many bison as horse. Caribou is fourth in abundance, with the exception of Fairbanks Creek, where it is tied for third, and Engineer Creek, where it is outnumbered twofold by musk-ox. The species generally ranking fifth is musk-ox. The two exceptions are the previously mentioned Engineer Creek and Fairbanks Creek. The relatively high frequency (4.7%) of moose at Fairbanks Creek is striking. At the other three localities moose comprise less than 1 percent of the fauna, about the

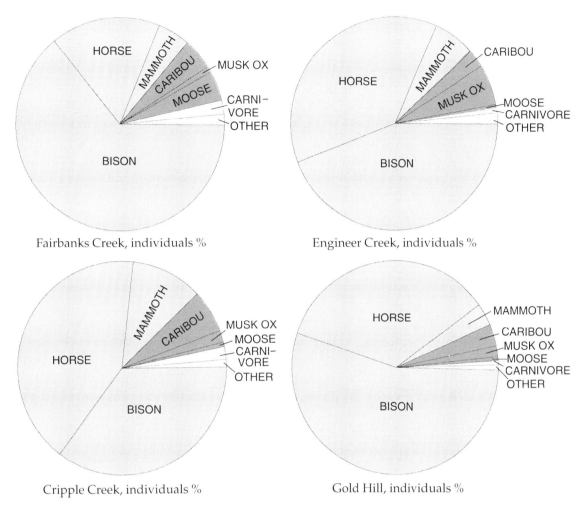

Fairbanks Creek, individuals %

Engineer Creek, individuals %

Cripple Creek, individuals %

Gold Hill, individuals %

Figure 2–28 The compositions of several large-mammal species at four fossil localities. The actual percentages are given in the I column in Table 2–5. The light areas represent grazers and the dark areas represent nongrazers.

same as wapiti. *Symbos* and *Cervalces* are the only other ungulates that occur in high enough frequencies to have been of much faunal significance—each averaging about 0.2 percent.

The number of wolf, lion, and bear are represented respectively in that order. The frequency of wolves averages slightly under 1 percent, lions half that, and bears less than half as frequent as lions. The average predator-prey ratio at these fossil localities is approximately one wolf per 130 ungulates and one lion per 250 or more ungulates. These ratios are not far removed from modern comparisons (Mech 1966; Wright 1960; Grzimek and Grzimek 1960), particularly when one considers that the wolf

and lion probably had some overlap in their prey species. The saber-toothed tiger, *Smilodon*, and the coyote, *Canis latrans*, also were present in the fauna for at least part of this depositional period. It is difficult to reconstruct the precise predator-prey relationships, as there are obviously no exact modern counterparts from which detailed comparisons can be drawn. But the ratio does not deviate greatly from the expected—given the predator species and the ungulate composition. Generally speaking, wolves are cervid predators, whereas lions prey more on equids and plains bovids. One might expect in this community that the bison, horse, and possibly young mammoth were the chief prey of the lion, and the

Pleistocene fossil exposure on Birch Creek, central Alaska. (*Source: University of Chicago Press. See Acknowledgments.*)

other forms prey of the wolf. This may account for the relatively high frequency of lions (see biomass information in Table 2–5), since the horse and bison were the most abundant ungulates.

One of the more difficult problems posed by this assemblage is the matter of density. How could such a complex and diversified fauna occupy a habitat which now supports so few individuals? Several estimates of large-mammal standing biomass have been made in different modern environments. The tundra environments are presently among the

lowest, with generally less than 300 kg per square km (estimated from the Nelchina basin, Alaska, by author), and some African savannas the highest, with 31,000 kg per square km (Bourlière 1963). Areas of faunal heterogeneity comparable to the Alaskan fossil community far surpass the present tundra in large-mammal biomass. Surely a fauna of such greater diversity would result in more efficient vegetational exploitation. Talbot (1966) has calculated that a replacement of the mixed native ungulate population of the African savannas by several introduced ungulates represents a loss in ungulate productivity of about 85 percent. The apparent high standards of living attained by Upper Palaeolithic man on the Eurasian loess-steppe (Butzer 1964), with virtually the same large-mammal community that existed in the Alaskan refugium, suggest relatively high densities of game. Rather than being comparable to the densities found on the present tundra, the ungulate production in the Alaskan refugium was perhaps several times as great.

The four fossil faunas point to a grazing community in the Fairbanks area during the Wisconsinan glaciation. Such a community is not unique to the interior of Alaska. Similar large-mammal faunas, with only minor variations, were present over much of northern Siberia (Vangengeim 1961). Northern Europe also had virtually the same species (Butzer 1964). Most of the large-mammal species present in late Pleistocene deposits in interior Alaska are also found in the University of Alaska's collections from north of the Brooks Range. Thus one might conclude that the climatic change accompanying the last glaciation (and earlier glaciations?), along with other factors, resulted in a widespread grassland environment in the northern Holarctic region and supported a rather extensive and complex large-mammal "grassland" community.

The differences between the past and existing large-mammal faunas of interior Alaska are not due to replacements but to eliminations. The Alaskan black bear is the one possible exception. It has not been found as a Pleistocene fossil in the Alaskan deposits and may be a recent southern immigrant. Of more than 20 fossil large-mammal genera, only six are now present in the area. The pattern of extinction involves chiefly the large grazers. Among the browsers, *Symbos*, *Mastodon*, and *Cervalces* also became extinct; however, these were only minor elements of the assemblage in the first place and may be interglacial representatives reworked into the glacial deposits. Although there is good circumstantial evidence from other areas that man was involved directly or indirectly in the widespread phenomenon of large-mammal extinctions in the late Pleistocene, the present paucity of the grazing niches in interior Alaska may provide evidence that at least some of these extinctions were due to a shift

in the climax plant species at the close of the last glaciation. The change in interior Alaska from grassland to coniferous forest and/or shrub tundra appears to have reduced the grazing habitat and increased interspecific competition until some species became extinct (Guthrie 1966). Since *Bison bison* lived in the interior after *B. priscus* (= *B. crassicornis*) had become extinct (Skinner and Kaisen 1947), at least *B. priscus* seems to have become extinct due to causes other than the elimination of the habitat. Viewing the extinction from this other side of the argument, one might visualize a situation where man was responsible for the extinction of these grassland species and, without the grazers to maintain it, the grassland habitat was gradually reforested. Although this latter hypothesis is possible, the former seems more plausible. This particular instance of vegetational causes of extinction may be an exception in the overall extinction pattern. Elsewhere, the grassland habitats continued to exist during and after the faunal extinctions (Martin 1963).

Although the major dietary divisions of these ungulate assemblages show striking similarities, there are definite differences among the four local faunas. The percentages of the chief grazers in the three assemblages dated as mainly Wisconsinan (Gold Hill cut has thick pre-Wisconsin sediments) form a pattern of increasing relative abundance of bison with increase in distance from the Tanana River floodplain and with higher altitude (Fig. 2–29). The percentages of horse and mammoth, on the other hand, decrease with increase in altitude and in distance from the floodplain. This complementarity may, of course, be an unavoidable artifact of dealing with percentages. The absolute numbers of bison may have increased directionally away from the floodplain, or the abundance of horse and mammoth increased toward the floodplain, or both. A comparison of the percentages of just horse and mammoth shows no clinal tendencies. Either altitude or distance could be the controlling variable or some factor related to one or both of these. It is impossible to separate the two effects, altitude and distance, since the distance of the alluvial deposits from the floodplain can almost be equated with accompanying changes in altitude. Taken at face

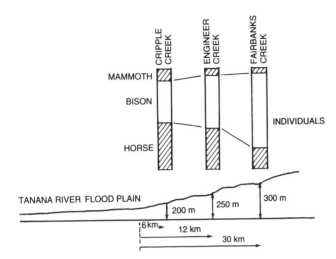

Figure 2–29 Percentages of the three chief grazers at three different localities representing different altitudes and distances from the valley bottom.

value, the data indicate that there was a tendency for bison to inhabit the higher areas while the mammoth and horse preferred the valley bottoms. Future data should provide more insight as to whether this correlation is directly causal, due to other causes, or is only spurious. Also note from Table 2–5 that there is a slight increase in the percentage of grazer biomass toward the floodplain.

The Alaskan Grassland

The most salient conclusion one can draw from the community reconstructed from these Alaskan mammalian fossil assemblages is that the vegetational patterns must have been radically different from those existing today in the same general region. The faunal composition (estimated from the counts conducted by the Alaskan Department of Fish and Game) of the living large mammals, within a 100-km radius of the four Pleistocene localities used in this study, is represented in Figure 2–30. Moose (*Alces*) and caribou (*Rangifer*) are the most common large mammals, with sheep present only in the high alpine meadows. The carnivores include wolves and two species of bears, the black bear (*Ursus americanus*) and the grizzly (*U. arctos*). It is a radically different fauna from the one portrayed in the fossil record. The chief difference can best be pointed out by ratios of the biomass of browsers and grazers. The four fossil faunas have biomass browser-grazer ratios of 5:100, 4:100, 3:100, and 1:100, whereas the ratio of the modern fauna from this same general area is higher than 100:1. By way of comparison, a grassland fauna in the Bison Range National Reserve in Wyoming has a biomass browser-grazer ratio of 28:100 (calculated from Bourlière 1963).

A ratio shift of this magnitude between the fossil and the recent mammalian faunas strongly supports the hypothesis that the area was a grassland during the Wisconsin glaciation, and has since changed to a spruce forest-shrub tundra environment. Both Colinvaux's (1964) profiles from the Seward Peninsula and Livingstone's (1955) profiles from north of the Brooks Range show a predominance of grass pollen during the last glaciation (see zones 0 and 1). Studies in Denmark (Hansen 1965) of fossil pollen, accompanying a similar assemblage of large mammal remains (dominated by horse and bison), also reveal a flora rich in grass and other herbs during late-glacial times (Younger and Older Dryas).

A comparison that may be meaningful in evaluating the grassland hypothesis is the mastodon-mammoth ratio. Apparently the mastodon was predominantly a browser and the mammoth a grazer. Green (1961) and Slaughter et al. (1962) have taken the ratios of mastodon to mammoth as a general indicator of the relative proportion of grassland and woodland. A contrast between these two proboscidian species, of similar size but different diets, avoids some of the difficulties encountered in a comparison of ungulates of dissimilar size and phylogeny. Generic identification of most proboscidian skeletal material is sometimes difficult. Cheek teeth, however, are easy to identify and are excellent for a comparison of this sort. Ratios of teeth are not directly equivalent to ratios of individuals, since individual mastodons have an average of about twice as many cheek teeth as do individual mammoths. Thus the correction factor is 2. If anything, the percentage of mastodon might be overestimated by this comparison as their teeth are bound entirely by thick enamel and would be more resistant to the

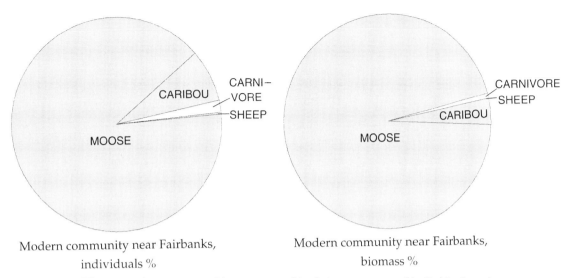

Modern community near Fairbanks,
individuals %

Modern community near Fairbanks,
biomass %

Figure 2–30   The approximate composition, expressed both in percentage of individuals and biomass, of the large mammals within a radius of 100 km of the four fossil localities.

rigors of time than the more friable lamellar teeth of mammoth. The numbers of teeth recovered from all four localities were: 13 mastodon and 677 mammoth. This corrects to a ratio of 1:104. Like the other ratios, this suggests a grassland environment.

With the cooler summer temperatures accompanying the Wisconsin glaciation the treeline was lowered more than 400 m (Repenning et al. 1964) in the Yukon–Tanana Upland of the Alaska refugium. A lowering of timberline of this magnitude on the southern exposure of the Tanana Hills would indicate an almost complete elimination of forested areas in the Alaskan refugium. There is even reason to believe that the Yukon–Tanana floodplains, although below timberline, were only sparsely forested or unforested. These extensive outwash floodplains were the sources for the heavy blanket of loess which now mantles the nearby high country, and as a consequence the floodplains were probably held in a disclimax condition. The periglacial climate and accompanying loess activity would no doubt have continually disturbed the adjacent vegetation, too. In Alaska the earliest successional stages are, as with most floras, grasses and other herbs.

Plant ecologists disagree concerning the various factors that favor maintenance of grasslands. However, it is the consensus that fires are one of the most important factors (Sauer 1950; Wells 1965). Carbon particles and pieces of charred wood are very frequently found in the frozen "muck" deposits in interior Alaska, so fire may have been important in maintaining the Alaskan grasslands. In his discussion of the effects of fire in interior Alaska, Lutz (1956) says that repeated burnings ultimately result in a stabilized fireweed-grass community. Hanson (1951) also illustrates how fires are frequently responsible for rich stands of grass in Alaska.

Other grass-generating forces may also have been present. Wind, in addition to increasing the effectiveness and extent of burns, also causes a direct selection for grasses—by its greater damage to other late successional forms—thus maintaining a continual disturbance situation or disclimax condition. The presence of the large glaciers may have produced more wind turbulence, as is suggested by the loess transport and the many presently inactive dune fields (Collins 1958) in what were unglaciated areas during Wisconsinan times.

Grasslands tend to be self-perpetuating since the many grazing mammals which they support continually redisturb the soil and trample the shrubs, selecting for grass regrowth. Murie (1944) and Bee and Hall (1956) have observed that concentrated grazing and trampling in the tundra result in an increased percentage of grass. One can conclude that there were a number of possible forces, aside from man, already present in the north that could have produced and maintained a grassland.

There are several grassland environments of limited extent in interior Alaska now. These have had a history of numerous fires and are subjected to winds far greater than average for interior Alaska. Interestingly enough, some of these are also areas where bison have been successfully reintroduced, but even after thirty-seven years, the bison have not expanded their distribution beyond the confines of these restricted grassland environments.

Rather than assuming a grassland in interior Alaska, one might postulate that the fossil animals being called grazers are not. But in the case of the horse and bison there is little room for doubt, although certainly all grazers rely on other herbaceous plants at times. The extant bison herds in the North (those reintroduced into Alaska and the indigenous herd in the Great Slave Lake area in northern Canada) and the once-existing feral horses in the Delta River, Alaska, area, utilized—and have been restricted to—limited grassland habitats. Pollen analyses of stomach contents of the frozen fossil mammoths have produced over 97 percent grass pollen (Tikhomirov 1958). That the mammoth was apparently more of a grazer than the modern elephant is indicated by the differences in tooth complexity. Studies of the food habits of modern wapiti (*Cervus elaphus*) (Murie 1951) also necessitate the classification of wapiti as primarily grazers, although they frequently shift to other plants during parts of the year. Wapiti, caribou, and musk-ox (*Ovibos*) are actually intermediate between browsers and grazers, but wapiti tend more toward the grazers and the living musk-ox more toward the browser preferences. Caribou also eat much grass, but are not generally classified as grazers. Although the saiga antelope feeds on a number of plants, grass is the major part of its diet also (Bannikow 1963). Sheep are almost strictly grazers.

Another possible explanation of the high percentage of fossil grazers is that the preservation biases favored the grazing forms. Quite to the contrary, the areas of deposition—the valley bottoms—were the habitat most likely to be occupied by one of the dominant elements of the modern fauna—the moose (*Alces*). Moose are as large boned (particularly the appendages) and would be just as likely to be preserved as either horse or bison, yet moose are one of the least common ungulates in the fossil assemblage. The woodland musk-ox, *Symbos*, is also rather rare in the collections. Apparently this genus was preferentially associated with a woodland situation (Semken et al. 1964) and, like *Alces*, could be expected to be rare in a grassland environment.

One of the chief limiting factors to grazers in the north during the winter is the condition of the snow. Flerow (1952), for example, states that snow line seems to be the most important factor influencing the northern perimeter of wapiti distribution. Depth and hardness seem to be the two main variables. If winter precipitation was less in the Alaskan refugium, as some suggest (Hopkins 1959), winter kill might not have been the obstacle to survival that it sometimes is today.

REFERENCES CITED

Bannikow, A. G. 1963. *Die Saiga-Antilope*. Wittenberg: A. Ziemsen.

Bee, J. W., and E. R. Hall. 1956. *Mammals of Northern Alaska on the Arctic Slope*. Misc. Pub. No. 8. Lawrence: University of Kansas Museum of Natural History.

Bourlière, F. 1963. Observations on the Ecology of Some Large African Mammals. In *African Ecology and Human Evolution*, ed. F. C. Howell and F. Bourlière, pp. 43–54. Chicago: Aldine.

Butzer, K. W. 1964. *Environment and Archeology*. Chicago: Aldine.

Colinvaux, P. A. 1964. The Environment of the Bering Land Bridge. *Ecological Monographs* 34:297–329.

Collins, F. R. 1958. Vegetated Sand Dunes in Central Alaska. *Geological Society of America Bulletin* 69:1752.

Flerow, C. C. 1952. *Fauna of the USSR. Mammals: Musk Deer and Deer*. Moscow: Nauka.

Green, F. E. 1961. The Monhans Dunes Area. In *Paleoecology of the Llano Estacado*, No. 1, ed. F. Wendorf. Santa Fe: University of New Mexico Press.

Grzimek, M., and B. Grzimek. 1960. Census of Plains Animals in the Serengeti National Park, Tanganyika. *Journal of Wildlife Management* 24:27–37.

Guthrie, R. D. 1966. The Extinct Wapiti of Alaska and Yukon Territory. *Canadian Journal of Zoology* 44:47–57.

Hansen, S. 1965. The Quaternary of Denmark. In *The Qua-*

ternary, Vol. 1, ed. K. Rankama, pp. 1–90. New York: Wiley Interscience.

Hanson, C. H. 1951. Characteristics of Some Grassland, Marsh, and Other Plant Communities in Western Alaska. *Ecological Monographs* 21:318–73.

Hopkins, D. M. 1959. Cenozoic History of the Bering Land Bridge. *Science* 129:1519–28.

Livingstone, D. A. 1955. Pollen Profiles from Arctic Alaska. *Ecology* 36:587–600.

Lutz, H. J. 1956. *Ecological Effects of Forest Fires in the Interior of Alaska*. U.S. Department of Agriculture Technical Bulletin 1133.

Martin, P. S. 1963. *The Last 10,000 Years*. Tucson: The University of Arizona Press.

Mech, L. D. 1966. *The Wolves of Isle Royale*. Fauna of the National Parks of the United States, Fauna Series, No. 7.

Murie, A. 1944. *The Wolves of Mt. McKinley*. Fauna of the National Parks of the United States, Fauna Series, No. 5.

Murie, O. J. 1951. *The Elk of North America*. Harrisburg, Pennsylvania: Stackpole.

Péwé, T. L. 1952. Geomorphology of the Fairbanks Area. Ph.D. dissertation, Stanford University.

———. 1965. Resume of the Quaternary Geology of the Fairbanks Area. In *Guidebook for Field Conference F, Central and South Central Alaska*, ed. T. L. Péwé. Seventh Congress, International Quaternary Association. Lincoln: Nebraska Academy of Sciences.

Repenning, C. A., D. M. Hopkins, and M. Rubin. 1964. Tundra Rodents in a Late Pleistocene Fauna from the Tofty Placer District, Central Alaska. *Arctic* 17:177–97.

Sauer, C. O. 1950. Grassland Climax, Fire, and Man. *Journal of Range Management* 3:16–21.

Semken, H. A., B. B. Miller, and J. B. Stevens. 1964. Late Wisconsin Woodland Musk Oxen in Association with Pollen and Invertebrates from Michigan. *Journal of Paleontology* 5:823–35.

Skinner, M. F., and O. C. Kaisen. 1947. The Fossil Bison of Alaska and Preliminary Revision of the Genus. *Bulletin of the American Museum of Natural History* 89:1–256.

Slaughter, B. H., W. W. Crook, Jr., R. K. Harris, D. C. Allen, and M. Seifert. 1962. The Hill-Schuler Local Faunas of the Upper Trinity River, Dallas and Denton Counties, Texas. Bureau of Economic Geology, University of Texas, Report of Investigations, No. 48.

Talbot, L. M. 1966. *Wild Animals as a Source of Food*. Bureau of Sport Fisheries and Wildlife. Special Scientific Report, Wildlife No. 98.

Tikhomirov, B. A. 1958. Natural Conditions and Vegetation in the Mammoth Epoch in Northern Siberia. *Problems in the North* 1:168–88.

Vangengeim, E. A. 1961. *Paleontological Basis of the Stratig-raphy of the Anthropogene Deposits of North East Siberia (According to the Mammal Faunas)*. Academy of Sciences of the USSR, Geological Institute. Vipusk 48.

Van Valen, L. 1964. Relative Abundance of Species in Some Fossil Mammal Faunas. *American Naturalist* 92:109–16.

Wells, P. V. 1965. Scarp Woodlands, Transported Grassland Soils, and Concept of Grassland Climate in the Great Plains Region. *Science* 148:246–49.

Wright, B. S. 1960. Predation on Big Game in East Africa. *Journal of Wildlife Management* 24:27–37.

A PALAEOENVIRONMENTAL RECONSTRUCTION OF THE "MAMMOTH EPOCH" OF SIBERIA*

Valentina V. Ukraintseva, Larry D. Agenbroad, and Jim I. Mead

The discovery and analyses of frozen carcasses of Siberian herbivores, ranging in radiocarbon ages from 9730 ± 100 to $53{,}170 \pm 880$ BP (Table 2–6), provide reconstructions of the vegetative cover and palaeoenvironment of the late Pleistocene and early Holocene "Mammoth Epoch" of Siberia. Analyses of the stomach and/or intestinal contents, as well as the enclosing sediments, serve as a contrast and comparison to the modern environments at the discovery sites. Six mammoths, one bison, and one horse (Fig. 2–31) provide the faunal basis for the botanical analyses of the ingested vegetation. The enclosing sediments provide a cross-check, and both ancient data sets are contrasted with modern environmental information.

Frozen fauna from the late Pleistocene of Siberia have been documented for more than 100 years (Chersky 1891, Pavlova 1910, Popov 1948, Sher 1971, Vereshchagin 1981, Lazarev 1982). More than fifty complete, or nearly complete, frozen speci-

*Adapted from *Vegetation Cover and Environment of the "Mammoth Epoch" in Siberia* by V. V. Ukraintseva. 1993. L. Agenbroad, J. Mead, and R. H. Hevly, eds. Published by The Mammoth Site of Hot Springs, South Dakota.

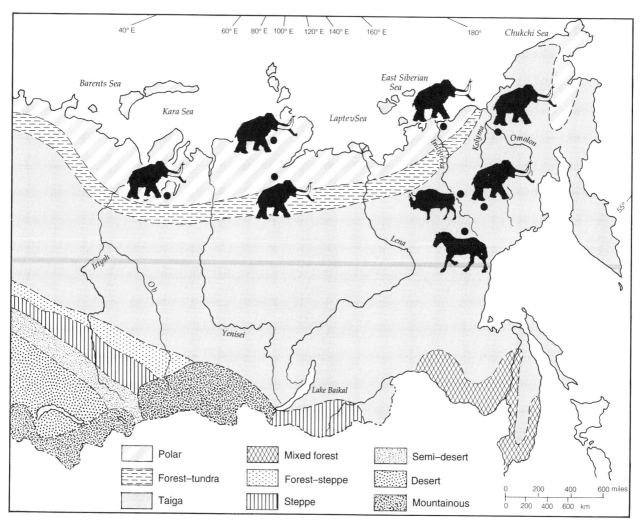

Figure 2–31   Map of Siberia, showing major vegetation zones and the locations of the fossil discoveries.

mens have been recovered. Rare instances provide frozen carcasses with well-preserved gastrointestinal tracts filled with plant matter reflecting vegetative conditions that were present in that area at the time of the animal's death.

Attempts at examination of food remains from fossil animals began as early as 1849, when investigators such as Brandt, Meyer, Marklin, and Schmalgausen analyzed plant remains stuck in the teeth of a woolly rhinoceros discovered on the Vilyuy River in Yakutia. The possibility of investigating vegetation from frozen gastrointestinal tracts began with the discovery of the Berezovka mammoth in 1900.

Analyses of the plant materials from this discovery were undertaken by several investigators (Sukachev, Gerz, Szafer, Tikhomirov, Popov, Zaklinskaya, Zenkova, Savich-Lyubikkaya, Abramova, Kupriyanova, and others) between 1900 and 1960. In addition to the gastrointestinal plant remains, studies of the enclosing stratigraphic units were also undertaken. Similar studies on other specimens revealed that pollen and spores were also well preserved and recognizable.

In the period 1968–1973, several well-preserved specimens of horse, bison, and mammoth were discovered. The gastrointestinal contents, as well as

Table 2–6 Radiometric Chronology of the Mammoth Fauna from Siberia, Some of Which Provided Associated Gastrointestinal Remains

	Indigirka River Basin (Eastern Siberia)	
Shandrin Mammoth (5)	Shandrin River, tributary to lower Indigirka River	$40,350 \pm 880$
Mammoth	Berelekh River, left tributary of the Indigirka River	$39,590 \pm 870$ $13,700 \pm 400$ $12,240 \pm 160$ $11,830 \pm 110$
Selerikhan Horse (3)	Selerikhan gold mine, upper Indigirka River (on Balkhan Creek)	$38,590 \pm 1120$
Mylakhchin Bison (4)	Middle Indigirka River	$29,500 \pm 1000$
	Kolyma River Basin (Eastern Siberia)	
"Dima" Mammoth (6)	Kirgilyakh Creek, left tributary of the Kolyma River	$41,000 \pm 1100$
Mammoth (1)	Berezovka River	$44,000 \pm 3500$
	Khatanga River Basin (Western Siberia)	
Vereshchagin's Mammoth (7)	Bolshaya Lesnaya Rassokha, right tributary of the Novaya River	53,170
*Mammoth	Terekhtiyk River	$44,540 \pm 1840$ $44,170 \pm 1870$
*Taimyr Mammoth (2)	Mamontovaya River	$11,450 \pm 450$
*Yuribey Mammoth (8)	Yuribey River, Gydan Peninsula	$10,000 \pm 70$ 9730 ± 100

*No plant remains within the gastrointestinal tract.

the enclosing sediments, were examined for palaeo-vegetational clues. In 1973 a program of botanical studies of fossil fauna was initiated by B. A. Tikhomirov and V. V. Ukraintseva. This program included the following steps: (1) macrovegetation analysis of the gastrointestinal tract; (2) carpological analyses of the material; (3) palynological analyses of various portions of the tract; and (4) radiocarbon dating of the remains. Geological and botanical investigations of the burial sites were also taken, where feasible, making the investigations multidisciplinary in nature. These combined data allowed a vegetational reconstruction for the period in which the animal lived (and died). This also provided a data base for the characteristics necessary for a palaeoclimatic reconstruction of the burial site.

The results of many of these investigations were issued in Russian in many publications of limited distribution. Little of these data were available to a non-Russian audience. Dr. Ukraintseva undertook the task of summarizing earlier reports, as well as her own research, in an English-language version of these investigations. In addition to synthesizing the ecological information amassed in the period 1900–1979, Dr. Ukraintseva used these data in an attempt to explain the cause of extinction of some members of the fauna of the late Pleistocene "Mammoth Epoch" of Siberia.

Of the late Pleistocene fossil animals depicted in Figure 2–31 and Table 2–6, analyses of the gastrointestinal contents of six of these animals (four mammoths, one horse, and one bison) was possible. The gastrointestinal contents are interpreted as vegetation taken in by the animals shortly prior to their

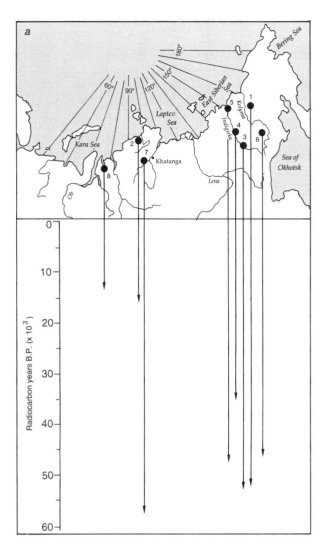

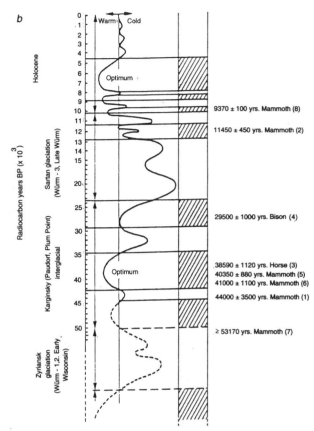

Figure 2–32b Schematic displaying the causes of relatively fast extinction of mammoth and some of its associates. Warm intervals are shown by hachures; ¹⁴C dates of death of some representatives of the faunal complex are designated by numbers.

Figure 2–32a Locations of frozen carcasses and skeletons of animals which perished in Siberia in the Pleistocene and early Holocene. Arrows designate their radiocarbon dates; date of discovery in parentheses. *(1)* Mammoth, Berezovka River (1900); *(2)* Taimyr Mammoth, Mamontovaya River (1948); *(3)* Horse, Elga River (Selerikhan Creek) (1968); *(4)* Bison, middle Indigirka River (1971); *(5)* Mammoth, Shandrin River (1971); *(6)* Mammoth ("Dima"), Kirgilyakh Creek (1977); *(7)* Vereshchagin's Mammoth, Bolshaya Lesnaya Rassokha River (1978); *(8)* Yuribey Mammoth, Yuribey River Valley (1979).

death and burial. Therefore, they represent unique samples of the vegetation existing at a time and place just before the animal's death. Each of the fossil floras represented has a "core" flora of plant species which still grow in the area of the fossil locality. They also contain a "zonal" component which is comprised of flora peculiar to, and distinct to, the discovery area at the time of sampling.

SYNOPSIS OF FOSSIL LOCALITIES AND THEIR ANALYSES

Vereshchagin (Khatanga) Mammoth The head with trunk and tusks, a femur, crus, and foot, plus hide, ligaments, and two ribs were found on the west bank of the Bolshaya Lesnaya Rassokha River in 1977 (Vereshchagin and Nikolaev 1981) (Fig.

Vicinity of Bolshoi Selerikhan River. Larch taiga association. (*Photo courtesy of N. V. Lovelius.*)

2–32*a* [7]). The remains were found in permafrost sand 105 cm above a peat bed. The locality is presently in a shrub tundra.

At the time of death (± 53,000 BP), the area was a meadow-like grass-forb community with sedge-grass communities near the river. Pollen gives evidence of cold conditions resulting in a polar-desert flora prior to the animal's death. Postmortem conditions created polygonal bogs and peat with invasion of larch forests.

Kirgilyakh Mammoth ("Dima") Gold miners found the remains of the baby mammoth, "Dima," on a tributary of the Berelekh River in 1977 (Shilo et al. 1983) (Fig. 2–32*a* [6]). The remains were unusu-

ally complete and well preserved and allowed anatomical and biochemical studies.

The site is in the Upper Kolyma highlands, which currently have a sparse larch forest with a birch-alder-mountain pine understory. At the time of the animal's death (± 41,000 to 40,000 BP), the environment was very similar to the present one, with sparse larch forests in river valleys and moist to dry tundra upland communities. Sedge marshes and meadows occupied the floodplain.

Pollen analyses suggest a warming period at about 38,000 BP in which tundras and herb communities decreased and larch forests spread.

Shandrin Mammoth A mammoth skull and tusks

were found on the east bank of the Shandrin River in 1972 (Fig. 2–32*a* [5]). Excavation revealed a skeleton at the base of the second terrace (Vereshchagin 1975). Internal organs were well preserved by permafrost and the entire gastrointestinal tract was transported in a frozen state to the Novosibirsk Biological Institute. Radiocarbon dating of the forage mass from the gastrointestinal tract provided an age of 40,350 ± 880 BP. Botanical analyses indicate the mammoth lived in larch forests and open woodlands with low shrub breaks. The Shandrin River valley was occupied by grass communities with herb-moss bogs and shrubby moss meadows. The fodder within the gastrointestinal tract indicated sedges, cotton grass, other grasses, plus sprigs and leaves of shrubs such as alder, willow, low birch, and larch shoots; mosses were also included in the diet.

At present the Shandrin River valley is treeless, an area of subarctic tundra. Modern plant communities, which would be similar to those at the time the mammoth died, occur 100–250 km south of the burial site.

Selerikhan Fossil Horse   Found in a gold mine, 8 to 9 m below the surface, this Chersky horse died with its stomach full around 38,590 ± 1120 BP (Fig. 2–32*a* [3]). It died in an ancient valley corresponding to a 110–120-m-high terrace of the Bolshoi Selerikhan River. Modern vegetation is comprised of floodplain forests of larch and poplar, with a thick willow and alder understory; hillsides are covered with arid larch forests and steep hillsides by steppe vegetation. There are also sedge marshes and sedge-cotton grass bogs.

The plants in the horse's stomach indicate it died suddenly after having eaten (Vereshchagin 1977). Residence time of food in a horse's gastrointestinal tract is usually less than ten hours; therefore, the plants in the stomach represent the vegetation of the burial site area at the time of the horse's death.

Dating of the horse and its fodder indicates it died during the Karginsky interglacial, possibly by having fallen in a thermokarst scour. That its death occurred in late July or early August is suggested by ripened fruits of sedges but unripened grasses.

The food indicates the animal grazed on a variety of habitats such as dry steppe, moist meadows, riparian aquatic vegetation, plus minor amounts of forest habitat plants. The conclusions reached were that the present vegetation is less rich than that of the time of the horse's death. The Lena–Aldan forests are regarded as modern equivalents.

Mylakhchin Bison   In 1971 a hunter found a fossil bison on the middle Indigirka River (Fig. 2–32*a* [4]). It was buried in a loess-loam at the base of the loess-ice terrace, about 40 m above the modern river. The bison died 29,500 ± 1000 BP in an area currently known as the Aluyi lowland. The modern Indigirka River is forested but contains many swamps and lakes. Larch forests alternate with dwarf birch communities and bogs. It is a relatively poor modern flora. Sedges and grasses dominated the bison's food plants with some low birch, willow, alder, and mosses. It is believed the animal died in early summer in the late Pleistocene (middle Würm/middle Wisconsin). The landscape at the time of the animal's death was interpreted as consisting of open, treeless plant communities such as tundra and forest-steppe.

Yuribey Mammoth   The youngest representative of the studied fauna, this mammoth died 10,000 ± 70 BP (Arslanov et al. 1980) (Fig. 2–32*a* [8]). The tightly filled gastrointestinal tract was preserved.

The modern vegetation community is comprised of dwarf willow and birch tundra. The animal died in an early Boreal period with stable vegetative cover being expanded due to a warming climate. The Yuribey River valley was occupied by sedge-grass communities and some sedge bogs. Bushy tundra may have been sparsely present. It appears the climatic conditions at the time of death were similar to modern. The mammoth apparently drowned in the river, or a lake.

Berelekh Mammoth Population   A large "cemetery" of mammoths was discovered on the Berelekh River (Vereshchagin 1977; Mochanov and Fedoseeva, this volume). The bone bed formed between 11,870 ± 60 and 10,260 ± 150 BP (Lozhkin 1977). A

Dyuktai culture archaeological site was found 200 m downstream. The "cemetery" contained an estimated 140 mammoths, 4 wolverines, 1 cave lion, 1 woolly rhinoceros, 3 Chersky horses, 4 reindeer, and 2 bison.

The cemetery thus began to develop in late Sartan. The locality was one of grass and low bush-grass tundra, with sedge-grass low areas. Study of the deposits suggests an interval of fluctuating wet-cold to dry-warm episodes.

SUMMARY AND CONCLUSIONS

Based on the data amassed from the gastrointestinal tracts, the enclosing sediments, and the radiocarbon dates, Ukraintseva proposes the following model of extinction of members of the mammoth fauna of Siberia.

1. In the past 75,000 years (equivalent to the Wisconsinan of North America) there were cycles of relatively cold (glacial) conditions and relatively warm (interstadial) intervals. The cold cycles favored the shrinkage of forest habitats with an increase in favorable mammoth fauna habitat, i.e., steppes and tundra. During the interstadials, forests advanced rapidly and swampy areas and boggy regions increased, causing segregation and breakup of favorable habitats, as well as increasing opportunity for fatal accidents and burials of representatives of mammoth fauna (Fig. 2–32b).

2. More than 221 taxa of plants are known from the analyses of gastrointestinal contents of large herbivores which died within the interval of 53,000–10,000 years ago. The plants represented in various specimens change due to selectiveness by the animal, and to minor differences in vegetational patterns of the areas of death. Dr. Ukraintseva concludes that wet ground fodder was of less nutritional value than cold, dry ground fodder. This supports a concept of greater mortality in interstadial intervals.

The duration of warm and cold intervals had an important effect on animal populations. Fluctuation toward cooling caused the emergence of cold-adapted organisms. A study of rates of loess accumulation led to the conclusion that separate cold, glacial epochs did not exceed 50,000 years, and that even in those epochs, warming or warm intervals totaled up to half of a given interval. This led to the conclusion that cold-adapted fauna necessarily had to exist in unfavorable climatic intervals.

Population size was governed by these oscillations. In cold (glacial) intervals, vast areas of treeless landscapes led to large interbreeding herbivore populations. Warm (interglacial) periods accompanied by forest and swamp/bog expansion were dominated by small and medium herbivore populations. These warm periods caused reductions of suitable habitat and food reserves, as well as causing deterioration of food nutrients. These combined factors led to lower birthrates and higher mortalities. Large interbreeding populations with high birthrates were divided into medium- and small-sized intrabreeding groups.

Certain elements of the mammoth fauna were adapted to cold, dry climates. Their food sources were also more nutritious. As rates of climatic change increased, small segments of formerly large animal populations were unable to adapt and became extinct.

Some members of the mammoth fauna such as bison, reindeer, musk-oxen, etc., were less specialized to cold environments than were mammoths, woolly rhinoceros, and others, and they, therefore, survived to the present. The Siberian model presents no crucial role for human predation in the extinction of mammoths and woolly rhinoceros. Human involvement at the Berelekh "cemetery" is suggested. Climate change is seen as sufficient explanation for the extinction of cold-adapted members of the mammoth epoch fauna.

The Siberian climate-change model viewed from a temperate North American perspective finds important differences. First, considering only the mammoths, the Siberian model is concerned solely with arctic and subarctic *Mammuthus primigenius*. In North America, by contrast, there were contemporary temperate zone species of mammoths (*M. imperator* and *M. columbi*), which were probably more numerous and widespread than *M. primigenius*. The environmental change depicted in the model driving the extinction of Siberian woolly mammoths would have had no effect on New World temperate zone mammoths—in fact, warming should have enhanced their environment, by expansion of suitable habitat.

Whereas human predation is not considered to be an effective element of extinction in Siberia, it has a more visible aspect in temperate North America.

REFERENCES CITED

Arslanov, Kh. A., N. K. Vereshchagin, V. V. Lyadov, and V. V. Ukraintseva. 1980. On the Dating of the Karginsky Interglacial Period and Siberian Landscape Reconstruction as Resulting from the Investigation of Mammoth Corpses and Their "Satellites." In *Geochronology of the Quaternary Period*, ed. I. K. Ivanova and N. V. Kind, pp. 208–13. Moscow: Nauka. (In Russian)

Chersky, I. D. 1891. Description of the Collection of Post-Tertiary Mammals Gathered by the Novosibirsk Expedition. Appendix I. *Zapiski Akademii Nauk* 65:706. St. Petersburg. (In Russian)

Lazarev, P. A. 1982. Fauna of Mammals and Biostratigraphy of the Cenozoic of Yakutia. XI Congress INQUA, Moscow, August 1982. Abstracts of reports 2:138. (In Russian)

Lozhkin, A. V. 1977. Living Conditions of Berelekh Mammoth Population. In *Mammoth Fauna of the Russian Plain and East Siberia*, ed. O. A. Skarlato, pp. 67–68. Leningrad: Nauka. (In Russian)

Pavlova, M. V. 1910. Description of the Fossil Remains of Mammals in the Troitskosavsku-Kyakhta Museum. *Trudy Troitskosavsk–Kyakhta Ord. Praimyrskogo Otdela Russdog Geograficheskogo Obshchestva* 13(1):21–59. (In Russian)

Popov, Y. N. 1948. Find of Fossil Mammal Carcasses in the Pleistocene Permafrost Layers of the Siberian Northeast. *Byulleten Komissii po Izucheniyu Chetvertichnogo Perioda* 13:74–75. (In Russian)

Sher, A. V. 1971. *Mammals and Stratigraphy of the Pleistocene in the Far East of the USSR and North America.* Moscow: Nauka. (In Russian)

Shilo, N. A., A. V. Lozhkin, E. E. Titov, and Y. V. Shumilov. 1983. *Kirgilyakh Mammoth.* Moscow: Nauka. (In Russian)

Ukraintseva, V. V. 1993. *Vegetation Cover and Environment of the "Mammoth Epoch" in Siberia.* ed. L. Agenbroad, J. Mead, and R. Hevly. Mammoth Site of Hot Springs, South Dakota.

Vereshchagin, N. K. 1975. On the Mammoth from the Shandrin River. *Vestnik Zoologii* 2:81–84. (In Russian)

———. 1977. Berelekh Mammoth Cemetery. In *Mammoth Fauna of the Russian Plain and East Siberia*, ed. O. A. Skarlato, pp. 5–50. Leningrad: Nauka. (In Russian)

———. 1981. Morphological Description of the Baby Mammoth. In *Magadan Baby Mammoth*, ed. N. K. Vereshchagin and V. M. Mikhelson, pp. 52–80. Leningrad: Nauka. (In Russian)

Vereshchagin, N. K., and A. I. Nikolaev. 1981. Remains of Other Mammals from the Baby Mammoth Burial Site. In *Magadan Baby Mammoth*, ed. N. K. Vereshchagin and V. M. Mikhelson, pp. 238–41. Leningrad: Nauka. (In Russian)

PALAEOECOLOGY OF THE PALAEOLITHIC OF THE RUSSIAN FAR EAST

*Yaroslav V. Kuzmin**

In Siberia and the Russian Far East, voluminous chronological, palaeogeographical, and faunal data were collected from the 1960s to the 1980s (Ravsky 1972; Tseitlin 1979; Tseitlin and Aseev 1982; Bazarova 1985; and others). The new archaeological discoveries introduced data on the ages and environments of sites. They also permitted palaeoecological reconstructions in which all components of the human environment could be appraised as they relate to the possibility of human existence in the vast and cold territory of Siberia (Arkhipov 1991).

Archaeological and palaeogeographical investigations in the Russian Far East have been summarized in several monographs (Krasnov 1984; Krushanov 1989). Using these data as a base and adding the results of the writer's own geoarchaeological investigations, the concern here will be with

*I thank Ruslan S. Vasilievsky, Natalia B. Verkhovskaya, Sergei S. Ganzei, Sergei A. Gladyshev, Valery A. Lynsha, and Valery P. Stepanov for fruitful discussions during preparation of this manuscript. I also thank John F. Hoffecker for correction of style and grammar. Research was supported in part by a grant from the International Research and Exchange Board with funds provided by the U.S. Department of State (Title VIII). Neither of these organizations is responsible for the views expressed.

Figure 2–33 Key Palaeolithic sites of far eastern Russia.

Landscape areas

① Boreal middle and southern taiga (larch and dark coniferous forests), *mari* (bogs with larch trees)

② Boreal to sub-Boreal landscapes (larch and dark coniferous forests, larch *mari*)

③ Sub-Boreal broadleaved forests and forest-steppes

④ Sub-Boreal mountainous, broadleaved and dark coniferous forests

▲ Sites

〰 Elevations of mountain ridges and peaks
1589

a general view of Palaeolithic human ecology for the Russian Far East: the middle and lower Amur Basin, Sakhalin Island, and Primorye.

The Far Eastern Palaeolithic may be divided into three stages, based on archaeological data: (1) Lower Palaeolithic; (2) Upper Palaeolithic (two phases); and (3) Final Palaeolithic (Mesolithic). Figure 2–33 shows the locations for the currently known Palaeolithic sites as well as some elements of the modern landscape.

MATERIALS AND METHODS

Faunal remains and the palaeogeographical conditions affecting human existence in the Palaeolithic have been studied, primarily in Primorye (Vereshchagin and Ovodov 1968; Ovodov 1977; Kuzmin 1991). Geoarchaeological data are still scanty for the middle-lower Amur Basin and the Sakhalin territories.

Geoarchaeological methodology is described in the works of Gromov (1955), Brothwell and Higgs (1969), Tseitlin (1979), Renault-Miscovsky (1986), Lazukov and Stepanov (1987), among others. The reconstructions presented here follow those methods, here applied to key sections of late Quaternary sediments of the Russian Far East. Site ages based on archaeological data are also employed (Table 2–7).

Currently known data for three important climatic events are collated on Table 2–7 and palaeolandscape reconstructions are presented for the following: (1) the final Malaya Kheta warming, 33,000–35,000 years ago (Fig. 2–34); (2) the maximum cooling during the Sartan glaciation ca. 18,000–20,000 years ago (Fig. 2–35); and (3) the Boreal warming ca. 8,000–9,000 years ago (Fig. 2–36).

PRE-UPPER PALAEOLITHIC

In the Russian Far East the period preceding the Upper Palaeolithic is represented by the following sites and find spots—Filimoshki, Ust-Tu, Kumara 1, and Bogorodskoye (Derevianko 1978, 1983). The Filimoshki and Ust-Tu site artifacts lie within grav-

eliferous riverbed facies, low terrace deposits of the Zeya River (the terrace heights are 8, 10 to 12, and 20 m above the water's edge). At the Filimoshki site, peat lying at the same level as ancient tools was ^{14}C dated 20,350 ± 850 BP (SOAN-825). The tools were probably redeposited from older deposits. S. M. Tseitlin (1979) assumes that the age of the 10-to-12-m terrace of the Zeya River dates to Karginsky–Sartan time (50,000–10,000 years ago).

There are many Lower Palaeolithic sites in northern China and Korea (Yi and Clark 1983). On the basis of similarity of the natural environments and the absence of strong natural obstacles between North China–Korea and Far Eastern Russia in prehistoric times, A. P. Derevianko (1983) assumes that the Lower Palaeolithic sites appeared first in the Russian Far East during the Middle Pleistocene. However, currently there are no *in situ* Lower Palaeolithic sites in Far Eastern Russia, which prohibits estimation of the age and environment of the places where these finds have been made.

EARLY UPPER PALAEOLITHIC

A few sites represent the early stage of the Upper Palaeolithic in the Russian Far East: the Osinovka and Geographical Society Cave sites (Derevianko 1983) and some find spots where materials are mainly on the surface—Gorniy Khutor, Illyushkina Sopka, Gadyuchya Sopka, and Astrakhanka (Primorye); Kumara 2, Tambovka, and Polina Osipenko village (Amur Basin); Imchin 2 (pebble tools) and Ado-Tyimovo (Sakhalin). According to Derevianko (1983), the Primorye sites form the Osinovskaya cultural complex.

There are both geomorphological and palynological data from the lower level (N 4) of the Osinovka site (Ganeshin and Okladnikov 1956; Nikolskaya 1970). However, the palynological data are scanty—between 34 and 39 pollen-and-spores grains were recovered. Birch and *Artemisia* dominate a pollen spectrum that suggests a slightly cool environment.

More data were obtained from the Geographical Society Cave site. Ovodov (1977) studied the mammalian fauna in detail; palynological and ra-

Table 2–7 Results of Geoarchaeological Studies of the Russian Far East Palaeolithic

Time, Ky BP \ Region	Primorye	Lower–Middle Amur	Sakhalin	Palaeoclimatic curve
	N E O L I T H I C			C − + W
8 — HOLOCENE	Illistaya 1, 7,840 ± 60		Kadilanya	
10 —	Timofeevka 1		Takoe 2 (C), Odoptu	
	Suvorovo 3	Neolithic		
12 — Late Glacial		?	Sokol 2, Takoe 2 (B)	
	Gorbatka 3, < 13,500 ± 200			
14 — Nyapan stage				
16 — Interstadial	Suvorovo 4, 15,300 ± 140	Novorybachiy		
		Barkasnaya Sopka 1–2		
18 —		Kumara 3	Takoe 2 (A)	
Gydan stage	Ustinovka 1	Ust–Ulma, 19,350 ± 60		
20 —		Filimoshki, 20,350 ± 850		
22 —				
24 — Lipovsko-				
26 — Novoselovsky				
warming				
28 —			Ado–Tyimovo (?)	
			Imchin 2 (?)	
30 —				
Konoschelie				
32 — cooling				
Malaya Kheta	Geographical Society Cave			
34 — warming	Osinovka			

diocarbon data were also obtained (Alekseev and Golubeva 1973; Gerasimov et al. 1983). There are typical Upper Palaeolithic faunal complex representatives in the Palaeolithic level of the site—mammoth (*Mammuthus primigenius* Blum.) and woolly rhinoceros (*Coelodonta antiquitatis* Blum.)—although dominant fauna are sika deer (*Cervus nippon* Temm.), Manchurian deer (*Cervus elaphus* L.), roe deer (*Capreolus capreolus* L.), bison (*Bison priscus* Boj.), horse (*Equus caballus* L.), and goral (goat-antelope) (*Nemorhaedus goral* Hard.). Pollen data indicate that coniferous/broad-leaved forests existed close to the site position during the period of mammal and human occupation. Horse and mammoth bones have been radiocarbon-dated to 32,570 ± 1510 BP (IGAN-341).

Early Upper Palaeolithic sites existed during the second Late Pleistocene interglacial period (Q^3_{III})—the

Karginsky of Siberia/Chernoruchye of Primorye. Palaeogeographical data allows one to establish the age of the Osinovskaya culture sites and the Geographical Society Cave site as being at the end of the Karginsky interglacial climatic optimum—the final stage of the Malaya Kheta warming, ca. 33,000–35,000 years ago (Kind 1974). Figure 2–34 suggests the palaeoenvironment of this period.

In the foothills of the middle and lower Amur Basin there were dense coniferous and larch forests with an admixture of birch and broad-leaved trees. Spruce-fir forests developed in the low-mountainous area (Markov 1978). Pollen spectra of the Bolshain Bira River alluvial deposits date to 34,600 ± 800 BP (GIN-712) and indicate birch-alder forest vegetation (Alekseev 1978). Birch-alder forests with an admixture of spruce, fir, hazel, and nut-tree existed in the mountains.

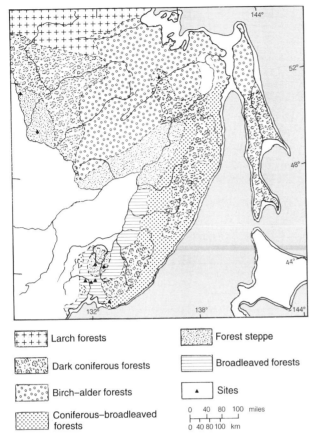

┌─┬─┐
│+ + + +│ Larch forests
│+ + + +│

Dark coniferous forests

Birch–alder forests

Coniferous–broadleaved forests

Forest steppe

Broadleaved forests

▲ Sites

0   40   80   100 miles
├──┼──┼──┼──┤
0 40 80 100 km

Figure 2–34 Palaeolandscapes of the early Upper Palaeolithic, 33–35 ka bp.

Spruce-fir forests with some broad-leaved trees covered most of the Sakhalin territory (Aleksandrova 1982); pollen spectra dated at younger than 37,950 ± 1000 BP (MGU-116) and 31,000 ± 1000 BP (GIN-415) (Svitoch et al. 1991). Birch-aspen landscapes probably had spread onto the northern Sakhalin plains.

In Primorye, the Osinovskaya culture corresponds in time to the climatic optimum of the Chernoruchye interglacial. In southern Primorye broad-leaved forests existed in the low-mountainous area; cedar (*Pinus koraiensis*)/broad-leaved forests in the intermediate-mountainous belt; and spruce-fir taiga vegetation in the higher mountainous area. In northern Primorye cedar/broad-leaved forests dominated (Korotky et al. 1980, 1988). This period has been ^{14}C-dated to 40,040 ± 1600 BP (MAG-274); 36,973 ± 750 BP (TIG-DVGU-16); and

29,000 ± 250 BP (MGU-325). Pollen spectra, dated to 37,500 ± 500 BP (MGU-407), indicate coniferous/broad-leaved forest vegetation near Khanka Lake (Bolikhovskaya et al. 1980).

During the Chernoruchye climatic optimum, the Sea of Japan was 20 meters below its present position (Korotky 1985). The natural environment was generally favorable for occupation by the Osinovskaya populations. Hunters would have found abundant game in the broad-leaved and cedar/broad-leaved forests of the Sikhote–Alin mountains, as well as in the vast forest-steppe in the Khanka and Amur–Zeya lowlands, where mammoth, woolly rhinoceros, horse, bison, and cave hyena (*Hyaena* sp.) were to be found (Ovodov 1977). There was also present a species rare in Siberia and the Far East—the Baikal wild yak (*Poephagus baicalensis* N. Ver.).

THE LATE UPPER PALAEOLITHIC

Many sites throughout the Russian Far East are assigned to the Late Upper Palaeolithic. The Amur Basin localities are Kumara 3 and Kumara Grotto, Novorybachiy, Barkasnaya Sopka 1 and 2 (Derevianko 1983; Derevianko et al. 1987), Ust-Ulma (Zenin 1988), Busse, and Novoaleksandrovka (Bolotsky et al. 1988; Sapunov 1990). On Sakhalin the known localities are Takoe 2 (site A) and Sokol 1 and 2 (Golubev and Lavrov 1988). Most materials are redeposited or strongly turbated. There is an obsidian date for the Sokol site of 16,300–11,800 BP (Vasilievsky et al. 1982). In Primorye, the Late Upper Palaeolithic sites are located primarily in the Zerkalnaya (Tadushi) River valley: Ustinovka 1 and 2, Suvorovo 3 and 4 (Vasilievsky and Gladyshev 1989), and Ustinovka 4 (Dyakov 1982). These sites form the Ustinovka cultural complex.

Archaeological cross-dating and radiocarbon dates show Late Upper Palaeolithic sites to be 10,000 to 20,000 years old (Derevianko 1983; Golubev and Lavrov 1988; Vasilievsky and Gladyshev 1989; Kuzmin 1991, 1992). Figure 2–35 presents the palaeoenvironmental reconstruction for the beginning of this period, ca. 18,000–20,000 years ago.

There were birch-larch open forests on the plains of the southern part of the middle-lower Amur Basin. Vast marshes and meadows and some dense coniferous forests were in the Amur Valley. In mountainous regions above 300–400 meters, vegetation was low-bush tundra (Makhova and Ter-Grigorian 1973; Markov 1978; Korotky et al. 1988), while tundra and forest-tundra had spread along the Sea of Okhotsk coast. Cirque glaciers—and, in some places, mountain and valley glaciers—had developed in the upper-mountainous belt.

In northern Sakhalin, and the intermontane plains, tundra landscapes prevailed; tundra and barren ground existed in the middle- and upper-mountainous belts. In the low-mountainous belt and on the southern Sakhalin ridges, open birch-larch forests dominated (Aleksandrova 1982; Korotky et al. 1988; Svitoch et al. 1991).

Throughout most of Primorye there were open birch-larch forests (Korotky et al. 1980). Middle- and upper-mountainous belts were occupied by barren ground and tundra (the upper limit of forest vegetation in southern Sikhote–Alin occurred at an elevation of 900 to 1,000 meters; in northern Sikhote–Alin, at 700 to 800 meters). In southern Primorye, local dense coniferous forests existed. Bogs and birch-larch forests spread over the Khanka Lowland. The beginning of the Sartan cooling is marked by birch-alder and birch–open coniferous forests, dated to $22,200 \pm 500$ BP (GIN-745) (Alekseev 1978); $22,556 \pm 516$ BP (TIG-DVGU-4); and $21,100 \pm 300$ BP (Ki-1885) (Korotky et al. 1989). The dates for the cooling maximum (open coniferous–birch forests with shrub birch) are $19,973 \pm 285$ BP (TIG-DVGU-3) and $19,939 \pm 199$ BP (TIG-DVGU-7). During this time small cirque glaciers existed on the central and northern Sikhote–Alin ridges.

The Sea of Japan, 18,000–20,000 years ago, was 120 to 130 meters below present (Korotky 1985). Sea ice conditions were severe; the southern limit of floating pack ice was in the latitude of 38–39°N (Pletnev 1985).

The conditions of periglacial environment did not prevent ancient people from living in the Zeya and Zerkalnaya river valleys throughout the period of maximal cooling—18,000–20,000 years ago. Evidence of those conditions shows in traces of strongly developed permafrost (ice-wedge casts) observed at the Ust-Ulma site (dated $19,360 \pm 65$ BP [SOAN-2619]) and at the Ustinovka 1 site (Zenin 1988; Kuzmin 1991).

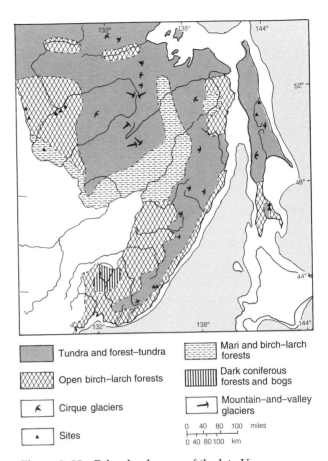

Figure 2–35   Palaeolandscapes of the late Upper Palaeolithic, 18–20 ka bp.

Dense coniferous forest components (spruce, fir, cedar, with a slight admixture of oak, elm, and nut tree) expanded with the end of cooling 18,000–20,000 years ago in Primorye. Pollen spectra indicating birch-larch and spruce-birch vegetation date to $17,400 \pm 150$ BP (Ki-1301) and $15,300 \pm 140$ BP (Ki-1130) (Korotky et al. 1989). Patches of cedar/broadleaved forests subsequently increased in southern Primorye; fir-spruce forests expanded in northern Primorye. In the Late Glacial, three expansions of birch-larch forests have been identified and have persisted since the end of the Pleistocene. The Late Glacial has been dated at $11,500 \pm 130$ BP (SOAN-188) to $10,780 \pm 50$ BP (SOAN-628).

At the Suvorovo 4 site, pollen spectra with a predominance of birch and hazel are dated to $15,300 \pm 140$ BP (Ki-3502) (Kuzmin 1990, 1992). The presence of single pollen grains of linden and oak reveal the increase of broad-leaved trees during the Pleistocene/Holocene transition at the Suvorovo 3 site, ca. 10,300 BP (Kuzmin and Chernuk 1990).

The results of palaeogeographical analysis of long cores from Late Glacial deposits in regions adjacent to the Russian Far East indicate that on southern Hokkaido the coldest climatic conditions occurred 19,790 to 17,440 BP; marked warming began ca. 13,040 BP (Sakaguchi 1989). The first trace of warming in the Late Glacial was dated to 16,000–17,000 years ago in the Pacific coastal area of Japan (Morley et al. 1986; Morley and Heusser 1989). The occurrence of spruce and pine pollen with an admixture of a little oak at ca. 14,000 years ago is noted in the southwestern Sea of Okhotsk area (Morley and Heusser 1991).

THE FINAL PALAEOLITHIC (MESOLITHIC)

The concluding stage of Palaeolithic cultural development in the Russian Far East is termed the Final Palaeolithic (previously called the Mesolithic) (Okladnikov 1969, 1977). Derevianko (1983) has subdivided it into two cultural complexes within the area observed: a blade culture—represented in Primorye by the second layer at Osinovka, the upper layer at Ustinovka 1, and the lower layer at Oleniy 1—and a pebble culture, represented by the Kumara 3 and Osipovka sites in the Amur Basin. Other Final Palaeolithic Primorye sites in the valleys of the Illistaya and Razdolnaya rivers are Illistaya 1, Gorbatka 3, and Timofeevka 1 (Kuznetsov 1980, 1986, this volume). On Sakhalin there are the Final Palaeolithic sites of Takoe 2 (sites B and C), Kadilanya, and Odoptu (Golubev and Lavrov 1988). An obsidian date of 9,000 BP was obtained for the Takoe 2 site (Vasilievsky et al. 1982).

Final Palaeolithic sites date broadly to the Pleistocene/Holocene boundary (10,000–11,000 years ago) and into the early Holocene (8,000–10,000 years ago). The main palaeoenvironmental characteristics of Boreal period landscapes (8,000–9,000 years ago) are shown in Figure 2–36.

In the southern part of the middle-lower Amur Basin there were coniferous/broad-leaved forests with an admixture of birch. Fir-spruce forests had

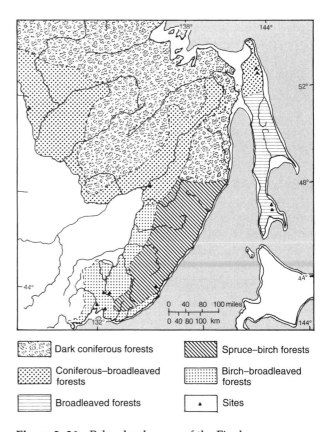

Dark coniferous forests		Spruce–birch forests	
Coniferous–broadleaved forests		Birch–broadleaved forests	
Broadleaved forests		Sites	

Figure 2–36 Palaeolandscapes of the Final Palaeolithic, 8–9 ka bp.

spread in the mountain area (Markov 1978). There were dense coniferous forests in the northern part of the Amur Basin; the ^{14}C date for these pollen spectra is 8730 ± 80 BP (DVGU-TIG-15-A) (Mikishin et al. 1987).

Throughout Sakhalin, oak- and elm-dominated broad-leaved forests, with an admixture of dense coniferous species, grew during the Boreal period (Khotinsky 1977; Aleksandrova 1982; Svitoch et al. 1991) (^{14}C dates of 8370 ± 120 BP [Vs-33] and 8370 ± 1000 BP [MGU-604]).

Birch/broad-leaved forests, with predominance of oak and elm, existed in the southern part of Primorye during the Boreal period; in northern Primorye there were spruce-birch forests (Korotky et al. 1980, 1989; Shumova and Klimanov 1989). These are radiocarbon-dated to 8020 ± 280 BP (MUG-63); 8180 ± 80 BP (MAG-332); and 8750 ± 110 BP (Ki-1920).

The Sea of Japan at the end of the Boreal period, ca. 8,000 years ago, was 20–25 meters below present (Korotky 1985).

The suite of game animals in the early Holocene was poor compared with the Late Pleistocene, with mammoth, woolly rhinoceros, and Baikal wild yak having become extinct (Ovodov 1977). Bison and cave hyena may have continued into the early Holocene. There are many hoofed animals in the Holocene faunal collections—Manchurian deer, elk (*Alces alces* L.), goral (goat-antelope), roe deer, Siberian musk deer (*Moschus moschiferus* L.), wild boar (*Sus scrofa* L.), horse, and sika deer. Carnivores consisted of brown bear (*Ursus arctos* L.), Asiatic black bear (*Ursus tibetanus* G. Cuv.), fox (*Vulpes vulpes* L.), raccoon-like dog (*Nyctereutes procionoides* Gray), and tiger (*Panthera tigris* L.). Among the mustelids, the badger (*Meles meles* L.) played an important role in hunting prey. Middle Holocene radiocarbon dates for this faunal complex (Alekseeva 1991) are 6825 ± 45 BP (SOAN-1212) to 5890 ± 45 BP (LE-4181).

Environmental conditions were relatively favorable for ancient inhabitants in the Final Palaeolithic. Birch-hazel forests with an admixture of broad-leaved species (oak, elm) (Kuzmin 1991, 1992) were present when the Illistaya 1 site was occupied (7840 ± 60 BP [Ki-3163]), and throughout this period nomadic hunters and gatherers occupied the Sikhote–Alin mountain-taiga territories (Stepanov 1984).

DISCUSSION

Initially, N. K. Vereshchagin and N. D. Ovodov (1968) assigned the Palaeolithic layer in the Geographical Society Cave to the last glaciation (Sartan time, $Q^4{}_{III}$), i.e., 15,000–20,000 years ago. According to the early data, the vegetation near the site was represented by forb steppe with birch groves. Subsequently, N. D. Ovodov (1977) recognized the existence of vast forest areas during site occupation. The cave has been dated by Tseitlin (1979) to 18,000–19,000 years ago and to 20,000–25,000 years ago by Derevianko (1983). I believe that the best agreement of pollen and radiocarbon data from the Geographical Society Cave with the general palaeo-

geographical conditions during the Karginsky (Chernoruchye) interglacial in Primorye suggests a date of 33,000–35,000 years ago.

The fission-track method was used to obtain obsidian dates for the Sokol and Takoe 2 sites (Vasilievsky et al. 1982). It is important to emphasize that this technique can estimate only the age of the *obsidian*—not that of the tools. With the younger sample dates (9,000–16,000 years ago), the measurement error is very high and the dates do not make sense (personal communication, S. S. Ganzei). Results of the dating of the Sokol and Takoe 2 sites must be considered as only preliminary.

L. V. Golubeva and L. P. Karaulova (1983) date the Ustinovka 1 site to the first Late Pleistocene cooling ($Q^2{}_{III}$), which corresponds with the Ermakovo glaciation in Siberia, approximately 50,000–100,000 years ago (Arkhipov 1987). That dating is associated with the Mousterian, not Upper Palaeolithic, and hence is unacceptable. V. A. Lynsha (1989) relied on a ^{14}C date from the lower level of the Ustinovka 1 site to establish the early Holocene age of the site (7800 ± 500 BP [GIN-2503]). The extent to which this date agrees with other geological data has been discussed previously (Kuzmin 1990). Clear evidence of a periglacial environment prohibits dating Ustinovka 1 younger than 10,000 years ago; most likely it is between 10,000 and 20,000 years old. Archaeological data indicate that the lower level at Ustinovka 1 most probably dates between 16,000 and 20,000 years BP, but the lack of adequate pollen and fossil data prevents narrowing this interval.

R. S. Vasilievsky (1986) and S. A. Gladyshev (Vasilievsky and Gladyshev 1989) assume that the Ustinovka cultural complex sites in the Zerkalnaya River valley were closely associated with coastal resource exploitation (initially, fishing). At the end of the Pleistocene the low floodplain of the Zerkalnaya had been occupied by a lagoon in which salmon could spawn. Because of lower sea level, only on the now-submerged shelf would there have been lagoons in the vicinity of the 10,000-to-15,000-year-old sites described. That is, they were much farther seaward than now. The Sea of Japan, 10,000–15,000 years ago, was 40 to 90 meters below present (Korotky 1985). Probably during occupation by the Ustinovka culture, the erosional downcutting of the Zerkalnaya River was much greater than today, and the valley had a V-shaped form (with narrow, flat, and steep sides). The shoreline, with lagoons and depositional geomorphological forms, had lain much farther away and nearer to the shelf 10,000–15,000 years ago than it does today.

S. M. Tseitlin (1979) assumes that mammoth fauna (mammoth, woolly rhinoceros) could have migrated into southern Primorye only during the maximal cooling during the Sartan period, ca. 18,000–19,000 years ago. However, remains of mammoth (*Mammuthus primigenius* Blum.) bones are known from the Russian Far East (the Amur–Zeya lowland and the lower Amur Basin) dating throughout the Karinsky–Sartan period (Alekseev 1978). Geographical Society Cave mammoth bones are of Karginsky age (Gerasimov et al. 1983). Representatives of the Late Pleistocene European–Siberian faunal complex are known in Manchuria and, more southerly, in the Hebei province of China (Kahlke 1976). The tusks of the Khorol mammoth (From the Khanka Lowland, Primorye) are radiocarbon-dated to 13,000–17,000 years ago. Thus, the mammoth fauna had existed in southern Primorye for a long time (10,000–40,000 years ago).

P. M. Dolukhanov (1989) described sea hunting in the Mesolithic economy in Primorye. The reason for this deduction is unclear because sites earlier than the Early Neolithic (^{14}C age: ca. 7500 BP) are unknown on the Primorye coast today. Sea mammal bones are known only from early Iron Age sites (^{14}C age: 2500 to 3000 BP).

SUMMARY

The results of geoarchaeological studies of the Russian Far East Palaeolithic are shown in Table 2–7. One may assume that the most favorable palaeoecological conditions were at the end of the Malaya Kheta warming, within the Karginsky interglacial, and in the early Holocene. Nevertheless, in the cold Sartan period, human activity was comparatively high. Drainage of part of the Sea of Japan and Sea of Okhotsk shelves, open vegetation, and the absence of large glaciers made human migrations comparatively easy.

REFERENCES CITED

Aleksandrova, A. N. 1982. *The Pleistocene of Sakhalin.* Moscow: Nauka. (In Russian)

Alekseev, M. N. 1978. *The Anthropogene of East Asia.* Moscow: Nauka. (In Russian)

Alekseev, M. N., and L. V. Golubeva. 1973. New Stratigraphic Data Concerning the Pleistocene of Southern Primorye. In *Stratigraphy, Palaeogeography and Lithogenesis of the Anthropogene of Eurasia*, pp. 12–24. Geological Institute AN SSSR, Moscow. (In Russian)

Alekseeva, E. V. 1991. Animal Remains from the Chertovi Vorota Grotto. In *The Neolithic of the Southern Far East: Ancient Settlement in Chertovi Vorota Cave*, ed. Zh. V. Andreeva, pp. 205–12. Moscow: Nauka. (In Russian)

Arkhipov, S. A. 1987. Pleistocene Chronostratigraphy in Northern Siberia. In *Proceedings of the First International Colloquium on Quaternary Stratigraphy of Asia and the Pacific Area*, ed. M. Itihara and T. Kamei, pp. 163–77. Osaka: Commission on Quaternary Stratigraphy.

———. 1991. Environment and Migration of Prehistoric Man into Siberia. In *The Evolution of Climate, Biota and Human Environment in the Late Cenozoic of Siberia*, ed. V. A. Zakharov, pp. 63–72. Institute of Geology and Geophysics, SO AN SSSR, Novosibirsk. (In Russian)

Bazarova, L. D. 1985. Palaeogeographical Reconstructions of Prehistoric Occupation in Southwestern Trans-Baikal. Ph.D. dissertation, synopsis. Institute of Geology and Geophysics, Novosibirsk. (In Russian)

Bolikhovskaya, N. S., T. N. Voskresenskaya, and M. V. Muratova. 1980. On the Stratigraphy and Palaeogeography of Late Pleistocene and Holocene Deposits of Primorye. In *Geochronology of the Quaternary*, ed. I. K. Ivanova and N. V. Kind, pp. 254–58. Moscow: Nauka. (In Russian)

Bolotsky, Y. L, V. V. Kolesnikov, and B. S. Sapunov. 1988. Concerning Mammoth Finds in the Upper Amur Basin. In *Stratigraphy and Correlation of Quaternary Deposits of Asia and the Pacific Area.* Papers of International Symposium V. 1, 93. DVO AN SSSR, Vladivostok. (In Russian)

Brothwell, D., and E. Higgs, eds. 1969. *Science in Archaeology: A Survey of Progress and Research*, 2d ed. New York: Praeger.

Derevianko, A. P. 1978. The Problem of the Lower Palaeolithic in the Southern Soviet Far East. In *The Early Palaeolithic in South and East Asia*, ed. F. Ikawa-Smith, pp. 305–15. Paris: Mouton.

———. 1983. *The Palaeolithic of the Far East and Korea.* Novosibirsk: Nauka. (In Russian)

Derevianko, A. P., P. V. Volkov, and A. V. Grebenshchikov. 1987. Palaeolithic Complexes of Barkasnaya Sopka on the Selemdga River. In *Antiquities of Siberia and the Far East*, ed. V. E. Larichev, pp. 73–82. Novosibirsk: Nauka. (In Russian)

Dolukhanov, P. M. 1989. The Palaeoenvironment of the Mesolithic in the Territory of the USSR. In *The Mesolithic of the USSR*, ed. L. V. Koltsov, pp. 11–17. Moscow: Nauka. (In Russian)

Dyakov, V. I. 1982. Ustinovka 4: A New Archaeological Site in the Far East. In *The Problems of Siberian Archaeology and Ethnography*, pp. 32–33. Irkutsk: Irkutsk University. (In Russian)

Ganeshin, G. S., and A. P. Okladnikov. 1956. Concerning Archaeological Sites in Primorye and Their Geological Significance. In *Proceedings of the All-Union Geological Scientific Institute*, new ser., part 1, pp. 50–57. (In Russian)

Gerasimov, I. P., O. A. Chichagova, A. E. Cherkinsky, V. L. Afonsky, V. M. Alifanov, and V. G. Tsiganov. 1983. Radiocarbon Findings of the Radiometric Laboratory of the Institute of Geography AN SSSR. *Bulletin of the Commission on Quaternary Studies* 52:205–11. (In Russian)

Golubev, V. A., and E. L. Lavrov. 1988. *Sakhalin in the Stone Age*. Novosibirsk: Nauka. (In Russian)

Golubeva, L. V., and L. P. Karaulova. 1983. *The Vegetation and Climatostratigraphy of the Pleistocene and Holocene of the USSR Far East, Southern Part*. Proceedings of the Geological Institute AN SSSR. Vol. 366. (In Russian)

Gromov, V. I. 1955. Archaeological Method. In *A Guide to Methods for the Study and Geological Mapping of Quaternary Deposits*, Part 2, pp. 127–33. Moscow: Gosgeoltechizdat. (In Russian)

Kahlke, H.-D. 1976. The Southern Boundary for the Late Pleistocene Eurosiberian Faunal Complex in East Asia. In *Beringia in the Cenozoic*, ed. V. L. Kontrimavichus, pp. 263–72. DVHZ AN SSSR, Vladivostok. (In Russian)

Khotinsky, N. A. 1977. *The Holocene of Northern Eurasia*. Moscow: Nauka. (In Russian)

Kind, N. V. 1974. *Geochronology of the Late Anthropogene According to Isotopic Data*. Proceedings of the Geological Institute AN SSSR. Vol. 275. (In Russian)

Korotky, A. M. 1985. Quaternary Sea-Level Fluctuations on the Northwestern Shelf of the Sea of Japan. *Journal of Coastal Research* 1(3):293–98.

———. 1991. Main Features of the Development of the Natural Environment in the Late Pleistocene–Holocene (Southern Soviet Far East). *CCOP Technical Publications* 22:45–54. Bangkok.

Korotky, A. M., L. P. Karaulova, and T. S. Troitskaya. 1980. *The Quaternary Deposits of Primorye: Stratigraphy and Palaeogeography*. Proceedings of the Institute of Geology and Geophysics, SO AN SSSR. Vol. 429. (In Russian)

Korotky, A. M., N. N. Kovalukh, and V. G. Volkov. 1989. Radiocarbon Dating of Quaternary Deposits (the Southern Far East). Preprint. DVO AN SSSR, Vladivostok. (In Russian)

Korotky, A. M., L. P. Karaulova, E. V. Alekseeva, and N. N. Kovalukh. 1981. Concerning the "Khorol" Mammoth Find (Primorye). In *Natural Development in the Pleistocene (the USSR Far East, Southern Part)*, pp. 29–39. DVNZ AN SSSR, Vladivostok. (In Russian)

Korotky, A. M., S. P. Pletnev, V. S. Pushkar, T. A. Grebennikova, N. G. Razhigaeva, E. D. Sakhebgareeva, and L. M. Mokhova. 1988. *The Development of the Natural Environment of the USSR Far East, Southern Part (Late Pleistocene–Holocene)*. Moscow: Nauka. (In Russian)

Krasnov, I. I., ed. 1984. *Stratigraphy of the USSR*. Vol. 2: *The Quaternary*. Moscow: Nedra. (In Russian)

Krushanov, A. I., ed. 1989. *History of the USSR Far East from Prehistoric Times until the Seventeenth Century*. Moscow: Nauka. (In Russian)

Kuzmin, Y. V. 1990. The Radiocarbon Chronology of Archaeological Sites of the USSR Far East, Southern Part. In *Chronostratigraphy of the Palaeolithic of Northern, Central, and East Asia and America (Palaeoecological Aspect)*, pp. 204–9. Papers of the International Symposium. Institute of History, Philology, and Philosophy SO AN SSSR, Novosibirsk. (In Russian)

———. 1991. Palaeogeography of the Ancient Settlements of Primorye (Palaeolithic–Neolithic). Ph.D. dissertation, synopsis. Institute of Geology and Geophysics, Novosibirsk. (In Russian)

———. 1992. *Palaeogeography and Chronology of the Ancient Cultures of the Stone Age in Primorye*. Novosibirsk: Institute of Geology and Geophysics. (In Russian)

Kuzmin, Y. V., and A. V. Chernuk. 1990. The Palaeoenvironment and Chronology of the Palaeolithic and Mesolithic of Primorye. In *Abstracts*, International Symposium on Quaternary Stratigraphy and Events of the Eurasian and Pacific Area, Vol. 1, pp. 118–19. YNZ SO AN SSSR, Yakutsk. (In Russian)

Kuznetsov, A. M. 1980. The Stone Age of Southwestern Primorye. Ph.D. dissertation, synopsis. Institute of Archaeology, Leningrad. (In Russian)

———. 1986. New Data Concerning the Stone Age of Primorye. In *The Palaeolithic and Neolithic*, pp. 138–42. Leningrad: Nauka. (In Russian)

Lazukov, G. I., and V. P. Stepanov. 1987. The Use of Anthropological and Archaeological Data for Studying the Pleistocene. In *Manual for the Study of Quaternary Deposits*, 2d ed., ed. P. A. Kaplin, pp. 194–204. Moscow: Moscow State University. (In Russian)

Lynsha, V. A. 1989. The Problems of the Age of the Ustinovka Culture in Light of Modern Studies of the Mesolithic in Southwestern Primorye. In *The Problems in Studying Stone Age and Bronze–Iron Age Sites in the Far East and Siberia*. Preprint, 3–7. DVO AN SSSR, Vladivostok. (In Russian)

Makhova, Y. V., and E. V. Ter-Grigorian. 1973. The History of the Development of Flora and Vegetation in the Northern Amur–Zeya Plain from the Late Oligocene to the Holocene. In *Geomorphology of the Amur–Zeya*

Plain and Low Mountains of Malyy Khingan, Part 1, ed. S. S. Voskresensky, pp. 83–104. Moscow: Moscow State University. (In Russian)

Markov, K. K., ed. 1978. *Sections of Quaternary Deposits of the Lower Amur Basin.* Moscow: Nauka. (In Russian)

Mikishin, Y. A., T. I. Petrenko, I. G. Gvozdeva, and G. G. Razova. 1987. The Stratigraphy of the 5 m Terrace Deposits of Chlia Lake. In *Palynology of the Eastern USSR,* ed. V. S. Markevich, pp. 94–101. DVNZ AN SSSR, Vladivostok. (In Russian)

Morley, J. J., and L. E. Heusser. 1989. Late Quaternary Atmospheric and Oceanographic Variations in the Western Pacific Inferred from Pollen and Radiolarian Analyses. *Quaternary Sciences Review* 8:263–76.

———. 1991. Late Pleistocene/Holocene Radiolarian and Pollen Records from Sediments in the Sea of Okhotsk. *Palaeoceanography* 6(1):121–31.

Morley, J. J., L. E. Heusser, and T. Sarro. 1986. Latest Pleistocene and Holocene Palaeoenvironment of Japan and Its Marginal Sea. *Palaeogeography, Palaeoclimatology, Palaeoecology* 53:349–58.

Nikolskaya, V. V. 1970. Palaeogeographical Data Concerning the Natural Environment of a Palaeolithic Site on the Osinovka River Terrace (Primorye Region). In *Siberia and Its Neighbors in Ancient Times,* pp. 60–62. Novosibirsk: Nauka. (In Russian)

Okladnikov, A. P. 1969. An Ancient Settlement on the Tadushi River at Ustinovka and the Problem of the Far Eastern Mesolithic (in Light of the 1964 Excavations). *Arctic Anthropology* 6(1);134–49.

———. 1977. The Mesolithic of the Far East: Preceramic Sites. Brief Communications of the Institute of Archaeology, AN SSSR 149:115–20. (In Russian)

Ovodov, N. D. 1977. Late Anthropogenic Mammalian Fauna of the Southern Ussuri Region. Proceedings, Biological Institute SO AN SSSR 31:157–77. (In Russian)

Pletnev, S. P. 1985. Stratigraphy of the Bottom Sediments and Palaeogeography of the Sea of Japan in the Late Pleistocene (on Planctonic Foraminifera). DVNZ AN SSSR, Vladivostok. (In Russian)

Ravsky, E. I. 1972. *The Sedimentation and Climates of Inner Asia in the Anthropogene.* Moscow: Nauka. (In Russian)

Renault-Miscovsky, J. 1986. *L'environment au temps de la préhistoire. Methodes et modeles.* Paris: Masson.

Sakaguchi, Y. 1989. Some Pollen Records from Hokkaido and Sakhalin. *Bulletin of Department of Geography, University of Tokyo:* 21:1–17.

Sapunov, B. S. 1990. Geology, Geomorphology, and Stratigraphy of the Upper Palaeolithic Locality Busse in the Upper Amur Basin. In *Chronostratigraphy of the Palaeolithic of Northern, Central, and East Asia and America (Palaeoecology Aspect).* Papers of the International Symposium, pp. 273–74. Institute of History, Philology, and Philosophy, SO AN SSSR, Novosibirsk. (In Russian)

Shumova, G. M., and V. A. Klimanov. 1989. The Late Glacial and Holocene Vegetation and Climate of the Coastal Zone in Middle Primorye. In *Palaeoclimates of the Late Glacial and Holocene,* ed. N. A. Khotinsky, pp. 154–60. Moscow: Nauka. (In Russian)

Stepanov, V. P. 1984. The Prehistoric Environment and Geography of Palaeoeconomy Types in the Far East. In *Man and Nature in the Far East* (Abstracts of conference), pp. 28–30. DVNZ AN SSSR, Vladivostok. (In Russian)

Svitoch, A. A., N. S. Bolikhovskaya, V. A. Bolshakov, T. N. Voskresenskaya, V. S. Gunova, O. V. Denisenko, O. B. Parunin, and G. M. Shumova. 1991. Quaternary Key Sections of Sakhalin Island. *Bulletin of the Commission on the Study of the Quaternary* 60:88–100. (In Russian)

Tseitlin, S. M. 1979. *Geology of the Northern Asian Palaeolithic.* Moscow: Nauka. (In Russian)

Tseitlin, S. M., and I. V. Aseev, eds. 1982. *Geology and Culture of Ancient Sites in Western Trans-Baikal.* Novosibirsk: Nauka. (In Russian)

Vasilievsky, R. S. 1986. The Upper Palaeolithic Cultures of the Southeast Maritime Regions. In *Pleistocene Perspective* 1:301–12.

Vasilievsky, R. S., and S. A. Gladyshev. 1989. *The Upper Palaeolithic of Southern Primorye.* Novosibirsk: Nauka. (In Russian)

Vasilievsky, R. S., E. L. Lavrov, and S. B. Chan. 1982. *The Stone Age Cultures of Northern Japan.* Novosibirsk: Nauka. (In Russian)

Vereshchagin, N. K., and N. D. Ovodov. 1968. Faunal History in Primorye. *Priroda* 9:42–49. (In Russian)

Yi, S., and G. A. Clark. 1983. Observations on the Northeast Asian Lower Palaeolithic. *Current Anthropology* 24(2):181–202.

Zenin, V. N. 1988. On the Question of Dating the Palaeolithic Site of Ust-Ulma in the Amur Basin. In *Historiography and Sources on the Study of the Opening up of Siberia* 1:13–14. Abstracts of Conference. Institute of History, Philology, and Philosophy SO AN SSSR, Novosibirsk. (In Russian)

Part Two

ARCHAEOLOGICAL
EVIDENCE

INTRODUCTION TO THE
ARCHAEOLOGY OF BERINGIA

John F. Hoffecker

IT NOW APPEARS certain that, excepting Antarctica, the New World was the last major land mass on earth to be colonized by a human population. It also appears more than likely that the explanation for this lies in the high latitude of the land that had to be traversed to reach this hemisphere, i.e., Beringia. The particular variable(s) that precluded movement across Beringia until the end of the Pleistocene remain(s) a matter of conjecture and debate. The problem is a baffling one, because it is apparent that humans had learned to cope with cold environments in Europe and northern Asia in much earlier times (Chard 1974; Gamble 1986; Fagan 1990).

A review of Beringian archaeology must begin with a delineation of spatial and temporal boundaries, which clearly affect questions of when and why the region was first settled. The term Beringia was originally proposed by the botanist Eric Hultén (1937) and included only those portions of the continental shelf between the present shores of northeast Asia and Alaska that were exposed during periods of lowered sea level. David Hopkins (1967) suggested a broader definition of Beringia that included northeastern Siberia and western Alaska, and Boris Yurtsev (1976) proposed an even greater area ("Megaberingia") encompassing much of northwestern Canada, Kamchatka, and the Lena Basin. Archaeologists have leaned towards the broader definitions of Beringia (e.g., West 1981; Morlan 1987; Hoffecker et al. 1993), which typically extends from the Lena River (or adjacent Verkhoyansk Range) to the Mackenzie River (or margins of the Laurentide and Cordilleran ice masses).

Beringia is a geographic concept in four dimen-

sions, and its temporal boundaries, which are a function of climate and sea level, are also fundamental. According to the previously established chronology, rising sea levels flooded the Bering Land Bridge (at a depth of 50 meters below present sea level) approximately 15,000–14,000 years BP (Hopkins 1982; McManus and Creager 1984). However, the analysis of new data from Barbados indicates that sea level did not exceed −50 meters until after 10,000 years BP (Fairbanks 1989). Fossil insects recovered from submerged peat deposits in the Chukchi Sea dating to 11,330–11,000 years BP confirm the existence of the land bridge at this time (Elias et al. 1992; see this volume). Thus, the temporal boundaries of Beringia, since the last glaciation, extend from roughly 25,000 to 10,000 years BP.

In the overview of archaeological sites assembled for this volume, Beringia has been defined in the broadest possible geographic and temporal terms, and regions and time periods normally considered to lie beyond its boundaries have been included to provide an adequate context for issues in Beringian prehistory. Sites in northeast Asia (or western Beringia) are described from the Lena Basin, eastern Yakutia, Chukotka, and Kamchatka, as well as the Russian Far East, which is generally excluded from formal definitions of Beringia. Sites in northwestern North America (or eastern Beringia) are described from all of Alaska and the Yukon. The temporal boundaries are extended beyond the submergence of the land bridge into the early Holocene (up to 6,000 years BP).

Beringian archaeology remains in an early phase of development. Archaeologists on both sides of the Bering Strait are still building cultural chronologies for Beringia, as well as the early Holocene (i.e., post-Beringia) prehistory of northeast Asia and Alaska and the Yukon. The immature state of Beringian archaeology is a function of limited modern settlement and industrial development. The extensive ground-disturbing activities—especially building, highway, and dam construction—that typically produce great quantities of archaeological remains have affected a comparatively small portion of these remote areas. This has been offset to some extent by large-scale surveys of undeveloped areas by both the Russian and American governments and by the clear potential of the region to contribute to the resolution of a major question in human prehistory.

Surveys have been conducted along portions of all of the major river systems of northeast Asia, including the Lena, Aldan, Indigirka, Kolyma, Omolon, Anadyr, and others (e.g., Mochanov 1977; Kiryak 1993), and many large federal land holdings in Alaska have been subject to inventories (e.g., Davis et al. 1981). Field projects specifically designed to investigate Native American origins have been undertaken in various parts of Alaska and the Yukon, including the Old Crow Basin, Middle Kuskokwim region, Porcupine River, and Central Alaska Range (Morlan 1980; West 1981; Powers 1983; Ackerman 1985; Dixon et al. 1985). This field research has inevitably yielded some early archaeological remains, although the number of reliably dated early sites—especially sites older than 10,000 years BP—remains small, given the immense size of the region.

The distribution of known sites reflects the bias of investigation, but also reflects local and regional historical geomorphology. Thus, for example, most early sites in central Alaska are concentrated along the major highway corridors, but they also occur in a region where deep aeolian deposits accumulated during the late Glacial and early Holocene (Hoffecker et al. 1993). The concentration of sites in the upper Aldan area apparently reflects both field survey emphasis and the presence of relatively accessible alluvial deposits dating the Last Glaciation (Mochanov 1977; Tseitlin 1979). Cave and rockshelter occupations are confined to only four major sites dispersed widely across Beringia: Dyuktai Cave (Aldan River), Trail Creek caves (Seward Pen-

insula), Lime Hills Cave (Middle Kuskokwim region, Alaska), and Bluefish caves (Bluefish Basin, Yukon). One cave site (Geographical Society Cave) is also known from the Russian Far East. East of the Lena Basin, most Beringian sites are buried in aeolian deposits that overlie river terraces or bedrock ridges and knobs. Sites in the Lena Basin and Russian Far East also are common along river terraces, but most are reportedly buried in stream or slope deposits (Tseitlin 1979).

Sites in the Russian Far East are concentrated in two areas: (a) Middle Amur River, and (b) southern Primorye. Sites located in the Middle Amur area are found in the Zeya-Bureya Lowland, and include one possible early locality along the Zeya River (Filimoshki) and several sites along the Amur River (Kumara 1–3). These sites yielded artifacts on the surface or buried in riverine gravels representing low terraces (Powers 1973: Chard 1974). Sites in the southern Primorye occur on both sides of the Sikhote–Alin Mountains, which extend along the coast of the Sea of Japan. They include Geographical Society Cave on the Suchan River and several open-air localities situated along small rivers, most notably Osinovka and Gorbatka 3 on the Illistaya River, and Ustinovka 1 and Suvorovo 3 on the Zerkalnaya River. The open-air sites are all located along low or medium terraces in slope sediments containing redeposited artifacts; the assemblages are thought to date to the late Upper Palaeolithic (Vasilievsky and Gladyshev 1989; Vasilievsky, this volume; Kuznetsov, this volume).

The largest group of sites in northeast Asia lies in the Lena Basin, which contains a massive river system that drains the Central Siberian Plateau and higher upland areas and mountains to the south and east. Many of these sites provide an important context for discussions of Beringian prehistory. With the exception of Dyuktai Cave on the Aldan River, these sites all represent open-air occupations located on low or medium river terraces near stream confluences. Artifacts (and sometimes faunal remains) at these sites are buried in channel or floodplain alluvium of the Last Glaciation, and have been dated to various phases of the Upper Palaeolithic (35,000–10,000 years BP) and post-Palaeolithic. Localities on the Aldan River include Ust-Mil 2, Ikhine 1–2, Verkhne-Troitskaya, Ust-Timpton 1, Ez-hantsy, and Tumulur. Several important sites are also located along the Olekma River, including Kurung 2 and Leten Novyy 1 (Mochanov 1977 and this volume; Tseitlin 1979; Alekseev 1987).

East of the Verkhoyansk Mountains lie areas—eastern Yakutia, Chukotka, and Kamchatka—that are especially critical for understanding the initial settlement of Beringia and its role in the peopling of the New World. Much of this region is characterized by low mountainous or upland terrain, although a major basin (Kolyma Lowland) occupies the north-central part of this region. The sites, which are confined to open-air localities dating to the late Upper Palaeolithic (less than 15,000 years BP) and post-Palaeolithic, are widely scattered and occur in a variety of geomorphic settings.

The majority of sites in this part of Beringia are found along river terraces in the Upper Kolyma area, and contain remains buried in aeolian sediment. These sites include Siberdik and Kongo on the Kolyma River, Uptar and Kheta on small tributary rivers, and Bolshoi Elgakhchan 1 on the Omolon River see (Slobodin and King, this volume; Dikov 1977 and this volume; Kiryak 1993 and this volume). The remaining sites represent more dispersed and isolated occurrences. On a small river near the Sea of Okhotsk lies Kukhtuy 3, where an assemblage of probable early Holocene or late Pleistocene age is apparently buried in alluvial sediment on a 25-meter terrace (Mochanov 1977 and this volume). An important complex of sites has been found in central Kamchatka along the shore of a lake (Ushki Lake) connected to the Kamchatka River (Ushki 1–5). These sites contain remains in a deep sedimentary context that is largely of aeolian origin (including volcanic ash), but may also comprise some fluvial and colluvial deposits (Dikov and Titov 1984). Near the Arctic Sea, on a tributary of the Indigirka River, is the Berelekh site, where a large concentration of mammal bones (primarily mammoth) and an assemblage of stone and bone artifacts are buried in fluvial deposits on a low terrace (Mochanov 1977; Tseitlin 1979; Mochanov and Fedoseeva, this volume).

Sites on the Seward Peninsula, which represents the westernmost remnant of eastern Beringia,

are confined to the Trail Creek caves. Only caves 2 and 9 yielded traces of early occupation in the form of lithic artifacts and bone deeply buried in cave deposits of sand, silt, and gravel. The lithic artifacts are dated to the early Holocene, while older radiocarbon dates (roughly 13,000–16,000 years BP) were obtained on underlying bones without associated lithics (Larsen 1968; C. West, this volume).

Most remaining sites in eastern Beringia have been found in central Alaska. They are open-air localities concentrated along major rivers, and invariably situated on terraces or bedrock ridges or knobs. These sites typically contain remains buried in aeolian sediment, which varies in thickness from a few centimeters to several meters and dates from the terminal Pleistocene through the Holocene (12,000 years BP to the present). They range from sites in the valleys of the Alaska Range (deglaciated at the end of the Pleistocene), to sites at intermediate elevations in the foothills zone, to sites in the low-lying Tanana Basin. Sites in the Alaska Range include Teklanika West in the upper Teklanika Valley and the Tangle Lakes sites (Phipps, Whitmore Ridge, and others) in the upper reaches of the Delta River. At least one site in the Alaska Range (Carlo Creek), which is located on the upper Nenana River, appears to be buried in stream deposits dating to the early Holocene (Bowers 1980). Many sites are in the valleys of the foothills, such as Owl Ridge (Teklanika River), Dry Creek, Walker Road, Moose Creek, and Nenana River, and Donnelly Ridge (Delta River). Sites in the Tanana Basin include Campus, Chugwater, Broken Mammoth, and Healy Lake.

Several early sites are known from southwestern Alaska, primarily in upland areas along tributaries of the Kuskokwim River. These sites are typically found on bedrock ridges buried in aeolian deposits, which are generally shallow in depth and often difficult to date (although deep loess deposits occur in the region [see Waythomas, this volume]). Nevertheless, on the basis of typological comparisons with well-dated remains from other parts of Alaska, these sites also appear to span the last 12,000 years BP. They include Ilnuk on the Holitna River and Spein Mountain and Nukluk Mountain on the Kisaralik River. Lime Hills Cave was recently discovered on a Kuskokwim River tributary in the Lime Hills (see Ackerman, this volume). Some early Holocene sites are concentrated along the margins of the numerous lakes of this region.

Northern Alaska has been of particular interest to American archaeologists as the only region in Beringia to produce fluted points, suggesting a link to the oldest firmly dated remains in mid-latitude North America. Most fluted points have been discovered in the northern Brooks Range, and typically in the form of surface finds. A major site here is Onion Portage, located along the Kobuk River on the southern margin of the western Brooks Range, where early Holocene and younger occupation layers are buried in a sequence of slope deposits.

Evidence of early occupation in unglaciated portions of the Yukon Territory now appears to be confined to the Bluefish caves, which contain mammal bones and a small quantity of artifacts buried in aeolian sediment. Although the association between the bones, which are dated to 25,000–12,000 years BP, and the artifacts remains problematic, the latter probably date at least to the terminal Pleistocene or early Holocene. Previously reported discoveries of modified bones dating to the earlier late Pleistocene and before are no longer accepted as sufficient evidence of human occupation.

The retreat of glaciers at the end of the Pleistocene opened up new areas of Beringia for colonization, not only at higher elevations in the Alaska Range and elsewhere, but also along low-lying coastal areas of southern Alaska that had been completely inundated by ice during the late Wisconsin. Sites in these areas date to the earliest Holocene and later (10,000–9,500 years BP and younger) and include localities along the shore of Cook Inlet (Beluga Point) and on the Kenai Peninsula (Round Mountain), as well as on the mainland coast and islands of southeastern Alaska (Ground Hog Bay 2 and Hidden Falls.

BIBLIOGRAPHY

Ackerman, R. E. 1985. Southwestern Alaska Archeological Survey. *National Geographic Society Research Reports* 19:67–94.

Ackerman, R. E., T. Hamilton, and R. Stuckenrath. 1979. Early Cultural Complexes on the Northern Northwest Coast. *Canadian Journal of Archaeology* 3:195–210.

Ackerman, R. E., K. C. Reid, J. D. Gallison, and M. E. Roe. 1985. Archaeology of Heceta Island: A Survey of 16 Timber Harvest Units in the Tongass National Forest, Southeastern Alaska. Project Report No. 3. Pullman: Center for Northwest Anthropology, Washington State University.

Alekseev, A. N. 1987. *Kamennyi vek Olekmy. (The Stone Age of the Olekma.)* Irkutsk: Irkutsk University Press.

Alexander, H. L. 1987. *Putu: A Fluted Point Site in Alaska.* Publication No. 17, Department of Archaeology. Burnaby, B.C.: Simon Fraser University.

Anderson, D. D. 1988. Onion Portage: The Archaeology of a Stratified Site from the Kobuk River, Northwest Alaska. *Anthropological Papers of the University of Alaska* 22(1–2).

Bowers, P. M. 1980. The Carlo Creek Site: Geology and Archeology of an Early Holocene Site in the Central Alaska Range. Occasional Paper 27. Fairbanks: Cooperative Park Studies Unit, University of Alaska.

Chard, C. S. 1974. *Northeast Asia in Prehistory.* Madison: University of Wisconsin Press.

Cinq-Mars, J. 1979. Bluefish Cave I: A Late Pleistocene Eastern Beringian Cave Deposit in the Northern Yukon. *Canadian Journal of Archaeology* 3:1–32.

Cook, J. P. 1969. The Prehistory of Healy Lake, Alaska. Ph.D. dissertation, University of Wisconsin.

Davis, C. W., D. C. Linck, K. M. Schoenberg, and H. M. Shields. 1981. *Slogging, Humping, and Mucking through the NPR-A: An Archaeological Interlude.* Vols. 1–5. Fairbanks: University of Alaska, Cooperative Park Studies Unit.

Dikov, N. N., 1977. *Arkheologicheskiye pamyatniki Kamchatki, Chukotki i Verkney Kolymy. (Archaeological Sites in Kamchatka, Chukotka, and the Upper Kolyma.)* Moscow: Nauka.

Dikov, N. N., and E. E. Titov. 1984. Problems of the Stratification and Periodization of the Ushki Sites. *Arctic Anthropology* 21(2):69–80.

Dixon, E. J. 1975. The Gallagher Flint Station, an Early Man Site on the North Slope, Arctic Alaska, and Its Role in Relation to the Bering Land Bridge. *Arctic Anthropology* 12(1):68–75.

Dixon, E. J., R. M. Thorson, and D C. Plaskett. 1985. Cave Deposits, Porcupine River, Alaska. *National Geographic Society Research Reports* 20:129–53.

Elias, S. A., S. K. Short, and L. Phillips. 1992. Paleoecology of Late-Glacial Peats from the Bering Land Bridge, Chukchi Sea Shelf Region, Northwestern Alaska. *Quaternary Research* 38:371–78.

Fagan, B. M. 1990. *The Journey from Eden: The Peopling of Our World.* London: Thames and Hudson.

Fairbanks, R. G. 1989. A 17,000-Year Glacio-Eustatic Sea

Level Record: Influence of Glacial Melting Rates on the Younger Dryas Event and Deep-Ocean Circulation. *Nature* 342:637–42.

Gamble, C. 1986. *The Palaeolithic Settlement of Europe.* Cambridge: Cambridge University Press.

Hoffecker, J. F., W. R. Powers, and T. Goebel. 1993. The Colonization of Beringia and the Peopling of the New World. *Science* 259:46–53.

Hopkins, D. M. 1982. Aspects of the Paleogeography of Beringia during the Late Pleistocene. In *Paleoecology of Beringia,* ed. D. M. Hopkins, J. V. Matthews, Jr., C. E. Schweger, and S. B. Young, pp. 3–28. New York: Academic Press.

Hopkins, D. M., ed. 1967. *The Bering Land Bridge.* Stanford: Stanford University Press.

Hultén, E. 1937. *Outline of the History of Arctic and Boreal Biota during the Quaternary Period.* Stockholm: Bokforlags Aktiebolaget Thule.

Irving, W. N., A. V. Jopling, and B. F. Beebe. 1986. Indications of Pre-Sangamon Humans near Old Crow, Yukon, Canada. In *New Evidence for the Pleistocene Peopling of the Americas,* ed. A. L. Bryan, pp. 49–63. Orono, Maine: Center for the Study of Early Man.

Kiryak, M. A. 1993. *Arkheologiya Zapadnoi Chukotki. (The Archaeology of Western Chukotka.)* Moscow: Nauka.

Kunz, M. L., and R. E. Reanier. 1994. Paleoindians in Beringia: Evidence from Arctic Alaska. *Science* 263:660–62.

Larsen, H. 1968. Trail Creek: Final Report on the Excavation of Two Caves on Seward Peninsula, Alaska. *Acta Arctica,* fasc. 15, Copenhagen.

McManus, D. A., and J. S. Creager. 1984. Sea-Level Data for Parts of the Bering–Chukchi Shelves of Beringia from 19,000 to 10,000 ^{14}C yr BP. *Quaternary Research* 21:317–25.

Mobley, C. M. 1991. *The Campus Site.* Fairbanks: University of Alaska Press.

Mochanov, Y. A. 1977. *Drevneishie Etapy Zaseleniya Chelovekom Severo-Vostochnoi Azii. (The Most Ancient Stages in the Settlement by Man of Northeast Asia.)* Novosibirsk: Nauka.

Morlan, R. E. 1980. *Taphonomy and Archaeology in the Upper Pleistocene of Northern Yukon Territory: A Glimpse of the Peopling of the New World.* Mercury Series 94, Archaeological Survey of Canada. Ottawa: National Museum of Man.

———. 1987. The Pleistocene Archaeology of Beringia. In *The Evolution of Human Hunting,* ed. N. H. Nitecki and D. V. Nitecki, pp. 267–307. New York: Plenum Press.

Powers, W. R. 1973. Paleolithic Man in Northeast Asia. *Arctic Anthropology* 10(2):1–106.

———. 1983. The 1977 Survey for Pleistocene Archaeological Sites. In "Dry Creek: Archeology and Paleoecology of a Late Pleistocene Alaskan Hunting Camp," by W. R. Powers, R. D. Guthrie, and J. F. Hoffecker, pp. 399–447. Report to the National Park Service.

Powers, W. R., and J. F. Hoffecker. 1989. Late Pleistocene Settlement in the Nenana Valley, Central Alaska. *American Antiquity* 54(2):263–87.

Tseitlin, S. M. 1979. *Geologiya paleolita Severnoi Azii. (The Geology of the Palaeolithic of Northern Asia.)* Moscow: Nauka.

Vasilievsky, R. S., and S. A. Gladyshev. 1989. *Verkhnii paleolit Yuzhnogo Primor'ya. (The Upper Palaeolithic of Southern Primorye.)* Novosibirsk: Nauka.

West, F. H. 1967. The Donnelly Ridge Site and the Definition of an Early Core and Blade Complex in Central Alaska. *American Antiquity* 32(3):360–82.

———. 1981. *The Archaeology of Beringia.* New York: Columbia University Press.

Yurtsev, B. A. 1976. Pozdnekainozoiskoi paleogeografii Beringii v svete botaniko-geograficheskikh dannykh. In *Beringiya v kainozoe (Beringia in the Cenozoic),* ed. V. L. Kontrimavichus, pp. 101–20. Vladivostok: Nauka.

WESTERN BERINGIA

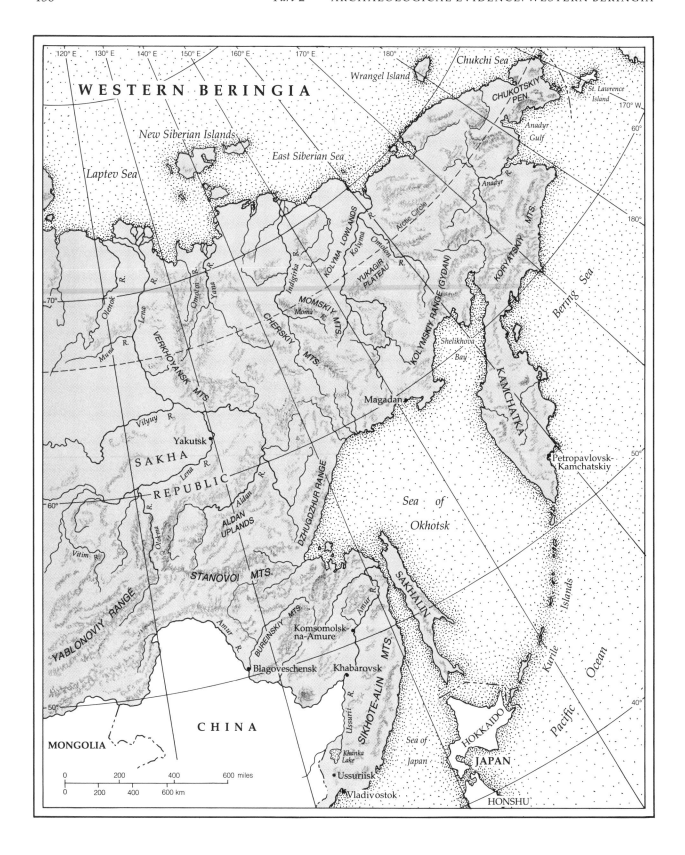

Chapter 3

ALDANSK:
ALDAN RIVER VALLEY,
SAKHA REPUBLIC

INTRODUCTION

In July 1960, during excavations of the ancient site of Kondon in Priamurye, Y. A. Mochanov found the first flint blade core of Gobi type to be discovered in northern Asia. The find was made to the north of the Amur River and was obviously ancient. Lying beneath strata containing remains of various Holocene cultures of the last 10,000 years, this level was uncovered in a number of test excavations. Here were revealed, among other things, wedge-shaped cores (frequently of Gobi type), bifacially worked tips of projectile points, knives, burins, scrapers, large and small core-struck blades, and other lithic artifacts. (Some of this material from Kondon is shown in Figure 4–7.) Further work by A. P. Okladnikov during 1962–1972 failed to identify clear Palaeolithic strata, although tools comparable to the *in situ* finds of 1960 were recovered mixed with artifacts of Neolithic age and younger.

Since 1960—over the past one-third of a century—the authors' archaeological paths have often led them to the search for traces of the ancient people of northern Asia, those people who, at different times, may have crossed into America by way of Be-

ring Strait. No matter where the authors' research has taken them over these long years, the question of the settlement of America has remained always one of their brightest guiding stars.

Before 1994, the authors worked on various expeditions on the Angar, Vilyuy, Yenisei, Amur, and Ussuri rivers, and in a number of other regions of northern Asia. In 1964, the Institute of Linguistics, Literature, and History at Yakutsk Scientific Center of the Academy of Sciences of USSR established the Prilensk Archaeological Expedition (PAE), and the authors were appointed to guide its operations. The Aldan was chosen as the first region of concentration by the PAE. This region was selected because the Aldan, with its right tributary, the Maya, happens to be a clear boundary between the Lena River in the west and the Sea of Okhotsk in the east. At the beginning of the expedition's operations in 1964, the following was noted: ''If, in fact, in ancient times, population was migrating from the south (i.e., from the Amur River basin) to the north—to the Chukotka, Kamchatka, and America—then, eminently, traces of these movements should have been left on the banks of the Aldan and its tributaries.''

The area is part of the Lena–Aldan plateau with elevations ranging from 200 to 500 m above sea level. To the east of the Aldan River, two low ridges, Digdiy-sis (up to 800 m) and Killyakh (up to 1,400

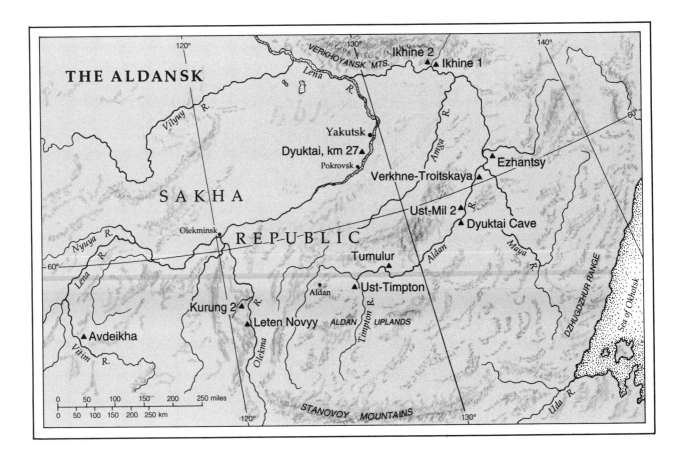

m), are the piedmont of the Sette–Daban range. The climate is strongly continental due to the winter Siberian anticyclone, with very cold winters and very hot summers. Permafrost covers almost all the region, with rare islands of thawing ground in the river valleys. The territory of the Aldan River basin is covered by taiga vegetation, characterized mostly by Dahurian larch (*Larix dahurica*) with an admixture of light-loving pines, birch, and aspen. Among the animals, the most common carnivores are brown bear, wolf, wolverine, fox, and some important fur species such as polecat (*Konocus sibiricus*) and sable (*Martes zibellina*). The hoofed species include elk and musk deer.

From 1964 through 1967, PAE discovered a number of stratified, multicomponent sites, associated with the floodplain facies of upper terraces and the higher terraces I–III of the Aldan. Here, in permanently frozen culture-bearing deposits, were undertaken vast excavations—on a scale unprecedented in northern Asia and America. Excavations

at Belkachi 1, with a deposit of 6.5 meters average depth, covered an area of over 600 square m. The experience gained and methodology developed in the Aldan proved invaluable in locating similar sites in other regions.

Complex research of such sites, in collaboration with geologists, frost and soil researchers, palaeontologists, and geochronologists, has allowed us to delineate a continuous sequence of hitherto unknown archaeological cultures: Dyuktai ($35,000 \pm 500$–$10,500/9,500$ years) and Sumnagin ($10,500/9,500$–6500 ± 100), of late and final Palaeolithic; Sialakhskaya (6500 ± 100–5200 ± 100), Belkachin (5200 ± 100–4100 ± 100), and Imyakhtan (4100 ± 100–3300 ± 100) of the Neolithic; Ust-Mil (3300 ± 100–2400 ± 100) of the Bronze Age; and, finally, various culture complexes of the early Iron Age (2400 ± 100–500 ± 100 years).

These cultures, which mostly relate to the Stone Age, were found all over Yakutia, Taimyr, and Chukotka. A number of cultures (especially Dyuktai,

The lower Aldan River Valley. (*Photo courtesy of Yaroslav V. Kuzmin.*)

Sumnagin, Belkachin, and Imyakhtan) were, at certain stages, extended into Kamchatka, Priokhotye, and Sakhalin, and were otherwise making a significant impact on the formation of ancient cultures of these regions.

The greatest value in the delineation of the Dyuktai lies, in our opinion, in the fact that its defining artifacts clearly rebutted a notion, current until the mid-1960s, about the atypicality of certain kinds of bifaces in the Palaeolithic of northern and central Asia, and, equally importantly, the similarly mistaken view that wedge-shaped blade and microblade cores found here—particularly, of the Gobi type—were late phenomena. Comparison of the Dyuktai remains with other Palaeolithic materials from northern, central, and eastern Siberia have allowed us to suggest a theory about the existence of a particular late Palaeolithic bifacial tradition that extended over the vast territory of northeast Asia. This tradition probably includes diverse remains of mammoth hunters, as well as those of hunters for other Pleistocene animals of Siberia, the Far East, Mongolia, China, and Japan. Remains of these animals occur with various bifacially worked projectile heads (perhaps including arrowheads), knives, and wedge-shaped cores of the Gobi type.

Possibly, local variations of this tradition could be recognized if our data approximated that of ethnography, but such distinctions are most often impossible to extract from archaeological data.

Presumably, the peculiarity of local environmental conditions, which would determine the availability of specific food resources, could result in producing some differentiation in various local Dyuktai populations. In order to see these particularities, one must know very well the fauna of the various sites. However, in most cases faunal remains are either meager or absent. In those sites where faunal remains are adequately represented, there are found, as a rule, bones of mammoth as well as those of woolly rhinoceros, bison, horse, musk-ox, and other late Pleistocene forms, which are typical mammals of the late Quaternary Euro-Siberian faunal complex.

Techno-typological attributes of the stone tools, the main criteria in determining similarities and differences between various Palaeolithic complexes, show that Dyuktai tradition assemblages, no matter where they are located, are always characterized by the presence of identical types of bifacially worked heads of piercing weapons, knives, and wedge-shaped nuclei. Differences are noted only on the level of subtypes of stone implements. At this point it is difficult to say what is concealed behind these differences (e.g., territorial or chronological peculiarities).

Most current knowledge on northeast Siberia derives from sites in the middle Lena Basin, the Aldan, the Vitim, the Olekma, Lena, and Vilyuy rivers. Individual Palaeolithic sites are known from the Olenek, the Indigirka, Kolyma, northwestern Priokhotye, northern Priamurye, and in Kamchatka. Almost all of these sites, except Ushki in Kamchatka (N. N. Dikov) and Kurung 2 and Leten Novyy 1 on the Olekma (A. N. Alekseev), were discovered and researched by PAE.

The late Palaeolithic remains of northeastern Asia, as known currently, are far from being of equal value. Some of them, such as Dyuktai Cave, Ust-Mil 2, Verkhne-Troitskaya, Ezhantsy, Ikhine 1, Ikhine 2, and Berelekh, lie in clear stratigraphical context and contain impressive archaeological, faunal, and palynological materials. Radiocarbon dates have been obtained for most.

Other sites—Ust-Timpton 1, Tumulur, Avdeikha 1, Kondon, Kurung 2, and Leten Novyy 1— have clear stratigraphy, but their archaeological complement is not accompanied by representative faunal remains. PAE recovered approximately 10,000 stone artifacts in a 2,000-sq-m area at the heavily researched site of Avdeikha. Since this evidence is important for comparison of late Palaeolithic materials of the northern Lena basin with that of Pribaikalye and ZaBaikalye, some of this material from Avdeikha is shown in Figure 3–36.

Many sites, particularly Ust-Chirkuo, Tuoy-Khaya, Olenek, Kyra-Krestyakh, Mayorych, Andreyevskaya, Kyunko 2, Malaya Zhikimda, Markhachan, Allakha, Tatil, Bolshoi Elgakhchan, and others, may be represented only by solitary surface finds or by an impressive tool kit, found lying mixed with elements of different cultures of the Holocene period. Their Pleistocene age and relation to

the Dyuktai culture are discovered only on the basis of their main techno-typological attributes.

Finally, it must be noted that, as of today, not a single late Palaeolithic site of northeastern Asia has been researched completely and in its entirety. Remains of Dyuktai age usually lie at great depth, in permafrost, and require a significant expenditure of time, labor, and finances.

Field archaeology in northeastern Asia, a region exceeding in extent the entire territory of Europe beyond the former USSR, is completely different from that of southern Siberia, Europe, or other regions where permafrost is absent. Even when we succeed in reaching the culture stratum, data retrieval is a major difficulty, especially in zones in which there are ice lenses. Thus, it has not, so far, been possible to reconstruct the original appearance of the Berelekh site (where, in our opinion, there were dwellings of mammoth bones) in the same way as was done in the Mezherich, Kosten, and Malta sites. It is all the more a pity, because we are convinced that in the permafrost of northeastern Asia are Palaeolithic evidences no less important than those of, say, Europe.

For the last ten years, since the discovery of the ancient Palaeolithic site of Diring, PAE appears to have practically terminated the quest and research of late Palaeolithic remains and to have abandoned the problem of the settlement of America. Such is not the case. Actually, this problem of human settlement of America is receiving a powerful new impulse for research.

In conclusion, it is necessary to emphasize that questions relating to the peopling of America will be resolved only by the efforts of archaeologists researching the Palaeolithic of Asia and America in field expeditions and in the laboratory. Not even the most sophisticated scholarly critique of Palaeolithic remains of Asia and America will produce anything positive towards elucidating the problem of the settlement of America, lacking a basis in concrete research. This is well illustrated by some of the "critical" accounts which find fault with various aspects of the Dyuktai culture. Those criticisms are probably explained by the fact that none of those "critics" has ever bothered to visit the subjects of their "critiques," which is to say, the sites of north-

eastern Asia, especially the Aldan (Powers 1973;
Derevianko 1975; Dikov 1977; Abramova 1979,
1981, 1984, 1989; Dolitsky 1985; Yi and Clark 1985).
All of their critical arguments about the stratigraphy of one or another site (e.g., Verkhne-Troitskaya,
Ezhantsy, Ikhine 1, Ikhine 2, Ust-Mil 2, etc.) pro-
voke nothing more than admiration for their irrelevant erudition. Those who argue about techno-
typological indices of Dyuktai assemblages have
not, as a rule, worked with them. It is apparent that
they have only a vague idea of the nature of the
Dyuktai materials. In light of all this, the present
publication seems very relevant and important.

The limited number of pages has forced us to
provide descriptions for only the most important
and representative relics of the late Palaeolithic.
Considering that the elaboration of a generally accepted typological nomenclature of Palaeolithic implement categories is still in the future, the majority
of critical specimens are shown in the accompanying illustrations.

Y. A. M., S. A. F.

BIBLIOGRAPHY

Abramova, Z. A. 1979. Regarding the Age of the Aldan Palaeolithic. *Soviet Archaeology* 4:5–14. (In Russian)
———. 1981. G. P. Sosnovsky and the Problems of the Northern Asian Palaeolithic. *Soviet Archaeology* 1:109–17. (In Russian)
———. 1984. The Late Palaeolithic of the Asian USSR. In *Palaeolithic of the USSR*, pp. 302–46. Moscow. (In Russian)
———. 1989. Palaeolithic of Northern Asia. In *Palaeolithic of Caucasus and Northern Asia*, pp. 145–240. Leningrad. (In Russian)
Aksenov, M. P. 1980. Archaeological Stratigraphy and Stratum-by-Stratum Description of the Tools of Verkholenskaya Gora 1. In *Mesolithic of Upper Priangarye*, pp. 45–93. Irkutsk. (In Russian)
Alekseev, A. N. 1980. Palaeolithic of the Lower Olekma. In *News in the Archaeology of Yakutia*, pp. 28–34. Yakutsk. (In Russian)
———. 1981. The Stone Age in the Lower Olekma. Ph.D. dissertation, Leningrad. (In Russian)
———. 1987. *The Stone Age of the Olekma*. Irkutsk. (In Russian)
Chard, C. S. 1974. *Northeast Asia in Prehistory*. Madison: University of Wisconsin Press.
Chen, C., and X. Wang. 1989. Upper Palaeolithic Microblade Industries and Their Relationships with Northeast Asia and North America. *Arctic Anthropology* 26: 127–56.
Derevianko, A. P. 1975. *The Stone Age in Northern, Eastern and Central Asia*. Novosibirsk. (In Russian)
———. *The Palaeolithic of the Far East and Korea*. Novosibirsk. (In Russian)
———. 1984. *The Palaeolithic of Japan*. Novosibirsk. (In Russian)
Dikov, N. N. 1977. *Archaeological Remains in Kamchatka, Chukotka, and the Upper Kolyma*. Moscow: Nauka. (In Russian)
Dolitsky, A. B. 1985. Siberian Palaeolithic Archaeology: Approaches and Analytic Methods. *Current Anthropology* 26:361–78.
Drozdov, N. I., V. P. Chekha, S. A. Laukhin, et al. 1990. Chronostratigraphy of Palaeolithic Remains of Mid-Siberia (Yenisei River Basin). Guide to the International Symposium on Chronostratigraphy of the Palaeolithic in North, Central, and East Asia and America. Novosibirsk. (In Russian)
Fedoseeva, S. A. 1968. *Ancient Cultures of the Upper Vilyuy*. Moscow. (In Russian)
———. 1970. Basic Stages of the Ancient History of the Vilyuy in the Light of New Archaeological Discoveries. In *Following the Traces of Ancient Cultures of Yakutia*, pp. 128–42. Yakutsk. (In Russian)
———. 1971. Archaeological Study of Yakutia for the Past Fifty Years. In *Questions of Historiography and Source Studying in Yakutia*, pp. 18–32. Yakutsk. (In Russian)
———. 1972. New Data on the Site Ust-Chirkuo (Upper Vilyuy) and Booroolgino (Lower Indigirka). In *Archaeological Discoveries of 1971*, pp. 260–61. Moscow.
———. 1980. Archaeological Relics of the Mid-Vilyuy. In *News in the Archaeology of Yakutia*, pp. 46–54. Yakutsk. (In Russian)
Gai Pei. 1990. Microlithical Complexes in the Chinese Palaeolithic. *Abstracts*, International Symposium on Chronostratigraphy of the Palaeolithic of North, Central, and East Asia and America, pp. 107–14. Novosibirsk. (In Russian)
Kalke, N. D. 1976. Southern Border of the Late Pleistocene Euro-Siberian Faunistic Complex in Eastern Asia. In *Beringia in the Cenozoic*, pp. 263–72. Vladivostok. (In Russian)
Kiryak, M. A. 1992. The Upper Palaeolithic Site Bolshoi Elgakhchan 1, in the Upper Reaches of the Omolon River. In *Palaeoecology and Settlement of Ancient Man in Northern Asia and America*, pp. 114–19. Krasnoyarsk. (In Russian)
Larichev, V. E. 1976. The Palaeolithic of Korea. In *Siberia, Central, and Eastern Asia in Ancient Times (the Palaeolithic)*, pp. 25–83. Novosibirsk. (In Russian)
Medvedev, G. I., A. M. Georgiyevsky, G. N. Mikhnyuk, and N. A. Savelyev. 1971. Sites of the Angaro–Belsky Region (Bajdaj Complex). In *The Neolithic of Upper Priangarye*, pp. 31–110. Irkutsk. (In Russian)

Michael, H. N. 1984. Absolute Chronologies of Late Pleistocene and Early Holocene Cultures of Northeastern Asia. *Arctic Anthropology* 21(2):1–68.

Mochanov, Y. A. 1966a. The Early Palaeolithic of the Aldan. *Soviet Archaeology* 2:126–36. (In Russian)

———. 1966b. Regarding the Initial Stages of Settlement of the New World. Geographical Society of the USSR, *Reports on Ethnography* 4:22–60. Leningrad. (In Russian)

———. 1966c. The Most Ancient Cultures of America. *Soviet Ethnography* 4:83–99. (In Russian)

———. 1969a. *The Multi-Stratified Site of Belkachi 1 and the Periodization of the Stone Age in Yakutia.* Moscow: Nauka.(In Russian)

———. 1969b. The Most Ancient Stages of the Settlement of Northeastern Asia and Alaska. *Soviet Ethnography* 1:79–86. (In Russian)

———. 1969c. The Earliest Neolithic of the Aldan. *Arctic Anthropology* 6(1):95–103.

———. 1970. Archaeological Reconnaissance along the Amga River and Lake Chukchagir. In *Following Traces of Ancient Cultures of Yakutia*, pp. 154–82. Yakutsk. (In Russian)

———. 1972a. New Data on the Bering Sea Route of American Settlement: Mayorych, the First Upper Palaeolithic Site in the Kolyma Valley. *Soviet Ethnography* 2:98–102. (In Russian)

———. 1972b. Palaeolithic Research on the Indigirka and Kolyma Rivers and on the West Coast of the Sea of Okhotsk. In *Archaeological Discoveries of 1971*, p. 251. Moscow. (In Russian)

———. 1973a. Human and Geographical Environment in the Palaeolithic and Neolithic of Siberia. In *Primitive Man and the Natural Environment*, pp. 72–73. Moscow. (In Russian)

———. 1973b. The Palaeolithic of Northern Eurasia and the Initial Stages of Human Settlement of America. In *Beringia and Its Significance for the Development of Holarctic Flora and Fauna in the Cenozoic*, pp. 14–15. Khabarovsk. (In Russian)

———. 1973c. Northeastern Asia 9,000–5,000 Years BC (the Sumnagin Culture). In *Problems of Archaeology of the Urals and Siberia*, pp. 29–44. Moscow. (In Russian)

———. 1973d. The Earliest Stages of the Peopling of America in Light of the Inquiry into the Dyuktai Palaeolithic Culture of Northeastern Asia. In *Reports of the Ninth International Congress of Anthropological and Ethnographical Sciences (Chicago)*. Moscow. (In Russian)

———. 1975a. Chief Results of Decades of Work by the Prilensk Archaeological Expedition. In *Yakutia and Its Neighbors in Ancient Times*, pp. 5–8. Yakutsk.

———. 1975b. Stratigraphy and Absolute Chronology of the Palaeolithic of Northeastern Asia Based on the data of 1963–1973. In *Yakutia and Its Neighbors in Ancient Times*, pp. 9–30. Yakutsk. (In Russian)

———. 1976. The Palaeolithic of Siberia (Research Results). In *Beringia in the Cenozoic*, pp. 540–65. Vladivostok. (In Russian)

———. 1977. *The Most Ancient Stages in the Settlement by Man of Northeastern Asia.* Novosibirsk: Nauka. (In Russian)

———. 1978. Stratigraphy and Absolute Chronology of the Palaeolithic of Northeast Asia. In *Early Man in America*, ed. A. L. Bryan, pp. 54–67. Occasional Papers of the Department of Anthropology, University of Alberta, No. 1.

———. 1982a. Initial Stages of the Human Settlement of Priokhotye, Kamchatka, and Chukotka. In *Problems of the Archaeology and Ethnography of Siberia*, pp. 34–36. Irkutsk. (In Russian)

———. 1982b. The Earliest Stages of Human Settlement of Yakutia (History and Research Results). In *Problems of Researching Ancient Cultures of Siberia and the Far Eastern USSR*, pp. 97–99. Yakutsk. (In Russian)

———. 1982c. The Palaeolithic of Northern Asia. In *Abstracts*, 11th International Congress of Research on the Quaternary Period, pp. 192–93. TPM.

———. 1986. Palaeolithic Finds in Siberia (Resumé of Studies). In *Beringia in the Cenozoic Era*, pp. 694–724. Rotterdam.

———. 1988. *The Most Ancient Palaeolithic of Diring (Archaeological Age of the Site) and the Question of Humanity's Foreland.* Yakutsk. (In Russian)

———. 1992a. Initial Research on the Northeast Asian Palaeolithic (Concepts of A. P. Okladnikov before the Discovery of the Dyuktai Palaeolithic). In *Archaeological Research in Yakutia*, pp. 3–20. Novosibirsk. (In Russian)

———. 1992b. *The Earliest Palaeolithic of Diring and the Question of the Non-Tropical Cradle of Humanity.* Novosibirsk. (In Russian)

———. 1992c. The Late Palaeolithic of North Asia 35,000–10,500 Years Ago. *Papers of the 45th Annual Northwest Anthropological Conference, Simon Fraser University*, pp. 1–11.

Mochanov, Y. A., and S. A. Fedoseeva. 1968. The Palaeolithic Site Ikhine in Yakutia. *Soviet Archaeology* 4:244–48. (In Russian)

———. 1973. Archaeology of the Arctic and Ethnocultural Links between the Old and the New Worlds via the Bering Sea during the Holocene. In *Beringia and Its Significance in the Development of Holarctic Flora and Fauna during the Cenozoic*, pp. 196–99. Khabarovsk. (In Russian)

———. 1974. The Basics of Correlation and Synchronization of Archaeological Remains of Northeastern Asia. *Ancient History of Peoples from the South of Eastern Siberia* 2:25–34. Irkutsk. (In Russian)

———. 1975a. Absolute Chronology of Holocene Cultures of Northeastern Asia (According to Evidence from the Multilayered Site Sumnagin 1). In *Yakutia and Its Neighbors in Ancient Times*, p. 38–49. Yakutsk. (In Russian)

———. 1975b. Periodization and Absolute Chronology of Yakutian Archaeological Remains. In *Correlation between the Ancient Cultures of Siberia and Cultures of Bordering Territories*, pp. 51–59. Novosibirsk. (In Russian)

———. 1976. Basic Stages of the Ancient History of Northeastern Asia. In *Beringia during the Cenozoic*, pp. 515–39. Vladivostok. (In Russian)

———.1980. Chief Results of the Archaeological Research of Yakutia. In *News in Yakutian Archaeology*, pp. 3–13. Yakutsk. (In Russian)

———. 1982a. Basic Stages of the Human Settlement of Yakutia. In *Geology of the Cenozoic in Yakutia*, pp. 157–67. Yakutsk. (In Russian)

———. 1982b. The Multilayered Site Ushki in Kamchatka and Its Place in the Ancient History of Northeastern Asia. In *Problems of the Archaeology and Ethnology of Siberia*, pp. 36–38. Irkutsk. (In Russian)

———. 1986. Main Periods in the Ancient History of Northeast Asia. In *Beringia in the Cenozoic Era*, pp. 669–93. Rotterdam.

Mochanov, Y. A., and G. M. Savvinova. 1980. The Natural Environment of Human Dwelling in the Palaeolithic and Early Neolithic (According to Archaeological Evidence). In *News of Yakutian Archaeology*, pp. 14–27. Yakutsk. (In Russian)

Mochanov, Y. A., S. A. Fedoseeva, S. P. Kistenev, and V. I. Ertyukov. 1980. Works of the Prilensk Archaeological Expedition in Chukotka and Northern Priokhotye. In *Problems of Archaeology and Ethnography of Siberia and Central Asia*, pp. 58–60. Irkutsk. (In Russian)

Mochanov, Y. A., S. A. Fedoseeva, E. N. Romanova, and A. A. Sementsov. 1970. The Multistratified Site of Belkachi 1 and Its Significance in Building an Absolute Chronology of the Ancient Cultures of Northeastern Asia. In *Following the Traces of Ancient Cultures of Yakutia*, pp. 10–31. Yakutsk. (In Russian)

———. 1972. Absolute Chronology of Ancient Cultures of Northeastern Asia in Light of Radiocarbon Dating (Following the Materials of Multilayered Sites of the Aldan River). In *Abstracts*, Reports of Sections, Dedicated to the Results of Field Researches in 1971, pp. 296–97. Moscow. (In Russian)

Mochanov, Y. A., S. A. Fedoseeva, I. V. Konstantinov, N. V. Antipina, and V. G. Argunov. 1991. *Archaeological Remains of Yakutia: The Vilyuy, Anabar, and Olenek River Basins*. Moscow. (In Russian)

Mochanov, Y. A., S. A. Fedoseeva, A. N. Alekseev, V. I. Kozlov, N. N. Kochmar, and N. M. Scherbakova.

1983. *Archaeological Remains in Yakutia: The Aldan and Olekma River Basins*. Novosibirsk. (In Russian)

Ohyi, H. 1990. Reconsideration of the Significance of the Dyuktai Culture in the Prehistory of Beringia. In *Chronostratigraphy of the Palaeolithic in North, Central, and East Asia and America: Papers of the International Symposium*, pp. 107–15. Novosibirsk. (In Russian)

Okladnikov, A. P., and I. I. Kirillov. 1980. *Southeastern Za-Baikalye in the Stone and Early Bronze Ages*. Novosibirsk. (In Russian)

Okladnikov, A. P., and R. S. Vasilievsky. 1980. *Northern Asia at the Dawn of History*. Novosibirsk. (In Russian)

Powers, W. R. 1973. Palaeolithic Man in Northeast Asia. *Arctic Anthropology* 10: 1–106.

Savvinova, G. M. 1980. Regarding the Palaeography of the Site Kukhtuy 3. In *Problems of the Archaeology and Ethnography of Siberia and the Far East,* pp. 51–52. Irkutsk. (In Russian)

Slobodin, S. B., and O. Y. Glushkova. 1992. The Kheta Site: The First Stratified Upper Palaeolithic Complex on the Kolyma River. In *Palaeoecology and Ancient Human Settlement in Northern Asia and America*, pp. 225–28. Krasnoyarsk. (In Russian)

Smith, J. W. 1974. The Northeast Asian–Northwest American Microblade Tradition (NANAMT). *Journal of Field Archaeology* 1:347–64.

Tomskaya, A. I., and G. M. Savvinova. 1975. Spore-Pollen Spectra of Accumulations from the Aldan River Valley, Including Palaeolithic Remains. In *Yakutia and Its Neighbors in Ancient Times*, pp. 31–37. Yakutsk. (In Russian)

Tseitlin, S. M. 1979. *Geology of the Palaeolithic of Northern Asia*. Moscow: Nauka. (In Russian)

Turner, K. G. 1983. Synodontia and Sundadontia: Origin, Microevolution and Settlement of Mongoloids in the Basin of the Pacific Ocean, Siberia and America, from the Odontological Point of View. In *Late Pleistocene and Early Holocene Cultural Links between Asia and America*, pp. 72–76. Novosibirsk. (In Russian)

West, F. H. 1979. The Beringian Tradition. Fourteenth Scientific Congress on the Pacific Ocean, *Abstracts*, vol. 2 (Khabarovsk), p. 215. Moscow.

———. 1981. *The Archaeology of Beringia*. New York: Columbia University Press.

Yeliseev, E. I., and P. S. Solovyov. 1992. New Evidence of the Dyuktai Culture in the Middle Lena River. In *Archaeological Researches in Yakutia*, pp. 42–47. Novosibirsk. (In Russian)

Yi, S., and G. Clark. 1985. The "Dyuktai Culture" and New World Origins. *Current Anthropology* 26:1–20.

DYUKTAI CAVE

Yuri A. Mochanov and Svetlana A. Fedoseeva

The cave is located on the right bank of the Dyuktai River 112 m from its confluence with the Aldan River at km 1031 in the Ust-Mayaskiy region of the Sakha Republic (formerly Yakutia) (59°18′N, 132°36′E). The site was discovered in 1967 by Y. A. Mochanov and was researched under his leadership by the Prilensk Archaeological Expedition in 1967–70, 1980, and 1982.

The cave is a karst cavity formed in Cambrian dolomitic limestone. It was cut into and enlarged by the Dyuktai River. The cave is subtriangular in plan with a surface area of about 60 sq. m. Its maximum length is 12.5 m; width at entrance is 10.5 m; height to ceiling is 2.7 m. The cave front is a smooth expanse extending along the river for 30–35 m; the distance between the cave entrance and the river's edge is 5–10 m. The maximum depth of sediment within the cave was 2.3 m and at the cave front up to 5.2 m. Three hundred and seventeen sq. m of deposit were excavated.

No artifacts were found on the surface either inside or outside the cave. In September of 1967 a test excavation was started in the fore part of the cave. This resulted in significant artifact recovery and ultimately the definition of a distinctive Palaeolithic culture called *Dyuktai* (Mochanov 1969a, 1969b, 1970a, 1970b). In 1968 through 1970 excavations were carried out in the cave and in the cave front area. The deposits within the cave differ significantly from those in the cave front.

STRATIGRAPHY

The following stratigraphic units were identified (Fig. 3–1):

1. A sod zone (strata 1 and 2) consists of dark grey, loosely cemented sandy loam. This unit ranges from 2 cm inside the mouth of the cave to 15–30 cm at the cave front. Gravels occur within it and limestone slabs were found lying on the surface inside the cave. Items of early Iron Age and late Neolithic were found here, as well as some Holocene faunal remains.

2. An underlying unit, subdivided as 3a and 3b, consists of seams of light brown sandy loam 0.5–6 cm thick and a dark brown, organic sandy loam of 1.5–8 cm thickness. Gravels are more abundant in the lower part of the section. A radiocarbon date of 740 ± 40 BP (LE-829) was obtained on charcoal from the middle part of the stratum. Bedding, stratigraphic location, and date indicate this as a late Holocene alluvium laid down over a period of approximately 1,000 years. Within it were found artifacts of the early Iron Age (upper segment) and late Neolithic (lower segment).

3. The cave front deposit underlying stratum 3b consists of a heavy grey graveliferous loam—stratum 5—20 to 40 cm thick. Discontinuous stringers of dark grey, slightly organic loam 2–3 cm thick are revealed in some areas. Stratum 5 is taken to represent the soil developed on the second terrace and is overlain by Holocene alluvium. The upper and middle segments of stratum 5 contained remains of the late Palaeolithic (Holocene) Sumnagin culture.

4. Stratum 7—subdivided a, b, c—is the thickest of the several units (390 cm maximum) and represents alluvium of the Terrace II floodplain facies deposited at the cave front. It is overlain discontinuously by a grey loam (stratum 5) and a yellow-grey graveliferous loam (stratum 6). Unit 7b overlies the more restricted Pleistocene deposit (stratum 8) at the mouth of the cave. Excavation of stratum 7 down to bedrock was restricted to an area of 84 sq. m. The overlying loams produced occasional fragments of mammoth tusk and chert flakes. The loams are probably a diluvium derived from the cave interior. Subdivision of stratum 7 is based upon soil color, character of bedding, and proportion of gravel.

Unit 7a consists of alternating bands of poorly sorted yellow sand (0.2–3 cm) and greyish, silty sandy loam containing cave detritus (0.5–2 cm). The overall thickness of 7a is generally 30–40 cm, occasionally as much as 60 cm. Evidence of solifluction is frequently encountered, but rarely evidence of ice wedges. Gravel, when found, is mostly fine. Various

Dyuktai Cave viewed from the river. (*Photo courtesy of Yuri A. Mochanov.*)

elements of Dyuktai culture and numerous bones of late Pleistocene fauna were found in unit 7a. Two radiocarbon dates were obtained, respectively, on wood and charcoal. These are 12,100 ± 120 BP (LE-907) and 13,200 ± 250 BP (GIN-405). Another date, 12,520 ± 250 BP (IM-462), was obtained on mammoth bone.

Unit 7b is composed of alternating stringers of grey, poorly sorted sands (0.1–1 cm) and grey-blue, silty sandy loam (0.3–4 cm). The average thickness of 7b is 100–120 cm but ranges from as little as 65 cm to over twice that thickness. An insignificant amount of gravel is present. Frost action is evidenced in the form of small soil veins and solifluction. Cultural and faunal deposits are similar to those in 7a. Two radiocarbon dates were obtained

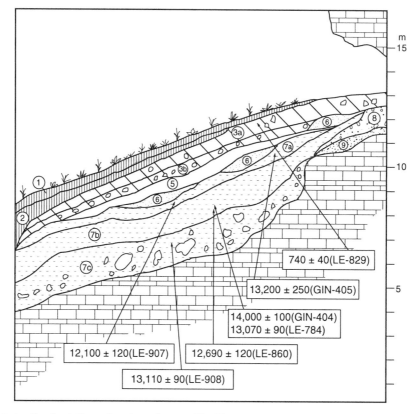

Figure 3–1 Dyuktai Cave. Stratigraphic profile. Elevation in meters above the Dyuktai River. See text for strata descriptions.

on charcoal from an eroded hearth: 13,070 ± 90 BP (LE-784) and 14,000 ± 100 BP (GIN=404). Another assay, 12,690 ± 120 BP (LE-860), was obtained on charcoal from a point 30 cm below the preceding. These, as well as its stratigraphic position, would place unit 7b in the range of 15,000–13,000 years.

Underlying unit 7b, the convoluted unit 7c consists of grey-blue sandy loam with lenses of light grey fine sand. The lower portions of the sandy loam are frequently graveliferous. Thickness of this unit varies from 130 cm to 210 cm. There is evidence of frost action and solifluction. The unit lies directly on bedrock. Materials of the Dyuktai culture and bones of Pleistocene fauna were found here. The

one radiocarbon date, 13,110 ± 90 BP (LE-908), obtained on wood from the upper portion of this unit, appears too young when compared with the assays from the overlying 7b, as well as with the radiocarbon dates from the comparable alluvium at Verkhne-Troitskaya. The age of level 7c can be estimated in the range of 23/22,000–15,000 years, pending receipt of new radiocarbon dates.

5. Stratum 8, a loosely consolidated grey-yellow sandy loam, heavily graveliferous, lies within the cave under the Holocene strata 3a and 4. This deposit ranges from 30 to 105 cm in depth. In it were large quantities of Pleistocene faunal remains and elements of Dyuktai culture.

6. Beneath stratum 8, overlying bedrock, and lying almost exclusively within the cave, is the yellow sandy loam designated stratum 9. Gravel is rare in this unit. Its thickness is on average 20–40 cm except for a small area of depression in the bedrock where it attains a thickness of 90 cm. Bedding indicates this deposit is a floodplain facies. The archaeological and faunal materials suggest that it is contemporaneous with stratum 7c.

THE CULTURAL STRATA

Faunal remains are not treated separately but are discussed in this section in order to provide a more comprehensive depiction of the cultural components. Finds from geological strata 7a, 7b, 7c, 8, and 9 are designated as Cultural Strata VIIa, VIIb, VIIc, VIII, and IX. Artifact specimen counts are presented as a ratio (15/331), i.e., 15 implements and fragments, 331 debitage.

Stratum VIIa Artifacts (121/1,572)   Blade cores are present in three forms: wedge-shaped (16, of which 12 are preforms); 3 flat unifacial subprismatic; and 2 prismatic. Also present are 2 subdiscoidal flake cores. Three subcategories of wedge-shaped cores are identified: 1 boat-shaped (Fig. 3–2: *a*); 2 triangular (Fig 3–2: *b*); 1 high subrectangular (Fig. 3–2: *c*). Core preforms are shown in Figure 3–2: *d, f*. Also recovered were 60 blades, complete and fragmentary, and 5 ski spalls.

Other tools include 2 angle burins (Fig 3–2: *e*); 2 knives on blades (Fig. 3–2: *g*); 10 bifacial curvate-base knives, preforms, and fragments of these (Fig. 3–3: *a, e*); 3 bifacial projectile points (Fig. 3–3: *b–d*); 3 backed blades; 4 blade inserts (Fig. 3–3: *f*); 2 side scrapers; 4 scrapers on flakes and blades (Fig. 3–3: *g*); and 1 coarse sandstone abrader.

Osseous artifacts are represented by 8 specimens, of which 4 are long bone fragments showing incisions and cut marks. Others include a small burned ivory fragment in the form of a fish fin; an antler fragment showing transverse incisions; and 2 fragments of implements—a sharpened awl-like tool (Fig. 3–3: *h*) and a hammer-like tool of reindeer antler (Fig. 3–3: *i*).

Stratum VIIa Faunal Remains (1,631)   Identifiable animal bones include mammoth, *Mammuthus primigenius* Blum. (24); bison, *Bison priscus* Boj. (2); horse, *Equus caballus* L. (1); reindeer, *Rangifer tarandus* L. (32); moose, *Alces alces* L. (33); snow sheep, *Ovis nivicola* (4); wolf, *Canis lupus* L. (4); red fox, *Vulpes vulpes* L. (9); arctic fox, *Alopex lagopus* L. (12); hare, *Lepus* sp. (38); ground squirrel, *Citellus* sp. (7); various rodents (32); various birds (2); various fish (2).

Stratum VIIb Artifacts (115/1,484)   Two types of blade cores are present: 2 prismatic rectangular showing irregular blade facets (Fig. 3–4: *a*) and 4 wedge-shaped cores (Fig. 3–4: *b*) and 6 wedge-shaped core preforms. The former are related to the high or vertical form. Also present is 1 subdiscoidal flake core (Fig. 3–4: *g*). In addition to 82 blades, there are also 2 crested blades (Fig. 3–4: *c, d*). Two blades for insetting fabricated on medial segments show very fine edge retouch; there are also 2 ski-spalls or flakes.

Burins were made on blades (2 angle burins) and on fragments of thin vein chert: 1 angle, 1 transverse (Fig. 3–4: *e*), and 1 dihedral (Fig. 3–4: *f*). Knives are present, 1 on a blade and 4 on thin fragments of vein chert on which one edge is bifacially retouched, the edge opposite left straight and unaltered, perhaps as a kind of backing (Fig. 3–4: *h*). Of 2 end scrapers, one is made on a blade-like flake, the other on a fragment of vein chert. Figure 3–4: *i* shows one of 2 side scrapers found, this on a pebble. Another specimen is crescentic, made on thin vein chert.

Stratum VIIb Faunal Remains (627)   Species represented include mammoth, *Mammuthus primigenius* Blum. (4, plus 3 fragments of tusk); bison, *Bison priscus* Boj. (1); horse, *Equus caballus* L. (4); moose, *Alces alces* L. (3); snow sheep, *Ovis nivicola* L. (6); large feline, *Panthera* sp. (3); red fox, *Vulpes vulpes* L. (5); arctic fox, *Alopex lagopus* L. (12); ground squirrel, *Citellus* sp. (17); marmot, *Marmota* sp. (1); beaver, *Castor fiber* L. (2); hare, *Lepus* sp. (11); and various small rodents (12). Of the mammoth remains, one was a large fragment of humerus, weighing 5.5 kg. The one radiocarbon date obtained for this stratum, $12,520 \pm 250$ BP (IM-462), was run on mammoth long bone, of which a large number were found.

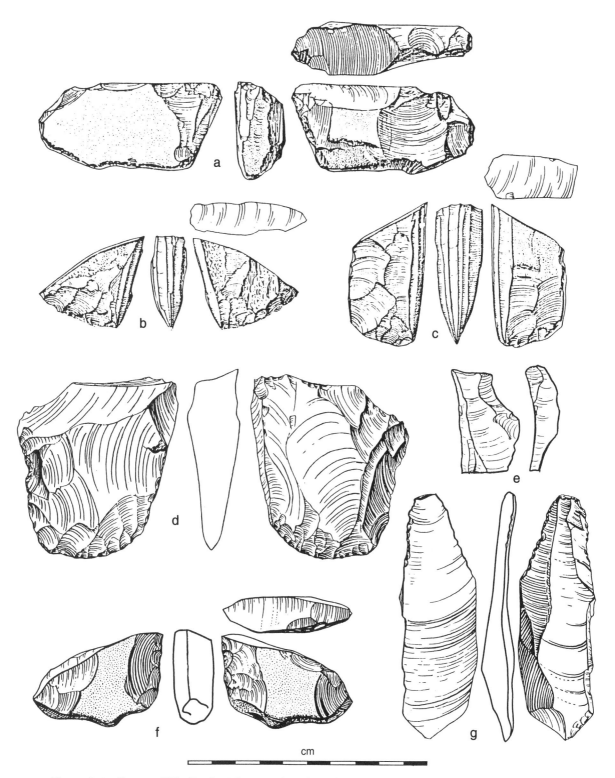

Figure 3–2   Stratum VIIa, Dyuktai Cave: wedge-shaped microblade cores (*a–c*); wedge-shaped core preforms (*d, f*); angle burin (*e*); knife on a blade (*g*).

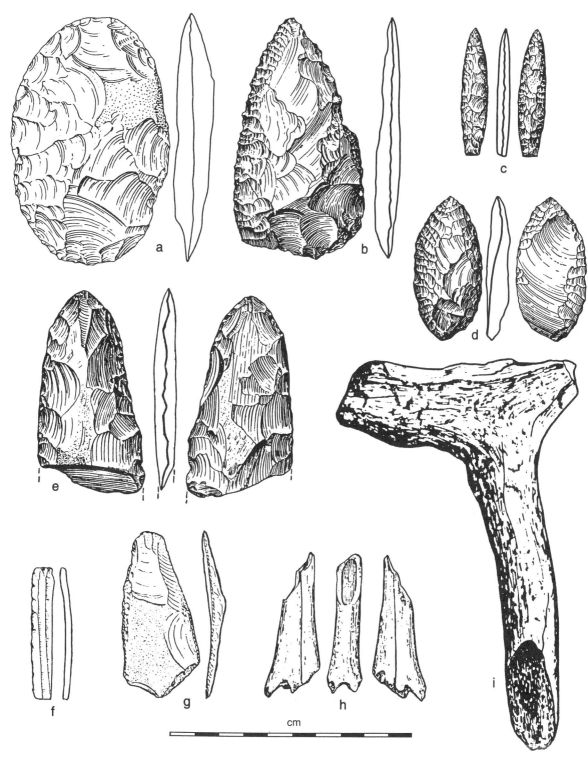

Figure 3–3 Stratum VIIa, Dyuktai Cave: biface knives (*a, e*); projectile points (*b–d*); blade insert (*f*); scraper on blade (*g*); bone awl-like tool (*h*); antler hammer-like tool (*i*).

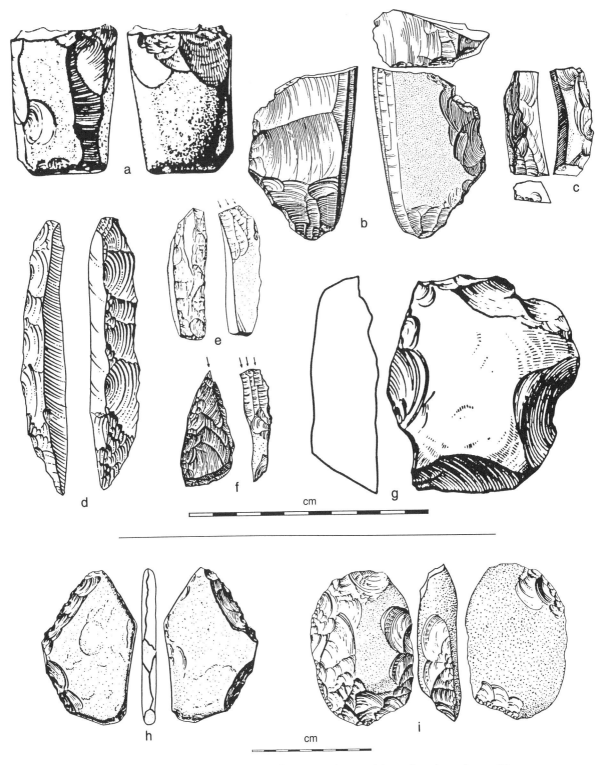

Figure 3–4   Stratum VIIb, Dyuktai Cave: rectangular prismatic core (*a*); wedge-shaped core (*b*); crested blades (*c, d*); burins (*e, f*); subdiscoidal flake core (*g*); backed knife (*h*); side scraper (*i*).

Stratum VIIc Artifacts (80/725) There are 2 wedge-shaped core preforms of the high vertical variety (Fig. 3–5: *b, c*) and a massive spall off the faceted end of a blade core (Fig. 3–5: *e*). Core products include 2 ski spalls, 62 blades, and 9 microblade insets (Fig. 3–5: *d*). Other tools include a lenticular biface of grey flint (Fig 3–5: *f*), 2 angle burins on blades, and a possible scraper on a blade.

Stratum VIIc Faunal Remains (2,570) The faunal assemblage consisted of mammoth, *Mammuthus primigenius* Blum. (7, all tusk fragments [according to O. V. Yegorov, it is probable that many unassignable long bone fragments are mammoth]); bison, *Bison priscus* Boj. (3); horse, *Equus caballus* L. (9); reindeer, *Rangifer tarandus* L. (2); arctic fox, *Alopex lagopus* L. (1); hare, *Lepus* sp. (20); ground squirrel, *Citellus* sp. (93); lemming, *Dicrostonyx torquatus* (7); various small rodents (98); and various birds (4).

Stratum VIII Artifacts (222/2,745) Blade core forms are represented by preforms for wedge-shaped cores (5) and a fragment of the faceted end of a wedge-shaped core. The shapes of the preforms suggest two completed forms: the Gobi and the high vertical. A subdiscoidal flake core was found in the lower horizon of this stratum (Fig. 3–5: *g*). In addition, 192 blades and fragments were found, as well as 2 ski spalls and 2 inset blades.

Burins consist of 2 transverse specimens on flakes, 1 dihedral on tabular chert (Fig. 3–5: *h*), and 2 multifacet forms on blades (Fig. 3–5: *i*).

There are 7 knives; five are made on tabular chert, of which 2 are preforms, 3 are finished tools, and 1 is a fragment. Another specimen is made on a large primary blade. Finished bifaces show fine, flat pressure flaking; on one of these the working edge is retouched, the edge opposite retains its cortex (Fig. 3–6: *a*). The medial section of a bifacial knife or projectile point is noteworthy. It is refitted from two pieces found in different excavation units. The larger specimen was found in the lower section of Stratum VIII, the second in the same quadrant but in the underlying Stratum IX (Fig. 3–6: *d*). End scrapers (3) were made on blades and flakes. Side scrapers (2) made on fragments of tabular chert

were found in lower Stratum VIII. One specimen is trapeziform with two working edges; the other, semilunar, is made on a large primary blade-like flake (Fig. 3–6: *b, c*).

Osseous artifacts consist of a laurel-leaf, unifacial, projectile point made on a massive flake of mammoth tusk. The specimen displays along the edge fine, flat, pressure retouch (Fig. 3–6: *e*). Modifications are seen in other mammoth tusk fragments in the form of splitting and edge retouch, in a large vertebra showing incisions, a long bone fragment with cut-marks, and half of a large canine tooth with deep incisions.

Stratum VIII Faunal Remains (3,771) Identifiable bones are those of mammoth, *Mammuthus primigenius* Blum. (all 621 fragments are of tusk, but according to O. V. Yegorov many long bone fragments are also probably mammoth); bison, *Bison priscus* Boj. (3); musk-ox, *Ovibos moschatus* Zimm (1); reindeer, *Rangifer tarandus* L. (3); moose, *Alces alces* L. (5); snow sheep, *Ovis nivicola* (3); large feline, *Panthera* sp. (6); wolf, *Canis lupus* L. (4); arctic fox, *Alopex lagopus* L. (16); arctic or red fox (14); red fox, *Vulpes vulpes* L. (3); hare, *Lepus* sp. (61); ground squirrel, *Citellus* sp. (6); lemming, *Dicrostonyx torquatus* (1); Ob lemming (1); various birds (24); and various fish (7).

Stratum IX Artifacts (38/650) Cores are subdivided into 1 subprismatic (a small fragment) and 3 wedge-shaped cores. All the latter are in the preform stage (Fig. 3–6: *h*). Eighteen blades and fragments were found as well as 5 blade insets. Possible retouch is seen on some specimens. Two transverse burins were found: one on a blade, one on a flake. Of 5 flake knives, 3 were on tabular chert (Fig. 3–6: *g*). Other artifacts include 1 fragmentary biface or projectile point, 1 end scraper on a blade, the distal end of a convergent side scraper on a flake (Fig. 3–6: *f*), and 1 fragment of a diabase abrader.

A number of modified animal bones complete the small inventory from Stratum IX. These include 1 fragment of mammoth tusk, 2 long bone fragments showing cut-marks, and 1 bone fragment with edge retouch and some polish.

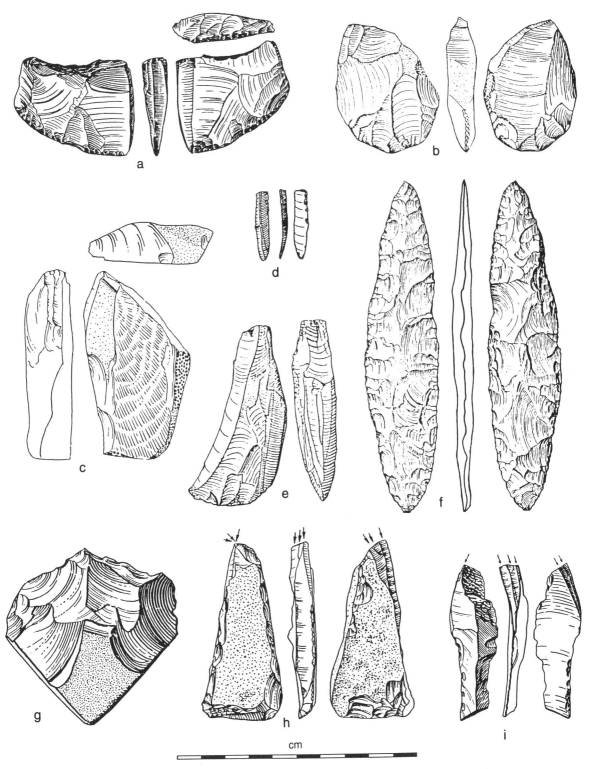

Figure 3–5   Stratum VIIc (*b–f*) and Stratum VIII (*g–i*), Dyuktai Cave (stratum of *a* not known): wedge-shaped microblade core (*a*); wedge-shaped core preforms (*b, c*); microblade (*d*); massive face spall from blade core (*e*); lenticular biface (*f*); subdiscoidal flake core (*g*); burins (*h, i*).

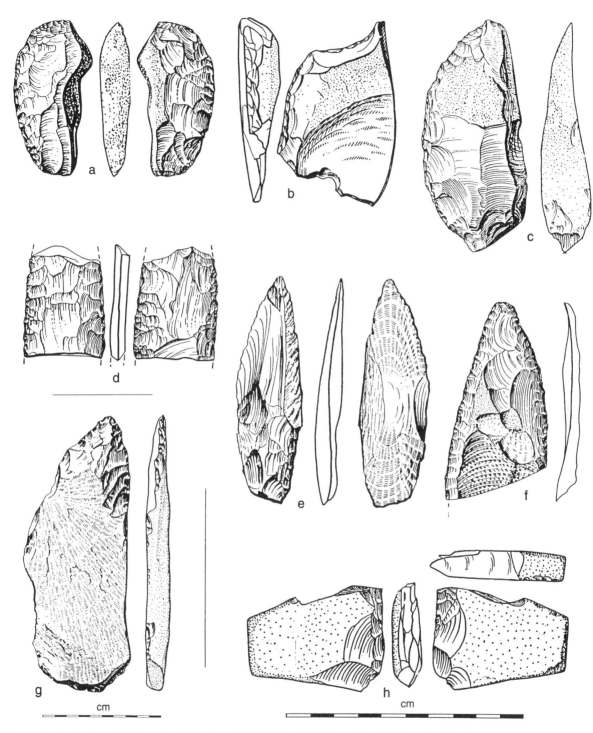

Figure 3–6 Stratum VIII (*a–e*) and Stratum IX (*f–h*), Dyuktai Cave: bifacial knives (*a, d*); side scrapers (*b, c*); laurel leaf point, mammoth ivory (*e*); convergent scraper, distal end (*f*); large flake knife of tabular chert (*g*); wedge-shaped core preform (*h*).

Stratum IX Faunal Remains (1,346)   The faunal assemblage consists of mammoth, *Mammuthus primigenius* Blum. (132—which includes 131 tusk fragments and 1 large humerus fragment); bison, *Bison priscus* Boj. (6); horse, *Equus caballus* L. (2); reindeer, *Rangifer tarandus* L. (4); red fox, *Vulpes vulpes* L. (1); arctic fox, *Alopex lagopus* L. (6); hare, *Lepus* sp. (66); ground squirrel, *Citellus* sp. (10); Ob lemming (1); various small rodents (11); various birds (23); various fish (3).

SUMMARY

Three Pleistocene horizons are identified in Dyuktai Cave, based on their stratigraphic relationships. Horizon A consists of Stratum VIIa and the upper unit of Stratum VIII; horizon B is composed of Stratum VIIb and the lower unit of Stratum VIII; horizon C consists of strata VIIc and IX. The apparent discrepancies are accounted for by the differential deposition within and without the cave (see Fig. 3–1). Horizon A seems to be appropriately dated in the range of 13,000–12,000 years BP. Horizon B is given an age of 13,000–15,000 years BP, while horizon C is dated at 15,000–16,000 years BP.

The great significance of the Dyuktai Cave research is that it was here, for the first time in northeast Asia, that there was delineated a new and distinctive culture dominated by a characteristic core-and-blade and biface industry in a context that included mammoth and other large, late Pleistocene fauna. Here, and subsequently elsewhere, these findings and associations have been confirmed and amplified. There is little question but that Dyuktai culture people were successful hunters of these large animals, and that eventually the Dyuktai people found their way to America.

BIBLIOGRAPHY

Mochanov, Y. A. 1969a. A New Upper Palaeolithic Culture of Northeastern Asia. In *Archaeological Discoveries of 1968*, pp. 214–15. Moscow. (In Russian)

———. 1969b. The Upper Palaeolithic Dyuktai Culture and Several Aspects of Its Genesis. *Soviet Archaeology* 4:235–39. (In Russian)

———. 1970a. The Most Ancient Stages of the Stone Age in Northeastern Asia (Yakutia and Chukotka). *News from the East-Siberian Branch of the Geographical Society of the USSR* 67:60–64. Irkutsk. (In Russian)

———. 1970b. Dyuktai Cave: A New Palaeolithic Site of Northeastern Asia. In *Following Traces of Ancient Cultures of Yakutia*, pp. 40–46. Yakutsk. (In Russian)

UST-MIL 2

Yuri A. Mochanov and Svetlana A. Fedoseeva

The Ust-Mil 2 site is located on the left bank of the Aldan River two km below the confluence of the Aldan and Mil rivers, 971 km from the mouth of the Aldan, in the Ust-Mayaskiy region of the Sakha Republic (Yakutia) (59°38′N, 133°07′E). The site was discovered by the senior author in 1966 and was researched under his leadership in 1966, 1968, 1970–73, 1980, and 1982.

The site is associated with the deposits of the third (16–18 m) terrace of the Aldan River. The area of excavation and test trenching extends over 312 sq. m.

STRATIGRAPHY

The following stratigraphic units were identified in excavation (Fig. 3–7):

1. Turf: 20–15 cm thick. Post-Neolithic age materials were recovered from the lower horizon of this unit.

2. A brownish red topsoil loam: 40–60 cm. Remains from various Neolithic cultures were recovered from the upper section of this unit; late Palaeolithic Sumnagin remains occurred in the middle section.

3. A yellow-grey sandy loam stratum interbedded with distinct seams of buried soil: 110–120 cm. A radiocarbon date, 12,200 ± 170 BP (LE-953), was obtained from wood taken from 25 cm above

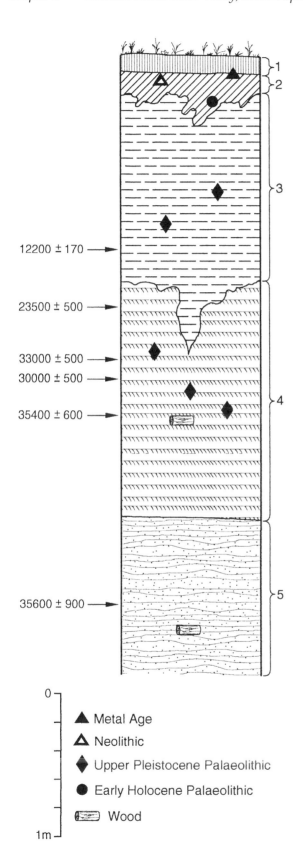

12200 ± 170

23500 ± 500

33000 ± 500
30000 ± 500

35400 ± 600

35600 ± 900

0

▲ Metal Age

△ Neolithic

◆ Upper Pleistocene Palaeolithic

● Early Holocene Palaeolithic

▱ Wood

1m

the base of this unit, indicating that this stratum was a third terrace, late Sartan alluvium.

4. Brownish grey, bedded alluvial loams: 150–160 cm. These underlie the third terrace alluvium. The upper section of the stratum bears traces of scouring. A radiocarbon date of 23,500 ± 500 BP (LE-999) on wood was obtained for the upper section of these deposits. Further radiocarbon dates on wood samples from the middle section of the deposit are 30,000 ± 500 (LE-1001); 33,000 ± 500 (LE-1000); and 35,400 ± 600 (LE-954) years.

5. Bedded yellowish sands with fine greyish seams of loam: more than 200 cm. A radiocarbon date on wood of 35,600 ± 900 years (LE-955) was obtained for the upper part of the unit at a depth of 55 cm from its upper contact. The character of the deposits would indicate that this unit represents an overbank sand facies on the third terrace.

CULTURAL STRATIGRAPHY

Cultural materials of Pleistocene age were recovered from within geological strata 3 and 4. These are related to cultural horizons A, B, and C. These materials are described separately from those collected on the surface.

Pleistocene Cultural Horizon A   Cultural horizon A, in the middle of stratum 3, yielded a small collection of 25 late-Pleistocene stone artifacts and 37 bones, mostly of extinct fauna. The lithic assemblage consists of 1 pebble core from a diabase boulder, 3 chert blade fragments, and 21 flakes of chert, quartzite, and diabase. Faunal remains include bones of mammoth, bison, horse, reindeer, and possible musk-ox. Although the A horizon sample is small, stratigraphic position and radiocarbon dates suggest it is synchronous with horizon A at Dyuktai Cave.

Pleistocene Cultural Horizon B   Cultural horizon B, in the upper to middle deposits of stratum 4, con-

Figure 3–7   Ust-Mil 2. Stratigraphic profile. Strata numbers at right. See text for descriptions.

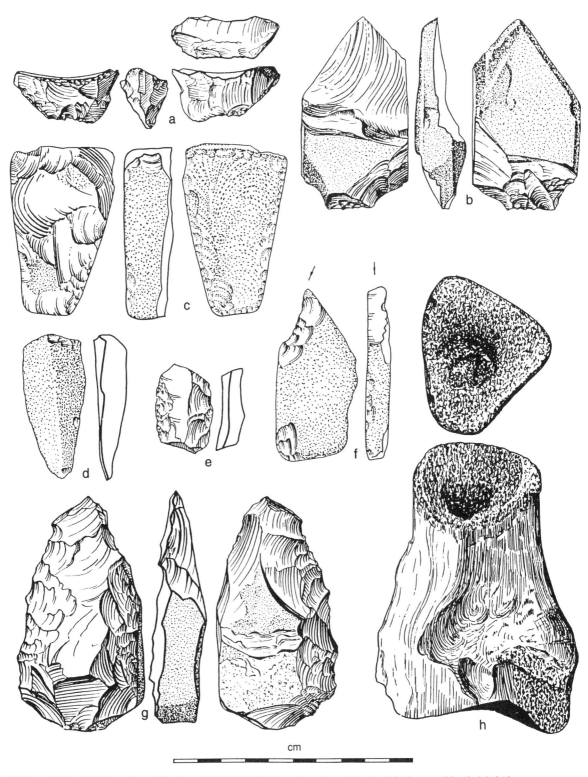

cm

Figure 3–8   Artifacts from Horizon C at Ust-Mil 2: wedge-shaped microblade core blank (*a*); biface blanks (*b, g*); burin (*f*); modified mammoth bone (*h*).

tained only 7 chert flakes, along with split bones of mammoth and woolly rhinoceros. These materials were recovered from between the wood samples dating 23,500 years (LE-999) and 33,000 years (LE-1000). The flakes are thus important for proving the presence of humans during this period.

Pleistocene Cultural Horizon C   Cultural horizon C, in the middle section of the alluvial loam deposits of stratum 4, with dates ranging from 30,000 to 35,400 years (specimens LE-1000, LE-1001, and LE-954), contained the most important cultural finds: 12 stone artifacts, 1 modified mammoth bone (Fig. 3–8: *h*), and 32 bones, some of which are split. The bones include those of mammoth, woolly rhinoceros, bison, and horse.

Distinctive elements of the artifact assemblage are 1 wedge-shaped core blank on a large chert flake (Fig. 3–8: *a*), an angle burin on a chert blade (Fig. 3–8: *f*), 1 end scraper on a blade-like flake of chert, 1 bifacial knife or speartip blank on a piece of vein chert (Fig. 3–8: *b*), 2 chert blades, and 5 chert flakes.

Unprovenienced Artifacts   Among the materials eroded from the site were 26 chert, wedge-shaped cores and blanks (Fig. 3–9: *a–e, g, j*), 11 unifacial subprismatic chert cores and blanks (see Fig. 3–10), 2 transverse burins (Fig. 3–9: *i, k*), 4 scrapers (2 of chert, 2 of diabase), 1 chert, chisel-like tool, 15 bifacial knives and blanks (see Fig. 3–10). Almost all of these artifacts have clear analogs, in terms of shape, material, and manufacturing technique, among the materials from Dyuktai Cave, as well as those from Ezhantsy and Ikhine 1 and 2. Unfortunately, due to the circumstances of their occurrence, these artifacts cannot be related to the specific cultural horizons.

(Figures 3–9 and 3–10 follow on pp. 178–179.)

SUMMARY

Good stratigraphy and a series of radiocarbon dates from clearly identified strata date this site at 30,000 to 35,000 years ago. The occurrence of Dyuktai Palaeolithic complex wedge-shaped cores and bifacial chert knives and speartips here in cultural horizon C indicates that these traits were present early in the Dyuktai cultural development and that the Dyuktai were hunters of mammoth, woolly rhinoceros, bison, and horse.

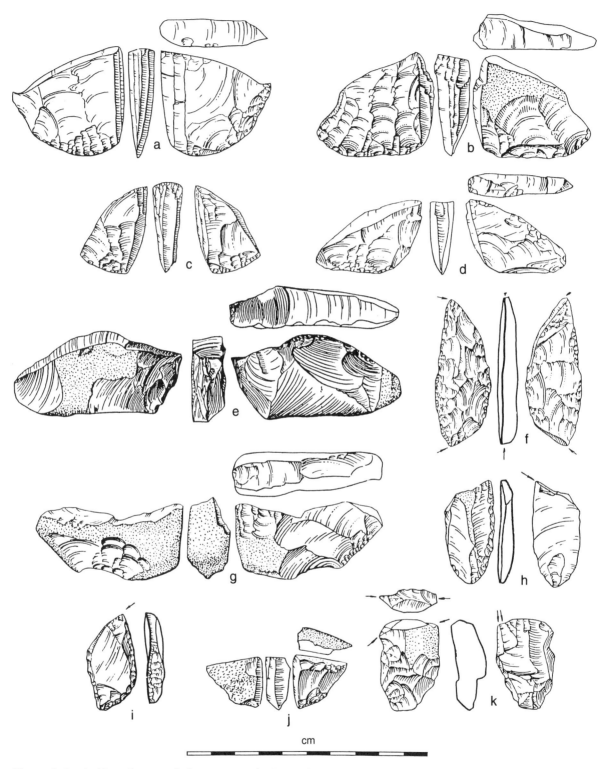

cm

Figure 3–9   Artifacts from eroded contexts at the Ust-Mil 2 site: wedge-shaped microblade cores (*a–e, g, j*); burins (*f, h, i, k*).

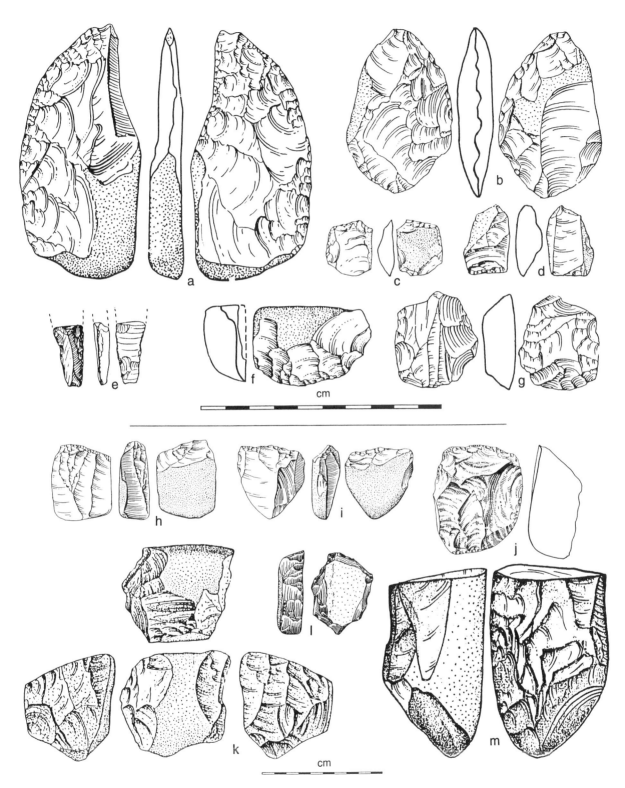

Figure 3–10   Artifacts from eroded contexts at the Ust-Mil 2 site.

VERKHNE-TROITSKAYA

Yuri A. Mochanov and Svetlana A. Fedoseeva

Verkhne-Troitskaya is located on the right bank of the Aldan River, 862 km from its mouth, in the Ust-Mayaskiy region of the Sakha Republic (Yakutia). It is 3 km upriver from the settlement of Troitskaya (60°21′N, 134°27′E). Discovered in 1969 by Y. A. Mochanov, the site was researched by him in 1969 to 1971, 1980, and 1982.

The site is associated with the second terrace deposits of the Aldan River, which vary from 12–13 m to 9–10 m, the latter elevations occurring at the eroded outer edge. The site has undergone extensive destruction since the time of its occupation due to lateral erosion by the Aldan and by the melting of exposed ice wedges. At the present time a sandy-pebbly shore about 50 m wide exists below the bank on which the site is found. During times of low water this surface is exposed. At the beginning of the Prilensk Archaeological Expedition work it was literally covered with bones of mammoth, woolly rhinoceros, bison, horse, and other animals, as well as numerous lithic artifacts.

When the terrace exposures were cleaned, significantly fewer remains were found than occurred on the shore below. The main part of the site obviously was gone, with only peripheral areas being preserved. The area that was excavated is 350 sq. m; much displaced material was collected from the riverbank.

STRATIGRAPHY

The following stratigraphic units have been identified (Fig. 3–11), of which three are cultural:

1. Turf: 30–40 cm thick.
2. A unit consisting of alternating stringers of dark brown loam (1–4 cm) and yellow-brown aleurites (0.5–2 cm). These deposits occur only on the sloping outer terrace edge. Their deposition began no earlier than 6,000 years ago. The unit's

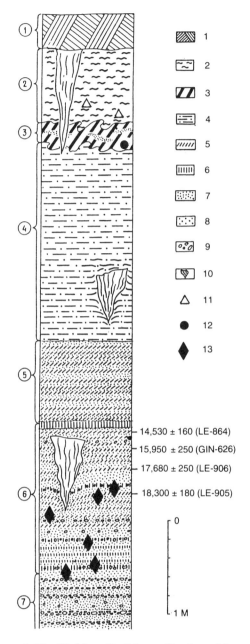

14,530 ± 160 (LE-864)
15,950 ± 250 (GIN-626)
17,680 ± 250 (LE-906)
18,300 ± 180 (LE-905)

Figure 3–11 Verkhne-Troitskaya. Stratigraphic profile. Major stratigraphic units numbered left. Descriptions in text. Soil units and artifact types (numbered right): (1) turf; (2) bedded sandy loam; (3) humic loam with traces of solifluction; (4) interbedded sandy loams and sands; (5) isolated stringers of silt-encrusted aleurites; (6) isolated stringers of loam; (7) fine-grained sands; (8) unsorted sands; (9) pebbles and gravel; (10) ice wedges; (11) Neolithic artifacts; (12) Early Holocene Palaeolithic artifacts; (13) Upper Pleistocene Palaeolithic artifacts.

lower strata yielded Neolithic remains (Cultural Stratum I).

3. Dark grey humic loam with thin lenses of poorly sorted sand (20–40 cm thick). This probably represents redeposition, by solifluction, of eroded, buried, second-terrace soil. Numerous diabase flakes, perhaps Sumnagin remains, were found. (This is Cultural Stratum II.)

4. An interbedded unit, 1.8–2 m thick, of greyish brown sandy loam (0.5–3 cm) and yellowish brown poorly sorted sands (0.2–3 cm). In the lower strata were numerous buried ice wedges. This alluvial unit represents a second-terrace floodplain facies. In it were rare, isolated bones of mammoth, bison, and horse.

5. Greyish purple silty aleurites (0.5–3 cm) interbedded with greyish brown poorly sorted sands (0.5–5 cm). Total thickness of the stratum is 0.8–1.0 m. Isolated bones of mammoth and bison were found.

6. A 160–180-cm unit of alternating seams of greyish purple silty aleurites (0.5–3 cm), loam stringers (1–4 cm) which increase in density with depth, and greyish brown poorly sorted sands (0.3–5 cm) thick. These deposits represent the lower strata of the second-terrace alluvium. The upper strata contain ice wedges.

Four radiocarbon dates were obtained on wood samples in geological stratum 6. These were collected from 12, 31, 56, and 84 cm below the upper boundary of this geological unit. They are 14,530 ± 160 years (LE-864); 15,950 ± 250 years (GIN-626); 17,680 ± 250 years (LE-906); and 18,300 ± 180 years (LE-905). Bones of mammoth, woolly rhinoceros, bison, and reindeer were found throughout the entire deposit. A loamy stringer, 5 cm above one dated at 18,300 years BP, yielded stone artifacts and an ivory needle (Palaeolithic Cultural Stratum III).

7. Cemented pebbles and gravel with grey poorly sorted sand. The stratum extends beneath the level of the Aldan. According to radiocarbon dates, the unit represents second-terrace riverbed alluvium that began to accumulate in the vicinity of Verkhne-Troitskaya at the end of the Lipovsko-Novoselovsky stage of the Karginsky interstadial, approximately 23,000–22,000 years ago.

CULTURAL STRATUM III

This Palaeolithic stratum yielded altogether 52 stone artifacts, 1 ivory needle (Fig. 3–12: *j*), and 49 split animal bones. The stone artifacts include 2 wedge-shaped chert cores (Fig. 3–12: *a, b*); 5 chert blades; 11 tools; and 34 chert flakes. The stone tools are represented by a midsection of a blade inset (Fig. 3–12: *c*); 2 burins on flakes (1 dihedral, Fig. 3–12: *d*; 1 lateral); 2 end scrapers on blades (Fig. 3–12: *e, f*); 3 knives on blades (Fig. 3–12: *g, i*) and on a flake (Fig. 3–12: *h*); 1 chisel-like chert flake tool, and 1 diabase pebble pick.

Cultural Stratum III faunal remains include bison (*Bison priscus* Boj.) (23); horse (*Equus caballus* L.) (9); mammoth (*Mammuthus primigenius* Blum.) (8); woolly rhinoceros (*Coelodonta antiquitatis* Blum.) (3); reindeer (*Rangifer tarandus* L.) (2); wolf (*Canis lupus* L.) (1), and 3 unidentified fragments.

ARTIFACTS DISPLACED BY EROSION

There are 87 stone artifacts from the eroded riverbank material that are distinctive techno-typologically. They have no analogs in any Holocene cultural complexes; they are comparable to Dyuktai Upper Palaeolithic material of the Aldan and other regions. These items include 21 blade cores, 4 ski-shaped spalls or crested blades (Fig. 3–12: *k*), and 62 tool forms. The blade cores include 10 wedge-shaped specimens on chert pebbles and tabular chert (2 blanks, 6 complete, and 2 fragments). Of the complete wedge-shaped cores, 2 are Gobi-type boat-like (Fig. 3–12: *l, m*) and 4 are tall triangular forms (Fig. 3–12: *n, o*). Among other cores are 1 black chert discoidal core (Fig. 3–13: *c*) and 10 sub-prismatic cores on chert and diabase pebbles (Fig. 3–13: *d, e*).

Stone tools include 33 knives, 12 scrapers, 4 burins, 1 perforator (Fig. 3–13: *f*), and 1 wedge chisel (Fig. 3–13: *g*). All of the burins are manufactured on flat flakes. There are two forms: lateral (1) (Fig. 3–13: *a*) and transverse (4) (Fig. 3–12: *p–r*; Fig. 3–13: *b*).

Knives may be separated into three categories based on blank form and manufacturing technique:

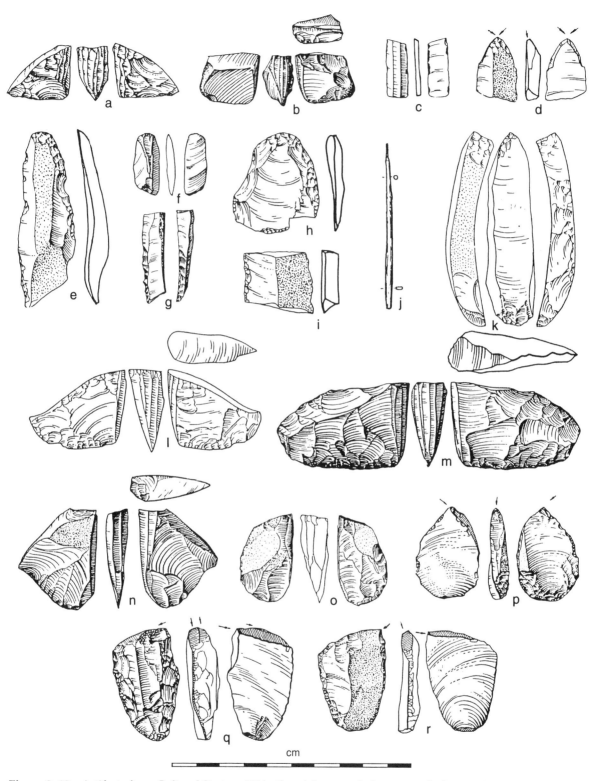

Figure 3–12 Artifacts from Cultural Stratum III (*a–j*) and from eroded contexts (*k–r*) at Verkhne-Troitskaya: wedge-shaped microblade cores (*a, b, l–o*); blade inset (*c*); dihedral burin (*d*); end scrapers on blades (*e, f*); knives on blades (*g, i*); knife on a flake (*h*); ivory needle (*j*); ski-shaped spall or crested blade (*k*); transverse burins (*p–r*).

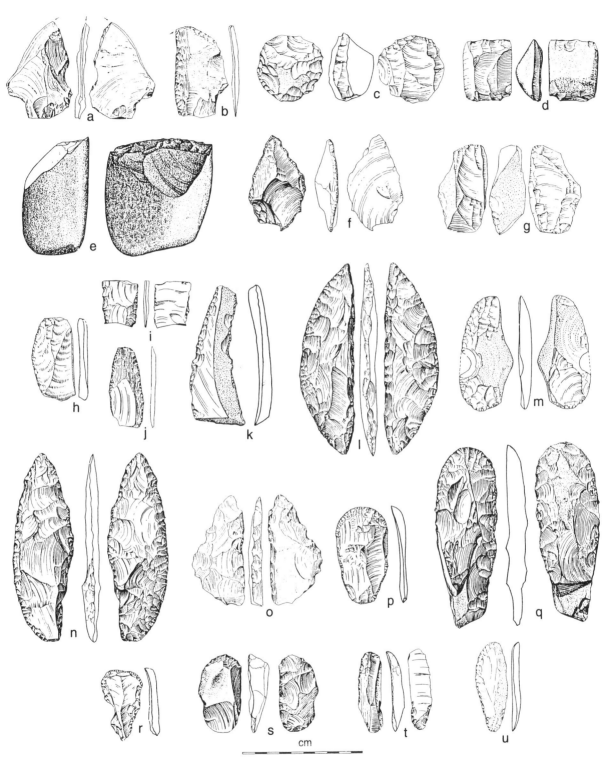

Figure 3–13   Artifacts from eroded contexts at Verkhne-Troitskaya: lateral burin (*a*); transverse burin (*b*); discoidal pebble core (*c*); subprismatic pebble cores (*d, e*); perforator (*f*); chisel-like tool (*g*); edge-retouched flakes (*h–j*); edge-retouched blade (*k*); crescentic knives (*l, m*); laurel leaf-shaped knife (*n*); subtriangular knife (*o*); flake scrapers (*p, r, s, u*); scraper on tabular chert (*q*); scraper on blade (*t*).

edge-retouched flakes (11) (Fig. 3–13: *h–j*); edge re-touched blades (2) (Fig. 3–13: *k*); and those bifacially prepared (6 complete and 10 broken). Among the biface fragments are not only knives but also spear and dart points. The complete bifacial knives are crescentic (plano-convex in outline) (Fig. 3–13: *l, m*); laurel-leaf–shaped (1) (Fig. 3–13: *n*); and subtriangular (2) (Fig. 3–13: *o*).

There are 9 chert end scrapers on flakes (Fig. 3–13: *p, r, s*); 1 end scraper on a blade (Fig. 3–13: *t*); and 1 bifacial side and end scraper (Fig. 3–13: *q*). Other scrapers of various forms are made on thick chert flakes (2) and thick flakes and fragments of diabase (10).

In addition to the lithic artifacts, an engraved rib fragment of woolly rhinoceros (?) has been attributed to the Dyuktai complex.

Faunal remains collected from the riverbank include horse (*Equus caballus* L.) (202); bison (*Bison priscus* Boj.) (201); woolly rhinoceros (*Coelodonta antiquitatis* Blum.) (60); mammoth (*Mammuthus primigenius* Blum.) (46); moose (*Alces alces* L.) (42); reindeer (*Rangifer tarandus* L.) (25); tiger (*Panthera* sp.) (3); and wolf (*Canis lupus* L.) (2). The remaining crushed bones were unidentifiable.

SUMMARY

The Verkhne-Troitskaya site occupies an important position in the Palaeolithic of northeastern Asia. Its materials represent a quite early Dyuktai culture stage, dating within the chronological range of 23,000/22,000 to 18,000 years. The most pronounced feature of this stage is the extensive use of bifaces in combination with wedge-shaped cores.

EZHANTSY

Yuri A. Mochanov and Svetlana A. Fedoseeva

The Ezhantsy site is located on the right bank of the Aldan River 794 km from its mouth in the settlement of Ezhantsy in the Ust-Mayaskiy region, Sakha Republic (Yakutia) (60°28′N, 135°08′E). Discovered in 1970 by Y. A. Mochanov, the site was investigated in 1970–72 and 1980 by the Prilensk Archaeological Expedition under his direction.

STRATIGRAPHY

The site is related to the third (16–18 m) terrace of the Aldan. The excavated area, including stratigraphic trenches, is 112 sq. m. The following stratigraphic sections have been identified at the site (Fig. 3–14):

1. Turf (5–10 cm).
2. Reddish brown loam (50–60 cm). Charcoal fragments, which have been moved by frost action from humic lenses within this unit into frost cracks, date 9000 ± 100 years (LE-997), $10,500 \pm 300$ years (LE-964), and $10,940 \pm 100$ years (GIN-737). These dates allow one to assume that the frostcracking of the third-terrace surface took place during the Norilsk stage of the Sartan glaciation, i.e., about 11,400–10,300 years ago.
3. Greyish beige alluvial loam (50–60 cm) containing weakly developed lenses that represent the third-terrace floodplain alluvium. At the lowermost contact of this unit with the underlying stratum there were recovered bones of Pleistocene animals and stone artifacts. This is the cultural stratum.
4. Greyish yellow alluvial sands and aleurites (up to 150 cm), associated with overbank sand deposits of the third terrace.

CULTURAL STRATUM

The presence of buildings over most of the site, together with large frost wedges, has made difficult a

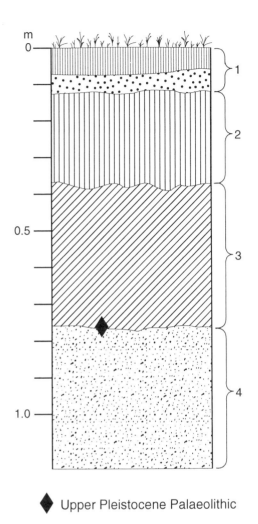

m

0

0.5

1.0

1

2

3

4

◆ Upper Pleistocene Palaeolithic

Figure 3–14   Ezhantsy. Stratigraphic profile. See text for strata descriptions.

Six hundred stone artifacts were recovered. These include 32 cores and core blanks: 14 subprismatic blade cores on large pebbles of diabase and chert (Fig. 3–15: *e, f*), 16 tall wedge-shaped cores of hornfels and chert (Fig. 3–16: *a–d*), 2 turtle cores (discoidal flake cores) of hornfels (Fig. 3–15: *a, b, d*); 2 microblades (Fig. 3–16: *k, l*), 6 ski-like spalls (crested blades) (Fig. 3–16: *i*), 33 retouched blades, 2 burin spalls, and 517 unmodified flakes. In addition, 30 objects could be classed as tools. Among them are 3 side scrapers, 1 end scraper (Fig. 3–16: *m*), 4 bifacial knives, 21 burins (Fig. 3–17), and 1 awl. Burins form the largest tool group. They are made both on flakes and blades. Formally, the burins may be divided into 4 categories: angular (7), lateral (6), central (4), and transverse (4). One central burin is formed on a fragment of a broken ovate biface, the broad surfaces of which retain evidence of fine pressure retouch (Fig. 3–16: *h*). There is also a fragment of a hammer-like tool made of antler (Fig. 3–15: *c*). On the transverse surface are traces of wear.

Summary

The occurrence of the cultural stratum in the lower section of the floodplain facies of the third terrace of the Aldan is important in estimating the site's age. Settlement had to have occurred when this alluvial terrace was beginning to form.

Elsewhere the third-terrace alluvial and overbank deposits have been radiocarbon dated 35,400 ± 600 years (LE-954) and 35,600 ± 900 years (LE-955), suggesting a site age here of approximately 35,000 years. It is, thus, one of the most ancient Dyuktai sites currently known, with an impressive stone tool assemblage.

(Figures 3–15 to 3–17 follow on pp. 186–188)

delineation of exact extent of the site. Frost heaves have displaced 50–60 cm upward some stone artifacts, bones, and alluvial loam, almost to the base of stratum 2. Conversely, some have been thrust downward into sand deposits as deep as 1 m.

Of 415 animal bones collected from the cultural deposit—almost all of which are split—55 have been identified by E. A. Vangengeim as to species. These were predominantly young animals. This faunal assemblage consists of bones of mammoth (*Mammuthus primigenius* Blum.) (8), bison (*Bison priscus diminutus* W. Grom.) (8), woolly rhinoceros (*Coelodonta antiquitatis* Blum.) (9), horse (*Equus caballus*) (20), and reindeer (*Rangifer tarandus*) (10).

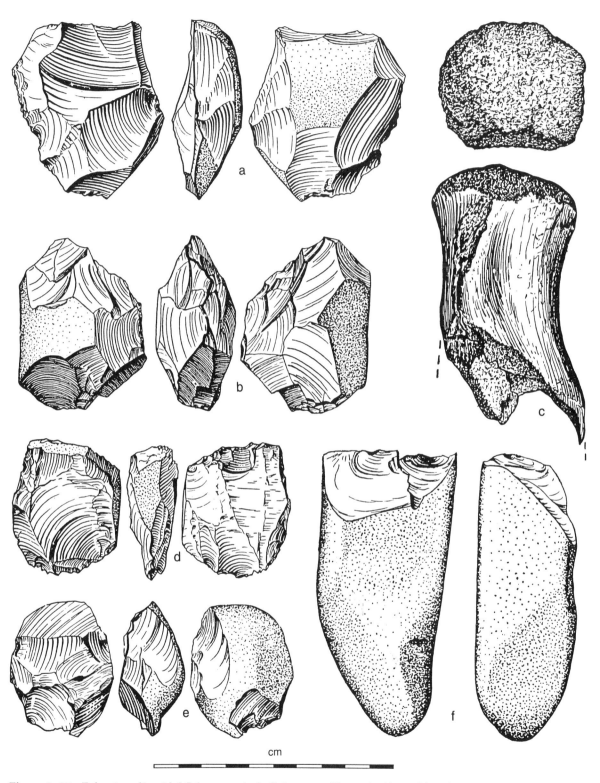

Figure 3–15 Ezhantsy: discoidal flake cores (*a, b, d*); hammer-like tool of horn (*c*); subprismatic cores on large pebbles (*e, f*).

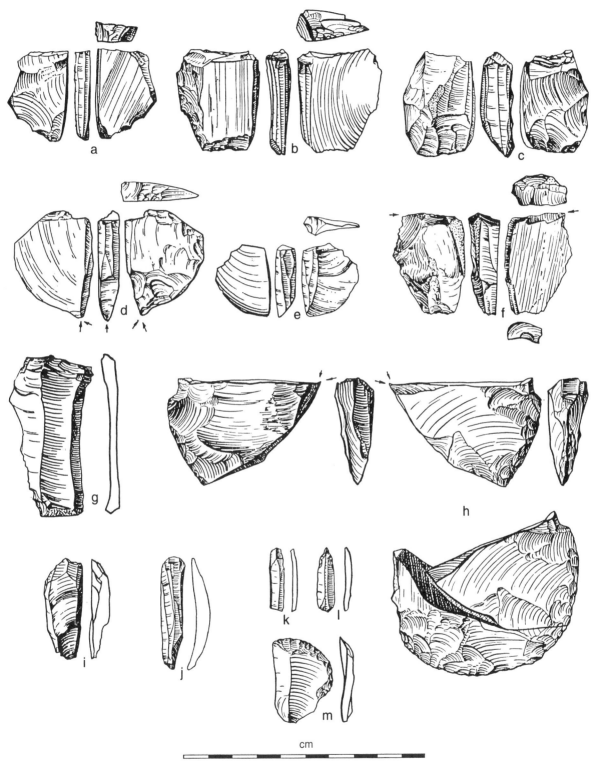

Figure 3–16   Ezhantsy: wedge-shaped cores (*a–d*); wedge-shaped core-burins (*e, f*); large blade (*g*); split biface modified as large burin or wedge-shaped core preform (*h*); crested blade (*i*); knife-like blade (*j*); microblades (*k, l*); scraper (*m*).

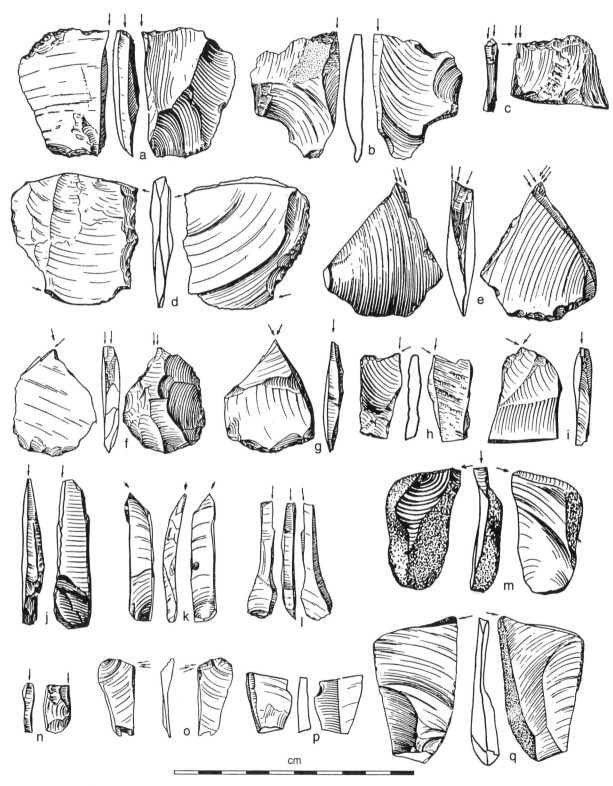

Figure 3–17 Ezhantsy: burins (*a–q*) from the cultural stratum.

IKHINE 1 AND 2

Yuri A. Mochanov and Svetlana A. Fedoseeva

The Ikhine sites 1 and 2 are located at the confluence of the Ikhine Stream and the Aldan River in
the Tattin region of the Sakha Republic (Yakutia)
283 km from the mouth of the Aldan (63°07′N,
133°36′E).

IKHINE 1

The Ikhine 1 site is located high on the right bank
of the Aldan River, just downstream from the point
where the Ikhine Stream enters it. Discovered in
1963 by Y. A. Mochanov and S A. Fedoseeva, it was
investigated by PAE under the direction of Y. A.
Mochanov during 1964–66, 1968, 1973–74, and
1980. Ikhine 1 is the first Palaeolithic site found in
Yakutia yielding *in situ* traces of Pleistocene
hunters.

Stratigraphy

The site is associated with deposits of the third
(16–18 m) terrace of the Aldan River. A 178 sq. m
excavation area revealed 4 cultural strata (3 Palaeolithic and 1 early Iron Age).

The following stratigraphic units have been
identified (Fig. 3–18):

1. Turf, 5–10 cm thick. Yielded isolated early
Iron Age ceramic fragments.

2. A 30–40 cm thick unit of reddish brown, di-
luvially derived loam. Palaeolithic stone artifacts
and Pleistocene animal bones were recovered from
the middle and lower sections (Palaeolithic Stratum
I, 15–20 cm).

3. An alluvial unit (75–90 cm) of interbedded
greyish blue and yellowish brown loam stringers
(2–5 cm), alternating with stringers of grey, poorly
sorted sand 0.1–4.0 cm thick. Some loam and sand
streaks contained isolated pebbles and gravel. The
two lowermost stringers of greyish blue loam
yielded bones of Pleistocene animals and stone artifacts (Palaeolithic Stratum II, 5–15 cm).

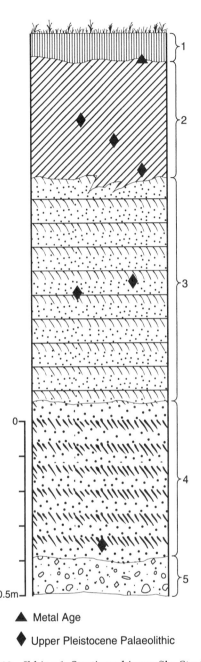

▲ Metal Age

◆ Upper Pleistocene Palaeolithic

Figure 3–18   Ikhine 1. Stratigraphic profile. Strata
numbered right. See text for descriptions.

4. An alluvial unit of grey sand with isolated
pebbles and gravel (40–45 cm). There are two merging stringers of greyish blue loam, each 3 to 4 cm
thick, in the lower part of the unit. The lowermost,
143 cm below surface, yielded bones of Pleistocene
animals, one of which had been worked (Palaeolithic Stratum III).

5. A riverbed alluvium of pebbles and gravel cemented by sand (>1.5–1.6 m).

Depositional characteristics suggest that Palaeolithic Cultural Strata II (geological unit 3) and III (geological unit 4) lie in the alluvial facies of the third terrace of the Aldan. The latter stratum may be associated with the lowest alluvial strata or with sandy overbank alluvium (Fig. 3–18).

Cultural Stratigraphy

Cultural Stratum I   Palaeolithic Stratum I yielded 5 stone artifacts and 8 animal bones. The nondiagnostic stone artifacts are represented by a flake of diabase and chert and 1 spalled chert pebble. Stratigraphically, this level could date to the end of the Pleistocene. The faunal assemblage does not contradict this assumption: it is comprised of bones of mammoth (*Mammuthus primigenius* Blum.) (1), bison (*Bison priscus diminutus*) (3), horse (*Equus caballus*) (1), reindeer (*Rangifer tarandus* L.) (2), and 1 unidentifiable fragment bearing a cut-mark.

Cultural Stratum II   Palaeolithic Stratum II yielded 10 stone artifacts and 30 animal bones well scattered through the stratum. Artifacts include 1 chert "Gobi-type" wedge-shaped core, 1 chert blade, 2 tools (1 chert chisel-like tool and 1 end scraper), 4 tool blanks, and 13 flakes of chert, argillite, and diabase. Two blanks are for diabase bifacial knives and two are indeterminate in form (see Figure 3–19.)

The faunal assemblage consists of 30 animal bones, including 3 mammoth (*Mammuthus primigenius* Blum.), 8 bison (*Bison priscus diminutus*), 1 woolly rhinoceros (*Coelodonta antiquitatis* Blum.), 4 horse (*Equus caballus*), and 4 reindeer (*Rangifer tarandus* L.).

Cultural Stratum II produced Dyuktai-like artifacts. Pending radiocarbon dates, the estimated age for this component is within the range of 30,000–25,000 years.

Cultural Stratum III   Palaeolithic Stratum III contains 1 culturally modified humerus of a young

mammoth (Fig. 3–19: *h*) and bones of 3 bison (*Bison priscus diminutus*), 2 horse (*Equus caballus*), 1 moose (*Alces alces*), and 3 unidentifiable fragments. These materials are at the contact of the third-terrace alluvium and underlying riverbed shingle. Considering the radiocarbon dates for the lower strata, this cultural component should date approximately 35,000 years.

Summary

Ikhine 1 is a multicomponent, stratified Dyuktai site related to other early Upper Palaeolithic remains dating to 35,000–25,000 years. It has considerable potential for future research.

IKHINE 2

Ikhine 2 is located on the high right bank of the Aldan immediately upstream from the Ikhine Stream's confluence with that river. Discovered two years after Ikhine 1, it was researched by PAE in 1965–66, 1968, 1973–74, 1980, and 1992.

Stratigraphy

The site is located on the third (15–16 m) terrace of the Aldan. An excavated area of 216 sq. m revealed two Palaeolithic cultural strata with the second stratum divisible into 5 horizons: IIa, IIb, IIc, IId, and IIe.

The following stratigraphic units have been identified (Fig. 3–20):

1. Turf, 10–15 cm.

2. A 5–50 cm unit of light grey, coarse alluvial sand. This represents late Holocene alluvial deposits no older than 1,000 years, occurring on the terrace slope as high as 12 meters.

3. A reddish-brown diluvial loam (35–45 cm). Early Iron Age ceramic fragments occurred in its upper section. In the lower section were isolated Pleistocene animal bones and stone artifacts (Palaeolithic Stratum I).

4. A complex 150–170 cm unit consisting of interbedded stringers of greyish blue and yellowish brown loams (0.5–2.0 cm). Within these loams are

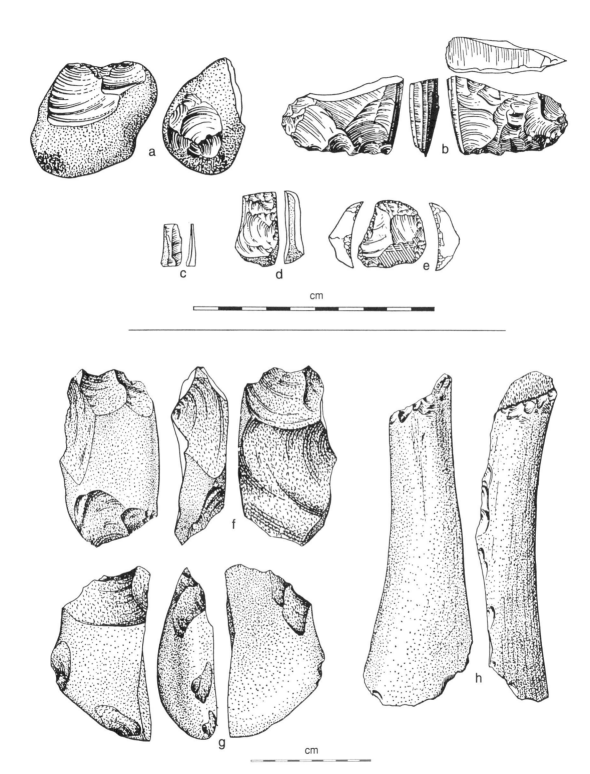

Figure 3–19 Artifacts from Cultural Strata I (*a*), II (*b–g*), and III (*h*) of the Ikhine 1 site: flaked pebble (*a*); wedge-shaped core (*b*); blade (*c*); chisel-like tool (*d*); rounded scraper (*e*); biface blanks (*f*, *g*); modified mammoth humerus (*h*).

The Aldan River near the site of Ikhine 2. (*Photo courtesy of Yaroslav V. Kuzmin.*)

embedded occasional strands of poorly sorted sand and gravel (0.1–0.3 cm). Quite clear streaks of light grey loam 1.0–5.0 cm thick occur at depths of 75, 110, 145, and 195 cm. These have allowed division of this unit, Palaeolithic Cultural Stratum II, into five distinct horizons: IIa, IIb, IIc, IId, and IIe, although the latter horizon has yielded only animal bones—mammoth, horse, bison, and reindeer. This unit represents the third-terrace floodplain alluvium.

     5. Beige-grey loam seams (1.0–2.0 cm) interbedded with poorly sorted sands (0.5–1.0 cm). (Total thickness 25–45 cm.) This facies is related to overbank alluvial sand deposits or the basal third-terrace alluvium.

     6. Pebbles and gravel cemented by alluvial sand. The visible portion is 11–14 m.

Cultural Strata

Palaeolithic Stratum I   Recovered from Palaeolithic Stratum I (5–10 cm thick) were 2 small diabase flakes, 6 bison bones (*Bison priscus diminutus*), and 5 horse bones (*Equus caballus*). They may be provisionally attributed to the end of the Pleistocene and correlated to finds in the Palaeolithic Stratum I at Ikhine 1.

Palaeolithic Stratum IIa   Palaeolithic Stratum IIa, 25–30 cm thick, contained 11 stone artifacts and 254 animal bone fragments. Stone implements included 1 wedge-shaped core blank on a black hornfels pebble (Fig. 3–21: *c*), 1 hornfels blade (Fig. 3–21: *a*), 2 tools, 4 diabase and chert flakes, and 3 pebbles (granite, diabase, and chert) with work traces. One of the tools is a hornfels knife on a blade midsection (Fig. 3–21: *b*); the other is a scraper on a black argillite pebble fragment (Fig. 3–21: *d*).

Identifiable animal bones include 7 mammoth (*Mammuthus primigenius* Blum.), 4 woolly rhinoceros (*Coelodonta antiquitatis* Blum.), 50 bison (*Bison priscus diminutus*), 71 horse (*Equus caballus*), 16 reindeer (*Rangifer tarandus* L.), and 1 moose (*Alces alces*).

Based on radiocarbon dates for underlying deposits and its stratigraphic position, Cultural Horizon IIa is estimated to be about 25,000–23/22,000 years old.

Palaeolithic Stratum IIb   Palaeolithic Stratum IIb, 35–40 cm thick, yielded 6 artifacts and 202 animal bones. The stone assemblage includes 1 wedge-shaped core of black hornfels (Fig. 3–21: *e*), a possible pebble core blank of diabase, 1 diabase pebble scraper formed on a longitudinally split pebble (Fig. 3–21: *g*), and 3 flakes of chert, argillite, and diabase (Fig 3–21: *f*).

Identifiable animal bones include 4 mammoth (*Mammuthus primigenius* Blum.), 5 woolly rhinoceros (*Coelodonta antiquitatis* Blum.), 58 bison (*Bison priscus diminutus*), 50 horse (*Equus caballus*), 6 reindeer (*Rangifer tarandus*), 1 polar fox (*Alopex lagopus* L.), 1 fox (*Vulpes vulpes* L.), and 1 fish.

Five radiocarbon dates were obtained from a wooden specimen that was collected from the middle of Cultural Stratum IIb: 24,330±200 (LE-1131); 24,500±480 (IM-203); 24,600±380 (IM-153); 27,400±800 (IM-205), and 30,200±300 years (GIN-1019). Tripling the average statistical error provides an age range of 23,000–31,000 years for these samples. When compared with dates from the middle alluvium at the Ust-Mil 2 site, Horizon IIb appears likely to date between 30,000 and 25,000 years, with 30,200 years (GIN-1019) the most acceptable date here.

Palaeolithic Stratum IIc   Palaeolithic Stratum IIc contained 1 black hornfels flake and half of a diabase pebble split longitudinally and partially retouched, possibly a scraper blank. Of the 64 animal bones recovered, 2 were identified as woolly rhinoceros (*Coelodonta antiquitatis* Blum.), 14 as bison

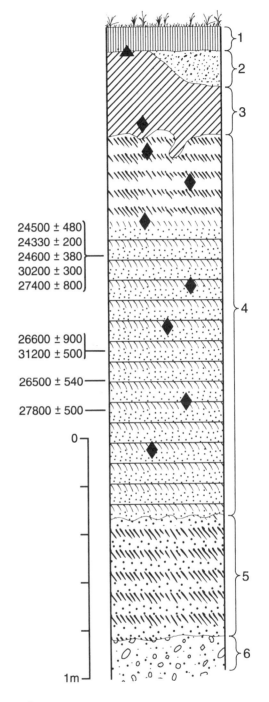

▲ Metal Age

◆ Upper Pleistocene Palaeolithic

Figure 3–20   Ikhine 2. Stratigraphic profile. Strata numbered right. See text for descriptions. (Note indistinct bedding upper stratum 4.)

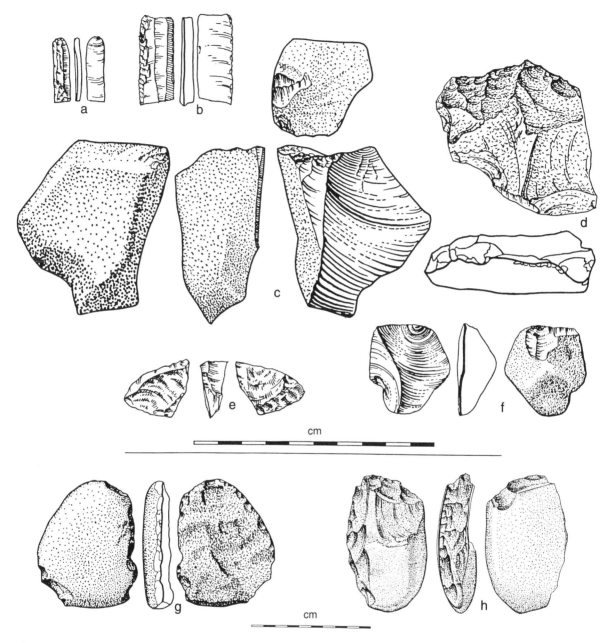

Figure 3–21   Artifacts from Cultural Strata IIa (*a–d*), IIb (*e–g*), and IId (*h*) of the Ikhine 2 site: blade (*a*); retouched blade knife (*b*); wedge-shaped core blanks on pebbles (*c, e*); pebble scrapers (*d, g, h*); flake (*f*).

(*Bison priscus diminutus*), 14 as horse (*Equus caballus*), and 2 as reindeer.

Two radiocarbon dates on wood from the middle section of this stratum, 26,600 ± 900 (IM-201) and 31,200 ± 500 years (GIN-1020), yield a statistical age range of 25,700–31,700 years. While tripling the

average statistical error provides an age range of 23,900–32,700 years, stratigraphic position suggests a date no younger than 30,000 years ago.

Palaeolithic Stratum IId   Palaeolithic Stratum IId contained 2 stone artifacts: an oval double-edged scraper (Fig. 3–21: *h*) on a fine crystalline diabase pebble and a chert pebble with a flake removal. There is also a tool made of a mammoth rib. Judging by use wear traces, it could be part of a boring tool.

The identifiable bones belong to mammoth (*Mammuthus primigenius* Blum.) (9), woolly rhinoceros (*Coelodonta antiquitatis*) (1), bison (*Bison priscus diminutus*) (29), horse (*Equus caballus*) (43), noble deer (*Cervus elaphus* L.) (1), reindeer (*Rangifer tarandus*) (10), and wolf (*Canis lupus* L.) (1). Wood from the upper part of Stratum IId dates $27,800 \pm 500$ years (IM-206), which appears young. Stratigraphic position suggests a date of 31,500–35,000 years, as for Horizon C at Ust-Mil 2.

Summary

Archaeological, stratigraphic, and faunal indicators suggest that Ikhine 2 site materials were left by Pleistocene hunters within the range of 35,000 to 25,000 years ago. Both the archaeological and faunal collections have been enriched substantially by finds from the 1992 field season. They confirmed the earlier proposal that the Ikhine site is a Dyuktai cultural component.

TUMULUR

Yuri A. Mochanov and Svetlana A. Fedoseeva

The Tumulur site is on the left bank of the Aldan River at km 1,456 in the Ust-Mayaskiy region (58°47′N, 128°11′E) of the Sakha Republic (Yakutia). The site was discovered in 1964 by the authors and was investigated by them and the PAE in 1964–66, 1969, 1971, 1974–75, 1980, and 1982.

The site is associated with deposits of the second (13-m) terrace of the Aldan River. Five hundred and twenty-five sq. m of deposits have been excavated.

STRATIGRAPHY

The following stratigraphic units were identified (Fig 3–22).

1. Turf: 5–10 cm.
2. A 20–35-cm thick unit of sand stringers (5–7 cm) interbedded with humic sandy loam (1–3 cm). This is a recent floodplain alluvium, disturbed by modern plowing over most of the excavation area. Frost-related cracks, leading to underlying deposits, occur frequently.
3. A unit of orange-yellow sandy loam (soil stratum II of a buried terrace, 12–20 cm), in which were found Neolithic, Bronze Age, and early Iron Age materials lying above a late Palaeolithic Sumnagin assemblage. The latter assemblage was asso-

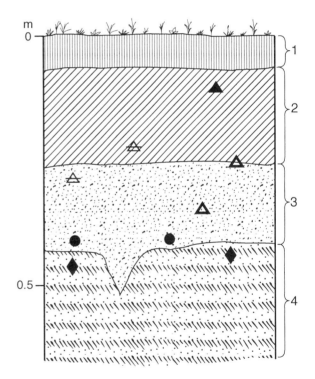

▲ Iron Age   △ Bronze Age
△ Neolithic   ● Early Holocene Palaeolithic
◆ Upper Pleistocene Palaeolithic

Figure 3–22 Tumulur. Stratigraphic profile. Strata numbered right. See text for descriptions.

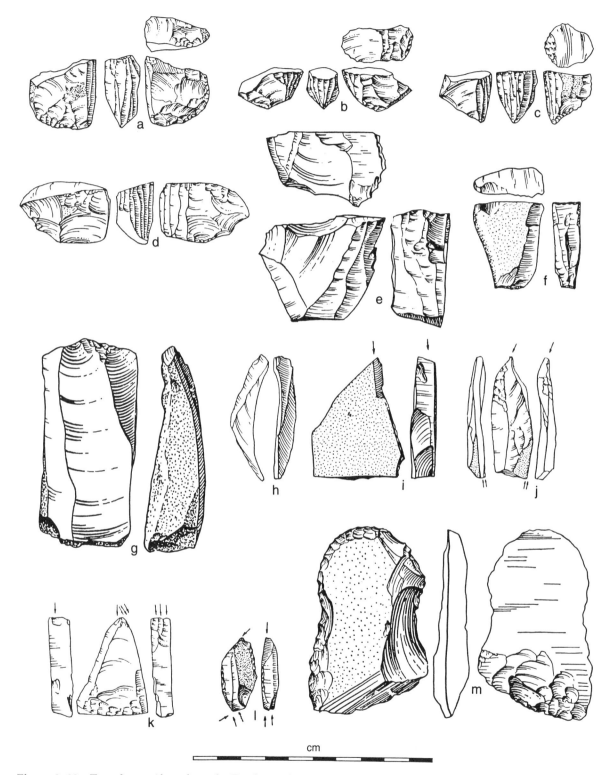

Figure 3–23 Tumulur: artifacts from the Dyuktai cultural stratum: wedge-shaped cores (*a–d*); wedge-shaped core blanks (*e, f*); blade core face removals (*g, h*); burins (*i–l*); end/side scraper (*m*).

ciated with the lower layers of sandy loam. The bottom of the stratum revealed frost cracking and ice wedges penetrating to underlying strata.

4. Stratified alluvium of the second terrace, extending to 250 cm below ground surface, in which there are two distinct horizons. Within a 25–30-cm-thick, upper orange-red sand and loam horizon are seams of yellow-brown, poorly sorted sands. These deposits mark the transition from overbank sand deposits to deposits of an episodically inundated high floodplain of Sartan time. Dyuktai cultural remains were recovered from an underlying reddish, dense sandy loam seam (0.5–3.0 cm) which in some areas separated to become 2 to 4 seams. The lower grey, poorly sorted sand (1–7 cm), an overbank facies, contained only isolated artifacts, introduced by the agency of frost cracks. Interbedded with it were bluish silty sandy loams 0.1–2 cm thick.

THE CULTURAL STRATUM

In the Dyuktai complex at Tumulur there were 37 worked artifacts of 94. There were 9 wedge-shaped cores and blanks of chert (Fig. 3–23: *a-f*) and 16 tools. Among the tools are 8 lenticular bifacial knives/spearpoints made of siliceous schist (Fig. 3–24: *a-e*), 2 bifacial round based knives (Fig. 3–24: *h*), 1 knife or spearpoint on a thick, faceted blade with extensive unifacial retouch on the ventral surface (Fig. 3–24: *f*), 4 burins (Fig. 3–23: *i-l*) and 1 end/side scraper (Fig. 3–23: *m*). There were also 57 flakes and 12 modified pieces of cortex.

The distribution of bifaces at the site is unusual. All were in a 20 x 25 cm cluster, although the total area of excavation was 460 sq. m. Many of these re-pieced bifaces had been broken and exfoliated. No dwelling or hearth remains were evident near the grouping. Since all of the bifaces were finished specimens, they may represent a cache.

SUMMARY

Since alluvial deposition of the second terrace ended circa 13,000–12,000 years ago and there are typological similarities to cultural materials from the upper alluvial deposits of Dyuktai Cave (radio-carbon-dated 12,100 ± 120 [LE-907] and 13,200 ± 250 [GIN-405] years), we may consider an age of 13,000–12,000 years as acceptable for the Dyuktai site of Tumulur, thus attributing it to the final stage of the Dyuktai culture.

(Figure 3–24 follows on p. 198.)

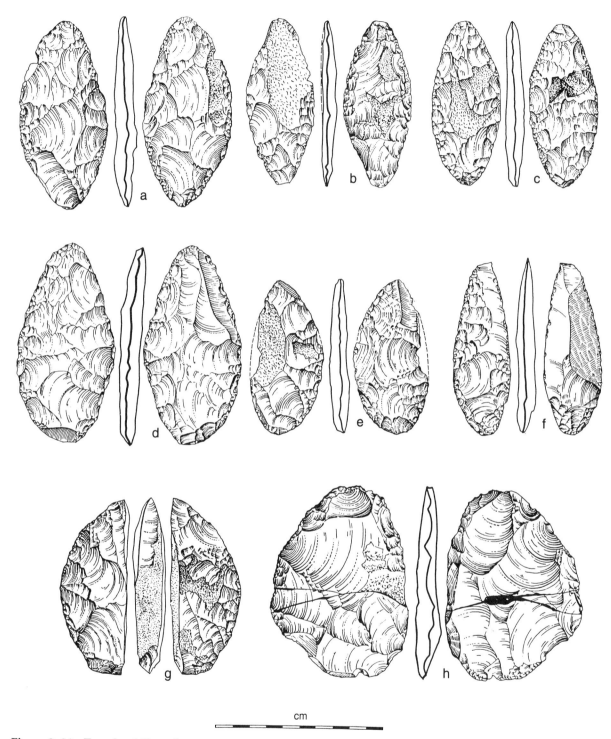

cm

Figure 3–24 Tumulur: bifaces from a possible cache in the Dyuktai cultural stratum: lenticular knives/spear points (*a–e*); knife/spear point on thick blade (*f*); bifacial butt knives (*g, h*).

UST-TIMPTON (STRATA Vb–X)

Yuri A. Mochanov and Svetlana A. Fedoseeva

The site is located on a promontory on the left bank of the Timpton River, at its confluence with the Aldan River at km 1,547 (58°42′N, 127°07′E). The Timpton is the largest right-hand tributary in the upper reaches of the Aldan. Administratively, this territory is included in the Aldanskiy region of the Sakha Republic (Yakutia). The authors discovered the site in 1964, and it was investigated by PAE in 1964–66, 1969–71, 1974–77, and 1979–82.

Ust-Timpton 1 is located on the second (12 m) terrace of the Aldan, which may be traced for 350–400 m along the Aldan and for 120–150 m along the Timpton. The terrace sequence is clear here: T 1 (6 m), 2–10 m wide, joins T 2, which in turn joins the 18-m T 3. On the downslope but associated with T 3 is Ust-Timpton 2.

Test excavations at Ust-Timpton 1 indicate that the site extends 110–120 m on the Aldan and 45–50 m on the Timpton. The site occupies approximately 5,500–6,000 sq. m. Thirteen hundred sq. m were excavated.

STRATIGRAPHY

Three geological units are recognized, based on age, genesis, lithology, stratigraphy, color, and cryogenic features (Fig. 3–5).

1. Turf, overlying alluvial sandy loam of the modern floodplain. Most of this unit is plowzone; surviving remnants of the original structure indicate a thickness of 60–65 cm. Early Iron Age artifacts (Cultural Stratum I) are associated with the two lower humic, sandy loam lenses here.

2. A sandy loam (8–25 cm) overlies alluvial deposits of the buried first terrace (70–95 cm). Cultural Stratum II, containing Imyakhtan late Neolithic evidences, is associated with the middle and lower sections of the sandy loam.

In Cultural Stratum III are Belkachin Neolithic

remains. These occurred in organic lenses 2 and 3 of the T 1 deposit. Late Palaeolithic (or Holocene) Sumnagin materials were found in the organic lenses 10–16 (40–60 cm thick) in the lower portion of the alluvium.

3. This unit consists of two sections. Above is a yellow-brown sand (6–10 cm) in which are two organic stringers. This is interpreted as an eroded buried soil possibly related to downcutting of the Aldan in the Pleistocene–Holocene transition. Cultural Stratum Va derives from the upper organic stringer; Cultural Stratum Vb, from the lower. Organic stringers were found in several excavation squares but it is almost impossible to interpret them owing to permafrost phenomena—frost cracking, solifluction, and erosion. Because Va and Vb cannot be differentiated definitively in these situations, finds here were assigned simply to Cultural Stratum V. Underlying these yellow-brown sands are overbank sand deposits (60–80 cm) interpreted as a facies of the first terrace. Cultural Strata VI, VII, VIII, IX, and X are associated with loamy stringers within this deposit (stringers 1–5, respectively). More of these may be discovered in future, but excavating here is exceptionally difficult due to the fact that these are all in permafrost.

RADIOCARBON DATING

Of the 26 radiocarbon dates obtained for these deposits, 13 relate to Cultural Strata IV–IX. The radiocarbon dates for the Sumnagin materials from Cultural Stratum IV (geological unit 2) are 9140 ± 150 and 8870 ± 90. The radiocarbon dates obtained for Cultural Stratum Va (geological unit 3) are 9400 ± 90 years (LE-896), 8900 ± 200 (IM-456), and 8900 ± 120. Radiocarbon dates for Cultural Stratum Vb, with materials transitional between Dyuktai and Sumnagin, are 9450 ± 300 (IM-455) and $10,740 \pm 100$ (LE-861). There is evidence of frost intrusion within Strata Va and Vb and evidence of artifact movement from level Vb into Va.

Radiocarbon dates were obtained from charcoal and charred wood in the Dyuktai Cultural Stratum VI: from a seam just overlying Stratum VI contact, $10,300 \pm 50$ (LE-920); from the upper horizon of

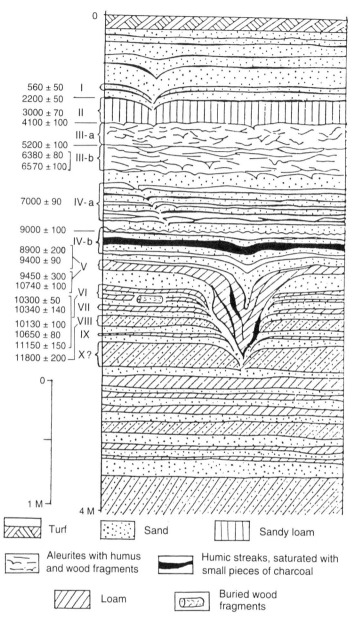

Figure 3–25 Ust-Timpton 1. Stratigraphic profile. Cultural strata numbered left. See text for descriptions.

Stratum VI, 10,340 ± 140 (LE-862); and from the underlying seam, 10,130 ± 100 (LE-897), 10,650 ± 80 (LE-898), and 11,150 ± 150 (IM-454). A radiocarbon date of 11,800 ± 200 (IM-453) was obtained on charcoal from Cultural Stratum VIII.

Strata VI–IX represent the final stage of Dyuktai culture and date in the range 11,000–10,500 yr BP.

FAUNAL REMAINS

In general, bone recovery was rare at Ust-Timpton 1. Only four animal bones were recovered from Cultural Stratum VI, probably due to unfavorable preservation conditions. According to A. V. Sher, one of these was caribou, another an exfoliated tooth of an elk (moose) recovered from a hearth, and two were unidentifiable fragments.

CULTURAL STRATA

Cultural Stratum V

Cultural Stratum V (a and b) are separable and associated with a buried soil in the upper section of geological unit 3. In further excavation squares, however, they merge and all those materials are assigned the designation "V" alone. It is probable, nevertheless, that the latter artifacts actually derive from Vb (based on techno-typological indices) and are to be included in Dyuktai. Cultural Strata VI, VII, VIII, IX, and X are associated with stringers of loam in the upper section of underlying permanently frozen deposits.

Stratum Vb Artifacts (27/133)   Upper Palaeolithic archaeological finds in Cultural Stratum Vb were found in place in two small concentrations. A selection of representative artifacts appears in Figures 3–26 through 3–28.

Blade cores are represented in two forms: 2 wedge-shaped "Gobi" (Fig. 3–26: *a, b*) and 2 subprismatic (Fig. 3–26: *c*). Also recovered were 11 complete blades and microblades (Fig. 3–26: *d, e*), 3 blades for insetting (Fig. 3–26: *f, g*), and one graver on a blade (Fig. 3–26: *o*). Also present is one trans-

verse burin made on a thin cortical flake (Fig. 3–26: *h*). Of 6 end scrapers, 3 are made on blades (Fig. 3–26: *i, j*) and 3 are on large flakes (Fig. 3–26: *k–m*).

Stratum V (combined a and b) Artifacts (8)   Blade cores are represented by 2 wedge-shaped cores, 1 subtriangular and 1 boat-shaped (Fig. 3–26: *n, p*). A discoidal flake core with a wedge-shaped longitudinal profile is shown (Fig. 3–26: *q*).

Two dihedral transverse burins on large blades were recovered (Fig. 3–27: *a, b*). The remaining tools included 1 end scraper with a high dorsal surface (Fig. 3–27: *d*), 1 thick scraper on a flake (Fig. 3–27: *c*), and 1 subrectangular wedge-like tool (Fig. 3–27: *e*). None of these forms is found in the later Sumnagin culture and are therefore included in the Dyuktai.

Cultural Stratum VI

Cultural Stratum VI (1–3 cm) contains 2–3 seams of loam, each 0.5–1.0 cm. An area of 468 sq. m was excavated along the Aldan side of the terrace. In the separate excavation units here Cultural Stratum VI is clearly divided into two horizons.

The upper horizon (VIa) contains only a few nondiagnostic finds and has a radiocarbon date of 10,340 ± 140 years BP. Horizon VIb yielded a date of 10,650 ± 80 years.

The major finds occurred in Horizon VIb. In an area 6 m by 2 m, extending north-south along the Timpton, were found remains of a probable skin dwelling (Fig. 3–28). These consisted of burned and cracked logs, of either pine or larch, compressed to 15–17 cm in width by 7–9 cm in thickness. These lay along the long sides of the rectangular structure. At right angles to them lay seven more logs. Below and above the logs were fragments of charred branches 3–5 cm in diameter stacked in 2 to 3 layers. Some branches were as long as 2 m. Fragments of similar-sized branches were found near the northern part of the dwelling, together with pieces of bark which probably represented remains of the roof. Found among the logs and branches were 1 burned granitic pebble and 8 small diabase flakes.

Nearby, approximately 1 m east of the dwell-

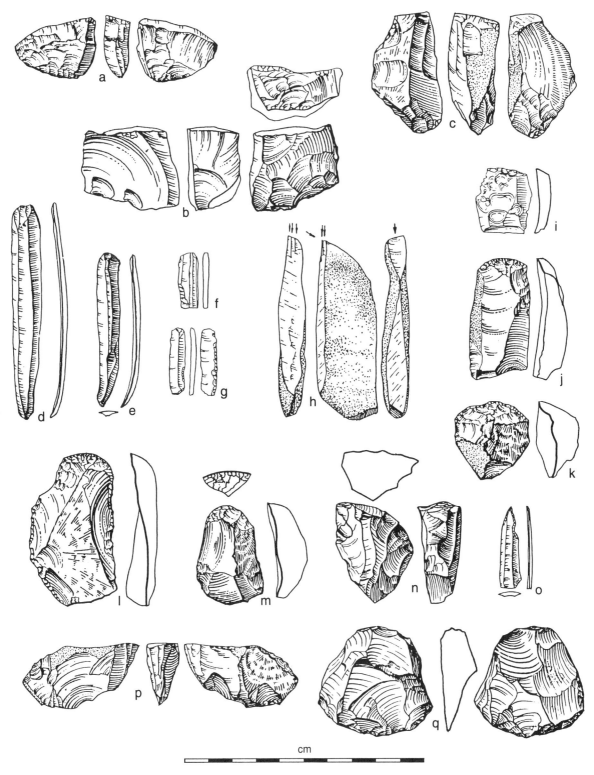

Figure 3–26   Artifacts from Stratum Vb (*a–m, o*) and combined Strata Va/b (*n, p, q*) of the Ust-Timpton 1 site: wedge-shaped microblade cores (*a, b, n, p*); subprismatic blade core (*c*); knife-like blades (*d, e*); blades for insetting (*f, g*); transverse burin (*h*); end scrapers (*i–m*); puncturer or perforator (*o*); discoidal flake core (*q*).

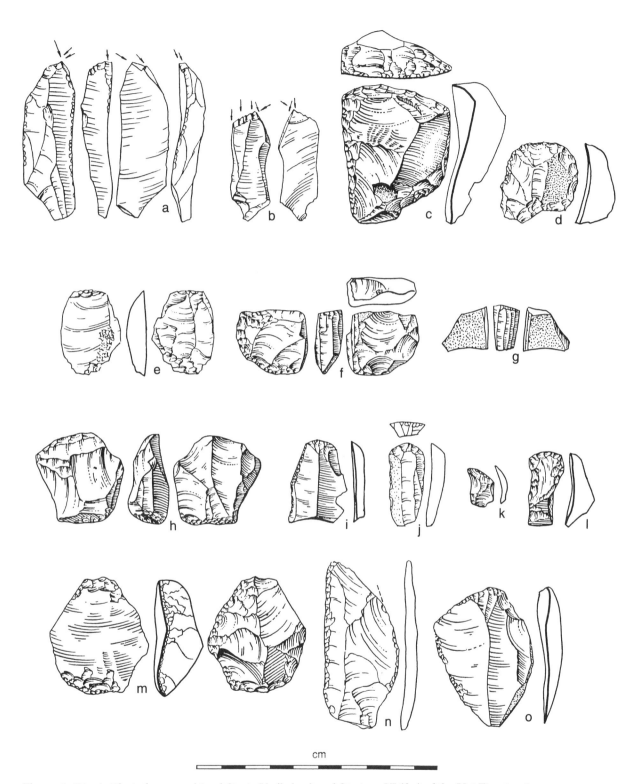

Figure 3–27   Artifacts from combined Strata Va/b (*a–e*) and Stratum VI (*f–o*) of the Ust-Timpton 1 site: transverse burins (*a, b*); scrapers (*c, d, i–m*), *pièce esquillée* or chisel-like tool (*e*); wedge-shaped microblade core (*f*); core fragment (*g*); discoidal core (*h*); knives on spalls (*n, o*).

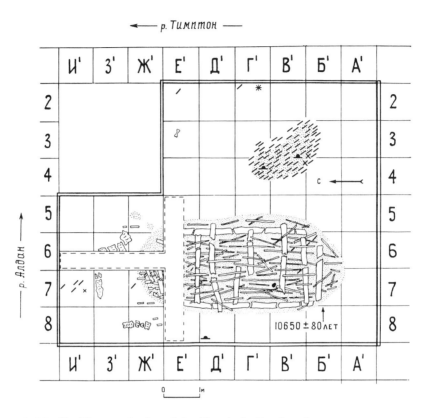

Figure 3–28   Ust-Timpton 1: plan of dwelling including hearth area.

ing, was an oval-shaped hearth area of 2 by 0.9 m (Fig. 3–8). In the hearth fill (1–2 cm in depth) were found 2 end scrapers, 2 chert blades, 5 chert flakes, and 675 flakes of diabase (some pressure removals). Four diabase flakes and 1 chert blade were discovered near the northern end of the dwelling. A chert scraper was found 80 cm west of the dwelling.

Stratum VI Artifacts (36/1031) In Stratum VI cores (5) are represented by 1 wedge-shaped microblade core (Fig. 3-27: *f*), 1 microblade core face fragment of tabular chert (Fig. 3-27: *g*), a prismatic core fragment, and 1 discoidal flake core (Fig. 3–27: *h*).

There are 4 end scrapers (2 on blades and 2 on flakes) (Fig. 3-27: *i–l*), 2 side scrapers (Fig. 3-27: *m*), 6 knives on spalls (Fig. 3-27: *n, o*) and on blades

(Fig. 3–29: *a*), 3 inset blades (Fig. 3–29: *b*), and 1 notched blade (Fig. 3–29: *c*).

There are 2 dihedral burins (Fig. 3-29: *d*) and 1 diagonal burin on a cortex spall (Fig. 3–29: *e*).

The remaining tools include 1 wedge- or chisel-like tool with a (possibly fortuitous) burin spall (Fig. 3–29: *f*) and 1 combination scraper/knife (Fig. 3–29: *g*).

Cultural Strata VII–IX

Stratum VII Artifacts (7/72) The number of finds from Cultural Strata VII–X is quite small; difficult excavating conditions restricted their discovery and definition.

In Stratum VII there are 5 blades (Fig. 3–29: *h*,

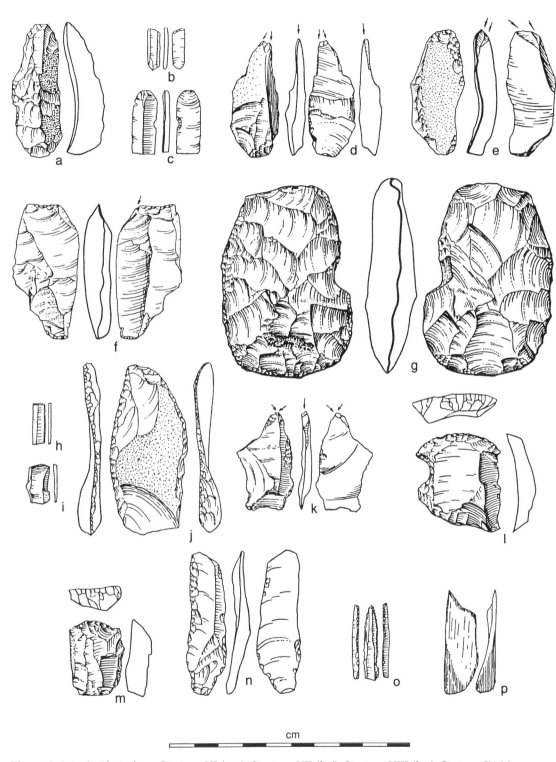

cm

Figure 3–29   Artifacts from Stratum VI (*a–g*), Stratum VII (*h–j*), Stratum VIII (*l–n*), Stratum IX (*o*) and Stratum X (*p*) of the Ust-Timpton 1 site: crested blade (*a*); blade for insetting (*b*); notched blade (*c*); dihedral burins (*d, k*); diagonal burin (*e*); *pièce esquillée* or chisel-like tool (*f*); scraper/knife (*g, j*); blades (*h, i*); end scraper (*l, m*); knife on blade-like flake (*n*); puncturer (*o*); ivory awl (*p*).

i), 1 dihedral burin made on a chert spall (Fig. 3–29: *k*), a combined scraper/knife on a large flake (Fig. 3–29: *j*), and 65 waste flakes.

Stratum VIII Artifacts (3/112) In Stratum VIII there are 2 end scrapers on flakes (Fig. 3–29: *l, m*), 1 knife on a blade-like flake (Fig. 3–29: *n*), and 109 waste flakes.

Stratum IX Artifacts (3/5) and Stratum X Artifacts (1) In Stratum IX there are two blades and 1 graver on a blade (Fig. 3–29: *o*), along with 2 waste flakes. In Stratum X there was found a single fragment of an ivory awl (Fig. 3–29: *p*).

SUMMARY

The archaeological materials from Strata Vb–X are uniquely distinctive from those of the Sumnagin Strata IVa–Va, marked particularly by the presence of wedge-shaped cores with bifacial retouched shaping. Tools on blades constitute 80–95% of the Sumnagin strata assemblage, while most of the tools of the lower levels (Dyuktai) are made on flakes.

A wedge-shaped microblade core from Stratum VI is almost identical to one from the Tumulur site, the materials of which related to the late stage of Dyuktai culture.

All of these observations, along with those of stratigraphy, palynology, and radiometry, allow us to assign the evidences of Strata Vb–X to the final stage of the Dyuktai culture (11,000–10,500 years BP).

The Ust-Timpton 1 site is one of the few sites representing the important transition at the Pleistocene/Holocene boundary. This unique site still needs additional detailed research and excavation, particularly in the lower horizons.

KURUNG 2 (STRATUM, VI)

Yuri A. Mochanov and Svetlana A. Fedoseeva

The Kurung site is located at km 298, on the right bank of the Kurung River, near its mouth. The Kurung is a tributary of the Olekma River (58°58′N, 121°32′E). Discovered in 1975 by A. N. Alekseev and I. E. Zeekov during fieldwork by the Yakut University expedition, the site was investigated by them in 1975–76 and 1978 (excavation area, 124 sq. m). Research was conducted by N. D. Arkhipov in 1977 (20 sq. m excavated) and in 1979–81 by A. N. Alekseev (276 sq. m). The estimated area of the site is 750–800 sq. m.

STRATIGRAPHY

The site is associated with the 14–15 m (second) terrace of the Olekma River. The following stratigraphic units were identified during excavation by A. N. Alekseev and I. E. Zeekov (Fig. 3–30).

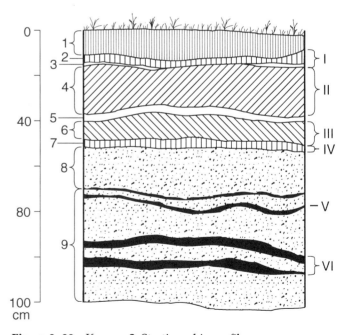

Figure 3–30 Kurung 2. Stratigraphic profile. Stratigraphic units numbered left; cultural strata numbered right. See text for descriptions.

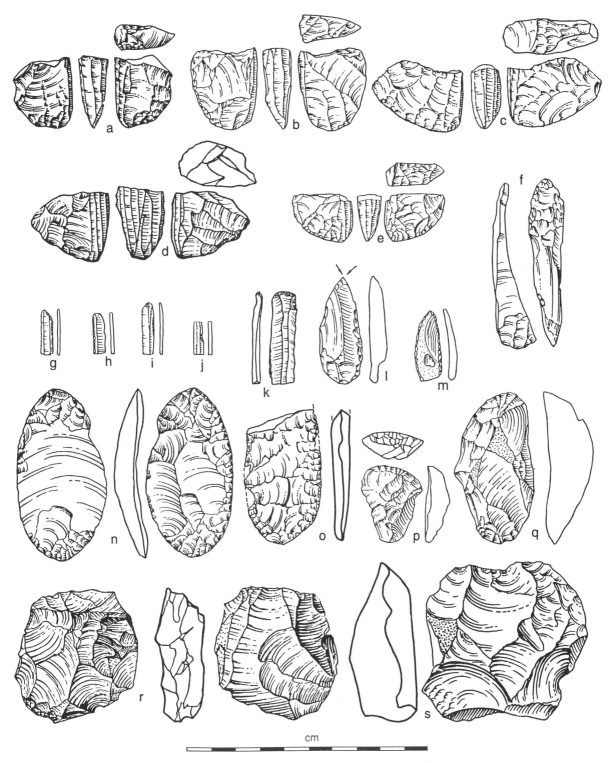

cm

Figure 3–31   Cultural Layer VI, Kurung 2: wedge-shaped microblade cores (*a–e*); microblade core tablet (*f*); microblade insets (*g–i*); notched microblade (*j*); blade (*k*); dihedral burin (*l*); modified flake (*m*); bifacial knives (*n, o*); scrapers (*p, q*); discoidal flake cores (*r, s*).

1. Turf (5–12 cm).

2. Humic sandy loam (2–3 cm thick), containing early Iron Age remains (Cultural Stratum I).

3. An intermittent seam of yellow sand (1–2 cm).

4. Reddish sandy loam with burned areas (15–24 cm). In its upper section is Cultural Stratum II (1–2 cm thick) containing late Neolithic Imyakhtan materials.

5. Fine-grained sand lens (1–3 cm).

6. Yellowish sandy loam (10–15 cm) containing Belkachin Neolithic materials (Cultural Stratum III).

7. Humic sandy loam (2–3 cm) containing early Neolithic Sialakhskaya remains (Cultural Stratum IV).

8. Light yellow, coarse-grained sand (15–20 cm).

9. Poorly sorted sands (35–40 cm) interbedded with four sandy loam stringers, each 4–5 cm thick and heavily impregnated with detritus. Sumnagin late Palaeolithic/Holocene materials were embedded within the second detrital seam (Cultural Stratum V). Dyuktai Upper Palaeolithic materials occurred in the third/fourth detrital seams of sandy loam (Cultural Stratum VI). In some areas these latter streaks merged into one.

Frost deformation of all strata including and below unit 6 (30–40 cm below surface, Cultural Stratum III) may have permitted mixing of some cultural materials (see Fig. 3–30).

ARTIFACTS

Cultural Stratum VI yielded 1,242 stone artifacts. The assemblage contains 9 microblade cores: 5 wedge-shaped (Fig. 3–31: *a–e*), 3 discoidal (Fig. 3–31: *r*, *s*), and 1 prismatic. Tools include 4 bifacial round based knives (Fig. 3–31: *n*, *o*), 1 dihedral burin on a blade-like flake (Fig. 3–31: *l*), 3 microblade insets (Fig. 3–31: *g–i*), and 5 scrapers on flakes (Fig. 3–31: *p*, *q*). There are also 273 blades and microblades and 947 waste flakes.

SUMMARY

A. N. Alekseev, who studied this stratified site over a number of years, suggests that the Cultural Stratum VI materials are related to the Upper Pleistocene strata of Dyuktai Cave (the final stage of the Dyuktai culture) dating 13,000–11,000 years ago.

LETEN NOVYY 1 (STRATUM IV)

Yuri A. Mochanov and Svetlana A. Fedoseeva

The Leten Novyy 1 site is located at km 281 on the right cape of Leten Novyy River, near its mouth. The river is a tributary to the Olekma River (58°58′N, 121°35′E). Discovered in 1970 by N. D. Arkhipov of Yakut University, the site was researched by him in 1970–72 (an excavation of 84 sq. m) and by A. N. Alekseev in 1974–75 (63 sq. m).

Arkhipov used only arbitrary excavation levels in his site excavations, rather than referring to geological horizons. As a result, his "cultural horizons" are a blend of materials from two and sometimes even three different epochs. The materials thus excavated must be treated as if they were from mixed strata, being sorted by typological comparison with other materials which were excavated according to geological strata.

STRATIGRAPHY

When excavations were resumed in 1975, Alekseev, using stratigraphic controls, plotted archaeological materials precisely within geological horizons. He identified the following stratigraphic units at the site (Fig. 3–32), which he relates to the 14-m (second) terrace:

1. Turf (4–10 cm).

2. Humic sandy loam (3–6 cm). This Cultural Stratum I contains Imyakhtan late Neolithic materials.

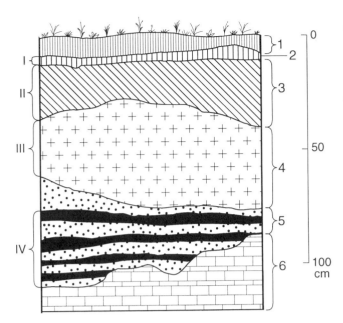

Figure 3–32 Leten Novyy 1. Stratigraphic profile. Cultural strata numbered left; stratigraphic units numbered right. See text for strata descriptions.

g). Other tools include 6 end scrapers on blades and flakes (Fig. 3–34: *a, b*), 5 lateral burins on blades (Fig. 3–34: *d–f*), 1 transverse burin on a flake (Fig. 3–34: *c*), 4 bifacial round based knives (Fig. 3–34: *g, j, l, m*), and 4 bifacial dart/spearpoints (Fig. 3–34: *h, k*).

Summary

Materials of Cultural Stratum IV are techno-typologically close to those from the Upper Pleistocene strata of Dyuktai Cave (Stratum VIIa), as well as the sites of Tumulur and Berelekh. They are thus related to the final stage of the Dyuktai culture and have an age of 13,000–10,500 years.

(Figures 3–33 and 3–34 follow on pp. 210-211.)

3. Grey sandy loam (15–30 cm); Cultural Stratum II. Contains Belkachin Neolithic remains.

4. Reddish brown alluvial sandy loam (25–50 cm). Contains Sumnagin Holocene (terminal) Palaeolithic materials (Cultural Stratum III).

5. A set of coarse-grained alluvial sands interbedded with humic sandy loam streaks (40–50 cm). Upper Palaeolithic Dyuktai materials associated with the three uppermost sandy loam streaks (each, 2–5 cm) (Cultural Stratum IV).

6. Cambrian limestone bedrock terrace base.

Artifacts

Cultural Stratum IV yielded 1,563 stone artifacts, as well as split bones of horse (*Equus caballus*). The assemblage consists of 21 blade cores of which 15 are wedge-shaped microblade cores (Fig. 3–33: *a–j*), 4 are subdiscoidal (Fig. 3–33: *k*), and 2 are subprismatic single platform cores. Also included are 197 retouched blades and 1,317 waste flakes. Of the wedge-shaped cores 7 are Gobi "stretched" cores (Fig. 3–33: *a–d*) and 9 are shortened cores (Fig. 3–33:

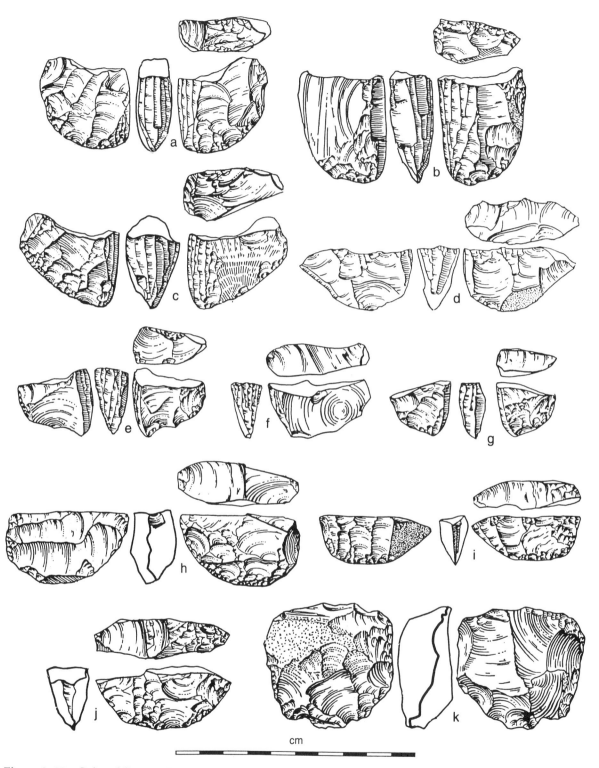

Figure 3–33 Cultural Stratum IV, Leten Novyy 1: wedge-shaped microblade cores (*a–j*); subdiscoidal flake core (*k*).

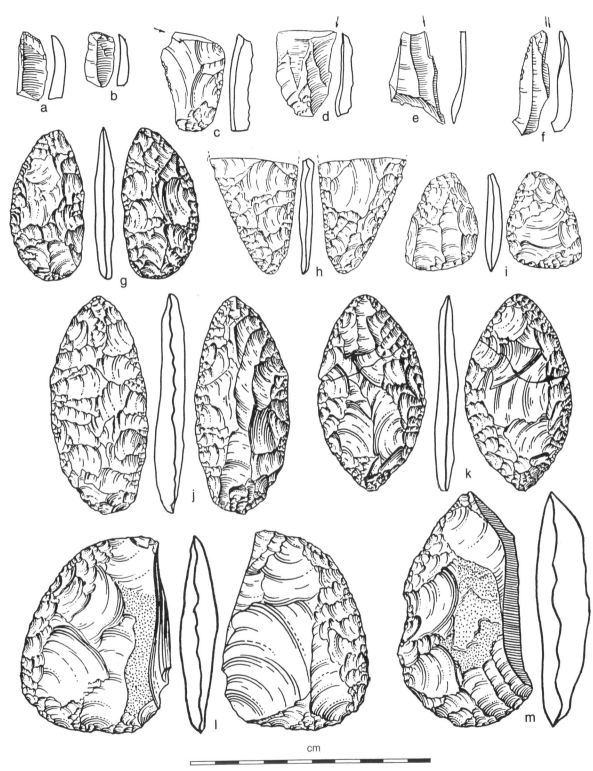

cm

Figure 3–34 Cultural Stratum IV, Leten Novyy 1: end scrapers on blades (*a, b*); burins (*c–f*); bifacial knives (*g, j, l, m*); dart/spear points (*h, k*).

THE DYUKTAI SITE AT KM 27 OF THE YAKUTSK–POKROVSK HIGHWAY

Yuri A. Mochanov and Svetlana A. Fedoseeva

The site is located on the left bank of the Lena River at km 27 of the Yakutsk–Pokrovsk highway. This road connects the city of Yakutsk and a regional center called Pokrovsk. The site lies on the border of the Yakutsk and Khangalazskiy regions of the Sakha Republic (61°52′N, 129°31′E). The site, in a highway borrow pit, was discovered by Y. A. Mochanov in 1985. The main part of the site was de-stroyed by bulldozers, therefore no proper excavations could be carried out. Instead, some displaced materials were collected and the eroded surfaces were cleaned and examined over the period 1985–1992.

This site is associated with the 100-m Tabagin terrace of the Lena River; it has been exposed by the slumping of its matrix. The site originally occupied one of the hills of which a small relic has been preserved. Here a horizon of brownish yellow buried soil 1–2 cm thick, within a layer of bright yellow sand, was traced out on the wall of one of the trenches at a depth of 20–30 cm below surface. This contained the cultural stratum.

ARTIFACT ASSEMBLAGE

The site assemblage consists of 142 stone artifacts; more than half are made of chert and the remainder of quartzite. Ancient high terrace exposures provided the material for manufacturing stone artifacts; there was no nearby riverbank to provide such pebbles and boulders.

There are 2 cores. One is a chert wedge-shaped "Gobi" microblade core with fine pressure retouch on the wide surfaces (Fig. 3–35: *a*). The platform was formed by removal from the core face. The 1 quartzite pebble core has a prepared striking platform, with a few flake removals. There are 4 biface

blanks whose wide surfaces are retouched but are rather amorphous. Of the remaining 8 tools there is 1 bifacial spear or dartpoint (Fig. 3–35: *e*), 1 bifacial knife and 1 edge fragment (Fig. 3–35: *c, d*), a large circular biface (Fig. 3–35: *b*), 2 lateral burins on small flakes (Fig. 3–35: *f, g*), 1 chisel-like tool on a flake, 1 flake knife with edge retouch on the working edge (Fig. 3–35: *h*). The remaining artifacts consist of 1 knife-like blade, 1 quartzite boulder, 122 waste flakes, and a number of fine, semi-transparent quartzite pebbles.

SUMMARY

The absence of fauna, stratigraphic indicators, and organic remains necessary for radiocarbon dating makes age determination difficult. However, the nature of the assemblage permits fairly accurate assessment of its cultural and chronological affiliations. Based on lithic techno-typological indicators (wedge-shaped and pebble cores associated with bifacial round base knives and projectile points), the site complex is clearly related to the Dyuktai culture. The presence of spear/dart points (analogous to those from Dyuktai Cave Stratum VII) allows us to indirectly date this site at 13,000–12,000 years.

(Figure 3–36 follows on p. 214.)

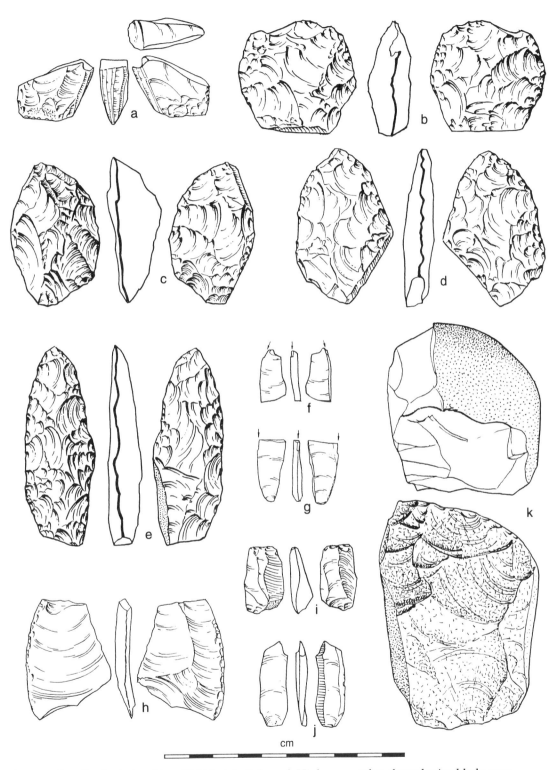

Figure 3–35 Dyuktai site at km 27, Yakutsk–Pokrovsk Highway: wedge-shaped microblade core (*a*); biface blanks (*b–d*); bifacial spear or dart point (*e*); burins (*f, g*); flake knife (*h*); chisel-like tool (*i*); knife-like blade (*j*); boulder chopper (*k*).

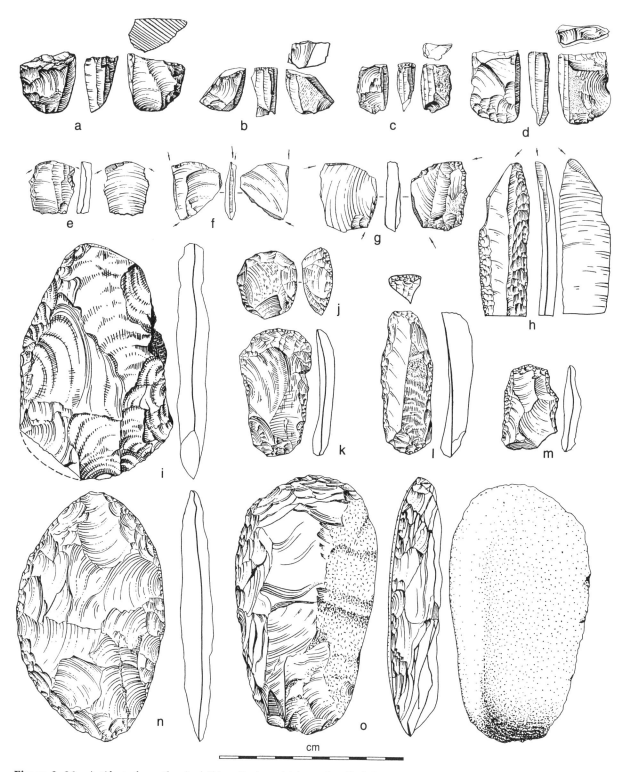

Figure 3–36 Artifacts from the Avdeikha site for which no detailed description exists: wedge-shaped microblade cores (*a–d*); burins (*e–h*); bifaces (*i, n*); scrapers (*j–m*); large uniface on split pebble (*o*).

Chapter 4

PRIOKHOTYE, KOLYMA RIVER BASIN, AND KAMCHATKA

INTRODUCTION

The area under consideration consists of two geographic regions. The first, which includes the northern coast of the Sea of Okhotsk (Priokhotye) and the basins of the Kolyma and Indigirka rivers, is the northeastern part of Siberia. The second, the Kamchatka Peninsula, belongs to the northern part of the Russian Far East.

The general relief of the Kolyma and Indigirka river basins comprises a complex system of mountain ridges and intermontane plateaus and depressions. There are two main mountain chains, the Cherskiy Range (with elevations up to 3,147 m above sea level) and the Gydan Range (up to 1,960 m). Between the ridges are located the Alazeya and Yukagir upland plateaus (both up to 1,000–1,500 m). Far to the north the Indigirka and Kolyma lowlands open onto the Arctic Ocean coast. The major rivers are the Kolyma (together with the Omolon), the Indigirka, the Yana, and the Alazeya (Arctic drainage); the Taui, the Yama, and the Gizhiga (Pacific drainage).

On Kamchatka, the Srednie Range (elevations up to 3,620 m) runs from the north down to the center of the peninsula. To the west is a broad, rather poorly drained coastal plain. To the east, beyond the Kamchatka River valley, lies a highly distinctive volcanic zone consisting of lava plateaus (up to 500–1,000 m) with dozens of volcanic cones (up to 4,750 m). The biggest river is the Kamchatka; other important water courses are the Tigil, the Icha, and the Bystraya.

CLIMATE AND PERMAFROST

There are three types of climate in the area. The Kolyma and Indigirka lowlands have an arctic climate which is distinguished more by cool summers than by the intensity of the winter cold. The whole area is underlain by thick (up to 500 m) permafrost.

The mountainous area, covering the main part of the Kolyma and Indigirka drainages and the Priokhotye, is characterized by a subarctic climate with extremely low winter temperatures coupled with short and relatively warm summers. The area is underlain by a continuous layer of permafrost (up to 200–300 m thick), with some scattered thawing areas in the Priokhotye.

Kamchatka has a Far Eastern monsoonal climate, with unpleasant moist and misty summers. The central portion of the peninsula, the Kamchatka River valley, has a warmer and drier climate, while the eastern shore receives much more precipitation,

with heavy snow in winter. In terms of permafrost, there are only some noncontiguous areas on the western coast.

Flora and Fauna

The vegetation of northeastern Siberia, in general, corresponds to the climate types. Tundra covers the Arctic coast and northern Kamchatka, down to 60°N latitude. The main type of vegetation is that of

bogs with abundant grasses, mosses, and dwarf birch (*Betula exilis*). To the south on the coast, the wooded-tundra vegetation is characterized by rare Dahurian larch (*Larix dahurica*) and dwarf birch. The typical animals are lemmings, reindeer (*Rangifer tarandus*), arctic fox, and birds (geese, ducks).

The mountains in the Kolyma and Indigirka drainages and in Priokhotye are covered by sparsely wooded larch taiga, with tundra on the higher elevations. The timberline lies at about 600–1,000 m above sea level. The forest consists mostly

In the Kolymskiy Mountains, Priokhotye. (*Photo courtesy of Paul A. Colinvaux.*)

of Dahurian larch and birches, with balsam poplar (*Populus suaveolens*) and Korean willow (*Chosenia macrolepis*) in the river valleys. Closer to timberline, the forests take on a wooded-tundra character of dwarf larch and dwarf stone pine (*Pinus pumila*). As for animals, bear, fox, weasel, and ermine are the most typical carnivores. Hoofed animals are represented mostly by elk (*Alces americanis phizenmayeri*) in the forests, and by snow sheep (*Ovis nivicola lydekkeri*) in the high mountains.

In river valleys on Kamchatka, the most common vegetation type is birch (*Betula japonica*) forests. They are park-like, while spruce and larch also occur on low ground. In the Kamchatka River valley, tall grass prevails. The forests on the mountainous east coast degenerate into dense growths of small, misshapen trees (mostly rock birch), and on the slopes they devolve into dwarf stone pine communities. The western coast is covered by hillocky peatbogs with dwarf rock birch and shrub alder.

Among the animals, bear (*Ursus piscator*), reindeer, snow sheep, wolverine, and weasel are most typical. The abundant fish resources of the Sea of Okhotsk and Bering Sea basins consist mainly of several kinds of salmon (Siberian salmon, humpback salmon, and silver salmon).

Y. V. K.

BIBLIOGRAPHY

Mellor, R. E. H. 1966. *Geography of the USSR.* London: Macmillan.

Shabad, T. 1961. *Geography of the USSR: A Regional Survey.* New York: Columbia University Press.

Suslov, S. P. 1961. *Physical Geography of Asiatic Russia.* San Francisco: W. H. Freeman.

Symons, L., J. C. Dewdney, D. J. M. Hooson, R. E. H. Mellor, and W. W. Newly. 1990. *The Soviet Union: A Systematic Geography.* New York: Routledge.

BERELEKH, ALLAKHOVSK REGION

Yuri A. Mochanov and Svetlana A. Fedoseeva

This site is located on the left bank of the Berelekh River, a tributary of the Indigirka, in the Allakhovsk region of the Sakha Republic (Yakutia) (70°24′N, 143°57′E). The site—regarded as a "mammoth cemetery"—had been well known to inhabitants since ancient times. In 1947, a permafrost specialist, N. F. Grigoryev, visited this site and made a small collection of fauna. Publication of his finds in 1957 introduced the site location into the scientific literature. In 1970, B. S. Rusanov of the Yakutian Branch of the Siberian Division, Academy of Sciences of USSR, directed a geological expedition to conduct research there. Professor N. K. Vereshchagin, a palaeontologist from Leningrad who joined the expedition, discovered the archaeological component at the site when using a hydropump to expose a bone-rich horizon of the "cemetery." Having had extensive experience with Palaeolithic sites, he searched for traces of human occupation in this horizon and discovered four man-made artifacts which he presented to Y. A. Mochanov in the spring of 1971. Mochanov saw in them ties to the Dyuktai culture (Vereshchagin and Mochanov 1972).

The discovery of a Dyuktai site north of the Arctic Circle led PAE to investigate the site from 1971 through 1973 and again in 1981. Use of the hydropump during the palaeontological work had resulted in extensive destruction of the site. Archaeological excavations were thus conducted on the areas that had survived erosion; sandpit exposures were brushed off and searched for any preserved cultural strata.

STRATIGRAPHY

The Berelekh site is associated with the 12-m terrace of the Berelekh River. The following stratigraphic sections have been identified (Fig. 4–1):

1. Turf (3–5 cm).
2. Yellowish brown, cemented, icy sandy loam with vertical undulate veins of blue-grey sandy loam (45–50 cm thick). Ice wedges extend from the base of this sandy loam to a depth of at least 5 meters. This unit, capping those underlying, was strongly affected by soil-forming and frost-related processes.

3. This 4.0–4.5-m unit consists of interbedded lenses of greyish brown sandy loam (0.1–2 cm) and poorly sorted grey sands (0.1–0.5 cm). It represents a floodplain facies of alluvium. Almost all sandy loam lenses contain plant remains. One lens, 250 cm below surface, consisted almost entirely of willow branches and small fragments of thin larch stems. In it were found stone tools as well as split and partially worked bones of mammoth and other animals. Several worked stone tools and bones were found in the underlying sandy loam lens at 252–253 cm. Since these two lenses merge in some units, the

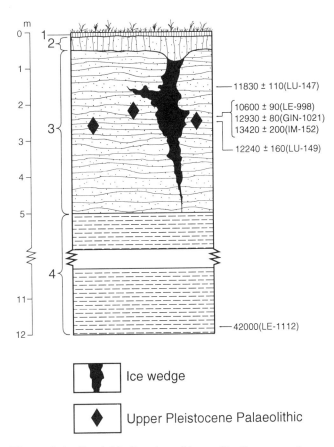

11830 ± 110(LU-147)

10600 ± 90(LE-998)
12930 ± 80(GIN-1021)
13420 ± 200(IM-152)

12240 ± 160(LU-149)

42000(LE-1112)

Ice wedge

Upper Pleistocene Palaeolithic

Figure 4–1 Berelekh. Stratigraphic profile. Strata numbers at left. See text for descriptions.

The Berelekh mammoth cemetery.

finds from both were attributed to the same cultural stratum.

4. An approximately 7.5-m unit of interbedded lenses of yellowish, poorly sorted sands (0.5–5.0 cm) and greyish blue sandy loam (1–3 cm). A radiocarbon date of 42,000 BP (LE-1112) was obtained on vegetative matter collected at a point 20 cm above the level of the Berelekh River. That dating, in conjunction with geological and geomorphological observations, suggests that this unit is composed of eroded alluvial deposits of a 20-m terrace (either T 3 or T 4) which incorporates deposits of the later 12-m terrace (T 2?).

The Cultural Stratum

In the process of cleaning the profile for examination, a number of artifacts without exact provenience were collected. These included 127 of stone, 49 of bone and ivory, and 1,003 fragments of animal bone. These latter may represent midden refuse. N. K. Vereshchagin examined the animal bones and identified 78 mammoth bones (*Mammuthus primigenius* Blum.), 3 bison or horse, 1 reindeer (*Rangifer tarandus* L.), 827 hare (*Lepus timidus* L.), 92 partridge, and 2 fish.

The two radiocarbon dates obtained from the cultural stratum, 12,930 ± 80 years (GIN-1021) and 13,420 ± 200 years (IM-152), agree well with palaeontological dates obtained from wood 160 cm below surface (11,830 ± 110 years [LU-147]) and from mammoth tusk found 255 cm below surface (12,240 ± 160 years [LU-149]).

Stone artifacts from the stratum include 1 flake core, 4 knife-like blades, 7 retouched blades and blade-like flakes (Fig. 4–2: *b, c*), 2 bifacial knives or spearpoint fragments (Fig. 4–2: *a, d*), 1 chisel-like tool on a chert flake, 4 pendants (Fig. 4–2: *f-i*), and 108 waste flakes. A fifth pendant, this of bright

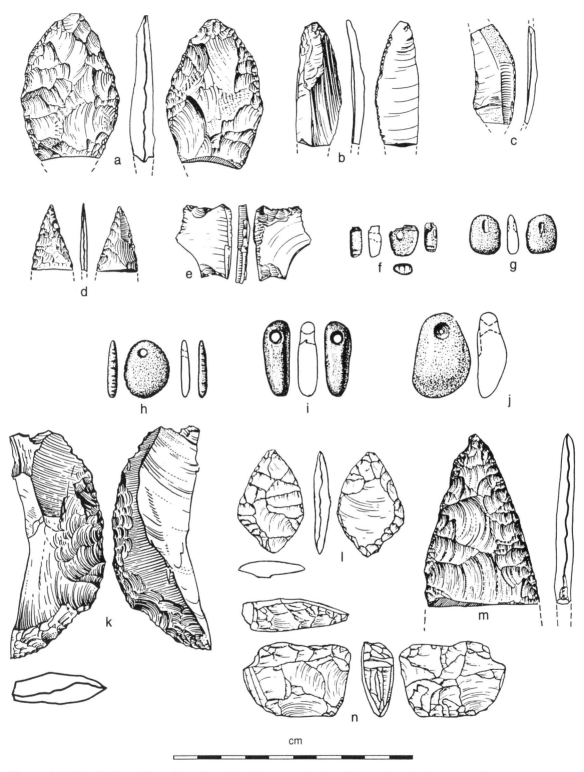

Figure 4–2 Berelekh: artifacts from the cultural stratum (*a–j*) and from disturbed contexts (*k–n*): bifaces (*a, d, k–m*); modified blade-like flakes (*b, c*); burin (*e*); pendants (*f–i*); wedge-shaped microblade core (*n*).

green jadeite, was found on the terrace slope. The 5 stone pendants are oval or rounded pebbles of various materials, each biconically pierced at one end. One specimen of white calcite bears 19 short parallel notches around the lateral edge. Another, of pyrophillite, shows 5 similar notches.

Bone and ivory artifacts include mammoth tusk and bone tools, blanks, and by-products. Among these are 4 mammoth tusk knives (oval, subtriangular, semilunar), one leaf-shaped spearpoint of mammoth tusk, and two scrapers. The tools of bone and tusk are prepared with the same fine pressure retouch seen on flint tools. Certain mammoth rib fragments have been smoothed, the traces suggesting possible use as polishers. Some of these objects are shown in Figure 4–3. A mammoth tusk spearpoint, with a broken tip, was discovered by N. K. Vereshchagin in the eroded terrace exposure. Some of these tools are engraved and incised with crossed lines (Fig. 4–3: *a*).

In 1965, local collectors gave to V. E. Flint, the biologist, a unique specimen from Berelekh—a tusk fragment bearing the engraved image of a mammoth (Fig. 4–3: *i*). This, the first such find in northeastern Asia, was published by Flint (1972), V. I. Gromov (1972), and O. N. Bader (1972, 1975).

Over the years members of PAE gathered 178 burned and split mammoth bones and 62 stone artifacts from the terrace slope and riverbank. The stone assemblage includes 1 black chert, unifacially worked, single platform, flattened core; 1 "Gobi" wedge-shaped microblade core (Fig. 4–2: *n*); 1 bifacial knife fragment (found by N. K. Vereshchagin); 1 complete bifacial knife or darthead; 3 biface fragments; 1 chisel-like tool; 1 pendant; and 54 waste flakes.

The Geological Institute of the Yakutia branch of the SO Academy of Sciences USSR expedition extruded 7,614 bone specimens from the soil matrix, using a hydropump in the bone-rich area. N. K. Vereshchagin identified mammoth (98.6% of the sample), woolly rhinoceros, bison, horse, reindeer, cave lion, wolf, glutton, and hare. The recovery of an entire hind leg of a mammoth, with flesh and wool, is now exhibited at the Geological Museum of the Yakutia Scientific Center. B. S. Rusanov believes it had been separated from the carcass in ancient times.

SUMMARY

The Berelekh site archaeological materials currently represent the northernmost occurrence of the Upper Palaeolithic in the world, representing the final stage of the Dyuktai culture (13,000–12,500 BP). It is possible that the numerous mammoth bones at the Berelekh site are the remains of prehistoric dwellings, rather than an accidental accumulation. They have been much disturbed by frost action, solifluction processes, and the palaeontologists' hydropump.

REFERENCES CITED

Bader, O. N. 1972. An Artist Was Drawing from Life. *Priroda* 8:95–96. (In Russian)

———. 1975. Palaeolithical Engraving from Indigir Area beyond the Polar Circle. In *Archaeology of Northern and Central Asia*, pp. 30–33. Novosibirsk. (In Russian)

Flint, V. E. 1972. Unique Find in Berelekh: Antiquity or Contemporaneity? *Priroda* 8:94. (In Russian)

Gromov, V. I. 1972. Drawing from Life: But Not Alive. *Priroda* 9:126. (In Russian)

zn2Vereshchagin, N. K., and Y. A. Mochanov. 1972. The Northernmost Traces of the Upper Palaeolithic in the World (the Berelekh Site in the Lower Reaches of the Indigirka River). *Soviet Archaeology* 3:332–36. (In Russian)

(Figure 4–3 follows on p. 222.)

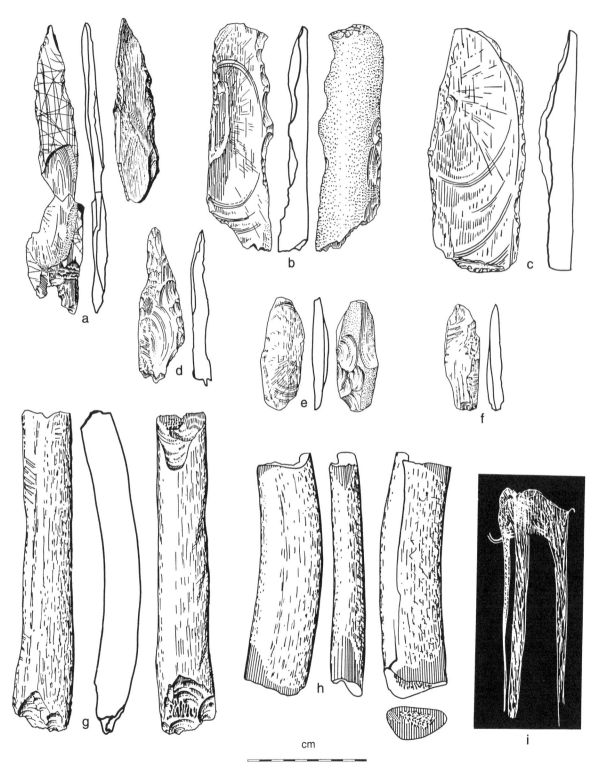

cm

Figure 4–3   Berelekh: ivory (*a–f, i*) and bone (*g, h*) artifacts. All are at the same scale except for the engraved mammoth.

MAYORYCH, YAGODINSKIY REGION, MAGADAN DISTRICT

Yuri A. Mochanov and Svetlana A. Fedoseeva

The site is located on the left bank of the stream Mayorych, near its mouth. Mayorych is an upper tributary of the Kolyma River, entering somewhat north of where the Kolyma cuts through the Cherskiy mountain range (62°31′N, 150°47′E). Mayorych was discovered in 1970 during an organized survey by PAE, under the direction of Y. A. Mochanov.

STRATIGRAPHY

The site is associated with the 14-m bedrock terrace of the Kolyma River. The following stratigraphic sections have been identified in the exposed sediments of the terrace's outer edge.

1. Turf (3–7 cm).
2. Beige-brown alluvial loam (25–40 cm).
3. Pebbles cemented by coarse-grained grey sand. This unit represents the riverbed facies of alluvium (7.5–8.5 m).
4. Terrace base composed of tufaceous mudstone at least 6–6.5 m thick.

CULTURAL STRATIGRAPHY

In numerous areas along the outer terrace edge facing the Kolyma River, turf and loam are loose, with shingle exposed on the surface. One such area, 12 sq m., yielded 3 chert flakes, 1 flattened wedge-shaped znchert microblade core (Fig. 4–4: *a*), and 2 tools. One of these is an end scraper/knife on a thick, faceted blade (Fig. 4–4: *c*); the second is a chert flake knife/chisel (Fig. 4–4: *b*).

SUMMARY

Even in the absence of clear stratigraphy, faunal remains, or radiocarbon dates, these artifacts can be attributed to the Dyuktai complex, where there are analogs in form and manufacturing technique (Ikhine 1, Verkhne-Troitskaya, Dyuktai Cave). Considering the wide distribution of wedge-shaped blade cores with bifacially formed bodies at the beginning of the Sartan Age (23/22,000–18,000 years ago) and their existence in the Dyuktai Cave deposits dated 13,000–12,000 years ago, the age of the Mayorych site must lie somewhere within the range of 23/22,000–12,000 years. (The Dyuktai culture ceased to exist about 11,000 years ago in northeastern Asia.) This estimate is supported by the fact that the 14-m Kolyma terrace alluvium could not have been deposited much earlier than the beginning of the Sartan.

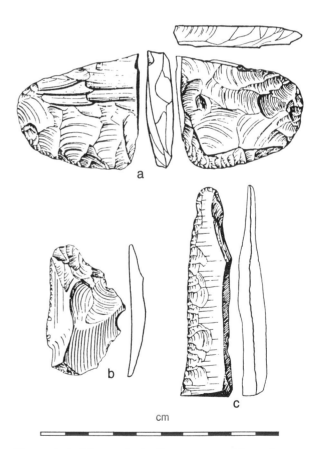

cm

Figure 4–4 Mayorych. Artifacts recovered from the eroded surface of the site: wedge-shaped microblade core (*a*); chert flake knife (*b*); knife on thick blade (*c*).

KUKHTUY 3

Yuri A. Mochanov and Svetlana A. Fedoseeva

The Kukhtuy site is located on the left bank of the Kukhtuy River about 1.5 km from the shore of the Sea of Okhotsk in the Okhotskiy region of Khabarovsk Territory, Russian Federation (59°22′N, 143°57′E). Y. A. Mochanov discovered the site in 1970 during a survey by PAE. Research on it was carried out in 1970–71 and 1981.

The site is associated with deposits of the 25-m terrace of the Kukhtuy River.

STRATIGRAPHY

The following stratigraphic units were identified (Fig. 4–5 *a*).

1. Turf: 3–7 cm thick.
2. Sandy loam, reddish brown, 30–35 cm. Early Iron Age materials were recovered from the top section of this loam, immediately beneath the turf; Neolithic remains were scattered throughout the remainder of the unit.
3. Stringers of yellowish brown sandy loam (0.5–3.0 cm) interbedded with poorly sorted yellowish brown sand (0.3–1.0 cm). The overall unit thickness ranges between 65 and 110 cm. Isolated Neolithic artifacts were found within 7–10 cm of the upper contact of this unit. A charcoal sample yielded a radiocarbon date of 4700 ± 100 years (LE-995). Lower sections of this unit contained Dyuktai-like stone artifacts.
4. A 6-to-7-m culturally sterile unit consisting of shingle and gravel cemented by greyish yellow coarse-grained sand.
5. Granitic bedrock at least 17–18 m.

POLLEN ANALYSIS

Eleven pollen-spore samples from various strata at Kukhtuy 3 provide the basis for a palynological characterization of the site deposits (Fig. 4–5*b*). Three phases of vegetational development were identified. In unit 4 the sample represents an open landscape, occupied by meadow-steppe associations. The lower and middle sections of the 25-m terrace alluvium contained spore plants for the most part. There were isolated seeds of arboreal/shrubby and herbaceous plants. The assemblage is indicative of cold, severe climatic conditions. The presence of Dyuktai artifacts in the lower part of the unit suggests that these deposits were forming at the end of the Pleistocene. The samples from the upper alluvial section and the covering sandy loam represent the beginning development of modern arboreal vegetation.

CULTURAL STRATUM

The Dyuktai cultural horizon yielded 30 stone artifacts—1 discoidal pebble core (Fig. 4–6: *a*), 10 implements, and 19 flakes of chert, hornfels, and siliceous limestone. Among the tools are 1 bifacial dartpoint (Fig. 4–6: *b*), 1 dartpoint tip fragment (Fig. 4–6: *c*), 2 bifacial knives (Fig. 4–6: *d, e*), 1 bifacial preform for a knife or spearpoint (Fig. 4–6: *f*), 1 primary blank for a bifacial knife (Fig. 4–6: *g*), 1 chisel-like tool (Fig. 4–6: *i*), 1 thick flake scraper (Fig. 4–6: *j*), and 2 flake knives (Fig. 4–6: *k, l*). There were two additional finds made while clearing the terrace slope. These may be attributed to the Dyuktai complex based on preparation techniques—spearpoint or knife fragments made on massive three-faceted blades (Fig. 4–6: *h*). A similar find was made at the stratified Dyuktai complex site of Tumulur.

SUMMARY

Although the assemblage here is closest to the Dyuktai complex in terms of techno-typological characteristics, stratigraphy, and vegetational associations, no wedge-shaped cores have been found at Kukhtuy. That absence may prove to be an indicator of the latest, or final, stage of Dyuktai tradition development, since wedge-shaped cores are absent in Holocene sites. Thus, there may be two Dyuktai subtraditions: the earlier one with wedge-shaped cores, the later lacking them.

Material from the lower cultural strata (Vb–VI)

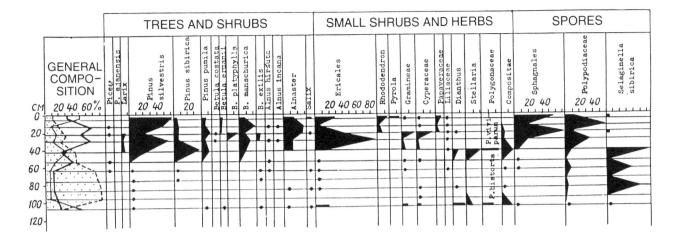

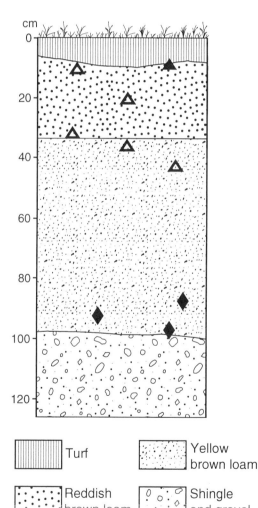

Figure 4–5 Kukhtuy 3: Stratigraphic profile and pollen profiles.

of the stratified Ust-Timpton site argues against such an interpretation, given the presence of wedge-shaped cores, an occasional bifacial knife, core-like subdiscoidal artifacts, and transverse burins among other artifacts in Strata Vb–VI. These strata are 11,000–10,500 years old and thus are presumed to relate to the final stage of the Dyuktai culture.

The Kukhtuy site, as one of the easternmost Dyuktai sites in continental northeastern Asia, is important not only for the study of the initial human settlement in Priokhotye and the western Pacific shore, but for indicating the paths and timing of human entry into America.

Figure 4–7 shows artifacts from the Kondon site, for which no detailed description exists.

(Figures 4–6 and 4–7 follow on pp. 226–227.)

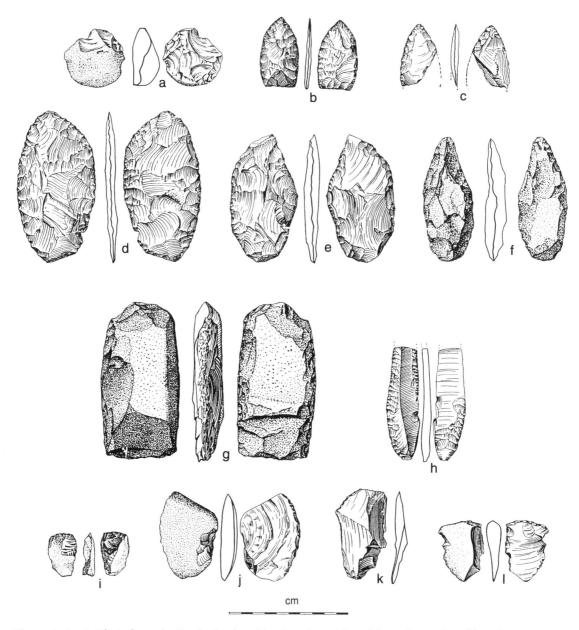

Figure 4–6   Artifacts from the Dyuktai cultural horizon (*a–g, i–l*) and from the surface (*h*) at the Kukhtuy 3 site: discoidal pebble core (*a*); bifacial dart point and tip (*b, c*); bifacial knives (*d, e*); biface blanks (*f, g*); spear point or knife on blade (*h*); chisel-like tool (*i*); scraper (*j*); flake knives (*k, l*).

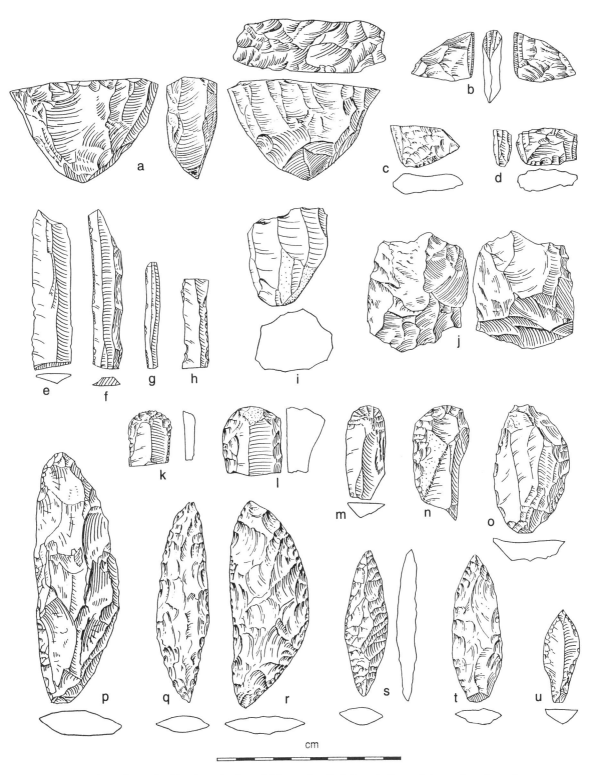

Figure 4–7   Artifacts from the Kondon site for which no detailed description exists: wedge-shaped cores (*a–d*); blades (*e–h*); blade core (*i*); discoidal flake core (*j*); scrapers (*k–o*); bifaces (*p–u*).

BOLSHOI ELGAKHCHAN 1 AND 2, OMOLON RIVER BASIN, MAGADAN DISTRICT

Margarita A. Kiryak

The sites Bolshoi Elgakhchan 1 and 2 are located at the mouth of the Bolshoi Elgakhchan River, a tributary to the Omolon River (see regional map, Priokhotye). Bolshoi Elgakhchan 1 is located on a cape 35.9 m above the present water level of the Omolon River (Fig. 4–8). Bolshoi Elgakhchan 2 lies 450 m to the east. The cape has been interpreted geomorphologically as a socle (erosive-denudational relict) of the river terrace. The summit of the socle measures 10 × 10 m and has a rounded, gently undulating contour. The ground surface is paved with large fragments of disintegrated weathered rock.

The archaeological group which visited this place in 1979, as a part of the Prilensk Archaeological Expedition (Yakutsk City), found cultural remains of various periods, dating from the late Neolithic to Mesolithic (Kistenev 1980:75).

In 1980 the West-Chukotka detachment of the Northeast Asian Complex archaeological expedition (SVKNII DVNC Academy of Sciences–USSR) inspected the whole area of the Bolshoi Elgakhchan Cape and laid out two trenches—one on the upper relict terrace and the other on a lower section 15 km to the northeast. The second trench revealed signs of human presence dating from the Upper Palaeolithic.

The surveyed terrain of the site has uneven turf. In the southern part the underlying socle sediments are exposed in almost all places. It is here that cultural remains of various time periods were discovered. The underlying sediments thicken to the north, following the surface slope. The first excavation here revealed cultural remains which appeared to date to the Upper Palaeolithic. To this time, 115 square m have been excavated; this report concentrates on the nature of the excavation and the archaeological material obtained in 1991. Up to that time, 97 square m had been excavated. In the 1992 season, the upper cultural layer, related to the be-

ginning of the late Palaeolithic, was discovered in clear stratigraphic context along the northern border of this site. This is a complex consisting of 265 artifacts of the Imyakhtan culture. Thus, Bolshoi Elgakhchan 1 may be considered at least a double-component site.

BOLSHOI ELGAKHCHAN 1

Geomorphology and Stratigraphy

Geomorphological research, conducted by Titov in 1991, has allowed us to conclude that the Bolshoi Elgakhchan Cape, in the third and final phase of Sartan ice of the Pleistocene (Q^4III), already constituted a normal alluvial river terrace (relative altitude 32–36 m) with loose alluvial cover no less than

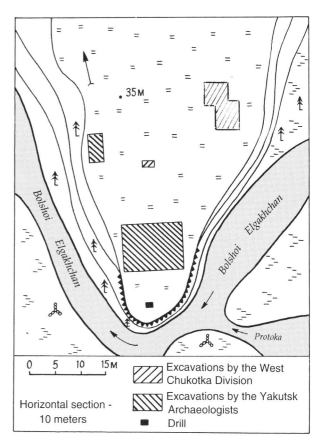

Figure 4–8　Bolshoi Elgakhchan 1 site plan.

View of Bolshoi Elgakhchan 1 on the Omolon River.

5 m thick. Remnants of this alluvium are still preserved today on the cape in the form of pebbles and granular earth. The Holocene saw the destruction of alluvium and the disintegration and weathering of native rocks of the socle. The upper horizons of alluvium could be younger geologically, but no younger than the beginning of the Holocene.

The stratigraphic cross section (see Fig. 4–9) at Bolshoi Elgakhchan 1 is:

1. Peaty turf, 4–14 cm thick.
2. Pale yellow sandy loam, 12–22 cm thick.
3. Greenish sandy loam with small black lenses, with traces of cryoturbation in the form of scallops, 25–58 cm thick.
4. Lumpy sandy loam, 4–56 cm thick (in the form of lenses).

5. Brownish grey humic soil with fine gravel, 4–56 cm thick.
6. A seam of yellowish brown poorly sorted sand, 5–46 cm thick.
7. Gravel with yellow deposits of sandy loam in fissures, 7–60 cm thick.
8. Socle bedrock.

The subturf layer over most of the area is hardened and reddened as a result of fire. Separate tongues of ash and charcoal extend into and occasionally through the yellowish sandy loam to the surface of the greenish sandy loam. Where turf is removed, the grooving of the ground shows through very sharply—a sign of frost deformation, which affected the location of cultural remains. Within the yellowish sandy loam and the upper horizon of greenish sandy loam, Upper Palaeolithic cultural remains are embedded within various levels, with some artifacts in a vertical position. The thickness of this cultural level reaches 27–30 cm.

Artifact Assemblage

The collection of Upper Palaeolithic cultural remains consists of core blanks, large blades and knife-like blades, burins, a ski spall, bifaces, arrowheads and fragments, large and small scrapers, gravers, a chopper, a chopper-abrader, a retoucher, flake knives, and other pebble tools (Fig. 4–10). Manufacturing waste includes flakes, shatter, and raw material debris. Altogether, 1,338 artifacts were collected; of these, 128 stone and bone artifacts could be considered tools. Siliceous schist, basalt, argillite, obsidian, chalcedony, and chert were used as raw materials. Large blades, scrapers, bifaces, and projectile points dominate the assemblage.

Core Blanks   No expended cores were found. However, frequently naturally subprismatic oblong pebbles were split to create blade core blanks (Fig. 4–10: *b*). A platform was prepared by pressure flaking or percussion, from which blade-like flake spalls and large blades were drawn (Fig. 4–10: *a, b*). For the second type of flake core blanks, the platform, flat or slanted, was prepared by removal of flat, wide spalls on a split pebble. The working face was trimmed and sometimes the longitudinal edge along the widest surface of the core was also struck (Fig. 4–10: *c*). Blades, knife-like blades, and spalls were thus produced to serve as tool blanks. Secondary processing included blank reduction and edge retouching, perhaps by means of a burin.

Blades   Most abundant at Bolshoi Elgakhchan 1 are large blades and blade segments, including knife-like blades. Blades (Fig. 4–10: *h, i*) vary in size from 5.8 × 3 to 2.1 × 1.6 cm. Basalt and siliceous schist knife-like blades (6.2 × 1.2 cm to 1.0 × 0.5 cm) were, as a rule, used without processing. Retouch and/or notched edges are characteristic, however, for knife-like blades made of obsidian (Fig. 4–10: *k–m*). It is important to point out that the obsidian blades are concentrated mainly in the

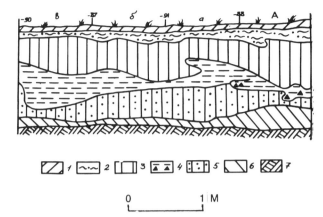

Figure 4–9   Bolshoi Elgakhchan 1. Stratigraphic profile. See text for descriptions.

southern part of the excavation, a socle projection covered by a thin layer of turf; therefore, the cultural remains in this section could be of mixed character. Blade knives made of obsidian were rarely found in the northern part of the excavation (only 4 specimens) and only in the upper horizon.

Burins   There are few independent burins in the collection (Fig. 4–11: *k, l*), which probably indicates the insignificance of the role given to this category of tool by the creators of this complex. Other tools—specifically, retouched and reworked small scrapers—could be used as burins, instead ("cutting edge," Fig. 4–11: *c*).

Scrapers   Numerically, the second biggest category of stone tools at Bolshoi Elgakhchan 1 is scrapers. Biface fragments, flake spalls (oval and sub-right-angular), and blade-like spalls served as blanks for these tools.

Large Scrapers   An example of a distinctive group of large (or side) scrapers with roughly trimmed working edges is seen on Figure 4–10: *o*. This end/side scraper has a sharpened distal end and a percussion-retouched dorsal edge on the left segment of the artifact. This trimming was most probably done to aid gripping the dulled edge of the artifact.

A second tool, with an asymmetrically curved working edge, has a small shaped haft widening

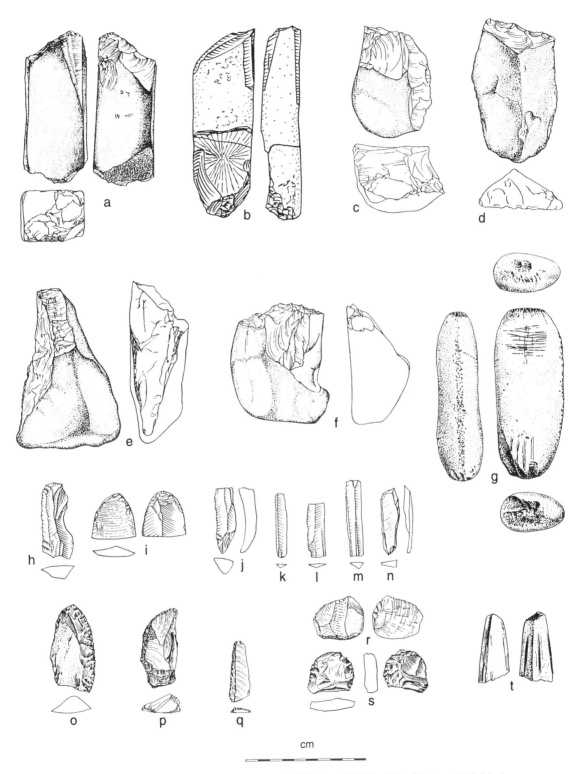

Figure 4–10 Artifacts from Bolshoi Elgakhchan 1 (*a–f, h–t*) and Bolshoi Elgakhchan 2 (*g*): blade cores (*a, b*); flake core (*c*); pebble tools (*d–f*); incised pebble hammerstone (*g*); blades (*h–m*); ski-like spall (*n*); large scrapers (*o, p*); small scrapers (*q–s*); abrasive stone (*t*).

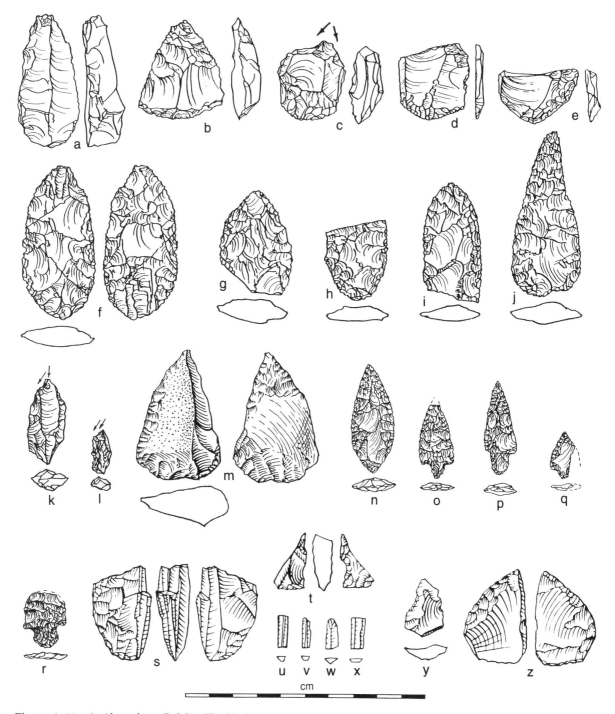

Figure 4–11   Artifacts from Bolshoi Elgakhchan 1 (*a–q*) and Bolshoi Elgakhchan 2 (*r–z*): small scrapers (*a–e*); bifaces (*f–j*); burins (*k, l*); graver (*m*); arrowheads (*n–r*); wedge-shaped microblade core (*s*); wedge-shaped core platform spall (*t*); microblades (*u–x*); flake tools (*y, z*).

near the base (Fig. 4–10: *p*). The convex-concave working edge might allow us to define this tool as a multifunctional shaver/scraper. The third large scraper with a thinned straight working edge has been narrowed by means of striking the base of the artifact; projections were formed by flaking the distal ends.

Small Scrapers   Among the group of smaller scrapers formed on blades, blade-like flakes, and flakes are end-of-blade scrapers manufactured on the distal segments of blades. Working edges are trimmed not only on the distal end of the blade but also on the edge (Fig. 4–10: *q*).

Other end scrapers include a steeply retouched specimen, on a thick blade-like flake (Fig. 4–11: *a*). The base has been spalled from both the ventral and dorsal sides, forming a working edge. There is use retouch all along the right lateral edge of the scraper's body.

There are 4 scrapers on sub-right-angular flakes of cream-colored siliceous schist. Two are illustrated here (Fig. 4–11: *d, e*). They were trimmed using flat unifacial pressure retouch on the working edges. A similar chert specimen (Fig. 4–11: *c*) exhibits small ventral retouch on the working edge. The rounded working edges of 3 semilunar scrapers are trimmed from the dorsal side. There appears to be deliberate breakage throughout this series of scrapers. Perhaps the intent was to recycle used tools. In the case of Figure 4–11: *c* there are 3 burin removals on the side opposite the working edge; the burin working edge is also trimmed. One end scraper was formed on a triangular flake (Fig. 4–11: *b*) and exhibits retouch traces on the lateral edges and a somewhat rounded working edge. The base was sharpened by spalling from the dorsal side.

Among the remaining scrapers is a small flake scraper and a biface fragment (Fig. 4–10: *s*) with scalar retouch on one of the working surface edges. The former has traces of previous spalls on the dorsal side, with a working edge probably retouched from both the dorsal and ventral sides. There are 2 other flake tools with outlines reminiscent of animal profiles, including one with scraper retouch (Fig. 4–10: *r*).

Bifaces   Bifaces, whole specimens and fragments, may have served the function of knives. Among these are distinctive wide (Fig. 4–11: *f–h*) and narrow (Fig. 4–11: *i, j*) leaf-shaped artifacts. Small flake retouching of the working edge is characteristic for bifaces. One specimen (Fig. 4–11: *f*) has marked traces of retouch toward the distal ends, following the longitudinal axis of the artifact; the working edge has deeply chipped segments.

Bifacially prepared artifacts include arrowheads made of siliceous schist. Of five specimens found at Bolshoi Elgakhchan 1, four are hafted or stemmed (Fig. 4–11: *o–q*) and one (Fig. 4–11: *n*) is leaf-shaped (haftless). Both surfaces of the specimen (Fig. 4–11: *n*) are thoroughly prepared with surface and edge retouch. There is very fine retouch on the point tip, with additional retouch on the body and rounded haft. The lower body of Figure 4–11: *o* ends in slightly drooping shoulders; the haft is triangular in shape. Specimen 4–11: *q* retains evidence of the original spall surface on one face; the other side has a secondarily retouched edge. This specimen, convex in side view, exhibits a narrowing haft.

Flake Tools

Knives   Flake knives were manufactured on cortex flakes and spalls.

Gravers   Two gravers were manufactured from cortex spalls. One was prepared by edge retouch on both the dorsal and ventral sides (Fig. 4–11: *m*); the second graver was prepared only on the dorsal side.

Other Tools

The collection also contains both a grooved crystalline stone and 2 flat pebbles with beveled cavities or notches that may have served as shaft straighteners.

A conical abrasive fine-grained sandstone fragment (Fig. 4–10: *t*), with five ground facets, exhibits grooved lines on its two flat facets. The grooved lines were possibly the result of sharpening bone tools.

A group of pebble tools retain pebble surfaces

except along the working edge. One is a chopper (9.5 × 8.45 cm [Fig. 4–10: *f*]) with a steep (45–50°) working edge; the pebble heel is very convenient for a hand grip. Three pebble tools with slanted or rounded working edges and high tops could be considered scrapers (Fig. 4–10: *d*). One pebble (Fig. 4–10: *e*) is a hacking tool with a sharpened distal edge and evidence of impact on its lateral working edge.

Manufacturing waste included some large pebble spalls modified for scraper use. Flakes, used as disposable tools, revealed grooves and notches which probably were formed during use.

Summary of Site 1

The Elgakhchan collection consists of a complex of stone tools in which certain categories of artifacts are found as a series: pebble core blanks, small scrapers, large scrapers, bifaces, and hafted arrowheads. Some tools are represented by single, but very impressive, specimens: a chopper, a graver, and a leaf-shaped point.

A peculiarity of the complex is the use of certain scrapers as burins. The stone industry of Bolshoi Elgakhchan 1, which consists of both micro and macro specimens, combines unifacial and bifacial technological traditions. This complex may be compared to the Ushki Upper Palaeolithic complex, since it is analogous in a number of categories.

While wedge-shaped cores are not typical for the Ushki culture, the production of flakes from amorphous cores and the production of knife-like blades from subprismatic cores form the basis for lithic production in the early Ushki Palaeolithic culture. Cores used to obtain flakes are present in the Elgakhchan 1 collection (Fig. 4–10: *c*), but naturally subprismatic pebble cores (some of which are faceted) are typical of the Elgakhchan 1 technology (Fig. 4–10: *a, b*). The deliberately chosen core shape consequently determined the products of flaking—blade-like flakes and blades.

The presence of one ski-like spall in the collection (Fig. 4–10: *n*) is hard to explain. Blades, blade-like flakes, and amorphous flakes were actually the blanks for Elgakhchan 1 tools. The techniques of

secondary processing included not only secondary edge retouch but also surface retouch. Both of these techniques are characteristic for the Ushki (Upper Palaeolithic layer) complex as well. The use of both unifacial and bifacial technological traditions binds these two complexes together.

The diagnostic elements of the Ushki complex are hafted arrowheads. Four such arrowheads were found at Bolshoi Elgakhchan 1 (Fig. 4–11: *o-q*) and one at Bolshoi Elgakhchan 2 (Fig. 4–11: *n*).

The morphological elements of proportion, technique of surface retouch, secondary retouching, trimming of cavities to form a haft, overlapping those cavities into the shoulder, and the prominent point of the arrowhead, together with trimming methods, are identical to the ones of Ushki. The leaf-shaped (haftless) points, present in both complexes (one specimen in each), are also analogous (Fig. 4–11: *n*).

Bifaces and certain types of small scrapers are also similar to ones found at Ushki. In both complexes burins are manufactured from other tools, particularly from scrapers (see Fig. 4–11: *c*). Obviously, there are differences as well. For example, the pebble technology is not typical for the Upper Palaeolithic layer of Ushki sites.

The techniques of primary splitting of core blanks, platform processing, and edge retouch at the Bolshoi Elgakhchan 1 site reveal elements which indicate a similarity with the complex of the Ezhantsy site on the Aldan River (Mochanov and Fedoseeva, this volume). This technology is typical in all details for the stone industry of the Upper Palaeolithic site of Ezhantsy, in which a whole series of such cores and blanks (13 specimens) were found (Mochanov 1977). However, wedge-shaped cores are inherent in the Aldan River Ezhantsy industry while absent in the Ushki and Bolshoi Elgakhchan 1 complexes. The Ezhantsy tool kit includes only a few burins, scrapers, and knives similar to the ones in Elgakhchan and, overall, the differences are quite significant.

There is a surprising similarity between the stone tools of Bolshoi Elgakhchan 1 and the Walker Road site complex in the valley of the Nenana River, Alaska. The analogies are evident in the presence of macro- and microtools in both complexes, in the technique of manufacturing pebble and chert tools, and in the combination of two traditions—unifacial and bifacial tool preparation. There are tool analogs as well. The difference between Walker Road and Bolshoi Elgakhchan 1 lies in the absence at Walker Road of hafted arrowheads and knife-like blades, which are characteristic of the Elgakhchan complex. A series of Walker Road complex radiocarbon dates range from $11,820 \pm 120$ to $11,010 \pm 230$ BP (Powers et al. 1990; Goebel et al., this volume).

Comparative typological analysis allows us to include the Elgakhchan 1 complex preliminarily in the same phase order as the Complex VII (Upper Palaeolithic) layer of the Ushki 1 site. It will be possible to date it only after conducting excavations at Bolshoi Elgakhchan 2, where a point analogous to the Ushki ones was found on an eroded surface (see Fig. 4–11: *r*). A wedge-shaped core was also found there (Fig. 4–11: *s*). Point tips are characteristic for the VII layer dated at $14,300 \pm 200$ BP (GIN-167), and wedge-shaped cores are characteristic for the VI layer of Ushki sites dated at $10,760 \pm 110$ BP (MAG-219). Finds from Bolshoi Elgakhchan 2 may provide additional materials that will allow a stronger argument for the genesis of these cultures and their correlation.

Palynology

Currently two palynological samples from a stratigraphic cross section at Bolshoi Elgakhchan 1 have been analyzed. Sample 1 was taken from a depth of 10–15 cm below the contemporary surface in the upper horizon of the Upper Palaeolithic cultural layer. Sample 2 was taken from a depth of 25–30 cm, the lower horizon of the cultural layer. Sample 1 contains pollen representative of flora of the Holocene period, excluding the phase of maximum warming. The pollen spectrum of the lower sample indicates a vegetational complex typical of the third and final phase of the Sartan Ice Age. Given previous radiocarbon dates for these assemblages, it seems that the age of the Upper Palaeolithic sites Bolshoi Elgakhchan 1 and 2 is not later than 10,200 years BP.

BOLSHOI ELGAKHCHAN 2

Artifact Assemblage

This site was discovered in 1980 by the West-Chukotka archaeological detachment (under the leadership of M. A. Kiryak). During a visual inspection of the 80-m terrace, which stretches from north to south for 2.5 km, a hafted arrowhead of the Ushki type (Fig. 4–11: *r*) was found on a partially eroded surface not far from the northern extremity of the terrace. A chopper was picked up 100 m from this location towards the south. In 1987, during a second survey of the same terrace surface, we found an oblong pebble with traces of battering and incisions on one side (Fig. 4–10: *g*). A wedge-shaped core (Fig. 4–11: *s*) was found 15 m southward. During a careful search of this section two microblades were found, made of the same raw material as the core (Fig. 4–11: *v*, *w*). An excavation of 2 × 2 m was begun 11 m northwest from these finds. The turf layer (1.5–2 cm thick) was removed, revealing the socle base of the terrace, covered with a thin seam of yellowish sandy loam near the eastern wall of the excavation. This seam, located beneath the humus stratum, was compositionally analogous to the second layer at Bolshoi Elgakhchan 1. The finds were embedded in the loose fill of the destroyed surface layer of the socle base.

The stone tool assemblage collected during this excavation numbered 305 items, of which 21 are tools. Both surface finds and excavated items are discussed below.

The wedge-shaped core found in the 1987 survey (Fig. 4–11: *s*), made of black obsidian, has a classic semi-segment-shaped form and a height of 2.9 cm, a length of 1.9 cm, and a thickness of 1.1 cm. The pressure-flaked striking platform is slanted; the lateral edge is sinuous and is retouched fully on one side and only partially on the other side (from the lower edge). Microblades with rectilinear facets, found near the core, were made of the same raw material (Fig. 4–11: *v*, *w*).

The chopper found in 1980 was made from a grey, flattened andesite-basalt pebble. Originally, the pebble was larger. During the process of primary splitting the small boulder most probably was split into four parts, resulting in the formation of a sub-quadrangular blank. The chopper was formed during secondary processing. The maximum height of the tool is 7.8 cm; width, 7.2 cm. A working edge (3.7 cm long) was struck on the narrowed distal end of the blank. The platforms formed as a result of the primary splitting remained unprocessed.

The pebble with incisions (Fig. 4–10: *g*) found during a surface survey in 1987 could have served as a hammerstone because both narrow distal ends bear traces of spalls.

The remaining artifacts discussed below were obtained from the 2 × 2-m excavation. Microblades (12 specimens) made of black, dull black, and greyish obsidian range in size from 0.7 × 0.2 cm to 1.3 × 0.6 cm and are triangular and trapezium-shaped in cross section (Fig. 4–11: *u, w, x*). A platform spall (Fig. 4–11: *t*) removed during retouch of a wedge-shaped core is of a black obsidian similar to the core collected from the surface (Fig. 4–11: *s*). There were also 2 burins on obsidian flakes.

There are 3 black obsidian flake tools with grooves: two have concave working edges (for example, Fig. 4–11: *y*) while the third has a convex-concave contour. In the collection there are also 2 flake knives of black obsidian: one of these has a bilaterally retouched blade edge (Fig. 4–11: *z*); the second one exhibits retouch only on the dorsal side.

Two other tools were manufactured on blanks of grey pebble analogous to the raw material from which massive tools were made at Bolshoi Elgakhchan 1. The raw material is found not far from here on the bank of the Omolon River. The first tool is a graver shaped like an oval leaf on a blade-like flake with a high dorsal ridge; it has a height of 4.3 cm, width of 2.5 cm, and thickness of 1.1 cm. One edge on the dorsal side is partly retouched. The second tool, a perforator, was formed on a pebble core platform that had striking retouch analogous to those recovered at Bolshoi Elgakhchan 1. The tip was formed by fine retouch, and other traces of retouching are found on the specimen.

Among the manufacturing waste are 2 large cortex flakes made from grey siliceous schist pebbles possibly intended as blanks for tools. There are 191 flakes made of obsidian, 53 of pebble material analogous to that utilized at Bolshoi Elgakhchan 1, 21 of siliceous schist, and 19 of flint. One-quarter of the total flakes are micro-specimens.

It is still early to draw any general conclusions based on these finds at Bolshoi Elgakhchan 2, since they are not numerous. More complex and extensive research is needed. Nevertheless, it is possible to ascertain that in the collection from Bolshoi Elgakhchan 2 there are elements which undoubtedly indicate its Upper Palaeolithic character. A few of the materials have their closest analogs in Layer VI of the Ushki site (Kamchatka) and in the Dyuktai culture (Yakutia).

BIBLIOGRAPHY

Dikov, N. N. 1979. *Ancient Cultures of Northeastern Asia (Asia Adjoining America)*. Moscow: Nauka. (In Russian)

Kistenev, S. P. 1980. New Archaeological Relics from the Kolyma Basin. In *News in the Archaeology of Yakutsk*. JF SO AN USSR (Academy of Sciences), pp. 74–87. Yakutsk. (In Russian)

Mochanov, Y. A. 1977. *The Most Ancient Stages of Human Settlement in Northeastern Asia*. Novosibirsk: Nauka. (In Russian)

Powers, W. R., F. E. Goebel, and N. H. Bigelow. 1990. Late Pleistocene Occupation at Walker Road. *Current Research in the Pleistocene* 7: 40–43.

UPTAR AND KHETA: UPPER PALAEOLITHIC SITES OF THE UPPER KOLYMA REGION

Sergei B. Slobodin and Maureen L. King

The relationships between northeastern Russia and Alaska have concerned archaeologists for many years (Dikov 1963; Anderson 1968; Mochanov 1977; West 1981). While substantial progress has been made in documenting terminal Pleistocene and early Holocene archaeological sites in eastern Beringia (e.g., Powers and Hoffecker 1989; Dixon 1993; Hoffecker et al. 1993; Kunz and Reanier 1994), the early prehistory of western Beringia is still under-

View of Kheta River from the Kheta site. (*Photo courtesy of Maureen L. King.*)

stood only very sketchily. Apart from the well-known Ushki sites in central Kamchatka (Dikov 1979) and the Berelekh site on the lower Indigirka River (Mochanov 1977; Mochanov and Fedoseeva, this volume), there are no sites east of the Verkhoyansk Range that can be dated reliably to the end of the Pleistocene or early Holocene (Dixon 1993; Hoffecker et al. 1993). Only a handful of assemblages in western Beringia have been assigned to the Upper Palaeolithic. These include Kukhtuy 3, Mayorych, and Urtychuk (Mochanov 1977; Dikov 1979; Dikov and Savva 1980; Derevianko 1990; Mochanov and Fedoseeva, this volume) among others. However, the assessment must be considered tentative since there is no independent means of establishing the dates of these materials (i.e., surface contexts, absence of datable materials).

Recently two sites have been discovered in the upper Kolyma region that can be assigned to the Upper Palaeolithic: the Uptar site and the Kheta site. Both sites occur in shallow stratigraphic contexts sealed by an early Holocene volcanic tephra and include assemblages that are Palaeolithic in character (Slobodin 1990, 1991). The apparent antiquity of these sites and their intermediate location between the Aldan basin and Kamchatka Peninsula make them especially important, not only for documenting the early prehistory of the region, but also for investigating the relationship of these assemblages to other early sites in Beringia.

The upper Kolyma region is located within the Magadan oblast that includes the upper Kolyma drainage basin and the area along the Okhotsk coastline. These two areas are separated by the Okhotsk–Kolyma upland. This upland forms the divide between river systems that drain into the Sea of Okhotsk and those that are tributaries of the upper Kolyma River, which flows northward into the East Siberian Sea. The region is mountainous, with elevations up to 2,500 m. The coast of the Sea of Okhotsk is tectonically active, and although much of the coastline is characterized by steep rock cliffs, tectonic depressions in other areas form flat coastal plains.

THE UPTAR SITE

The Uptar site is located in northern Priokhotye, about 40 km north of Magadan. The site is situated on the uppermost fluvial terrace on the right bank (north side) of the Uptar River, 4–5 m above the current floodplain at an elevation of 160 m. The Uptar River flows into the Khasyn River, a major tributary of the Arman river system that drains into the Sea of Okhotsk. At the Uptar site the surface of the terrace has very little relief and a larch (*Larix dahurica*) forest dominates, along with dwarf stone pine (*Pinus pumila*) and birch (*Betula exilis*), with fructose lichens (in particular, *Cladonia rangifer*) and moss as common plants of the forest understory.

The site was discovered by Slobodin in 1985 when lithic artifacts were found in an area that had been disturbed when a bulldozer moved trees along the terrace. Approximately 25 m north of the terrace edge is a gravel borrow area. Excavation of this gravel pit had destroyed a portion of the site. Archaeological investigations at Uptar began in 1986 and an area of 32 sq. m has since been excavated.

Stratigraphy

Surface deposits at the site vary in thickness, from 15 to 45 cm, and are underlain by massive alluvial deposits (Fig. 4–12). Beneath the surface organic layer is a thin (2–10 cm), well-defined tephra layer. This in turn is underlain by a reddish brown, poorly sorted, fine to medium sand. A radiocarbon date of 8260 ± 330 years BP (Lozhkin 1989) was obtained from charcoal collected within the tephra deposit and at the interface between the tephra and underlying sand deposit. The date conforms with the 8300 years BP date for the Elikchan tephra (Begét

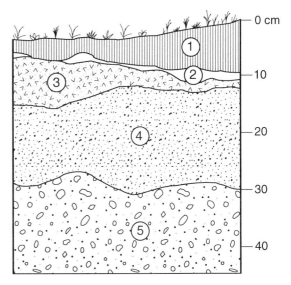

Figure 4–12   Uptar. Stratigraphic profile. Stratigraphic units as follows: (1) dark brown surface organic horizon; (2) A-2 eluvial horizon, greyish brown fine sand/silt with organic; (3) Elikchan tephra, light grey to grey fine sand/silt with glass shards (10x magnification); (4) reddish brown to brown, poorly sorted fine to medium sand with some gravel; (5) greyish brown poorly sorted coarse sand with gravel, massive alluvium over 4 m thick.

et al. 1991) and is likely associated with that tephra fall.

A single cultural component has been found at Uptar. Excavation revealed that artifacts were positioned within the upper extent of the sand deposit directly beneath, although in instances extending into, the tephra. Approximately half of the artifacts in the assemblage were collected from disturbed areas of the site but conform with those from the excavated areas in both typology and stone material type. Included in the assemblage are 45 tools and tool fragments, 6 blades/microblades, ca. 2,000 pieces of stone debitage, and a single stone pendant. No organic material was found.

Artifact Assemblage

Artifacts were manufactured from siliceous slate, chert, and basalt. Most of the tools in the Uptar collection are bifaces and biface fragments (80%). Bifaces are lanceolate to ovate in plan view with lenticular cross sections (Fig. 4–13: *a–h*). Examples

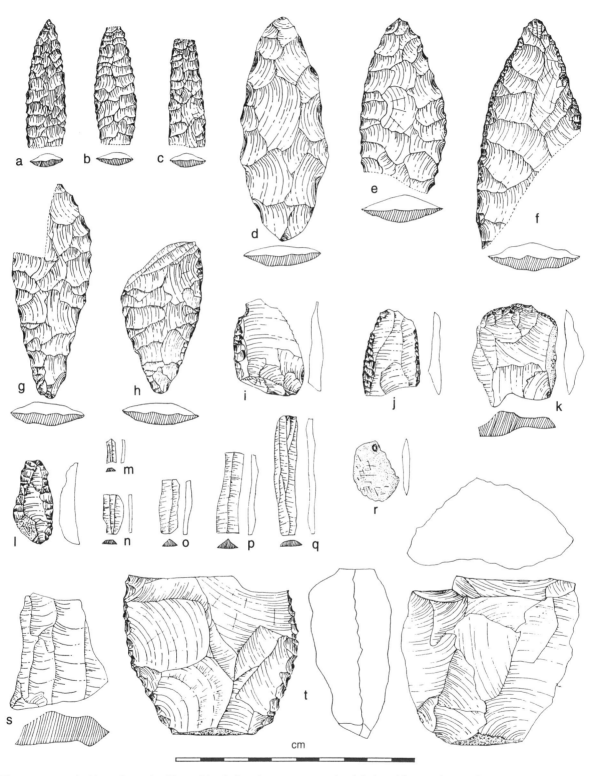

Figure 4–13   Artifacts from the Uptar Site: bifacial arrow points (*a–c*); bifacial knives/spear-dart points (*d–h*); utilized flakes (*i–l*); blades (*m–q*); pendant (*r*); possible subprismatic blade core fragment (*s*); biface core (*t*).

of two sizes were observed: bifacial arrow points (Fig. 4–13: *a–c*) and bifacial knives/spear-dart points (Fig. 4–13: *d–h*). A small number of the larger biface fragments exhibit fine bifacial, marginal retouch (Fig. 4–13: *f*).

Five utilized flakes have also been identified. All exhibit wear/retouch that is unifacial. With the exception of one artifact, the configuration of the edge damage is complex, indicating multiple use (Fig. 4–13: *i–l*). The blades/microblades represent a small component of the assemblage (Fig. 4–13: *m–q*). None of these specimens exhibit evidence of retouch or wear. Aside from an artifact that may represent a fragment of a subprismatic core (Fig. 4–13: *s*), no blade or microblade cores were found. Three cobble tools are also represented, one of which may be described as a biface core fragment with edge modification (Fig. 4–13: *t*). Finally, the pendant is a dark brown, oval stone that is lenticular in cross section and has a biconically drilled hole on one end (Fig. 4–13: *r*). Polishing is evident along the edges, and short, narrow incisions occur on the edge.

This description of the Uptar materials is preliminary. Originally, it was thought that a blade reduction strategy was represented here (Slobodin 1990). It is possible that the blades, sorted from thousands of pieces of debitage, are incidental byproducts of the stone technology practiced at Uptar. In addition, the bifaces may represent progressive stages of manufacture. A detailed analysis of the Uptar collection will provide important information concerning the stone technology and the number and kinds of tool classes.

Summary

While the precise date of the occupation of the Uptar site is not known, the tephra overlying the artifact level provides an important limiting date. We know that the site is at least as ancient as the early Holocene and possibly much older. The collection from Uptar is unique for western Beringia and does not compare readily with sites from either the Aldan basin (Dyuktai culture [Mochanov 1977; Mochanov and Fedoseeva, this volume]) or Kam-

chatka (Ushki sites [Dikov 1979, this volume]). The sites of Siberdik and Kongo located on the upper Kolyma River have been dated to the early Holocene (Dikov 1977, 1979) and contain bifacially formed tools, but blade reduction is also an important component of these assemblages.

THE KHETA SITE

The Kheta site is located near the divide of the Kolyma and Okhotsk watersheds, in the foothills of the Maymanzhinskii Range, 180 km northeast of Magadan. The site is situated on a terrace edge, at an elevation of 800 m, 15 m above the confluence of the left and right Kheta rivers. Vegetation on the terrace is an open larch forest with a dwarf stone pine and birch-shrub tundra as important components of the forest understory.

During the Sartan (late Wisconsin equivalent), glacial activity on the western slope of the Maymanzhinskii Range was restricted, leaving a large part of the area ice-free, especially deep watershed saddles and larger valleys, like those of the Kheta rivers. Along the upper course of the Kheta, glacial features are rare. Only a few glacial cirques have been identified at elevations from 1,000 to 1,300 m, and these likely predate the Sartan (Glushkova 1984). The palaeogeographic conditions of the Kheta valley during the Sartan glaciation do not, therefore, conflict with the possibility of a human occupation in this region during that time.

The Kheta site was discovered in 1990 by Slobodin, who observed that the site had been inadvertently disturbed during the installation of a telephone cable in the 1960s. Limited excavation and surface collection conducted in 1991 and 1992 within a 60-sq.-m area revealed that the site is multicomponent and that a limited area of the lower component remained undisturbed.

Stratigraphy

The deposits at Kheta represent an inset terrace relationship (Fig. 4–14). A stratigraphic section south of the site shows a surface organic layer underlain by sandy sediment. The basal unit is bedrock shale. Gravel on the bedrock surface suggests that the terrace was scoured prior to the deposition of the sand, an event that occurred during the Karginsky (middle Wisconsin equivalent) warming epoch (Glushkova 1983). No artifacts were found in the sandy sediments beneath the organic mat in this area. In the central portion of the site a layer of aeolian reworked alluvial sands overlain by a tephra was found. The tephra has been tentatively identified as the Elikchan tephra, a tephra that dates to ca. 8,300 years BP (Begét et al. 1991). The sand deposits seen in the southern section were eroded from the terrace some time prior to the deposition of the reworked sands. Artifacts occur directly beneath the reworked sand deposit and are lagged within the alluvial gravels and bedrock schist. The reworked sand, overlying the artifact level, is likely a terminal Pleistocene deposit (O. Y. Glushkova, personal communication 1994).

Artifact Assemblage

The Kheta site includes at least two components. Only the material associated with the lower component is discussed here. The assemblage includes 1 microblade core, microblades, a ski spall, 4 bifaces, 2 end scrapers, 4 flake cores, a burin, a perforator, 2 pendants, and over 500 pieces of lithic debitage. No organic material was observed during excavation or on the exposed surface of the site. All of the artifacts are manufactured from a "plate chert" material. While the source for this material has not been located, geologists who work in the region suggest that it is local.

The microblade core in the Kheta collection is of particular interest because it is similar to microblade cores from the Aldan basin that are associated with assemblages assigned to the Upper Palaeolithic occupation—the Dyuktai culture (Mochanov 1977; Mochanov and Fedoseeva, this volume). The microblade core is wedge-shaped (Fig. 4–15: *b*). It was manufactured from a biface, struck longitudinally (removing a flake termed a "ski spall" [Fig. 4–15: *a*]) to form a platform along one of the margins. Microblades were then removed

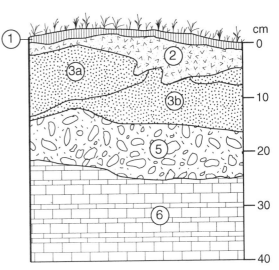

Figure 4–14   Kheta. Stratigraphic profile. Stratigraphic units as follows: (1) surface organic layer, dark brown, fine to coarse sand; (2) Elikchan tephra, light grey to grey, fine medium sand with glass shards (10x magnification); (3a) B-1 soil horizon, reddish brown, poorly sorted very fine with medium to coarse sand, aeolian-reworked alluvial sand; (3b) B-2 soil horizon, yellowish brown, very fine, aeolian-reworked sand. Artifacts found at contact with underlying unit 5. (Stratum 4 contained no cultural material and is not shown here.) (5) Brown, poorly sorted coarse to very coarse sand with gravel and tabular bedrock schist. Contains artifacts in areas where deposit is exposed or overlain by stratum 3a/3b. (6) Bedrock schist.

from one end of the biface. Blade scars are no more than 3 mm in width and 2.8 cm in length. The keel of the core exhibits bifacial flake scars likely associated with edge preparation.

Four bifaces/biface fragments have been found. Three of these appear to represent finished forms (Fig. 4–15: *g–i*), while one may be a preform for a microblade core (Fig. 4–15: *j*). The two end scrapers were manufactured on flakes that have a trapezoidal cross section (possibly blades) and exhibit steep unifacial retouch on the distal margins (Fig. 4–15: *k, l*). The burin was made from a biface with burin blows on opposing ends (Fig. 4–15: *m*). One of the blades in the collection exhibits retouch on the distal end that converges to form a point (perforator) (Fig. 4–15: *n*). Finally, the two pendants are made

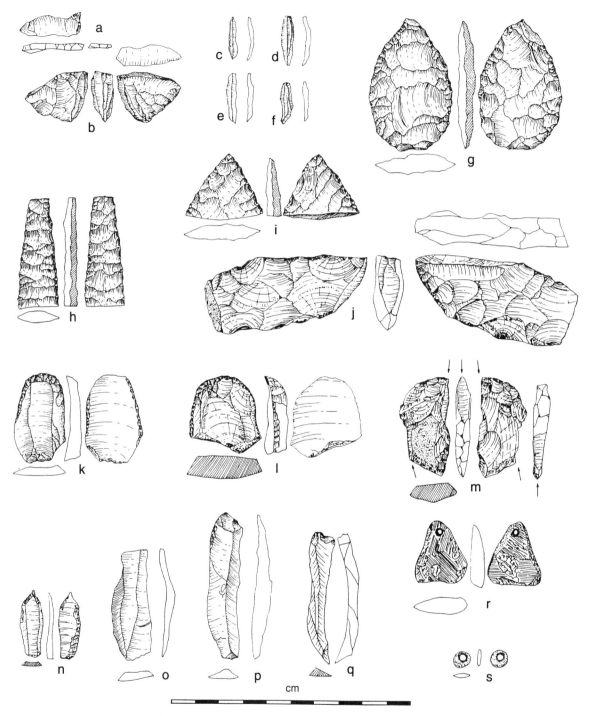

Figure 4–15   Artifacts from the Kheta site: ski spall (*a*); wedge-shaped microblade core (*b*); microblades (*c–f*); bifaces (*g–i*); probable bifacial microblade core preform (*j*); end scrapers (*k, l*); burin (*m*); retouched blade perforator (*n*); blades (*o–q*); pendants (*r, s*).

from agalmatolite pebbles. One is subtriangular with a biconically drilled hole and evidence of abrasion on the surface. The other is round and polished with a biconical hole in the center (Fig. 4–15: *r, s*).

Summary

The collection from the Kheta site is typologically similar to material from the Upper Palaeolithic Dyuktai culture (Mochanov 1977) and the material from Ushki 1 Level VI in Kamchatka (Dikov 1979). Based on the interpretation of the depositional context of the artifacts, the lower component at the Kheta site likely dates to the terminal Pleistocene. Ongoing investigations will, it is hoped, provide organic materials suitable for dating.

Discussion and Conclusions

The Uptar site and the Kheta site provide important new data about western Beringian prehistory. With the exception of the Berelekh site, the lower component at the Kheta site contains the only wedge-shaped microblade core industry found in a datable context in the vast area between the Aldan basin and the Kamchatka Peninsula. The Uptar site, where bifacial reduction is predominant, is not comparable to other early western Beringian collections. Both sites provide information relevant to questions concerning early connections between Asia and North America.

The identification of the Elikchan tephra in the upper Kolyma region holds promise for providing relative dates for archaeological assemblages. This is especially important for sites where there is little or no preservation of the organic material necessary for obtaining radiocarbon dates. In addition, this tephra is an important benchmark for archaeological investigations. Recent analysis of sediment cores from the Jack London Lake area provides a record of late Quaternary vegetation (Lozhkin et al. 1993). This record shows that modern vegetation associations (*Larix dahurica*, *Pinus pumila*, and *Betula* are predominant) were established by ca. 9,000 years BP. In the absence of more precise dating methods, the tephra is a general marker for terminal Pleistocene/

early Holocene assemblages that date to the transition from full-glacial to modern conditions.

References Cited

Anderson, D. D. 1968. A Stone Age Campsite at the Gateway to America. *Scientific American* 218:24–33.

Begét, J. E., D. M. Hopkins, A. V. Lozhkin, P. M. Anderson, and W. R. Eisner. 1991. The Newly Discovered Elikchan Tephra on the Mainland of Soviet Asia. *Geological Society of America Abstracts with Programs 1991*, A62.

Derevianko, A. P. 1990. *Paleolithic of North Asia and the Problem of Ancient Migration.* Novosibirsk: Nauka. (In Russian)

Dikov, N. N. 1963. Archaeological Materials from the Chukchi Peninsula. *American Antiquity* 28(4):529–36.

———. 1977. *Arkheologicheskie pamyatniki Kamchatki, Chukotki i Verkhnej Kolymy (Archaeological Monuments of Kamchatka, Chukotka, and the Upper Kolyma).* Moscow: Nauka.

———. 1979. *Drevnie kul'tury Severo-Vostochnoj Azii (Ancient Cultures of Northeast Asia).* Moscow: Nauka.

Dikov, N. N., and N. E. Savva. 1980. Novaya Vnutrikontinental'naya Stoyanka Kamennogo Veka v Severnom Priokhot'e: u istokov rush. Urtychuk (The New Continental Palaeolithic Site in the Northern Priokhotye: Around Urtychuk Creek). In *Noveishie Danniye po Arkheologii Severa Dal'nego Vostoka*, ed. N. N. Dikov, pp. 64–68. Magadan.

Dixon, E. J. 1993. *Quest for the Origins of the First Americans.* Albuquerque: University of New Mexico Press.

Glushkova, O. Y. 1983. Istoriy Razvitiya Rel'efa Severnogo Priokht'ya v Pozdnem Pleistosene i Golotsene (The History of the Northern Priokhotye Relief Development in the Pleistocene and Holocene). In *Stratigrafiya i Paleogeografiy Pozdnego Kainozoya Vostoka SSSR*, pp. 114–23. Magadan.

———. 1984. Morfoloiya i Paleogeografiya Pozdnepleistotsenovykh Oledenenii Severo-Vostoka SSR (Morphology and Palaeogeography of the Late Pleistocene Glaciations of East Asia, USSR). In *Pleistotsenovie Oledeneniya Vostoka Azii*, pp. 28–42. Magadan.

Hoffecker, J. F., W. R. Powers, and T. Goebel. 1993. The Colonization of Beringia and the Peopling of the New World. *Science* 259:46–53.

Kunz, M. L., and R. E. Reanier. 1994. Paleoindians in Beringia: Evidence from Arctic Alaska. *Science* 263:660–62.

Lozhkin, A. V. 1989. *Radiouglerodnoe datirovanie arkheologicheskikh pamyatnikov na territorii Magadanskoi oblasti (Radiocarbon Dating of Archaeological Monuments in Magadan Oblast).* Otchet. Fondy SVKNII. Report on file, USSR Academy of Science, Magadan.

Lozhkin, A. V., P. M. Anderson, W. R. Eisner, L. N. Rovako, D. M. Hopkins, L. B. Brubaker, and P. A. Colinvaux. 1993. Late Quaternary Lacustrine Pollen Records from Southwestern Beringia. *Quaternary Research* 39:314–24.

Mochanov, Y. A. 1977. *Drevnejshie etapy zaeseleniya chelovekom Severo-Vostocnoj Azii (Ancientmost Stages of the Settlement by Man of Northeast Asia).* Novosibirsk: Nauka.

Powers, W. R., and J. F. Hoffecker. 1989. Late Pleistocene Settlement in the Nenana Valley, Central Alaska. *American Antiquity* 54(2):263–87.

Slobodin, S. B. 1990. Issledovaniya Kontinental'noi Stoyanki Uptar I v Severnom Priokhot'e (Investigation of the Uptar 1 Continental Site in Northern Priokhotye). In *Drevnie Pamyatniki Severa Dal'nego Vostoka,* ed. N. N. Dikov, pp. 65–79. Magadan.

———. 1991. Kemmenyi vek Verknei Kolymy i Kontinental'nogo Priokhot'ya (The Upper Kolyma and Continental Priokhotye Stone Age). In *Problemy Arkheologii i Etnografii Sibiri i Dal'nego Vostoka,* pp. 15–16. Krasnoyarsk.

West, F. H. 1981. *The Archaeology of Beringia.* New York: Columbia University Press.

THE USHKI SITES, KAMCHATKA PENINSULA

Nikolai N. Dikov

The Ushki sites are located at several promontories on the southern shore of Great Ushki Lake, an oxbow, or cut-off meander, of the Kamchatka River. Of the five sites, the most productive have been Ushki 1, 2, 4, and 5. The lowest levels are dated to late Pleistocene–early Holocene. The banks of the Kamchatka in this vicinity formerly supported a dense population of Kamchadal people who made extensive use of the fishery.

HISTORY OF RESEARCH

These sites were discovered by the writer in 1961 and 1962. With various associates he has worked intensively at Ushki every year from the time of its discovery virtually to the present. Geological research intended to augment the archaeology was begun in the 1970s, as were pollen studies. Particular attention was given the tephrochronology, as late Pleistocene and Holocene volcanic ash layers are found widely in the area as well as in the sites themselves. The correlation of these layers has been especially important in cross-dating and in establishing the integrity of the archaeological levels. On the basis of geomorphological studies of the region correlated with the archaeology, it appears that the oxbow lake formed in the late Holocene, and thus the occupations discussed here were on an active channel of the Kamchatka River.

CHARACTERIZATION OF THE SITES

Of the five sites, only Ushki 1 and Ushki 5 contain all seven cultural levels, and of these Ushki 1 has provided the greatest data recovery. Seven components thus characterize the sites, occurring variously across the five. Level VII is the oldest. The stratigraphy is readily correlated from site to site, greatly aided by the tephrochronology. It is stressed that with the rare exception of house pits dug down into one of the earlier levels, allowing for a small amount of mixing, the levels are unmixed, sealed off from one another, and stratigraphically distinct. Sterile zones are interbedded, as are the tephra zones.

In cultural characterization, the upper levels—I through IV—are Neolithic in age; Levels V through VII are Palaeolithic. In terms of palaeoclimate, Level V is early Holocene; Levels VI and VII are late Pleistocene. The assemblages from VI and V are both classified as Late Ushki Palaeolithic and are assigned to the Beringian Tradition. Similarities to Dyuktai are quite clear. The Level VII assemblage pertains to an entirely different Palaeolithic culture, termed Early Ushki. Table 4–1 presents radiocarbon dates for these early levels. The stratigraphy at Ushki 1 and the burial pit associated with Level VII are shown in Figure 4–16.

Ushki Lake and the locality of the sites.

ARTIFACT ASSEMBLAGES

Cultural Level V   This is characterized by rather crude wedge-shaped microblade cores and their products, burins, foliate bifaces of variable size, unifaces of similar outline, end scrapers, large blades and blade-like flakes, and side scrapers.

Cultural Level VI   The stone assemblage is generally similar to that of the overlying level. "Gobi," or wedge-shaped, cores are present, as are microblades, burins, foliate bifaces, and various scrapers. The most remarkable discoveries of Level VI, however, were in areas other than conventional lithics. At Ushki 1 there were found over twenty house floors, apparently of wood-framed skin habitations. This may represent the largest complex of late

Palaeolithic dwellings in the world. The floors, formed by a shallow excavation, tend to be roughly circular, usually have an entry passage, and, typically, a stone-encircled hearth. Fish bones were frequently encountered in the hearths, along with charcoal.

In the floor of one house at Ushki 1, there was found the skeleton of a dog that appeared to have been deliberately buried. Vereshchagin sees wolf ancestry in this dog. Three lithic specimens found close by may have been interred with it. Considering the late Pleistocene date, this must be seen as very early evidence of domestication and certainly the earliest such evidence for northeastern Asia.

In another house floor was found a burial pit containing the remains of a child (Fig. 4–17). The body had been bound and placed in a sitting posi-

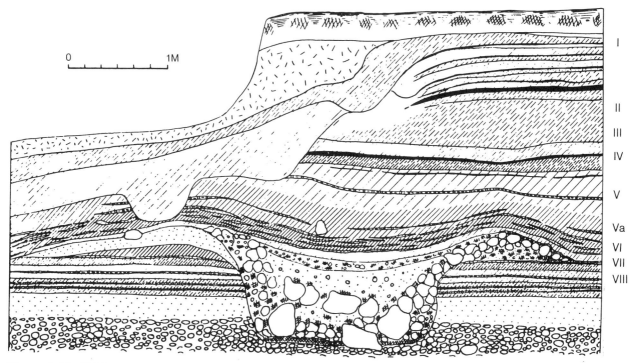

Figure 4–16   Ushki 1. Stratigraphic profile. Interbedded levels of sandy loam and humic sandy loam with a cross section of possible burial pit in Cultural Level VII. (*Source:* Arctic Anthropology. *See Acknowledgments.*)

tion. Beneath it was a kind of ritual mat made from the incisor teeth of lemmings. Amongst the bones were ground stone slabs, a fragment of a pendant, wedge-shaped cores, and a number of retouched microblades. The pit also contained a large quantity of red ochre.

Several specimens of ornamental stone work

were found in Level VI. These include several pendants drilled at one end and a sandstone slab with a cross-like design made by drilling holes; another slab, oval in form, bears incisions depicting tents. A number of whetstones were also found. A paddle-shaped object of bone also occurred in this level. Artifacts from Level VI are shown in Figure 4–18.

Level VI and Level V are considered constituents of the Dyuktai complex but with differences that may relate to different ancestry and to the Ushki occupations as representing the remains of people following a riparian economy.

Table 4–1   Radiocarbon Dates for Ushki Sites, Levels V, VI, VII

Level	Sites	Assays (yrs BP)
V (Late Ushki)	1, 2, 4, 5	8790 ± 150 (MAG-321)
VI (Late Ushki)	1, 2, 4, 5	10,860 ± 400 (MAG-400)
		10,790 ± 100 (MAG-518)
		10,760 ± 110 (MAG-219)
		10,360 ± 350 (MO-345)
VII (Early Ushki)	1, 5	14,300 ± 200 (GIN-167)
		14,200 ± 700 (MAG-550)

Cultural Level VII   The materials from this level are sharply different from those of Level VI. They are found more than 2 m below the surface. The level is best demonstrated at Ushki 1. Dwelling remains are present, variable in plan and dimension, unlike those of the level above. These are much larger (40–100 sq. m) and contain multiple hearths. Charcoal was abundant in these; they lacked stone

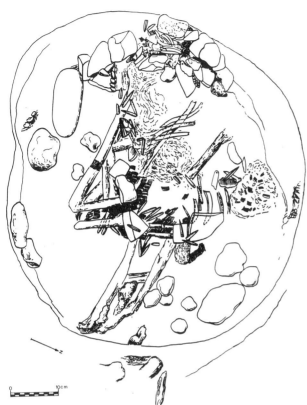

Figure 4–17 Burial pit in house floor. See text for description.

rings. In these concentrations were found traces of ochre, hematite pieces, and tools. Foliate bifaces in various sizes are characteristic, as are stone beads, pendants, but especially the stemmed projectile points, which are unique to Level VII. More than 50 of these have been found, invariably within the charcoal concentrations. This earliest occupation is radiocarbon-dated in the range of >13,000 to >14,000, but may be more ancient.

A possible burial pit was found in Level VII, filled with stones and red ochre. Only traces of bones were seen in the pit, but 800 beads of pyrophyllite were found, along with several burin-like points evidently used in fabricating them. Two stemmed arrowpoints were found at the edge of the pit (see Fig. 4–19).

The floor area of the largest house discovered was defined by the distribution of charcoal covering more than 100 sq. m. It lay near the burial pit. About its center were pieces of hematite, smears of red

ochre, and a number of lithic artifacts. These included 23 stemmed points, 17 leaf-shaped bifaces, various scrapers, burins, 3 stone pendants, and a small piece of elk antler. An adjacent house of about 75 sq. m contained a similar inventory as well as a piece of charred clay and duck gastroliths.

A series of house floors first found as an exterior row around the south side of the site were, again, of variable size. One of these, investigated in 1989, included two large hearth pits containing stemmed points and other lithics. In another nearby "external" house was revealed a hearth pit containing red ochre and hematite. The small hearth showed layers of charcoal, grey calcined bone, yellow calcined bone, and, at base, a brown sandy loam.

Other finds in the occupation floors of this stratum included leaf-shaped bifaces of variable size, stone beads and pendants, burin-like points, stemmed points, and many flakes and blades.

All of the 50 stemmed points were found *in situ*. The size and stem treatment of these points vary, as may be seen in Figure 4–19. There is no doubt of their ubiquity in Level VII or of the integrity of this level as there is no evidence whatever of mixture with others. Early Ushki culture is quite different from that of Levels VI and V.

INTERPRETATION

The Ushki sites are especially important because they provide a long, continuous record of human occupation and cultural change in a critical area of northeastern Siberia. Two significant Palaeolithic cultures are present: Early Ushki Palaeolithic, seen in Level VII, and Late Ushki Palaeolithic, seen in Levels VI and V, the latter dating to early Holocene. The integrity of the five site records is assured by the fact that, while relatively near to each other, they are separated by intervening areas in which there are no sites. Each occupies its own promontory on Ushki Lake and the stratigraphy may be seen as clearly unmixed by reference to the volcanic ash layers that seal off the strata from one another. A pollen analysis carried out here is shown as Figure 4–20.

Figure 4–18 Cultural Assemblage of Level VI at Ushki 1: bifacial arrowheads (*a–c*); bifacial knives (*d, e*); wedge-shaped microblade cores (*f–i*); ski-shaped spalls (*j, k*); burins (*l, m*); scraper (*n*); skreblo-like biface (*o*); slate skreblo-burnisher (*p*); slate knife (*q*); stone labrets (*r, s*); pendants (*t, u*); images of tent-like houses on sandstone slab (*v*).

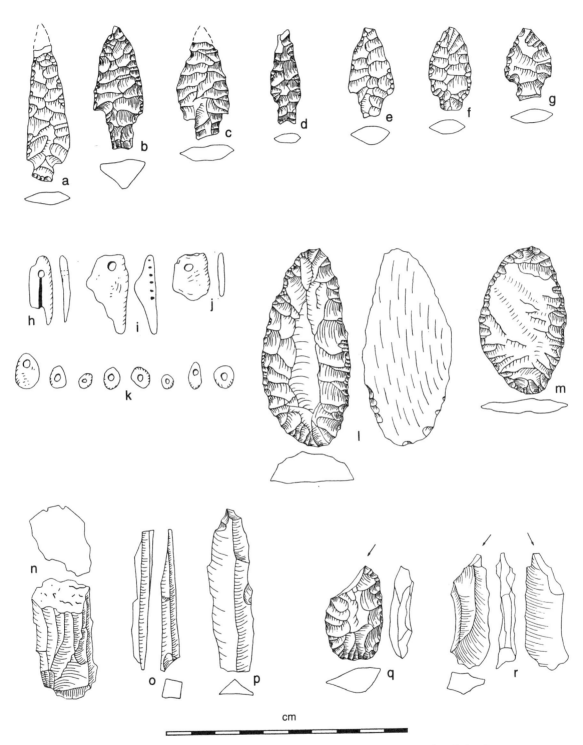

cm

Figure 4–19 Cultural Assemblage of Level VII at Ushki 1 and 5: stemmed arrowheads (*a–g*); pendants (*h–j*); stone beads (*k*); unifacial knife (*l*); bifacial knife (*m*); core (*n*); drills (*o, p*); burins (*q, r*).

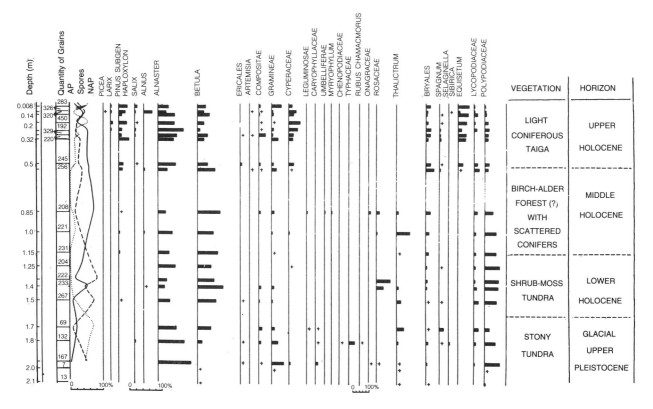

Figure 4–20 Ushki 1. Pollen diagram. (*Source:* Arctic Anthropology. *See Acknowledgments.*)

The stemmed points of Level VII—the earliest bifacial stemmed points of this culture in Eurasia—are quite similar to those of the Stemmed Point Tradition found in northwestern North America (Bryan 1980). The Ushki sites provide basic information for the study of the earliest relations of northeastern Asia and northwestern North America.

BIBLIOGRAPHY

Bryan, A. L. 1980. The Stemmed Point Tradition: An Early Technological Tradition in Western North America. In *Anthropological Papers in Memory of Early H. Swan-son, Jr.*, ed. L. B. Harten, C. N. Warren, D. R. Tuohy, pp. 77–107. Pocatello: Idaho Museum of Natural History.

Dikov, N. N. 1968. The Discovery of the Paleolithic in Kamchatka and the Problem of the Initial Occupation of America. *Arctic Anthropology* 5(1):191–203.

———. 1988. On the Road to America. *Natural History* 97(1):12–15.

———. 1989. The Palaeolithic of Kamchatka and Chukotka Related to the Problem of the Peopling of America via Beringia. Kamchatkan Palaeolithic. Circum-Pacific Prehistory Conference, Seattle.

Dikov, N. N., and E. E. Titov. 1984. Problems of the Stratification and Periodization of the Ushki Sites. *Arctic Anthropology* 21(2):69–80.

Chapter 5

SOUTHERN PRIMORYE

INTRODUCTION

The territory of Primorye includes the coast of the Sea of Japan, the Ussuriisk–Shankhayst plain, and the southeastern leg of the Sikhote–Alin mountain range. This territory is bounded on the east by the Sikhote–Alin mountains, which extend to the northeast, parallel to the coast of the Sea of Japan and the Straits of Tatar, for approximately 1,900 km. The entire region is part of Eurasia.

Environmental Research

During the Upper Pleistocene, the northeast part of Primorye adjoined the southern reaches of Beringia. The deposits of the Upper Quaternary period in the coastal zone of Primorye are represented by alluvial soils: shingle (gravels), sands, and loamy soils; coastal-maritime sediments: sands, aleurites, peat, and loam; and various slope-wash deposits. These cyclical deposits formed: (1) a floodplain terrace at 2–4 meters, (2) a floodplain terrace at 4–6 meters, and (3) a floodplain terrace at 10–12 meters. These are evident at the base of the Holocene section at river mouths on the coast of the Sea of Japan

(Korotky et al. 1988:63–65). The deposits are divided into five stratigraphic horizons, four of which are Upper Quaternary deposits: Q^1III, the Nakhodkinsky; Q^2III, the Lazovsky; Q^3III, the Chernoruchynsky; and Q^4III, the Partizansky. The fifth horizon represents the Holocene-South Seaside (Korotky et al. 1980:73–74).

The Nakhodkinsky horizon (Q^1III) is comparable to the midglacial periods of Riss–Würm in Europe and Kazantsevsky in eastern Siberia. Two ^{14}C dates have been obtained for these deposits: 54,000 BP (Kn-2406) and 57,000 BP (Ki-2407).

The pollen spectrum is characterized by three groups of pollen: (1) pollen of broad-leafed trees, abundantly and diversely represented by *Quercus, Ulmus, Juglans, Corylus, Acer, Carpinus, Oleaceae*; (2) pollen of coniferous trees (*Pinus koraiensis, P. n/p haploxyion, P. pumila, Abies, Picea* section (sect.) *Omorica, Picea* sect. *Eupicea, Larix*); and (3) pollen of small-leafed trees (*Betula manshurica, B. schmidtii, B. costata, B. dahurica*, and other *Betula* spp.).

The Lazovsky horizon (Q^2III) is comparable to the Würm glaciation period in Europe and the Zyrian in eastern Siberia. Sediment deposition in this horizon in the coastal zone occurred against the background of the regression of the Sea of Japan. The pollen of small-leafed species is predominant in the pollen spectrum of the Lazovsky horizon's deposits: *Betula* sect. *Nonal, Betula manshurica, Al-*

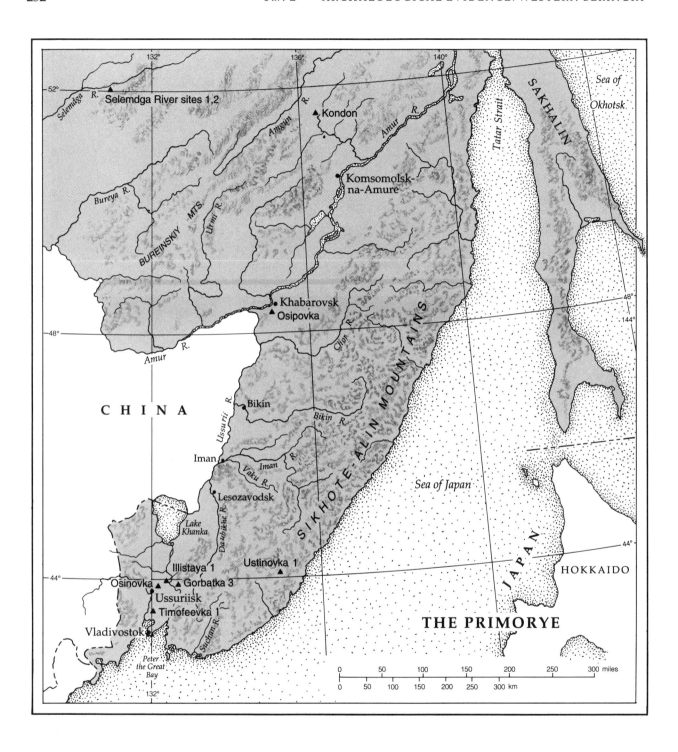

naster, et al. (^{14}C dates: 53,400 ± 1600 BP [Ki-2305], above 50,000 [SOAN AN-279]).

The Chernoruchynsky horizon (Q³III) is comparable to warming periods in the late Pleistocene: the mid-Würm in Europe and the Karginsky in eastern Siberia. Deposits of this period have been well stud-

ied at the base of the 6-to-8-meter terrace of the Venyukova River. Three pollen spectra are discerned in the section.

In the lowest spectrum, identified at the base of the terrace (^{14}C date: 40,000 ± 1200 BP), the pollen *Pinus koraiensis* (up to 40%) is predominant among

Vicinity of the Ustinovka site.

the coniferous species. Of the small-leafed species, the pollen *Betula* sp. *albae* ranges up to 20% and the *Betula* sp. *costate* extends to 10%. The broad-leafed species are represented by *Quercus* pollen (up to 10%), *Ulmus* (up to 4%), and *Fraxinus, Juglans,* and *Corylus* occasionally. In the second spectrum (3.15–4.50 m) the pollen of small-leafed species predominate (*Betula manshurica,* 4.2%; *B. dahurica,* 1.5%; *B. fruticosa,* 2%; *B. exilis,* 3%; *B. middendorffii,* 5%; *Betula* sp. 2–8%; *Alnus,* 4%; *Alnaster,* 26–80%). Coniferous species pollen content is moderate: *Picea* sect. *Omorica,* 3–22%; *Picea* sect. *Eupicea,* 3%; *P. koraiensis,* 3%. Pollen of broad-leafed species has been noted infrequently and only in scattered samples (*Quercus, Ulmus, Corylus*).

Two ^{14}C dates were obtained for the peat layer in this part of the core section (3.5–3.6 meters): 22,700 ± 800 (MAG-341) and 29,429 ± 475 BP (TIG-DVGU-17).

The third pollen spectrum (2.23–3.15 meters) differs from the second in the increased quantity of pollen from coniferous species (especially *Picea* sect. *Omorica,* 15–31%; *P.* sect. *Eupicea,* 5–27%; to a lesser extent, *Abies,* 2–9%; *Pinus koraiensis,* 3–6%; *Pinus* n/p *haploxylon,* 1–6%). In this spectrum *Larix* pollen is present constantly (0.5–4.0%). The small-leafed

species are represented by the pollen of arboreal birches. Also notable is a sharp decrease in the quantity of alder pollen (1–5%).

The Partizansky horizon (Q⁴III) is comparable to the late Würm in Europe and the Sartan period in eastern Siberia. The stratigraphic level representing the middle-to-late Würm boundary has been studied in detail in the upper section of the 6-to-8-meter terrace of the Maksimovka River, producing five radiocarbon dates in the range of 19,773 to 23,292 BP. The formation of ancient alluvium in this section (2.0–4.6 m) relates to the beginning of a cooling period at the end of the Karginsky period. The ubiquitous pollen from the base of this section (with ^{14}C dates 23,292; 21,932; 21,556 BP) is one of the dark coniferous species (*Picea* sect. *Omorica,* up to 6.3%; *Picea* sect. *Eupicea,* 2.1%). It is accompanied by pollen of small-leafed species, particularly shrub birches and alders. Upward in the section (^{14}C: 19,793) there is a complete predominance of small-leafed pollen, particularly of the frigid forms (*Betula exilis,* up to 20%; *B. nana,* up to 35%; *B. middendorffii,* up to 14.4%; *Alnaster,* up to 21%). The presence of the *Pinus* n/p *haploxylon* pollen (up to 9.3%) probably indicates the presence of cedar creeper. Such a spectrum corresponds with the climatic minimum of the late Würm (Korotky et al. 1988:64–77). At Venyukova the late Würm–early Holocene boundary has been identified in the 3-to-4-meter terrace section (^{14}C date: 10,317 ± 70 BP).

PALAEONTOLOGY

The most impressive late Pleistocene faunal complex was discovered during excavations of the palaeontological site called Geographical Society Cave. Large animals represented in the assemblage include mammoth, horse, woolly rhinoceros, roe deer, deer, cave lion (tiger), bison, bear, elk, and ounce (snow leopard) (Ovodov 1977). The ^{14}C date obtained for the layer that overlies this assemblage is 32,570 ± 1510 BP (Ovodov 1977; Gerasimov et al. 1983). Mammoth existed in Primorye in more recent times, as indicated by bone remains from the later Palaeolithic site in the Bliznets cave in the Partizanskaya River basin (^{14}C date: 11,965 ± 65 BP [SOAN-1550], Korotky et al. 1988).

PERIGLACIAL AND OTHER PHENOMENA

The character and extent of late Quaternary glaciation in the Far East and in Primorye is debatable. Most researchers consider it impossible for massive ice sheets to have developed, even in the mountain masses of Sikhote–Alin (Alekseev and Golubeva 1973; Korotky et al. 1988). In southern Primorye, nivation cirques developed with a spread of snow cover, mainly on eastern-exposure slopes of Sikhote–Alin, during the late Pleistocene cold maximum (Korotky et al. 1988). The cold and dry climate during the late Pleistocene cold maximum was conducive to the development of permafrost in Primorye. Permafrost evidence is seen in the relief of the terrace and deposits of various landscape/climate zones. Cryogenic formations, the result of deep freezing of mountain rocks, were found in sections I (Q^{3-4}_{III}) and II (Q^{1-2}_{III}) of super-floodplain terraces in the river valleys of Sikhote–Alin. Extensive permafrost has led to an active frost-stable fissure formation. In a number of cases the late Würm-phase fissure/polygonal soil formation preceded deformation related to solifluction processes. Intensive solifluction evidence has been found in river terrace sections of southern Primorye (Alekseev and Golubeva 1973; Korotky et al. 1980; Korotky et al. 1988). Reiterated patterns of frost-stable fissures and solifluction layers identified in these sections indicate a fluctuating climate.

ARCHAEOLOGY

The development of the Upper Palaeolithic of southern Primorye may be divided into three stages corresponding, respectively, to the Nakhodkinsky warming period, the Lazovsky cooling period, and the terminal/late Pleistocene. Osinovka, a stratified multicomponent site on the 20-to-25-meter terrace of the right bank of the Osinovka, near the city of Ussuriisk, should be considered the oldest known early Upper Palaeolithic site in southern Primorye.

Well-stratified deposits at Osinovka include three cultural horizons: early Upper Palaeolithic, late Palaeolithic–Mesolithic, and a find zone representing the age of metallurgy ("Palaeo-Metal" period).

The stone tools of the lower horizon are primitive, consisting of disk-like pebble cores, Levallois-like cores with radial and subparallel flaking techniques, hacking tools such as choppers, pebble scrapers, massive blade-like flakes, blades, and blade products. As a whole, the stone industry is characterized by the combination of Levallois lithic reduction and a "Soan"-like pebble-tool technology (Okladnikov and Derevianko 1973; Derevianko 1983).

Through geological evidence and pollen spectra data, the cultural remains of the lower horizon are attributed to one of the warm periods of the late Pleistocene (the Nakhodkinsky), when there were forest landscapes with broad-leafed tree species in the valley of Osinovka. The *absolute* chronology of these deposits is in the range of 35,000–40,000 years ago, although the age range for the corresponding geological horizon is estimated to be ca. 54,000–57,000 BP.

Currently in Primorye and Priamurye there are more than ten known locations with stone tools of the Osinovka type: these assemblages define the Osinovskaya complex. The stone industries of Osinovskaya sites are analogous to complexes of similar age—the middle Upper Palaeolithic—in the Amur basin, North China, the Korean Peninsula, and southeastern Asia (Derevianko 1983).

The second cultural stage includes those sites which correspond to the Lazovsky time period (ca. 40,000–54,000 BP). They developed during a cooling period and the spread of steppe landscapes. One of these, the Geographical Society Cave site, is located 25 meters above the floodplain of the Suchan River, near the village of Ekaterinovka, not far from the Nakhodka River. Cultural remains are contained within a layer of brown clay filled with limestone debris (layer 4 of the cave fill).

Stone tools are few, but characteristic finds are bipolar pebble cores and massive flakes with indications of preparation and utilization. The preparation technique for cores is analogous to the Osinovskaya technique. Among bone artifacts is a socket for a stone tool, made of elk horn (Derevianko 1983). Tools were found together with numerous remains of Pleistocene fauna, as described earlier. Many bones are broken and split. Ends of some tubular bones are slightly ground. As indicated previously, the overlying layer of light grey loam has a radiocarbon date of $32,570 \pm 1510$ years ago (Gerasimov et al. 1983). Obviously, the underlying layer with cultural remains should be older.

The third stage of the Upper Palaeolithic of Primorye includes sites from the final phase of the Pleistocene, whose stone industry is characterized by a blade technology. Those sites where excavations have been carried out include: Ustinovka 1, 2, and 4; Suvorovo 2 and 4, in the Zerkalnaya River valley; the lower horizon of the Oleniy site in the Artemovka River valley, and the sites described in this chapter by Kuznetsov—Gorbatka 3, Illistaya 1 and 2, and Timofeevka 1.

R. S. V.

REFERENCES CITED

Alekseev, M. N., and L. V. Golubeva. 1973. *New Data on the Stratigraphy of the Pleistocene of Southern Primorye: Stratigraphy, Palaeography, and Lithogenesis of the Eurasian Anthropogene.* Moscow. (In Russian)

Derevianko, A. P. 1983. *The Palaeolithic of the Far East and Korea.* Novosibirsk: Nauka. (In Russian)

Gerasimov, I. P., O. A. Chichigova, A. E. Cherkinsky, et al. 1983. Radiocarbon Results from the Radiometric Laboratory of the Institute of Geography, Academy of Science, USSR. *Bulletin of the Commission on Research of the Quaternary Period*, No. 52. (In Russian)

Korotky, A. M., L. P. Karaulova, and T. S. Troitskaya. 1980. *Quaternary Deposits of Primorye: Stratigraphy and Palaeography.* Novosibirsk. (In Russian)

Korotky, A. M., S. P. Pletnev, V. S. Pushkar, T. A. Grebennikova, N. G. Razhigaeva, E. D. Sakhebgareeva, and L. M Mokhora. 1988. *The Development of the Natural Environment of the Southern Far East.* Moscow: Nauka. (In Russian)

Okladnikov, A. P., and A. P. Derevianko. 1973. *The Distant Past of Primorye and Priamurye.* Vladivostok. (In Russian)

Ovodov, N. D. 1977. *Late Anthropogenic Mammalian Fauna in the Southern Ussury Land: Fauna and Taxonomy of Vertebrates of Siberia.* Novosibirsk: Nauka. (In Russian)

USTINOVKA 1

Ruslan S. Vasilievsky

The Ustinovka site is located on a bluff on the right bank of Bezymyannyy Stream near its entrance into the Zerkalnaya (formerly Tadushi) River. This stream enters the Zerkalnaya approximately 30 km from the point at which it empties into the Sea of Japan, 2.5 km northeast of the village of Ustinovka, and 60 km northwest of Olga Bay (approximately 44°03′N, 135°04′E). The site is confined to the second terrace of the Zerkalnaya River. The terrace is 10–12 meters above the contemporary river level. To the southeast and the northwest the site is bordered by the terrace-like spurs of Sikhote–Alin, which form a natural amphitheater.

The terrace on which the site is located slopes 2–3° in a northwest-to-southeast direction. Loose, friable terrace deposits are up to 4 m thick. Its cut-

Figure 5–1   Ustinovka site plan.

Excavation at Ustinovka 1.

bank, exposed to the Zerkalnaya River, has been destroyed by the Ustinovka–Bogopol highway (see Fig. 5–1).

Currently, the site area and its environment consist of dense, broad-leafed forest (Dahurian oak, Dahurian and white-stemmed birch, Manchurian nut, hazel, and alder trees), as well as bushes and lianas (lemon tree, various types of aralias). Combinations of elements of Manchurian and Okhotskiy flora are characteristic. The Zerkalnaya River floodplain is actually a water meadow. In the surrounding forest one may see wild boar, brown bear, musk deer, tiger, and Manchurian hare, as well as many birds. The insect fauna are quite diverse too. It includes some specific relict forms (among butterflies, e.g., *Luehdorfia*) (Ovodov 1977). In July and August each year the far eastern salmon migrate up the river.

The region of Ustinovka lies in a zone that is characterized by a continental climate with monsoon features. The average temperature in January is −13.7° C, and in June +20.6° C. Summer monsoon rains begin in July and continue until mid-August. At times precipitation reaches up to 150 mm per 24-hour period. Overall, the period from April to October sees 92 percent of the annual precipitation norm, leaving the remaining 8 percent for the rest of the year. The riverside area has a precipitation rate of 550–600 mm per year (Far East 1961).

HISTORY OF RESEARCH

Ustinovka was discovered in July 1954 by geologist B. F. Petrun, who made a geomorphological study of the site and gathered a small collection of artifacts. He related those artifacts temporally to the Verholenskaya Gora site in Pribaikalye and sites near Khabarovsk on the Amur River (Petrun 1956).

In the summer of 1961, J. V. Andreeva, while surveying coastal regions of the Sea of Japan, launched an excavation of a 6-sq.-m area at the Ustinovka site. Over 250 artifacts were collected here. In July 1963, full-scale excavations were conducted by A. P. Okladnikov. An area of 20 sq. m was excavated and 13,128 artifacts were collected. Okladnikov described the topography and stratigraphy of the site and analyzed the stone industry (Okladnikov 1964, 1966). In the summer of 1966, J. V. Andreeva, with the participation of Quaternary geologist T. I. Khudyakov, researched the site and conducted a series of excavations and tests in a 48-sq.-m area; 14,113 artifacts were collected (Andreeva and Khudyakov 1973). In 1968 excavations were conducted by A. P. Okladnikov in a 83-sq.-m area, and over 15,000 artifacts were collected. Okladnikov identified two cultural layers; the upper one is dated to the Mesolithic and the lower one to the late Palaeolithic (Okladnikov 1969). Excavations of Ustinovka were continued by R. S. Vasilievsky with the participation of V. A. Kashin in 1980–81 (Vasilievsky and Kashin 1983) and in 1984 and 1986 with S. A. Gladyshev (Vasilievsky and Gladyshev 1989). During these years an area of 192 sq. m was excavated, and a collection of 60,401 artifacts was made. Overall, the Ustinovka 1 collection currently includes nearly 100,000 artifacts, of which 15 percent are tools and 85 percent are by-products of tool production (flakes, spalls, cortex spalls, and core fragments).

Research of the stratigraphy and geology of the site was conducted by archaeologists and the following geologists: T. I. Khudyakov in 1966, M. N. Alekseev and L. V. Golubeva in 1971 (see Alekseev and Golubeva 1973), and A. M. Korotky in 1986 (Korotky et al. 1980, 1988).

Stratigraphy and Dating

Cultural remains are contained within multilayered, multicolored, and multicompositional flood deposit loams with sand lens inclusions. These loams represent covering deposits of a former floodplain (terrace II) of the Zerkalnaya River. They overlie pebble-boulder deposits left by the Bezymyannyy Stream (a left bank tributary of the Zerkalnaya River). Terrace formation took place either in the Karginsky-Sartan (Alekseev and Golubeva 1973; Tseitlin 1979) or the Kazantsevsky–Zyrian periods, the latter of which corresponds to the Nakhodkinsky–Lazovsky period in Primorye (Korotky and Karaulova 1975). The maximum thickness of the

flood deposits (1.4–1.5 m) was in the northwestern area of the site; the minimum in the southeastern parts (0.7 m).

The following is an example of the most typical stratigraphic section, obtained as a result of the 1980 excavations (see Figure 5–2):

1. Modern soil layer, 0.0 m–0.15 m thick.
2. Brown-grey loam, .26 m–.4 m thick. Humus, in the form of vegetable rootlets, penetrates the entire thickness. The upper boundary of the layer is even, but the lower one is corrugated and unclear. The layer reveals permafrost evidence in the form of frost wedges. The maximum width of these wedges sometimes reaches 0.27–0.29 m; depth is 0.45–0.48 m.
3. Fawn-colored loam with signs of buried soil and with inclusions of fine gravel. Thickness of this layer varies from 0.12 to 0.25 m. Its lower boundary is marked by seams of fine gravel; the upper archaeological complex is associated with this layer.

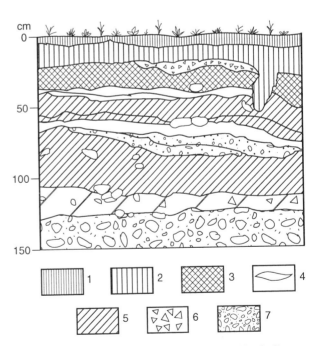

Figure 5–2 Ustinovka 1. Stratigraphic profile. Sediment symbols as follows: (1) modern soil; (2) loamy grey-brown soil; (3) fawn-colored loam; (4) seams of grey-blue clays; (5) brown to dark brown loams; (6) rock debris; (7) pebbles.

4. Dark brown loam of mixed composition. In the middle part of the layer there are seams of grey-blue clay, sand, and gravel. The clays are in the form of lenses, whose maximum thickness is 0.16 m. Layer thickness varies from 0.1 to 1.10 m. The lower archaeological complex is associated with this layer.

5. Medium-sized, lightly rounded pebbles. The space between separate pebbles and boulders is filled with loamy soil rich in tin. The unit is up to 1 m thick. No artifacts were found at this level.

The seams of grey-blue clay identified in the fourth layer are actually solifluction deposits. They denote the coldest phase of the late Pleistocene Partizansky period in Primorye (comparable to phase II of the Sartan in eastern Siberia). Cold temperature anomalies in the form of frost wedges, found in the second layer, formed during the period of climatic minimum at the Pleistocene–Holocene boundary (about 10,000 years ago).

The seams of grey-blue clay thus serve as a natural chronological boundary, separating the upper and lower layers. Therefore, if cryoturbated clays were indeed accumulating at the beginning of the Partizansky (Sartan) period, then the cultural remains lying beneath them must date to no less than 16,000 years ago. Artifacts of the upper archaeological complex contained within loam deposits are related to the last global temperature decrease in Primorye (11,780 ± 50 bp [SOAN-628]—the Norilsk stage of the Sartan Ice Age in Siberia, 10,300–10,800 BP) and thus date between 10,000 and 12,000 BP.

The dating of the archaeological complexes of Ustinovka 1 is confirmed by correlation with well-stratified and ^{14}C-dated site complexes on Hokkaido Island in Japan (Derevianko 1983:89–91; Vasilievsky and Gladyshev 1989:99–106).

FAUNA AND FLORA

No faunal remains of Pleistocene mammals were found at the site. Pollen spore and diatom samples were collected from profile 1 and profile 2 (in the trench adjacent to the 1980 excavation) to determine palaeoclimatic and palynological conditions during deposit formation.

From profile 1, a representative amount of pollen was found only in the surface sample. For profile 2, representative pollen spectra were identified for the depth interval of 0.55–0.90 m. Pollen of arboreal and herbaceous plants is present in approximately equal amounts. The pollen of arboreal species is represented mainly by birches, with a slight predominance of shrub forms as well (*Betula*, 36–76%). Poorly preserved scattered seeds of broad-leafed species were found. The predominant herbaceous pollen comprises varieties of *Artemisia* (up to 68%) and various grasses (up to 32%). Spores are represented mainly by Polypodiaceae.

The data indicate that forest-tundra had already developed in the site vicinity, interfingered with tundra-steppe (represented by the herbaceous pollen) during formation of these deposits. Climatic conditions were cold and dry. No diatoms were found in profiles 1 and 2.

ARTIFACT ASSEMBLAGE

Collections from three field seasons (1980–81, 1984) are in the process of analysis. The total number of artifacts is 60,398 specimens. The assemblage from the upper complex consists of 29,642 artifacts; the lower one, 30,756 artifacts.

The Upper Complex

Artifacts were distributed evenly within stratigraphic level 3. No cultural features have been identified. Siliceous tuff was the main raw material used for tool manufacturing. Outcrops of siliceous tuff appear in the rocky exposures near the site.

The assemblage represents functional artifact categories characterized by various preparation and reduction techniques.

Cores Core technology is the principal determinant of the character of the site's lithic industry. Cores are divided into eight classes based on platform and face reduction preparation techniques, orientation of reduction scars, and the quantity and placement of the striking platforms relative to these reduction faces. They consist of: single-platform,

Artifact Categories from the Upper Complex

Category	Quantity	Percentage
Cores	89	0.3
Core Blanks	21	0.1
Core Debris	11	—
Core Spalls	37	0.1
Tools	180	0.6
Blades	2,276	7.7
Waste Flakes	23,796	80.3
Macrospalls	42	0.1
Spalls	1,973	6.7
Microspalls	52	0.2
Cortex Spalls	1,165	3.9

single-faced (36); single-platform, double-faced (5); double-platform (bipolar), single-faced (9); double-platform, double-faced (6); multiple-platform, multifaceted (1); amorphous, with unsystematic preform reduction techniques (15); wedge-shaped (14); disk-shaped, with radial reduction techniques (3). The vast majority of cores are made of angular pieces of siliceous tuff (75 of 89).

Sizes of cores vary from very large (33 × 13 × 12 cm) to small (6 × 3 × 4 cm). Judging by the orientation of flake removal scars at Ustinovka 1, all systems of reduction were used: parallel, subparallel, convergent, radial, and unsystematic.

SINGLE-PLATFORM, SINGLE-FACED (36) This variable group consists of artifacts of different shapes, sizes, outlines, and stages of use wear. Of these, 10 have flat platforms and 26 have acute striking platforms (Fig. 5–3: *e*). Two types of cores are well defined: subprismatic, flattened with tapered base (Fig. 5–3: *c*)—and cubiform. Blades alone were drawn from subprismatic flattened cores; flakes and blades, from the cubiform ones.

SINGLE PLATFORM, DOUBLE-FACED (5) Three of these artifacts have two flat, laterally adjoining reduction surfaces. In the case of the other two specimens, the reduction surfaces are opposite each other (bipolar). Cores may be divided into two types: subprismatic and cubiform. Four of them were used to obtain blades and one for flake removal (average removal scar size is 3.5 cm × 2.3 cm).

DOUBLE-PLATFORM (BIPOLAR), SINGLE-FACED (9) All of these cores have alternate striking platforms (i.e., they are bipolar). The striking platforms are small in size and are shaped like irregular polygons; the platform angles vary. Generally, these cores constitute a discrete group with thoroughly prepared striking platforms, thinned surfaces, and use retouch. They form one type, the "subprismatic core," which was used to produce blades.

DOUBLE-PLATFORM, DOUBLE-FACED (I.E., 2 REDUCTION SURFACES) (6) The striking platforms of five cores are bipolar and one has adjoining striking platforms (two adjacent surfaces). In most cases the striking platforms have identical angles. Generally, these are close in shape and appearance to the cores of the third group and belong to the subprismatic class.

MULTIPLE-PLATFORM, MULTIFACETED (1) This core has three striking platforms and three reduction faces. All three platforms are adjoining; one is flat, the other two are angled, from the reduction surface to the counterface. The largest blade scar has a length of 11.7 cm and a width of 2.7 cm; the smallest one, 1.9 cm and 0.7 cm. This core belongs to the cubiform class.

AMORPHOUS, WITH UNSYSTEMATIC REDUCTION EVIDENCE, USED TO OBTAIN FLAKES (15) The blanks of these cores were fragments of thin slabs of siliceous stone. The number of striking platforms and reduction faces varies among the cores.

WEDGE-SHAPED (14) The reduction faces of these cores have a length-width ratio of over 3:1. The reduction face is short and intended for producing shortened microblades. The striking platform is always smooth, with no additional preparation. In most cases, wedge-shaped cores have straight bases.

There are two wedge-shaped core types, based on shaping techniques. The first one consists of wedge-shaped cores made on bifaces (4) (Fig. 5–3: *b*). These cores were discovered in the upper horizon of the first cultural layer. Wedge-shaped cores of the second type are made from massive flakes with a trapeziform cross section (Fig. 5–3: *a*).

DISK-SHAPED (BY A RADIAL REDUCTION TECHNIQUE) (3) Earlier, these specimens were interpreted as discoidal cores (Okladnikov 1966:355). However, in contrast to the classic oval discoidal cores, these are

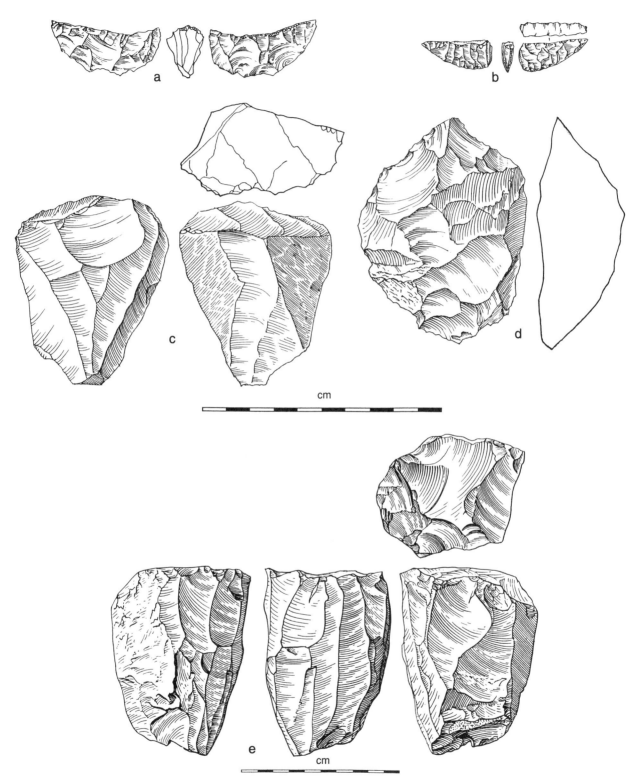

Figure 5–3   Cores from the upper complex of the Ustinovka 1 site: wedge-shaped microblade cores (*a*, *b*); single-face core with slanted platform (*c*); disk-shaped core (*d*); single-face, straight platform, subprismatic core (*e*).

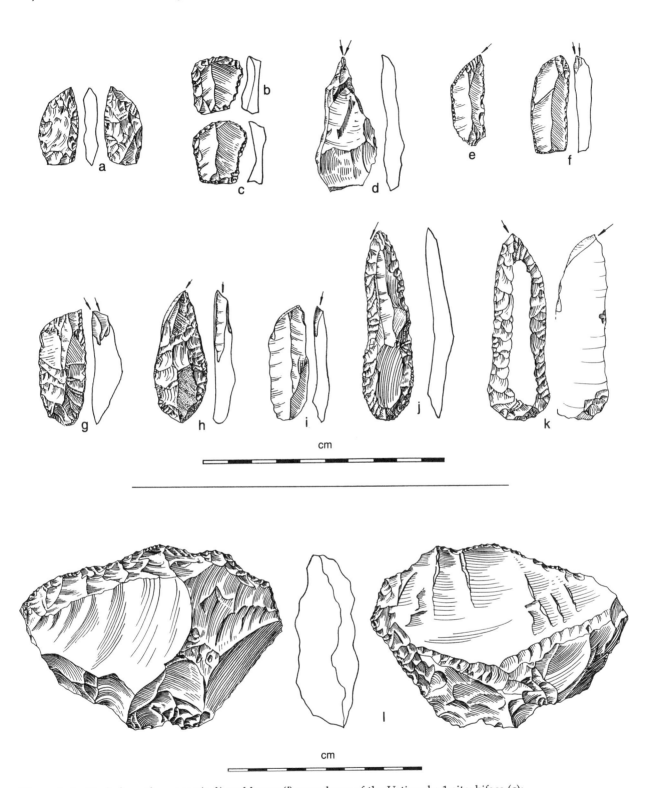

Figure 5–4 Tools from the upper (*a–k*) and lower (*l*) complexes of the Ustinovka 1 site: biface (*a*); small scrapers (*b, c*); dihedral burin (*d*); transverse burins (*e–k*); large scraper (*l*).

polyhedral; the cross section is polygonal. Preform removal scars occupy only part of the flaking surface. The ventral surface (back) of these cores is unworked (Fig. 5–3: *d*).

Tools   The upper complex includes 181 specimens categorized as tools. They constitute a typical late Palaeolithic tool assemblage and may be divided into the following types: large scrapers (41); small scrapers (43); knife-like tools (40); burins (transverse, dihedral, corner/angle, lateral) (36); bifaces (8); spokeshaves (5); perforators (5); gravers (3); adze-like tools (1).

LARGE SCRAPERS (41)   In most cases large, lenticular spalls and flakes of variously colored siliceous tuff were used as blanks; less frequently, fragments of blades were used. In plan-view large scrapers are predominantly oval, subtriangular, and subquadrangular. The scrapers are divided into seven classes based on the working edge: simple straight (10), simple convex (18), double straight (1), double biconvex (2), convergent (1), angular (4), and transverse straight (5).

SMALL SCRAPERS (43)   Small flakes and blades served as blanks for these scrapers. They are divided into five types: end (10), side (2), combined (2), amorphous (atypical) (28), and uniquely shaped (1) (Fig. 5–4: *b, c*).

KNIFE-LIKE TOOLS (40)   All are made on blades or blade fragments. There are four distinctive types based on the shape of the cutting edge as compared with the back, viz., convex-straight, straight-straight, etc.

BURINS (36)   Four types are distinguished: lateral (13), corner/angle (6), dihedral (10), and transverse (7). Blades, blade fragments, flakes, and blade-like spalls served as blanks (see Figure 5–4).

BIFACES (8)   Seven bifaces are made of greenish blue, thick, fine-grained flint; one is made of white siliceous tuff. Most bifaces are represented by fragments; there are only two whole specimens (Fig. 5–4: *a*). They are leaf-shaped. A characteristic specimen is a large oval. Both of its surfaces are thoroughly pressure-retouched. Imputing function is difficult. Possibly symmetrical specimens served as projectile points; asymmetrical bifaces may have been used as cutting tools.

SPOKESHAVES (5)   Spalls and, more rarely, flakes

were used as preforms. They are generally irregularly triangular and subquadrangular.

PERFORATORS (5)   Two perforators are made on small, thin blades, two are on medium-sized flakes, and one is on a thin slab. Tools of this group have a characteristic spur on the end, isolated by multifaceted retouch. Spurs of two specimens exhibit intensive polishing. Some of these artifacts may have been used as borers or drills.

GRAVERS—UNIFACIAL AND BIFACIAL (3)   Two are made on blade-like flakes, one on a flake. One of the former has a carefully prepared hafting projection on the base. This projection has been shaped by steep edge-retouch. This is the only known hafted graver in the collections of Ustinovka 1.

ADZE-LIKE TOOL (1)   The form of this piece and the location of the cutting edge suggests its placement in the adze category.

DEBITAGE   This constitutes the biggest category of artifacts (29,304). It is typical for sites of the late Palaeolithic. The most interesting group included in this category is the blades. The collection from the upper complex includes 2,276 blades and fragments; whole blades constitute only 28%. The predominant types are proximal (37%) and distal segments (26%). The classic prismatic blades are rare. Mainly represented are double- and triple-faceted blades; quadrangular ones are represented by only a few specimens. Nearly 25 percent of the blades are edge-retouched. Dimensions vary from large (13.8 × 6.2 cm) to small (2.5 × 0.8 cm), averaging approximately 8 × 3.5 cm.

The Lower Complex

The second stratigraphic concentration of artifacts (stratigraphic level 4) is lithologically heterogeneous; the thickness of the stratigraphic level is significantly greater than that of the first one. However, as shown in the mapping, the spatial distribution of cultural remains in the layer is generally not uniform. Two concentrations of artifacts have been observed. One of them is located in the northern part of the 1980 excavations and is related to an oval-shaped, cup-like depression. The other one is seen on the square section A-38 and consists of a complex of cores and flakes.

The lower complex comprises 30,756 artifacts.

Artifact Categories from the Lower Complex

Category	Quantity	Percentage
Cores	69	0.2
Core Blanks	74	0.2
Core Debris	32	0.1
Core Spalls	56	0.18
Tools	117	0.4
Blades	1,705	5.5
Waste Flakes	18,869	61.3
Macrospalls	9,325	30.3
Cortex Spalls	509	1.82

The primary raw material is siliceous tuff. Color varies from light grey to blue-grey-white (98%). Other rock types are rare; there are no obsidian artifacts.

Cores This group is divided into seven categories: single-platform, single-reduction faced (30); single-platform, double-faced (6); double-platform, single-faced (3); double-platform, double-faced (4); multiple-platform, multifaceted (8); wedge-shaped (16); and disk-like with radial reduction orientation (2).

The cores of the lower complex are very similar to those of the upper, but with some minor differences, e.g., striking platforms of the single-platform, single- and double-faced cores here are well prepared by removal of small thinning spalls. In most cases, these cores have angled faceted striking platforms (Fig. 5–5: *e, f*). Among them, there is a marked division by type: subprismatic, conical, and cubiform.

Wedge-shaped cores are made of massive spalls (triangular in cross section) and have a smooth striking platform. The lateral surfaces are only partly smoothed. On the core edges there are microblade removal scars.

Disk-like cores are made from large flakes. They are oval in plan with an irregular, lens-like cross section. The striking platforms are slightly angled, from the reduction surface to the counterface, and are retouched by small parallel spalls. Both broad surfaces of these cores are prepared by radial removal of spalls of different sizes, which

gives these specimens their disk-like appearance. However, the preforms for these cores were removed in parallel planes.

Tools There are 117 tools which may be divided into eight categories: large scrapers (43); small scrapers (10); knife-like tools (39); burins (13); spokeshaves (5); perforators (2); gravers (4); adze (1).

LARGE SCRAPERS (43) This is the most numerous category of the whole tool assemblage (37.1%). Primarily made on large flakes, they are divided into types based on the shape of the working edge(s): simple straight (4), simple convex (24), simple concave (3), double convexo-convex (4), double straight (2), angular (2), convergent (2), and simple transverse (2) (Fig. 5–4: *l*).

SMALL SCRAPERS (10) This collection, traditional in Upper Palaeolithic complexes, constitutes 8.6 percent of the tools in the lower layer. There are 5 end scrapers and 5 side scrapers. End scrapers are made on blades and blade-like flakes (Fig. 5–5: *a*).

KNIFE-LIKE TOOLS ON BLADES/FLAKES (39) Sixteen are flake knives and 23 are made on blades. Knives made on flakes have a subtriangular shape. The working edge is completely retouched. Working edges in the flake group tend to be straight; in the blade group, convex. The most distinctive form is made on an oblong blade of subtriangular outline. The tool is retouched all along the straight back edge of the blade. The opposing convex edge was retouched only at the upper end. Such tools are known in Japan as backed or "Moro" knives (Fig. 5–5: *b, c*) (Shimpei 1984).

BURINS (13) Burins are represented by two types: lateral (7) and dihedral (6). They are made on blades and blade fragments.

SPOKESHAVES (5) Four are made on flakes, one tool is on a large, blade-like spall. Two are combination tools: a spokeshave-scraper, with a steeply retouched convex working edge, and a spokeshave-knife, on which the edge opposite the shave has been retouched as a knife.

PERFORATORS (2) These are made on secondary flakes.

GRAVERS (4) All gravers are made from small blades. Three tools are worked on the dorsal surface and one on the ventral side. Two gravers have an

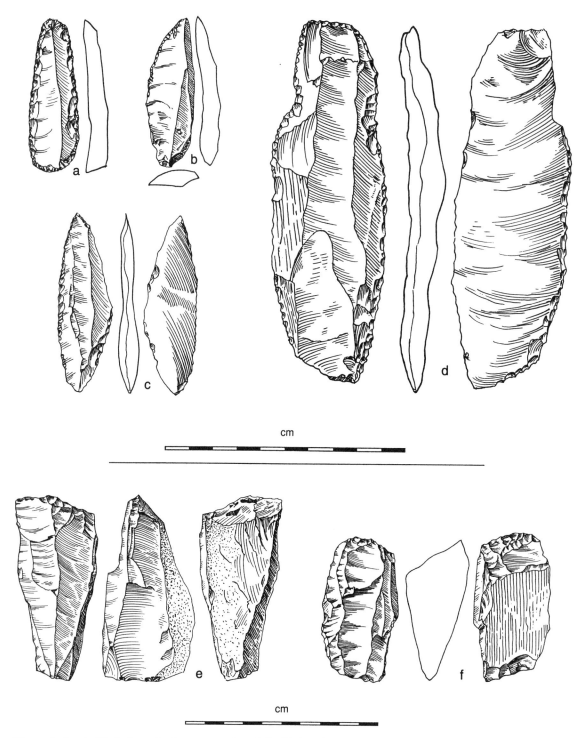

Figure 5–5   Artifacts from the lower complex of the Ustinovka 1 site: small end scraper (*a*); "Moro" type knives (*b, c*); knife with straight blade (*d*); subprismatic bipolar core with slanted striking platform (*e*); flattened bipolar core with slanted striking platform (*f*).

expanded diamond-shaped form in plan-view and an isolated hafting area–nozzle. One graver, leaf-shaped in form, is made from a triangular spall.

ADZE (1) This tool is made on a large, thick, blade-like spall. It is obovate in plan and triangular in cross section.

Summary

A comparison of the tool assemblages from the upper and lower complexes shows that bifaces, large- and small-oval end scrapers, and retouched transverse burins are not present in the lower.

In terms of form, the scrapers are more diverse in the lower complex; the distinctive artifacts, biconvex scrapers, have concave working edges. There are no such scrapers in the upper complex. The presence of such diagnostic elements as Moro-type knives in the lower complex and a hafted graver in the upper complex is also worth noting. Unifacial tool preparation is characteristic of the lower complex; in the upper one both uni- and bifacial techniques are present. However, these differences do not alter the overall pattern of genetic closeness of the complexes.

Interesting data about the functional purpose of tools and their correlation with manufacturing waste was obtained by T. F. Korobkova of the Institute of Archaeology, Leningrad, while researching the stone industry of Ustinovka I that resulted from the 1963 excavations. According to the results of this analysis, artifacts were distributed in the following way: 11,393 flakes of tuff and flint; 1,139 blades and their fragments; 395 cores and core debris; 51 spokeshaves for wood processing; 13 knives for cutting animal carcasses; 3 planing knives for bone and wood; 19 small and larger scrapers for processing skins; 8 pressure flakes; 2 borers for stones; 2 perforators; 1 fret-saw; 1 pick ax (?). In total, there were 13,128 specimens; of these, cores and manufacturing waste accounted for 13,028 and tools for 100. In addition, some flakes and blades were used without additional preparation. So, among the 1,176 flakes found in 1963, 133 had traces of utilization (Okladnikov 1977).

Such correlation of technical waste and tools is characteristic for the entire collection of Ustinovka

1. The abundance of manufacturing waste indicates that the initial preparation of raw materials was done at the site. At the same time, the presence in the collection of cores of finished and utilized specimens, along with flakes, indicates that at Ustinovka there was not only a workshop site but also a settlement. This was proven by the results of the 1968 and 1980 excavations, during which remains of possible dwellings (small cup-like depressions) were uncovered.

In 1980 the remains of one such feature were delineated in the northern part of the excavations: this was an oval depression 3.4 m long and 2 m wide, with the long axis oriented from northwest to southeast. The northeastern side of the hollow is quite steep; the other slopes are gentle. The greatest depth of the depression is 36 cm. The fill was composed of grey-blue clay with small seams of ochre clay. The ochre layer (thickness 2–5 cm) extended over all of the dwelling's floor, distinctively outlining the cup-like depression. In excavating this dwelling a concentration of large boulders lying close to each other was found distributed along the southern wall. This deposit represents the ruins of a rudimentary construction, possibly a foundation for plank beds. Remains of a circular stone hearth, 26 by 30 cm in diameter, were found along the southeastern wall of the depression. In the hearth fill were two distinct, thin (up to 1.5 cm) charcoal lenses.

The excavations at the workshop-habitation site of Ustinovka 1 yielded an impressive collection of artifacts. The stone industry of the site is characterized by a blade technology. The predominant core forms are single-platform and bipolar subprismatic cores, from whose one wide face long knife-like blades were drawn. Microblades were drawn from the wedge-shaped cores. Blades were mainly used as blanks in the manufacture of tools such as the knives and gravers. Blade fragments and blade-like flakes were used for small scrapers, spokeshaves, and burins, while large flakes were used for large scrapers. The bifacial flaking technique occurs in the upper horizon. The depth of the deposit and the numbers of stratified artifacts indicate that Ustinovka 1 was inhabited for a long period of time.

Even in the first years of research on Ustinovka 1, two culture-bearing stratigraphic layers were detected (Okladnikov 1966:353–54; Andreeva and Khudyakov 1973:16–25; Okladnikov and Derevianko 1973:75–76). According to geologists, these layers constitute two chronologically separate stages. The earlier one is "close to the end of the Upper Pleistocene" and the later one "relates to the late Sartan cooling period" (Tseitlin 1979:241–42).

The aforementioned geologists and palynologists also indicate that the cultural remains lying beneath the grey-blue clays must be no younger than 16,000 BP, and the artifacts of the upper archaeological complex must date within the limits of 10,000–11,000 BP.

The artifact collections of Ustinovka 1 accord with these geological and palynological conclusions. Basing his conclusions on comparison with materials from Japanese island sites, A. P. Okladnikov defined two cultural layers in the deposits of Ustinovka 1. He relegated the lower artifact-bearing unit to "17,000–15,000 years ago" and estimated the age of the upper layer—which contained leaf-shaped, bifacial gravers—as 12,000 years old (Okladnikov 1977:117).

A. P. Derevianko, noting the similarity of Ustinovka stone implements to preceramic complexes of Japan, provided a site age estimate of 10,000–20,000 BP (Derevianko 1983:91). Such conclusions were corroborated by comparison with Japanese stone tool types. The stone industries of Japan, with well-defined tool typologies and well-dated chronology, could indeed serve as indicators to the age of the Ustinovka 1 collections, whose artifacts are similar in a number of attributes to those of the stone industries of Japan. Of the upper Ustinovka 1 assemblage, end scrapers on blades, graver-bifaces and graver-unifaces, transverse burins with diagonal spalls, oblong wedge-shaped cores, and subconical cores with parallel scars on a single face find direct analogies in the complexes of northeastern Hokkaido–Shirataki 30 and 33, Tatikarusunai A (10,000–15,000 years ago), and Tatikov 2 and 3 (8,500–9,500 years ago) (Yosidzaki 1961). The "Moro" knives found in the Ustinovka 1 lower complex bear close resemblance to the Moro knives of Matsuyama on Kusu, which are dated between 14,000 and 15,000 years ago (Tatibana 1988:6–8). Comparison of Ustinovka's graver-unifaces with hafting elements (upper complex), subprismatic

and conical (lower complex) single-faced cores can also be made to artifacts from Japanese sites.

Based on the correlation with dated assemblages of Japanese artifacts, and considering the geological, palynological, and stratigraphic data, the age of the lower complex of Ustinovka 1 should date between 12,000 and 16,000 years ago, and the upper one between 10,000 and 11,000 years ago. The Ustinovka and northeastern Hokkaido complexes probably constitute parts of the same culture, synchronous or chronologically close in time, but extending over different geographic areas.

Correspondences to this blade technology culture are also found in the northern Pacific Ocean basin: in Kamchatka (Ushki sites) and on the Aleutian Islands (Anangula). The late Ushki Upper Palaeolithic culture of Kamchatka presents manufacturing traditions like those of the Ustinovka 1 collections (Dikov 1979:68–69): oblong, wedge-shaped cores; knives/bifaces; small end scrapers on blades; and small side scrapers with concave cutting edges. These correspondences are also present, as noted, in the stone tool assemblages of the Hokkaido sites (Shirataki, Tovarubetsu, and Tatikova).

Interesting analogies are also found in the comparison of materials from Ustinovka 1 and Anangula on the Aleutian Islands. Here the resemblance is evident in the technique of manufacturing oblong, wedge-shaped cores and the types of graver-unifaces on attenuated triangular blades; edge-retouched blades; transverse burins with diagonal burin spalls; and some types of small scrapers (Vasilievky 1973:42–44; Laughlin and Aigner 1966; McCartney and Veltre, this volume). However, assemblages from Anangula are younger than the analogous specimens in Primorye and Japan. Thus, we may speculate that cultural influences at the end of the Pleistocene probably were spreading from the south—from Primorye and Hokkaido—to the north, toward the outlying southwestern regions of Beringia.

REFERENCES CITED

Alekseev, M. N., and L. V. Golubeva. 1973. New Data on the Stratigraphy of the Pleistocene of Southern Primorye. In *Stratigraphy, Palaeography, and Lithogenesis of the Eurasian Anthropogene.* Moscow. (In Russian)

Andreeva, J. V., and G. I. Khudyakov. 1973. A Palaeolithic Memorial on the Zerkalnaya River. In *Materials on the History of the Far East.* Vladivostok. (In Russian)

Derevianko, A. P. 1983. *The Palaeolithic of the Far East and Korea.* Novosibirsk: Nauka. (In Russian)

Dikov, N. N. 1979. *Ancient Cultures of Northeastern Asia.* Moscow. (In Russian)

Far East. 1961. *Physical/Geographic Characteristics.* Moscow. (In Russian)

Korotky, A. M., and L. P. Karaulova. 1975. New Data on the Stratigraphy of the Quaternary Deposits of Primorye. In *Questions of Geomorphology and Quaternary Geology of the Southern Far Eastern USSR.* Vladivostok. (In Russian)

Korotky, A. M., L. P. Karaulova, and T. S. Troitskaya. 1980. Quaternary Deposits of Primorye: Stratigraphy and Palaeography. Proceedings of the Institute of Geology and Geophysics, SO AN SSSR, Vol. 429. Novosibirsk. (In Russian)

Korotky, A. M., S. P. Pletnev, V. S. Pushkar, T. A. Grebennikova, N. G. Razhigaeva, E. D. Sakhebgareeva, and L. M. Mokhova. 1988. *The Development of the Natural Environment of the Southern Far East.* Moscow: Nauka. (In Russian)

Laughlin, W., and J. Aigner. 1966. Preliminary Analysis of the Anangula Unifacial Core and Blade Industry. *Arctic Anthropology* 3:(2):41–56.

Okladnikov, A. P. 1964. Siberia in the Stone Age. In *Materials on the Ancient History of Siberia.* Ulan-Ude. (In Russian)

———. 1966. An Ancient Settlement on the Tadushi River, near the Village Ustinovka and the Problem of the Far-Eastern Mesolithic (in Connection with the Excavations of 1964). In *The Quaternary Period of Siberia.* Moscow: Nauka. (In Russian)

———. 1969. Excavations near Ustinovka. In *Archaeological Discoveries of 1968.* Moscow: Nauka. (In Russian)

———. 1977. The Mesolithic of the Far East. *Brief Reports of the Institute of Archaeology,* No. 149. Moscow. (In Russian)

Okladnikov, A. P., and A. P. Derevianko. 1973. *The Distant Past of Primorye and Priamurye.* Vladivostok. (In Russian)

Ovodov, N. D. 1977. Late Anthropogenic Mammalian Fauna in the Southern Ussury Land: Fauna and Taxonomy of Vertebrates of Siberia. *Proceedings, Biological Institute SO AN SSSR* 31:157–77. Novosibirsk: Nauka. (In Russian)

Petrun, B. F. 1956. Regarding the Age of River Terraces of Southern Primorye (Finding Stone Tools on the Coast of the Sea of Japan). In *Materials on Geology and Useful Fossils in Eastern Siberia and the Far East.* Moscow. (In Russian)

Shimpei, K. 1984. On a Backed Point from the Maritime Region, Far East. In *Human Culture and Environmental Studies in Northern Hokkaido,* No. 5. University of Tsukuba, Japan.

Tatibana, M. 1988. The Matoumaya Site. *Beppo University Herald,* Issue 1, No. 29. (In Japanese)

Tseitlin, S. M. 1979. *Geology of the Northern Asian Palaeolithic.* Moscow: Nauka. (In Russian)

Vasilievsky, R. S. 1973. Regarding the Role of Beringia in the Peopling of the Aleutian Islands. In *Beringia in the Cenozoic.* Vladivostok. (In Russian)

Vasilievsky, R. S., and S. A. Gladyshev. 1989. *The Upper Palaeolithic of Southern Primorye.* Novosibirsk: Nauka. (In Russian)

Vasilievsky, R. S., and V. A. Kashin. 1983. The 1980 Excavations of the Multi-layered Settlement Ustinovka I. In *The Palaeolithic of Siberia.* Novosibirsk. (In Russian)

Yosidzaki, M. 1961. The Settlement of Sirataki and Pre-Ceramic Cultures of Hokkaido. *Midzokuchaku Kenku* 16(1):13–23. (In Japanese)

LATE PALAEOLITHIC SITES OF THE RUSSIAN MARITIME PROVINCE PRIMORYE

Anatoly M. Kuznetsov

Before Y. A. Mochanov's discovery of the Dyuktai culture, there were close parallels with the Alaskan Palaeolithic at early sites in the southern part of the Russian Far East (the Maritime and Cis-Amur regions). Based on this evidence, in the 1950s A. P. Okladnikov supported Nelson's hypothesis of the peopling of the New World across northeastern Asia (Okladnikov 1959:42–43). During the 1960s, when Palaeolithic sites were found in the Yakutia and Kamchatka regions adjacent to Alaska, data from the Cis-Amur and Maritime provinces were no longer discussed in connection with New World origins. The obvious lack of Palaeolithic evidence from northeastern Asia, coupled with similarities of Palaeolithic assemblages of northern northwest America and those of the southern Russian Far East, make it possible to include these data in attempting to answer the Beringian question. Recent research in Primorye requires attention in this respect.

Favorable natural resources created like condi-

The Illistaya site.

tions for human life in Primorye. Many archaeological sites of various ages have been discovered here, but only a few are Palaeolithic. During the ninth to thirteenth centuries A.D. the province was a part of the Bokhay and Chyurdchenyan medieval states. There remained only small groups of Nanay and Udege hunters and fishers from the thirteenth-century Mongol invasion until the annexation of this region by Russia in 1860.

Palaeolithic evidence in Primorye was first found in the 1950s by the Far Eastern Archaeological Expedition of the Institute of Archaeology, Leningrad Branch of the USSR Academy of Sciences, led by A. P. Okladnikov. From 1953 to 1968 only six Palaeolithic localities were investigated: Osinovka, Ustinovka 1, Artemovka 1, Geographical Society Cave, Illyushkina Sopka, and Razdolnaya. That number is now augmented; to date, archaeologists from Novosibirsk, Ussuriisk, and Vladivostok have discovered about 40 sites with Palaeolithic-like stone artifacts, but only 11 have been excavated: Osinovka; Ustinovka 1, 4; Suvorovo 3, 4; the Geo-

graphical Society Cave; Artemovka 1; Gorbatka 3; Illistaya 1, 2; and Timofeevka 1. Many sites have been destroyed; there is no stratified multicomponent site, and Pleistocene fauna with rare artifacts were found only at the Geographical Society Cave site. The Palaeolithic sites in Primorye are distributed over three areas: (1) the Illistaya River basin, (2) the Razdolnaya River basin, and (3) the Zerkalnaya River basin.

The Illistaya basin in southwestern Primorye has approximately 15 sites (Osinovka; Gorbatka 2, 3, 5; Illistaya 1, 3; Ivanovka 2, 3; Khalkidon 2, 3; Firsanova Sopka; Chernigovka; Medvedya 1–3). In the Razdolnaya basin, also in southwestern Primorye, are approximately 13 sites (Razdolnaya 1, 2; Timofeevka 1–3; Illyushkina Sopka; Terekhovka; Novo-Nikolskaya; Krawnovka 2; Danilovka 2; Rakovka 7, etc.). The Artemovka 1 and the Geographical Society Cave sites are in the southeast. The Zerkalnaya River basin, in northeastern Primorye, has approximately 8 sites (Ustinovka 1–4; Suvorovo 3, 4; and others). (See regional map, Primorye.)

The assemblages of the lower levels of Osinovka and the Geographical Society Cave sites included rough pebble tools. At Ustinovka 1 and Razdolnaya 1, blade cores and large blades were obtained. True microblade remains were found primarily in the collections from Artemovka 1 and Firsanovaya Sopka. Thus, Okladnikov's first concept of the preceramic period of Primorye was based on the proposition of a pebble tool Palaeolithic and a blade-microblade Mesolithic here (Okladnikov 1959:26–40; 1966). But soon excavations revealed the core and blade component at Osinovka and new clay levels with blade cores under the Mesolithic layer at Ustinovka 1. The new data were interpreted by Okladnikov and A. P. Derevianko as evidence for a more ancient dating of the Osinovka lower level (from 20,000 to 30,000 BP) and for a Palaeolithic blade period for the Ustinovka culture (Okladnikov and Derevianko 1968).

The discovery of new sites in the 1970s (Gorbatka 3, Illistaya 1, Timofeevka 1, and Ustinovka 4) led Primoryan archaeologists V. I. Dyakov, A. M. Kuznetsov, and then V. A. Lynsha to suggest that there is no stratigraphic or archaeological evidence for separating the known Primoryan Palaeolithic sites into different cultures. They also rejected the ancient dates of Osinovka and Ustinovka 1 (Dyakov 1982; Kuznetsov 1988; Lynsha 1989). But the Novosibirsk investigators A. P. Derevianko, R. S. Vasilievsky, and S. A. Gladyshev in their latest works wrote that Osinovka represented the first stage of the Primoryan Palaeolithic and the Ustinovka sites represented the second stage. They also proposed a new age for Osinovka (at the Middle–Upper Palaeolithic boundary) and for Ustinovka 1 (about 10,000–24,000 BP) (Derevianko 1983, 1985; Vasilievsky and Gladyshev 1989). It may thus be said that there exist several interpretations for the Palaeolithic in Primorye. The foundation was laid down by A. P. Okladnikov and built upon by other Novosibirsk scholars. A second point of view was declared by the Primoryan archaeologists, based on their own data. In their view, the most firm evidence on stratigraphy, stone industry, and chronology has been obtained from the Palaeolithic sites of southwestern Primorye, which were excavated by an archaeological team of the Ussuriisk Pedagogical Institute led by this author.

STRATIGRAPHY

Most sites in Primorye, including Gorbatka 3 and Illistaya 1, are situated on terrace-like surfaces with basalt or granite bedrock at their base. Only the Timofeevka group of sites, Razdolnaya 1, and some Ustinovka sites were connected with terraces 8, 12, and 30 m in height. Typically, Palaeolithic assemblages were excavated from light loams, whitish-greyish-yellow in color, 8 to 30 cm thick. Stone artifacts at Gorbatka 3, Osinovka, Ustinovka 1, and Suvorovo 3 were also obtained from the next lower stratigraphic unit—brownish-blackish heavy loams from 30 to 60 cm thick. The Palaeolithic assemblages from light loams at Gorbatka 3, Illistaya 1, Timofeevka 1, Ustinovka 4, and Suvorovo 3 were mixed, with some polished stone implements and pot sherds of Bronze and early Iron ages. No remains of dwellings, hearths, or localized activity areas were discovered in the Primoryan sites. Shallow lenses of ash were found only at Ustinovka 1 and Gorbatka 3.

The most complete stratigraphic sequence was excavated in a central part of the Gorbatka 3 site. It is one of 12 Palaeolithic sites found in the middle area of the Illistaya River valley. This is a border territory between the Cis-Khanka plain and the western range of the Sinie Mountains where the river has carved out a vast basalt surface about 8 to 20 m above the Illistaya water level. Four major sedimentary units were exposed in the Gorbatka 3 area (Fig. 5–6). In order, from top to bottom, they are:

1. Humus, 8–15 cm thick, without cultural remains.
2. Greyish white light loam, 10–25 cm thick, with numerous Palaeolithic artifacts and a number of pot sherds of Bronze and early Iron ages.
3. Brown, brown-black, red-brown, and black heavy loams, 20–60 cm thick, with Palaeolithic artifacts.
4. Yellow loam, 40–60 cm thick, without cultural remains.
5. Detritus mantle-basalt bedrock.

There was a gradual slope (3–6°) toward the perimeter of the site surface and here the detritus mantle, yellow loam, and part of the dark-colored heavy loam band had eroded. It may be supposed

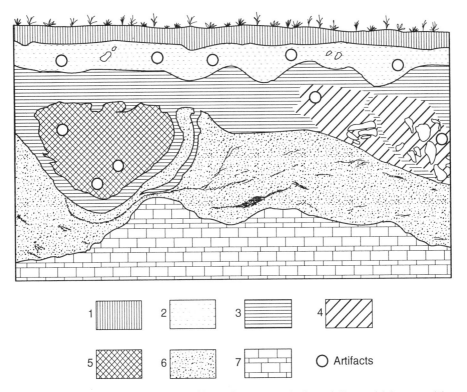

Figure 5–6 Gorbatka 3. Stratigraphic profile. Sediment symbols as follows: (1) humus; (2) greyish white light loam; (3) brown, brown-black, red-brown, and black heavy loams; (4) heavy loam with stones; (5) ice wedge; (6) yellow loam; (7) bedrock.

that the layers were destroyed by sheet erosion from rains and especially from typhoons, because a typhoon may bring from 350 to 400 mm precipitation in one or two days (*Physical Geography of the Primoryan Territory* 1990:49–50). Two levels of cryoturbation were discovered in the sedimentary units at Gorbatka 3. The upper one included ice wedges about 20 cm in height, dissecting the top of the heavy loam band. The next cryoturbations were ice wedges as well, but the height was about 60 cm and upper width was 1.2 m. These ice wedges dissected the entire dark heavy loam, crushing the top of the yellow loam (Fig. 5–6).

The Illistaya 1 profile is similar in part to the Gorbatka 3 stratigraphic units. Excavation pits at this site were dug on the upper section of the sloping (6–8°) terrace surface. Only three depositional bands remained:

1. Humus, 8–15 cm thick, with iron implements and pottery of the medieval epoch.
2. Yellowish white light loam, 8–30 cm thick, with Palaeolithic artifacts and Bronze Age pot sherds and stone implements.
3. Red-brown heavy loam, 30–50 cm thick, and overlapping basalt bedrock, without cultural remains.

The absence of yellow loam and detritus mantle in this profile may be explained by sheet erosion since Illistaya 1 is situated 5 km to the northeast of Gorbatka 3, in a similar topographic position. There were no cryoturbations similar to the Gorbatka 3 ice wedges at this site.

The Osinovka stratigraphic profile is very close to those of Gorbatka 3 and Illistaya 1. This site was 16 km away from the Gorbatka sites and was lo-

The Osinovka site.

cated on the 25-m terrace-like surface, with granite bedrock at the base. According to A. P. Okladnikov and V. V. Nikolskaya, there were several kinds of exposed deposits:

1. Humus, 8–15 cm thick, with dwelling pits and Iron Age implements.
2. Greyish straw-colored light loam, 13 cm thick, with Neolithic remains.
3. Greyish straw-colored heavier loam with granite detritus, Mesolithic horizon.
4. Dark greyish brown medium loam with granite detritus, Palaeolithic artifacts.
5. Clayed detritus overlapping bedrock, 50–52 cm.

The entire height of the Osinovka profile was about 152 cm (Okladnikov 1959:28–30; Nikolskaya 1970).

The three upper bands of the stratigraphic units at the Illistaya River Palaeolithic sites may be interpreted as brown forest soil. Primoryan pedologists have described three horizons in this soil:

A_1—humus, 7–12 cm thick.
A_2—eluvial, whitish horizon, 20–30 cm thick.
A_3—transitional, whitish yellow horizon, 0–10 cm thick.
B_1—illuvial horizon of brownish dark color, 30–40 cm thick (*Physical Geography of Primoryan Territory* 1990: 124–26).

The stratigraphy of the Razdolnaya River sites differed from those described previously because the investigated localities were located on river terraces. However, in this area some sites are situated on the Borisovka basalt plateau (Borisovka 2, Krawnovka 2) and their sediments are similar to brown forest soil. At the Timofeevka 1 site, the top of the 30-m terrace deposits were excavated:

1. Humus, 8–12 cm, without cultural remains.
2. Yellowish white light loam, 6–10 cm, with rare pot sherds.
3. Yellowish white light loam with a lot of small

Cryoturbation in the Gorbatka 3 excavation.

pebbles and detritus, 6–8 cm, with Palaeolithic artifacts.
4. Greyish white light loam with less clay, 10–22 cm, rare artifacts.
5. Reddish brown medium loam, about 65 cm.
6. Pebbles.

At Timofeevka 2, located on a 12-m terrace 0.5 km to the east of the previous locality, there was a band of sand and sandy loam with large ice wedges beneath the artifact-bearing yellowish white light loam and brown medium loam units. (According to Okladnikov, the stratigraphic profile most similar

to Gorbatka 3 was at Artemovka 1 [Okladnikov 1983]).

Composite deposits were excavated at Usti-novka 1, which is situated in the east range of the Sikhote–Alin Mountains. The stratigraphic profile here is described by R. S. Vasilievsky and S. A. Gladyshev:

Section I: light brown and dark grey loams with gravels and detritus subdivided into two hori-zons—(a) grey sandy loam, 3–10 cm; (b) greyish brown or fawn loams with some pebbles and small pieces of tuff;

Section II: two clay levels of different brown tints—(a) dark greyish brown and humic clay with a lens of light brown clay without pebbles and detritus; (b) dark brown ("chocolate") clay with tuff pieces;

Section III: boulder layer.

There was also a cryoturbated lens of grey-blue clay with sand between the layers in section II. Another cryoturbation level was described in section I of this profile. It consisted of ice wedges 20–25 cm long, dissecting the greyish brown loam. Except for the humus layer, all stratigraphic units at Ustinovka 1 (about 2 m in height) included stone artifacts. However, there were no clear differences among the items found in each of the layers in sections I and II. The genesis of Ustinovka 1 area deposits is suggested by its topography. The site is located on a 12-m-high, terrace-like surface, connected by a 95-m-long slope to a 30-m-high watershed stream surface. Palaeolithic artifacts were located at the slope and also on the upper surface (Vasilievsky and Kashin 1983; Vasilievsky and Gladyshev 1989:31–37). Evidence of sheet erosion was discovered in the sediments by the geologist G. I. Khudyakov (Andreeva and Khudyakov 1973).

Ustinovka 4 is situated on a flatter terrace surface not far from Ustinovka 1. Here was a simple stratigraphic sequence:

1. Humus.
2. Brown and yellow loam horizon.
3. Red-brown clay.
4. Pebbles with sand and clay lens.

Palaeolithic artifacts, according to V. I. Dyakov, were distributed in layer 2 and a part of them penetrated other horizons.

At the Suvorovo 3 site, a stratigraphic profile was excavated that is similar in part to the Ustinovka 4 one:

1. Humus, 4–12 cm thick.
2. Bright brown loam with some small pebbles, 6–18 cm thick.
3. Bright yellow heavy loam with pebbles and gravel, 2–24 cm thick.
4. Pebbles-gravel horizon, about 10 m thick.

Palaeolithic artifacts were in layers 2 and 3, but Vasilievsky and Gladyshev assume there is only one archaeological assemblage (Vasilievsky and Gladyshev 1989:79–81).

It is evident that there are unique stratigraphic sequences for each area where Palaeolithic sites occur in Primorye. However, the stone artifacts at each of these sites were obtained from shallow deposits at a depth of 20–30 cm to 150 cm from the current surface. These subaerial deposits were often destroyed by sheet erosion and were cryoturbated; artifacts contained in light, medium, or heavy loams were often compressed, redeposited, and mixed. Therefore, there are no primary contexts for artifacts of the Primoryan Palaeolithic.

LITHIC INVENTORY OF THE PRIMORYAN LATE PALAEOLITHIC

The most complete and characteristic artifact assemblages of this region were found at Gorbatka 3 and Illistaya 1. The main raw materials at these sites were big red diabase pebbles and small obsidian pebbles, taken from the Illistaya River banks. Because the raw materials were obtained not far from the sites, there were many unbroken and split pebbles, pebble fragments, and a large number of flakes. Other kinds of implements—cores, microblade cores, by-products of their manufacture, blades, microblades, tool preforms, and finished tools—were not as numerous (see Table 5-1), but were represented by tool groups ranging from several score to hundreds of artifacts.

Blade-core groups from these sites included large, crude specimens of red diabase and other rocks (Fig. 5–7: *a*) and small, but more diagnostic, implements made from obsidian pebbles (Fig. 5–7: *b, d*). All of these cores are characterized by their oblique platforms; preparation using one, two, or several blows without trimming; and blade detachment on usually one surface (sometimes two or three). Only 3.1 percent of the Gorbatka 3 cores and 0.4 percent of Illistaya 1 cores display the scars of previous processing. Blades from these sites were not numerous when compared to the number of cores. Blade form was not regular, and many blades were fragmented (Fig. 5–7: *g–j*).

Many obsidian flakes were removed from flat pebbles split on hard anvils without any previous

Table 5–1   Percentage of Artifact Classes from Palaeolithic Sites of the Russian Maritime Province

Sites	Excavated Area (sq. m)	Total Artifact Count	Debitage		Blade Cores	Micro-Blade Cores	Tech-nical Spalls	Micro-blade Core Parts	Blades	Micro-Blades	Flake Cores	Tool Pre-forms	Tools
			Flakes	Source Mat. or Frags.									
Gorbatka 3 (upper level)	300	38,272	84	9.0	1.0	0.30	0.04	0.90	0.7	0.2	3.3	0.1	0.9
Gorbatka 3 (lower level)	300	6,238	77	13.1	1.0	0.20	+	0.60	1.2	0.2	5.0	+	1.3
Illistaya 1	400	24,781	87	5.0	0.8	0.20	+	0.70	1.5	0.6	3.6	0.1	0.7
Timofeevka 1	100	3,602	90	2.8	0.4	0.30	0.4	0.10	3.4	0.1	—	1.9	1.1
Ustinovka 1 (upper level)	309	29,644	89	3.9	0.3	0.02	0.1	0.03	7.7	?	—	0.2	0.2
Ustinovka 1 (lower level)	309	30,757	92	1.8	0.2	0.01	0.2	0.02	5.5	?	—	0.2	0.1
Suvorovo 3	180	2,982	77	6.5	0.5	0.20	6.9	+	4.5	1.7	—	+	2.3

preparation. The distinctive attributes of such core-like artifacts are the absence of platforms, small scars on their tops and bases, and amorphous flake scars on the front and back surfaces (Fig. 5–7: *e, f*).

The Gorbatka 3 and Illistaya 1 assemblages also contained a microblade component: microblade cores, their blanks and by-products of manufacture, and microblades. The microblade cores may be subdivided into three main groups according to shape: wedge-shaped, conical/semiconical, and tall, wedge-shaped. The wedge-shaped cores include those made on flakes (23% at Gorbatka 3 and 23.8% at Illistaya 1); split bifaces (8.3% at the first site and 39% at the second); and boat-shaped blanks (35.6% at Gorbatka 3 and 23.8% at Illistaya 1). Platforms of some wedge-shaped cores on flakes were retouched, but usually there was no processing. Some of the wedge-shaped cores on boat-shaped blanks are very small (about 1 cm in height and 3 cm in length).

Conical and semiconical microblade cores were rare in these complexes (3% of Gorbatka 3 microblade cores and 3.4% at Illistaya 1). They are tall with rounded retouched platforms. Long microblade facets cover the whole surface of the cores or a large part of them. Pebble cortex may be seen on the back sides of the semiconical cores (Fig. 5–7: *k, l*).

The tall wedge-shaped cores share attributes with the previously mentioned microblade core forms. Their short platforms are oblique but retouched; microblade scars overlap on wide frontal facets, extending partially onto the lateral sides of the cores. Bases of such tall microcores are keel-like in form (Fig. 5–7: *m*). Microblade cores were associated with split bifaces (Fig. 5–7: *o*) and boat-shaped blanks (Fig. 5–7: *n*). A small number of ski spalls were present (Fig. 5–7: *p–r*); one implement had pebble cortex on the dorsal surface (Fig. 5–7: *p*). Microblades of these assemblages vary in length from 4 to 0.9 cm; many of them are irregular in shape (Fig. 5–7: *s–y*). There were no retouched microblades at the Gorbatka 3 and Illistaya 1 sites.

These assemblages yielded many end scrapers, retouched blades, and flakes. Not so numerous were bifaces, transverse burins, points, adzes, per-

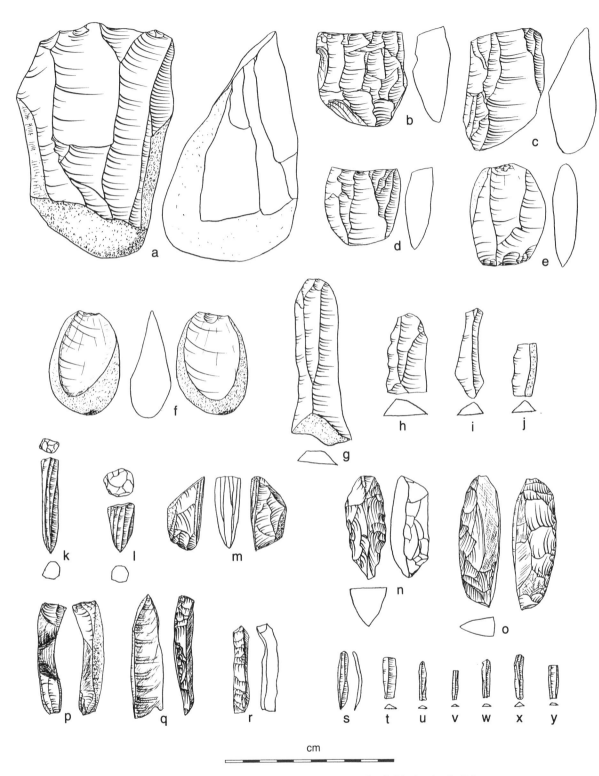

Figure 5–7   Artifacts from Gorbatka 3 and Illistaya 1: blade cores (*a–d*); blades (*g–j*); flake cores (*e, f*); conical and semiconical microblade cores (*k, l*); tall wedge-shaped microblade core (*m*); boat-shaped blanks for wedge-shaped microblade cores (*n*) and split biface (*o*); ski spalls (*p–r*); microblades (*s–y*).

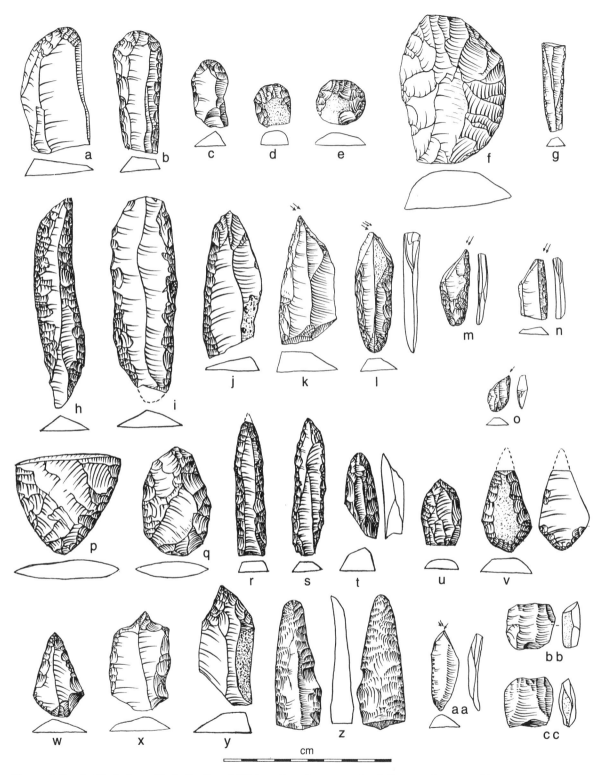

Figure 5–8   Artifacts from Gorbatka 3 and Illistaya 1: end scrapers (*a–d, f*); round scraper (*e*); retouched blades (*g–j*); burins (*k–o*); bifaces (*p, q*); unifacial points (*r–w*); perforators (*x, y*); water-worn artifacts (*z–cc*).

forators, and some other tools. End scrapers vary in dimensions and processing characteristics. Some of them were retouched only on the working end, others were processed on one or both side edges as well (Fig. 5–8: *a–d*). At Illistaya 1, Gorbatka 2, and Khalkidon 3, rounded scrapers were found (Fig 5–8: *e*). Retouched blades and flakes were heterogeneous in both shape and processing. Perhaps some of them were only preforms for burins and other kinds of tools (Fig. 5–8: *g–i*). The most impressive of these implements were big blades with a natural backing opposite a sharp edge (Figs. 5–8: *j*).

Many of the bifaces were fragments of elongate leaf-shaped points (Fig. 5–8: *p*), some of oval and semilunar shapes (Fig. 5–8: *q*). More bifaces were excavated at the Gorbatka 3 site than at Illistaya 1.

Various forms of burins were used at these sites: transverse, oblique, and central. The most common types were single and multifaceted transverse burins. These small tools were produced by blows directed from the right to the left side (Fig. 5–8: *m–o*). Burins on large blades were prepared by blows made from left to right (Fig. 5–8: *k, l*). The bases of some transverse burins were rounded by retouch which covered one or both surfaces; notches were not formed on these tools (Fig. 5–8: *l, m*).

Some symmetrical and asymmetrical unifacial points with convergent retouched sides were part of the tool kit of these sites (Fig. 5–8: *r–u*). Stemmed points with wide bases which were retouched on both surfaces (Fig. 5–8: *v, w*) should also be noted. Other tool categories from these sites included large scrapers (Fig. 5–8: *f*), elongate adzes, perforators (Fig. 5–8: *x, y*), and pebble tools (anvils, hammerstones).

The stone industries of Gorbatka 3 and Illistaya 1 are very similar in their raw materials, tool categories, and implement forms. Based on the percentage of raw material, debitage, cores, preforms, and finished tools, the assemblages could be interpreted as those of workshop and habitation sites (Table 5-1). Similar artifact collections were obtained from the excavated site of Illistaya 2 and the destroyed sites of Ivanovka 3, Gorbatka 2 and 5, Firsanovaya Sopka, and Khalkidon 3.

It must be noted that some water-worn artifacts were found at Gorbatka 3 and Illistaya 1 in the same stratigraphic position as other Palaeolithic implements. It is possible that these artifact groups—which included cores, flakes, blades, boat-shaped blanks, end scrapers, retouched blades, burins, and points—represent the early components of these sites (Fig. 5–8: *z–cc*).

The stratigraphy and typology of additional new sites investigated in the Illistaya River area resembled that of the Osinovka site. The Palaeolithic assemblage of Osinovka, according to Okladnikov and Derevianko, included over 400 artifacts (384 flakes, 64 blades and blade-like flakes). The implements of this site (Fig. 5–9: *a–i*) were represented by processed pebbles, unifacial cores, end scrapers, and retouched blades and flakes (Okladnikov 1959; Derevianko 1983). Because of the small size of this assemblage, one may suggest that the area excavated was a specialized activity area and that the tools are only a small portion of the total Osinovka site industry.

The excavated lithic inventory of the Timofeevka 1 site, in the Razdolnaya River valley, consisted of basalt, crystalline ignimbrite, and rare obsidian artifacts. Despite the difference in the raw materials employed, the core, microblade core, and other tool forms were quite similar to those from the Illistaya River basin. Wedge-shaped cores on flakes from Timofeevka 1 (Fig. 5–9: *j, k*) were larger than those at Gorbatka 3 and Illistaya 1. No conical or semiconical microblade cores were found. The finished tools of this site were represented by end scrapers, retouched blades and flakes, fragments of leaf-shaped bifaces, transverse burins, and points. An anvil and hammerstone were part of this complex, too. Artifacts similar to those of Timofeevka 1 were collected at the destroyed sites Timofeevka 2 and Razdolnaya (investigated by A. P. Okladnikov).

The Artemovka 1 site is located 50 km to the east of the Timofeevka sites. Here were excavated a small number of obsidian, crystalline ignimbrite, and other rock artifacts, including blade cores, wedge-shaped microblade cores on flakes and split bifaces, blades, microblades, retouched blades, end scrapers, and small transverse burins. The Artemovka 1 assemblage is similar to those of other Palaeolithic Primoryan sites (Derevianko 1983; Okladnikov 1983).

In the southeastern part of Primorye (the Parti-

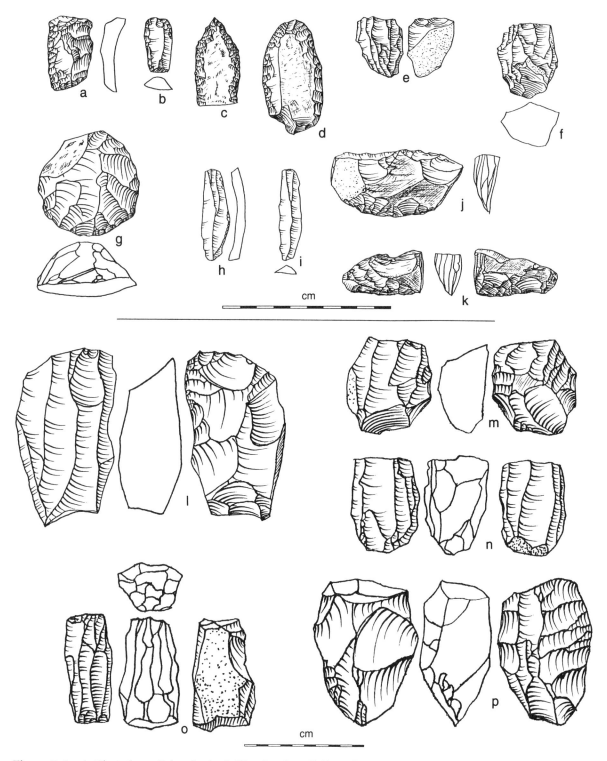

Figure 5–9 Artifacts from Osinovka (*a–i*), Timofeevka 1 (*j, k*), and Ustinovka 1 (*l–p*).

zanskaya River area) is located the Geographical Society Cave site. At this cave Pleistocene fauna (including mammoth) and some crude pebble cores and flakes were excavated. Because of the insufficiency of archaeological data this assemblage is not discussed here.

Primoryan Palaeolithic sites in the Zerkalnaya River area were investigated by V. F. Petrun, A. P. Okladnikov, R. S. Vasilievsky, S. A. Gladyshev, V. I. Dyakov and V. A. Kashin, and A. V. Tabarev. The sites were located near the villages of Ustinovka and Suvorovo. Ustinovka 1, one of the most well-known Primoryan sites, yielded about 100,000 stone artifacts. The main part of this assemblage, according to Okladnikov, Vasilievsky, Gladyshev, and Kashin, was composed of raw material—fragments and flakes of siliceous shale and tuff.

In addition to the blade cores resembling those of other sites (Fig. 5–9: *l, m*), there are a number of prismatic-shaped cores with horizontal platforms (Fig. 5–9: *n, o*). Many of the cores also have retouched platform edges. The bases of many cores share multiple small flake scars; the cores were obviously struck on hard anvils. In comparison with other sites there are many blades at Ustinovka 1, but many are broken, with pseudo-retouch scars on their edges. Microblades are rare, as are microblade cores. Many specimens are described as scrapers, knives, angle burins, etc. But the fragile blades and flakes on which they are made are redeposited. These questionable artifacts show no systematic edge retouch. The presence of pseudo-retouch on many blades was noted by Vasilievsky and Gladyshev (1989). At best, if they lack pseudo-retouch, they may be classified as tool preforms.

Another issue at Ustinovka 1 is the number of cultural layers. Artifacts from six geological horizons were integrated into two cultural layers, Mesolithic and Upper Palaeolithic, according to Okladnikov; Vasilievsky and Gladyshev suggested two Upper Palaeolithic layers (Okladnikov 1968; Vasilievsky and Gladyshev 1989). However, from a techno-morphological point of view, items from these levels were very similar. The compilation of artifact variations between the upper and lower Ustinovka 1 layers, made by Vasilievsky and Gladyshev, was based on questionable scrapers and

knives. They noted that the transverse burins and leaf-shaped bifaces were found only in the first layer at this site (Vasilievsky and Gladyshev 1989:75). The total quantity of unmistakable tools was small and the site was interpreted as a workshop.

To resolve this problem one must turn to the collection of artifacts from Ustinovka 4, located .5 km to the southwest of Ustinovka 1. According to Dyakov, raw materials used at this site were similar to those at Ustinovka 1, but the total number of nodules, their fragments and flakes, was less here than at the workshop area. At Ustinovka 4 some small blade cores were also excavated which were similar to Ustinovka 1 specimens and blades. The microblade component of this site included wedge-shaped cores on split bifaces, ski spalls, boat-shaped blanks, and microblades. Among other implements at Ustinovka 4 were end scrapers, retouched blades, leaf-shaped bifaces, transverse burins, an adze, and a small stemmed arrowhead retouched on both surfaces (Dyakov 1982, 1984).

The next site in this area, Suvorovo 3, is located 12 km from the village of Ustinovka. The Suvorovo 3 assemblage is characterized by a small quantity of artifacts (Table 5-1). Artifacts, including small flakes and several utilized blade cores, were made of siliceous shale, chert, and argillite. The bases of these cores had small flake scars. The microblade component of Suvorovo 3 is marked by wedge-shaped cores, similar to Ustinovka 1 items, and microblades. A number of other tools were discovered at this site: end scrapers, retouched blades and flakes, leaf- and oval-shaped bifaces, transverse and angle burins, adzes, unifacial points, and perforators. The Suvorovo 3 assemblage is similar to those of the previous sites (Vasilievsky and Gladyshev 1989: 81–89).

Table 5-1 summarizes the artifact class percentages for Palaeolithic sites of the Maritime Province. There were differences both between Palaeolithic sites in the same area of the province and between groups in different areas. For example, at Gorbatka 3 there were more bifaces but fewer wedge-shaped cores, compared to the Illistaya 1 assemblage. At Ustinovka 1 there were only a small number of cores and tools but many blades. The sites of each area also differed in raw materials used, and some of the assemblages' peculiarities may be understood on the basis of this factor. (Flaked core-like implements of the Illistaya River area were made from flat obsidian pebbles.) Other special features may be explained by the functional variation among these sites. But on the whole, all of the known Primoryan Palaeolithic sites are very similar in techniques of blade-microblade detachment and tool typology.

As was described earlier, many blade cores of these assemblages have common attributes: oblique platforms and blade facets on one plane surface of the core. On such implements the first blades/flakes removed during the beginning utilization stage were taken from the lateral sides of the cores, often including cortex removal. At the next stage of core utilization, blade detachment shifted to the frontal surface. The next row of blades was removed from the wide ovoid face of such cores. In the final utilization stage, flat cores were formed with one wide, faceted surface. This blade removal technique was described by A. P. Okladnikov; he considered it epi-Levalloisian (Okladnikov 1966). Nevertheless, such cores are widespread in the Upper Palaeolithic sites of Siberia (Makarovo 4, Dyuktai sites, Kokorevo, Afontova, and others). There were no tortoise cores or Levalloisian points in Primoryan or Siberian assemblages. Therefore, this technique is considered subprismatic in modern Palaeolithic studies (Suleimanov 1972:85–100). Isolated prismatic cores were also produced at Primoryan sites, especially at Ustinovka 1.

The common attribute of the Palaeolithic assemblages of this region was the use of several techniques for preparing microblade core blanks, some of which resemble methods described for Hokkaido (Morlan 1967:173–78).

Sometimes the wedge-shaped cores made from flakes are close in form to multifaceted transverse burins. Obviously these two classes of implements were produced in a common way.

In many cases, the shapes of Primoryan Palaeolithic tools were determined by the form of blade-and-flake preforms. Bifaces (including leaf-shaped ones), rounded scrapers, adzes, unifacial points, and perforators had intentional shapes, as did large

transverse burins with left-right orientation of the burin facets and stemmed points with bifacial basal retouch.

At present the Palaeolithic of the Maritime Province is represented by some sites without firm stratigraphic positions for the artifacts. However, assemblages of various habitation and workshop sites are similar in both technique and typological attributes. It may be said that all Palaeolithic sites of this region belong to the same stone industry with a microblade component. Despite the redeposited character of these assemblages, the Palaeolithic artifacts from these sites may be considered real complexes based on the commonality of assemblage composition and major implement forms.

The chronology of the Primoryan Palaeolithic is based on stratigraphic positions of artifacts and some radiocarbon dates. Two radiocarbon dates were obtained for the Gorbatka 3 site: 2590 ± 85 BP (SOAN-1921) for the light loam layer and $13,500 \pm 200$ BP (SOAN-1922) for a black loam lens from the bottom of the dark heavy loam unit. There is one date for the light loam of the Illistaya 1 site: 7840 ± 60 BP (Ki-3163) (Kuznetsov et al. 1988). The lower-level clay of Ustinovka 1 was dated 7800 ± 500 BP (GIN-2500) (Lynsha 1989). The date obtained for a loam horizon underlying humus at Suvorovo 4 was $15,300 \pm 140$ BP (Ki-3502) (Kuzmin 1990). All samples for radiocarbon dating were spatially discrete charcoal collected within the stratified layers but were not from ancient fireplaces or hearths. Therefore, it is not clear how these dates may be correlated with archaeological assemblages, the more so since radiocarbon dates of Ustinovka 1 and Suvorovo 4 contradict each other. The dates of light loams at Gorbatka 3 and Illistaya 1 are also very different. In conclusion, therefore, the chronological problem of the Primoryan Palaeolithic cannot be resolved on this evidence alone.

The dating of these assemblages is more firmly established on stratigraphic data. As was noted in the Gorbatka 3 profile, the Palaeolithic artifacts here, including water-worn ones, cannot be earlier than the time of formation of the big ice wedges, because all of these artifacts were found in or above the cryoturbated deposits (Fig. 5–6). In the opinion of A. M. Korotky and some other geologists, such ice wedges were formed in the Cis-Khanka area during the final stage of the late-glacial period of the Pleistocene (Korotky et al. 1980:189–91). The small ice wedges of the Gorbatka 3 profile may be dated to the early Holocene because they lay above the first ones and may be correlated with early Holocene cryoturbations in other profiles. Therefore, a date of about 13,000 BP for the lens crushed by the large ice wedge does not contradict the age of the cryoturbation formation. The date of about 2,500 BP must be connected with the stone tools and early Iron Age pottery that were also found in this layer. The fact that the upper part of the Holocene-dated ice wedges lay in the light loam layer suggests that this loam was deposited during the Holocene epoch. The radiocarbon date of Illistaya 1 provides good support for this proposition. The Palaeolithic artifacts of Gorbatka 3 can be dated to the final Pleistocene–early Holocene interval (about 11,000–9,000 BP) and the Illistaya 1 assemblage to the early Holocene period.

At Osinovka the Palaeolithic artifacts were found in dark heavy loam units, but it is not clear whether they are contemporary with this band or were deposited later, as at Gorbatka 3.

Artifacts from the Timofeevka sites were excavated in light loams, so they may also be dated to the final Pleistocene or early Holocene. The radiocarbon date from Ustinovka 1 has been discussed (Kuzmin 1990) but there is no firm stratigraphic or archaeological evidence for its late Pleistocene dating. The Suvorovo 3 and 4 sites, because of their stratigraphic and typological similarities with other Primoryan sites, must have the same chronology. Despite the possible early Holocene dating of these assemblages, they can be assigned typologically and technologically to the late Palaeolithic on the basis of close similarities with Palaeolithic assemblages from eastern Siberia and Hokkaido, which have been dated to between 15,000 and 11,000 BP.

The Maritime Province contains Palaeolithic sites of the final Pleistocene–early Holocene period. The technological and typological attributes of these assemblages are very similar, so it may be suggested that all of them belong to the same Primoryan stone industry. Other cultural features were not preserved at these sites.

REFERENCES CITED

Andreeva, J. V., and G. I. Khudyakov. 1973. Paleolitichesky pamyatnik na reke Zerkalnoy. (Palaeolithic Monuments on the Zerkalnaya River.) In *Materialy po istorii Dalnego Vostoka. (Historical Materials of the Far East.)* Vladivostok.

Derevianko, A. P. 1983. *Paleolit Dalnego Vostoka i Korei. (The Palaeolithic of the Far East and Korea.)* Novosibirsk: Nauka.

———. 1985. *Paleolit Tikhookeanskogo Basseina i problemu Tikhookeanskoi arheologii. (The Palaeolithic of the Pacific Ocean Basin and Archaeological Problems of the Pacific Ocean Region.)* Vladivostok.

Dyakov, V. I. 1982. Ustinovka IY: novyi paleoliticheskii pamyatnik Dalnego Vostoka. (Ustinovka 4: New Palaeolithic Monument of the Far East.) In *Problemy archeologii i etnographii Sibiri. (Archaeological and Ethnographic Problems of Siberia.)* Irkutsk.

———. 1984. Archeologicheskaya kartina zaceleniya zapadnogo poberehia Yaponskogo morya i Tatarskogo proliva v kamennom veke. (An Archaeological View of the Settling of the Western Seacoast of the Sea of Japan and Tatar Strait in the Stone Age.) In *Problemy issledovania kamennogo veka Evrasii. (Research Problems of the Stone Age in Europe and Asia.)* Krasnoyarsk.

Korotky, A. M., L. P. Karaulova, and T. P. Troitskaya. 1980. *Chetvertichnye otolohenia Primoria. (Quaternary Deposits in Primorye.)* Novosibirsk: Nauka.

Kuzmin, Y. V. 1990. *Radiouglerodnaia hronologiia arhelologicheskih pamiatnikov iuga Dalnego Vostoka SSSR i hronostratigrapnaia paleolita severnoi centralnoi i vostochnoi Azii i Ameriki. (Radiocarbon Chronology of Archaeological Monuments of Southern Far Eastern Russia and the Chronostratigraphy of the Palaeolithic of North, Central, and East Asia and America.)* Novosibirsk.

Kuznetsov, A. M. 1988. Sovremennoe sostoyanie problemy paleolita i mesolita v Primorie. (Current Status of the Palaeolithic and Mesolithic Problem in Primorye.) In *Stratigraphia i korrelyatsia Chetvertichnykh otlohenii Asii i Tikhookeanskogo regiona, T. 1. (Stratigraphy and Correlation of Quaternary Deposits of Asia and the Pacific Ocean Region, Vol. 1.)* Vladivostok.

Kuznetsov, A. M., V. A. Lynsha, V. A. Panychev, N. N. Kovaluh, and Y. V. Kuzmin. 1988. Stratigraphia i hronologia dokeramicheskih pamyatnikov Ugo-Zapadnogo Primoria. (Stratigraphy and Chronology of the Monuments of the Preceramic Period in Southwestern Primorye.) In *Stratigraphia i korrelatsia Chetvertichnykh otlohenii Asii i Tikhookeanskogo regiona, T. 1. (Stratigraphy and Correlation of Quaternary Deposits of Asia and the Pacific Ocean Region, Vol. 1.)* Vladivostok.

Lynsha, V. A. 1989. Problema vozrasta ustinovskoy kultury v svete noveishikh issledovanii mesolita v Ugo-Zapadnom Primorii. (Problem of the Age of the Ustinovka Culture, Taking into Account the Latest Mesolithic Research in Southwestern Primorye.) In *Problemy isuchenia pamatnikov kamennogo veka i paleometalla Dalnego Vostoka i Sibiri.* (*Research Problems Concerning the Stone Age and Palaeometal Period Monuments in the Far East and Siberia.*) Vladivostok.

Morlan, R. E. 1967. The Preceramic Period of Hokkaido: An Outline. *Arctic Anthropology* 4(1):164–220.

Nikolskaya, V. V. 1970. Paleogeographicheskie dannye o prirodnykh usloviah drevnih pocelenii cheloveka na terrace r. Osinovka. (Palaeogeographic Evidence Concerning the Natural Environment of Ancient Human Settlements on the Osinovka River Terrace.) In *Sibir i ee socedi v drevnosti.* (*Siberia and Her Neighbors in Ancient Times.*) Novosibirsk: Nauka.

Okladnikov, A. P. 1959. *Dalekoe proshloe Primoria.* (*The Ancient Past of Primorye.*) Vladivostok.

———. 1966. Drevnee pocelenie na r. Tadushi u d. Ustinovka i problema dalnevostochnogo mesolita. (Ancient Settlement on the Tadushi River near the Village of Ustinovka and the Problem of the Far Eastern Mesolithic.) In *Chetvertichnyi period Sibiri.* (*The Quaternary Period in Siberia.*) Moscow: Nauka.

———. 1968. Raskopki u s. Ustinovka na r. Tadushi. (Excavations near the Village of Ustinovka on the Tadushi River.) In *Arheologicheskie otkrytia 1968 goda.* (*The 1968 Archaeological Discoveries.*) Moscow: Nauka.

———. 1983. Artemovka 1, mesoliticheskii pamyatnik v Primorie. (Artemovka 1, Mesolithic Monument of Primorye.) In *Isyskania po mesolitu i neolitu SSSR.* (*Mesolithic and Neolithic Research in Russia.*) Leningrad: Nauka.

Okladnikov, A. P., and A. P. Derevianko. 1968. Paleolit Dalnego Vostoka. (The Palaeolithic of the Far East.) In *Problemy isuchenia Chetvertichnogo perioda.* (*The Problem of the Study of the Quaternary Period.*) Khabarovsk.

Phisicheskaya geographia Primorskogo kraya. (*Physical Geography of the Primoryan Territory.*) 1990. Vladivostok.

Suleimanov, R. H. 1972. *Statistcheskoe isuchenie kultury grota Obi-Rakhmat.* (*Statistical Research Concerning the Grotto Obi-Rakhmat Culture.*) Tashkent: Phan.

Vasilievsky, R. S., and S. A. Gladyshev. 1989. *Verkhnii paleolit Uhnogo Primoria.* (*The Upper Palaeolithic of Southern Primorye.*) Novosibirsk: Nauka.

Vasilievsky, R. S., and V. A. Kashin. 1983. *Raskopki mnogosloinogo poselenia Ustinovka 1 v 1980 godu, in Paleolit Sibiri.* (*The 1980 Excavations of the Stratified Settlement Ustinovka 1 in Palaeolithic Siberia.*) Novosibirsk: Nauka.

LATE PLEISTOCENE SITES OF THE SELEMDGA RIVER BASIN

Anatoly P. Derevianko

In 1982 through 1984 the author and A. I. Mazin discovered in the basin of the Selemdga River (the largest tributary of the Zeya) more than ten well-stratified, multi-layered sites of Upper Palaeolithic and Mesolithic ages (Derevianko et al. 1987). The valley of the Selemdga, where it enters the Zeya, is asymmetrical, characterized by a differential process of terrace formation. At the mouth of the Selemdga, only the first terrace above the floodplain is well marked. The geomorphology of the earlier part of terrace formation here is rather complicated, due to the fact that in the late Pleistocene the river did not change its bed.

GEOMORPHOLOGY AND STRATIGRAPHY

In the lower reaches of the Selemdga, the river is of the mountain-valley type. Its right bank is mostly high, with outcrops of rocks, whereas the left one—a vast floodplain 10 km wide—is characterized by the subtle transition into the second and third terraces. The process of terrace formation is best traced on the right bank of the river. All the terraces are bedrock. The height of the second above-riverbed terrace is 16–20 m; of the third, 22–30 m; of the fourth, 31–38 m; and of the fifth, 40–55 m. The width of the bedrock terrace surfaces overlain by unconsolidated deposits ranges between 30 and 100 m. The same deposits can be traced on all levels of the terraces (Fig. 5–10). The strata are as follows:

1. Modern soil, 0.03–0.1 m thick.
2. Pale yellow sandy loam (aleurite), 0.1–0.25 m thick.
3. Brown loam (BL) with inclusions of irregularly fractured rock, 0.7–1.2 meters thick.
4. Dark brown loam (DBL), 0.2–0.3 m thick.
5. Stratum containing tree macrofossils, 0.2–0.5 m thick.

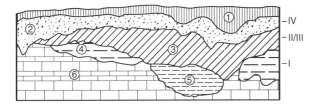

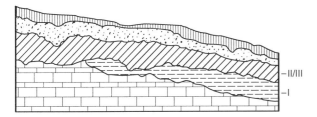

Figure 5–10   Ust-Ulma 1 (top) and Ust-Ulma 2. Stratigraphic profiles. Cultural Horizons (I–IV) shown at right. See text for strata descriptions.

With the exception of the dark brown loam, these lithological subdivisions of various thicknesses lay on all terraces. The loam was also found in the lens-like fillings of cavities in the layer containing tree macrofossils. The initial formation of unconsolidated deposits on the bedrock terraces of the Selemdga—the dark brown loam—was followed by solifluction, erosion, and other processes which promoted the destruction of this stratum. It is quite probable that there was a long interval of sedimentation after the first lithological horizon had been formed. The alluvial inclusions in the thick unconsolidated layers of the Selemdga are episodic. These are represented by numerous pebbles. Barkasnaya Sopka 3 is the only exception. Here, an alluvial stratum over 3 m thick was found on the third terrace. This alluvium is of Middle Pleistocene age. The intensive process of cryogenesis is traced elsewhere within the pale yellow loam, reflected in ice-wedge casts and polygonal fissures.

Cultural Horizons

During 1982–1984 ten sites, situated on the second to fifth terraces, were excavated in the lower Selemdga. Over 4,000 sq. m were excavated, and four cultural horizons were revealed. The first one is within the dark brown loam, the second and third are within the brown loam, and the fourth is associated with the pale yellow sandy loam. The brown loam, which lay everywhere except Barkasnaya 3, yielded the greatest amount of archaeological material. During the excavations, over 100,000 artifacts were collected. Tools constituted over 4,000 specimens (4% of the total number of finds).

Cultural Horizon I

This unit is characterized by wedge-shaped blade cores with narrow and wide fronts; by those of the Horoko type, of the single- or multiple-platform types; as well as by prismatic and discoidal varieties, monofrontal with single platforms. In the tool kit are burins (diagonal corner and core-like); retouched blades, including specimens with alternate and bifacial treatment; end scrapers made on blades and blade-like flakes, and side scrapers on edge-retouched flakes; skrebloes (corner and transverse, as well as those made from cores); notched tools on large blade-like flakes; bifaces ranging from core-like forms with one thick end to leaf-shaped ones; knives made on blades; and large blanks.

It is important to note the presence in the lower cultural horizon of the ski- and boat-like flakes and massive bifaces, which were struck in the process of making striking platforms on wedge-shaped cores and on the low wide-platform ones. The first technological tradition has been called *Yubetsu* and the second one *Horoko* by Japanese archaeologists.

In the lower cultural horizon, artifacts of archaic appearance take a special place. The specific technique of splitting stone (pebbles), giving backed flakes, is connected with choppers. Such flakes were obtained in the process of treatment of their working edge.

The finds of the lower cultural horizon were connected with the dark brown loam (DBL), traced at five of the ten excavated sites in the lower Selemdga basin. Some of the finds lay in a stratum preserving tree remains. The lower cultural horizon correlates with the beginning of sedimentation on the bedrock terraces of the river. The dark brown loam layer shows no traces of cryogenic processes.

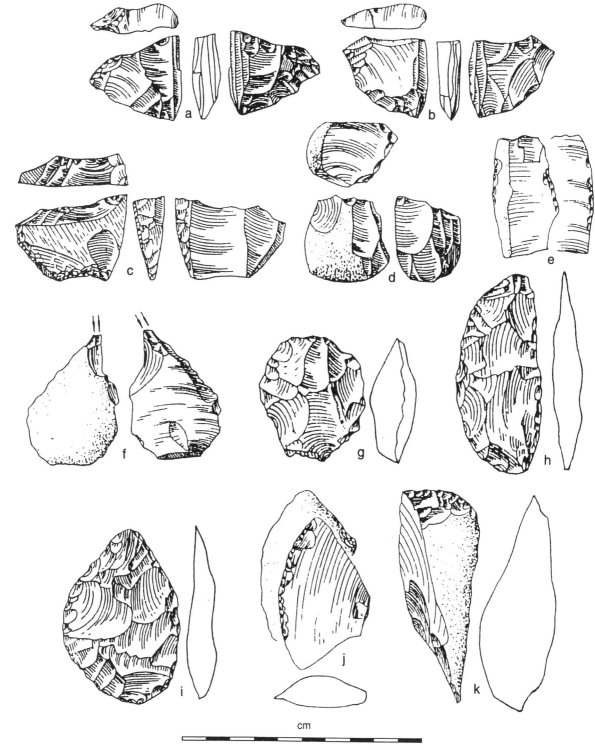

cm

Figure 5–11   Selemdga River basin sites, Cultural Horizon I: blade cores (*a*, *b*, *d*): blade core preform or scraper (*c*); retouched blade (*e*); burin (*f*); bifaces (*g–i*); retouched flake (*j*); scaled piece (*k*).

Formation of this loam evidently took place in the warm period preceding the last cold period (Partizansky according to the far eastern scheme, or Sartan in the Siberian scheme). If this was so, the finds ought to date not younger than 25,000–23,000 years. Artifacts of Cultural Horizon I are shown as Fig. 5–11.

Cultural Horizon II

This horizon is connected with the base of the brown loam (BL). The finds lay at the same depth in this 30-cm-thick layer. In the course of excavations several workshops were identified. Here, a number of heavily worn nuclei, technical flakes, and different preforms were revealed. At the Ust-Ulma 1 site, two hearths were exposed.

The primary technique, in the second cultural horizon, is represented by blade cores of single-platform monofrontal forms, disk and prismatic cores, and those with wide platforms. Wedge-shaped nuclei with narrow front or face are few here, whereas those with wide face for removing microblades are rather numerous. New types of nuclei appear—single-platform, bifrontal, and double-platform, uni- and bifrontal specimens, as well as those of the Levallois tradition.

Some other types of artifacts in the tool kit of the second cultural horizon are also more perfect examples of those represented in the first one: blades with inverse alternate and bifacial retouch; leaf-shaped symmetrical and asymmetrical bifaces; backed oval bifaces and core-like skrebloes made on cores and flakes with lateral convex edge; notched scaled pieces with inverse retouch, made on flakes; and also choppers, planes, and retouched flakes. In addition to these, a series of new types of artifacts appears: blades with obverse retouch; small and large end and side scrapers, corner ones and those on narrow-spined blades which are retouched around their perimeter; burins (double, dihedral, and asymmetric specimens with retouched edges); skrebloes made on flakes and blades with uni- and bifacial retouch (those with two alternate working edges, and also specimens made on pebbles and cores with bifacially flaked working edge); notched unifacial artifacts with inverse retouch;

drills on blades; massive bifaces resembling small hand-axes (triangular, round, and leaf-shaped asymmetrical ones); flakes with alternate inverse and obverse retouch; pebble tools with bifacially flaked transversal edge. For the second cultural horizon, the date 19,360 ± 65 years (SDAS-2019) was obtained on charcoal taken from a hearth. Artifacts of Cultural Horizon II are shown in Figure 5–12 (see page 286).

Cultural Horizon III

The archaeological finds lay here in the upper part of BL, the brown loam. The second and third horizons were found at all Selemdga sites. Artifacts of these two cultural layers are found in the same stratum separated from each other by a sterile zone 25–40 cm thick. Besides, at Zmeyenaya Sopka, the dark brown streak of 8–12 cm thickness lay between them. The heaviest cryogenic processes are connected with the upper BL layer: the frost wedges not only break through the brown loam, but in some cases penetrate into the stratum containing remains of trees.

The stone inventories of the second and third horizons have much in common in the technique of primary flaking and the tool kits. Thus, for instance, 35 percent of tools make analogous pairs. The industry of the third cultural horizon retains all the types of cores typical for the previous stage, and some new types of nuclei appear, too, such as the single-platform bifacially conjugated and the double-platform unifacial ones with flakes struck in the longitudinal-transverse direction. The tool kit contains retouched blades with inverse treatment, blades with bifacial retouch over one edge, side scrapers with a spine and perimeter retouch, burins (corner, diagonal, and core-like), double skrebloes on flakes, those of the side variety with inverse and obverse retouch, skrebloes on cores, notched artifacts, scaled pieces, bifaces, retouched planes, and flakes.

In the third cultural horizon, some new types of artifacts are represented: blades with inverse and alternate retouch, those with oblique obverse retouch of the working edge; a microblade with inverse retouch; microscrapers with retouched edges;

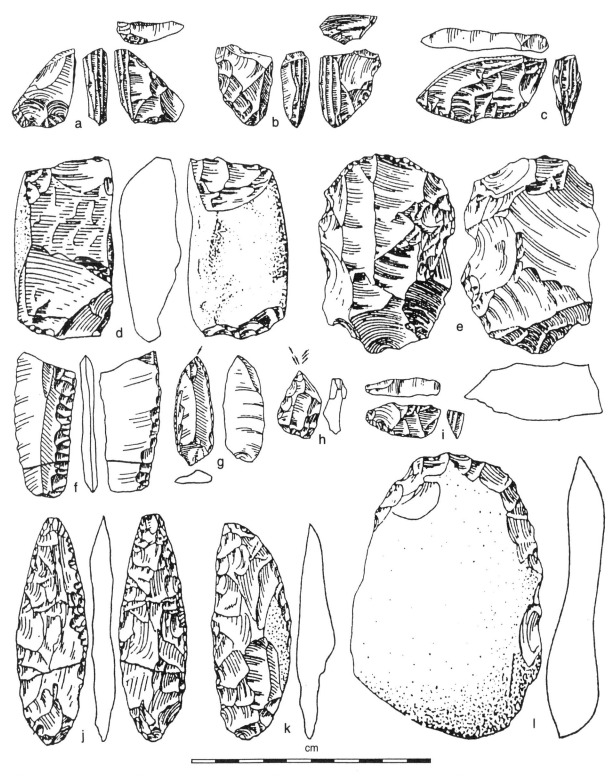

Figure 5–12   Selemdga River basin sites, Cultural Horizon II: blade cores (*a–c*); cores (*d, e*); retouched blade (*f*); burins (*g–i*); leaf-shaped bifaces (*j, k*); skreblo (*l*).

macroscrapers and medium-sized ones made on flakes; dihedral burins, asymmetrical and symmetrical, with retouched edges, and also *becs*; skrebloes on blades (side, corner, and double); skrebloes on flakes with transverse concave edge and lateral concave-convex obverse edge retouching and angular double convergent forms; notched artifacts with obverse retouch (made on blades); adze-like/skreblo-like tools; massive bifaces on primary flakes (leaf-shaped symmetrical and asymmetrical). Artifacts from Cultural Horizon III are shown in Figure 5–13.

The dating of the third cultural horizon can be described on the basis of the cryogenic evidences. As noted above, the frost wedges are connected with the top of the brown loam (BL) and they pierce through the second and third cultural horizons. These frost manifestations are found at all the Selemdga River sites. In Siberia, during the Sartan glaciation, the lowering of temperature is connected with the Gydan stage (20,900 ± 300 to 19,900 ± 500 BP [Kind 1974]) and the Nyapan (14,320 ± 300 to 13,300 ± 30 BP). Thus we have good grounds to consider the third cultural layer to be more than 13,000 to 14,000 years old.

Cultural Horizon IV

This horizon is found in the pale yellow sandy loam. Its industry is characterized by a decrease in the number of wedge-shaped cores, as well as by an increased percentage of prismatic cores and conical nuclei with flaking of blades all over the perimeter. There are no specimens of the Levallois tradition here.

The tool kit demonstrates that the inhabitants of the site went on using blades with obverse retouch, microblades with inverse (ventral) retouch, end and side macroscrapers and those on wide flakes retouched over their perimeter, corner and diagonal burins, *becs*, side skrebloes on flakes, and transverse ones. The number of skrebloes sharply increases as does that of notched tools. Adze-like/skreblo-like instruments, hammerstones, drills, planes, leaf-shaped symmetrical or asymmetrical bifaces are more numerous in this collection.

The new types of tools are: large blades with obverse (dorsal) retouch; blades with bifacially retouched edges and those with obverse retouch of their transversal edge; blades with inverse and alternate treatment and those with retouched corner; double macroscrapers; bifacially symmetrical burins, dihedral burins (symmetrical with straight or concave retouched edges); notched unifacially flaked tools on pebbles; perforators on flakes; combination tools—a knife-scraper, a burin-scraper, a burin-knife, and so on; arrowheads made on blades; irregular blades on schistose slabs. Artifacts of Cultural Horizon IV are shown in Figure 5–13.

The stratum of pale yellow sandy loam was formed in the final stage of the late Pleistocene. The final cold period in Siberia is connected with the Norilsk stage—11,450 ± 250 to 10,700 ± 200 years BP. The cryogenic processes in the third horizon took place in the Gydan stage, whereas the polygonal fissures and the ice wedges of the fourth cultural horizon are, possibly, of the Nyapan stage. But, in any case, the fourth horizon is no younger than 12,000 to 10,500 years. It is important to note that in the latter, at the Barkasnaya 3 and Ust-Ulma 2 sites, a small number of ceramics were found in association with wedge-shaped cores, bifaces, and adze-like/skreblo-like tools. Ceramics are also found in the late Pleistocene deposits of the Gasa Settlement (the Osipova culture), in Sakachi-Alan on the Amur River 60 km downstream from Khabarovsk. These finds are dated on charcoal to 12,960 ± 120 years BP (LE-1781). Such early appearance of ceramics in the southern Soviet Far East conforms well with finds in Japan, where over 20 localities have yielded ceramics more than 10,000 years old.

SUMMARY

Investigation of the multi-layered sites in the lower reaches of the Selemdga has yielded an enormous and unique body of data on the characteristics of the development of human culture in the late Pleistocene. The stone industries of their four cultural horizons exhibit the presence of wedge-shaped and prismatic cores, Levallois nuclei, bifaces, scrapers, skrebloes, burins, artifacts made on blades, and other tools belonging to a common morphotypological tradition.

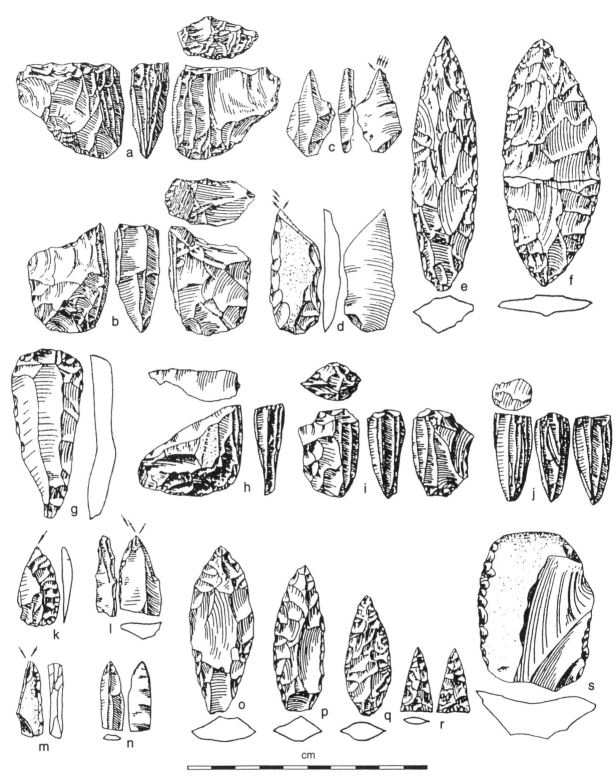

Figure 5–13   Selemdga River basin sites, Cultural Horizons III (*a–g*) and IV (*h–s*): blade cores (*a, b, h–j*); burins (*c, d, k–m*); leaf-shaped bifaces (*e, f, o, p*); scraper (*g*); drill (*n*); arrowheads (*q, r*); skreblo (*s*).

Similarities with the Selemdga materials may be traced over a vast area of eastern Siberia—in the Cis-Baikal in such sites as Verkholenskaya Gora 1 and 2, Horizon V of Sosnovyy Bor (Leshnenko and Medvedev 1983), and Igetey Ravine 1 (Medvedev 1983), as well as in the upper Lena. In the Trans-Baikal the appearance of the blade industry, wedge-shaped cores, and those of wide platform is also connected with the Sartan at the sites of Sannyy Mys, Oshurkovo, and others (Okladnikov and Kirillov 1980). In Yakutia, Mochanov (1977, this volume) has called attention to the "Dyuktai Culture" but the grouping, its dating, and its wide distribution are questioned by some (Abramova 1979). In our view, the main techno-complex of the Dyuktai tradition shows a genetic connection with that of the Selemdga sites.

Prismatic and wedge-shaped cores may be found as well in Mongolia, China, and Korea. In Japan similar types (Horoko and Yubetsu) appear no earlier than 19,000–18,000 years ago. In eastern Beringia, the earliest dates (ca. 13,000 and 15,000 years ago) have been obtained at Bluefish Cave (Morlan and Cinq-Mars 1982; Ackerman, this volume).

The origin of this early Asian archaeological phenomenon is unknown. The Selemdga sites are of great importance in this connection: the technical traditions seen in the wedge-shaped and prismatic cores, as well as in those with wide platform, were formed there not later than 25,000 to 24,000 years ago. These coexisted with the technology of radial, Levallois, and pebble flaking. At the earliest stage, various types of bifaces appear to be a specific indicator of the Selemdga complexes. There is undoubted continuity in the fabrication dynamics of many types of cores and tools in all cultural horizons. This allows us to single out the Selemdga localities as a special culture that existed no less than 15,000 years ago.

The unity apparent over these vast reaches must be seen as related to the sweeping ecological changes under way from 25,000 to 11,000 years ago. Those changes had to affect the evolution of man being dialectically connected in his life to his surroundings.

The complexes referred to no doubt have common roots but the historical significance is a matter of debate. It is important not to approach this matter simplistically. An ancient and common background must, in our view, account for some similarities. Changes occurred as a result of adaptations to specific ecological conditions. This matter, of great interest to archaeologists, will be resolved when scientists derive common methodologies and terminologies for the description of ancient industries.

REFERENCES CITED

Abramova, Z. A. 1979. K voprosu o vozraste aldanskogo paleolita. (On the Problem of the Age of the Aldan Palaeolithic.) *Sovetskaya arkheologiya* 3:5–15.

Derevianko, A. P., P. V. Volkov, and A. V. Grebenshchikov. 1987. Paleoliticheskiye kompleksi Barkasnoy Copki na p. Selemdzhe. (Palaeolithic Complexes of Barkasnaya Sopka on the River Selemdga.) In *Istoriya i kul'tur Vostoka Azii (The History and Culture of East Asia)*, pp. 73–83. Novosibirsk: Nauka.

Kind, N. V. 1974. *Geokhronologiya pozdnego antropogena po izotopnim dannim. (Geology of the Late Anthropogene According to Isotopic Data.)* Moscow: Nauka.

Leshnenko, I. L., and G. I. Medvedev. 1983. Issledovaniya paleoliticheskikh i mezoliticheskikh gorizontov stoyanki Sosnoviy Bor na peke Beloy v. 1966–1971 gg. (Investigations of the Palaeolithic and Mesolithic Horizons of the Sosnovvyy Bor Site on the Belaya River in 1966–1971.) In *Paleolit i mezolit yuga Sibiri (The Palaeolithic and Mesolithic of Southern Siberia)*, pp. 80–107. Izdatel'stvo Irkutskogo universiteta. Irkutsk.

Medvedev, G. I. 1983. Issledovaniye paleoliticheskogo mestonakhozhdeniya Igeteyskiy Log I. (Investigation of the Palaeolithic Locality Igetey Ravine I.) In *Paleolit i mezolit yuga Sibiri (The Palaeolithic and Mesolithic of Southern Siberia)*, pp. 6–34. Izdatel'stvo Irkutskogo universiteta. Irkutsk.

Mochanov, Y. A. 1977. *Drevneishiye etapi zaseleniya chelovekom Severo-Vostochnoy Azii. (The Ancient Stages of the Peopling of Northeastern Asia.)* Novosibirsk: Nauka.

Morlan, R., and J. Cinq-Mars. 1987. Ancient Beringia: Human Occupation in the Late Pleistocene of Alaska and the Yukon Territory. In *Paleoecology of Beringia*, ed. D. M. Hopkins, J. V. Matthews, Jr., C. E. Schweger, and S. B. Young, pp. 353–81. New York: Academic Press.

Okladnikov, A. P., and I. I. Kirillov. 1980. *Yugo-Vostochnoye Zabaykal'ye v epokhu kamn'a i ranney bronzi. (The Southeast of the Trans-Baikal Area in the Stone and Early Bronze Ages.)* Novosibirsk: Nauka.

EASTERN BERINGIA

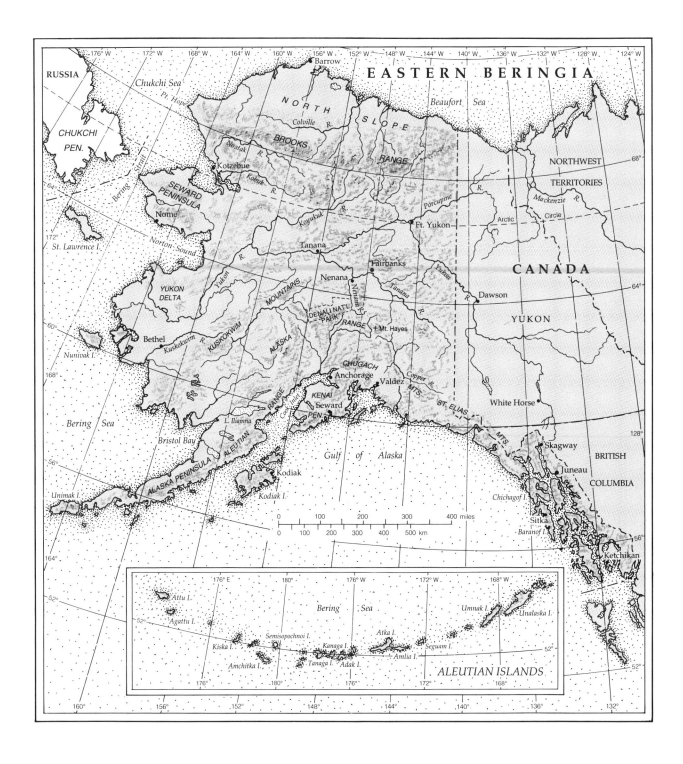

Chapter 6

CENTRAL ALASKA:
TANANA RIVER VALLEY

INTRODUCTION

The Tanana River is second in volume only to the Yukon, of which it is the major tributary. Its sources are in the Nabesna and Chisana rivers, which rise in the Wrangell Mountains (approx. 63°N, 142°W). From that point, the Tanana follows a northwesterly course where, in its braided middle and upper courses, it is constrained on the south by the Alaska

Range and on the north by the Yukon–Tanana upland. On leaving these bounds at about the latitude of Fairbanks, the river enters the extensive Tanana–Kuskokwim lowland in which it meanders broadly to its junction with the Yukon at the village of Tanana, a distance of approximately 600 km airline from its source.

The Yukon–Tanana highlands, although topographically rugged, is a region of relatively low relief. By contrast, the Alaska Range is a great arcuate grouping of mountains over 800 km long. In the several massifs making up the range there are many

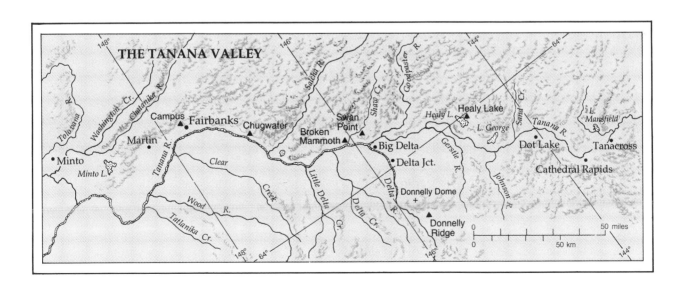

The Alaska Range across the Tanana River. Mount Hayes at right.

peaks of 3,000 to 4,000 meters height. Within it also, of course, is Mount McKinley, which, at somewhat over 6,000 meters, is the highest point in North America. The Precambrian Birch Creek Schist, which is the oldest formation of the Alaska Range, outcrops widely north of the range and composes the bluffs on which are located several of the Tanana Valley archaeological sites.

Although these mountains constitute a major barrier between most of the Tanana Valley and the Pacific littoral, there are several important rivers that, rising south of the range, pass northward through the range and allow relatively free movement north and south. Among these are the Nenana and Delta rivers. The higher portions of the mountains are perpetually ice covered. At least four major Quaternary glaciations are recorded in the Alaska Range as well as some minor Holocene readvances. The Yukon–Tanana upland is part of the vast interior of Alaska that was never glaciated.

The bed of the Tanana is very broad through most of its braided course and contains immense amounts of sands and silts which are continually replenished. These provide the source for the extensive aeolian deposits that characterize wide areas adjacent to the river. The archaeological assemblages discussed in this section are found in these late Quaternary-age silt and sand deposits.

Shielded by the Alaska Range from the ameliorative effects of northern Pacific circulation, the climate of the Tanana basin is strongly continental with intensely cold winters and relatively warm summers. Temperature extremes may range from a summer high of 30°C to a winter low of −50°C.

Interior Alaska is characterized by discontinuous permafrost, the presence or absence of which may have important effects, as in vegetation. This is, in general, a region of boreal forest in which the dominant tree species are white and black spruce (*Picea glauca, P. mariana*), the latter often associated with permafrost, paper birch (*Betula papyrifera*), aspen, and cottonwood (*Populus* spp.). The occurrence of these species is variable, however, with important determinants being altitude and summer temperatures. Thus, there are significant transitional, or ecotonal, areas in which the vegetation of the treed lowland is mingled with that of the alpine tundra. These areas would typically be dominated by dwarf or resin birch (*Betula glandulosa*) and might include important components of shrub alder (*Alnus crispa*). Between this shrub tundra and the high reaches of perpetual snow and ice, the alpine tundra is composed of low growth in which barren rocky areas may be interspersed with mats of mountain avens (*Dryas octopetala*), alpine bearberry (*Arctostaphylos alpina*), and such small plants as *Cassiopes*, moss campion, grasses, and recumbent willow.

The fauna of this region is that which characterizes all of interior Alaska. These are generally considered from the standpoint of human subsistence, which translates, in the main, to large mammals. Principal among these are the ungulates: caribou (*Rangifer tarandus*), moose (*Alces alces*), and to a lesser extent Dall sheep (*Ovis dalli*); carnivores such as the brown (grizzly) bear (*Ursus arctos*) and black bear (*U. americanus*). Other forms include wolf (*Canis lupus*), red fox (*Vulpes vulpes*), and wolverine (*Gulo gulo*); and fur bearers such as beaver (*Castor canadensis*), ground squirrel (*Citellus parryi*), and the snowshoe hare (*Lepus americanus*).

Before the American period (1880–), the Tanana Valley was occupied by small bands of Athapaskan speakers whose subsistence was largely dependent upon the hunting and trapping of those animals listed above as well as upon the taking of fish, especially the several varieties of salmon that, in spawning, reach some of the waters of this region.

The discovery of the Campus site preceded by thirty years the next such find. It remained, through that period, clear evidence—as emphasized by Nelson—of northeast Asian derivation. But it remained, as well, an enigma until discoveries of the 1960s and later revealed it to be one of a series of related sites that would provide definitive evidence on American origins.

F. H. W.

BIBLIOGRAPHY

Alaska Department of Fish and Game. 1973. *Alaska's Wildlife and Habitat*. Juneau.

Hartman, C. W., and P. R. Johnson. 1978. *Environmental Atlas of Alaska*. Fairbanks: Institute of Water Resources, University of Alaska.

Hultén, E. 1968. *Flora of Alaska and Neighboring Territories*. Stanford, Calif.: Stanford University Press.

Nelson, N. C. 1935. Early Migration of Man to America. *Natural History* 35(4):356. New York: American Museum of Natural History.

———. 1937. Notes on Cultural Relations between Asia and America. *American Antiquity* 2(4): 267–72.

Richardson and Glenn Highways, Alaska. 1983. In *Guidebook to Permafrost and Quaternary Geology, Guidebook 1*, ed. T. L. Péwé and R. D. Reger. International Conference on Permafrost, Fairbanks.

Viereck, L. A., and E. L. Little, Jr. 1972. *Alaska Trees and Shrubs*. Agriculture Handbook 410. Washington, D. C.: U. S. Department of Agriculture, Forest Service.

Williams, H., ed. 1958. *Landscapes of Alaska: Their Geologic Evolution*. Berkeley and Los Angeles: University of California Press.

CAMPUS SITE

Charles M. Mobley

The Campus site is located approximately 152 meters above mean sea level at the University of Alaska campus in Fairbanks, on a bluff overlooking the Chena and Tanana river floodplains to the south.

HISTORY OF RESEARCH

Between 1931 and 1933, students found several artifacts on the university campus. Students John Dorsh and Albert Dickey excavated at the site in 1934 and 1935 (*Farthest North Collegian* 1934). The stone artifacts they collected were studied by the American Museum of Natural History's Nels C. Nelson (1935, 1937). Froelich G. Rainey dug more of the site in 1936 (Rainey 1939:381–83). Most of the artifacts found in the 1930s remain in museums, but no detailed excavation notes survived.

The site was again excavated three decades later by H. M. Morgan (in 1966), E. Hosley and J. Mauger (in 1967), and J. P. Cook (in 1971) (Hosley 1966, 1968; Hosley and Mauger 1967; Aamodt et al. 1968). In the late 1960s, much of the site vicinity was covered by fill for a campus parking lot and perimeter road. About 300 square meters have been excavated over the years, producing over 9,000 artifacts. Pieces of the site likely remain intact.

Three museums have accessible collections from the Campus site: University of Alaska Museum, Canadian Museum of Civilization, and Arizona State Museum. Twenty-two specimens received by the University of Moscow in the late 1940s are known only by entries on the shipping list. Many researchers have examined specific aspects of selected Campus site artifacts, often comparing them with other collections (Irving 1953, 1955; Bandi and West 1965; West 1967, 1981; Cook 1968; Hayashi 1968; Sanger 1968; Mauger 1970, 1971; Morlan 1970, 1978; Smith 1974; Del Bene 1978, 1980; Aigner 1986). The author assembled and described the three available collections and obtained radiocarbon dates (Mobley 1983, 1984, 1985, 1991).

FLORA AND FAUNA

Soil samples, collected as part of the 1960s work, have not been analyzed for microflora or microfauna. No plant remains were noted during excavation. Over the years, excavators have recovered a little burned and calcined bone, representing bear (*Ursus* sp.), beaver (*Castor canadensis*), hare (*Lepus americanus, Lepus* sp.), canid (*Canis* sp.) (wolf-sized), ungulate (bison-sized), bird, and an intrusive skeleton of ground squirrel (*Citellus parryi*) (Mobley 1991:71).

STRATIGRAPHY AND DATING

Artifacts were found in subarctic brown forest soil—a stable matrix derived from unbedded Wisconsinan and Illinoian loess (Péwé 1965:14–15)—from the ground surface 40 cm down to the basal schist bedrock. The majority of specimens came from depths of 5–25 cm. No natural soil stratigraphy was discerned by the excavators.

Nelson (1935:356) compared the Campus site microblades and microblade cores with Old World palaeolithic industries and estimated the site to be 9,000–12,000 years old. Typological comparisons led Rainey (1939:388; 1953:43–44) to surmise a pre-Athabascan date for the site. West (1967:373, 378) used typological similarities to group the Campus site with three others—Teklanika East, Teklanika West, and Donnelly Ridge—as type sites for the Denali complex, which, based on comparison with Siberian data then available was estimated to date within the range of 15,000 to 10,000 BP. With radiocarbon dates from Tangle Lakes sites, West (1975) revised the dates for the Denali complex to 12,000–8,000 BP.

An obsidian hydration date of 8,400 BP for the Campus site was reported by Y. Katsui (Bandi 1969:52; West 1981:142). Five obsidian samples prepared by Cook (1975:129) suggested a date of about 3,300 BP, while C. E. Holmes (1978) used obsidian

The University of Alaska campus, College, Alaska, circa 1938. The Campus site lies just above bluff to right. *(Photo by Bradford Washburn. Courtesy of Boston Museum of Science.)*

hydration data from Lake Minchumina and Healy Lake to suggest a dating range of 4,500–1,000 BP for the Campus site.

Mobley (1991:73–83) took from the museum collections several charcoal and bone samples, chosen to obtain a broad horizontal and vertical sample throughout the site, and radiocarbon-dated them. Because features and natural stratigraphy were absent from the site, the excavators collected organic specimens from the soil and recorded their depth in arbitrary excavation levels. The uncalibrated results of the radiocarbon dating were: charcoal, depth 5–10 cm, modern (DIC–2793); bone, depth 10–15 cm, 650 ± 200 BP (Beta 10879); charcoal, 15–20 cm, 2860 ± 180 BP (Beta 4260); charcoal, 20–25 cm, 2725 ± 125 BP (Beta 7075); charcoal, 20–25 cm, 40 ± 110 BP (Beta 10878); charcoal, 20–30 cm, 240 ± 120 BP (Beta 7224); charcoal, 20–30 cm, 3500 ± 140 BP (Beta 6829).

ARTIFACT ASSEMBLAGE

No features were described during excavations. In the 1960s work, most artifacts were found near the center of the horizontal grid, with microblades and

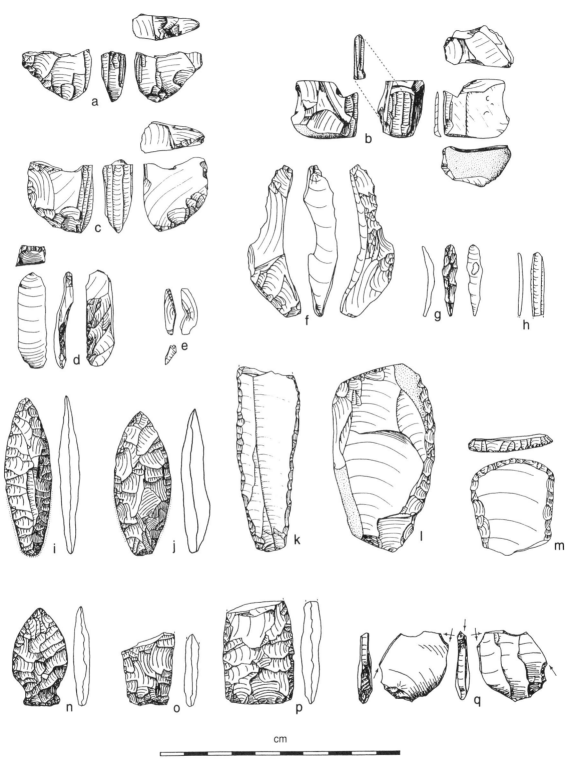

cm

Figure 6–1   Artifacts from the Campus site: microblade cores (*a–c*), one with attaching microblade (*b*); microblade core tablet (*d*); gull-wing flake (*e*); initial microblade removals (*f, g*); microblade (*h*); lanceolate bifaces (*i, j*); side-notched biface (*n*); flat-based biface fragments (*o, p*); retouched macroblade (*k*); side- and end-retouched flakes (*l, m*); Donnelly burin (*q*). *(Source: University of Alaska Press. See acknowledgments.)*

flakes having similar density distributions. Microblade, flake, and biface thinning-flake densities also correspond in terms of their vertical distribution. All were found in few numbers between 0–5 cm below surface, increasing to maximum density in the 10–15-cm level, then tapering to a few examples of each just above bedrock at about 40 cm.

The collections contain only prehistoric stone artifacts and historic trash. The total sample is biased by the selective artifact recovery technique used in the 1930s, when many microblades and flakes were apparently not saved. The relative frequencies of artifact types are meaningful only for the later collections. Chalcedony and other terms have been used to describe some of the siliceous stone, but Mobley (1991:23) grouped the materials into five categories: obsidian, chert, quartz, coarse material with conchoidal fracture, and coarse material with tabular fracture. Obsidian makes up 2.3% of the total collection; a few specimens have been confirmed as from the Batza Tena groups B/B′ (Mobley 1991:85). About 2.5% of the flakes are thermally altered, interpreted along with their absence of spatial patterning to be accidental.

Chipped stone artifacts were produced using microblade, macroblade, biface, and core/flake technologies. Products of these four basic technologies were sometimes refined by retouching or removal of burin facets. Microblade manufacture at the Campus site is represented by microblade cores, microblades, initial spalls, trifaced ridge flakes, rejuvenation flakes of two kinds, and gull-wing flakes. Both bifaces and biface thinning flakes (Fig. 6-1) represent the biface technology at the site, and flake cores, flakes, and shatter indicate the existence of a core-flake technology. Miscellaneous items collected from the site include 4 tabular slabs, 6 unmodified cobbles, 4 hammerstones, 3 cobble spalls, 39 unaltered pebbles, 3 groundstone specimens, and 48 pieces of historic trash. The groundstone items are a chipped siltstone biface with ground facets, a tabular rock with three small pits ground into it with a rotary motion, and a vesicular rock showing shallow grooves indicating use as an abrader. These artifact classes are listed in Table 6-1.

Microblade cores were usually made from a flake or biface, but an important exception is one microblade core made from a split cortical cobble (Fig. 6-1). Microblade removal was accomplished unidirectionally, with no core rotation. Each core has only one platform and a single blade face displaying from 2 to 12 microblade removal scars. Most often rejuvenation of the core platform was done using a single force executed head-on toward the blade face; less often the force was applied from other angles.

Except for the specimen made from a split cortical cobble, most microblade cores are small, thin, unifacially or bifacially flaked pieces with one straight truncated edge used as a platform for microblade removal. Two or more fluted facets emanating from the platform form the blade face, where microblades were removed. The third portion of the specimen's perimeter is flaked to form a keel, completing the basic morphology termed a "Campus-site core." Descriptive statistics for the 42 microblade cores are: range in core length, 15.0–45.6 mm; range in core height, 14.0–29.0 mm; range in core width, 4.3–17.3 mm; range in core weight, 0.8–16.0 g; range in chord length, 4.0–18.6 mm; range in face height, 12.0–28.1 mm.

Microblades are small and mostly broken, with descriptive statistics for the complete specimens as follows: range in length, 7.8–42.8 mm; range in width, 2.5–14.0 mm; range in thickness, 0.8–7.4 mm; and range in weight, 0.1–3.0 g.

Microblade core platform rejuvenation is indicated by three types of flakes. Gull-wing flakes reflect sequential forces applied from the side of the core (perpendicular to the long axis), with the point of force deliberately aligned below the negative bulb of the previous removal. The two other rejuvenation flake types are long slender spalls removed using force applied head-on towards the blade face—they differ from one another in that one shows remnants of the blade face and the other does not.

Macroblade technology is a minor element in the Campus site collection. No macroblade cores were recovered; specimens consist of 12 macroblades, half of which were subsequently retouched. Width and thickness easily differentiate the macroblades from the microblade sample, with descrip-

Table 6–1 Artifact Types Recovered from the
Campus Site*

Artifact Class	1966–1971 Samples Number	Total Sample Number
Microblade Technology		
Microblade cores	9	42
Microblades	591	604
Initial spall	9	9
Tri-faced ridged flake	0	1
Rejuvenation flake with blade face remnant	9	20
Rejuvenation flake with no blade face remnant	3	8
Gull-wing flake	9	15
Macroblade Technology		
Macroblades	1	6
Biface Technology		
Irregular biface	20	40
Battered biface	2	8
Oval biface	3	5
Triangular biface	2	4
Round basal fragment	1	9
Square basal fragment	3	5
Lanceolate point	2	7
Biface tips	9	16
Notched point	1	2
Biface thinning flakes	427	468
Core/Flake Technology		
Cores	8	17
Flakes	7,295	7,417
Nondiagnostic shatter	58	69
Retouched Specimens		
Retouched microblade	1	3

Artifact Class	1966–1971 Samples Number	Total Sample Number
Retouched macroblade	1	6
Flake: short axis retouch	18	43
Flake: long axis retouch	5	13
Flake: irregular retouch	20	79
Flake: miscellaneous retouch	0	2
Burinated Specimens		
Donnelly burin	6	22
Flake: no platform preparation	3	4
Biface: platform preparation	1	3
Biface: no platform preparation	0	1
Retouch: platform preparation	0	1
Retouch: no platform preparation	0	1
Burin spall	6	10
Miscellaneous Stone		
Tabular slab	0	4
Unmodified cobble	1	6
Hammerstone	0	4
Cobble spall	2	3
Ground stone	1	3
Pebbles	39	39
Historic Trash	48	48
Charcoal Samples	102	102
Bone	38	39
Shell	18	18
Totals	8,772	9,226

*Compiled from materials in Mobley 1991. *See Acknowledgments.*

tive statistics as follows: range in length, 29.9–72.8 mm; range in width, 13.4–30.2 mm; range in thickness, 3.1–14.3 mm; range in weight, 1.6–20.5 g.

Evidence for biface technology comes from the bifaces and biface thinning flakes (Fig. 6-1). The collection contains large oval bifaces, some with battered edges, and quite a few biface fragments. Two side-notched points (one of obsidian, one of chert)

are present, but the dominant point type is a lanceolate form. Seven relatively complete specimens are characterized by basal grinding, and in two cases the base is formed not by bifacial flaking but by a flat, unflaked facet. As much as one-third of the specimens' edges, where they form a contracting stem, are also ground. Nine other rounded biface fragments, with grinding, are believed to be the basal portions of similar lanceolate points. Biface thinning flakes comprise almost 5 percent of the total artifact collection.

Core-flake technology was also practiced at the Campus site. Nondiagnostic shatter, reflecting experimentation with prospective cores, is not common and displays a high proportion of quartz examples. Flake cores are not common either, but the 17 recovered show a wide range in size, from a weight of 8.8 to 721.1 g. Flake removal often employed core rotation, using more than one platform. Flakes are the most prevalent artifacts in the collection, ranging from 0.1 to 112.9 g.

Retouching (unifacial modification) was performed on macroblades, microblades, and flakes. Macroblades were retouched on both the distal end and the lateral edges, usually on the dorsal surface. Only three microblades were retouched, in each case on the dorsal surface and lateral edges. Flakes were retouched on both their long and short axes, usually on their dorsal surfaces, to create steeply beveled tools. Rarely, a flake was retouched to create a point or a notch.

Burin faceting was performed on flakes, retouched flakes, and bifaces. One long biface has a spectacular burin scar from a force which removed the entire edge. Most common are "Donnelly burins"—flakes on which one or more small notches have been flaked to serve as platforms for burin removal.

SUMMARY

The Campus site yielded the first recognized microblade assemblage from Alaska, and the collection still contains the largest sample (42) of Alaskan microblade cores. Consequently, it has long been a reference point for comparing other collections. Early characterizations have been largely qualitative and based on partial samples, in contrast to Mobley's 1991 study. Controversy centers on the dating of the microblade material, with West (1967, 1975, 1981) assigning the site to the Denali complex (12,000–8,000 BP) based on typological similarities with such sites as Dry Creek. In contrast, L. R. Owen (1988:118) saw morphological similarities between the Campus site microblade cores and those

of the Otter Falls site, dated by W. B. Workman (1978:186) to 4570 ± 150. This author (1991:95) attributes the radiocarbon dates (between 3500 and 2725 BP) from the Campus site to the occupation responsible for the microblade material because no older radiocarbon dates were obtained; no natural stratigraphy exists to suggest multiple occupations; the horizontal distribution of microblade and flake material is similar throughout the site; the vertical distribution of microblades, flakes, and biface thinning flakes is unimodal through the arbitrary 5-cm excavation levels; and the relative frequencies of microblades, flakes, and biface thinning flakes are very similar from level to level. Thus, Mobley (1991) suggests the Campus site reflects one major period of occupation around 3,000 years ago by people who practiced several lithic technologies, including microblade technology.

REFERENCES CITED

Aamodt, M. W., W. K. Hao, and K. Humphries. 1968. New Diggings from the University Slag Heap. *Denali: The University of Alaska Yearbook*: 24–29.

Aigner, J. S. 1986. Footprints on the Land. In *Interior Alaska: A Journey through Time*, pp. 97–146. Anchorage: Alaska Geographic Society.

Bandi, H.-G. 1969. *Eskimo Prehistory*. College, Alaska: University of Alaska Press.

Bandi, H.-G., and F. H. West. 1965. The Campus Site and the Problem of Epi-Gravettian Infiltrations from Asia to America. Paper presented at Field Conference F, Central and South Central Alaska, Fairbanks. International Association for Quaternary Research, 7th Congress, U.S.A.

Cook, J. P. 1968. Some Microblade Cores from the Western Boreal Forest. *Arctic Anthropology* 5(1):121–27.

———. 1975. Archaeology of Interior Alaska. *The Western Canadian Journal of Anthropology* V(3–4):125–33.

Del Bene, T. A. 1978. A Microscopic Analysis of Wedge-Shaped Cores from the Alaska Sub-arctic: A Preliminary Study. Paper presented to the Tenth I.C.A.E.S. Prehistoric Technology Session.

———. 1980. Microscopic Damage Traces and Manufacture Process: The Denali Complex Example. *Lithic Technology* (2):34–35.

Farthest North Collegian [A student publication of the Alaska Agricultural College and School of Mines, now the University of Alaska-Fairbanks]. 1934. Volume 12, no. 12, pp. 1, 8.

Hayashi, K. 1968. The Fukui Microblade Technology and Its Relationships in Northeast Asia and North America. *Arctic Anthropology* 5(1):128–90.

Holmes, C. E. 1978. Obsidian Hydration Studies: A Preliminary Report of Results from Central Alaska. Paper presented at the 31st Annual Northwest Anthropological Conference, Pullman.

Hosley, E. 1966. Interim Report: Campus Site Archaeological Investigations. Manuscript in possession of the author.

———. 1968. The Salvage Excavation of the University of Alaska Campus Site. Manuscript in possession of the author.

Hosley, E., and J. Mauger. 1967. The Campus Site Excavations—1966. Paper presented at the Annual Meeting of the Society for American Archaeology, Ann Arbor.

Irving, W. N. 1953. Evidence of Early Tundra Cultures in Northern Alaska. *Anthropological Papers of the University of Alaska* 1(2):55–85.

———. 1955. Burins from Central Alaska. *American Antiquity* 20(4):380–83.

Mauger, J. E. 1970. A Study of Donnelly Burins in the Campus Archaeological Collection. M.A. thesis, Washington State University.

———. 1971. The Manufacture of Campus Site Microcores. Manuscript in possession of the author.

Mobley, C. M. 1983. A Report to the Geist Fund for the Campus Site Restudy Project. Manuscript on file, University of Alaska Museum, Fairbanks.

———. 1984. A Report to the Alaska Historical Commission for the Campus Site Restudy Project. Manuscript on file, Alaska Historical Commission, Anchorage.

———. 1985. A Report to the Alaska Historical Commission for the Campus site Restudy Project. Manuscript on file, Alaska Historical Commission, Anchorage.

———. 1991. *The Campus Site: A Prehistoric Camp at Fairbanks, Alaska*. Fairbanks: University of Alaska Press.

Morlan, R. E. 1970. Wedge-Shaped Core Technology in Northern North America. *Arctic Anthropology* 7(2):17–37.

———. 1978. Technological Characteristics of Some Wedge-Shaped Cores in Northwestern North America and Northeast Asia. *Asian Perspectives* 19(1):96–106.

Nelson, N. C. 1935. Early Migration of Man to America. *Natural History* 35(4):356.

———. 1937. Notes on Cultural Relations between Asia and America. *American Antiquity* 2:264–72.

Owen, L. R. 1988. *Blade and Microblade Technology*. Oxford: BAR International Series 441.

Péwé, T. L. 1965. Resume of the Quaternary Geology of the Fairbanks Area. In *Guidebook for Field Conference F—Central and South Central Alaska*, ed. T. L. Péwé,

O. J. Ferrians, Jr., D. R. Nichols, and T. N. L. Karlstrom, pp. 6–36. International Association for Quaternary Research, 7th Congress, U.S.A. Lincoln: Nebraska Academy of Science.

Rainey, F. 1939. Archaeology in Central Alaska. *Anthropology Papers of the American Museum of Natural History* 36(4):355–405.

———. 1953. The Significance of Recent Archaeological Discoveries in Central Alaska. *Memoirs of the Society for American Archaeology* 9:43–46.

Sanger, D. 1968. The High River Microblade Industry, Alberta. *Plains Anthropologist* 12(41):190–208.

Smith, J. W. 1974. The Northeast Asian-Northwest American Microblade Tradition (NANAMT). *Journal of Field Archaeology* 1(3/4):347–64.

West, F. H. 1967. The Donnelly Ridge Site and the Definition of an Early Core and Blade Complex in Central Alaska. *American Antiquity* 32(3):360–82.

———. 1975. Dating the Denali Complex. *Arctic Anthropology* 12:76–81.

———. 1981. *The Archaeology of Beringia*. New York: Columbia University Press.

Workman, W. B. 1978. *Prehistory of the Aishihik-Kluane Area, Southwest Yukon Territory*. Paper 74, Mercury Series, Archaeological Survey of Canada. Ottawa: National Museum of Man.

DONNELLY RIDGE

Frederick H. West

The site Donnelly Ridge (Mt. Hayes 5) is located approximately 40 km south of Delta Junction, which is in turn about 161 km south of Fairbanks. It lies in the foothills of the Alaska Range and in the shadow of Donnelly Dome, a major landmark in this part of the state. Its coordinates are 63°46′N, 145°48′W.

PRESENT-DAY LANDSCAPE

The site is located on a high ridge trending north-south, one of a series which are identified as terminal moraines of the Donnelly glaciation (Péwé and Holmes 1964) of the Delta River valley. The Delta

River lies just west of the locality and joins the Tanana River at Delta Junction.

The site area itself lies some 788 meters above sea level and, while lower-lying areas will support some amount of sparse and irregular spruce growth, the vegetation here is primarily shrub tundra in which the dominant species are dwarf birch and low willows. Partly because of the high winds characteristic of this area, exposed ridgetops are frequently devoid of higher vegetation and support only alpine tundra forms or bare gravelly surfaces.

The climate of the Donnelly area is typical of interior Alaska except that even in winter very windy conditions are common.

Caribou and moose are fairly frequent in the vicinity of Donnelly Dome. Much rarer, of course, but an important predator, is the grizzly bear. Avian forms include ptarmigan and, during the summer, many ponds of the low flat areas to the east support large numbers of ducks and geese. The usual northern insects are present as well, their effectiveness often nullified by the high winds.

Until the advent in the early 1900s of American gold-seekers, there was little human habitation in this region. The first decade of this century saw roadhouses spotted along the important Valdez Trail, running from Valdez on tidewater to Fairbanks in the interior. This later became known as the Richardson Highway. One of these roadhouses was located at Donnelly.

The presence of archaeological sites at this locality was discovered by Michael Brady in the fall of 1963 in the course of a moose-hunting trip. His initial discovery (Mt. Hayes 1) produced one microblade core and was sufficient to excite interest in the possibility of further sites. That site was also a ridgetop occurrence. The most productive of the several sites discovered, that described here, was situated on the top of the highest of these morainal ridges. The denuded areas upon which artifacts were first discovered are termed "blowouts" and are quite familiar to Alaskan archaeologists. Fortunately, adjacent to these pockets were found areas in which artifacts still lay *in situ.*

The Donnelly Ridge site was excavated in its totality in the summer of 1964 (West 1967). In 1976 the Alaska Pipeline was routed directly over the site.

Donnelly Ridge was a single-component site, the assemblage lying in a shallow, unstratified matrix of loess of which the maximum thickness was about 65 cm. Most of the assemblage was found at depths between 15 and 30 cm from surface.

ARTIFACT ASSEMBLAGE

The Donnelly Ridge collection, by its typological coherence, provided the key by which a new complex came to be defined for the interior of Alaska. It was, and remains, a straightforward Denali complex assemblage. A total of 1,513 artifacts were found, of which slightly over one-third showed evidence of use or working (407). As originally described, this is a core and blade assemblage and that technology fairly dominates the collection.

The microblade cores from Donnelly Ridge conform well to the wedge-shaped, single-edge-faceted form termed, variously, *Denali, Gobi,* or *Campus.* The American occurrence was first brought to notice by Nelson (1935, 1937). The cores were first described in detail in the original account of this site (West 1967). In the typical example, the core body was bifacially shaped. Less frequently, a thick flake was unifacially worked to the required form. In either case, the core body is relatively thick, providing the girth required for the removal of multiple parallel adjoining microblades off the face. Viewed from above—once more the ideal case—the form is wedge-shaped. The base, sometimes termed the "keel," retains its bifaciality. Viewed from the faceted end, or face, the form is, once more, wedge-shaped. The opposite edge (i.e., the top) was worked into an irregularly straight, flat surface, at one end of which the striking platform would be created. Typically, this was accomplished by the removal of a single short spall struck burin-fashion from the core face. This spall then hinged up about one-fourth to one-third of the distance back from the face. A stop notch on the top at that point ensured the termination of the core tablet spall there and prevented the entire core top being carried away. The result was a smooth platform for blade removal. The characteristic stop notch, or side blow, flakes are relatively common in Denali assem-

View to the east from the Donnelly Ridge site.

blages. Primary core tablets, bearing traces of the original core body shaping and having no facets, are sometimes found (see the example from the Reger site, this volume) but much commoner, obviously, are the platform renewal spalls (core tablets) which bear on the upper surface the negative of a previous removal and show truncated facets at the struck end. At some Denali sites as many as five renewal spalls can be refitted to a particular core (*viz.* Dry Creek, Component II). They are common elements of Denali assemblages. The platform angle of these Denali cores approximates 75°.

Let it be noted that blade cores (*not* microblade cores) technologically and formally quite similar to these occur not only in Siberia and northern Japan but also, sparingly, in the Aurignacian of France.

Microblades were probably removed by pressure rather than by punch, as originally supposed by this writer (West 1967:368). First removals from the core face, or more rarely from the top, produced these characteristic specimens termed *"lames a crêtes,"* "ridge flakes," or "crested blades." The cross section of a typical ridge flake will approximate an isoceles triangle and bear elements of the original bifacial shaping on the two equal sides.

The blade-producing portion of the Denali core is a partial cone. It may well be that this form evolved to provide better control of microblade width. It would, as well, have made pressure flaking easier which would, in turn, have ensured the desired symmetry in these small blades. The blades, or segments thereof, were presumably used almost exclusively as insets for edging bone or antler weapon heads.

The Denali microblade core is certainly one of the most complex and efficient devices to have developed in the realm of lithic technology. It can only be assumed that its evolution is fairly strictly a product of time and place—an extraordinarily demanding and peculiar environment of the late Ice Age. Following is a breakdown by class of the major artifact forms.

Blade Cores (10)   The specimens are small. In this collection height of the faceted end averages 2.41 cm; total length (face to tail), 2.63 cm; and width, or thickness, 1.06 cm. Several of these specimens are shown (Fig. 6-2: *a–d*). Two very interesting preforms are shown in the Ravine Lake article, this volume.

Microblades (320)   Expectably, most of these specimens are broken, but perhaps of some interest the longest measured intact specimen at 33 mm is a bit longer than the longest (or highest) microblade core face in the collection. Average width is 4.8 mm. Virtually none of these reveal any indication of edge usage.

Other evidences of this industry found at Don-

nelly Ridge include core tablets—the first description of this form—corner platform spalls, and other distinctive reduction debris associated with the manufacture of the core and its products.

Large Blades and Blade-like Flakes (6)   While no large blade cores were discovered, their former presence is well attested by the 2 specimens illustrated here (Fig. 6-2: *m, n*). As may be judged by the scale, the larger of these is about 10 cm in length and over 5 in maximum width. Portions of three previous removals show on the dorsal face. The third specimen shown (Fig. 6-2: *o*) once more displays three facets across the face of this proximal blade fragment. The widest of these facets measures just under 3 cm and the specimen itself may, in fact, represent a portion of a former core. However, it is technically a large (6 cm in width) blade. One edge has been carefully retouched.

Burins (8)   The Donnelly Ridge collection provided the first recognition of this artifact form in Alaska. Two of these are shown in Figure 6-2: *e, f*. These specimens are at one with those that have been found in further Denali complex sites and have been as a class termed Donnelly burins. They are made on flakes and seldom show any surface modification. Restricted areas of edge retouch are usually, but not invariably, associated with shallow notches created as platforms or as points of arrest ("stop notches"). Figure 6-2: *e* shows one of the common forms, a notched burin which in this case serves as platform. In other instances there are stop notches. Once more in common with the subsequent finds of these burins, this specimen has facets on several edges. These were termed in the Donnelly report "multiple burins." It should be noted here and throughout that these are burins in the archaeological sense, i.e., they display that distinctive spall removal technique. Almost none of them in any collections give evidence of having been used as a true burin or graving tool.

Burin Spalls (56)   While their dimensions and cross sections—rectangular or triangular—are usually decisive, there are intergrades between these and microblades which may cause problems of identification. Triangular spalls are often removals of outer edges and show traces of fine retouch.

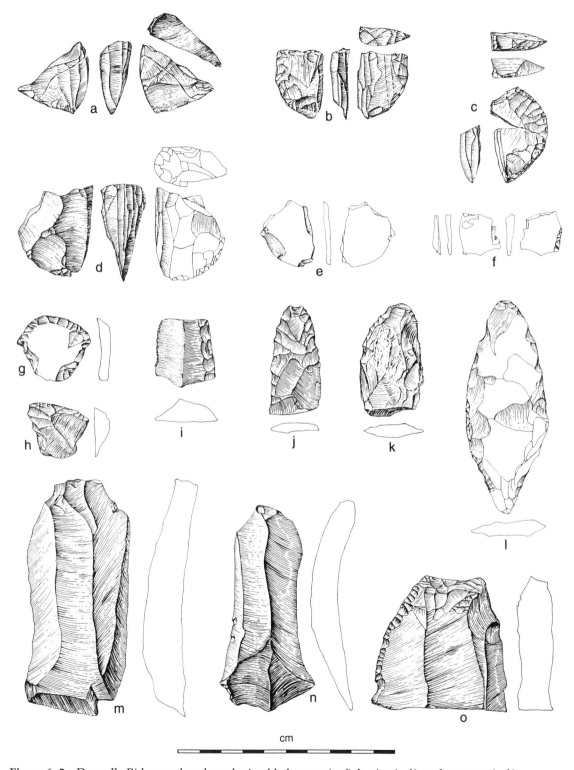

Figure 6–2   Donnelly Ridge: wedge-shaped microblade cores (*a–d*); burins (*e, f*); end scrapers (*g, h*); bifaces (*j–l*); large blades (*m–o*), *o* laterally retouched.

Bifaces (4)   Only one of these specimens was complete (Fig. 6-2: *l*). It is bifacial, lenticular in form, and shows irregular working. The other two pieces shown likely conform to this simple biface form, i.e., their outlines are composed of two segments. They are shown as *j* and *k*.

End Scrapers (3)   Of the two pieces shown, one exhibits a spur to the left and is essentially flat (Fig. 6-2: *g*). The other specimen (Fig. 6-2: *h*) is irregular in form and rather crudely worked. Earlier it was thought that spurred scrapers might have diagnostic value but this has not been borne out.

Summary

There were no hearths found at Donnelly nor was there any significant spatial patterning observed in the excavation. There were no organic remains of any description. Two radiocarbon dates were run on samples obtained from shallow depths in the soil column. These both dated well into the Christian era and clearly had no bearing on the actual age of the site, obviously referring to a tundra fire.

It was the analysis of the Donnelly collection that led the writer and his then student assistant at the University of Alaska to search out the debitage from the Campus site. At that time it was not even housed in the university museum but was found in a rather obscure place ensconced in two 16-gallon wooden boxes (''Blazo'' boxes to the initiated) that had not been touched since the 1930s. It was certain that amongst the ''waste flakes'' of those excavations there would be found some of the newly recognized artifact forms from Donnelly. That proved to be the case. Most important of these were the first true burins recorded for Campus and the first core tablets. Many of the latter could be refitted to cores in the museum.

Direct comparison of the Donnelly and Campus collections left no doubt as to their relatedness. As that study proceeded, two sites on the Teklanika River (Teklanika West and Teklanika East) were drawn in and the upshot of all was the formulation of the Denali complex as described in 1967. It was but a small step to enlarge on Nelson's (1935, 1937)

and Rainey's (1939, 1940) conjectures on Gobian and Siberian relationships. However, available comparative materials from Siberia were still sparse and rather far from Bering Strait, hence the parallels drawn (to Verkholenskaya Gora and Afontova Gora, among others) were rather attenuated. Shortly thereafter, the research of Mochanov and Fedoseeva, Dikov, and others began to appear and quite specific comparisons became possible.

In view of the subsequent firm datings derived for Denali sites, it is of no small interest to recall Nelson's estimation of the age of the Campus site: 7,000–10,000 BP.

References Cited

Nelson, N. C. 1935. Early Migration of Man to America. *Natural History* 35(4):356. New York: American Museum of Natural History.

———. 1937. Notes on Cultural Relations between Asia and America. *American Antiquity* 2(4): 267–72.

Péwé, T. L., and G. W. Holmes. 1964. Geology of the Mt. Hayes D-4 Quadrangle, Alaska. *Miscellaneous Geologic Investigations Map I-394*. Washington, D.C.: U.S. Geological Survey.

Rainey, F. 1939. Archaeology in Central Alaska. *Anthropology Papers of the American Museum of Natural History* 36(4):355–405.

———. 1940. Archaeological Investigation in Central Alaska. *American Antiquity* 5(4):299–308.

West, F. H. 1967. The Donnelly Ridge Site and the Definition of an Early Core and Blade Complex in Central Alaska. *American Antiquity* 32(3): 360–82.

CHUGWATER

Ralph A. Lively

The Chugwater site (FAI-35) is located on Moose Creek Bluff, 35 km southeast of Fairbanks and 0.5 km north of the village of Moose Creek, at approximately 64°41′30″N, 147°13′5″W. Moose Creek Bluff is a kidney-shaped extension of the Yukon–Tanana upland that lies roughly 5 km west of the uplands in the Tanana River valley. The bluff measures 1.5 km from east to west and 0.8 km from north to south, and comprises an eastern and western summit connected by a lower saddle area. The location provides a panoramic view of the Tanana valley, Alaska Range, and Yukon–Tanana upland. The eastern summit, on which the main site is located, rises to an elevation of 67 m above the river floodplain and 224 m above mean sea level. Moose Creek flows along the southern margin of the bluff and empties into the Tanana River approximately 3 km to the west.

Table 6–2　Radiocarbon Dates, Chugwater Site*

Lab. No.	Square	Area	Years BP
Beta-7567	N19E73	NE	(modern)
Beta-7636	S0W85	W	160 ± 170
Beta-9247	S3E59	SE	500 ± 100
Beta-7637	S0W85	W	720 ± 60
Beta-9249	S2E60	SE	840 ± 90
Beta-7565	S2E68	SE	870 ± 50
Beta-9248	S3E59	SE	950 ± 105
Beta-9245	N11W70	NW	1000 ± 95
Beta-9250	S2E60	SE	1120 ± 90
Beta-7566	S7E64	SE	1120 ± 90
Beta-9246	S5E62	SE	1320 ± 80
Beta-15115	S5W38	W	1720 ± 90
Beta-15116	S2W43	W	2020 ± 100
Beta-9252	S3E64	SE	2300 ± 70
Beta-9251	S2E59	SE	2370 ± 80
Beta-9253	S3E64	SE	2530 ± 110
Beta-7570	S19W10	SW	6260 ± 390
Beta-7569	S19W10	SW	7760 ± 130
Beta-18509	N12W26	N	8960 ± 130
Beta-19498	N11W26	N	9460 ± 130

*Adapted from Erlandson et al. 1991. *See Acknowledgments.*

The Chugwater site, view westward toward Tanana River.

HISTORY OF RESEARCH

J. L. Giddings was the first to document cultural material at Moose Creek Bluff. In 1941, he made drawings of a group of pictographs found on a rock face at the western end of the bluff during construction of the highway between Fairbanks and Delta Junction. The eastern summit, which was subsequently identified as Chugwater Main, was initially surveyed by Cook and White in 1976. Additional surveys were conducted by C. E. Holmes and L. Yarborough in 1978, and by J. S. Aigner in 1979. Subsequent research at the site was undertaken by the U.S. Army Corps of Engineers during 1982–83 (Maitland 1986), and the University of Alaska–Fairbanks during 1984–87 (Aigner and Lively 1986; Lively 1988). A total area of approximately 400 sq. m has been excavated, yielding approximately 25,000 artifacts.

STRATIGRAPHY AND DATING

Moose Creek Bluff is composed of fragmented Birch Creek schists and late Cretaceous to early Tertiary intrusive granites and gneiss (Mark-Anthony and Tunley 1976). Numerous veins of highly weathered quartz can be seen cross-cutting these materials in rock quarries near the eastern and western ends of the bluff. The bedrock is overlain by shallow deposits of loess and, in protected areas, by sand. The sand is found in test pits across the saddle area and in bedrock fractures on the eastern summit of the bluff. Solid bedrock was encountered in only a few of the units excavated on the eastern summit; in most of these units, loess and sand overlay hard, pea-sized gravel and colluvium. Deposition of the sand is thought to date to approximately 16,000–13,500 years BP (Thorson 1983; D. M. Hopkins, personal communication 1984).

Loess deposits across the bluff vary in depth, with the deepest deposits in areas protected from the prevailing southwest winds. The loess is generally shallow, averaging 30 cm and rarely exceeding 50 cm in depth. On the western summit, less than 25 cm is usually present. A generalized stratigraphic profile and stratigraphic features (Fig. 6-3) have

been described by Thorson (1983). In most places, a layer of colluvium (stratum 4), comprising rubble, sand, and silt, unconformably overlies the sand (stratum 5) or bedrock (stratum 6) and rests at the base of the loess. The loess is divided into lower (stratum 3) and upper (stratum 2) silt layers, which are separated by a buried soil (comprising many incipient A horizons). The upper portion of the upper silt layer is commonly oxidized by soil formation processes. The sequence is capped with a modern soil that contains charcoal with some finely divided humus (stratum 1). All buried artifacts were recovered from the loess, and, despite the shallow sedimentary context, the writer distinguished three archaeological components on the basis of horizontal and vertical separation and typological comparison with assemblages from better-dated sites (Lively 1988).

The dating of the artifact assemblages at Chugwater has been problematic due to shallow sediment, cryoturbation, root penetration, and prehistoric and historic human activity at the site. The origin of the dated charcoal samples is questionable due to the ephemeral nature of the possible hearths

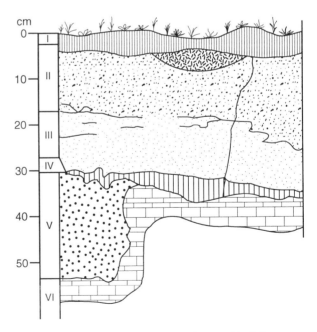

Figure 6–3 Chugwater. Stratigraphic profile (after Thorson 1983). See text for description.

and evidence of past forest fires in the area (Aigner and Lively 1986). A total of 20 ¹⁴C samples, collected by researchers from the U.S. Army Corps of Engineers and the University of Alaska, produced dates ranging from the earliest to the latest Holocene (Lively 1988; Erlandson et al. 1991). However, no firm correlation can be made between these dates and any of the diagnostic artifacts recovered from the site. The dates (in uncorrected radiocarbon years before present) are shown in Table 6–2.

Although no dates are associated with the lowermost component (Component I), two dates are tentatively associated with Component II, suggesting an age of approximately 9,500–9,000 years BP, which is consistent with the typological character of the assemblage (i.e., Denali complex).

FAUNA AND FLORA

A small quantity of bone and several gastroliths have been recovered from test excavations at the site. Most of the bones are too fragmented for taxonomic assignment, although several of them may be assigned to birds or small mammals.

FEATURES

The cultural material recovered *in situ* from Moose Creek Bluff represents numerous small, often discrete, activity areas extending from the eastern summit across the saddle area to the west side of the western summit. Cultural material has also been identified on two lower ridges extending north and south from the main ridge line (Aigner and Lively 1987). In selected areas, usually with southeast to southwest aspects and slopes of 4–6°, these areas overlap both vertically and horizontally, representing perhaps hundreds of occupations of the bluff through time (Maitland 1986; Lively 1988).

ARTIFACTS

Three components have been identified on the basis of vertical and horizontal artifact distributions from two adjoining areas (SW1 and SW3) occupying a

total of 49 sq. m on the southwest side of the eastern summit (Lively 1988). These areas yielded 3,239 artifacts, including 89 tools. Among the total number of artifacts recovered from Chugwater (approximately 25,000), a wide variety of lithic raw materials is represented, including approximately 100 colors and textures of chert, chalcedony, obsidian, moss agate, quartzite, rhyolite, siltstone, slate, sandstone, and miscellaneous metamorphic and granitic materials (Maitland 1986; Lively 1988).

Component I A total of 189 artifacts recovered from the lowermost component were found only in areas SW1 and SW3. The tools include 3 points, 1 bifacial knife fragment, and 7 small end scrapers (Fig. 6–4). All of the artifacts assigned to this component were recovered from depths of 29–22 cm below the surface. One additional point that may belong to this assemblage was found on a road surface on the eastern summit. The artifacts are similar to those found in the earliest assemblages from the Nenana Valley (Powers and Hoffecker 1989) and

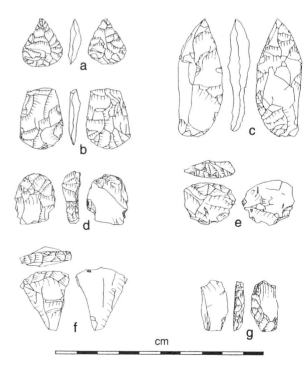

Figure 6–4   Artifacts from Component I (Nenana complex), Chugwater site: projectile points (*a*, *b*); bifacial blade or knife fragments (*c*); end scrapers (*d–g*).

Healy Lake (Cook 1969, this volume), and this component has been classified as part of the Nenana or Chindadn complexes.

Component II   The assemblage from this component (1,223) contains a wedge-shaped microblade core, 22 microblades, and 20 tools, including small bifacial knives, gravers, scrapers, point fragments, and retouched flakes. Artifacts assigned to this component were recovered from depths of 21–12 cm below the surface in areas SW1 and SW3. Component II is also present in many of the test excavations on the eastern summit, and in shovel tests dug across the saddle area and on the western summit. The character of the assemblage, tentative associated dates, and relative position of the component in the loess unit indicate that it should be assigned to the Denali complex as defined by West (1967, this volume).

Component III   A total of 1,827 artifacts is assigned to the uppermost component. The assemblage contains a flake core, 32 microblades, and 57 tools, including points (leaf-shaped, notched, triangular, and lanceolate), burins (?), spokeshaves, large end and side scrapers, unifacially and bifacially retouched blades, retouched flakes, hammerstones, and a shaft straightener (?). Artifacts were recovered from depths of 1–11 cm below the surface, and occasionally at depths of more than 14 cm below the surface. Side- and corner-notched points were found only in test units on the southeast side of the eastern summit, while all the lanceolate points were recovered from the surface or from berms on the northeast side of the summit. This assemblage is similar to those classified as late Denali or Tuktu, and includes point types of the Northern Archaic Tradition.

Summary

Despite the heavy disturbance to many areas on Moose Creek Bluff and the shallow sedimentary setting, Chugwater provides a significant record of human activity in the northern Tanana valley over the past 11,000 years. Component I extends the geographic range of the Nenana or Chindadn com-

plexes, and supports its identity as separate from the younger Denali complex. Although testing at the site was undertaken over a six-year period, controlled excavations have taken place at only a small portion of the known activity areas, and much of the bluff remains untested. The results of shovel tests conducted across the saddle area in 1986–87 (Aigner and Lively 1987) suggested that the deeper sediment found in this area may provide a clearer stratigraphic record of human occupation of the bluff.

References Cited

Aigner, J. S., and R. A. Lively. 1986. Proposal for an Extended Program of Archaeological Research at Moose Creek Bluff, the Chugwater Site (FAI-035), Alaska 1986–1990. Unpublished manuscript on file, Department of Anthropology, University of Alaska, Fairbanks.

———. 1987. A Report on Archaeological Research Conducted at the Chugwater Site (FAI-035) in 1987. Unpublished manuscript on file, Department of Anthropology, University of Alaska, Fairbanks.

Cook, J. P. 1969. The Early Prehistory of Healy Lake, Alaska. Ph.D. dissertation, University of Wisconsin.

Erlandson, J., R. Walser, H. Maxwell, N. Bigelow, J. Cook, R. Lively, C. Adkins, D. Dodson, A. Higgs, and J. Wilber. 1991. Two Early Sites of Eastern Beringia: Context and Chronology in Alaskan Interior Archaeology. *Radiocarbon* 33(1):35–50.

Lively, R. A. 1988. A Study of the Effectiveness of a Small Scale Probabilistic Sampling Design at an Interior Alaska Site, Chugwater (FAI-035). Unpublished report on file, U.S. Army Corps of Engineers, Alaska District, Anchorage.

Maitland, R. E. 1986. The Chugwater Site (FAI-035), Moose Creek Bluff, Alaska. Final Report, 1982 and 1983 Seasons. Unpublished report on file, U.S. Army Corps of Engineers, Alaska District, Anchorage.

Mark-Anthony, L., and A. T. Tunley. 1976. *Introductory Geography and Geology of Alaska.* Anchorage: Polar Publishing.

Powers, W. R., and J. F. Hoffecker. 1989. Late Pleistocene Settlement in the Nenana Valley, Central Alaska. *American Antiquity* 54(2):263–87.

Thorson, R. M. 1983. Stratigraphic Reconnaissance of the Chugwater Site, Alaska. Unpublished report on file, U.S. Army Corps of Engineers, Alaska District, Anchorage.

West, F. H. 1967. The Donnelly Ridge Site and the Definition of an Early Core and Blade Complex in Central Alaska. *American Antiquity* 32(3):360–79.

Section of Shaw Creek Bluff containing the Broken Mammoth site.

BROKEN MAMMOTH

Charles E. Holmes

The Broken Mammoth site (XBD-131) is located near the confluence of Shaw Creek and the Tanana River in east-central Alaska, approximately 22 km northwest of Delta Junction and 97 km southeast of Fairbanks (64°16′N, 146°07′W).

HISTORY OF RESEARCH

The Broken Mammoth site was discovered in 1989 by C. E. Holmes and D. McAllister. Research began in 1990 and excavations continued through 1993. This new site (XBD-131) was given the name Broken Mammoth because of the recovery of mammoth-tusk fragments, associated with other faunal re-

mains and lithic tools, from the eroding bluff face. Initial testing in 1989 of a 1 m × 2 m pit 5 m from the bluff face revealed the presence of two distinct occupation zones separated vertically by almost a meter of sterile loess containing a thin sand layer. Although the site suffered considerable damage by highway construction undertaken in the 1970s, preliminary investigation indicated that substantial portions of the site remained undisturbed.

The 1990 testing program was a joint project between the Office of History and Archaeology (Holmes) and the University of Alaska, Anchorage (D. Yesner), designed to answer basic questions critical for evaluating site significance and research potential (Holmes 1990). About 6 percent of the site was excavated in 1990 to define limits, locate artifact concentrations, and study the stratigraphy. Work continued by Holmes in 1991 focused on better defining the various archaeological components. Continuation of the project in 1992 and 1993 was again a joint undertaking by the Office of History and Archaeology and the University of Alaska, Anchorage, directed at data recovery. Approximately 20,000 items (ca. 10,000 lithic and ca. 10,000 faunal elements) have been recovered from 200 sq. m. A conservative estimate of the area left undisturbed by the highway construction at the Broken Mammoth site is 650 square meters. Thus far, approximately 24 percent of the site area has been excavated.

STRATIGRAPHY AND DATING

The stratigraphy at the Broken Mammoth site consists of a series of aeolian sediments (sand and loess) overlying a frost-shattered and weathered felsic gneiss bedrock of the Yukon–Tanana crystalline terrace (Table 6–3 and Fig. 6–5). Unit A, a grey sand, is the oldest sedimentary unit. Péwé (1975) correlates this grey sand with the Delta glaciation of Illinoian age. However, at the Broken Mammoth site the lower sand seems more likely to have been deposited after full late Wisconsinan deflation-formed ventifacts and before about 12,000 BP. More precise dating of unit A must wait until there is direct dating of regional sand units in the area.

Massive, compact aeolian silt (loess) makes up units B and D. Unit B contains three palaeosol complexes. Loess deposition began sometime prior to 12,000 BP and was interrupted by that of a (overlying) thin sand layer (unit C) sometime after 9,000 BP. The palaeosol complexes represent soil formation due to decreasing loess accumulation. Dilley (1991) has characterized the two oldest palaeosol complexes (units B1 and B2) as cryepts and the upper palaeosol complex as an entisol. The lower palaeosol complex is dated between 11,800 BP and 11,200 BP. The middle palaeosol complex is radiocarbon-dated between 10,300 BP and 9300 BP. No direct dating is available for the upper palaeosol complex. The upper sand (unit C) appears to be a local change in sediment source. It may relate to nearby changes of the Tanana River and floodplain or possibly reworking of exposed sand of unit A.

The upper loess, unit D, represents silt deposition over approximately the last 8,000–9,000 years and shows a gradual fining of sediments upwards.

Table 6–3   Major Geologic Units*

Unit D—Upper Loess
Sandy silt (up to 120 cm thick), pale brown in lower portion to yellowish brown in middle portion and brown in upper portion. Massive bedding containing occasional sand lenses up to 20 cm long and 3 cm thick and diffuse sandy zones. The upper 35–40 cm contains modern sub-arctic brown forest soil (with up to 5 cm very dark grey organic layer) overlying oxidized (yellowish brown) silt. Sharp undulating lower contact.

Unit C—Upper Sand
Very fine sand (light brownish grey), 1–5 cm thick, continuous across bluff. Abrupt smooth lower contact.

Unit B—Lower Loess
Silt (50–65 cm thick), pale brown in upper portion to light yellowish brown in lower portion. Massive bedding containing 3 palaeosol complexes and occasional organic lenses between complexes. Lower portion of unit more indurated, containing abundant carbonate features in the form of vertical root casts and discontinuous thin, horizontal laminations. Lower contact intercalated with underlying sand and can be gradational, sharp, or interfingered.

> *Unit B3—Upper Palaeosol Complex*
> 8–13-cm–thick complex consisting of 1 to 7 thin, continuous to discontinuous, undulating organic layers. Each organic layer is less than 1 cm thick, often difficult to distinguish, and separated from others by no more than 5 cm. Complex B3 is separated from complex B2 by at least 10 cm of loess.
> *Unit B2—Middle Palaeosol Complex*
> 5–10-cm-thick complex consisting of 2 to 3 undulating organic layers. Middle layer, continuous across bluff, consists of a 1–5-cm–thick, highly organic layer which shows strongest palaeosol development of all the palaeosols at the site. No layer within complex B2 is separated by more than 5 cm. B2 is separated from B1 by 5–15 cm of loess.
> *Unit B1—Lower Palaeosol Complex*
> 3–10-cm-thick complex consisting of 2 to 3 continuous to discontinuous undulating organic-rich layers, each up to 4 cm thick and separated from others by less than 6 cm.

Unit A—Lower Sand
Fine sand (light brownish grey) interbedded with silty sand, up to 1 m thick. Horizontal bedding with 1–10-cm–thick beds of silt intercalated with sand, both gradational and with sharp contacts between beds. Vertical carbonate root casts in upper portion. Lower contact abrupt and irregular, with quartz ventifacts occurring at or near contact with weathered bedrock.

Weathered Bedrock (Birch Creek schist)

*Based on T.E. Dilley 1991.

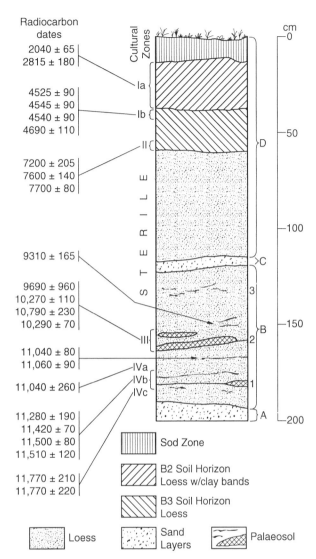

Figure 6–5 Broken Mammoth. Stratigraphic profile. Geologic units (right of profile) described in text.

At the top of unit D is the modern incipient soil classified as a cryochrept, which is the typical sub-arctic brown forest soil common in interior Alaska. Radiocarbon dates indicate an age of less than 5000 BP for this modern soil.

Twenty-three radiocarbon dates on charcoal and bone are available for the four cultural zones (Fig. 6–5). Cultural Zone Ia, the youngest, is associated with dates between 2000 and 2800 BP. For Cultural Zone Ib there are four dates that center on

about 4500 BP, which includes dates on two hearths. Charcoal from two hearth features, defining Cultural Zone II, yielded dates between 7200 and 7700 BP. Cultural Zone III, which is associated with the middle palaeosol complex, has provided four dates on two hearths centering on 10,300 BP. The oldest cultural material, Cultural Zone IV, is associated with the lower palaeosol complex. Cultural Zones IVb and IVc produced six radiocarbon dates ranging from 11,800 to 11,200 BP. Table 6–4 is a list of the available radiocarbon dates.

FAUNA AND FLORA

One of the distinctive characteristics of the Broken Mammoth site is the well-preserved faunal remains associated with Cultural Zones II, III, and IV. (Cultural Zone I has yielded only a few unidentified burned bone fragments.) Cultural Zone II produced a substantial amount of fauna associated with a large hearth. Species represented include large and small mammals (small rodents, ground squirrel, hare, beaver, caribou, moose, and bison) and unidentified birds.

Preliminary analysis indicates that Cultural Zones III and IV are not very different in faunal composition. Species represented include: bison, elk, caribou, small rodents, ground squirrel, snowshoe hare, possible otter, swans, geese, ducks, other birds, numerous burned bone fragments, and fish (salmonoids). Proboscidian tusk fragments (cf. mammoth) were uncovered during the initial investigation and are associated with either Cultural Zone III or Cultural Zone IV. A cache of three ivory artifacts associated with Cultural Zone IV yielded an AMS date of 15,800 BP, indicating that "old" ivory was collected for use in artifact manufacture. Waterfowl are more numerous in Cultural Zone IV. Analysis is continuing for the fauna recovered from the excavations completed to date.

Preliminary examination of flotation samples indicates the presence of plant macrofossils, insects, and mammal hairs. Initial pollen scans indicate that although pollen is present in the loess, the pollen may be too poorly preserved to be of analytic value.

Table 6–4   Radiocarbon Dates for the Broken Mammoth Site

WSU-4267	2040 ± 65	CZ-1A
UGA-6255D	2815 ± 180	CZ-1A
WSU-4458	4525 ± 90	CZ-1B, west charcoal concentration associated with WSU-4457, 4545 ± 90, probable hearth
WSU-4457	4545 ± 90	CZ-1B, west charcoal concentration associated with WSU-4458, 4525 ± 90, probable hearth
WSU-4456	4540 ± 90	CZ1B, east charcoal concentration
WSU-4350	4690 ± 110	CZ1B, central hearth
UGA-6281D	7200 ± 205	CZ-2, north hearth, split with WSU-4264, 7600 ± 140
WSU-4264	7600 ± 140	CZ-2, north hearth split with UGA-6218D, 7201 ± 205
WSU-4508	7700 ± 80	CZ-2, west hearth?
WSU-4266	9310 ± 165	CZ-3A, upper stringer/middle palaeosol complex, 3 samples combined
UGA-6256D	9690 ± 960	CZ-3, central hearth, middle/middle palaeosol complex, split with WSU-4263, 10,270 ± 110
WSU-4263	10,270 ± 110	CZ-3, central hearth, middle/middle palaeosol complex, split with UGA-6256D, 9688 ± 961
WSU-4019	10,790 ± 230	CZ-3, charcoal scatter ca. 1.5 m below surface in eroded bluff face, probably from same hearth as WSU-4263, 10,270 ± 110
CAMS-5357	10,290 ± 70	CZ-3, east hearth, middle/middle palaeosol complex (AMS)
CAMS-7203	11,040 ± 80	CZ-3/4A? (lg. mammal bone), between palaeosol complexes (AMS)
CAMS-7204	11,060 ± 90	CZ-3/4A? (wapiti bone), between palaeosol complexes (AMS)
UGA-6257D	11,040 ± 260	CZ-4A?, upper/lower palaeosol complex
WSU-4265	11,280 ± 190	CZ-4B, middle/lower palaeosol complex, 2 samples combined
CAMS-5358	11,420 ± 70	CZ-4B, east hearth, middle/lower palaeosol complex (AMS)
CAMS-8261	11,500 ± 80	CZ-4B (swan bone), middle/lower palaeosol complex (AMS)
WSU-4262	11,510 ± 120	CZ-4B, central hearth, middle/lower palaeosol complex
WSU-4351	11,770 ± 210	CZ-4C, lower/lower palaeosol complex
WSU-4364	11,770 ± 220	CZ-4C, lower/lower palaeosol complex
CAMS-9898	15,830 ± 70	CZ-4C? (Beta-67690) ivory artifact from lower/lower palaeosol complex, "old" ivory used as raw material for artifact manufacture (AMS)

FEATURES

A number of former hearths were uncovered during excavations. A shallow lenticular pit hearth was excavated in Cultural Zone Ib, associated with a few flakes and obsidian microblades. Cultural Zones Ia and Ib artifact distributions suggest occupation of the area near the present bluff edge. Cultural Zone II consists of one large hearth smear (with some evidence for hearth stones) suggestive of multiple fires built on the former ground surface in a radius of 4 m away from the bluff and another hearth area near the bluff edge. Two, and possibly a third, large hearth smears associated with hearth stones, similar to Cultural Zone II, were uncovered in Cultural Zone III. Lithics and fauna were scat-

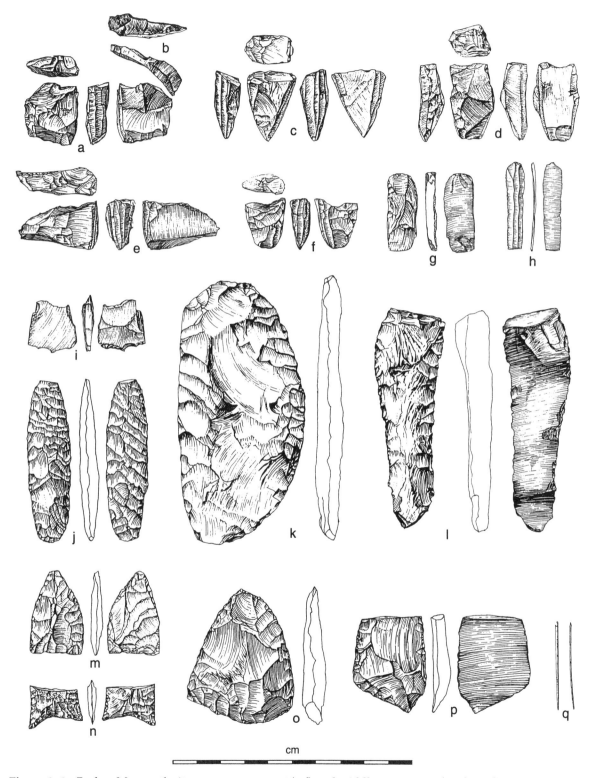

Figure 6–6   Broken Mammoth site, upper component (*a–l*) and middle component (*m–q*): wedge-shaped microblade cores (*a, c–f*); microblade core tablets (*b, g*); microblade (*h*); burin (*i*); projectile points (*j, m, n*); bifaces (*k, o*); modified blade (*l, p*); eyed bone needle (*q*).

tered around and in the hearths. Similarly, two large hearth smears were also excavated for Cultural Zone IV. The artifact, fauna, and hearth distributions for both Cultural Zones III and IV indicate occupation near to the bluff edge, more so than do those in Cultural Zone I.

ARTIFACTS

Several lithologies (rhyolite, chalcedony, chert, basalt, obsidian) are represented in the Cultural Zone I assemblage. Artifacts include retouched flakes, end scrapers, side scrapers, point fragments, flake burins, burin spalls, microblades, and small, wedge-shaped microblade cores. A nearly complete lanceolate point (Fig. 6–6: *j*) is thick, lenticular in cross section, and has a convex base with basal edge grinding. The flaking pattern is diagonal/parallel. A flake with platform preparation (Fig. 6–6: *i*) has been burinized on two edges. Figure 6–6: *c* shows a rhyolite microblade core. Thus far, only a few flakes, fire-broken rocks, and hearthstones have been recovered from Cultural Zone II away from the bluff. However, there is a likelihood that microblade technology is also part of Cultural Zone II near the bluff edge loci.

Artifacts found in Cultural Zone III include numerous tiny flakes (primarily rhyolite and chert), retouched flakes, large biface fragments, point fragments, hammers (made of quartz ventifacts), and anvils (made of Birch Creek schist). Two chert points are thin lenticular in cross section and have concave bases with basal edge grinding. One almost-complete point (Fig. 6–6: *m*) is trianguloid in outline, exhibits slight basal edge grinding, and appears to have been resharpened. The other point base (Fig. 6–6: *n*) is well made and also shows slight basal thinning. A small, eyed, bone needle (Fig. 6–6: *q*) was associated with a hearth dated to 10,300 BP.

Cultural Zone IV has produced chipping detritus of rhyolite, basalt, obsidian, chert, and quartzite. Finished stone tools are rare. Artifacts include retouched flakes, various scrapers, and a large quartz chopper/scraper/plane (Fig. 6–7: *a*). The presence of biface thinning flakes indicates that bifacial tools were part of the tool kit. Several ivory tusk frag-

ments have scratches that could have been produced by stone tools. There are two sizes of scratches, a narrow set approximately 1 mm across and a wider set about 3 mm across. One tusk piece has a stone microchip embedded in one of the wide-scratch channels. Experimentation with a replicated flake burin suggests that a tiny chip broke away during the ivory grooving process and became embedded in the channel. Other ivory pieces include a cache of tools that include two points (Fig. 6–7: *d, e*) and a possible handle (Fig. 6–7: *c*). A radiocarbon date of 15,800 BP was obtained by the AMS method on a small fragment of ivory from this cache. This age suggests that the ivory was collected from an older context for artifact use. It should be noted that there is no evidence of a microblade industry in either Cultural Zone III or IV artifact assemblage.

REFERENCES CITED

Dilley, T. E. 1991. Late Quaternary Stratigraphy and Soils of the Broken Mammoth Archaeological Site, Central Alaska. Unpublished manuscript in possession of author.

Holmes, C. E. 1990. The Broken Mammoth Site: Evaluation for Significance and Determination of Site Boundary. Unpublished report, Alaska Department of Natural Resources, Division of Parks, Anchorage.

Péwé, T. L. 1975. *Quaternary Geology of Alaska*. U.S. Geological Survey Professional Paper 835.

(Figure 6–7 follows on p. 318.)

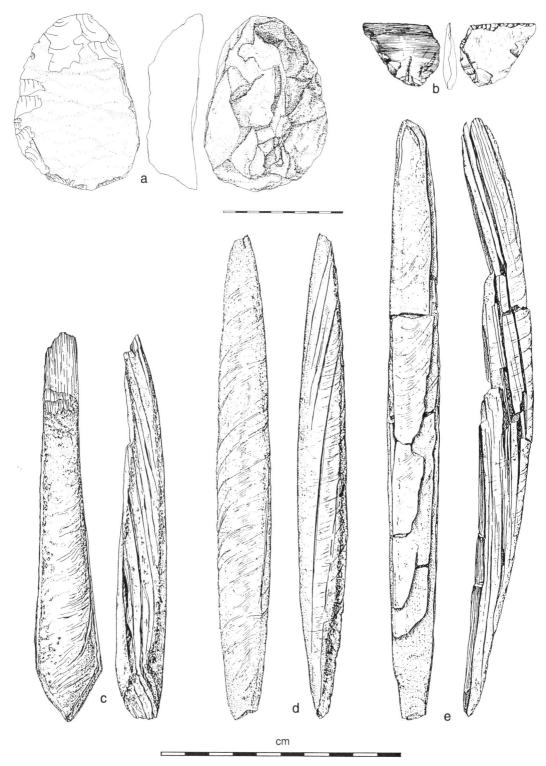

Figure 6–7   Broken Mammoth site, lower component: large quartz core/scraper (*a*); modified flake (*b*); mammoth ivory tools (*c–e*). Note lithic artifacts are at half the scale of ivory artifacts.

Swan Point at left rear. *(Photo courtesy of Richard VanderHoek.)*

SWAN POINT

Charles E. Holmes, Richard VanderHoek, and Thomas E. Dilley

The Swan Point site (XBD-156) is located on the northern edge of the Shaw Creek Flats in the central Tanana valley (63°18′N, 146°02′W), approximately 90 km southeast of Fairbanks. The site occupies a small, dome-shaped knoll at the eastern edge of a km-long ridge, rising 25 meters above the surrounding lowlands at 322 meters above mean sea level.

HISTORY OF RESEARCH

The site was discovered by R. VanderHoek and T. E. Dilley in August, 1991, working under the direction of C. E. Holmes. A single square meter was dug in 1991, exposing multiple components that included mammoth ivory in the basal cultural materials. Four square meters were excavated to bedrock in 1992 with perimeter testing, and a similar amount excavated to bedrock in 1993. Geologic studies of the Swan Point site were undertaken in 1993 by T. E. Dilley. Artifactual analysis is ongoing, with a multi-year excavation planned for the future.

STRATIGRAPHY AND DATING

The site stratigraphy consists of up to one meter of late Quaternary aeolian sediments overlying the frost-shattered bedrock of the knoll (Holmes et al. 1994). The stratigraphic sequence is firmly dated by eight radiocarbon dates spanning the last 11,700 years (Table 6-5 and Fig. 6-8). The upper surface of the bedrock has very irregular topography with nu-

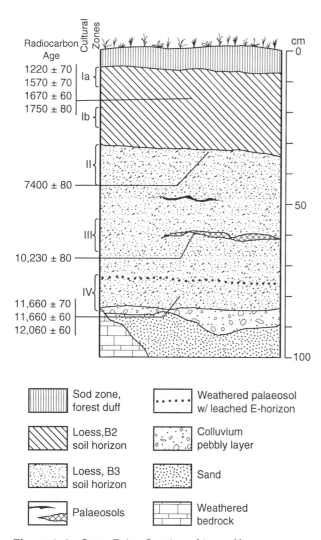

Figure 6–8   Swan Point. Stratigraphic profile.

merous large cracks (possible frost wedges) and en-larged joints. Angular gneiss pebbles are common along this contact, with the more quartz-rich fragments exhibiting slight ventifaction and wind polish.

A greyish brown coarse aeolian sand overlies the frost-shattered bedrock and in places fills cracks and joints. The mineralogy of the sand appears to indicate two possible sources. The brown sand appears to have been derived from the weathered gneissic bedrock. The grey sand is similar to sand from the nearby large late Pleistocene dune field in the Shaw Creek Flats (Péwé 1975; Lea and Waythomas 1990) which partially surrounds the site. The

sand is absent on the higher subsurface bedrock ir-regularities, indicating a period of deflation and/or downslope movement.

Overlying the aeolian sand and in places lying directly on the bedrock is a pebbly layer. This unit is a 3–5-cm-thick layer of small angular pebbles in a grey silty sand matrix. The unit has a sharp lower contact over the aeolian sand and can be traced back to the higher subsurface bedrock irregulari-ties. This seems to indicate that the pebbly layer was deposited by downslope frost movement or sheet-wash from the higher bedrock irregularities over a frozen or semifrozen sand.

A tan loess up to 80 cm thick overlies the pebbly layer. Numerous cultural levels and palaeosols are found in the loess and the modern subarctic brown forest soil that has developed at its surface. The palaeosols are thin, discontinuous organic string-ers, composed of small charcoal fragments and dif-fuse organic material. The lowest palaeosol lies directly on the pebbly layer at the base of the loess and contains the oldest cultural material at the site. Two AMS radiocarbon dates of 11,660 BP are from this lowest palaeosol. At 50–55 cm below the sur-face is a set of palaeosols with abundant cultural material, hearth charcoal, and an associated radio-carbon date of 10,230 ± 80 BP. A radiocarbon date of 7400 ± 80 obtained at 38 cm beneath the surface (just below the B horizon of the modern soil) provides a reference to the earlier Holocene loess which forms the C horizon of the soil. The modern surface soil is a typical subarctic brown forest soil (cryochrept) with a reddish brown B horizon down to about 35 cm beneath the surface. A thin (0.5–1.0 cm) discon-tinuous red clay layer is found near the base of the modern B horizon. Four radiocarbon dates indicate the upper 15 cm of the loess spans the last 1,700 years. Eight radiocarbon dates are from charcoal samples and one is from ivory. All dates in Table 6-5 are in radiocarbon years BP.

The lateral continuity and integrity of the palaeosols and other strata indicate there has been little or no cryoturbation or bioturbation at the site since the beginning of loess accumulation. The overall stratigraphy at Swan Point is similar to the stratigraphy at the two nearby sites, Broken Mammoth and Mead (Yesner et al. 1992; Holmes, this volume). However, the stratigraphy at Swan Point

Table 6–5   Radiocarbon Dates for Swan Point

Component	¹⁴C Age	Laboratory Number	Material
SP-2	1220 ± 70 BP	WSU-4523	charcoal (*Betula* spp.)
SP-2	1570 ± 70 BP	WSU-4524	charcoal (*Picea* spp.)
SP-2	1670 ± 60 BP	WSU-4522	charcoal
SP-2	1750 ± 80 BP	WSU-4521	charcoal/resin
SP-5	7400 ± 80 BP	WSU-4426	charcoal
SP-6	10,230 ± 80 BP	Beta-56666, CAMS-4251	charcoal
SP-7	11,660 ± 70 BP	Beta-56667, CAMS-4252	charcoal (*Populus/Salix* group)
SP-7	11,660 ± 60 BP	Beta-71372, CAMS-12389	charcoal (*Salix* spp.)
SP-7	12,060 ± 70 BP	NSRL-2001, CAMS-17045	collagen/ivory (*Mammuthus* spp.)

is thinner due to its being farther from the Tanana River, the probable main source of the loess.

FAUNA AND FLORA

Faunal remains from Swan Point are poorly preserved in comparison to those from the nearby Broken Mammoth site. Still, a few preliminary identifications have been made. Goose (*Branta* spp.) and large cervid remains have been identified from the terminal Pleistocene component. Moose (*Alces alces*) have been recovered from the mid-Holocene and historic components.

Charcoal samples associated with the oldest cultural material have been identified with *Salix* and *Populus/Salix* groups.

FEATURES

Hearth remains were observed during excavation, with the most distinct hearth features observed at the early Holocene cultural level. In this horizon the hearth charcoal is abundant enough to make the palaeosols predominantly cultural in origin.

ARTIFACTS

Swan Point is still in the early stages of excavation and analysis, yet at least five components have been identified to date.

Terminal Pleistocene   Artifacts at the oldest cultural level, at the base of the loess, are firmly dated to 11,660 radiocarbon years BP and include worked mammoth tusk fragments (one over 50 cm in length), microblades, microblade core preparation flakes, blades, dihedral burins, red ochre, pebble hammers, and split quartz pebble tools such as chopper/planes (Fig. 6–9: *k–o*).

Latest Pleistocene   Artifacts in this zone, at the 50–55 cm level, date to 10,230 ± 80 BP. They include a variety of bifacial forms: small lanceolate points with convex to straight bases and thin triangular points, as well as graver spurs made on broken points and quartz pebble choppers or hammers (Fig. 6-9: *d–j*).

Mid- to Mid-Late Holocene   Artifacts in this level include converging-based lanceolate points with heavy edge grinding, subconical microblade cores, microblades, and various scrapers. The upper part of this level included notched points, lanceolate points, flake burins, microblades, a tabular microblade core, and a graver spur on a flake (Fig. 6–9: *a–c*).

Late Holocene   The upper 15 cm of loess spans the last 1,700 years. Representative artifacts include pecked and ground stone fragments, scrapers, *tci thos*, straight-based lanceolate points, and microblades.

Historic   At the top of the deposit, within the modern forest litter, were found examples of tin

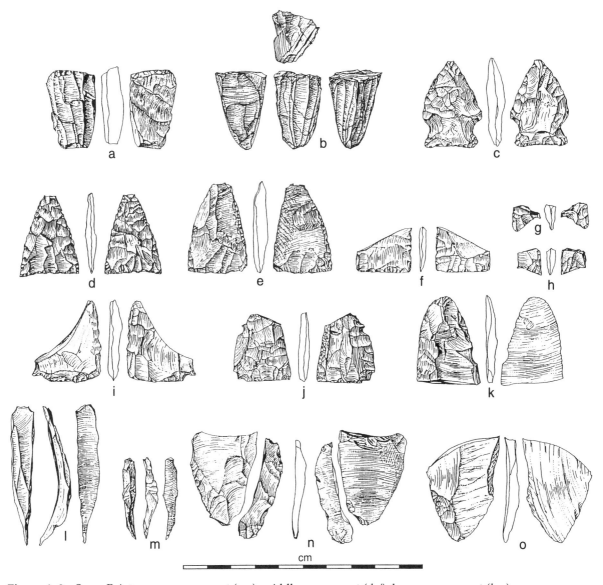

Figure 6–9 Swan Point, upper component (*a–c*), middle component (*d–j*), lower component (*k–o*): microblade cores (*a, b*); projectile points (*c–h*); perforators on projectile point fragments (*i, j*); large blade (*k*); microblade core preparation flakes (*l, m*); burins, one with an attaching spall (*n, o*).

cans, 30-30 rifle cartridges, an iron knife tang, an early historic glass bottle, and a moose bone flesher.

SUMMARY

The Swan Point site is a multicomponent site whose periodic use spans the entire Holocene. Two items of considerable importance from Swan Point are the presence of culturally worked mammoth ivory in association with microblades, both of late Pleistocene age. These microblades appear to be the oldest solidly dated microblades in eastern Beringia. Microblade technology at Swan Point is found in cultural horizons to late Holocene times. The worked mammoth ivory at Swan Point, in conjunction with the ivory artifacts found at the Mead and Broken Mammoth sites (Yesner 1994; Holmes, this volume) suggest a late Pleistocene organic technology based in part on the utilization of scavenged tusk material from extinct species (Holmes and VanderHoek 1994).

References Cited

Holmes, C. E., and R. VanderHoek. 1994. Swan Point: A Multi-Component, Late Pleistocene/Holocene Site in the Tanana Valley, Central Alaska. Paper presented at the 59th annual meeting of the Society for American Archaeology, Anaheim, California.

Holmes, C. E., R. VanderHoek, and T. E. Dilley. 1994. Old Microblades in the Tanana Valley: The View from Swan Point. Paper presented at the 21st annual meeting of the Alaska Anthropological Association, Juneau.

Lea, P. D., and C. F. Waythomas. 1990. Late Pleistocene Eolian Sandsheets in Alaska. *Quaternary Research* 34:269–80.

Péwé, T. L. 1975. *Quaternary Geology of Alaska*. U.S. Geological Survey Professional Paper 835.

Yesner, D. R. 1994. Human Adaptation at the Pleistocene–Holocene Boundary (Ca. 13,000 to 8,000 yr BP) in Eastern Beringia (Alaska and the Yukon Territory). Paper presented at the 59th annual meeting of the Society for American Archaeology, Anaheim, California.

Yesner, D. R., C. E. Holmes, and K. J. Crossen. 1992. Archaeology and Paleoecology of the Broken Mammoth Site, Central Tanana Valley, Interior Alaska. *Current Research in the Pleistocene* 9:53–57.

HEALY LAKE

John P. Cook

The Healy Lake Village site is located at the tip of a westward-trending ridge that projects into Healy Lake, directly opposite the lake outflow through the Healy River. The Healy River flows about two miles, entering the Tanana River, which in turn empties into the Yukon River. The Village site produced the greatest evidence of late Pleistocene–early Holocene occupation from at least 12 sites located on Healy Lake (Cook 1969:18). (Coordinates 64°N, 144°45'W).

The potential of prehistoric resources at Healy Lake was recognized by Robert McKennan, and subsequent test excavations were conducted by McKennan, J. P. Cook, W. B. Workman, and A. D. Shinkwin in 1966; detailed excavations were directed by Cook between 1967 and 1972. The work through 1968, constituting about 24 percent of total excavation (ca. 400 sq. m) has been reported in detail (Cook 1969). Subsequent reports cover different aspects of the culture history and chronology (Cook 1975, 1989, 1994; Erlandson et al. 1991). The final report, for work after 1968, has not been completed.

Stratigraphy

Stratigraphy was relatively consistent across the site, although depths of different soil horizons varied to some degree (Cook 1969:31). Three major stratigraphic units above the degraded Birch Creek Schist bedrock include:

1. A humus and sod layer that contained historic period artifacts.
2. A relatively thick layer of loess (60–120 cm) that included excavation layers 1–10.
3. A thicker layer of coarser sandy silt (also windblown) overlying bedrock.

The loess (unit 2) is subdivided into an interrupted A2, B2, a faint gray A2b (lower podzol), and a lower B2b soil horizon which starts about 25–30 cm below the sod level and is 15–20 cm thick (Fig. 6–10). Within the upper B horizon are thin reddish brown, vertically oriented bands of clay forming a reticulate pattern. Similar bands, although thicker and both horizontal and vertical, are characteristic of the buried soil horizons. These are interpreted as actively accumulating sesquioxides. The loess itself is increasingly coarse downward, grading to a sandy silt, indicating an early rapid deposition and much slower deposition in the last few thousand years.

The site was excavated by five-foot squares in two-inch (5-cm) levels beginning at the base of the historic period sod (the "Upper Level"); screening was generally not used. Artifacts were recovered

Aerial view of the Healy Lake Village site.

throughout the upper eight levels of excavation—only five squares produced material below level 9—or to approximately 50 cm below the base of the sod. The lower levels (6–10) generally correspond to the lower A2b and B2b horizons, but an exact correlation is generally not possible, given that the datum level (base of the sod) may not have paralleled prehistoric surfaces, soil zones varied in thickness, and their boundaries were often diffuse.

FEATURES AND FAUNAL REMAINS

The most frequent features at the Village site were areas of burned earth, charcoal, and calcined bone. Other concentrations included clusters of flakes, fire-cracked rocks, or unburned bone (Cook 1969:235). Large quantities of burned bone were more frequently encountered in lower levels. One feature from level 7 (NW quadrant of N45/W0) contained much burned bone, a concentration of blade-like flakes, four microblades, and a biface fragment; an AMS date of 11,100 ± 60 was obtained. An oval, bone-filled hearth (28 × 45 cm) was found in the eastern half of the same unit, level 8. Calcined bone from this feature produced the original early

date of 11,090 BP (Cook 1969:251), although the apatite fraction that was dated may bring the date into question. A microblade and a "wedge" were considered to be contemporary with this hearth. Calcined bone fragments from the lower levels have been only tentatively identified due to their small sizes. These appear to be mostly bird and small mammal bones (rabbit/squirrel) and some larger mammal (caribou/sheep).

ARTIFACT ASSEMBLAGES

In 1969, the artificial 5-cm levels were grouped into a four-stage cultural scheme; this was later simplified (Table 6–6).

Of the various artifact forms, the greatest numbers occurred in the upper three levels, including a variety of notched and lanceolate points, several biface forms, Donnelly burins, and more than 1,000 microblades. This strong microblade technology is very similar to that of the Campus site, as are the burins and points.

The early levels (6–10) are considered as a group because of similar technology and artifact types as well as the apparent rapid deposition of

the coarser silts. This is the type site for the Chindadn ("ancestor") complex. Quartz crystal and a
red-yellow agate are characteristic, as is the presence of large, thin percussion flakes. These flakes
make up 93 percent of the total Chindadn collection
which is nearly the same ratio of flakes to tools as
the upper levels (91.9%). Flake size is somewhat
larger but not quantified. Obsidian is found in
slightly less frequency in Chindadn levels as in the
upper levels (0.9% vs. 1.5%) but the obsidian
sources are the same—Groups B and A—with one
major exception. One sample of Group H obsidian
was found in level 7; the source of this material is
unknown, and only one other sample, from the Fish
Creek site near Paxson Lake, has been identified
(Cook 1995:96).

The diagnostic artifact is the Chindadn point or
knife, which is very thin, bifacially flaked, with a
teardrop outline. Some are definitely projectile
points, while others are larger, and not so pointed,
knives. Some small triangular points are associated
with these (Fig. 6–11). These two artifact types have
lately been associated with the Nenana complex
(Hoffecker et al. 1993). One basally thinned, con-
cave-based point, which appears to have been heat-
treated, has sometimes been cited as a "fluted"
point; it is not.

While not as strongly present as in the upper
levels, the microblade industry is represented by
nearly 100 microblades and two cores. One of the
cores is wedge-shaped—although not quite like
Campus cores—and very near, but not directly associated with, a date of 9,885 BP (an average of four
dates from the same feature). The other core is a

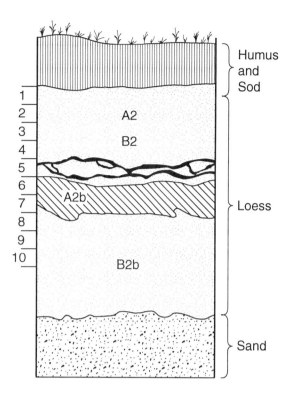

Figure 6–10   Representative profile from the Healy Lake
Village site with approximate excavation levels (5 cm)
below the base of the humus.

very small wedge-shaped core of obsidian; there are
no obsidian microblades in the Chindadn levels. A
summary of the microblades appears in Table 6–7.

An end scraper with very definite graver spurs
at both corners of the working edge is present, as is
a fine-grained quartzite graver with two well-defined spurs. An unusual, and so far unique, tool is a

Table 6–6   Cultural Stages, Associated Artifact Totals, and Radiocarbon Dates

Stage	Levels	Artifacts Total	Tools	Number of Dates	Average BP
I. Athapaskan	1	11,260	785	2	678
	2	10,176	797	8	2359
	3	7,028	681	3	2600
II. Transitional	4	4,877	471	3	3060
	5	2,700	269	1	5000
III. Chindadn	6–10	6,998	489	17	9700

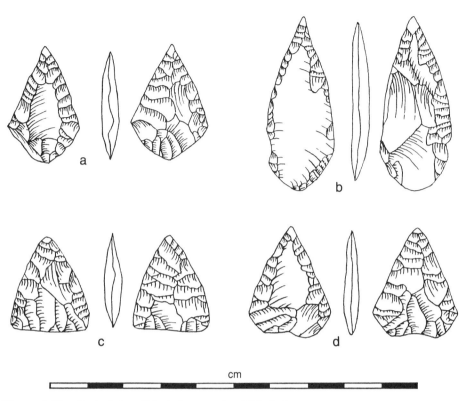

Figure 6–11   Healy Lake, Chindadn complex: Chindadn points (*a, b, d*) and small triangular point (*c*).

Table 6–7   Healy Lake Village Site Microblades

	Level	Count		Length (mm)	Width (mm)	Thickness (mm)	Width/ Thickness
Athapaskan	1	148		14.2	5.9	1.4	4.08
	2	366		14.0	5.7	1.4	4.08
	3	318		13.6	5.6	1.4	3.94
Transition	4	212		13.5	5.5	1.4	3.94
	5	99	1143	13.3	5.3	1.4	3.91
Chindadn	6	54		15.1	5.5	1.3	4.15
	7	29		16.1	6.2	1.6	3.93
	8	9	92	15.4	6.6	1.9	3.45
Total		1235					

doubly concave implement, probably a spokeshave, looking like a projectile point with two bases. The more than 20 other end scrapers have a variety of shapes and sizes.

RADIOCARBON CHRONOLOGY

A large series of radiocarbon analyses (42) have been run; 12 of these were AMS dates (Table 6–8).

Table 6–8   Radiocarbon Dates from the Healy Lake Village Site, Ordered by Artificial Levels

Lab No.	AMS No. CAMS	Level	Radiocarbon Age BP	Material
GX-2166		1	455 ± 130	charcoal
Gak-1886		1	900 ± 90	charcoal
GX-1945		1	modern	charcoal
GX-2160		2	905 ± 90	charcoal
Gak-1887		2	1360 ± 80	charcoal
GX-2168		2	1655 ± 180	charcoal
GX-2169		2	2875 ± 140	charcoal
GX-2165		2	3580 ± 140	charcoal
AU-4		2	3655 ± 426	charcoal
Beta-76058	16521	2	380 ± 50	plant
Beta-76068	16524	2	4460 ± 60	plant
GX-2176		3	2660 ± 100	charcoal
Beta-76059	16522	3	1790 ± 50	plant
Beta-76061	15915	3	3350 ± 50	charcoal
GX-2161		4	2150 ± 80	charcoal
GX-2163		4	4010 ± 110	charcoal
GX-1340		4	8960 ± 150	bone
Beta-76063	15916	4	3020 ± 50	charcoal
GX-2162		5	(modern)	charcoal
Beta-76069	16525	5	5000 ± 60	plant
GX-2173		6	10,250 ± 380	charcoal
Beta-76062		6	7920 ± 90	soil
Beta-76060	15914	6	11,410 ± 60	charcoal
Beta-76064		6	5110 ± 90	charcoal
Beta-76067	15918	6	11,100 ± 60	charcoal
Beta-76071	15920	6	10,410 ± 60	charcoal
GX-2171		7	8655 ± 280	charcoal
GX-2170		7	8680 ± 240	charcoal
Beta-76066	15917	7	10,290 ± 60	charcoal
Beta-76070	15919	7	8990 ± 60	charcoal
Beta-76065	16523	7	11,550 ± 50	plant
AU-1		7[1]	9245 ± 213	charcoal
GX-2174		7[1]	9895 ± 210	charcoal
SI-737		7[1]	10,150 ± 210	charcoal
GX-1341		8	11,090 ± 170	bone
AU-2		9[2]	9401 ± 528	charcoal
GX-2159		9[2]	6645 ± 280	charcoal
SI-738		9[2]	8210 ± 155	charcoal
AU-3		10[3]	10,434 ± 279	charcoal
GX-2175		10[3]	8465 ± 360	charcoal
SI-739		10[3]	10,040 ± 210	charcoal
GX-1944		10	10,500 ± 280	charcoal

[1,2,3] Dates run on split portions of the same sample.

Most of the samples (34) were charcoal; 2 of these came out "modern." The two bone dates are from the apatite fraction and are therefore suspect. Any such series is going to have some anomalies and the Village site is no exception. However, the dates generally support the stratigraphic and superpositional interpretations. Briefly, the (uncalibrated) range of dates from the upper levels (1–5) span between 380 and 8960 radiocarbon years (excluding modern dates). The oldest of these was run on calcined bone and is probably wrong, while the next oldest date is 5000 BP. Lower levels (6–10) produced dates ranging from 5110 to 11,550 BP. Some of these can be discarded for technical reasons (Erlandson et al. 1991:48). Considering only dates on charcoal (as opposed to calcined bone, plant material, and soil— one of each), only 2 dates of 19 are younger than 8210; both of these might be excluded for technical reasons. Five dates fall between 8000 and 9000 BP; three between 9000 and 10,000 BP; seven between 10,000 and 10,500 BP; and two between 11,100 and 11,410 BP. Given the arbitrary levels and uncertainty of precise superposition over the extent of the site (125 × 130 feet), the greatest resolution of the Chindadn complex at the Healy Lake Village site, at present, is a range of 8210 to 11,410 radiocarbon years, with an average of 9700 BP.

REFERENCES CITED

Cook, J. P. 1969. Early Prehistory of Healy Lake, Alaska. Ph.D. dissertation, University of Wisconsin.

———. 1975. Archaeology of Interior Alaska. *Western Canadian Journal of Anthropology* 5(3–4):125–33.

———. 1989. Historic Archaeology and Ethnohistory at Healy Lake, Alaska. *Arctic* 42(2):109–18.

———. 1994. The Chindadn Complex. Paper presented at the 21st Alaska Anthropological Association Conference, Juneau.

———. 1995. Characterization and Distribution of Obsidian in Alaska. *Arctic Anthropology* 32(1):92–100.

Erlandson, J., R. Walser, H. Maxwell, N. Bigelow, A. Higgs, and J. Wilber. 1991. Two Early Sites of Eastern Beringia: Context and Chronology in Alaskan Interior Archaeology. *Radiocarbon* 33(1):35–50.

Hoffecker, J. F., W. R. Powers, and T. Goebel. 1993. The Colonization of Beringia and the Peopling of the New World. *Science* 259:46–53.

Chapter 7

NORTH CENTRAL ALASKA RANGE: NENANA AND TEKLANIKA VALLEYS

INTRODUCTION*

The central portion of the northern Alaska Range is occupied by a broad foothill belt that stretches from the Kantishna Hills in the west to the Johnson River in the east, and extends up to 30 km beyond the range front. The foothills are represented by a series of parallel ridges that range from 900 to 1,500 m above sea level and valleys that range from 300 to 800 m above sea level. The valleys have been incised by northward-flowing rivers that empty into the Tanana basin. These rivers possess steep gradients and braided floodplains and have, through alternating phases of aggradation and downcutting, created sequences of terraces in the major valleys. Most archaeological sites of the late Pleistocene and early Holocene occupy the surfaces of these terraces.

The foothill belt lies today in the transitional zone between the spruce-hardwood forest of the Tanana lowland and the herbaceous tundra of the range. On the upland interfluves, moist herbaceous

tundra dominates, with cotton grass (*Eriophorum* spp.), sedges (*Carex* spp.), dwarf arctic birch (*Betula nana*), and other species. The valleys support a spruce-hardwood forest characterized by mixed stands of white spruce (*Picea glauca*) and deciduous trees, including balsam poplar (*Populus balsamifera*), paper birch (*Betula papyrifera*), and quaking aspen (*Populus tremuloides*), on south-facing slopes. Stands of black spruce (*Picea mariana*) and deciduous species are often found on north-facing slopes and other areas of poorly drained soil.

The foothills are composed primarily of an uplifted mass of poorly consolidated Tertiary rock, but they also contain Quaternary deposits in the form of glacial till, outwash, alluvium, and aeolian sediment. The older and less extensively exposed formation (late Oligocene–Miocene) comprises sandstone, claystone, and coal, and crops out along the range front and central foothills. The younger Nenana Gravel represents the most abundant formation in the foothills, consisting chiefly of moderately well-sorted conglomerate and sandstone pebbles and cobbles. The Nenana Gravel appears to have been deposited as a coarse-grained alluvial apron following a mid-Tertiary orogeny in the Alaska Range (Capps 1940; Wahrhaftig 1958).

Traces of ancient Pleistocene glaciation are visible in many parts of the foothills. Isolated deposits of till have been identified in the Nenana, Teklanika,

*This text is a revised version of material originally presented in Applied Geomorphology and Archaeological Survey Strategy for Sites of Pleistocene Age by J. F. Hoffecker. 1988. *Journal of Archaeological Science* Vol. 15, pp. 683–713. Published by Academic Press, London.

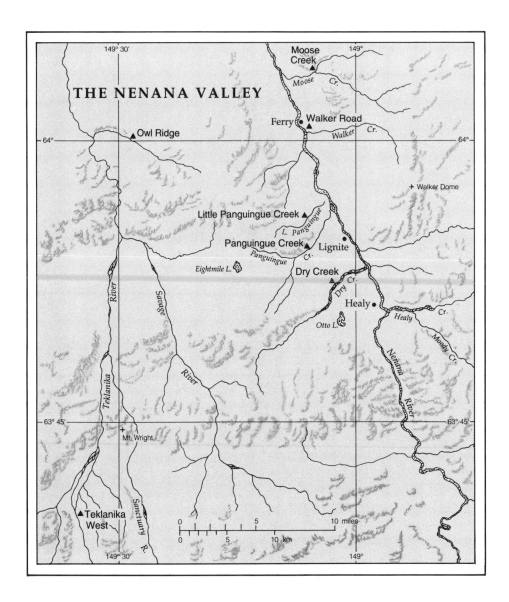

and Delta valleys and tentatively assigned to an extensive glaciation of early Pleistocene age (Wahrhaftig 1958; Péwé 1965). However, the earliest well-documented glaciation is probably of late Middle or early late Pleistocene age and is represented by the Healy moraine (or Delta moraine in the eastern foothills) and terraces composed of outwash and alluvium. In the Nenana Valley, both a depositional and erosional terrace is present, and side-valley streams have built large alluvial fans over the outwash surfaces (Ritter 1982; Ritter and Ten Brink 1986).

The late Wisconsinan is represented in the

northern Alaska Range by extensive moraine systems, but only in the eastern foothills do they extend beyond the mountain front into the foothills. The main advance is dated to 25,000–17,000 BP, followed by smaller readvances during 15,000–13,500 BP; 12,800(?)–11,800(?) BP; and 10,500(?)–9,500 BP (Ten Brink and Waythomas 1985). The main late Wisconsinan advance generated a major episode of outwash deposition, and terraces can be identified in all of the larger valleys of the foothill belt; distinct outwash terraces that correspond to the first and second readvances can also be recognized in many valleys, but outwash terraces of the final readvance

View of Nenana River valley.

are either absent or minimally developed (Wahrhaftig 1958,1zn 1970a, 1970b; Ten Brink 1983; Ten Brink and Waythomas 1985). During the interstadial periods, side-valley streams created alluvial fans on the outwash plains (Ritter and Ten Brink 1986).

Although deep accumulations of loess and aeolian sand of late Wisconsinan age (and earlier) occur in the Tanana and Kuskokwim lowlands, aeolian deposits appear to be rare in the foothills until the late Glacial. Their scarcity may reflect especially strong wind conditions in the foothills region prior to 14,000–13,000 BP (Hoffecker 1985; Thorson and Bender 1985). During the interstadial phase that followed, loess began to accumulate on both the Pleistocene terraces and older Tertiary surfaces in the valleys (Hoffecker 1988). Deposition of loess and aeolian sand continued throughout the Holocene, which has been characterized by a series of minor alpine glacial readvances (Ten Brink and Waythomas 1985). Buried soil horizons are observable in

many aeolian sections, apparently reflecting intervals of milder climates and reduced loess deposition. Archaeological remains of the terminal Pleistocene and Holocene are typically buried in these aeolian deposits, which often provide an excellent stratigraphic context (West 1967; Thorson and Hamilton 1977; Powers and Hoffecker 1989).

Sources of palynological data for the late Quaternary are rare in the north-central Alaska Range, although several cores have been recovered from lakes in the foothills that span the terminal Pleistocene and Holocene (Ager 1980, 1983). During the late Wisconsinan (25,000–14,000 BP), the region probably supported a herbaceous tundra; the absence of aeolian deposits in the major foothill valleys may indicate especially harsh conditions at this time. Shrub tundra vegetation, comprising dwarf birch, willows, ericads, and herbaceous plants (sedges, grasses, and forbs) began to invade the foothills after 14,000 BP. However, corridors of herbaceous tundra (with abundant sedge, grass, and *Artemisia*) may have survived in the valleys for several millennia (Ager, 1980).

J. F. H.

REFERENCES CITED

Ager, T. A. 1980. An Outline of the Quaternary History of Vegetation in the North Alaska Range. *In* North Alaska Range Project Final Report on 1978–82 Geoarcheological Studies, ed. N. W. Ten Brink. Unpublished report to the National Geographic Society and the National Park Service, Washington, D.C.

———. 1982. Vegetational History of Western Alaska during the Wisconsin Glacial Interval and the Holocene. In *Paleoecology of Beringia*, ed. D. M. Hopkins, J. V. Matthews, Jr., C. E. Schweger, and S. B. Young, pp. 75–94. New York: Academic Press.

———. 1983. Holocene Vegetational History of Alaska. In *Late-Quaternary Environments of the United States*, ed. H. E. Wright, Jr. Vol. 2 *The Holocene*, ed. H. E. Wright, Jr., pp. 128–41. Minneapolis: University of Minnesota Press.

Capps, S. R. 1940. Geology of the Alaska Railroad Region. *U.S. Geological Survey Bulletin* 907.

Hoffecker, J. F. 1985. Archaeological Field Research 1980. *National Geographic Society Research Reports* 19:48–59.

———. 1988. Applied Geomorphology and Archaeological Survey Strategy for Sites of Pleistocene Age: An Example from Central Alaska. *Journal of Archaeological Science* 15:683–713.

Péwé, T. L. 1965. Delta River Area. In *Guidebook to Field Conference F, Central and South Central Alaska*, ed. T. L. Péwé, pp. 55–93. INQUA Seventh Congress. Lincoln: Nebraska Academy of Sciences.

Powers, W. R., and J. F. Hoffecker. 1989. Late Pleistocene Settlement in the Nenana Valley, Central Alaska. *American Antiquity* 54(2):263–87.

Ritter, D. F. 1982. Complex River Terrace Development in the Nenana Valley near Healy, Alaska. *Geological Society of America Bulletin* 93:346–56.

Ritter, D. F., and N. W. Ten Brink. 1986. Alluvial Fan Development and the Glacial-Glaciofluvial Cycle, Nenana Valley, Alaska. *Journal of Geology* 94:613–25.

Ten Brink, N. W. 1983. Glaciation in the Northern Alaska Range. In *Glaciation in Alaska: Extended Abstracts from a Workshop*, ed. R. M. Thorson and T. D. Hamilton, pp. 82–91. Occasional Paper No. 2. Fairbanks: University of Alaska Museum.

Ten Brink, N. W., and C. F. Waythomas. 1985. Late Wisconsin Glacial Chronology of the North-Central Alaska Range: A Regional Synthesis and Its Implications for Early Human Settlements. *National Geographic Society Research Reports* 19:15–32.

Thorson, R. M., and G. Bender. 1985. Eolian Deflation by Ancient Katabatic Winds: A Late Quaternary Example from the North Alaska Range. *Geological Society of America Bulletin* 96:702–709.

Thorson, R. M., and T. D. Hamilton. 1977. Geology of the Dry Creek Site: A Stratified Early Man Site in Interior Alaska. *Quaternary Research* 7:149–76.

Wahrhaftig, C. 1958. *Quaternary Geology of the Nenana River Valley and Adjacent Parts of the Alaska Range*. U.S. Geological Survey Professional Paper 293-A.

———. 1970a. *Geological Map of the Fairbanks A-5 Quadrangle*. U.S. Geological Survey.

———. 1970b. *Geologic Map of the Healy D-5 Quadrangle*. U.S. Geological Survey.

West, F. H. 1967. The Donnelly Ridge Site and the Definition of an Early Core and Blade Complex in Central Alaska. *American Antiquity* 32(3):360–82.

TEKLANIKA WEST

Frederick H. West

The site is located in the central Alaska Range within the confines of Denali National Park. The region lies on the north slope of the range, south of the broad Tanana lowlands and east of the mining district of Kantishna. Coordinates: 63°40′35″N, 149°45′20″W.

PRESENT-DAY LANDSCAPE

Teklanika West occupies a high knoll directly over the Teklanika River. The region is hilly to mountainous in relief. The Teklanika River is one of several sizeable rivers rising in the range and flowing northward to join the Tanana. The Teklanika River contributes thus to the Bering Sea drainage. The river is broad and complexly braided. Seasonally it carries a heavy flow of water and, importantly, heavy loads of silt and gravel, some of which are

The Teklanika West site atop the bluff overlooking the river.

deposited in the site vicinity. The further transport of that silt is greatly abetted by the high winds that are frequent in these valleys. Also within the confines of the park is Mount McKinley ("Denali" to the Ahtna people), which, at 6,000 m, is the highest point in North America.

Teklanika West and the nearby related, but much smaller, Teklanika East are situated atop a discontinuous igneous dike. The elevation of Teklanika West is approximately 788 m.

Some stunted spruce are found directly on the site. Treeline lies at about 848 m here. Otherwise, the landscape could be characterized as having a forest tundra aspect with shrub vegetation (chiefly willow) relatively dense and high adjacent to the river but thinning considerably away from there. Alder is the other, higher, constituent. Ground cover consists of extensive areas of Labrador tea

and *Vaccinium*. The faunal assemblage includes those species expectable in this part of the sub-Arctic—moose, caribou, grizzly bear, lynx—with the addition of alpine tundra forms such as ground squirrel and mountain sheep.

The climate is deeply continental with long, intensely cold winters and short, cool, and, often, rainy summers. Snowfall values are high due to the mountainous setting. The area was almost certainly within the hunting territory of the Western Ahtna and perhaps others who came upstream from the Tanana flats.

The knoll occupied by Teklanika West shows some degree of microrelief including a boggy area just north of the site. The land falls off in all directions from the site proper. At the west the drop-off is precipitous to the river below. Erosion at the bluff face led to the site's discovery.

HISTORY OF RESEARCH

The site was discovered in 1960 by Richard D. Reger. Extensive test excavations were conducted in 1961 for the University of Alaska by Ronald Boyce and Burle Beard with two high school student assistants; this was done under the indirect supervision of the writer. In 1963 Adan Treganza conducted a general survey in the park, in the course of which he excavated several squares at Teklanika West. That material is incorporated here (Treganza 1964). In subsequent summers, groups of various sizes, under the writer's supervision, undertook further, but always restricted, research at the site. This involved short periods in 1964, 1967–1971, and 1974. Partial results of that work have appeared in several publications, including West 1965, 1967, 1974, 1975, 1976, 1979, 1980, 1981.

CHARACTERIZATION OF SITE

The soil matrix here was originally misinterpreted (West 1965, 1967) but was partially corrected in a 1974 publication (West 1974). The site is stratified and appears to contain two components. The primary and major occupation is a Denali one. A second, very recent component, consisting of bone fragments and a few flakes, occurs sporadically just below the surface. Its identity is unclear. Unless otherwise stated, all discussion here concerns the Denali occupation exclusively. Teklanika West was one of the four constituents in the original definition of the Denali complex.

Soil depths range from about 50 cm near the bluff edge to almost twice that value away from that edge. Attention is called to Figure 7–1, taken from Schweger (1971). It was his research in that year that resulted in the recognition of the buried soil and a detailed depiction of the site stratigraphy. The soils are loessial in origin except the upper 10–15 cm, which is also highly organic, becoming peaty, especially downslope where the deposit grades into a boggy area. According to Schweger, the uppermost portion of this unit becomes the modern soil (intermittently identifiable) while the greater part of it is identified as "a (multiplying

downslope) series of azonal soil units." He estimates that "at a maximum, there may be as many as twenty of these azonal units" (Schweger 1971). In the lower part of this unit a thin tephra layer occurs. This is assumed to be Jarvis ash (of the Hayes tephra set) (Péwé 1965; Riehle et al. 1990; Begét et al. 1991). Whether this should be interpreted as a buried soil would seem to be a matter of definition.

The lower palaeosol (Ab and Bb) is a more conventional buried soil. By far the greater part of the material recovered came from here. This unit was "irregular and discontinuous becoming darker [away from the bluff] with the addition of organic matter, particularly rootlets and charcoal" (Schweger 1971). The small quantity recovered from the unit just above, B(?), is interpreted to result from the extensive bio- and cryoturbation seen throughout the site.

Two radiocarbon assays on charcoal from the presumed A1b horizon of the palaeosol were clearly too recent to pertain to the assemblages. These results were 3465 ± 120 BP (UGa-253) and 3820 ± 115 BP (UGa-527). A date of 3630 ± 150 (I-5710) was obtained from the overlying unit here shown as B(?).

In view of its setting the site has probably been subject to repeated cycles of deposition and partial denudation. Still, the presence of the palaeosol would seem to suggest a period of some stability, which, in return, may have some palaeoclimatic implications.

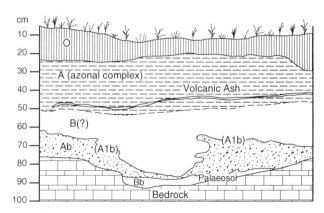

Figure 7–1   Teklanika West soil profile.

An earlier excavator at the site. *(Photo courtesy of Ronald Boyce.)*

ARTIFACT ASSEMBLAGE

The Teklanika West collection consists of over 6,000 artifacts. This is a core and blade assemblage that fits conformably into the definition of the Denali complex. There are, nonetheless, certain elements of the collection that are somewhat unusual and which, as has been previously suggested, might indicate a placement at one end or the other of the Denali continuum (West 1981: Table 33). One of these is the relatively large number of irregular blade cores—tabular and "pillar" forms. However, the site is situated on an igneous dike in which one of the constituents is a fairly coarse-grained chert-like rock. This material was extensively used and its abundance may have encouraged the seemingly inefficient blade production technique. Likewise, of

course, the occurrence of that chert may have been at least partly determinative in the use of this site. Certain it is that there are few archaeological sites in the vicinity yet many other spots that would have provided an equally good game lookout.

Bifaces (53) Of this total, only 10 are complete specimens. This is the largest class of tools in the collection. They may, in general, be characterized as *simple* forms, i.e., the outline consisting of 2 or 3 segments. In almost all cases the flaking is random. The size variation is relatively wide, ranging from about 4 cm to about 13 cm. In some cases use wear is present. The dominant form is the lenticular or biconvex (two segment) biface found in virtually all Denali assemblages. Several of these may be seen in Figure 7–2. Repeatedly here, as in other parts of this

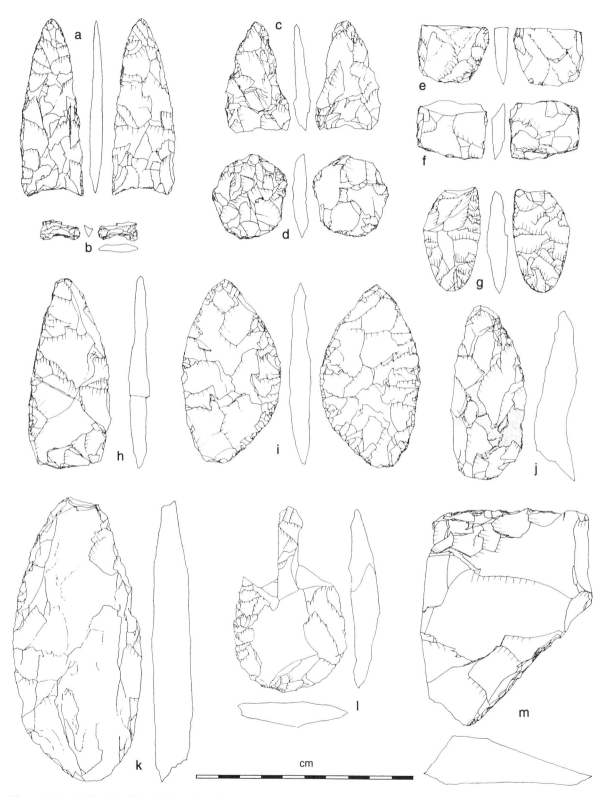

Figure 7–2   Teklanika West: bifaces (*a–m*).

collection, the control exerted by the local chert is evident: there are pieces that might normally be identified as blanks that, in fact, give evidence of use (e.g., Fig. 7–2: *j*). Apparent crudeness, then, appears to result from mechanical constraints imposed by the somewhat intractable local material.

Figure 7–2: *a* is well thinned, subtriangular, and has a concave base. One edge shows light retouch, perhaps from use. This piece was found in the lower section of Palaeosol 1. Figure 7–2: *b* is a basal fragment of indeterminate original form. It may have resembled the preceding.

The specimen Figure 7–2: *c* is triangular, under 5 cm in length, and shows heavy, irregular retouch originating from one side only and fairly high on the sides. This suggests the possibility of its having been hafted. Its appearance is somewhat crude. The mended specimen (Fig. 7–2: *h*) is likewise of the local rock and reveals unidirectional retouch along the median section of one edge. As may be seen, specimens Figure 7–2: *e, f, h* are essentially straight-based but, like 7–2: *a* and 7–2: *c*, are three-segment, simple bifaces. Figure 7–2: *d* is a small, ovate biface on which the upper edge, in the illustration, has been reworked as an end scraper.

Figure 7–2: *g* is—probably—the bottom portion of a small biface. Figure 7–2: *i* is decidedly asymmetrical, while the reconstructed Figure 7–2: *l* is a cordate form. The large rectangular biface (Fig. 7–2: *m*) may have functioned as a knife or scraper. Another, smaller flake artifact in the collection displays the same straight, bidirectionally worked edge.

Tabular Blade Cores (16)   These have been referred to as "pillar" cores. They are mainly made from the local chert and it is frequently the width of the chert vein that determines the dimension and form of these tools. They show little or no preparation and often employ the unmodified edge of the tabular (vein) chert as platform. In some cases it appears only a single removal was accomplished. Several of these tabular cores are shown in Figure 7–3: *e, f, g* along with a blade derived from one (Fig. 7–3: *m*). They are clearly tools of expediency.

Microblade Cores (4)   Relatively few microblade cores were recovered. These are all typical Denali complex forms (Fig. 7–3: *a–d*). Only one of these

specimens is made from the local rock (Fig. 7–3: *c*). Its form is irregular. In addition to the specimens illustrated, however, four core tablets were found, none of which matches any one of those cores and each of which is a different rock type. Additional evidence of "normal" microblade technology is found in the form of core fragments and one possible preform.

Blades (40) and Microblades (91)   This size dichotomy is mirrored in materials preference. Those specimens definitely within the microblade category are mostly of finer-grained, lustrous silex. There is one of obsidian. Those clearly in the "true" blade category (width >6 mm) are of coarser-grained rock (including the local) and often are irregular to the extent that the distinction between a "core" and a "blade" is not straightforward. Deliberate edge retouch is mainly restricted to blades over 6 mm in width. (See Fig. 7–3: *j, k, l*.)

Burins (12)   Two of the items here may, in fact, be microblade cores (Fig. 7–3: *n* and *o*). Another specimen (Fig. 7–3: *p*) has the appearance of a small end scraper on a thick flake from which transverse removals were made. The burin body and one articulable spall were found in the eroded area of the site, while the second articulating spall was found in a test pit at a level corresponding to Schweger's buried palaeosol. Although not set up as a formal type, this piece has been referred to as the "Teklanika burin" in recognition of its distinctive form.

The remainder of this grouping conforms to the Donnelly burin category. Several specimens are shown in Figure 7–3.

End Scrapers (8)   Two of those shown could be termed "keeled" (Fig. 7–4: *a, b*); two others resemble each other (Fig. 7–4: *c* and *g*) in showing two blade-like removals from the dorsal surface and with similar edge retouch. The last specimen could be termed a crude blade end scraper (Fig. 7–4: *e*).

Unifacial Points (2)   The small, pointed, and carefully trimmed uniface (Fig. 7–4: *i*) is matched by the adjacent basal fragment (Fig. 7–4: *j*). They are unique at this site and occurred in the lower section. When reported in 1965 they were termed "unifacial

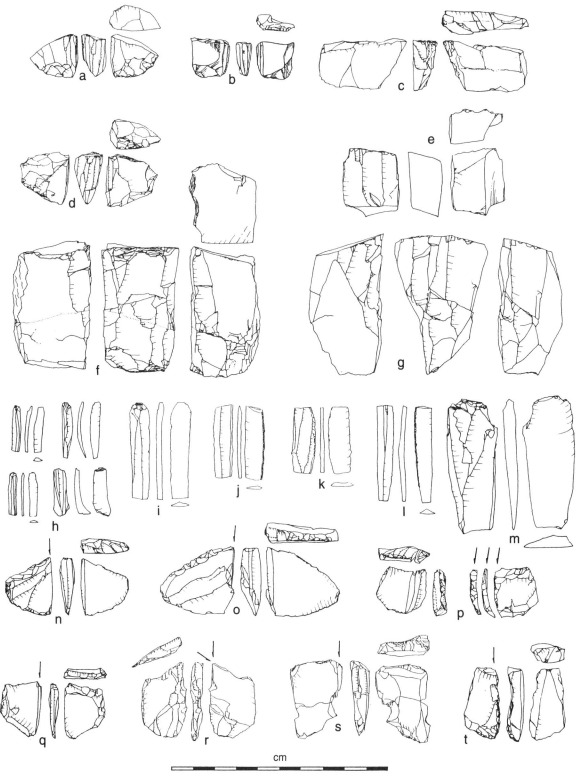

Figure 7–3   Teklanika West: wedge-shaped microblade cores (*a–d*); tabular or pillar blade cores (*e–g*); microblades (*h*); small blades (*i–l*; *j, k,* and *l* are unifacially retouched on one side); large blade (*m*); burins (*n–t*).

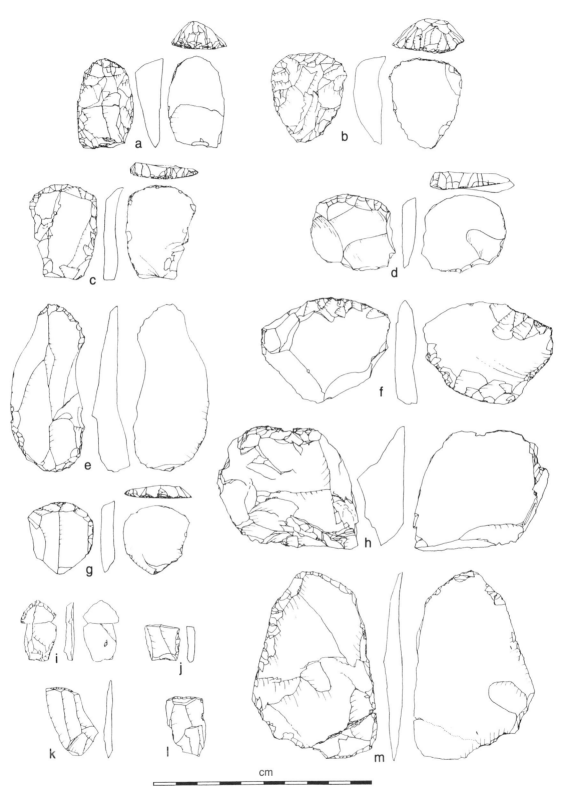

Figure 7–4   Teklanika West: end scrapers (*a–c, e, g*); side scrapers (*d, f, m*); scraper plane or carinate scraper (*h*); unifacial points (*i, j*); end retouched blades (*k, l*).

points" and perhaps their transverse fracture lends support to that identification.

Side Scrapers and Other Unifaces (22)   Included here are rough scrapers on large flakes, retouched flakes, etc. Three objects are identified as *tci thos* or boulder chip scrapers. While these specimens occur in most Denali sites, they are often so nondescript that, found alone, they would be passed over.

Scraper Plane or Carinate Scraper (1)   This large specimen (Fig. 7–4: *h*) was originally termed a "scraper plane burin" to accommodate the apparent burin facet across the lower edge. The undersurface is distinctly smooth and flat. The pattern of extensive step fracturing around the bottom edge suggests heavy usage. A similar specimen was found in Component I of Dry Creek and in the lower component of Whitmore Ridge.

SUMMARY

When compared with other Denali assemblages, it is apparent that certain discrepancies characterize this collection. While it is an obvious truism that no two sites are ever exactly alike, in this case it is argued—as above—that there is reason to suppose that this site may have been chosen for occupation because of the ready availability of chert there. Many, perhaps most, of the variations from the more usual Denali pattern may result from the extensive use of this material.

It must be supposed that what may be termed—here, at any rate—an "archaeological occupation" is more likely to represent intermittent occupations of variable and indeterminate duration over many generations. The earliest of these may have taken place just as the first loess deposits were being laid down. Subsequent occupations may have continued through the time of the—supposed—reduced loess fall, and the formation of a mature soil, to sometime beyond. Thus, the relatively few materials occurring between the palaeosol and the organic horizon may be interpreted in one of two ways: a continued use by later Denali people but on a reduced scale, or simply the result of churning of

the entire deposit by the action of frost and burrowing animals. It cannot be overemphasized that throughout the soil column there is significant evidence of cryo- and bioturbation. That condition appears to bear heavily upon the dating of this site.

Organic Remains

Some small quantity of bone was recovered from the A horizon of the modern soil. These pieces were identified by Dr. Arthur E. Spiess as bones of modern mammals—mountain sheep and caribou as well as several small mammals (Spiess 1982).

Apart from evidence of burning—presumably natural—in the upper segment of the palaeosol layer, no hearths or significant spatial patterning were observed.

Dating of Teklanika West

Several radiocarbon assays have been run on samples from the lower palaeosol. All have been too recent to have anything to do with the Denali occupation at Teklanika. All, in fact, have dated in with or later than the Jarvis ash which lies some 30 to 40 cm above the main find zones. The causes are not far to seek in the vertical displacement emphasized above. Recent inspection of extant charcoal samples from Teklanika West revealed that all samples contain variable amounts of partially burned wood. It seems most probable that the movement of pieces of charcoal and partially reduced wood is mostly the work of animals but conceivably also due to the penetration of roots. Perhaps one of the most instructive photographs taken at the site was that of a ground squirrel peering out of its burrow in the exposure of an excavation.

More help in dating may be provided by several obsidian hydration measurements made on a small series of samples from the occupation zone. Five samples (Lab. nos. 2616–2620) were submitted to the Pennsylvania Obsidian Dating Laboratory. The rind measurements were 2.63 microns, 2.72 microns, 2.74 microns, 2.76 microns, and 2.92 microns (Marshall 1976).

While being entirely cognizant of the difficulties that surround obsidian dating, nevertheless attention may be directed to the results of Smith's 1977 study of a long series from Component II (Denali) at Dry Creek. The peculiar significance of this comparison resides in the fact that these two sites are within 20 km of one another. The Dry Creek samples are quite consistent both within the Component II group and when contrasted with the results from the overlying Component III samples. (There apparently was no obsidian in Component I.) For purposes of his study, Smith (1977) averaged the results for each of the five populations (discrete pods of obsidian flakes) at Dry Creek II. These results were 2.20, 2.61, 1.60, 1.82, and 2.05.

While the fundamental results of Smith's pioneering study were less clear than he had wished—for reasons that may reflect the complexities of the hydration process—nevertheless the comparability of the Teklanika measurements with those of Dry Creek Component II is striking and, at least, suggestive.

REFERENCES CITED

Begét, J. E., R. D. Reger, D. A. Pinney, T. Gillispie, and K. Campbell. 1991. Correlation of the Holocene Jarvis Creek, Tangle Lakes, Cantwell, and Hayes Tephras in South-Central and Central Alaska. *Quaternary Research* 35(2):174–89.

Marshall, N. M. 1976. Report to author on obsidian rind measurements from site Teklanika West, Alaska.

Péwé, T. L. 1965. *Guidebook to the Quaternary Geology of Central and South-Central Alaska*. INQUA Guidebook for Field Conference F. Lincoln: Nebraska Academy of Sciences.

Riehle, J. R., P. M. Bowers, and T. A. Ager. 1990. The Hayes Tephra Deposits, an Upper Holocene Marker Horizon in South-Central Alaska. *Quaternary Research* 33:276–90.

Schweger, C. E. 1971. Stratigraphy of Site Teklanika West. Unpublished field report.

Smith, T. A. 1977. Obsidian Hydration as an Independent Dating Technique. M.A. thesis, Department of Anthropology, University of Alaska, Fairbanks.

Spiess, A. E. 1982. Report to author on faunal identifications.

Treganza, A. E. 1964. An Archaeological Survey of Mount McKinley National Park. Report to the National Park Service, Anchorage.

West, F. H. 1965. Excavations at Two Sites on the Teklanika River, Mount McKinley National Park, Alaska. Report to the National Park Service, University of Alaska, College.

———. 1967. The Donnelly Ridge Site and the Definition of an Early Core and Blade Complex in Central Alaska. *American Antiquity* 32(3):360–82.

———. 1974. Amphitheatre Mountain Complex. In *Proceedings of the International Conference on the Prehistory and Palaeoecology of the Western Arctic and Sub-Arctic, November 1972*. Calgary: University of Calgary Press.

———. 1975. Dating the Denali Complex. *Arctic Anthropology* 12:76–81.

———. 1976. Arkeologicheskii kompleks Tangl-leiks (tsentral'naya Alyaska) i ego svyaz' so starym svetom. (Old World Affinities of Archaeological Complexes from Tangle Lakes, Central Alaska). In *Beringiya v Kainozoe (Beringia in the Cenozoic)*, ed. V. R. Kontrimavichus. Materialy vsesoyuznogo simpoziuma "Beringiiskaya Sucha i ee znachenie dlya Razvitiya Golarkticheskikh flor i faun Kainozoe" (Khabarovsk). Vladivostok: Nauka.

———. 1979. The Beringian Tradition. Fourteenth Pacific Science Congress. *Abstracts of Papers*, Vol. 2 (Khabarovsk). Moscow: Pacific Science Association.

———. 1980. Late Palaeolithic Cultures in Alaska. In *Early Native Americans*, ed. D. L. Browman, World Anthropology. The Hague: Mouton.

———. 1981. *The Archaeology of Beringia*. New York: Columbia University Press.

ADDENDUM
RECENT RESEARCH AT TEKLANIKA WEST: SITE STRATIGRAPHY AND DATING

Ted Goebel

[*The following notes were prepared by Dr. Goebel at the editor's request. On July 29, 1992, Goebel, in company with Nancy H. Bigelow and Sergei Slobodin, visited Teklanika West. Investigations carried out that day were reported to the National Park Service and are described below.*]

In 1992, Goebel, Bigelow, and Slobodin revisited Teklanika West to better establish the stratigraphic context and age of the Denali complex occupation. Walls of previously excavated units were cleaned and three stratigraphic sections were drawn. One of these, called profile A-A' (presumably excavated by Treganza in 1963), yielded a fairly clear section con-

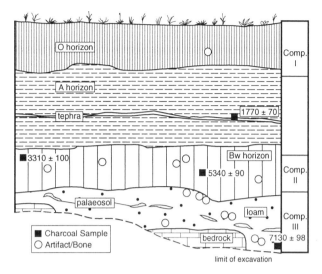

Figure 7–5 Teklanika West. Stratigraphic profile.

taining artifacts, bone, and wood charcoal *in situ*. It is shown in Figure 7–5 and is described briefly below.

Profile A-A' is a 90–100-cm-thick heavily weathered loam (or loess) deposit. The top of the section is dominated by a 70-cm-thick loam containing distinct O, A and Bw horizons of the modern soil. The A horizon is gleyed and contains a distinct but thin band of volcanic ash near its base (ca. 40 cm below the surface). Underlying the Bw horizon is approximately 20 cm of unweathered loam; no change in texture or grain size was noted at this contact. Within the basal loam is a thin, intermittent palaeosol (Palaeosol 1) (ca. 75–85 cm below the surface). Bedrock occurs 80–100 cm below the modern surface.

The loams appear to be aeolian in origin. Closer to the bluff edges, changes in grain size are apparent (loams and sandy loams occur). The modern soil has been identified as a sphagnic-borofibrist (or peat). Nearer the bluff edge this peat is replaced by what appears to be a moderately weathered inceptisol. These differences in modern soil are likely due to differences in micro-relief; the bluff edge is well drained, while the area surrounding Profile A-A' is along the edge of a wet bog. The ash band contained within the modern soil could be either the

Devil or Watana tephra, independently dated elsewhere in south-central Alaska to ca. 1,400–1,600 BP and 1,900–2,700 BP, respectively (Dixon and Smith 1990:394). We await results of microprobe analysis by J. Begét. The palaeosol identified low in the section appears to extend across most of the site, as it is visible in other profiles examined.

Lithic artifacts (small flakes) in Profile A-A' occur in what appear to be three distinct horizons, referred to here as Components I, II, and III (Fig. 7–5). Component III, the uppermost component, consists of two flakes and a retouched blade fragment exposed at the top of the section in the O horizon of the modern soil (10–15 cm below the surface). Component II includes 5 flakes, 1 blade fragment, and 3 bone fragments (presumably large mammal). These were encountered at the base of the modern soil within the Bw horizon, approximately 70 cm below the surface. Component I is represented by 6 flakes situated immediately above or below Palaeosol 1 (80–90 cm below the surface).

Four wood charcoal samples collected from the newly exposed face of Profile A-A' were submitted for radiocarbon dating. A tiny lump of charcoal encountered near the top of the tephra band yielded an AMS date of 1770 ± 70 BP (Beta-59592), and two small samples of charcoal recovered from the Bw horizon of the modern soil in stratigraphic association with Component II artifacts yielded an AMS date of 5340 ± 90 BP (GX-18517) and an extended counting time date of 3310 ± 100 BP (Beta-59591). The fourth charcoal sample was collected from the very base of the section, about 5 cm below Palaeosol 1. It produced an AMS date of 7130 ± 98 BP (GX-18518).

These results indicate that Teklanika West is a stratified, multicomponent site. Three components appear to exist, the first perhaps dating to the early Holocene (ca. 7,000 BP), the second to the mid-Holocene (3,000–6,000 BP), and the third to the late Holocene (less than 1,500 BP). The Teklanika West assemblage described above should not be treated as a single industry, but one containing a mix of artifacts likely spanning the Holocene. Additional excavations will be necessary to identify the diagnostic elements of each of these components.

REFERENCES CITED

Dixon, E. J., and G. S. Smith. 1990. A Regional Application of Tephrochronology in Alaska. In *Archaeological Geology of North America,* ed. N. P. Lasca, and J. Donohue, pp. 383–98. Centennial Special Vol. 4. Boulder, Colorado: Geological Society of America.

Goebel, T. 1992. Geoarchaeological Research at Teklanika West, Denali National Park, Alaska. Preliminary report submitted to the National Park Service, Anchorage.

DRY CREEK

John F. Hoffecker, W. Roger Powers, and Nancy H. Bigelow

The Dry Creek site (HEA-005) is located in the north-central foothills of the Alaska Range near the town of Healy, approximately 125 km southwest of Fairbanks (63°53′N, 149°02′W). The site occupies a southeast-facing bluff on the north side of Dry Creek in the Nenana Valley, approximately 470 m above mean sea level.

HISTORY OF RESEARCH

The site was discovered by C. E. Holmes in May 1973. During the late summer of 1973, three small test excavations were undertaken along the exposed bluff edge (Holmes 1974). In 1974, a 2 m × 15 m test trench was excavated; three to four components were defined, and a total of 2,827 artifacts was recovered. Four separate areas were uncovered during 1976; 12,951 artifacts were recovered. An additional 172 sq. m and 19,033 artifacts were excavated in 1977, bringing the total excavated area to 347 sq. m and the total number of artifacts to 34,811 (Powers et al. 1983). Geologic studies were undertaken at the Dry Creek site during the period from 1973 to 1976 by T. D. Hamilton and R. M. Thorson (Thorson and Hamilton 1977). Faunal remains from the site were analyzed by R. D. Guthrie (Guthrie 1983).

STRATIGRAPHY AND DATING

The artifacts, faunal, and other remains are buried in loess and aeolian sand deposits that cap a Pleistocene terrace. The terrace is composed of cobbles and pebbles in a sandy matrix, and represents glaciofluvial outwash deposited by the Healy glaciation, which appears to be of pre-Wisconsinan age (Wahrhaftig 1958, 1970). The aeolian deposits unconformably overlie the Healy outwash and span the last 12,000 years; they achieve a depth of up to 2 m (Thorson and Hamilton 1977) (see Fig. 7–6).

The lowermost unit (Loess 1) is a silt with minor sand, devoid of archaeological remains. It is overlain by a sandy silt with minor clay (Loess 2) that contains the lowermost archaeological component (Component I) and is radiocarbon-dated to 11,120 ± 85 years BP. A thin, discontinuous sand lens separates Loess 2 from overlying Loess 3, which is another sandy silt with minor clay unit containing archaeological Component II (10,690 ± 250 years BP). Loess 3 also contains a series of thin, organic lenses distributed throughout the unit (Palaeosol 1), and is capped with two more strongly developed and continuous organic bands (Palaeosol 2). Palaeosol 2 appears to date to the earliest Holocene; two significantly older dates on this buried soil apparently reflect contamination from nearby coal-bearing rocks (Thorson and Hamilton 1977:166–67).

The upper portion of the aeolian sequence at Dry Creek comprises a series of alternating loess (sandy silt) and aeolian sand (sand and silty sand) units that span the remainder of the Holocene. An important marker horizon that has been recognized elsewhere in the Nenana Valley is Palaeosol 3, which is composed of a thick, continuous organic band overlain by eight less well-developed organic lenses (Thorson and Hamilton 1977:161). Palaeosol 3 is dated to roughly 6,250–8,600 years BP and may be correlated with the Altithermal; it contains scattered artifacts. The uppermost part of the sequence contains a buried forest soil complex (Palaeosol 4a

Dry Creek site during 1992 excavtion by N. H. Bigelow and W. R. Powers. *(Photo courtesy of Frederick Hadleigh West.)*

and 4b) dated to the middle Holocene. The lower part of the complex (Palaeosol 4a) yielded an assemblage of several thousand artifacts (Component IV).

A total of 24 radiocarbon dates are available from the site; six of them are AMS dates produced by the University of Arizona (AA) (Table 7–1). All of the dates were obtained on charcoal from buried soils or charcoal lenses with the exception of SI-1933B (peat and roots). Thorson and Hamilton (1977:166–67) regard some dates as anomalously old (e.g., SI-1544, SI-1938, SI-1936) and probably contaminated by airborne lignite derived from nearby coal-bearing rock; other anomalies (e.g., AA-11730, AA-11731) may be due to post-depositional movement of charcoal (Bigelow and Powers 1994). Despite these anomalies, the chronology of the aeolian sequence at Dry Creek appears to be under control, especially when correlated with other, more recently dated profiles in the Nenana

Valley (e.g., Walker Road, Panguingue Creek) (Hoffecker 1988:694, Fig. 4; Powers and Hoffecker 1989:270, Table 1; Goebel et al., this volume; Goebel and Bigelow, this volume). In addition to the radiocarbon dates, a thermoluminescence date of 11,350 ± 1000 years was obtained on a sediment sample from Loess 1 (Hoffecker 1988:693); however, while consistent with most of the other absolute dates, this is a preliminary estimate on a sample that was subsequently destroyed and should not be accorded substantial weight (C. F. Waythomas, personal communication).

FAUNA AND FLORA

A small quantity of poorly preserved mammalian faunal remains was recovered from Dry Creek during the period from 1974 to 1977; in addition, avian gastroliths were excavated from the site in 1977 (Guthrie 1983).

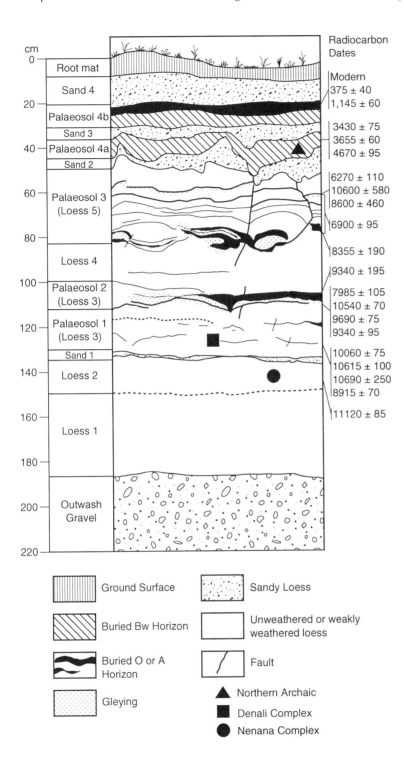

Figure 7–6 Dry Creek. Stratigraphic profile. Note: dates with standard deviations greater than 1,000 years not shown. *(Source:* Current Research in the Pleistocene. *See Acknowledgments.)*

Mammalian remains excavated from Component I (Loess 2) were identified as Dall sheep (*Ovis dalli*) and wapiti (*Cervus canadensis*). Remains from Component II (Loess 3/Palaeosol 1) were assigned to steppe bison (*Bison priscus*) and Dall sheep. Identifiable specimens are confined to fragments of tooth enamel, and Guthrie (1983) cautions that the specific (but not generic) assignments for wapiti

Table 7–1   Radiocarbon Dates from the Dry Creek Site*

SI-1933A	modern	Palaeosol 4b
SI-1933B	375 ± 40	Palaeosol 4b
SI-2333	1145 ± 60	Palaeosol 4b
SI-2332	3430 ± 75	Palaeosol 4a/Component IV
SI-1934	3655 ± 60	Palaeosol 4a/Component IV
SI-1937	4670 ± 95	Palaeosol 4a/Component IV
SI-2331	6270 ± 110	Palaeosol 3
SI-1935C	6900 ± 95	Palaeosol 3
SI-2328	7985 ± 105	Palaeosol 2(?)
SI-1935B	8355 ± 190	Palaeosol 3
SI-2115	8600 ± 460	Palaeosol 3
AA-11730	8915 ± 70	Palaeosol 1 (lower member)
SI-2329	9340 ± 195	Palaeosol 2/Component II
AA-11733	9340 ± 95	Palaeosol 2 (lower member)
AA-11732	9690 ± 75	Palaeosol 2 (middle)
AA-11727	10,060 ± 75	Palaeosol 1 (upper member)
AA-11731	10,540 ± 70	Palaeosol 2 (upper member)
SI-1935A	10,600 ± 580	Palaeosol 3
AA-11728	10,615 ± 100	Palaeosol 1 (lower member)
SI-1561	10,690 ± 250	Palaeosol 1/Component II
SI-2880	11,120 ± 85	Loess 2/Component I
SI-1936	12,080 ± 1025	Palaeosol 2
SI-1544	19,050 ± 1500	Palaeosol 3
SI-1938	23,930 ± 9300	Palaeosol 2

*After Thorson and Hamilton (1977:166, Table 4) and Bigelow and Powers (1994). Dates are given in uncalibrated radiocarbon years BP and are arranged in chronological order as per Thorson and Hamilton.

and bison are tentative. The sheep specimens (not subdivided according to occupation level) consist of three isolated M^2s, one $P^4–M^3$ sequence, and one $M_2–M_3$ sequence. A minimum of five individuals appears to be represented (one juvenile, one juvenile/adult, and three adults) (Guthrie 1983:220, Table 6.1). The bison remains include two $M_2–M_3$ sequences, one M_3 fragment, one $P_3–M_3$ sequence, and one unerupted crown fragment. A minimum of five individuals is represented (three young and two older/old) (Guthrie 1983:242, Table 6.2). Wapiti remains are limited to one $M_2–M_3$ sequence and one $P_4–M_3$ sequence; two individuals are represented (one adult and one old) (Guthrie 1983:252, Table 6.3). The two modern taxa present (i.e., Dall sheep and wapiti) exhibit larger than modern body size (Guthrie 1983:285).

Gastroliths (i.e., gizzard stones) were recovered from both occupation levels and archaeologically sterile units (Guthrie 1983:274–81). The mean size of the gastroliths is 2.14 mm, which suggests that they were derived from birds in the ptarmigan size range. The angular morphology of most specimens indicates that they were deposited during the summer, fall, or early winter. Their association with occupation episodes at Dry Creek remains problematic.

During 1977, T. A. Ager (personal communication) collected pollen/spore samples from the site, but their contents were too sparse for meaningful analysis. However, Guthrie (1983:282–83) reports that opaline phytoliths are abundant in the sediment (including former hearth lenses); although not studied in depth, many specimens can be assigned to festucoid grasses.

FEATURES

The broad-scale excavations at the site (347 sq. m) uncovered large occupation areas and documented numerous features in the form of debris concentrations and remnants of former hearths in Components I and II (Powers et al. 1983; Powers and Hoffecker 1989:276–81).

Component I   Three large artifact concentrations were mapped on the Component I occupation floor, accounting for approximately 50 percent of the artifacts in the component (Powers and Hoffecker 1989:282, Fig. 10). The debris concentrations range in size from ca. 2 sq. m to 5 sq. m and comprise from 110 to 1,160 items. Smaller accumulations of debris, including a concentration of faunal remains, are scattered across the occupation area. Isolated faunal remains are associated with all three concentrations, and charcoal fragments were recovered within at least 1 m of their margins. An analysis of lithic reduction sequences by Smith (1985) indicates that the Component I concentrations reflect manufacture of bifaces and possibly scrapers, along with the reduction of several large river cobbles.

Component II   Fourteen large debris concentrations were uncovered in Component II, accounting for roughly 70 percent of the total artifacts in the component (Powers and Hoffecker 1989:277, Fig. 7). These concentrations range from ca. 4 to 13 sq. m and contain from less than 350 to over 3,500 items. Ten of the concentrations are associated with faunal remains, and charcoal fragments were found within 1 m of all but one (partially excavated). The concentrations vary significantly in terms of contents. Five of them contain microblade cores, microblade core parts, and microblades; these concentrations also contain numerous burins and burin spalls. The remaining nine concentrations lack evidence of microblade manufacture, with the exception of an isolated microblade core in one and a single microblade in another; burins and burin spalls also are either rare or absent in these concentrations (Fig. 7–7).

ARTIFACT ASSEMBLAGE

A total of 34,811 artifacts was recovered from the site between 1973 and 1977. Although four separate components were defined during the earlier excavations (Powers and Hamilton 1978), Component III was eventually combined with Component II.

Component I The lowermost assemblage contains a total of 3,517 artifacts, including 4 cores, 3,474 flakes and fragments, and 39 tools (1%). The most common raw materials are brown chert, grey chert, light rhyolite, and a degraded quartzite. The cores represent large, elongate river cobbles of degraded quartzite (3) and rhyolite (1). The tools include 1 projectile point, 2 projectile point bases, 1 biface base, 1 biface tip, 3 small bifaces, 3 transverse scrapers, 2 side scrapers, 11 end scrapers, 1 double end scraper, 1 end scraper/burin, 1 quadrilateral uniface, 6 retouched flakes, and 3 split cobble tools. The isolated whole point is small and triangular in form (31 × 16 × 4 mm) with excurvate sides and a straight base (except for a small spur at one corner). At least one of the point bases appears to be derived from a similar point, while the other exhibits parallel sides and lacks a corner spur. The small bifaces are larger (55 to 64 mm in length), but also triangular with excurvate sides and a convex base. The end

scrapers were manufactured on flakes or blade-like flakes (Powers et al. 1983; Powers and Hoffecker 1989:281) (Fig. 7–8).

Component I has been assigned to the Nenana complex on the basis of the presence of small triangular/teardrop-shaped points, a high proportion of end scrapers, and the absence of wedge-shaped microblade cores, microblades, and polyfaceted burins (Powers and Hoffecker 1989:277–81).

Component II The overlying assemblage contains a total of 28,881 artifacts, including 126 cores and core fragments, 28,529 flakes and blades, and 194 tools. Among the wide variety of raw materials represented are degraded quartzite, light and dark rhyolite, various cherts (grey, brown, black, and green), chalcedony, obsidian, quartzite, argillite, diabase, pumice, and sandstone (Figs. 7–9, 7–10).

The cores and core fragments include 21 wedge-shaped microblade cores, 8 aberrant microblade cores, 4 subprismatic cores, 3 microcore preforms, 21 miscarried microcore preforms, 45 core tablets, and 24 miscellaneous wedge-shaped core parts. The wedge-shaped microblade cores are small, averaging approximately 30 mm in length, 25 mm in height, and 14 mm in thickness; they possess 3 to 8 microblade flutes each (mean = 5.6). Unretouched flakes and blades include 1,772 microblades. Most of the microblades (ca. 90%) are incomplete and manifest a mean width of 3.8 mm.

The tools in this assemblage include 29 burins, 8 core/burins, 44 bifaces, 47 heavy bifaces, 21 scrapers, 21 retouched flakes, 18 blade-like retouched flakes, and 3 hammerstones. The burins include 3 dihedral, 2 angle, 13 transverse, and 10 burins-on-a-snap. The core/burins represent polyhedral forms that resemble the wedge-shaped microcores, but exhibit macroscopic traces of heavy edge wear. A total of 35 burin spalls was recovered. The bifaces include a small lanceolate projectile point and 2 projectile point tips and 6 bases; the point bases are apparently derived from lanceolate or stemmed forms. Most of the 26 remaining bifaces represent spatulate, ovate, triangular, and lanceolate knives.

Component II is assigned to the Denali complex, as defined by West (1967, 1981, this volume), primarily on the basis of the presence of wedge-

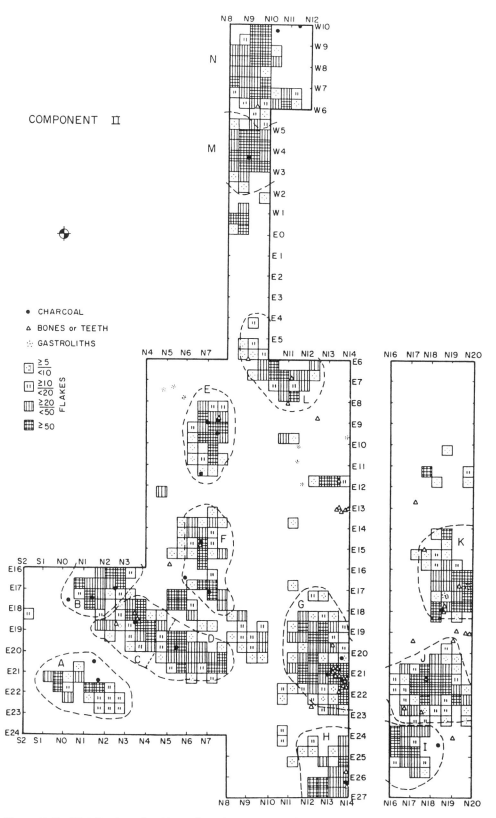

Figure 7–7   Distribution of artifacts, faunal remains, and charcoal fragments in Component II (Loess 3). *(Source: American Antiquity. See Acknowledgments.)*

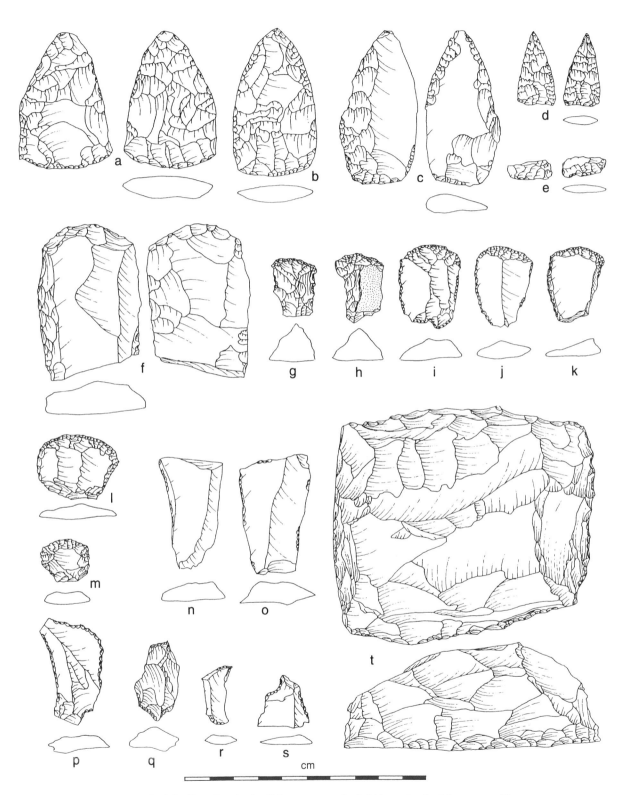

Figure 7–8   Component I of the Dry Creek site (Nenana complex): bifaces (*a–e*); side scraper (*f*); end scrapers (*g–m*); bilaterally retouched blades (*n, o*); retouched flakes (*p–s*); quadrilateral uniface or plane (*t*).

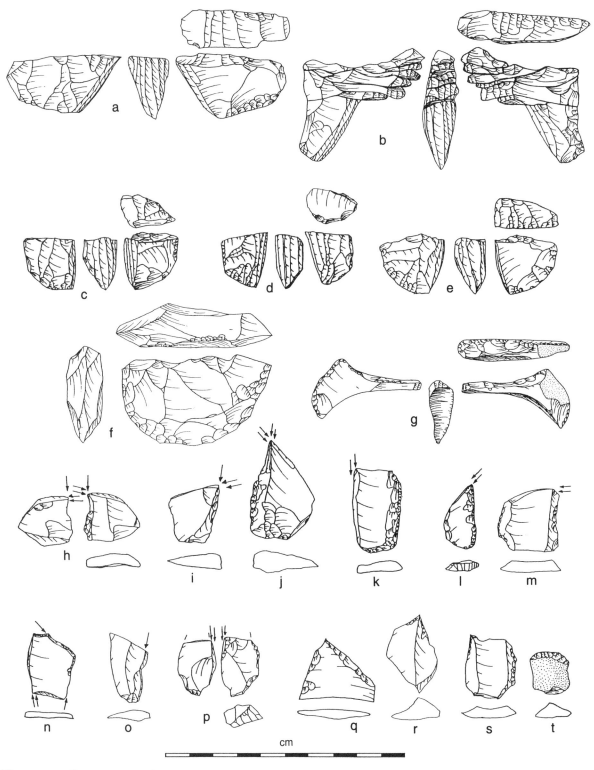

Figure 7–9   Component II of the Dry Creek site (Denali complex): wedge-shaped microblade cores (*a–e*, *b* with attaching core parts); wedge-shaped core preform (*f*); wedge-shaped core tablet (*g*); dihedral burins (*h–j*); angle burin (*k*); transverse burins (*l*, *m*); burins-on-a-snap (*n–p*); retouched flakes (*q–s*); end scraper (*t*).

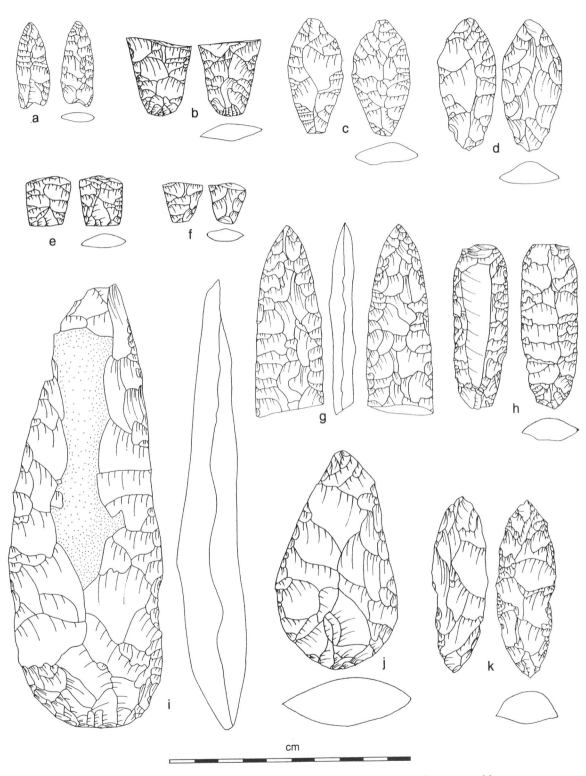

Figure 7–10   Bifaces from Component II of the Dry Creek site: complete projectile point and base fragments (*a, b, e, f*); biface knives and preforms (*c, d, g–k*).

shaped microblade cores, microblades, and polyfaceted burins (Powers et al. 1983; Powers and Hoffecker 1989:272–76). Bifacial lanceolate points, including projectile points, although less common in other Denali assemblages (West 1981), are also characteristic of this complex (e.g., Panguingue Creek, Broken Mammoth).

Component IV The uppermost assemblage contains a total of 2,372 items, which includes 2,115 flakes, 241 rocks and bone fragments, and 16 tools (0.8%). The raw materials comprise rhyolite, quartzite, degraded quartzite, obsidian, chert, and siltstone. The tools include 4 stemmed/side-notched points, 1 point base, 8 end scrapers, and 2 retouched flakes. The assemblage is assigned to the Northern Archaic tradition on the basis of the point typology.

Summary

Dry Creek remains one of the most important sites in Beringia, and its investigation has played a significant role in the development of ideas about Beringian prehistory and the settlement of the New World. It was the first deeply stratified site in the Alaskan interior to yield assemblages radiometrically dated to the Pleistocene. The well-studied site stratigraphy became a type section for aeolian deposits in other parts of the north-central Alaska Range (Powers and Hoffecker 1989). Dry Creek provided important corroboration of the terminal Pleistocene/early Holocene age of the Denali complex, which had been previously based on a small number of radiocarbon dates from shallow sedimentary contexts (West 1975). The Dry Creek Denali assemblage contributed significantly to our knowledge of this complex, yielding a large and diverse sample of artifacts, occupation floor patterns, and associated faunal remains. The Dry Creek site also provided the first glimpse of a possible earlier complex in dated stratigraphic context (i.e., Nenana complex). Although no major unresolved issues remain to be addressed at Dry Creek, the results of its investigation continue to influence research in the foothills region and other parts of Alaska.

References Cited

Bigelow, N. H., and W. R. Powers. 1994. New AMS Dates from the Dry Creek Paleoindian Site, Central Alaska. *Current Research in the Pleistocene* 11:114–16.

Guthrie, R. D. 1983. Paleoecology of the Dry Creek Site and Its Implications for Early Hunters. In *Dry Creek: Archaeology and Paleoecology of a Late Pleistocene Alaskan Hunting Camp*, ed. W. R. Powers, R. D. Guthrie, and J. F Hoffecker, pp. 209–87. Report to the National Park Service, Washington, D.C.

Hoffecker, J. F. 1988. Applied Geomorphology and Archaeological Survey Strategy for Sites of Pleistocene Age: An Example from Central Alaska. *Journal of Archaeological Science* 15:683–713.

Holmes, C. E. 1974. New Evidence of a Late Pleistocene Culture in Central Alaska: Preliminary Investigations at Dry Creek. Paper presented at the 7th annual meeting of the Canadian Archaeological Association, Whitehorse, Yukon.

Powers, W. R., R. D. Guthrie, and J. F. Hoffecker. 1983. Dry Creek: Archaeology and Paleoecology of a Late Pleistocene Alaskan Hunting Camp. Report to the National Park Service, Washington, D.C.

Powers, W. R., and T. D. Hamilton. 1978. Dry Creek: A Late Pleistocene Human Occupation in Central Alaska. In *Early Man in America from a Circum-Pacific Perspective*, ed. A. L. Bryan, pp. 72–77. Occasional Papers No. 1, Department of Anthropology. Edmonton: University of Alberta.

Powers, W. R., and J. F. Hoffecker. 1989. Late Pleistocene Settlement in the Nenana Valley, Central Alaska. *American Antiquity* 54(2):263–87.

Smith, T. A. 1985. Spatial Analysis of the Dry Creek Site. *National Geographic Society Research Reports* 19:6–11.

Thorson, R. M., and T. D. Hamilton. 1977. Geology of the Dry Creek Site, a Stratified Early Man Site in Interior Alaska. *Quaternary Research* 7:149–76.

Wahrhaftig, C. 1958. *Quaternary Geology of the Nenana River Valley and Adjacent Parts of the Alaska Range*. U.S. Geological Survey Professional Paper 293-A:1–68.

———. 1970. *Geologic Map of the Healy D-5 Quadrangle*. U.S. Geological Survey.

West. F. H. 1967. The Donnelly Ridge Site and the Definition of an Early Core and Blade Complex in Central Alaska. *American Antiquity* 32(3):360–82.

———. 1975. Dating the Denali Complex. *Arctic Anthropology* 12(1):76–81.

———. 1981. *The Archaeology of Beringia*. New York: Columbia University Press.

OWL RIDGE

*John F. Hoffecker, W. Roger Powers,
and Peter G. Phippen*

The Owl Ridge site (FAI-91) is located in the north-central foothills of the Alaska Range, west of the town of Healy and roughly 130 km southwest of Fairbanks. Owl Ridge occupies a south-facing bluff on the north side of an unnamed creek in the Teklanika Valley, approximately 80 m above the modern floodplain. It is situated at approximately 64°00'45"N, 149°33'25"W and 470 m above mean sea level. In early unpublished reports, Owl Ridge was referred to as the First Creek site.

HISTORY OF RESEARCH

The site was discovered in August 1976 by D. C. Plaskett and R. M. Thorson during a survey of the Teklanika Valley (Plaskett 1976; Thorson 1977). At that time, a 1 m × 1 m test pit was excavated approximately 5 m from the bluff edge. The site was briefly revisited in August 1977 by W. R. Powers, N. W. Ten Brink, and several field assistants, who excavated several additional test pits (Powers 1983). In July 1978, Ten Brink, Thorson, and J. F. Hoffecker recorded the aeolian stratigraphy and collected the first radiocarbon sample (Hoffecker 1978, 1988). P. G. Phippen and a field assistant worked at Owl Ridge in August 1982, excavating a total of six 1-sq.-m units. Phippen revisited the site with two assistants in August–September 1984, and excavated an additional 20 sq. m (Phippen 1988).

STRATIGRAPHY AND DATING

The artifacts and associated features are buried in aeolian sand and silt that overlies a glaciofluvial outwash terrace. The outwash is composed of rounded-to-angular gravels in a clay-to-sand matrix, and is assigned to the Healy glaciation (Péwé et al. 1966), which probably dates to the late middle Pleistocene. The aeolian deposits unconformably overlie the outwash and are dated to the terminal Pleistocene and Holocene (see Fig. 7–11).

The aeolian deposits average approximately 1 m in thickness (Hoffecker 1988:694, Fig. 4*d*; Phippen 1988:76, Table 1). At the base of the sequence lies a silt layer (unit 2) containing the lowermost archaeological component (Component I), which has yielded ^{14}C dates of 11,340 and 9060 years BP. This unit is capped with a sand horizon that varies between 0 and 15 cm in thickness. Above the sand horizon lies a well-developed buried soil complex (unit 4) that contains Component II and has been ^{14}C-dated to 9325–7230 years BP. A bed of fine-to-medium sand overlies the buried soil; Component III is deposited in the uppermost 5–10 cm of this unit. The upper portion of the sequence comprises silty sand (unit 6) and a series of grey and reddish brown horizons that represent middle and late Holocene soil formation, capped with the modern forest soil.

Thirteen radiocarbon dates from Owl Ridge are reproduced in chronological order in Table 7–2 (Phippen 1988:92, Table 2). All dates are given in (uncorrected) radiocarbon years BP.

The dates on unit 4 (with the exception of B-11081) were obtained on soil organics; the remaining dates were obtained on samples of charcoal. Phippen (1988:91–93) regards four of the dates (B-5416, B-11079, B-11081, and B-11082) as stratigraphically discordant and suggests that they reflect introduction of younger materials from subsequent erosion, exposure, and redeposition of overlying layers.

The Owl Ridge aeolian sequence is very similar to dated profiles in the neighboring Nenana Valley (e.g., Moose Creek, Walker Road). Correlation with these other localities provides a consistent picture of aeolian deposition and soil formation in the valleys of the northern foothill region during the past 12,000 years (Powers and Hoffecker 1989).

FAUNA AND FLORA

Carbonized remains of spruce (*Picea* sp.) were recovered from a charcoal concentration (former

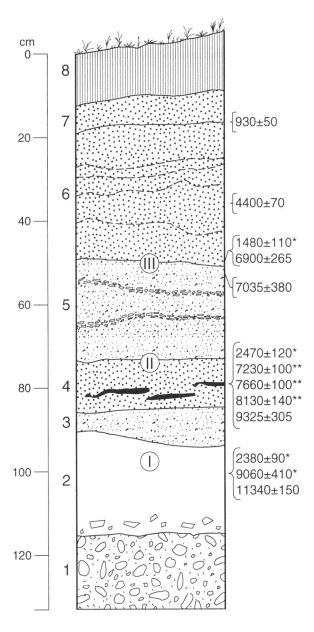

* Dates regarded as discordant
** Dates on organic rich silt

Figure 7–11 Owl Ridge. Stratigraphic profile. Strata units to left: (1) gravel; (2) silt; (3) sand; (4) silty sand with palaeosols; (5) sand with lag schist fragments; (6) silty sand with forest beds; (7) silty sand; (8) surface organics. Archaeological components circled. *(After Phippen 1988.)*

hearth?) associated with Component III and dated to 7,035–6,900 years BP. (Phippen 1988: 70, Fig. 17). No faunal remains have been reported from the site.

FEATURES

On the lowermost occupation level (Component I), Phippen (1988:36–71) reports two diffuse debris concentrations. Cluster A occupies an area of less than 1 m × 2 m and comprises 2 retouched pieces, 25 flakes, and several charcoal fragments. Cluster B occupies approximately 1.5 sq. m and contains 2 tools, 4 utilized flakes, 1 hammerstone, and 25 flakes, associated with a small quantity of charcoal and a fragment of red ochre.

No debris concentrations were encountered in Component II, which yielded few artifacts, but Plaskett (1976) reported a concentration of seven large cobbles subsequently assigned to this occupation level. An additional cobble was recovered by Hoffecker (unpublished notes), and Phippen (1988:43) encountered a ninth cobble located 3 m from the main concentration. Phippen (1988) suggested that the cobble concentration might represent a former tent ring.

A large debris concentration (feature 1), occupying an area of approximately 2 square m and containing 7 tools, 674 flakes and rock fragments, and a small quantity of charcoal, was excavated from Component III (Phippen 1988:66).

Table 7–2 Radiocarbon Dates from Owl Ridge

D-3071	930 ± 50	Unit 6/7
B-11082	1480 ± 110	Unit 5/6
B-11079	2380 ± 90	Unit 2
B-11081	2470 ± 120	Unit 4
B-11080	4400 ± 70	Unit 6 (middle)
D-3070	6900 ± 265	Unit 5a/6c
GX-13009	7035 ± 380	Unit 5a
B-11437	7230 ± 100	Unit 4
B-11436	7660 ± 100	Unit 4
B-5418	8130 ± 140	Unit 4
B-5416	9060 ± 410	Unit 2
GX-6283	9325 ± 305	Unit 4
B-11209	11,340 ± 150	Unit 2

ARTIFACTS

The lowermost component at Owl Ridge (Component I) contains a total of 151 artifacts, including 11 retouched or utilized pieces (7%); additional items include a hammerstone and a red ochre fragment (Phippen 1988:104–18). The most common lithic raw materials are rhyolite, chert (black, blue-grey, and brown), and chalcedony (brown and translucent). No cores or core fragments are reported; waste flakes include 133 flakes or flake fragments and 6 blade-like flakes and flake fragments. The retouched items are limited to 3 bifaces and 2 biface fragments. The complete specimens include thin ovate, triangular, and large crude forms; the fragments represent a lanceolate point base and point base fragment (Fig. 7–12). The assemblage is tentatively assigned to the Nenana complex, primarily on the basis of its lack of diagnostic Denali elements (microblades, burins) and its stratigraphic/geochronologic position (Phippen 1988:137).

Component II contains only seven artifacts and associated rocks and rock fragments (some of which are classified as manuports or fire-cracked rock) (Phippen 1988:118–23). The artifacts (rhyolite, chert, and degraded quartzite) include a thick, ovate biface, sub-triangular biface, blade fragment, utilized flake, and 3 waste flakes. Phippen (1988:137–38) suggested that this assemblage was probably affiliated with the Denali complex on the basis of its age (i.e., 9,325–7,230 years BP).

The uppermost component (Component III) contains 743 artifacts and associated lithic debris (Phippen 1988:123–32). Raw materials consist primarily of andesite, rhyolite, and argillite. The artifacts include 3 unifacial cobble tools, 1 retouched flake, 2 large crude bifaces, 1 flake knife, 1 burnisher, 1 hammerstone, and 724 waste flakes and fragments. The culture-stratigraphic affiliation of this assemblage (which dates to approximately 7,000 years BP) is unclear.

SUMMARY

Owl Ridge is an important locality in the northern foothills of the Alaska Range because of the strati-

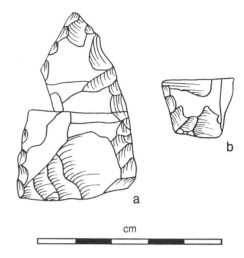

Figure 7–12   Owl Ridge: thin triangular biface (*a*); biface base corner fragment (*b*).

graphic and geochronologic context of the site. However, ambiguities remain regarding the culture-stratigraphic affiliations of the archaeological components due to the scarcity of diagnostic artifacts. The artifactual contents and stratigraphic context of the lowermost component is very similar to Component I at Moose Creek in the Nenana Valley, which, although formerly assigned to the Nenana complex (Powers and Hoffecker 1989), now appears of uncertain affiliation. This uncertainty is due in large part to the problematic status of bifacial lanceolate points in this complex. Component II is almost certainly of comparable age to Denali assemblages in other parts of central Alaska, such as Panguingue Creek, but has yet to produce any diagnostic materials.

REFERENCES CITED

Hoffecker, J. F. 1978. On the Potential of the North Alaska Range for Archaeological Sites of Pleistocene Age. Report to the National Geographic Society and the National Park Service, Washington, D.C.

———. 1988. Applied Geomorphology and Archaeological Survey Strategy for Sites of Pleistocene Age: An Example from Central Alaska. *Journal of Archaeological Science* 15:683–713.

Péwé, T. L., C. Wahrhaftig, and F. R. Weber. 1966. *Geologic Map of the Fairbanks Quadrangle, Alaska.* U.S. Geological Survey Miscellaneous Geological Investigations, Map I-455.

Phippen, P. G. 1988. Archaeology at Owl Ridge: A Pleistocene–Holocene Boundary Age Site in Central Alaska. M.A. thesis, University of Alaska, Fairbanks.

Plaskett, D. C. 1976. Preliminary Report: A Cultural Resource Survey in an Area of the Nenana and Teklanika Rivers. Report to the Alaska Division of Parks, Anchorage.

Powers, W. R. 1983. The 1977 Survey for Pleistocene Age Archaeological Sites. *In* Dry Creek: Archeology and Paleoecology of a Late Pleistocene Alaskan Hunting Camp, ed. W. R. Powers, R. D. Guthrie, and J. F. Hoffecker. Report to the National Park Service, Washington, D.C.

Powers, W. R., and J. F. Hoffecker 1989. Late Pleistocene Settlement in the Nenana Valley, Central Alaska. *American Antiquity* 54(2):263–87.

Thorson, R. M. 1977. Reconnaissance Surficial Geology of the Savage–Teklanika Rivers Region with Particular Emphasis on Prehistoric Archaeological Remains. Unpublished manuscript.

WALKER ROAD

Ted Goebel, W. Roger Powers, Nancy H. Bigelow, and Andrew S. Higgs

---

The Walker Road site (HEA-130) is located 12 km north of the village of Healy, central Alaska, in the northern foothills of the Alaska Range (63°58′N, 149°05′W). It is situated on a south-facing bluff overlooking the confluence of the northward-flowing Nenana River and a small side-valley stream locally known as Cindy and James Creek, at an elevation of approximately 430 m above sea level.

HISTORY OF RESEARCH

J. F. Hoffecker and S. M. Wilson discovered Walker Road in August 1980, uncovering a cluster of artifacts near the base of a 1-m-thick loess deposit mantling terrace alluvium (Hoffecker 1985). In 1984, Hoffecker and W. R. Powers excavated a 1-sq.-m test pit, revealing a small, unlined hearth feature and associated artifacts and bone fragments, also near the base of the loess. Charcoal from the hearth was conventionally ^{14}C-dated to 11,820 ± 200 BP (Beta-11254) (Powers and Hoffecker 1985). Full-scale excavations were carried out during 1985–90. In 1985–86 Powers and T. Gillispie exposed 50 square m of the site, discovering a second hearth feature (Powers et al. 1990). In 1987, Bigelow (1991) conducted geoarchaeological tests along the Walker Road terrace, and in 1988–90, Powers and T. Goebel resumed full-scale excavations, uncovering an additional 149 sq. m (Goebel 1990). Approximately 200 sq. m have been excavated at Walker Road to date.

STRATIGRAPHY AND DATING

The Walker Road site is found along the south-facing bluff edge of a Healy-aged strath terrace, 15 m below the main Healy terrace surface and 60 m above the modern river floodplain. The strath terrace, which dips 2.5° westward toward the Nenana River along the same plane as Cindy and James Creek and 1° southward, was formed by the downcutting of the creek after the deposition of the Healy outwash. The outwash consists of well-rounded gravel in a coarse, sandy matrix. The cobbles are capped with a layer of silt 1 m thick. This late Quaternary stratigraphic profile consists of heterogeneous loesses containing varying amounts of aeolian sand, two major palaeosol horizons, and the modern soil (Fig. 7–13).

The basal loess units (Loesses 1 and 2) are gleyed, weakly oxidized, and unweathered silt loams. They are stratigraphically sealed by Sand 1, a nearly continuous band of sandy loam reaching 10 cm in thickness. It is virtually continuous across the site and exhibits clear upper and lower contacts. Overlying Sand 1 is Loess 3, a silt loam containing a series of small gravel lenses as well as laterally discontinuous, oblique organic stringers (Ob horizons). Loess 4 is a silt loam with a higher clay fraction than the lower-lying loesses. It contains Palaeosol 1, a highly contorted, organic-rich, greyish brown pedocomplex with distinct Ab, Bgb, and Cgb horizons.

Overlying Palaeosol 1 is a series of alternating loams (Sands 2, 3, 4) and silt loams (Loesses 5, 6) of

Excavations at Walker Road. *(Photo courtesy of W. Roger Powers.)*

aeolian origin. Loess 5 and Sand 3 contain Palaeosol 2. This is a reddish brown, heavily oxidized palaeo-sol consisting of distinct Bwb and Ab horizons. The modern soil is developed upon Sand 4 and Loess 6, and bears O, A, AE, and Bw horizons. Near the bluff edge, Palaeosol 2 and the modern soil are stratigraphically separated by a discontinuous band of unweathered loam (Sand 3a). Across most

of the site, however, this stratum is absent, and distinguishing between Palaeosol 2 and the modern soil is difficult.

Two stratigraphically separate cultural components have been identified at Walker Road. The lower one (Component I) is situated within Loess 2 and 1. On some parts of the site this component was partially deformed by solifluction processes. In ad-

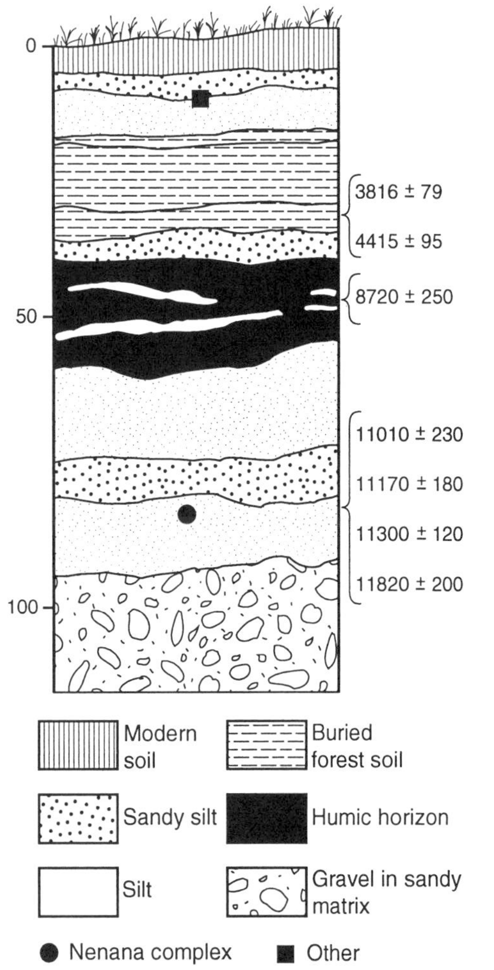

Figure 7–13   Walker Road. Stratigraphic profile. *(Source: Science. See Acknowledgments.)*

The small lithic assemblage ascribed to Component II was recovered from near the top of the loess profile within the modern soil. These artifacts were found nearly 70 cm above Component I and clearly represent a second, stratigraphically separate occupation of the site.

The initial conventional ¹⁴C date from the basal loess at Walker Road (11,820 ± 200 BP [Beta-11254]) was obtained on charcoal from Hearth Feature 1. Subsequent AMS dating of wood charcoal from this hearth resulted in younger dates of 11,170 ± 180 (AA-1681) and 11,010 ± 230 BP (AA-1683) (Powers et al. 1990; Goebel et al. 1991). Wood charcoal from Hearth Feature 2 was dated by AMS ¹⁴C to 11,300 ± 120 BP (AA-2264) (Powers et al. 1990; Goebel et al. 1991). This date is statistically contemporaneous to the AMS dates from Hearth 1. Several additional dates have been obtained from higher in the loess profile. Natural wood charcoal from Palaeosol 1 was AMS-dated to 8720 ± 250 BP (AA-1692). Furthermore, natural charcoal from Palaeosol 2 was AMS-dated to 3816 ± 79 BP (AA-1693) and conventionally dated to 4415 ± 95 BP (GX-12875) (Goebel et al. 1991). These dates assign the two palaeosols to the early and mid- to-late Holocene, respectively.

FLORA AND FAUNA

Numerous small fragments of charred bone have been recovered from Component I at Walker Road, but no identifiable faunal remains have been found to date (Powers and Hoffecker 1989:281; Powers et al. 1990). The majority of these remains were found associated with Feature 1.

FEATURES

The majority of cultural remains from Component I at Walker Road can be ascribed to four horizontally separate activity areas (1–4). Areas 1 and 2 are major concentrations of lithic and bone debris with centrally located hearths, while Areas 3 and 4 are relatively small, activity-specific areas. Refitting of artifacts demonstrates that on numerous occasions

dition, isolated lithic artifacts were found slightly higher in the loess profile in Sand 1 and Loess 3. The majority of these artifacts were oriented vertically and appear to have been secondarily displaced upward by seasonal freezing and thawing. Artifact refitting studies, spatial reconstructions, and ¹⁴C dates indicate that a single component, perhaps reflecting one or two contemporaneous occupations, is represented. Although some artifacts have been secondarily displaced (both vertically and horizontally), the majority of cultural remains were recovered either in their primary contexts or in positions not far removed from their original contexts.

blanks were manufactured in one activity area and then transported to another activity area for further reduction, use, resharpening, and finally disposal.

Activity Area 1 This activity area, situated along the terrace edge, is one of two larger clusters of cultural debris. Numerous artifacts were found displaced along the terrace slope in front of this cluster, indicating that it has been partially destroyed by slope erosion. The hearth associated with this cluster, Hearth Feature 1, was circular in shape, unlined, and composed of charcoal, several unidentifiable bone fragments, and bits of red ochre. Two anvil stones were recovered 1.5–2 m north of the hearth feature. In addition, a dense concentration of over 50 lithic implements and 1,000 pieces of lithic debris were found surrounding the hearth; this concentration forms a roughly semicircular outline. Tools from this activity area include end scrapers, side scrapers, gravers, cobble tools, retouched blades, and retouched flakes.

Activity Area 2 This represents the second major activity area of the site. Its southern border was approximately 8 m north of the modern terrace edge. Near the center of this concentration of cultural remains was uncovered a circular, bowl-shaped hearth (Hearth Feature 2) that was excavated into the outwash gravels at the base of the loess mantle. This hearth was filled with abundant wood charcoal, unidentifiable bone fragments, and isolated flecks of red ochre. One large anvil stone was uncovered 1.5 m northwest of the hearth. Surrounding the hearth were over 150 lithic implements and about 2,000 pieces of lithic debris.

This activity area can be further broken down into two distinct subconcentrations, each of which is distributed in the form of an arc across either the northern or southern side of the hearth feature. The northern cluster consisted almost exclusively of end scrapers and hundreds of tiny retouching chips removed from the distal ends of these scrapers, while the lithic concentration south of the hearth was much more dense and heterogeneous. Besides flakes, chips, and splinters, it consisted chiefly of end scrapers, side scrapers, wedges, planes and other cobble tools, and Chindadn points—small, bi-

facially worked, teardrop-shaped implements (Goebel 1992).

Activity Area 3 This concentration consisted of nearly 1,000 splinters, flakes, and blade fragments of brown chert found in an area of less than 4 sq. m; no tools were recovered. The lithic scatter formed an elongate oval, which fanned out in a southwesterly direction, and represented a single flaking event in which one flint knapper prepared a core and removed a series of blades and flake-blades from this core. Many of the blades manufactured here were carried to other areas of the site and used as tools.

Activity Area 4 This activity area included a diffuse concentration of about 15 tools and 100 flakes. Most of the flakes are of a greyish black siltstone, and appear to be by-products of the manufacture of a single, unfinished Chindadn point. Other tools include two notches and retouched flakes, a hammerstone, retoucher, end scraper, side scraper, and chopping tool. Several cores and core fragments were also found in this concentration.

Artifacts

The Walker Road artifact assemblage from Component I presently consists of 4,980 lithic pieces. Lithic raw materials include cherts (40%), rhyolite (29%), and chalcedony (16%), followed by smaller percentages of basalt, black siltstone, argillite, quartzite, and obsidian. Nearly all stone may have been procured in the form of cobbles from the alluvium underlying the site and at other localities along the Nenana River floodplain and its terraces, and only the obsidian (0.4% of all raw materials) appears to be from a nonlocal source.

Primary Working Techniques

The majority of the assemblage is made up of splinters (lithic pieces not bearing platforms or bulbs of percussion), small flakes, and retouch chips. The majority of these are clearly by-products of core preparation and tool manufacture. In addition, cortical spalls occur in relatively high numbers, mostly in the form of small primary (>50% cortex on dorsal surface) and secondary (<50% cortex) spalls. Among the cores in the assemblage there is no single, predominating form. Two cores were prepared for the removal of blades and blade-like flakes. One of these was made on a black siltstone cobble and bears two platforms and three sub-prismatic fronts. The second blade core is "flat-faced," not prismatic, and was manufactured on an argillite cobble. It bears two opposing fronts, one flaked unidirectionally (from a single smooth platform), and the other flaked bidirectionally (from two opposing cortical platforms). Eight flake cores vary in terms of size, number of fronts, and number of platforms. Among them is a monofrontal radial core manufactured on a cobble of tan-colored chert, which exhibits a cortical platform surface. Prismatic blade cores as well as wedge-shaped microblade cores are absent from the assemblage. Striking platforms are for the most part smooth and unfaceted. Dihedral and faceted platforms are rare. Most blanks bear large striking platforms and prominent ventral bulbs of percussion, suggesting that they were removed from cores through hard-hammer, direct percussion flaking. Over 300 chips and small flakes, however, bear small, often imperceptible, retouched platforms, suggesting that they were removed from tool edges through soft hammer percussion or through pressure flaking. Blades, flake-blades, and their fragments (18.3%) make up a minor, yet significant, element of the assemblage. Blade widths are relatively wide, ranging from 5 to 52 mm and averaging 20.47 mm. Only 4 blades are less than 9 mm wide, and nearly 76% fall between 10 and 25 mm in width. Microblades are absent.

Secondary Working Techniques

Tools were manufactured on flakes (51.4%), blades (36.2%), and cobbles (12.4%). The frequency of blades used as tools (36.2%) is twice as high as the frequency of blades noted in the entire assemblage (18.3%). This may indicate a propensity toward the selection of blades over flakes as a suitable blank form. Analyses of the debitage have also shown that a significant amount of retouching occurred at

Walker Road. The widespread occurrence of chips and small flakes with retouched platforms resulted chiefly from end scraper resharpening and bifacial point manufacture. Bifacial technology is weakly represented. Only 5 tools (2.4%) are bifacially worked, and 23 (11%) are bimarginally worked. Four tools display invasive retouching on the ventral surface, and 2 display alternate, bimarginal retouching. The majority of the assemblage (176 tools), however, was retouched unifacially. Degrees of retouch intensity have been categorized into four ordinally scaled values. Only 53 tools (25.7% of those measured) were retouched facially; these include unifacially worked side scrapers and cobble tools and bifacially worked points and wedges. Most tools (51.9%) were shaped or resharpened through simple marginal retouching; these include end scrapers, gravers, and some retouched blades and flakes. Use wear without retouch is seen 20.8 percent of the time, typically on utilized, "ad hoc" flakes and blades.

Tools The tools comprise 218 lithic artifacts (4.3%), the majority of which fall into the following tool classes: retouched/utilized flakes and blades, end scrapers, side scrapers, and cobble tools. Retouched flakes and blades together make up nearly 50 percent of the entire tool assemblage. Retouched blades consist mainly of unilaterally and bilaterally retouched blades (>5 cm in length) (Fig. 7–14: *x, y*), whereas bladelets (<5 cm in length) and flake-blades occur less frequently. End scrapers (18.3%) make up the largest portion of the shaped tools found at Walker Road and have been classified into the following seven forms (excluding fragments): on blades (Fig. 7–14: *f–k*), on flakes (Fig. 7–14: *l–o*), pan-shaped (Fig. 7–14: *p–r*), double (Fig. 7–14: *s*), circular (Fig. 7–14: *t*), carinated (Fig. 7–14: *u*), and spurred. Side scrapers (9.2%) also make up a considerable portion of the Walker Road tool kit. Single-convex and double-convex occur most frequently, as do convergent, transverse, and angle (*déjeté*) scrapers. Other forms include single- and double-straight, single-concave, and scraper fragments.

Bifaces, projectile points, and preforms (4.6%) are weakly represented in the assemblage. The four

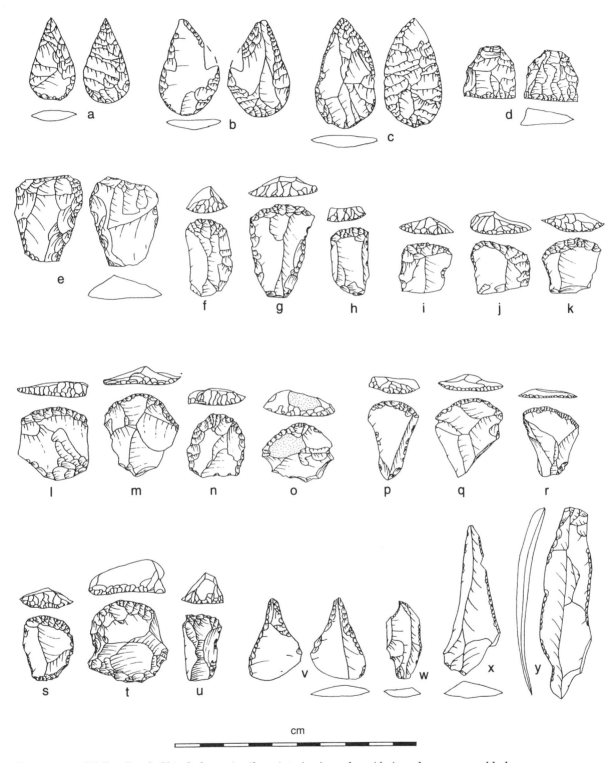

Figure 7–14 Walker Road: Chindadn projectile points (*a–c*); wedges (*d, e*); end scrapers on blades (*f–k*); end scrapers on flakes (*l–o*); pan-shaped end scrapers (*p–r*); double-end scraper (*s*); circular end scraper (*t*); steeply keeled end scraper (*u*); perforators (*v, w*); bilaterally retouched blades (*x, y*).

complete points are Chindadn points (Fig. 7–14: *a–c*). The three preforms are also bifacially worked and may represent Chindadn points in a preliminary stage of manufacture. Bifaces are fragmentary and unexpressive. Cobble tools (7.8%) constitute a major component of the Walker Road tool kit. Six of the 17 cobble tools recovered have been called plano-convex tools, or planes, which display smooth, sometimes cortical, ventral surfaces and extremely steep flaked dorsal surfaces. The worked margins of these planes display direct percussion flaking, as well as some heavy crushing. Other cobble tools found at Walker Road include three choppers and one chopping tool, as well as two hammerstones, one retoucher, and three anvil stones. The hammerstones and retoucher display crushing and flaking along otherwise cortical surfaces.

Wedges (4.1%) (*pièces ésquillées*) exhibit bipolar crushing which is difficult to interpret (Fig. 7–14: *d, e*). On the one hand, these may have been tools perhaps used for splitting apart bone or wood. On the other hand, they may be cores produced during the removal of flakes from small cobbles via the block-on-block, or anvil, technique of lithic reduction. Wear studies are needed to resolve this issue. Eight gravers (3.7%) have been recovered, two of which are multiple gravers, displaying more than one worked tip. Two single perforators (Fig. 7–14: *v, w*) are bimarginally retouched and symmetrical in form. The others are unifacial, usually with single hooked or curved spurs. Other tool types that occur infrequently in the Walker Road assemblage include notches (or spokeshaves) (4.1%), denticulates (0.9%), and knives (0.5%). Burins and retouched/ utilized microblades are absent.

The Component I assemblage is assigned to the Nenana complex of the Palaeoindian Tradition (Haynes 1987; Powers and Hoffecker 1989; Goebel et al. 1991).

Discussion

In terms of primary working techniques, the Walker Road lithic industry is characterized by core-and-blade as well as core-and-flake technologies. Tool blank and debitage analyses demonstrate that both of these reduction strategies were stressed somewhat equally. Cores were manufactured on small- and medium-sized cobbles, for the most part locally available in the Nenana River alluvium. All segments of primary reduction are represented, from the initial flaking and preparation of cores, to the removal and selection of blanks for use as tools. A major portion of primary working involved direct, hard-hammer percussion flaking. Microblades and microblade cores are absent.

The high frequency of retouching chips indicates that much secondary working of blanks and resharpening of tools occurred on the site. The vast majority of tools were retouched on one face, usually only marginally. Even those tools that display retouch on both faces were more often retouched marginally than facially. Furthermore, about 20 percent of all tools display only use wear, with no retouch, a sign of rapid use and discard. These features indicate that little effort was expended in the curation and resharpening of the majority of the Walker Road tools, most likely because of the close proximity to sources of raw material. However, this does not appear to have been the case with end scrapers, which were often resharpened to the point of exhaustion. Most of these end scrapers were manufactured on high grades of cryptocrystalline chert, which is harder to find in local alluvium.

The Component I assemblage is an important part of the Nenana complex sample, both in terms of geochronologic context and lithic inventory.

REFERENCES CITED

Bigelow, N. H. 1991. Analysis of Late Quaternary Soils and Sediments in the Nenana Valley, Central Alaska. M.A. thesis, University of Alaska.

Goebel, T. 1990. Early Paleoindian Technology in Beringia. M.A. thesis, University of Alaska.

———. 1992. The Chindadn Point: A New Type Fossil for the Paleolithic of Beringia. In *Paleokologiia i rasselenie drevnego cheloveka v Severnoi Azii i Amerike*, pp. 277–79. Krasnoyarsk: Nauka.

Goebel, T., W. R. Powers, and N. Bigelow. 1991. The Nenana Complex of Alaska and Clovis Origins. In *Clovis Origins and Adaptations*, ed. R. Bonnichsen and K. Turnmire, pp. 49–79. Corvallis, Oregon: Center for the Study of the First Americans, Oregon University.

Haynes, C. V. 1987. Clovis Origins Update. *The Kiva* 52:(2):83–93.

Hoffecker, J. F. 1985. Archaeological Field Research 1980. *National Geographic Society Research Reports* 19:48–59.

Powers, W. R., and J. F. Hoffecker. 1985. Nenana Valley Cultural Resources Survey 1984. Unpublished report submitted to the Alaska Division of Parks, Anchorage.

———. 1989. Late Pleistocene Settlement in the Nenana Valley, Central Alaska. *American Antiquity* 54:263–87.

Powers, W. R. , F. E. Goebel, and N. H. Bigelow. 1990. Late Pleistocene Occupation at Walker Road: New Data on the Central Alaskan Nenana Complex. *Current Research in the Pleistocene* 7:40–43.

MOOSE CREEK

John F. Hoffecker

The Moose Creek site (FAI-206) is located in the north-central foothills of the Alaska Range, approximately 18 km north of Healy and 120 km southwest of Fairbanks (64°04′ N, 149°06′ W). The site occupies a southwest-facing bluff on the north side of Moose Creek in the Nenana Valley. The bluff edge lies at 210 m above the modern river floodplain (ca. 500 m above mean sea level) and represents the highest terrace level in the valley.

History of Research

The site was discovered by J. F. Hoffecker and C. F. Waythomas during July 1978. At the time of discovery, one test pit was excavated along the edge of the bluff, yielding 75 artifacts from an occupation layer near the base of the aeolian sediment that overlies the terrace gravels. A second pit (archaeologically sterile) was excavated approximately 10 m from the bluff edge, from which a stratigraphic profile was recorded by R. M. Thorson. During August 1979,

15 sq. m were excavated by Hoffecker and a field assistant, and over 900 artifacts were recovered, including some from a younger component (Hoffecker 1982, 1985). Further work was undertaken in August 1984 by W. R. Powers, Hoffecker, and two field assistants, who excavated an additional 4 sq. m and recovered 1,250 new artifacts (Powers and Hoffecker 1985). In 1987 Moose Creek was briefly revisited as part of a broader geoarchaeological study of the valley (Hoffecker et al. 1988). To date, approximately 20 sq. m of the occupation area have been excavated.

Stratigraphy and Dating

The remains are buried in loess and aeolian sand deposits that overlie an ancient terrace. The terrace is composed of unconsolidated conglomerate and sandstone pebbles, and is assigned to the Nenana gravel formation of late Tertiary age by Wahrhaftig (1958:11–12). At the site location, Wahrhaftig recorded a rare deposit of till or glacial outwash (comprising coarse sand and gravel with scattered erratics) that caps the Nenana gravel and apparently represents an extensive and ancient glaciation in the North Alaska Range (1958:25). The aeolian sand and silt that contain the artifacts overlie this glacial deposit.

The aeolian sediments vary in thickness from 55 cm at the bluff edge to 180 cm at 10 m east of the bluff edge. A detailed stratigraphic profile was recorded where the sediment achieves its maximum depth, and buried organics are best preserved (see Fig. 7–15). A silt unit lies at the base of the sequence and contains a buried soil complex. The lower archaeological component is associated with this unit. A fine-medium sand layer (archaeologically sterile) overlies the basal silt; it contains at least one buried soil horizon. Above the sand lies a silty sand unit containing several weakly developed buried soil horizons and the upper archaeological component. The sequence is capped with a modern forest soil.

The following radiocarbon dates (radiocarbon years BP) were obtained on soil organics from the basal silt unit (Hoffecker 1985:38, Table 2):

The Moose Creek site. *(Photo courtesy of W. Roger Powers.)*

A-2168	8160 ± 260	lower palaeosol complex
A-2144	8940 ± 270	lower palaeosol complex
I-11227	10,640 ± 280	lower palaeosol complex
GX-6281	11,730 ± 250	lower palaeosol complex

These dates, in conjunction with stratigraphic correlations (primarily based on buried soils) with other dated profiles in the region (Hoffecker 1988; Powers and Hoffecker 1989), indicate that the basal silt may be assigned to the terminal Pleistocene and/or earliest Holocene (i.e., ca. 8,000–12,000 years ago). Sediment samples for thermoluminescence dating were collected in 1987, but have yet to be processed (Hoffecker et al. 1988). The absence of former hearths precludes direct dating of the lower occupation level, which is presently assumed to be roughly contemporaneous with occupation horizons in comparable stratigraphic positions elsewhere in the north-central Alaska Range. The upper component may be tentatively assigned to the mid-

dle Holocene on the basis of correlation of associated buried soil units with dated soil horizons at Dry Creek (Powers and Hoffecker 1989; Hoffecker et al., this volume).

FAUNA AND FLORA

No vertebrate faunal remains were recovered at the site. Sediment samples have not been analyzed for pollen and spores; palynological studies at other Nenana Valley sites in similar sedimentary contexts (e.g., Dry Creek) suggest that Moose Creek is unlikely to yield adequate samples.

FEATURES

No former hearths or other features were encountered during excavation. In terms of spatial pattern-

ing, most artifacts are grouped into small lithic
concentrations of 1 m or less in diameter.

ARTIFACTS

The artifact assemblage from the lower occupation
level comprises approximately 2,250 items. The following breakdown with respect to raw material was
determined from the 1978–79 collections (44% of
total assemblage): light rhyolite (68%), dark rhyolite
(23%), grey chert (7%), black chert (1%), green chert
(1%), and other (<1%). No cores or core fragments
are present, and most of the assemblage is composed of unretouched flakes. A single microblade-
like fragment was recovered in 1979 from a mass of
flakes and fragments of the same raw material (light
rhyolite) and apparently represents an incidental
by-product of core reduction (Powers and Hoffecker 1989:278).

Retouched items (n = 11) account for .5 percent
of the total assemblage, and include point fragments (2), biface fragments (4), side scrapers (1),
and retouched flakes (4). The point fragments represent basal portions of bifacial lanceolate points;
they exhibit concave bases and expanding sides (see
Fig. 7–16). The biface fragments are relatively undiagnostic and appear to be derived from small (ca.
6–10 cm long) ovate bifaces (see Fig. 7–16). The side
scraper and retouched flakes are similarly undiagnostic (not shown in Fig. 7–16).

Artifacts from the upper component are confined to 13 items, including an ovate biface and 12
unretouched flakes (rhyolite and chert).

DISCUSSION

Uncertainties remain concerning both the dating
and cultural affiliation of the lower component at
Moose Creek. Stratigraphic correlation with other
dated profiles in the Nenana Valley suggests that
this component, which is buried in the basal silt
horizon, is contemporaneous with or older than the
assemblages dating to 11,000–11,500 BP at Dry
Creek and Walker Road (Powers and Hoffecker
1989; Hoffecker et al., this volume; Goebel et al., this

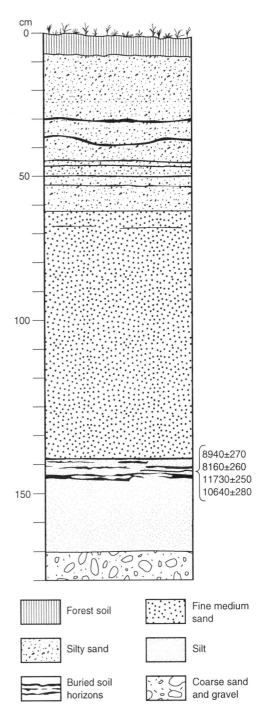

Figure 7–15 Moose Creek. Stratigraphic profile.
(Adapted from Hoffecker 1985. See Acknowledgments.)

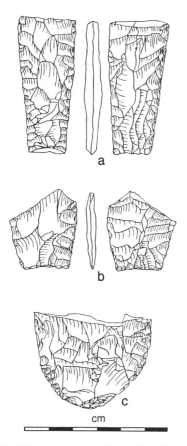

Figure 7–16   Biface fragments from the Moose Creek site. *(After Hoffecker 1985. See Acknowledgments.)*

REFERENCES CITED

Hoffecker, J. F. 1982. The Moose Creek Site: An Early Man Occupation in Central Alaska. Unpublished report to the National Park Service and the National Geographic Society, Washington, D.C.

———. 1985. The Moose Creek Site. *National Geographic Society Research Reports* 19:33–48.

———. 1988. Applied Geomorphology and Archaeological Survey Strategy for Sites of Pleistocene Age: An Example from Central Alaska. *Journal of Archaeological Science* 15:683–713.

Hoffecker, J. F., C. F. Waythomas, and W. R. Powers. 1988. Late Glacial Loess Stratigraphy and Archaeology in the Nenana Valley, Central Alaska. *Current Research in the Pleistocene* 5:83–86.

Phippen P. G. 1988. Archaeology at Owl Ridge: A Pleistocene–Holocene Boundary Age Site in Central Alaska. M.A. thesis, University of Alaska.

Powers, W. R., and J. F. Hoffecker. 1985. Nenana Valley Cultural Resources Survey 1984. Unpublished report to Alaska Department of Natural Resources, Anchorage.

———. 1989. Late Pleistocene Settlement in the Nenana Valley, Central Alaska. *American Antiquity* 54(2): 263–87.

Wahrhaftig, C. 1958. Quaternary Geology of the Nenana River Valley and Adjacent Parts of the Alaska Range. *U.S. Geological Survey Professional Paper* 293-A:1–68.

West, F. H. 1981. *The Archaeology of Beringia*. New York: Columbia University Press.

volume). The Moose Creek radiocarbon dates provide age estimates for the lower buried soil complex, but not for the artifacts which lie within and below the soil horizons and necessarily antedate them. The assemblage is assigned to the Nenana complex on the basis of its stratigraphic position, apparent lack of microblade technology, and presence of bifacial lanceolate points (Powers and Hoffecker 1989). However, while lanceolate points are present in Nenana complex assemblages at Dry Creek and Owl Ridge (Phippen 1988; Hoffecker et al. this volume), they are also present (although not common) in the Denali complex (West 1981, this volume; Powers and Hoffecker 1989). Further research, including absolute dates on the occupation level (^{14}C and TL), may warrant reclassification of Moose Creek as a Denali site.

PANGUINGUE CREEK

Ted Goebel and Nancy H. Bigelow

The Panguingue Creek site (HEA-137) is located in the north-central foothills of the Alaska Range, approximately 4.5 km northwest of the town of Healy. The site is found on a narrow promontory along the north side of Panguingue Creek in the Nenana Valley, roughly 490 m above sea level and 200 m above the modern river level.

The Panguingue Creek site. *(Photo courtesy of W. Roger Powers.)*

Panguingue Creek was discovered in September 1976 by T. A. Smith and J. F. Hoffecker. Test excavations were undertaken in 1977 and 1985, and revealed three occupation horizons ranging in age from 10,000 to 4,000 BP (Powers et al. 1983; Powers and Maxwell 1986a). Broad-scale excavation was undertaken in 1991. A total of approximately 100 sq. m has been uncovered to date.

STRATIGRAPHY AND DATING

The artifacts and associated remains are buried in deposits of aeolian sand and silt that overlie a Pleistocene river terrace composed of glacial outwash (Fig. 7–17). The outwash is assigned to the Healy glaciation, which appears to be of late middle Pleistocene or earlier late Pleistocene age (Wahrhaftig

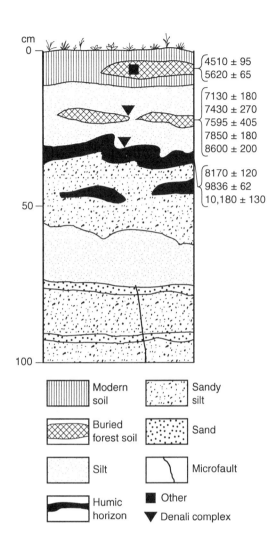

Figure 7–17 Panguingue Creek. Stratigraphic Profile. *(Source:* Science. *See Acknowledgments.)*

Capping this basal silt is a 1-m-thick unit of medium sand intercalated with 2-cm-thick bands of coarse sand, pebbles, and granules. These bands are horizontally continuous, although difficult to see when dry. Vertical cracks, occurring about 10 m apart, run from the top of the sand unit to the top of the underlying silt. These cracks are nearly 1 cm wide and filled with well-sorted medium sand distinct from the surrounding matrix.

Overlying the bedded sand is approximately 10 cm of unweathered silt loam. This is, in turn, capped by 30 cm of loam with intermittent lenses of granules and small pebbles. Two buried soils occur within this unit. Palaeosol 1 consists of one, and in some places two, levels of discontinuous, nearly horizontal organic stringers. Each stringer is about 0.5 cm thick and at most 25 cm long. Palaeosol 2, which represents the dominant weathering horizon of this loam unit, is characterized by a nearly continuous, although contorted, Ob horizon (1–10 cm thick), often with an associated Bwb horizon. This palaeosol has yielded two ¹⁴C dates that average 9951 ± 56 BP; a third date is highly discordant and has been discounted. The lowermost occupation level (Component I) is associated with Palaeosol 2.

The uppermost portion of the sequence comprises a homogeneous silt loam 30 cm thick. Two buried soil horizons lie immediately below and within the modern soil layer. They are represented by discontinuous, organic-rich lenses. The lower palaeosol (Palaeosol 3) has produced five ¹⁴C dates that average 7711 ± 97 BP and is associated with Component II. The uppermost occupation level (Component III) occurs in Palaeosol 4, which has yielded two ¹⁴C dates that average 5250 ± 54 BP. The modern soil comprises distinct Bw and gleyed BC horizons, with little evidence of organometallic leaching.

The complete sequence of dates (expressed in uncorrected radiocarbon years before present) is shown in Table 7–3.

The aeolian sequence reflects changing depositional environments through the terminal Pleistocene and Holocene. The basal silt loam capping the outwash gravel appears aeolian in origin, while the overlying bedded sand is probably a low-energy colluvial deposit. The absence of pebbles and granules in the overlying silt loam suggests a brief inter-

1958; Ritter 1982). The aeolian deposits, which reach depths of up to 1.8–2.0 m, are radiocarbon-dated to the terminal Pleistocene and Holocene (Goebel and Bigelow 1992).

At the base of the aeolian deposits lies a 10-cm-thick silt loam that rests unconformably on the glacial outwash. This unit is gleyed, with minor oxidation mottles, some of which contain preserved rootlets. It recently yielded a conventional ¹⁴C date of 13,535 ± 400/380 BP (A-6744) on a bulk sample of organic matter.

val of renewed aeolian activity, followed by the formation of a loam deposit possibly derived from sheetwash colluviation. The discontinuous stringers of Palaeosol 1 within this loam may simply represent detrital organics. The silt loam capping the sequence is aeolian in origin, and the three buried soils within this unit probably formed during brief intervals of reduced aeolian activity.

FAUNA AND FLORA

Numerous small burned/calcined bone fragments were recovered from a debris concentration in Component II, but none of the fragments was identifiable to taxon (Powers and Maxwell 1986a).

FEATURES

No features were encountered in the lowermost and uppermost occupation levels (Components I and III). Excavation of the middle occupation level (Component II) revealed two separate activity areas, 8–10 m in diameter. One of these areas contained a hearth feature characterized by a dense concentration of charcoal and burned bone.

ARTIFACTS

Component I The lowermost component comprises 1 subprismatic blade/flake core, 60 flakes, and 6 tools. The tools include 2 transverse scrapers on short, wide flakes, 1 *tci tho*, and 2 lanceolate projectile points (Fig. 7–18: *b, c, i*). The small size of the assemblage precludes firm assignment to a defined complex, although it appears to fall within the time range (roughly 10,000 years BP) generally prescribed for the Denali complex.

Component II The assemblage from this component contains 9 cores, >150 microblades, 10 burin spalls, >5,000 unretouched flakes, and 60 tools. Among the cores are 7 wedge-shaped and subconical microblade cores. The tools include 5 lanceolate and 2 ovate bifacial knives, 7 side scrapers, 5 end scrapers, 5 burins (transverse, angle, and dihedral),

Table 7–3 Panguingue Creek Radiocarbon Dates (Uncorrected)

GX-13011	4510 ± 95	Component III
SI-3237	5620 ± 65	15–20 cm below surface
Beta-15094	7130 ± 180	34–38 cm below surface
AA-1688	7430 ± 270	Component II
GX-13012	7595 ± 405	Component II
Beta-15093	7850 ± 180	35–38 cm below surface
AA-1689	8600 ± 200	Component II
AA-1687	8170 ± 120	Component I
GX-17457	9836 ± 62	Component I
AA-1686	$10,180 \pm 130$	Component I
A-6744	$13,535 \pm 400/380$	Basal silt loam

16 retouched flakes, 2 chopping tools, 2 *tci thos*, 2 hammerstones, and 2 anvil stones (Fig. 7–18: *a, d–h*). This assemblage is assigned to the Denali complex and appears to date to approximately 8,500–7,500 BP (Powers and Maxwell 1986b).

Component III The uppermost occupation level yielded 20 waste flakes and 3 tools. The tools include 2 end scrapers and 1 side scraper. This component cannot be assigned to a defined complex or tradition, because it has thus far failed to yield diagnostic artifact types.

DISCUSSION

Although Component I was assigned to the Denali complex in an earlier publication (Goebel and Bigelow 1992), its cultural affiliation remains ambiguous due to the scarcity of diagnostic items in the component and to continuing uncertainties about the sequence of industries in early Holocene Alaska. The closest parallel to this assemblage may be Moose Creek, which also remains problematic. By contrast, Component II exhibits all of the characteristic features of the Denali complex (e.g., wedge-shaped microblade cores, microblades, burins, bifacial knives), although the subconical microblade cores are not common in other Denali assemblages. This occupation appears to represent a relatively short-term hunting camp dating to a poorly under-

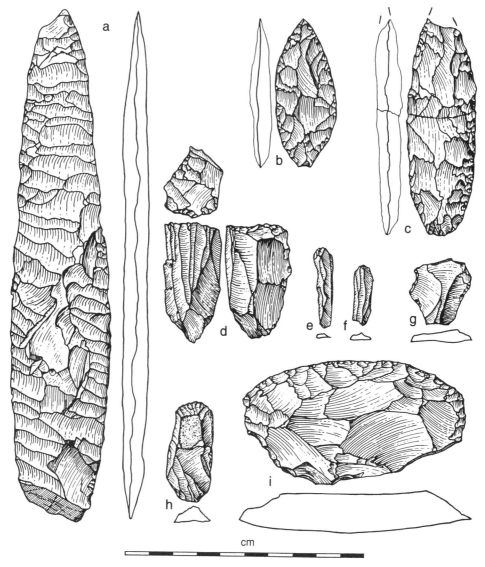

Figure 7–18 Panguingue Creek Component II (*a, d–h*) and Component I (*b, c, i*): biface, parallel to obliquely flaked (*a*); lenticular bifaces (*b, c*); microblade core (*d*); microblades (*e, f*); transverse burin (*g*); end scraper (*h*); side scraper (*i*).

stood period of archaeology of early Holocene central Alaska. It warrants further technological and spatial analysis.

REFERENCES CITED

Goebel, T., and N. Bigelow. 1992. The Denali Complex at Panguingue Creek, Central Alaska. *Current Research in the Pleistocene* 9:15–18.

Hoffecker, J. F., W. R. Powers, and T. Goebel. 1993. The Colonization of Beringia and the Peopling of the New World. *Science* 259:46–53.

Powers, W. R., R. D. Guthrie, and J. F. Hoffecker. 1983. Dry Creek: Archaeology and Paleoecology of a Late Pleistocene Alaskan Hunting Camp. Report to the National Park Service, Washington, D.C.

Powers, W. R., and H. E. Maxwell. 1986a. Alaska Range Northern Foothills Cultural Resources Survey 1985. Unpublished report submitted to the Alaska Division of Parks and Outdoor Recreation, Anchorage.

————. 1986b. *Lithic Remains from Panguingue Creek, an Early Holocene Site in the Northern Foothills of the Alaska Range.* Studies in History 189. Anchorage: Alaska Historical Commission.

Ritter, D. F. 1982. Complex River Terrace Development in the Nenana Valley near Healy, Alaska. *Geological Society of America Bulletin* 93:346–56.

Wahrhaftig, C. 1958. *Quaternary Geology of the Nenana River Valley and Adjacent Parts of the Alaska Range.* U.S. Geological Survey Professional Paper 293-A.

The fan alluvium comprises gravels and cobbles in a sandy matrix and is associated with outwash deposition of the Healy glaciation (Wahrhaftig 1958; Ritter and Ten Brink 1986). The aeolian sediment unconformably rests on the alluvium and is unlikely to antedate 12,000–11,000 years (Hoffecker 1988:693–94).

The aeolian deposits achieve depths of up to 200 cm and are primarily composed of silt and sandy silt (Fig. 7–19). At the base of the sequence, as at other localities in the north-central foothills of

LITTLE PANGUINGUE CREEK

John F. Hoffecker and W. Roger Powers

The site (HEA-038) is located in the north-central foothills of the Alaska Range, north of the town of Healy and approximately 125 km southwest of Fairbanks. The site occupies a bluff on the north side of Little Panguingue Creek in the Nenana valley. It is situated at approximately 445 m above mean sea level at 63°56′N, 149°06′W.

HISTORY OF RESEARCH

The Little Panguingue Creek site was discovered in 1976 by D. C. Plaskett during a survey of the Nenana valley (Plaskett 1976). Initial testing (1 sq.-m) during the summer and early fall of 1976 by Plaskett, T. A. Smith, and others revealed the presence of a microblade assemblage buried less than 50 cm below the surface. The site was revisited in 1979 by J. F. Hoffecker, who recorded a complete stratigraphic profile (Hoffecker 1979). In 1984, W. R. Powers and Hoffecker excavated four 1 m × 1 m units and located a lower component (Powers and Hoffecker 1985).

STRATIGRAPHY AND DATING

The artifacts are buried in aeolian sand and silt that overlie coarse fan alluvium of pre-Wisconsinan age.

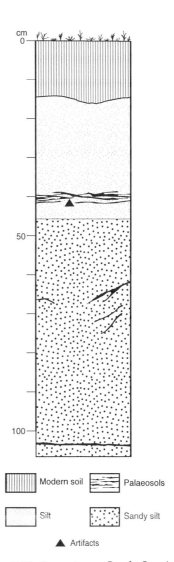

Figure 7–19 Little Panguingue Creek. Stratigraphic profile.

View from Little Panguingue Creek site. *(Photo courtesy of W. Roger Powers.)*

the Alaska Range, lies a silt unit capped with a thin sand horizon. The overlying 150–175 cm comprises sandy silt containing scattered organics and several buried soils. The lower palaeosol is represented by a dark brown organic horizon (1–2 cm thick) buried 100 cm below the surface. It is associated with the lower archaeological component. Several thin, discontinuous organic lenses occur in the overlying sandy silt (between 60 and 100 cm in depth). A heavily disturbed buried soil (upper palaeosol), comprising lenses of yellowish red, light grey, and dark brown silty clay, lies only 30–45 cm below the modern surface; it is associated with the upper archaeological component. In certain parts of the site, this buried soil appears to have been largely obscured by modern soil formation.

Comparison with other profiles in the Nenana valley indicates that the basal silt unit at Little Pan-

guingue Creek antedates 11,000 years BP, and that the lowest buried soil probably dates to the terminal Pleistocene or early Holocene (10,000–8,000 years ago?). However, this remains to be confirmed by absolute dating. Although the upper palaeosol exhibits an appearance similar to palaeosol complex 3 at the Dry Creek site, which is dated to 8,600–6,270 years BP, an AMS radiocarbon date of only 1825 ± 68 years BP (AA-1699) was obtained on charcoal from a former hearth associated with the upper archaeological assemblage (Powers and Hoffecker 1989:270, Table 1).

FAUNA AND FLORA

No faunal or floral remains have been recovered from the site.

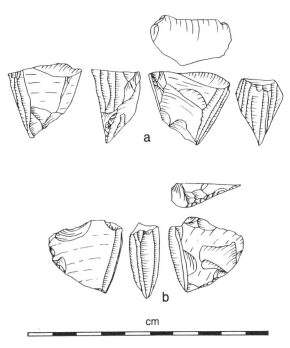

Figure 7–20   Wedge-shaped microblade cores from the Little Panguingue Creek site.

FEATURES

Despite the limited area exposed to date, a well defined former hearth (38 × 40 cm) was encountered in the upper component. The charcoal lens was associated with a large cobble that may represent a hearthstone and a concentration of flakes and microblades.

ARTIFACTS

The artifact assemblage from the lower component is confined to approximately 15 unretouched flakes of rhyolite and chert. No microblades or diagnostic retouched pieces have been recovered to date.

Artifacts from the upper component include 5 wedge-shaped microblade cores (Fig. 7–20), 3 microblade core tablets, ca. 70 microblades, 1 burin, 2 bifaces, 1 end scraper, and >100 unretouched flakes (Powers and Hoffecker 1989:276). Although detailed analyses of the artifacts have not been undertaken, it is apparent that the assemblage exhibits close metric and morphologic affinities to the Denali complex.

DISCUSSION

The dating of both components at Little Panguingue Creek remains somewhat problematic. The Denali complex affinities and apparent association with an early Holocene soil of the upper component suggest an age in the range of 10,000–7,000 years ago. However, the single radiocarbon date from the level indicates that it may represent a late Holocene microblade industry, which has been reported from other localities in interior Alaska (e.g., Campus, Broken Mammoth). If the upper component dates to the late Holocene, the age of the lower component is unclear. If the upper palaeosol actually contains an early Holocene Denali assemblage, then the lower component almost certainly represents a terminal Pleistocene assemblage (possibly Nenana complex).

(References follow on p. 374.)

REFERENCES CITED

Hoffecker, J. F. 1979. The Search for Early Man in Alaska: Results and Recommendations of the North Alaska Range Project. Unpublished Report to the National Geographic Society and the National Park Service, Washington, D.C.

———. 1988. Applied Geomorphology and Archaeological Survey Strategy for Sites of Pleistocene Age: An Example from Central Alaska. *Journal of Archaeological Science* 15:683–713.

Plaskett, D. C. 1976. Preliminary Report: A Cultural Resource Survey in an Area of the Nenana and Teklanika Rivers of Central Alaska. Unpublished report to the Alaska Division of Parks, Anchorage.

Powers, W. R., and J. F. Hoffecker. 1985. Nenana Valley Cultural Resources Survey 1984. Unpublished report to the Alaska Department of Natural Resources, Anchorage.

———. 1989. Late Pleistocene Settlement in the Nenana Valley, Central Alaska. *American Antiquity* 54(2): 263–87.

Ritter, D. F., and N. W. Ten Brink. 1986. Alluvial Fan Development and the Glacial-Glaciofluvial Cycle, Nenana Valley, Alaska. *Journal of Geology* 94:613–25.

Wahrhaftig, C. 1958. Quaternary Geology of the Nenana River Valley and Adjacent Parts of the Alaska Range. *U.S. Geological Survey Professional Paper* 193-A:1–68.

Chapter 8

SOUTH CENTRAL ALASKA RANGE: TANGLE LAKES REGION

INTRODUCTION

The lakes that dominate this region are of moderate size, forming a chain trending generally north-south. There are six major named lakes. From south to north, they are: Landlock, Upper Long Tangle Lake, and Butcher's Pond (a local name), which connects via Tangle River to Round Tangle Lake. North of this is Lower Long Tangle Lake and, finally, Lower Tangle Lake. At about this point the lakes attenuate, develop a current, and shortly they become the upper Delta River. Many smaller lakes and ponds are found in the vicinity of the larger lakes, especially around those to the south. Most of these contribute to the Delta drainage and are considered parts of the Tangle Lakes. The coordinates 63°N, 146°W lie less than 4 km east of a midpoint on Landlock Lake. The Denali Highway runs through the district, connecting Paxson and the Richardson Highway on the east with Cantwell, about 250 km distant, to the west. Round Tangle Lake lies just north of the road; Butcher's Pond, Upper Long Tangle Lake, and Landlock Lake lie south. The latter two are the largest lakes in the system, Landlock being almost 4 km in length and slightly under 1 km in greatest width. Depths may range over 25 m.

The Tangle Lakes region occupies a broad divide between three major drainages. The Delta River, of which these are headwaters, flows northward through the Alaska Range, joins the Tanana River, and becomes part of the Bering Sea drainage. Just south and southeast, the Gulkana River rises out of Dicky and Swede lakes and flows southeast to join the Copper River, which empties into Prince William Sound. To the west, the Maclaren River has its source in the Maclaren Glacier of the Alaska Range. It flows south, then west, to join the Susitna River, which ultimately enters Cook Inlet near Anchorage.

The northern portion of the Tangle Lakes region is dominated by the imposing facade of the Amphitheater Mountains, with elevations to 2,400 m. These form a front range to the main Alaska Range, the white summits of which are visible through the several passes in the Amphitheaters.

Most of the Tangle Lakes region is mapped on the U.S. Geological Survey Mt. Hayes quadrangle, and that peak, about 80 km northwest of the Tangle Lakes, is the highest on the quadrangle. Its elevation is well over 4,000 m.

South of the Amphitheaters the region about the lakes is marked by very irregular topography, most of which is composed of unconsolidated sediments—many of them ice-contact features—resulting from late Pleistocene glaciation. Some of these features are intimately tied to the interpreta-

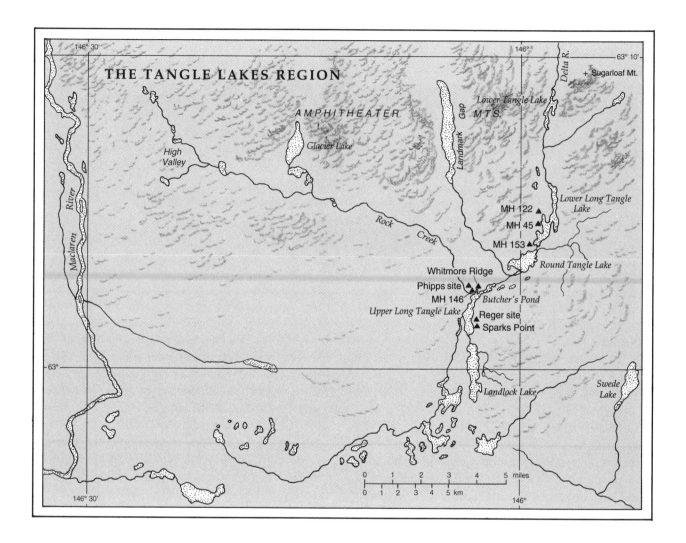

tion of Tangle Lakes prehistory. Of particular note is the evidence, ubiquitous around the lakes south of the Denali Highway, of a high lake stand—a former proglacial lake.

The dominant vegetation association is an alpine shrub tundra in which the dominant species is dwarf birch (*Betula glandulosa*). Dwarf birch in many places is quite dense and may reach heights of six feet and more. Willow (*Salix* spp.) bushes may intermingle with it. To the distant southeast in the lowlands occupied by the west fork of the Gulkana River, spruce (*Picea*) forest dominates. In the Tangle Lakes spruce trees do occur spottily in favorable locations and, because this represents an altitudinal apparent limit to their distribution, are reported to reproduce vegetatively. Other higher forms also of irregular distribution include willows (*Salix*) and,

in the fairly extensive delta of Rock Creek where it enters Upper Long Tangle Lake, occasional stunted cottonwood (*Populus*) trees. Small copses of *Populus* may also occur on warm hillsides. On the lower southeastern slopes of the mountains, fairly dense stands of alder (*Alnus* spp.) will be encountered. Generally, though, the appearance of the landscape is dominated by low tundra vegetation, the visual effect of which is to accentuate rather than conceal the architecture of the region. In higher areas, depending upon conditions of slope and drainage, these associations will be characterized by alpine bearberry (*Arctostaphylos uva-ursi*), cranberry (*Vaccinium vitis idaea*), blueberry (*V. uliginosum*), ground willow (*Salix* spp.), mountain avens (*Dryas octopetala*), and juniper (*Juniperus* spp.), which may take a creeping or low bush form, among others. Unvege-

Mount Hayes and Maclaren Glacier viewed from High Valley, vicinity of Tangle Lakes.

tated areas are common on these gravel ridgetops which, once denuded for whatever cause, revegetate very slowly. As most of the sites occur on such glacial features, the underlying soil matrix is char-acteristically graveliferous. Overlying the gravels in most places is a mantle of silt, generally of inconsiderable depth.

The Tangle Lakes region is part of the normal

range of the Nelchina caribou herd so that, depending upon the circumstances directing their movement and presence, *Rangifer tarandus* may be found almost anywhere in the Tangle Lakes. Moose (*Alces alces*) occur sparingly throughout the district. The major predator is the grizzly bear (*Ursus arctos*), which is not uncommonly encountered. The wolf (*Canis lupus*) is present but rarely seen. Among the smaller mammals are ground squirrels (*Citellus undulatus*), pikas (*Ochotna*), and hairy mammoths (*Marmota marmota*). Common avian forms of the summer include loons, both common and red-throated (*Gavia*), three species of jaegers, arctic terns, shore birds, and a variety of small passerine birds. Ptarmigan (*Lagopus*) are year-round residents.

Lake trout (*Salvelinus namaycush*) are found in the larger lakes and arctic grayling (*Thymallus arcticus*) in the streams and lakes. Salmon spawn in the Gulkana but do not occur here. Biting insects—blackflies, mosquitoes, and black gnats (''no-see-ums'')—are abundant in the summer, demanding so much attention of higher forms that it may almost seem the evolutionary order of dominance has been reversed.

There is no official recording station in the near vicinity, but characterization of the climate is fairly straightforward. As in all of the interior, winters are extremely cold but here, because of the mountains, there is an additional factor in the form of frequent high winds and much greater amounts of snow. Breakup on the lakes usually occurs in the first week or so of June, and not infrequently the snow pack is still present through that same period. Summers are short and cool, and occasional snowfalls at any time in the summer are not uncommon. Windy conditions tend to prevail, with directions being generally out of the south or from the north.

Until the opening of the Denali Highway through the region in 1956, there was very little human traffic in the Tangle Lakes. Today the transient population of people camping or simply traveling through on the Denali Highway may be relatively heavy and in fall there will be a fairly large number of hunters also present. As far as can be determined, of native peoples the only users of the Tangle Lakes were the Ahtna of Copper River,

who made occasional visits to the region as far up as Maclaren Glacier. The appearance is, however, that in immediate pre-contact time—in contrast to more distant times—the region saw relatively little use.

HISTORY OF RESEARCH

The first archaeological discoveries in the Tangle Lakes were made by William Hosley in the course of a hunting expedition in 1956. These were reported to his University of Alaska colleagues Ivar Skarland and Charles J. Keim, who made a brief survey the following summer. Their findings, especially those at the site they called Hosley Ridge, were reported in 1958 (Skarland and Keim). Sporadic forays of brief duration were carried out in subsequent years by Skarland and Keim, and by the writer (in early fall and late spring) with small classes from the University of Alaska. In 1964 a slightly more serious, but brief, reconnaissance was conducted by the writer and some of the crew who were, that summer, working at Donnelly Ridge. These included John V. Matthews and Jeffrey Mauger. Another brief trip in 1965, this time with Douglas R. Reger as assistant, provided convincing proof that intensive survey was in order. In 1966 a month was spent conducting survey in the region, once more with Douglas Reger as field assistant. New sites were found almost every day, and the first certain Denali complex site, *in situ*, (the Reger site, Mt. Hayes 92) was found that season. This survey was continued in the following summer, again with Douglas Reger, and aided by the late Bond Whitmore, Jr., then a student at Alaska Methodist University. By the end of 1967 more than 100 sites had been catalogued. Laboratory assistance in 1966–67 was under Douglas Reger, and in 1967–68 under Linda C. Ellanna. Field schools were conducted by Alaska Methodist University in 1968, 1969, 1970, 1971, and 1973, with studies focusing on archaeology, geology, and, at various times, biology. Those directly involved with these programs were Charles E. Schweger, Ross L. Schaff, Douglas R. Reger, Jeffrey Mauger, John V. Matthews, William Frohne, Leonard Freese, Robson Bonnichsen

View of Landmark Gap across Upper Long Tangle Lake.

(faculty) and, as researchers, Constance F. West, R. Greg Dixon, James Hamilton, and Peter M. Bowers.

Between 1964 and 1978, 204 archaeological sites were recorded. Most of these are flake scatters yielding little information, but a significant number of productive localities are included. Among these are 20 Denali and related late Pleistocene–early Holocene sites, of which 8 are *in situ*.

A number of other researchers in other fields have worked here as well: Thomas A. Ager, Vera Alexander, Robert J. Barsdate, Catherine Campbell, Robert Hinton, David M. Hopkins, Donald R. Nichols, Troy L. Péwé, Richard D. Reger, and John D. Sims, among others.

Since 1980 most archaeology fieldwork has been undertaken by, or for, the Interior Department's Bureau of Land Management. Among these field researchers are Brian and Teresa Zinck, Charles E. Mobley, Peter M. Bowers, Thomas E. Gillispie, John L. Beck, Robert E. King, and Patricia

McCoy. These surveys have thus far more than doubled the number of sites recorded in the Tangle Lakes Archaeological District.

THE LATE PLEISTOCENE PROGLACIAL LAKE AND BURIED SOILS

The wave-cut foreshore of the former lake stands at 880 m, or 30 m above present lake levels. The sites that will be discussed here are *all* found above this elevation, from which it is inferred that those peoples who inhabited these sites were in the region when this old high lake was still in existence. The downstream end of the lake was dammed by an ice terminus at Round Tangle Lake which stagnated there, leaving distinctive collapse features. This proglacial lake incorporated practically all of Landlock and Upper Long Tangle lakes and a great deal of territory to the west and southwest. It was about 10 km in length and over 3 km in greatest width. It existed for some time following regional deglaciation.

Two major loess falls appear to be recorded in the Tangle Lakes. It is important to note that there are no sources for loess here today. The first evidently occurred upon deglaciation as areas marginal to the ice were briefly bare as the ice withdrew. This loess (Loess 1) is found uniquely above 880 m and occurs over an area of some 23 sq. km (Robert Hinton, personal communication). That deposit lay subaerial for a long time and there formed on it a fully developed podzol (or spodosol).

In contrast to those problematical isolated organic-appearing stringers often identified as palaeosols, this buried soil is preserved intact and may be found virtually anywhere above 880 m elevation. It is well preserved, with all three major soil horizons normally easily discernible—an exceedingly rare condition. It is in this soil that *in situ* Denali materials are found. It is shown for the several sites discussed below.

That this Lower Holocene Podzol, as it was formally designated (West 1981), is preserved intact evidences a sudden, rapid, short-term event. That event must surely have been the lowering of the old lake to approximately present-day levels. It becomes thus an invaluable chronostratigraphic marker: depending upon the interpretation, this soil (and any included archaeology) was buried by a second loess fall (Loess 2) ca. 6,000–8,000 years ago. Moreover, it is assumed that within this podzol itself a relative chronology may be read based upon the position within it of the assemblage in question. As will be seen, this becomes particularly interesting in respect to the relative placement of Denali assemblages vis-à-vis the assemblages from Whitmore Ridge (Mt. Hayes 72) and Mt. Hayes 122.

The indication is that the old lake dropped

quite rapidly. There is little or no evidence of intermediate stillstands. The newly exposed silts were picked up by the winds—which, because of the continued presence of glacial ice in the near vicinity, probably well exceeded present normal velocities—and deposited over a broad area. This loess sheet, on which is developed the modern podzol, overlies all of the occupations discussed here. Thus, the time of its deposition provides a "stop date" for all those assemblages. The draining of the lake has been estimated at >8,000 BP by Schweger (1981) and at 7,000–6,000 BP by the present writer (West 1981).

In a back eddy of a small cove on the 880-m lake there was accumulated over time a sizeable organic deposit which included many plant macrofossils. (This has been informally referred to as a "fossil beaver dam.") The plant remains include a great deal of beaver-gnawed wood, sections of small *Populus* tree trunks, a mass of unmodified driftwood, *Alnus* catkins, spruce cones, and pollens. Donald R. Nichols discovered this site, secured a radiocarbon date on it, and made the initial observations of its relationship to the old strandline. His findings were greatly augmented by Schweger (1981), some of whose data are reproduced here. The present writer carried out some work here as did, latterly, Peter Bowers.

F. H. W.

References Cited

Schweger, C. E. 1981. Chronology of Late Glacial Events from the Tangle Lakes, Alaska Range, Alaska. *Arctic Anthropology* 18(1):97–101.

Skarland, I., and C. J. Keim. 1958. Archaeological Discoveries on the Denali Highway. *Anthropological Papers of the University of Alaska* 6(2):79–88.

West, F. H. 1981. *The Archaeology of Beringia.* New York: Columbia University Press.

Phipps site left of center in middle ground. (*Photo courtesy of R. Greg Dixon.*)

PHIPPS SITE

Frederick H. West, Brian S. Robinson, and Mary Lou Curran

The Phipps site (Mt. Hayes 111) is located approximately 1 km west of Butcher's Pond and south of the Denali Highway (coordinates: 63°02′30″N, 146°03′20″W). The Whitmore Ridge site (Mt. Hayes 72) is about 250 meters east and south. The feature on which the Phipps site is located appears to be a sandspit associated with the old high lake (Schweger 1981)—perhaps a remnant of a former (late Pleistocene) Rock Creek distributary channel. The site was discovered by F. H. West and D. R. Reger in 1966. Test excavations were made in 1967 and 1968. Profiles were redrawn in 1978. The site consists of a single stratified Denali complex component within the buried soil, the matrix of which is a shallow, sandy loess column lying on coarse alluvial sands and silts (Fig. 8–1). Some portion of the sand component may be windblown and perhaps colluvial as well. In any case, the resultant lower horizons of the podzol that developed are characteristically sandy. The buried soil is remarkably well preserved at the Phipps site, showing clearly its podzol or spodosol profile. Apart from several charcoal samples, no organics were found here.

Artifact Assemblage (587/1,041)*

The Phipps site produced a Denali complex assemblage unusual in that it was restricted almost entirely to wedge-shaped microblade cores, burins, and their products. The assemblage may represent a single episode of occupation.

Microblade Cores (22) and Parts (67)   The 22 microblade cores from the Phipps site are typical Denali complex wedge-shaped cores (Fig. 8–2: *a–f*). The core parts consist of 21 tablets, 37 notch flakes, and 9 other miscellaneous core parts. Two of the wedge-shaped cores (Fig. 8–2: *a–f*) have removals from both ends, which is interpreted only as a variation on initial core preparation. Two probable preforms are made of the local weathered chert, one of which was fitted together from relatively large

*Artifact tabulations are presented as a ratio of deliberately worked specimens to debitage.

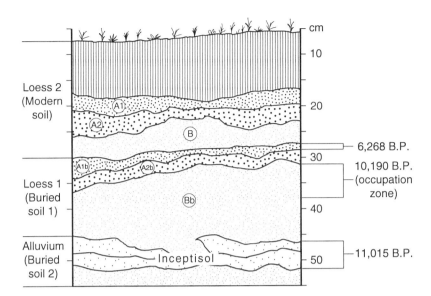

Figure 8–1 Phipps site. Stratigraphic profile. Modern soil (O–B) underlain by the buried soil developed on Loess 1 (A1b–Bb), which at this site alone is underlain by a third, incipient, soil developed in alluvium. Denali materials were found in middle–upper Loess 1, i.e., in Bb–A2b soil horizons. Base is coarse sand. Dates are averages.

reduction flakes and fragments. The latter specimen represents the only large, early-reduction-stage tool recognized at the site.

Raw materials employed include grey chert (10/45%), weathered green/tan chert (8/36%), and sard (4/18%). Two yellow-brown sard or jasper specimens (Fig. 8–2: *a, f*) are made from well-thinned biface blanks reminiscent of the sard specimens at Mt. Hayes 146. Biface technology is apparently limited to microblade core reduction at the Phipps site.

Only 1 of 21 core tablets (5%) is a complete removal of the top (Fig. 8–2: *e*). The top of this specimen is flat rather than notched, and the resulting microblade core was discarded after the perhaps inadvertent full-top removal. Two partial tablet removals (Fig. 8–2: *h, i*) were refitted with a broad notch flake (Fig. 8–2: *g)* forming a particularly long core top (Fig. 8–2 reconstruction 6.3 cm) even though the specimen is incomplete.

Microblades (357) and Burin Spalls (81) Seven of 438 microblades and burin spalls were retouched after removal. Two of the microblades have retouched spurs or points on the ends (Fig. 8–3: *a, b*), with one of these having points at both ends (Fig. 8–3: *b*). One burin spall (Fig. 8–3: *c*) shows a completely retouched edge as well as a steeply retouched distal end. The remaining retouched specimens have more or less irregular lateral retouch or lateral notches.

Burins (23) Twenty-three typical Donnelly burins were recovered at the Phipps site, made from grey/black chert (16/73%), yellow-brown sard (5/22%), and weathered chert (2/9%). Fourteen specimens (61%) have visible nicking (probable use wear) on the spall facet edge. The burins are made on small unifacial flakes in most cases modified by marginal retouch applied on the dorsal side. A few, such as the equivocal piece shown as Figure 8–3: *g*, show short platforms created by a removal like that of the microblade core tablet. Others have straight platforms, some very short and retouched or, in one instance, based on a snap fracture. The preponderance, however (13/57%), exhibit notched platforms,

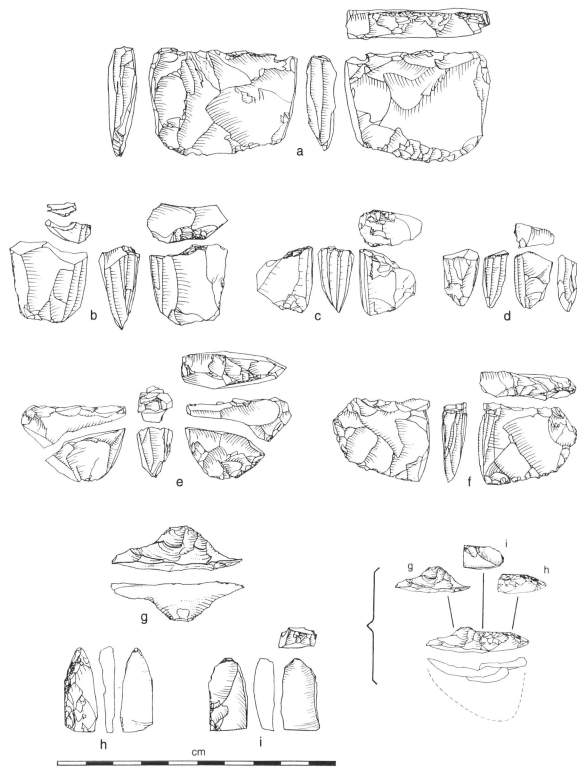

Figure 8–2   Phipps site: wedge-shaped microblade cores (*a–f*), some with articulating core tablets and one complete top (*e*). Three core top elements from the same core represent three distinct steps in core top preparation: a notch flake (*g*); a primary platform removal (*h*); and a secondary removal (*i*). Reconstructed core top elements *g, h,* and *i,* shown at lower right, are not to scale.

and five of these show paired opposing notches at approximately right angles to the long axis of the body (Fig. 8–3: *h–k*). While these may have served in some instances as hafting devices, it seems clear that on certain specimens (e.g., Fig. 8–3: *i, j*) they functioned as stop notches. Two burins were made on core tablets, one of them shown as Figure 8–3: *e*.

Snapped-Flake Gravers (4) Snapped-flake gravers were first identified by Robinson in the Mt. Hayes 146 assemblage. After a search of the Phipps site assemblage, four examples were identified here as well, made of grey chert (3) and sard (1) (Fig. 8–3: *l, m*). Use wear is relatively light on all specimens although the form is similar to those at Mt. Hayes 146.

Modified Flakes (23) and Blade-like Flakes (9) The majority of the modified specimens are small, snapped flakes reminiscent of the snapped-flake gravers. Snapped edges have use retouch and nicking similar to that on burins. Thus, many of the modified flakes and blade fragments seem to be part of the general "snapped-flake industry" proposed for Mt. Hayes 146. A small number of large blade-like flakes or large blades occur (e.g., Fig. 8–3: *d*) that may or may not be from prepared cores. No well-formed end scrapers were recovered at Phipps although a few flakes were lightly modified or utilized at the distal end.

Flakes (1,041) Debitage raw material types include: weathered chert (69%), grey chert (26%), and sard (2.4%). Local weathered chert dominates the debitage counts, although not the respective tool counts. Burins show the strongest contrast from the debitage material counts, indicating the greatest degree of selection.

Hammerstone (1) One well-used cobble hammer was recovered.

SUMMARY

The inordinately high number of microblade cores and burins, to the virtual exclusion of other classes,

attests a specialized activity at the Phipps site. The most reasonable explanation is that weapons points were being produced here in quantity—surely slotted antler or bone points armed with microblade insets: the burins employed in shaping and slotting the antler; the microblades inserted to provide the highly effective cutting edges.

A number of radiocarbon dates have been obtained for the Phipps site. These are presented in Table 8–1.

The draining of the old lake was, of itself, a major event in the Tangle Lakes region. It was probably related to other changes occurring in the environment. As such, its timing is of considerable interest. But the dates derived from the Buried Soil 1, A horizon, need to be read cautiously. The A horizon occupies the uppermost segment of the Loess 1 depositional unit. Even though preserved

Table 8–1 Radiocarbon Dates from the Phipps Site

Buried Soil 1: A Horizon (above occupation)
Charcoal:
     6735 ± 390 BP (UGa-974)
     6610 ± 80 BP (Beta 62224; CAMS-6408)
     5460 ± 60 BP (Beta 63671; CAMS-7658)
Soil Humic Acid NAOH insoluble fraction:
     5740 ± 110 BP (SI-2171B-1)
     5230 ± 65 BP (SI-2171A-1)
Soil Humic Acid NAOH soluble fraction:
     5100 ± 60 BP (SI-2171B-2)
     3920 ± 65 BP (SI-2171A-2)

Buried Soil 1: Lower A2b Horizon (occupation zone)
Charcoal:
     8155 ± 265 BP (UGa-927) [Lower A2]*
     10,150 ± 280 BP (UGa-572) [Upper Bb]
     10,230 ± 70 BP (Beta-63672; CAMS-7659) [Upper Bb]

Buried Soil 2: Incipient A Horizon? (below occupation)
Twigs:
     11,000 ± 70 BP (Beta-63673; CAMS-7660)
     11,030 ± 80 BP (Beta-63674; CAMS-7661)

*This date was obtained from various small pieces of charcoal contained in a bulk soil sample from the lower portion of the buried A horizon. This may have been a hearth, but the paucity of charcoal in the deposit renders that interpretation unsure. There were various artifacts in association. The assay is clearly too recent to pertain to the occupation, however.

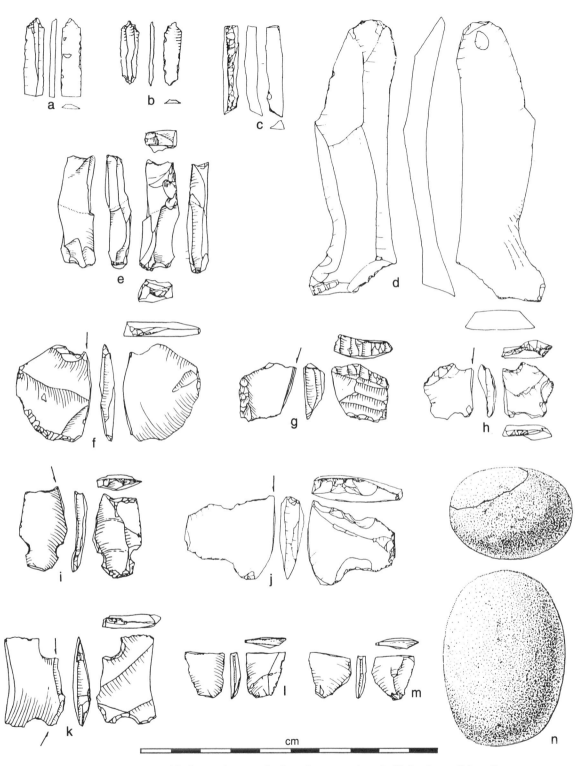

Figure 8–3   Phipps site: microblades with retouched perforator points (*a, b*); burin spall heavily retouched prior to removal (*c*); large blade (*d*); core tablet with burin facet (*e*); burins (*f–k*); snapped-flake gravers (*l, m*); hammerstone (*n*).

by burial, this cannot be construed as the A horizon that formed when this soil first developed. It is, rather, the amalgam of an A horizon dynamic, a continuum that spanned an unknown interval from its earliest appearance until its burial. More importantly for the dating, as brought out by Thorson in his discussion of Bowers's Carlo Creek site, being relatively unstable, humus "decays more rapidly [than charcoal] increasing the probability of sampling young humus" (Thorson 1990:407). As pointed out by Holliday in the same volume, samples derived from buried A horizons are subject to "contamination by younger organic compounds moving in from overlying soils" (Holliday 1990: 536). This stricture obviously applies to any organics so collected, including charcoal. These considerations would suggest that Schweger's older estimate of 9,000 BP might be a better approximation of the true dating of this early Holocene event. This would, in turn, allow speculation that there may have been a larger correlative—the warming that introduced the Holocene.

REFERENCES CITED

Holliday, V. T. 1990. Pedology in Archaeology. In *Archaeological Geology of North America*, ed. N. P. Lasca and J. Donahue, pp. 525–540. Boulder: Geological Society of America.

Schweger, C. E. 1981. Chronology of Late Glacial Events from the Tangle Lakes, Alaska Range, Alaska. *Arctic Anthropology* 18(1):97–101.

Thorson, R. M. 1990. Geologic Contexts of Archaeological Sites in Beringia. In *Archaeological Geology of North America*, ed. N. P. Lasca and J. Donahue, pp. 399–420. Boulder: Geological Society of America.

WHITMORE RIDGE

Frederick H. West, Brian S. Robinson, and Constance F. West

In the long series of late Quaternary sites in the Tangle Lakes, Whitmore Ridge (Mt. Hayes 72) is unique and presents certain special interpretive problems. These are detailed below. The site occupies the crest of a generally north-trending ridge. The ridge is part of an esker complex that is truncated just to the south by Rock Creek. It is on the face of this truncation that the fossil beaver dam is exposed. The site lies about 600 m west of Butcher's Pond. It was discovered in 1966 by Douglas R. Reger. Coordinates: 63°02′46.1″N, 146°02′52.5″W.

Most of the Whitmore Ridge collection occurred *in situ* and within the buried soil (Lower Holocene Podzol, Fig. 8–4). On this high exposed ridge the soil column is quite shallow and in a few places the double soil is either absent or cannot be definitively identified. High winds and chronic deflation undoubtedly contributed to this condition.

Occupation was found distributed continuously over a distance of about 40 m, and within that distribution were three notable artifact concentrations. The northernmost of these, Locus 1, was restricted in extent and produced a small artifact series composed entirely of wedge-shaped microblade cores and their products. These were manufactured of fine-grained lustrous chert. About 15 m south, in Locus 2, there were found examples of a distinctive conoidal blade core technology. Here also were biface fragments and one burin. The conoidal cores are predominantly of a black marine chert with a few specimens of red sard. The technology is quite unlike that seen in Denali.

Locus 3 lies about 8 m south of Locus 2. It was marked by a dense concentration of debitage, mostly large flakes indicative of bifacial thinning. The rock type involved is that which dominates the entire site, an argillite ("chert," "welded tuff") derived from outcroppings at the base of the Amphitheater Mountains about 3 km directly north. (This was probably not the immediate source; many artifacts of this material were clearly made from cob-

View south from Whitmore Ridge. Alphabet Hills in background. (*Photo courtesy of Frederick Hadleigh West.*)

bles.) It is the sheer localized density of this material within the context of its broader distribution throughout the site that distinguishes Locus 3. Loci 1 and 2 differ from the preceding in that what is seen in both instances is a preferential use of particular lithics applied to specific core and blade technologies.

Loci 1 and 2 are taken to represent relatively short-term activities. The question is whether the loci represent different activities by the same people over a relatively brief span of time (such as is seen in the contrasting microblade and biface loci in Component II of Dry Creek) or whether one or more of them may be rather widely spaced in time from the others. This problem, unfortunately, cannot be unequivocally resolved on the basis of stratigraphic context.

ARTIFACT ASSEMBLAGE (760/13,964)

Core and blade technology characterizes this entire collection. In addition to cores and products of different forms and dimensions—some of which are rare in Alaskan archaeology—there are other unusual forms: large, well-made bifaces, some showing parallel flaking, unifacial and bifacial side scrapers, end scrapers, as well as other objects.

Blade Cores (16) and Blade Core Parts and Fragments (104) To take first the most divergent of the three blade core technologies, three of the conoidal specimens are shown here (Fig. 8–5: *a-c*). These were recovered from Locus 2. The two complete specimens (Fig. 8–5: *a, b*) are 4.0 and 4.4 cm in height respectively, each with a platform diameter

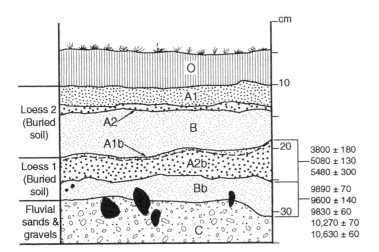

Figure 8–4   Whitmore Ridge. Stratigraphic profile. Both components are contained in Loess 1. Component 1 primarily in the Bb soil; Component 2 in the A2b. The first four assays in the Bb horizon are interpreted as cultural. The lowest assay (10,630 BP) may pertain to Component 1 as well, since some Component 1 materials occurred at the base of Loess 1.

of 2.8 cm. Blades were removed from virtually the entire perimeter. The products were small blades. Whether these should be called microblades or not is a matter of definition. Robinson's plotting of these blade widths against those from Locus 1—which are typically Denali in dimension—finds a clear bimodality. Blades derived from wedge-shaped cores average 4.6 mm in width, while the average for the conoidal core blades is 6.8 mm. The latter are correspondingly longer as well. The preferred platform rejuvenation was achieved by removal of the entire top in a single blow (see Fig. 8–5: *a, c*). Initial platform preparation is illustrated by the primary core tablet (Fig. 8–5: *c*). Figure 8–5: *a* is a secondary (any removal after primary) core tablet. The degree of core reduction can be judged by the overhang of the core tablet (Fig. 8–5: *a*). Similarly, it is clear that the resultant almost-smooth top served as the renewed platform. Conoidal cores were found only in Locus 2.

Denali-style cores are shown as Figure 8–5: *f, g*. Despite the discreteness of this concentration at Locus 1, a typical Denali core tablet and articulating notch flakes were found in Locus 3. Further linkage is provided by the distribution of microblades. An unusual feature of the small Locus 1 series is that all of these appear to be exhausted cores.

Examples of large, irregular, subconoidal blade cores are seen in Figure 8–5: *d, e*. The presence of still larger cores is attested by the crested blades (Fig. 8–5: *h, i*) and other large specimens. Some large blades, of course, may have been struck from unprepared, expedient cores, possibly in the process of cobble or block reduction. The larger cores from Locus 3 shown in Figure 8–5: *d, e* do not yield readily to typological classification. But despite their formal infirmities, it may be noted that their size range is similar to that found in the small but well-characterized assemblage from Mt. Hayes 122.

In his study of these evidences Robinson finds a clear association between core form and rock type preference. The microblades deriving from Denali-style cores are almost exclusively of fine-grained, lustrous chert. The conoidal cores are mainly of black marine chert (or indurated shale) which frequently shows speckled bedding planes and fossil bivalve inclusions. The large core evidences tend to be in the local argillite.

Small Blades (264), Microblades (95), and Large Blades (109)   The derivation of the first two has already been noted. Blades less than 5 mm in midpoint width are termed micro (Fig. 8–6: *b*); 5–11 mm

Figure 8–5   Whitmore Ridge: conoidal cores of black marine chert from Locus 2 (*a–c*), *a* and *c* with articulating core tablets; weathered chert subconoidal cores from Locus 3 (*d, e* [reconstructed]); wedge-shaped microblade cores from Locus 1 (*f, g*); large crested and corner removal blades from Locus 3 (*h, i*).

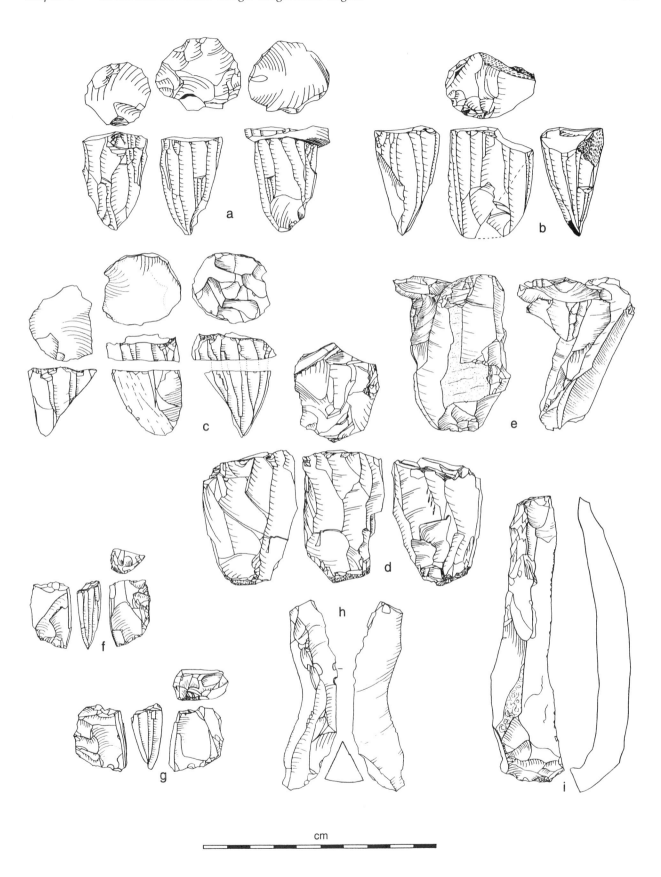

cm

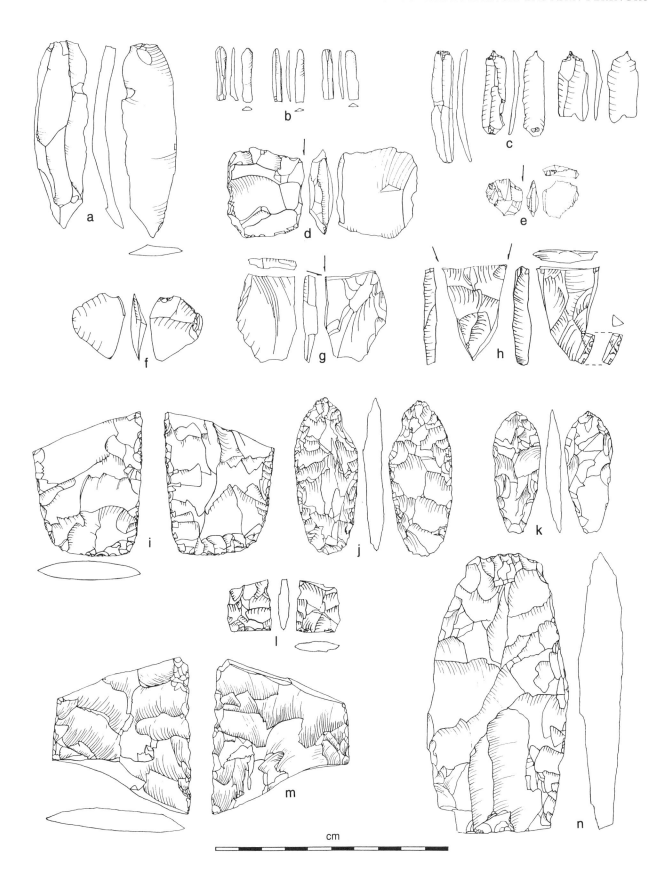

cm

are small blades (Fig. 8–6: *c*). The distribution of these finds would tend to support the contemporaneity of Locus 1 and Locus 3 (Table 8–2). The 35 microblades found in Locus 1 are all of lustrous chert. Of 10 microblades in Locus 2, 8 are of black chert, and 2 of lustrous chert. In Locus 3, however, 9 of 13 specimens are of lustrous chert. A number of the small blades show careful retouch, suggesting functions quite different from the microblades. (Site totals above include surface finds.)

Burins (7) and Burin Spalls (7)   Six of these specimens are complete (Fig. 8–6: *d–h*); the seventh is badly fire-spalled. As may be seen, their forms vary. That shown as Figure 8–6: *d* shows a clear burin facet. The removal may have been across the working end of an end scraper. Figure 8–6: *h* shows spall removals from both edges of what was part of a well-thinned biface. Figure 8–6: *e, f* conform reasonably closely to the notched burins of the Denali complex. The fifth specimen (Fig. 8–6: *g*) may be termed a dihedral burin, an apparent facet serving as platform or spall removal surface. The sixth and seventh specimens are separate burins that were refitted to form a snapped, modified blade. The heterogeneity of this sample and its small size would indicate that burin use was not important. That observation, in turn, may suggest that the use of osseous materials was relatively rare, and that may correlate with the relatively high proportion of bifaces found here and the small number of microblades.

Bifaces (60)   Nine of these are complete; perhaps twice that number allow a reasonable inference as to their original form. Attention is called to the two large specimens, Figure 8–6: *i, m*. The latter shows collateral flaking along one face; the former, also very well made, shows a straight to slightly convex base. The small biface shown as Figure 8–6: *l* is straight based, with one corner missing, and appears to have been subtriangular in form. Finally,

Table 8–2   Whitmore Ridge Distribution, All Blades

Artifact Concentration	Locus 1	Locus 2	Locus 3	Total
Micro (<5 mm)	35	10	13	58
Small (5–11 mm)	17	91	77	185
Large (>11 mm)	0	12	63	75

the large biface shown as Figure 8–6: *n* may have served as a scraper (there is wear along the upper right edge) or may represent a blank. The two blade-like removals on one face were struck from a fracture.

Apart from the preceding suggestions, it is not possible to ascribe uses to these particular specimens. Figure 8–6: *j* shows heavy edge grinding from just below the shoulder on both edges almost to the bottom of the specimen. Figure 8–6: *k* shows a pattern of possible use retouch along one edge, lower left in the illustration. (It is possible that this piece is inverted.) The ovate specimen in Figure 8–7: *a* , while bifacial, is markedly plano-convex in cross section.

As may be seen, there is considerable variation within this general category of simple biface, ranging from lenticular to lanceolate to that of a small cordiform (Fig. 8–7: *b*).

End Scrapers (15)   Eight of the items in this category are complete specimens. The remainder are fragmentary or questionable. Four end scrapers are shown as Figure 8–7: *d, e, g, h*. Figure 8–7: *e*, a double-ended specimen, is the only such found in any collection in the Tangle Lakes. It is very cleanly executed and both ends are retouched as end scrapers. The large blade-like flake (Fig. 8–7: *d*) could perhaps better be termed a side and end scraper or perhaps a beaked scraper. The working end of 8–7: *g* shows edge wear in the form of smooth beveling. Figure 8–7: *h* is made from the same black marine chert as the conoidal cores.

Figure 8–6   Whitmore Ridge, including artifacts from Locus 1 (*b*), Locus 2 (*c–e, i*), and Locus 3 (*a, g, l–n*): large blade (*a*); microblades (*b*); small blades (*c*), two of which have retouched perforator points; burins (*d–h*); bifaces (*i–n*).

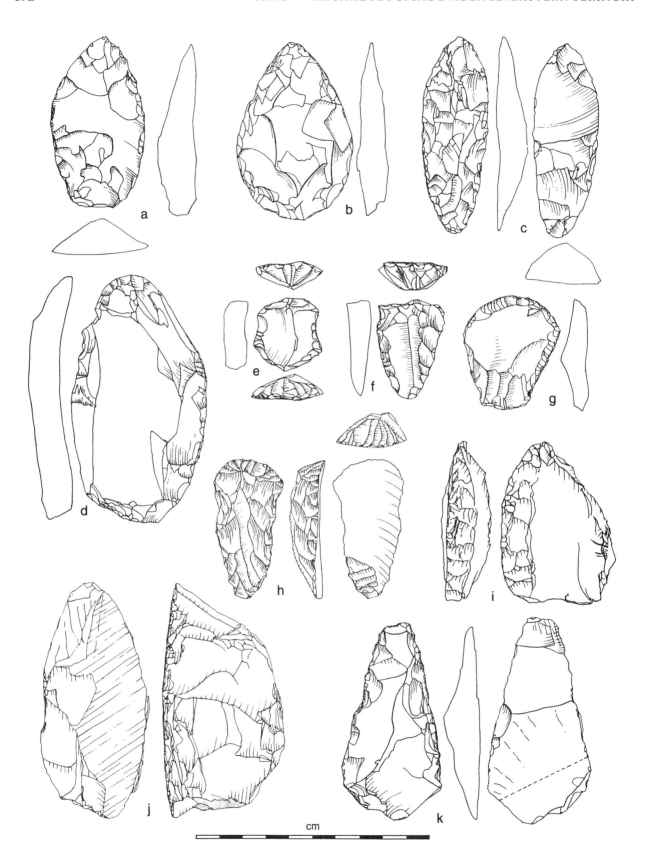

cm

Scraper Plane or Carinate Scraper (1) This large, turtle-shaped uniface (Fig. 8–7: *j*) is quite similar to one in the Teklanika West collection except that this is more distinctly oval in plan. A pattern of step flaking occurs around all edges. (In the figure, the right-hand representation is a side view.) These specimens resemble generally the carinate scraper found in some Aurignacian collections. A similar find was made in Dry Creek I.

Limace (1) This specimen (Fig. 8–7: *c*) is markedly plano-convex in cross section but about one-half of the ventral surface has been worked. Similar pieces have been found in some other Alaskan collections. They resemble the Old World limaces.

Side Scrapers (8) and Other Unifacial Tools (6) This is expectably a variable category as may be seen in the illustration (Fig. 8–7: *f, i, k*). Specimen *i* in Figure 8–7 exhibits edge wear compatible with heavy use. Figure 8–7: *f* appears to be carefully formed but its function is unclear.

Other Tools (67) In addition to the foregoing, about 64 flakes show evidence of modification either deliberately or by use and may probably be categorized as tools of expediency. Only one form is noted that suggests a specialized use, this having unifacial notches or modifications at the corners of snap fractures suggesting use of the corner for scraping. Included here also are 3 irregular flake cores made from argillite cobbles. That hammerstones were not recovered even from the area of intensive bifacial reduction may imply that soft hammer was used. Several possible boulder chip scrapers (*tci thos*) were found.

Debitage (13,964) Waste flakes are separated into five general lithologic categories: argillite (96%); lustrous chert (1%); black marine chert (0.5%); sard (0.2%); and other types (2%).

SUMMARY

The materials at this site all occurred within the lower loess unit, Loess 1. The double soil development was present across most of the site but with some significant disjunctions. These variations in depth and development are ascribable to the exposed nature of this high esker ridge. It has been subject to wind scouring from the time of its first exposure to the present day. Locus 1 was at the highest point of the ridge and here soil was thinnest. Fuller development (i.e., both soils present) was seen at Locus 2. Here the distinctive conoidal cores and related objects were found in the lower Ab horizon. Soil depth was greatest at Locus 3, which is situated in a swale.

A provisional interpretation of Whitmore Ridge identifies two components which are *not* superpositionally arrayed. Both are exclusively within Loess 1. Component 1, the older, occurs virtually from the contact with the esker deposits up into the lower A2 horizon of the buried soil. It is spatially restricted to Loci 1 and 3 and is interpreted as a Denali assemblage. Component 2 is found only at Locus 2, in the A2b soil horizon, and is taken to represent a distinctive early Holocene core and blade complex that derives, in some manner, from the Sumnagin of northeast Siberia. An alternate hypothesis is allowed: this would find the entire assemblage to be one component, representing multiple occupations over a fairly long period of time. This grouping would then be interpreted as transitional from Denali to that early Holocene complex. The radiocarbon dates derived from the A2b horizon, as shown in Table 8–3, were obtained on natural charcoal and should not be taken to pertain to the occupation of this level. They are too recent. The Component 2 (or Locus 2) material compares well with that from Birch Lake (Skarland and Giddings 1948), Girl's Hill (R. Gal personal communication, 1977), and somewhat less well

Figure 8–7 Whitmore Ridge, including artifacts from Locus 2 (*d, f, h*) and Locus 3 (*b, c, j*): simple bifaces (*a, b*); limace-like implement (*c*); end and side scraper (*d*); end scrapers (*e, g, h* [*h* of black marine chert]); flat-faced uniface (*f*); side scraper (*i*); scraper plane or carinate scraper (*j*); uniface (*k*).

Table 8-3 Radiocarbon Dates from Whitmore Ridge

Buried A Soil Horizon (Ab)
    5480 ± 300 (UGa-530)
    3800 ± 180 (Beta-64575)
    5080 ± 130 (I-4231)

Buried B Soil Horizon (Bb)
    9890 ± 70 (Beta-62222; CAMS-6406)
    9600 ± 140 (Beta-64578; CAMS-8300)
    9830 ± 60 (Beta-70240; CAMS-11255)
    $10,270 \pm 70$ (Beta-77286; CAMS-16834)

Microstratum (Below Bb)
    $10,630 \pm 60$ (Beta-77285; CAMS-16833)

with Panguingue Creek (Powers and Maxwell 1986; Goebel and Bigelow, this volume), and the Kagati complex of southwest Alaska (Ackerman 1992). Component 2 at Whitmore should be roughly contemporaneous with these sites. Radiocarbon dates from Panguingue Creek are in the range of 7000–8000 years BP (Goebel and Bigelow, this volume). As noted above, this early Holocene Beringian tradition is surely related to the Sumnagin of Siberia.

The four radiocarbon dates shown for the lower B soil horizon are considered cultural, as all were within distinct flake concentrations. Two of the dates (Beta-70240; Beta-77286) are well associated with Locus 3. Beta-77286 consisted of charcoal tightly bound with small flakes. The date of $10,630 \pm 60$ (Beta-77285) came from a possibly intrusive "microstratum" just below Beta-77286 (Table 8-3).

The unusual emphasis on bifaces and the presence of certain other rarely seen artifacts, as well as its dating, could suggest that Component 1 is relatively late in the Denali sequence. But perhaps another line of reasoning might better approach this problem. The site location, on a high exposed ridge, is quite unlike other Denali sites in the Tangle Lakes and this may be an indication of seasonality. Certainly, the shoreline sites are better protected from the wind and they may have been so sited primarily to facilitate the hunting of large game. Based upon the analogy of winter caribou behavior in the recent past, the wind-scoured lakes, frozen over for eight

months of the year, would have served as the primary avenues for the movement of large game. From these sites their movements could have been followed from long distance, and access to such game animals would have been swift and relatively easy. The somewhat variant character of this Denali assemblage may be ascribable to variant hunting practices of the short summer season.

REFERENCES CITED

Ackerman, R. E. 1992. Earliest Stone Industries on the North Pacific Coast of North America. *Arctic Anthropology* 29(2):18–27.

Powers, W. R., and H. E. Maxwell. 1986. Lithic Remains from Panguingue Creek, an Early Holocene Site in the Northern Foothills of the Alaska Range. *Alaska Historical Commission Studies in History* 189.

Skarland, I., and J. L. Giddings, Jr. 1948. Flint Stations in Central Alaska. *American Antiquity* 14(2):116–20.

SPARKS POINT

Frederick H. West, Brian S. Robinson, and R. Greg Dixon

Sparks Point (Mt. Hayes 149) is a small, single-component, stratified Denali complex site occurring in Loess 1. It occupies an eminence overlooking the northeast corner of Landlock Lake near the portage between this and Upper Long Tangle Lake. Site coordinates are 63°09'50"N, 146°01'30"W. The site is approximately 10 m above the shoreline of the old proglacial lake and lies about 300 m south and east of the Denali site Mt. Hayes 95. (The latter, although virtually entirely surficial, appears to be an uncontaminated representative of the culture.) The buried soil here (Lower Holocene Podzol) also preserves all of its horizons (Fig. 8-8).

The greater part of the Sparks Point assemblage was found in the lower (buried) A to B horizon. Some, at least, of the material recovered in the bur-

View west from Sparks Point looking across Upper Long Tangle Lake. (*Photo courtesy of R. Greg Dixon.*)

ied B horizon is attributed to extensive bioturbation, in this case burrowing by ground squirrels.

Test excavations at Sparks Point were carried out by S. and L. Sparks, R. G. Dixon, D. Reger, and F. West. In the course of his work there, Dixon uncovered a ring of stones or small cobbles resting on the uppermost Bb horizon. Their upper edges protruded almost to the top of the A1b horizon. The ring was about 55 cm in diameter (Fig. 8–9). This has been referred to as a hearth although no indication was found of ash or burning, nor of the characteristic profile often associated with such features. It is assumed that if this was a hearth such organic traces have leached away. Waste flakes and several small charcoal samples were recovered from within the circle. Whatever its use, the stones are clearly manuports and the ring is very similar to that found previously at the Reger site (Mt. Hayes 92).

ARTIFACT ASSEMBLAGE (281/305)

Sparks Point produced a Denali complex assemblage that includes wedge-shaped microblade cores, burins, a lenticular biface, and several large

cores that appear to have been used to produce boulder chip scrapers (*tci thos*). The relatively small number of specimens recovered suggests a single episode of occupation.

Microblade Cores (8) and Parts (24)   The 8 microblade cores from Sparks Point are typical Denali

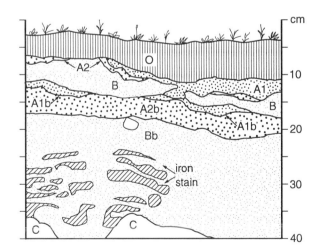

Figure 8–8   Sparks Point. Stratigraphic profile. The Denali assemblage occurred in the upper Loess 1, primarily associated with the A2b.

Figure 8–9   Hearth feature, Sparks Point. (*Photo courtesy of Frederick Hadleigh West.*)

complex, wedge-shaped cores (Fig. 8–10*a–e*). Five of the cores are produced from a variety of lustrous silicates; the remaining three are of weathered tan and grey chert. The core parts include 6 tablets, 2 full top removals, and 16 notch flakes. A lustrous black chert core top appears to have been used as an end and side scraper (Fig. 8–10: *f*). This specimen and three others can be articulated to two microblade cores.

Microblades (115)   Microblades constitute 24 percent of all artifacts including waste flakes. At least four microblades were retouched, one with a single lateral notch, one with a straight retouched distal end, and two with well-formed perforator or graver points (Fig. 8–10: *m, n*).

Large Blades (2)   The Sparks Point assemblage has only a few large blades. One medial blade fragment of lustrous chert (43 × 13 × 3 mm) and a proximal fragment of a wider (21 mm) blade-like flake were recovered. No evidence of prepared core reduction was found.

Burins (10)   Four of the burins are typical notched-platform specimens (Fig. 8–10: *g, k, l*). Two have straight retouched platforms (Fig. 8–10: *h*) and three have short "tablet" removals struck into unifacially retouched platforms, allowing their designation as dihedral burins (Fig. 8–10: *i, j*). The tenth specimen has the platform spalled off. Two of the specimens are weathered chert and the remainder are lustrous grey, brown, and yellow chert.

Burin Spalls (58)   The high number of burin spalls includes a variety of material types that are not included in the burin sample. The most notable attribute of the spalls is that 18 specimens (31%) had znheavy unifacial—probably use—retouch on the lateral edge (Fig. 8–10: *p*), representing edge-dulling or, perhaps, planing use prior to removal. Another 9 specimens have lighter, but still pronounced, nicking and edge stepping that probably also represents prior use retouch. Three specimens were retouched after removal, two show retouch on the distal end. One exhibits a well-formed point (Fig. 8–10: *o*), recalling the microblade noted above.

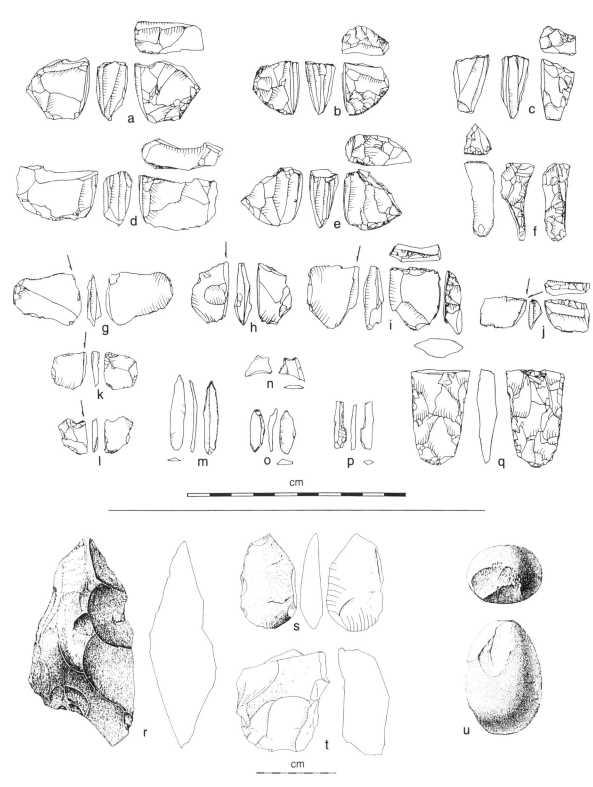

Figure 8–10   Sparks Point: wedge-shaped microblade cores (*a–e*); core tablet reused as end and side scraper (*f*); burins (*g–l*); microblades (*m, n*) and burin spall (*o*) with retouched perforator points; burin spall heavily retouched prior to removal (*p*); biface fragment with well-ground basal edges (*q*); large coarse porphyry flake core (*r*); *tci tho* or boulder chip scraper that fits core *t* (*s*); large cobble flake core or *tci tho* core (*t*); large hammerstone (*u*).

Bifaces (4)   A fragment of a lenticular biface shows basal grinding (Fig. 8–10: *q*). The specimen is well thinned and may have been either a projectile point or knife. A large, crudely flaked biface is made of a coarse igneous rock which resembles that of the large flake cores described below. Two biface fragments may have been produced in the fabrication of microblade cores. In general, little evidence was found of bifacial reduction at the Sparks Point site.

Unifaces (3)   Uniface technology is limited to two edge fragments and one laterally retouched flake that appears to have been a deliberate manufacture.

Denticulate (1)   One small flake of lustrous chert has multiple unifacial notches producing pronounced spurs. Its purpose may have been similar to that of the microblade and burin spall tool.

Snapped-Flake Gravers (2)   This category of modified flake was first described by Robinson for Mt. Hayes 146 and consists of corner use on snapped flakes, probably for incising or engraving. These two specimens were made of weathered chert.

Modified Flakes, Miscellaneous (46)   Forty-one (89%) of the modified flakes are made of lustrous chert with the remainder of weathered chert (4) and a single specimen of yellow-brown sard. As with other tool classes at Sparks Point, grey chert dominates the assemblage while weathered chert and sard occur as a minor proportion of the raw materials.

***Tci thos* (3)**   Three *tci thos* or boulder chip scrapers are identified, primarily on the basis of their regularity of form. One of the specimens (Fig. 8–10: *s*) has edge battering or retouch subsequent to its removal from the core. All are made of dense, coarse igneous rock that is characteristic of this form of scraper. ("Boulder chip" follows Alaskan convention. The rocks involved are actually more nearly cobbles in size.)

Large Flake Cores (3) and Derived Flakes (58)   Three large flake cores were produced from two water-worn igneous boulders. The largest specimen

(26 × 13 × 8.3 cm, Fig. 8–10: *r*) is an elongate biface core of coarse porphyry bearing an eroded cortex. Thirty-six flakes of similar material were recovered. At least two flakes attach to the core. The two other cores (e.g., Fig. 8–10: *t*) fit together with six flakes into a large, uniform, water-worn cobble that originally measured 23 × 15 × 9.5 cm. The dense granitic stone was split in two and each portion further reduced, yielding large flakes with cobble cortex. Two of the attaching flakes (e.g., Fig. 8–10: *s*) are included in the *tci tho* category and it is possible that all three of these specimens were intended for the production of *tci thos*. One of the two conjoining cores (Fig. 8–10: *t*) was found within or directly adjacent to the stone ring or hearth.

Hammerstone (1)   This specimen is exceptionally large (14 × 9.7 × 7.8 cm, 1,640 grams) and shows an exceptional degree of battering at either end (Fig. 8–10*u*). It is made from a dense, igneous cobble, and likely served in the reduction of the large flake cores.

Debitage (Siliceous) (247)   This constitutes 42% of the assemblage. Materials include lustrous chert (55%), weathered grey chert (17%), weathered tan chert (22%), sard (1%), and miscellaneous materials (4%). The comparatively low proportion of waste flakes is characteristic of microblade and burin assemblages with minimal biface reduction.

Summary

As may be seen, this is a rather typical Denali site assemblage, the exceptions consisting of the large flake or boulder chip cores (Fig. 8–10: *r–t*) and the stone ring or hearth. The relatively high proportion of microblades and cores, burins, and spalls is also notable here.

Four radiocarbon dates have been obtained. That of 4310 ± 150 BP (UGa-950) is clearly too recent. A combined sample from several points in the upper Bb horizon yielded an assay of 9060 ± 425 BP (UGa-941). Dates of 9200 ± 60 BP (Beta-62773; CAMS-6772) and 9110 ± 80 BP (Beta-64577; CAMS-8299) come from the upper buried B horizon on which the hearthstones rested. It is the authors' impression that the latter dates are rather too recent.

View is north, Reger site indicated by arrows. Below it, the old lake foreshore can be traced, trending laterally here. Landlock Lake to right, Upper Long Tangle Lake to left. High bank at upper left is truncated esker on which is Whitmore Ridge.

REGER SITE

Frederick H. West

The Reger site (Mt. Hayes 92) is located about 200 m south and east of the southeast shore of Upper Long Tangle Lake, close to the portage into Landlock Lake to the south. The view from the site is broad and largely to the north and west. Much of Long Tangle Lake, the lower course of Rock Creek, and, beyond, Landmark Gap and the Amphitheater Mountains to either side, may be seen. The site lies about 20 m above the shoreline of the old high lake. That old foreshore is marked here as elsewhere by a lag deposit consisting of cobbles and coarse gravels. Approximate site coordinates are 63°01′00″N, 146°28′00″W.

HISTORY OF RESEARCH

This was the first *in situ* Denali complex assemblage recognized and tested in the Tangle Lakes. It was also the first where there was seen the double soil horizon in which all subsequent *in situ* Denali assemblages were found to occur. It was, as well, the first site discovered in a clear relationship just above the old lake shoreline. The initial surface finds were made in a blowout of the sand that is the soil matrix here. The very first find was a typical Denali complex microblade core (Fig. 8–11: *a*). Its platform had been renewed but no blades had been struck from the new surface. A scatter of flakes lay nearby. Tests were made and subsequently expanded into a small excavation in the course of which the core tablet fitting the surface find core turned up (Fig. 8–11: *a*). The site was discovered late in the 1966 season by

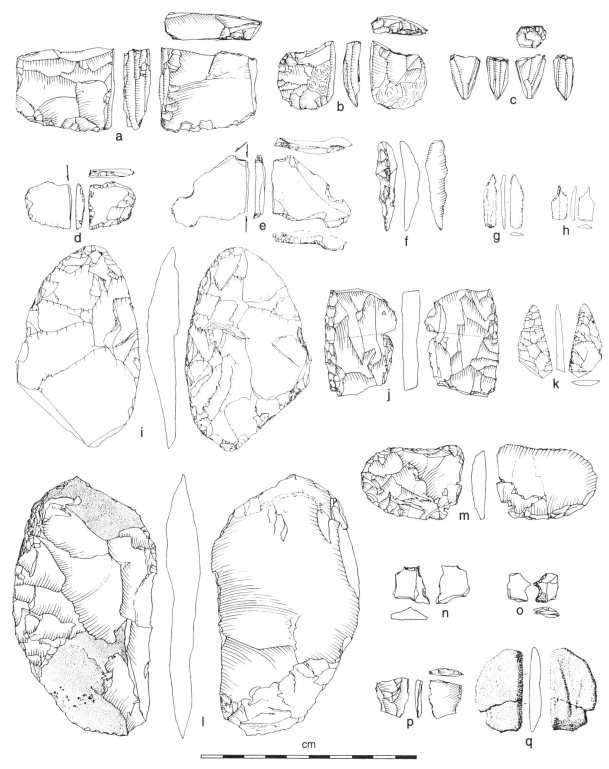

Figure 8–11 Reger site: wedge-shaped microblade cores (*a–c*); notched burins (*d, e*); crested microblade (*f*); retouched microblades (perforators) (*g, h*); bifaces (*i–k*); side scrapers (*l, m*); denticulates (*n, o*); snapped-flake graver (*p*); abrader (*q*).

the writer and Douglas Reger, and most of the work there was carried out in the following summer and later in restricted confirmatory tests.

In the summer following the discovery of the Reger site, Robert Hinton of the U.S. Soil Conservation Service in Palmer visited the site and shared his knowledge of the nature and distribution of the buried soil. Its full significance archaeologically would await further research and radiocarbon dating.

STRATIGRAPHY

This is a single-component site. The assemblage occurred in the buried soil at the contact of the A2b and the underlying Bb horizons (Fig. 8–12). The matrix is silty sand. In 1968, John V. Matthews identified two depositional units below the upper loess in which the modern soil is developed. The upper member, which is capped by the buried soil, consists of reworked sand and silt. Beneath this is an apparent dune sand (approximately 1–5 m depth) showing well-defined cross-bedding. No attempt was made to reach the bottom of this deposit.

One of the most interesting finds here was a small circle of stones. This, as with the later find, has been termed a "hearth," but it revealed no evidence of a typical hearth profile (see Fig. 8–13). The other occurrence to which it is quite similar was found at Sparks Point.

ARTIFACT ASSEMBLAGE (256/808)

The Reger site produced a Denali complex assemblage dominated by wedge-shaped microblade cores, burins, and their products, with a smaller number of biface and uniface tools. The complete assemblage may represent a single occupation.

Microblade Cores (10), Parts (27), and Miscellaneous Fragments (11)   The 10 microblade cores from the Reger site are typical Denali complex, wedge-shaped cores (Fig. 8–11: *a–c*). One very small specimen was reduced from both ends, resulting in a conoidal form (Fig. 8–11: *c*). Four of the cores are made of lustrous grey cherts and 5 are of burned,

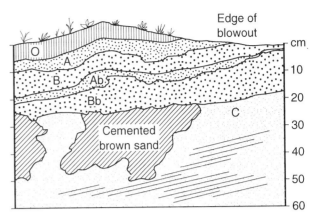

O - Moss and moss roots
A - Ashy grey to black, upper A horizon
B - Red, upper B horizon, sandy loess
Ab - Grey, lower A horizon, occupation zone, sandy loess
Bb - Red, lower B horizon, sandy loess
C - Bedded grey sand, C horizon

Figure 8–12   Reger site. Stratigraphic profile showing the buried Lower Holocene Podzol (Ab–Bb). The artifact assemblage occurred in the Ab horizon.

and often heat-spalled, red sard or jasper. Microblade core parts include 11 tablets and 16 notch flakes. Most of the 11 miscellaneous fragments are related to wedge-shaped cores.

Microblades (134)   Microblades constitute 12 percent of all artifacts including waste flakes. Proportions of material types (37% sard, 37% lustrous chert, and 26% weathered cherts and argillites) reflect the narrow range of materials used in the cores. Thirty percent of the sard microblades are yellow-brown in color (unburned), indicating that burning did not necessarily represent heat treatment. At least 6 microblades were retouched, including 3 possible perforators (Fig. 8–11: *g, h*).

Large Blades (4)   Four large blades were recovered, three of which revealed edge utilization. Two of the utilized blades can be refitted against each other and represent consecutive removals. Both are somewhat irregular and the core form is not apparent. The largest of these is 96 mm long × 36 mm wide × 8 mm thick. The conjoining specimens are red/grey nonlustrous chert and the remaining two are of lustrous grey chert and chalcedony.

Figure 8–13   Hearth feature at Reger site.

Burins (22) and Burin Spalls (59)   There were over twice as many burins as any other worked artifact form except modified flakes. Platform preparation techniques included unifacial notching or, less frequently, straight retouch (13) (Fig. 8–11: *d, e*); double burin removal or dihedral burins (7, including one "core burin"); and unretouched snapped platforms (2). Materials include lustrous grey chert (7), yellow-brown and red sard (6), and weathered tan and grey cherts (9). At least 2 specimens were double-notched at either end of the burin removal, representing either a notch-and-snap technique or alternating platforms. One of these (Fig. 8–11: *e*) had four distinct notches, with removals in three directions. The two opposed notches were probably intended as stop notches or platforms though not used. It is similar to 2 specimens from the Phipps site. A high number of burin spalls were recovered (59), of which at least 10 had lateral edge retouch prior to removal while 5 had some form of modification after removal, including 1 possible perforator.

Bifaces (6)   Three of the five bifacial specimens are small fragments that could represent either microblade core by-products or biface fragments. Three specimens are considered bifacial implements (Fig. 8–11: *i–k*). A large ovoid biface is unifacially retouched on opposing faces of the two edges (Fig. 8–11: *i*). A poorly preserved, highly weathered midsection is shown as Figure 8–11: *j* and a very thin tip fragment as Figure 8–11: *k*. These specimens are of weathered tan chert or argillite. The remaining fragments are of lustrous chert (2) and sard (1).

Unifaces (7)   Only one end scraper fragment was found. Four of the remaining unifaces (Fig. 8–11: *l, m*) are of comparatively large size. The remainder are modified flakes and fragments.

Perforators (6) These deliberately pointed specimens were produced on microblades, burin spalls, and flakes.

Denticulates (2) Two finely worked denticulate tools were made from red sard and weathered chert (Fig. 8–11: *n, o*). These tools may overlap in function with perforators. Taken together (6) they constitute a relatively high number of such specimens compared to other Denali complex sites here.

Snapped-Flake Gravers (3) In contrast to Mt. Hayes 146, where it was first recognized, only a small number of this expedient tool class were identified (Fig. 8–11: *p*). They may have served, with burins, as bone-working tools.

Modified Flakes, Miscellaneous (59) Modified flakes excluding the preceding categories were produced from lustrous chert (25), weathered chert or miscellaneous materials (25), and sard (9).

***Tci Thos* (4) and Abraders (2)** One utilized *tci tho* (boulder chip scraper) and three other possible specimens were recovered. A small sandstone abrader bears an incised groove across its length. One edge is well beveled from use (Fig. 8–11: *q*). No unequivocal hammerstones were recovered.

Waste Flakes (808) The dominant rock types used were weathered chert or argillite (43%), lustrous chert (20%), sard (10%), and miscellaneous lithologies (27%).

SUMMARY

The artifact assemblage from Reger is dominated by microblade and burin technology, suggesting an emphasis on the production of composite bone or antler weapons heads. The assemblage also includes a number of large bifaces and unifaces, as well as delicate denticulate tools and perforators.

As in virtually every Denali assemblage, there occurred here some seemingly unique artifact forms (e.g., the distinctive double-notched burin [Fig. 8–11: *c*]). These, it is thought, can be attributed

to the partial sampling of a single manifestation of a culture—a factor that must constitute an inbuilt bias of all site interpretations. Properly seen, they may provide the researcher with a useful cautionary device in the reading of a given archaeological record. In any event, they are here interpreted as normal constituents of a Denali assemblage—which is not to rule out the possibility that some of these may represent chronological markers *within* Denali that may come to be recognized in future. For the present it seems safest to accept these as being all coeval forms from one Denali site to another. Certainly within the Tangle Lakes sequence the appearance is primarily one of unity—despite the possibility that there may well be represented here Denali occupations spanning as much as 1,500 years.

OTHER SITES IN THE TANGLE LAKES

Frederick H. West

There is a remarkable density of sites in the Tangle Lakes district. A gratifyingly large number are ones that appear to relate to the earliest movement of people into this region. Brief mention will be made here of three more of these.

Mt. Hayes 146

This is a small Denali site in the same stratigraphic context as the others but situated at the south edge of a remnant ancient Rock Creek delta. About 700 m north and east lies the Phipps site, also on this topographic feature. When this delta formed, Rock Creek emptied into the old high lake. Despite its size, the site produced a fairly large number of microblade cores (14) together with by-products and fragments related to them. The only biface fragment found was refitted by Brian Robinson to a microblade core. The presence of 13 burins further at-

Region of lower Tangle Lakes and Amphitheater Mountains, in which Mt. Hayes 45, 122, 153, and other sites occurred.

tests the specialized nature of this site, a character shared with the larger Phipps site.

It was in this assemblage that Robinson discovered and delineated the existence of a distinctive artifact form he terms a "snapped-flake graver." These small tools were made from blade-like flakes, and show a pattern of edge rounding, stepping, and, occasionally, "micro-burin spall" removals at the working corner, the latter presumably also from use. They appear to have been employed for grooving bone (or antler) or wood. Examples have also been found in the Sparks Point site. Interestingly, it is these tools that show textbook burin use. The technological burins, throughout Denali, appear to have been used as shavers, presumably to shape antler tools (projectile points) after the fashion of Bordes's demonstration (Bordes 1965). Figure 8–14 shows some elements of this assemblage.

MT. HAYES 122

An unusual small assemblage, this was found well to the north of most of the sites discussed, away from the lakes and near the base of one of the Amphitheater Mountains. While the soil context was not as strongly developed as some, the assemblage, nevertheless, quite clearly occurred in the buried soil. Essentially one class of manufacture only is represented here. These are large blade cores (4), two of which are technologically reasonably close to the Denali form and, in the case of the best characterized example, reasonably close to the conoidal cores from Whitmore Ridge. Their size may result from proximity to a quarry, and certainly that showing the least work resembles quarry block reduction. Some of these specimens are shown in Figure 8–15.

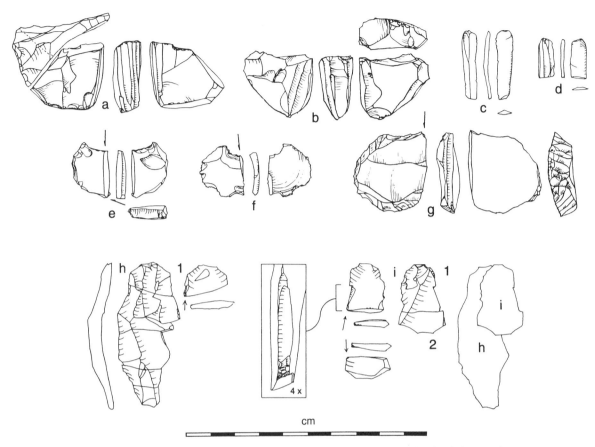

Figure 8–14 Mt. Hayes 146: wedge-shaped microblade cores (*a, b*); microblades (*c, d*); burins (*e–g* [*g* is also an end scraper on a large blade]); snapped and utilized blade-like flakes that attach to each other (*h, i*); *h1, i1,* and *i2* are snapped-flake gravers (inset shows use-stepping and wear).

THE AMPHITHEATER MOUNTAIN SITES

In 1974 the existence of another complex termed "Amphitheater Mountain" was proposed (West 1974). It was composed of a series of small sites near the base of the Amphitheater Mountains. All sites were surficial and, in fact, in this particular area there is little to—often—no soil development. The technology was described as "Mousteroid" but it was acknowledged that these could represent quarry facies. In 1981 they were listed among sites of "unclear standing" and the definitional difficulties were enumerated (West 1981: Table 3.1). The archaic appearance of some of these specimens is apparent, aided significantly by the extremely heavy weathering throughout. It may well be that the "Mousteroid" character reflects nothing more than the not-very-elaborate techniques that were involved in reducing quarry materials to usable size. The examples shown in Figure 8–16 came from the site Mt. Hayes 45, with three additional pieces from Mt. Hayes 153 shown in Figure 8–17.

REFERENCES CITED

Bordes, F. 1965. Utilisation possible des côtés des burins. *Fundberichte aus Schwaben*. Neue Folge 17.

West, F. H. 1974. Amphitheater Mountain Complex. In *Proceedings of the International Conference on the Prehistory and Palaeoecology of the Western Arctic and Sub-Arctic, November 1972.* Calgary: University of Calgary Press.

———. 1981. *The Archaeology of Beringia.* New York: Columbia University Press.

(Figures 8–15 to 8–17 follow on pp. 406–408.)

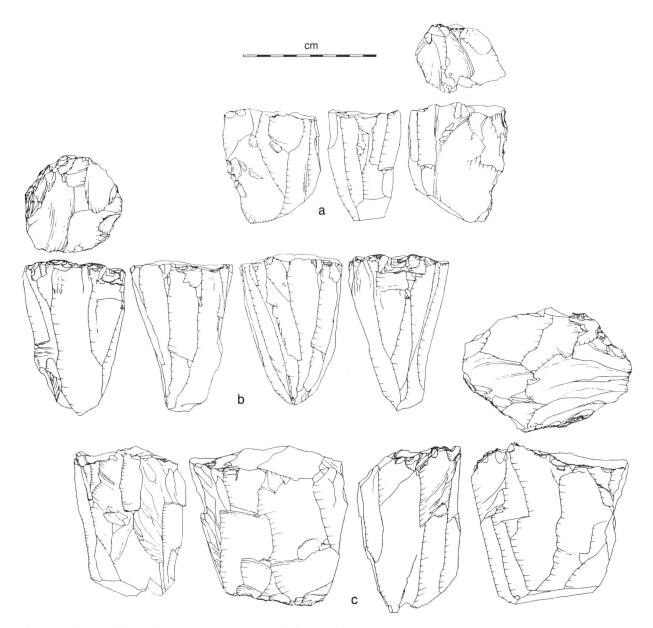

Figure 8–15   Mt. Hayes 122: large wedge-shaped blade core (*a*); large conoidal blade core (*b*); large blade core (*c*).

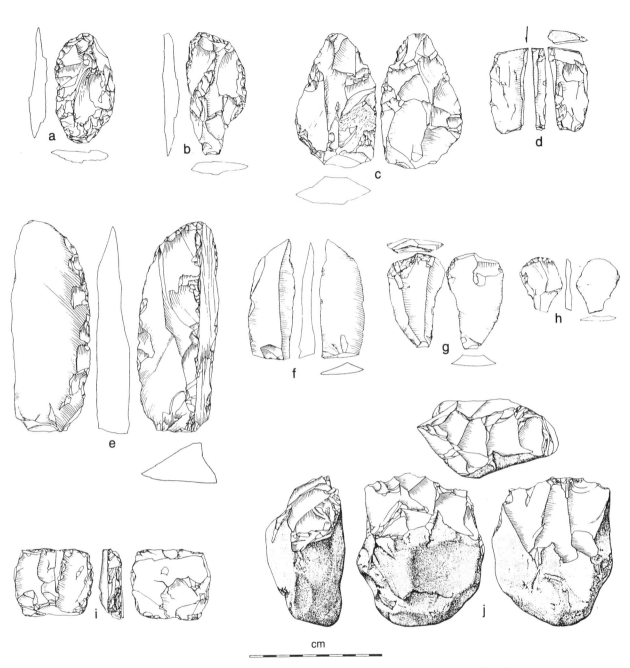

Figure 8–16   Mt. Hayes 45: bifaces (*a–c*); large blade with burin-like removals (*d*); large utilized blades (*e–g*); end scraper (*h*); tabular core (*i*); flake core (*j*).

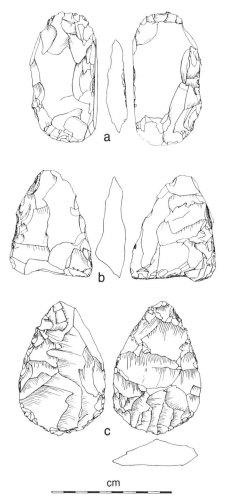

cm

Figure 8–17　Mt. Hayes 153: bifaces (*a–c*).

Chapter 9

NORTH PACIFIC
LITTORAL, ALASKA

INTRODUCTION

All portions of this large region either are on or are proximate to the ocean. As a result, maritime and near maritime climates are characteristic throughout. The contrast with the climate of the interior is striking. Winter temperatures on the northern Pacific littoral are generally mild for their latitude while those of summer are quite cool. Thus, where the wide range of temperatures may be said to be the most obtrusive aspect of climate in the interior, here precipitation dominates. Cloudiness is commonplace as are strong surface winds, especially in the winter. Rainfall values range from abundant to copious. Snowfall mirrors summer precipitation with, in both cases, the highest values being recorded on and near to the southern and southeastern coasts. As much as 10 to 15 meters of snow may be recorded on the higher slopes of these coastal mountains. Much reduced values are found in the piedmont and along the coasts proper; the same is true of the Susitna and Matanuska valleys. These latter areas are, in fact, intermediate in character between the maritime regimes of the coast and the more distinctly subarctic climates of the interior.

Nevertheless, snowfall and rainfall are abundant. Average annual precipitation for those valleys ranges from 16 inches in Upper Cook Inlet–Susitna Lowland to 12 inches in the vicinity of Long Lake. The Alaska Peninsula likewise has high annual precipitation rates similar to those found on Kodiak Island and in the eastern Aleutian Islands. Most of this occurs as rain and may aggregate as much as 200 centimeters per annum. In these areas freezing weather is uncommon. Windy and foggy conditions prevail throughout.

Two quite monumental factors combine to produce this situation, unusual for the far north. The first of these is the Alaska Current, an eddy of the Japanese Current, which, flowing from the south, brings warm waters to the coast of southeastern and southern Alaska. One immediate effect of this warming is that sea ice cannot form except under exceptional circumstances, usually in sheltered embayments of restricted size. A second result is that fog and humid air are normal occurrences. Prevailing winds from the west and southwest transport this moisture landward where the other great factor comes into play: the entire arcuate southeastern-southern coast is composed of mountains, including some of the highest peaks in Alaska and adjacent Canada. Here also are some of the most extensive icefields and glaciers in the world—outside of Greenland and Antarctica.

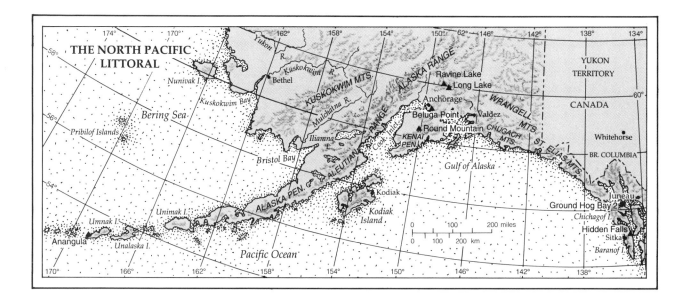

The heavily fiorded nature of these northern Pacific coastlines—especially marked in the Alexander Archipelago—results in there being several thousand kilometers of coastline in the northern Pacific littoral. One exception to that characterization occurs in the archipelago where the strongly linear Chatham Strait and Lynn Canal preserve the morphology of a great fault of Tertiary times. The mountains of this region are relatively low, ranging from 1,500 to 2,000 meters. Just to the north, however, lies the massif of Mount Fairweather where, among peaks of over 3,000 meters, Mount Fairweather itself rises to almost 4,000 meters. Snowfields and glaciers form a virtually unbroken chain from midway of the Alexander Archipelago westward to Prince William Sound. Malaspina Glacier, about 200 kilometers west of Mount Fairweather, is the largest of the piedmont glaciers on this glaciated coast. It is approximately 48 kilometers across and from its source in the St. Elias Mountains descends some 45 kilometers to tidewater in the Gulf of Alaska. It is estimated to be over 500 meters thick. Mount Logan, in this group, is 5,090 meters in height; Mount St. Elias, slightly over 4,600 meters. They are, respectively, the second and fourth highest peaks in North America. Mount St. Elias is thought to have been first sighted—and named—by Vitus Bering in 1728.

The great arc described by the coastal system continues its westward trend with the Chugach Mountains, then turns abruptly south, becoming the much more subdued Kenai Mountains. This segment of the arc covers some 800 kilometers. The highest points here are in the eastern Chugach and rise to over 3,000 meters. Prince William Sound, with its many islands and bays, lies in the northernmost part of the curve. Hinchenbrook Entrance is the major passage between the sound and the Gulf of Alaska.

To the west of the Kenai Peninsula is Cook Inlet, which is enclosed on its western side by the Alaska Peninsula. Cook Inlet is one of the great structural features of the Alaskan coast. From its head, southwestward to Kennedy Entrance is a distance of about 240 kilometers. From its upper end, Turnagain Arm bears off to the southeast. The constricted character of the inlet and Turnagain Arm give rise to extraordinarily high tides—maximum high tides of ca. 8 meters in the lower inlet and over 10 meters in the upper inlet. In Turnagain Arm tidal bores are frequent. The upper inlet is, in addition, subject to heavy siltation and constant shifting of channels such that navigation is hazardous. It is a measure of Captain James Cook's navigational skill that the name "Turnagain Arm" was applied by him to that point where he judged it prudent to return southward.

The Susitna River flows into Cook Inlet from

Dundas Bay in Glacier Bay National Park. (*Photo courtesy of Robert E. Ackerman.*)

the north. It is the largest tributary. Just east of here the Matanuska River enters via Knik Arm. It transports a heavy load of detritus derived from its source, the nearby Matanuska Glacier as well as other glacial streams that add to its flow.

This region of Alaska is considered to be one of the world's most seismically active. In 1964 an earthquake of between 8.3 and 8.6 on the Richter scale struck here. Its epicenter was in Prince William Sound. It is considered the strongest earthquake recorded since the advent of modern monitoring equipment.

Cook Inlet is bounded on the west by the Alaska Peninsula, the narrow backbone of which is the Aleutian Range. The peninsula and the Aleutian Islands mark the southern limits of the Bering Sea. The great chain of volcanoes which constitute the Aleutian Range begins with Mount Spurr at the north—west of the upper inlet—continues down the peninsula and thence, as the Aleutian Islands, far to the west. From Mount Spurr to Cape Wrangell on Attu Island is a distance of slightly over 2,500 kilometers. The Alaska Peninsula portion of this

great volcanic arc contains 32 volcanoes, about half of which have been active—in some instances repeatedly—in historic time. The most famous modern eruption was that of 1912 involving mounts Katmai and Novarupta when ejecta from the latter buried 75 square kilometers of an adjacent valley under as much as 215 meters of pumice. The great caldera of Mount Katmai was created in this eruption. About 47 volcanoes occupy the islands of the Aleutian chain. Shishaldin, which dominates Unimak Island, is often compared to Mount Fuji because of its perfect symmetry. The topography of the islands is mountainous, rugged, and on many islands there exist great vertical sea cliffs.

The climate of the Aleutians is decidedly maritime, and freezing temperatures are rare. Fog, high winds, and storminess are common. The southern sides of the islands are exposed to the full sweep of the northern Pacific weather and there is no counterpart of the Alaska Current nor topographic barriers to ameliorate its effects. The surf resulting from northern Pacific storms is often very destructive. One result of this chronic condition is that most permanent settlements face the Bering Sea. That condition is peculiar to the Aleutian Islands setting. Here also, though, as elsewhere in the north Pacific littoral, the areas chosen for habitation are close to the shore, this encouraged by the topography and the littoral economies found throughout.

The first European explorers of this region were the Russians, beginning with the expeditions of Vitus Bering in 1728 and 1741. They, and later Russian, British, French, and American explorers and traders, were to discover a great diversity of native cultures. The Aleut, speakers of a distinctive Eskimoan language, occupied the Aleutian chain and the lower half of the Alaska Peninsula. They were maritime hunters but also used intensively the rich resources of their shorelines. Dialects of the Yupik Eskimo language were spoken in the great region including Prince William Sound, the western Kenai Peninsula, Kodiak Island, the upper Alaska Peninsula, and the southwestern coast as high up as Norton Sound. These Eskimos followed substantially the same way of life as the Aleut with observable differences largely attributable to the presence or absence of different marine resources. In the southwest proper, land hunting played an important role. Upper Cook Inlet and the valleys of the Susitna and Matanuska rivers were the territory of Athapascan speakers whose subsistence was equally dependent upon fishing and the hunting of moose and caribou. In most respects these folk, the Tanaina, were far more like their interior cousins than like their coastal neighbors. East of the Tanaina lay the territory of the Chugachigmiut and beyond that the realm of Northwest Coast culture. The dominant peoples here in terms of numbers, size of territory, and power were the Tlingit. While their economies were assuredly geared to the exploitation of maritime and littoral resources, in their rich material culture, their ceremonialism, and their elaborated

social organization, they were completely unlike the people just discussed. The Eyak, a small group located at the mouth of the Copper River, between the northernmost Tlingit and the Chugach Eskimos, were related culturally and linguistically to the Tlingit and were a remnant of a somewhat larger group partially absorbed by the latter. Although archaeological sites are on record that can be ascribed to the Tlingit, to date Ground Hog Bay and Hidden Falls are the first, and only, early sites reported from the area of southeastern Alaska.

F. H. W.

BIBLIOGRAPHY

Barnes, F. F. 1958. Cook Inlet–Susitna Lowland. In *Landscapes of Alaska, Their Geologic Evolution*, ed. H. Williams, pp. 43–47. Berkeley and Los Angeles: University of California Press.

Emmons, G. T. 1991. *The Tlingit Indians.* ed. and with additions by F. de Laguna. New York: American Museum of Natural History; Seattle: University of Washington Press.

Hartman, C. W., and P. R. Johnson. 1978. *Environmental Atlas of Alaska.* Fairbanks: Institute of Water Resources, University of Alaska.

Miller, D. J. 1958. Gulf of Alaska Area. In *Landscapes of Alaska, Their Geologic Evolution*, ed. H. Williams, pp. 19–29. Berkeley and Los Angeles: University of California Press.

Powers, H. A. 1958. Alaska Peninsula–Aleutian Islands. In *Landscapes of Alaska, Their Geologic Evolution*, ed. H. Williams, pp. 61–75. Berkeley and Los Angeles: University of California Press.

Reed, J. C. 1958. Southeastern Alaska. In *Landscapes of Alaska, Their Geologic Evolution*, ed. H. Williams, pp. 9–18. Berkeley and Los Angeles: University of California Press.

Rieger, S., D. B. Schoephorster, and C. E. Furbush. 1979. *Exploratory Soil Survey of Alaska.* Washington, D.C.: Soil Conservation Service, U.S. Department of Agriculture.

HIDDEN FALLS

Stanley D. Davis

The Hidden Falls site is located on the northeastern shore of Baranof Island, thirty km northeast of Sitka, Alaska, at latitude 57°13′N, 134°52′W. Baranof Island is one of the many islands that make up the Alexander Archipelago of southeastern Alaska. The site is at the head of Kasnyku Bay; it lies in a low saddle on a small point that extends northward into the bay approximately 8 m above sea level. Just to the west of the site is Hidden Falls Lake and waterfall, which drain into a geologically recent tidal lagoon. Chatham Strait lies to the east (Fig. 9–1). The Baranof Mountains lie to the west of the site.

Hidden Falls Lake occupies a cirque in the lower end of a hanging valley south and west of the archaeological site. The valley is connected by a sequence of higher cirques to highlands which rise to over 1,220 m at the center of the island. The main glacial activity that flowed over the site and into Kasnyku Bay in Pleistocene and early Holocene times originated in the cirque–hanging valley highland area (Davis 1989; Swanston 1989).

Geological evidence suggests that mainland ice flows covered much of the Alexander Archipelago to a depth of more than 1,000 m (3,200 ft.) during the late Pleistocene (Mann 1986; Swanston 1989); precisely when retreat of the late Wisconsinan ice began in southeastern Alaska is unknown.

Ice streams exited the archipelago by way of such now-submerged valleys as Chatham Strait, Cross Sound, Summer Strait, and Dixon Entrance onto the Continental Shelf. Deglaciation within southeastern Alaska can only be inferred, but was well under way by at least 16,000 years ago. By 13,000 years ago southeastern Alaska would have been dominated by valley glaciers and a rising shoreline (Mann 1986; Swanston 1989).

Raised marine terraces are recognized throughout the Alexander Archipelago, dating between ca. 13,000 and 10,000 BP. By the beginning of the Holocene low-level terracing was evident along extensive areas of coastline in southeastern Alaska.

Southeastern Alaska's climate is cool, moist, and maritime, due, in part, to the interaction of the weather circulation patterns and the mountainous topography. The waters of southeastern Alaska are warmed by the circulation of the Kuroshio Current, which produces abundant moisture. The mountains lift the air mass, producing yearly rains up to 400 cm (13.2 ft.) (Davis 1989).

Soil development in southeastern Alaska is relatively young, beginning after the recession of the Pleistocene ice. Early Holocene ice fluctuations and subsequent neoglacial advances delayed or impeded development in localized areas (Davis 1989).

The predominant plant community is the coastal western hemlock–Sitka spruce forest. The local plant communities include alpine tundra, muskeg, meadows, beach fringe, and mature conifer forest (Davis 1989).

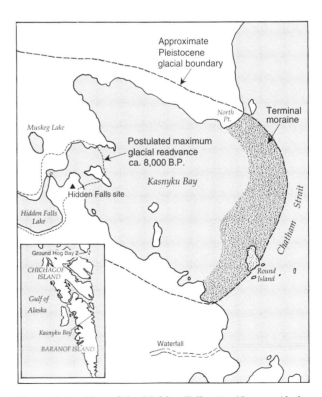

Figure 9–1   Map of the Hidden Falls site. (*Source: Alaska Anthropological Association. See Acknowledgments.*)

Area of the Hidden Falls site.

FAUNAL HISTORY

There is a consensus among researchers that there were ice-free areas (refugia), possibly extending onto the Continental Shelf, during the last maximum glaciation between 25,000 and 10,000 years ago (Mann 1983, 1986; Molina 1986). Two major mammalian faunal refugia were present during the Pleistocene: one in the interior Alaska–Bering Sea region north of Baranof Island and one south of the Cordilleran ice sheet that covered the mainland of the northwestern coast.

Small isolated coastal refugia supported a very limited biota and were inundated by the rising sea level as the late Pleistocene came to an end between 10,000 and 16,000 years ago. There were small isolated refugia on the mainland (Lituya Bay [Mann 1983, 1986]) and on the outer islands (the Queen Charlotte Islands) that predate 13,000 BP (Clague et

al. 1982). These small biotic communities were probably limited to small rodents, plants, and insects.

Klein (1975) suggests that large mammals are postglacial arrivals from the larger refugia located in either the south or north.

Two species of bear, the earliest mammals to arrive in the archipelago, exhibit limited territorial overlap, suggesting that the brown bear (*Ursus arctos*) originated in the northern refugia and the black bear (*Ursus americansis*) originated in the southern refugia. While both species exist on the islands of southeastern Alaska, the black bear is located primarily south of Frederick Sound and the brown bear north of the sound. Similarly, there is a division between large moose (*Alces alces*) from the north and the smaller moose (*Alces alces andersoni*) from the south.

The early presence of black bear is indicated

by a sample dated 10,745 ± 75 BP (AMS-AA7793); brown bear remains from the El Cap cave on northern Prince of Wales Island, dated to 9760 ± 75 BP (AMS-AA7794) (Autrey, J., personal communication 1992), indicate that the Alexander Archipelago had, at least in part, a stable environment by 11,000 to 10,000 BP (at the beginning of the Holocene). The early date for black bear is consistent with the theory that the southern section of the northwestern coast may have been deglaciated first. Smaller species, such as wolverine and mink, arrived soon after the bear.

SUBSISTENCE RESOURCES

The environs of Baranof Island and adjacent southeastern Alaska include ecozones and organisms that have provided a resource-rich subsistence base since the late Pleistocene and early Holocene. The resources include abundant sea mammals, birds, fishes, intertidal shellfish, terrestrial mammals, and diverse plants.

RECENT OCCUPATION HISTORY

While the Tlingit are the one principal indigenous population occupying southeastern Alaska today, they arrived after ca. 1,500 BP. Three additional native groups are protohistoric/historic arrivals to the southern archipelago—the Tsetaut, Tsimshian, and Kaigan Haida—and along the northern boundary at Yakutat were Eyak and Athabaskan. The earliest evidence for human occupation in greater southeastern Alaska is associated with remains that are recognized as belonging to the Palaeomarine tradition (Moss 1989; Davis 1991, 1992), a coastal cousin of the American Palaeo-Arctic tradition (Davis 1980, 1989, 1990).

The Palaeomarine tradition (11,000 to 6,500 BP) was represented by a coastal Na-Dene people who used a microblade technology. They produced microblades primarily from wedge-shaped microblade cores on flakes; they had a predominantly unifacial tool industry, with rare evidence of a bifacial component; and they had a marine subsistence pattern (Davis 1989, 1990, 1991).

HISTORY OF RESEARCH

The Hidden Falls site was first identified in February 1978 by the USDA Forest Service. The site was being adversely affected by construction of a salmon hatchery. At the time of discovery, the site had been bisected by an access road from the beach to the construction area. The initial investigation in that year by the writer (Davis 1989) was limited because of cold weather, frozen ground, and two feet of snow. The only observable surfaces were the 2-to-5-m-high stratigraphic profiles along the road cut. Identified within these profiles were fourteen distinct strata. Charcoal-rich anthropogenic sediments containing artifacts, shell, and fish bone were exposed within the upper portion of the profile.

Additional investigations were conducted at various times in the spring of 1978 to determine the archaeological character of the deposit and delineate the stratigraphy revealed along the road-cut profile (Davis 1979, 1989). Several lithic artifacts were collected near the bottom of the profile immediately above and within a palaeosol that was situated between what seemed to be two glacial deposits. The information gained would be correlated with the regional and local geomorphic setting to permit reconstruction of the prehistoric landscape (Davis 1979, 1989; Swanston 1989). Additional studies focused on biotic environmental reconstructions. The site was sampled for plant macrofossils and pollen. Pollen and plant macrofossils were also collected from nearby dated sections of undisturbed sediments (Davis 1989; Holloway 1989).

THE SITE

Hidden Falls is a stratified, multicomponent site, containing three archaeological components. In Component I, the oldest (ca. 10,000 BP), there is no evidence of a transition from the microblade technology to a ground-stone-and-bone industry.

Component II (ca. 4620 ± 110 BP [Beta-7442] to ca. 3010 ± 40 BP [SI-4505]), can be placed within the early phase of the Developmental Northwest Coast Stage (Davis 1989, 1990). The tool assemblage here consists of a ground-stone-and-bone industry, including single-edged tools, abraders, hammer-stones, ground-stone-and-bone points, adzes, beads, labrets, and a flaked-stone industry produced predominantly by bipolar reduction (Davis 1989; Lightfoot 1989). Microblades and microblade cores and evidence of a bifacial industry are absent from this assemblage.

The most recent prehistoric occupation at Hidden Falls, Component III, appears to have placed more of an emphasis on ground-bone technology and may be placed in the middle phase of the Developmental Northwest Coast Stage dating between ca. 3015±55 BP (SI-3974) and ca. 1315±105 BP (Davis 1990).

STRATIGRAPHY

Fourteen stratigraphic units were identified at Hidden Falls. Seven units, A–G, are associated with Components II and III and will be discussed only briefly. The earliest occupation at the site, Component I, is found in units G–I. Units E (E1 and E3) and I are buried soils. Units J1 and J2 are Pleistocene glacial deposits (Fig. 9–2).

Unit A is the modern organic accumulation with plant remains. It and unit B have incorporated the historic refuse from a modern sawmill.

Units C, D, and E2 are anthropogenic sediments in which Component III materials are incorporated. Volcanic ash is present in small quantities throughout unit C and near the contact with unit B.

Unit D is highly organic, with shell concentrations, as much as 1 m thick, and shell lenses making up 15 to 20 percent of the deposit.

Unit E (E1 and E3) is a buried organic soil horizon that is continuous over the entire site. Interposed between E1 and E3 is the discontinuous unit E2. Units E1 and E2 are buried organic soil that represents a significant break in the occupation of the site.

Units F and G are anthropogenic sediments that have incorporated within them Component II artifacts. Component II is primarily contained in unit F. Unit G is the upper level of an early Holocene fluvioglacial deposit that exhibits prolonged weathering, soil formation, and subareal water move-

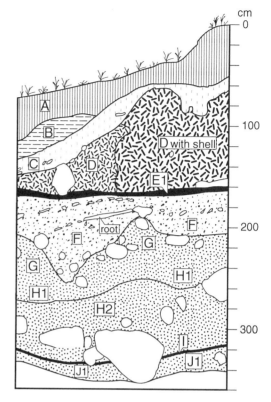

Figure 9–2   Hidden Falls. Stratigraphic profile. See text for strata descriptions. Not all substrata are present in this section. (*Source: Alaska Anthropological Association. See Acknowledgments.*)

ment. The weathering corresponds in time to the Hypsithermal Interval in southeastern Alaska. This event may have increased chemical weathering characteristics of the site. Heusser (1960) places the Hypsithermal between 8,000 and 3,500 BP, and Goldthwaite (1966) places it between 7,000 and 4,000 BP. Both preburial dispersal and postdepositional disturbance are responsible for the occurrence of Component II materials within the upper horizon of unit G.

Stratigraphic units G, H1, H1a, H2, and I contain the oldest archaeological remains at Hidden Falls, Component I. Units G, H1, and H1a are interpreted as a Holocene ablation till.

Unit G is the weathered upper unit and unit H1 the lower unweathered part of the ablation deposit.

Fluvioglacial in origin, unit H1a is a localized area of moderately stratified, poorly sorted, medium to very coarse sand and pebbles. Unit H2 is a discontinuous, poorly sorted, high density, very compact, massive structure deposit that represents a lodgement till underlying unit H1a and overlying unit I. Units G, H1, H1a, and H2 constitute the deposit of a minor early Holocene ice advance and retreat across the Hidden Falls site. Component I artifacts were incorporated in part into this 8,000–9,000 BP Holocene deposit by ice and water movement across the surface of unit I, and by differential compaction of the substrata. This resulted in the vertical and horizontal displacement of some of the archaeological residues.

The buried organic soil, unit I, made up of 3 to 7 percent medium-to-very-coarse sand, is discontinuous over the site. This palaeosol varies in thickness from a very thin (1 to 3 cm) horizon with no recognizable plant remains to areas of a thick (75 cm) organic palaeosol with plant remains that are identified as wood fragments, branches, twigs, roots, needles, and cones of *Tsuga mertensiana* (mountain hemlock). This artifact-bearing stratum containing flaked lithic debitage, unifacial tools, microblades, and microblade cores comprises the Component I assemblage and dates from ca. 10,000 BP.

Radiocarbon dates place this palaeosol as beginning ca. 10,300 BP and indicate that it was overridden by an early Holocene ice advance between 8,000 and 9,000 BP. Swanston (1989) suggests that Baranof Island was depressed at least 61 m, based on high-level marine terracing at Sitka. This depression allowed for the inundation of most of Baranof's lowland area still uncovered by ice. At present no direct evidence of any high-level marine deposits has been found at Hidden Falls, but the area-wide occurrence of these marine deposits indicates that the site may have been covered by the sea, possibly for as long as 2,500 years after the retreat of Pleistocene ice. Later emergence of the site area by isostatic rebound occurred prior to the establishment of vegetation. This was determined by radiocarbon dating of wood samples (e.g., 10,345 ± 95 BP) obtained from the unit I buried soil, which overlies the Pleistocene glacial deposit.

Unit J is the underlying stratum that can be divided into two units, J1 and J2. (J2 is not shown in Figure 9–2.) Unit J1 is an ablation till and unit J2 a

lodgement till that are associated with the last advance of Pleistocene ice over the area (Fig. 9–1). When the retreat of Pleistocene ice took place is unknown, but the establishment of an early Holocene forest within the Hidden Falls area was well under way by 10,500 years ago.

COMPONENT I ASSEMBLAGE

Analysis of the Pleistocene–Holocene artifact assemblage of Component I revealed a unifacial tool, microblade, and microblade core industry. Six hundred and twelve lithic artifacts and debitage belonging to Component I were recovered during the 1978 and 1979 excavations. Artifacts modified by use comprised 32.4 percent (n = 198) of the total collection; the remaining 67.6 percent (n = 414) are waste flakes. Numerical breakdown by class/type is presented in Table 9–1.

The artifact assemblage of Component I consists of microblade cores, microblades, split-cobble and split-pebble tools, choppers, hammerstones, a variety of scraping tools, gravers, an abrader, burins, a unifacial blade or point, and utilized flakes. [Author's note: In 1989 the term *burinized flakes* was incorrectly used to describe an artifact class. The correct terminology for this class of artifacts is *burin*.]

One additional point of clarification is necessary concerning the presence of a biface fragment recovered at the site. The exact provenance of the biface tip is unknown. It was recovered from mixed sediments that were associated with the removal of a tree stump. It cannot be determined whether the biface fragment was associated with Component I or with one of the later components. A description of each artifact class/type in Component I follows.

Microblade Cores and Core Fragments (14) Two types of microblade cores are identified: four wedge-shaped microblade cores on flakes (Fig. 9–3: *a–d*) and two split-pebble microblade cores. Wedge-shaped cores are produced on flakes with little or no evidence of flake scars on the lateral edges. Core platforms are prepared by removing flakes across the platform from one edge. There is

Table 9–1 Summary of Component I Artifacts

Artifact Class/Type	Number
Split-Cobble Tools	4
Split-Pebble Tool	1
Cobble Choppers	2
Block Chopper	1
Core Scrapers	2
Flake Cores and Core Remnants	26
Gravers	31
Hammerstone	1
Abrader and Incised Stone	2
Unifacial Point or Blade	1
Primary Spall	1
Scrapers	36
Notched Scrapers (N-11)	
End Scrapers (N-5)	
Side Scrapers (N-19)	
Narrow Bit Scraper/Scraper Plane (N-1)	
Burins	9
Blade-like Flakes	5
Microblades	9
Microblade Cores	14
Wedge-Shaped Cores (N-9)	
Pebble Core (N-2)	
Fragmented Cores (N-3)	
Utilized Flakes	53
Nonutilized Flakes	414
Total	612

little or no evidence of flake scars on the cores' lateral sides. The three obsidian cores (Fig. 9–3: *a-c*) have platform edge-crushing along one lateral side. Microblades had been removed from the wide ends of the flakes, and on one face only. A fourth core had facets that extended around to a second face (Fig. 9–3: *d*). This core had been rotated, due to a flaw in the rock. Based on observation of eight cores and fragments, keel modification was unifacial on four cores, bifacial on only one, and unmodified on two others. Keel modification on the remaining fragment could not be determined. Crushing was also evident along the keel on several specimens, probably from rebound concussion during blade removal.

Two split-pebble microblade cores were recovered. These were produced by splitting small pebbles in halves. In one case, the two lateral edges

were removed, forming a triangular outline. The platform was the unmodified cortex. One microblade had been detached from one core and two microblades had been removed from the other. The blades' scars indicated that these were wider microblades with less curvature than those that had been detached from cores of obsidian or chert. These microblade cores give the impression of being expedient tools; the knapper could quickly produce one or two blades which could then be discarded. On the other hand, these cores may have been used as practice pieces for inexperienced knappers.

Microblades (9) and Blade-like Flakes (5) Microblades (Fig. 9–3: *e–i*) are defined as longitudinally struck flakes from a microblade core that has been prepared specifically to produce parallel edges on an arris (dorsal ridge). Microblade edges showed some variation, but all microblades were struck from the same direction. If complete, microblades would have been at least three times longer than the widths. Only two of the specimens are complete (Fig. 9–3: *f, g*) and only one shows use wear along one edge (Fig. 9–3: *f*). The other specimens are broken at their distal or proximal ends. In contrast, blade-like flakes have multiple arrises and nonparallel edges. One of the blade-like flakes had been retouched along one lateral edge.

Primary Spall (1) This ski-shaped artifact, detached from the face of a microblade core, has a dorsal ridge that has been bifacially retouched. Retouching appears to have facilitated the removal of the spall.

Unifacial Blade or Point (1) This artifact is a thin, tabular piece, leaf-shaped in outline, with a straight base. The lateral edges of the dorsal and ventral sides have been flaked alternately (Fig. 9–3: *l*). This may be an unfinished piece; no wear patterns are evident to indicate its use as a blade or a point.

Burins (9) Three types of flake burins identified in Component I were classified according to shared morphological features and whether they were lateral, transverse, or notched. Both lateral and trans-

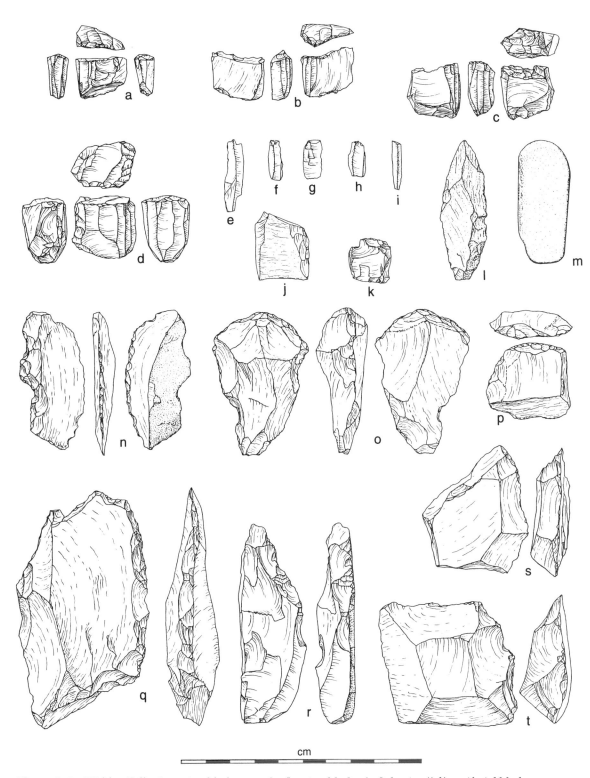

Figure 9–3 Hidden Falls site: microblade cores (*a–d*); microblades (*e–i*); burins (*j, k*); unifacial blade or point (*l*); abrader (*m*); notched scrapers (*n, o*); end scraper (*p*); side scrapers (*q, r*); core scrapers (*s, t*).

verse types have either unprepared platforms prepared by bifacial retouch (Fig. 9–3: *j*), by notching (Fig. 9–3: *k*), or by a transverse burin blow. One specimen, a burin on a microblade (Fig. 9–3: *e*), has two burin facets, one at each end of the right lateral edge. There is evidence of use wear or retouching along the lateral edge just below the platform. Of core burins, the majority have had multiple edges used for spall removal. Several of the burins show use as gravers, as well as wear along the lateral edges of the burin facet.

Scrapers (36)   This artifact class is made up of four types. Notched scrapers (11) (Fig. 9–3: *n, o*) have a well-defined notch or a gorge-like groove. Specimens with grooves exhibit abrading and polish along the depression. Tools with notches have multiple step-fractures, striations, and polish within the notch. In addition, they have been retouched along one or more edges and striations show that they have been used for scraping.

A second type is the narrow-bit scraper or scraper plane (1). This specimen has a narrow bit placed at a right angle to the long axis on a narrow flake. The convex tip, which has been formed by retouching, shows step-fracturing, and striations run along the long axis of the bit.

The third type is the end scraper (5) (Fig. 9–3: *p*). As the name implies, the working edge of the scraper is at one end of the long axis, usually the distal end. This edge has been retouched at a very steep angle.

The remaining specimens can be classed as side scrapers (19) (Fig. 9–3: *q, r*). These tools have been retouched along one edge of the long axis.

Gravers (31)   These tools have a variety of shapes, all having a utilized beaked tip, point, or points. Some have been intentionally formed by pressure retouching; others have a natural point that was used with little or no modification. Wear patterns include rounding, polish, striations, and microchipping along the graver tips.

Flake Cores and Core Remnants (26)   These core types are multidirectional, with multiple platforms on previous flake scars or on a cortex surface that

is prepared by chipping. Morphological categories vary in size and include small, angular fragments, flakes, faceted blocks, and tabular fragments.

Hammerstone (1)   This cobble-size implement shows extensive battering and pitting on one end indicating use.

Choppers (3)   Two types of chopping tools were identified within Component I, cobble and block. The cobble type was made from locally available beach or stream cobbles. These tools were bifacially flaked along one edge by removing flakes alternately from side to side. The block type of chopping tool (1) was produced on meta-quartzite blocks from vein deposits. Modification is similar to that of the cobble tool.

Core Scrapers (2)   Core scrapers (Fig. 9–3: *s, t*) are specialized cores, at least in the unique way in which flakes were removed. Flake shape is predetermined by the preparation of the core before the flake is removed, suggestive of the Levallois core preparation technique. During the initial core reduction, flakes are detached from the core edges in a patterned manner—flakes are removed alternately from one face and then the opposing face. The exhausted specimen resembles a truncated-pyramid–shaped core. Figure 9–3: *s* may be such a primary flake rather than a core remnant.

Split-Cobble and Split-Pebble Tools (5)   Cobbles and pebbles were split or reduced by a blow, producing a cortex fragment. Additional spalls, removed to reduce the thickness at the platform end, allowed the implement to be more easily held in the hand. These split-cobble and split-pebble tools were probably used for scraping, as indicated by limited-use retouch, striations, and polish along one lateral edge.

Abraded and Incised Stone (2)   One specimen is a tabular piece, exhibiting polish on one face and grinding on both faces and one edge (Fig 9–3: *m*). The other specimen is a narrow fragment, having a shallow, incised line across one surface. The incised line is U-shaped in cross section, with the inside of the groove exhibiting striations and polish.

Utilized (53) and Nonutilized Flakes (414) Modification of the utilized flakes was in the form of
edge-rounding, striations, irregular edge-chipping,
hinging, and step-fracturing. Use was probably
based on availability, and the user's need for a
sharp edge or a specific shape determined which
flake was selected.

Three obsidian samples were selected from
nonutilized flakes and subjected to neutron activa-
tion analysis studies conducted by Dr. John Cook
(Davis 1989). Two of the samples were recovered
from unit H2 and one was from the contact zone
between units F and G. One sample from the contact zone between units F and G and one sample
from unit H2 were identified as coming from the
Mount Edziza flows in northern British Columbia.
A remaining sample from unit H2 was identified as
being from an obsidian source on Suemez Island, at
the southern end of the Alexander Archipelago,
west of Prince of Wales Island. The long distances
between the obsidian sources and the Hidden Falls
site suggest that the users either traveled long distances, traded with intermediate groups, or both.

RADIOCARBON DATING

Radiocarbon determinations from Component I
were obtained from 14 samples. Eleven wood samples were from stratigraphic unit I and one charcoal
and two wood samples were from unit H2 (Table
9–2). Radiocarbon determinations provide an estimate for the formation of the palaeosol (unit I) and
the establishment of a mountain hemlock–spruce
forest after the last glacial advance and retreat of
Pleistocene ice. Forest development was terminated
by an early Holocene glacial readvance that over-
rode the site area between 7,900 and 8,600 BP. Radiocarbon ages determined from wood samples range
in age from 10,345 ± 95 (SI-4360) to 7900 (SI-4340) BP.
The one radiocarbon determination from charcoal
provided a date of 9060 ± 230 BP. The sample, which
was associated with the scattered remains of a
hearth, was collected from the base of stratigraphic
unit H2 at the contact with unit I. The hearth consisted of fire-altered rocks that were spatially distributed in an easterly direction along an elliptical

outline. Two of the radiocarbon dates (7175 ± 155 BP
[SI-3777] and 7900 ± 90 BP [SI-43400]) are considered
too recent to date the Holocene glacial readvance.
Based on radiocarbon determination associated
with the pollen cores taken from Muskeg Lake
(Holloway 1989), it is suggested that the occupation
occurred ca. 9,500 BP.

FLORAL AND FAUNAL REMAINS

The only vertebrate faunal remains recovered from
Component I consisted of one unidentified fish
bone from the contact zone between units H1 and I
(Davis 1989; Moss 1989). The bone was recovered
from the general area of the scattered hearth remains. Also recovered at the contact between unit
H1 and I, in the same quadrants as the fish bone
and the scattered hearth remains, were two occurrences of shell (Davis 1989; Erlandson 1989). One
of several heavily weathered fragments of clamshell
was identified as *Thais lamellosa*. A single fractured
valve of clam was identified as possibly *Saxidomus
giganteus*.

Palynological studies were conducted on both
pollen and plant macrofossil samples recovered
from the Hidden Falls site and from selected areas
surrounding the site (Davis 1989; Holloway 1989).
Four sedimentary columns were taken from three
muskegs located within the Kasnyku Bay area. In
addition, sediment samples in the form of columns
and bulk samples from features were collected for
pollen extraction within the archaeological site.

Plant colonization of the archaeological site and
the Kasnyku Bay area was well under way by 10,300
BP with the establishment of a *Tsuga mertensiana-
Picea* forest, including an understory composed of
Alnus and a lush growth of Polypodiaceae. This is
inferred from the microbotanical remains and the
pollen assemblage recovered from unit I. Identifiable plant remains from unit I include intact female
cones identified as those of *Tsuga mertensiana*, several varieties of conifer needles, and a specimen of
Rhytidiadelphus loreus. *Rhytidiadelphus loreus* is widespread in cold temperate regions, associated with
moist, coniferous woodlands. Its presence at ca.
10,300 BP argues for a cool-moist climate and a well-

Table 9–2 Radiocarbon Dates for Component I

Zone	Material Sample	Date in Years *BP*	Date in Years *BC*	Laboratory Number
I	Wood/Diluted	7175 ± 155	5225 ± 155	SI-3777
I	Wood	7900 ± 90	5950 ± 90	SI-4340
I	Wood	8640 ± 70	6690 ± 70	SI-4357
I	Wood	8750 ± 65	6800 ± 65	SI-4356
H2	Charcoal	9060 ± 230	7110 ± 230	Beta-7440
I	Wood	9085 ± 70	7135 ± 70	SI-4353
I	Wood	9290 ± 70	7340 ± 70	SI-4358
I	Wood	9405 ± 75	7455 ± 75	SI-4355
H2	Wood	9410 ± 70	7460 ± 70	SI-3778
I	Wood	9690 ± 70	7740 ± 70	SI-4359
H2	Wood	9860 ± 75	7910 ± 75	SI-3776
I	Wood	$10,005 \pm 75$	8055 ± 75	SI-4352
I	Wood	$10,075 \pm 75$	8125 ± 75	SI-4354
I	Wood	$10,345 \pm 95$	8395 ± 95	SI-4360

Source: Alaska Anthropological Assn. See Acknowledgments.

developed forest (Davis 1989; Holloway 1989). The early Holocene glacial readvance (between 7,900 and 8,600 BP) effectively scoured the unit I soil development except in isolated pockets and in a discontinuous thin palaeosol.

Pollen recovery from unit H1 was unexpected, since the deposit was a glacial till. The distribution of pollen in unit H1 was dominated by *Alnus* and secondarily by *Picea* and *Tsuga mertensiana*. *Tsuga heterophylla* was absent from the lower levels of the unit. Also present, but in lower frequencies, were *Pinus*, Gramineae, *Salix*, *Rosaceae*, and *Ericaceae*. This has been interpreted by Holloway (1989) as resulting from the downward percolation of water through the till during and after the unit was deposited. The pattern of taxa represented in unit H1 is similar to that of early successional communities found either on morainal deposits or growing on stagnant ice covered with a thin layer of soil. The vegetational surface of unit G/H1 corresponded to the 7,400 BP level at Muskeg Lake. The pollen record at Muskeg Lake and the deeply weathered condition of unit G/H1 are interpreted as corresponding to the onset of the Hypsithermal, which began about 7,400 BP with successional forest dominated by *Tsuga mertensiana*, *Picea*, *Alnus*, and Polypodiaceae. Rapid replacement of the earlier forest

species by *Tsuga heterophylla* and *Picea* occurred due to the improving climate during the Hypsithermal (Holloway 1989).

The warming climate of the Hypsithermal coincides with the oscillating occupation patterns at Hidden Falls between 5,000 and 3,200 BP. These intermittent occupations effectively disturbed the understory habitat, yet the forest canopy continued to mature, forming an increasingly more dense stand. The oscillating occupation at the site aided the diversification of the area's herbaceous and shrub understory. Pollen evidence from Kasnyku Bay does not show any clear marker between Hypsithermal and post-Hypsithermal vegetation. The core from Muskeg Lake indicates that there was slightly more moisture between 3,500 and 1,400 BP. There were also periods of local drying conditions, as indicated by the presence of such taxa as *Selaginella* (Holloway 1989). The latest prehistoric occupation at Hidden Falls coincides with the onset of neoglacial conditions (3,000 BP).

OTHER SITES

In addition to Hidden Falls, Ground Hog Bay 2, another early occupation within southeastern Alaska,

is also a multicomponent site on an 8- to 12-m terrace. The site is located at Ground Hog Bay on the
Chilkat Peninsula, at the confluence of Chatham
and Icy straits, at the north end of the archipelago.
The site has been dated to ca. 10,000 BP. Ground Hog
Bay 2, Component II, has similarities with Component I at Hidden Falls in the use and technology
of microblade cores (Ackerman 1968, this volume;
Ackerman et al. 1979). Chuck Lake, now an interior
site, on Heceta Island, west of Prince of Wales Island, has also yielded evidence of a microblade and
microcore technology that dates to 8220 ± 125
(WSU-3241) and 8180 ± 130 BP (WSU-3243). Of great
importance at the Chuck Lake site is a faunal assemblage including shellfish, sea mammals, and
terrestrial mammals (Ackerman et al. 1985; Davis
1990). The lowest levels at the Thorne River site, on
Prince of Wales Island, have produced dates as
early as 7650 ± 160 BP (WSU-3618) and lithic artifacts similar to those at Hidden Falls, Component I
(Holmes et al. 1989). While it can be assumed that
such early sites may occur elsewhere in southeastern Alaska—some in association with uplifted-ter-
race environments—not enough data exist to
describe the frequency or type of settlement pattern.

REFERENCES CITED

Ackerman, R. E. 1968. *Archaeology of the Glacier Bay Region, Southeastern Alaska*. Laboratory of Anthropology Report of Investigations No. 44. Pullman: Washington State University.

Ackerman, R. E., T. D. Hamilton, and R. Stuckenrath. 1979. Early Culture Complexes on the Northern Northwest Coast. *Canadian Journal of Archaeology* 3:195–208.

Ackerman, R. E., K. C. Reid, J. D. Gallison, and M. E. Roe. 1985. *Archaeology of Heceta Island: A Survey of 16 Timber Harvest Units in the Tongass National Forest, Southeastern Alaska*. Center for Northwest Anthropology Project Report No. 3. Pullman: Washington State University.

Clague, J. J., J. R. Harper, R. J. Hebda, and D. E. Howes. 1982. Late Quaternary Sea Levels and Crustal Movements, Coastal British Columbia. *Canadian Journal of Earth Science* 19:597–618.

Davis, S. D. 1979. Hidden Falls, a Stratified Site in South-

——— . 1980. Hidden Falls: A Multicomponent Site in the Alexander Archipelago of the Northwest Coast. Paper presented at the 45th annual meeting of the Society of American Archaeology, Philadelphia.

——— . 1990. Prehistory of Southeastern Alaska. In *The Handbook of North American Indians*, ed. W. Sturtevant. Vol. 7: *Northwest Coast*, ed. W. Suttles, pp. 197–202. Washington, D.C.: Smithsonian Institution.

——— . 1991. A Hypothesis on Prehistoric Migration and Cultural Relationships in Southeastern Alaska. Paper presented at the 18th Alaska Anthropological Association meeting, Anchorage.

Davis, S. D., ed. 1989. *The Hidden Falls Site, Baranof Island, Alaska*. Aurora, Alaska Anthropological Association Series, No. 5. Anchorage.

——— . 1992. Cultural Resources Baseline Study: Polk Inlet Environmental Impact Statement. On file, U.S. Forest Service, Tongass National Forest, Ketchikan, Alaska.

Erlandson, J. M. 1989. Analysis of the Shellfish Assemblage. In *The Hidden Falls Site, Baranof Island, Alaska*, ed. S. D. Davis, pp. 131–58. Aurora, Alaska Anthropological Association Monograph Series, No. 5. Anchorage.

Goldthwaite, R. P. 1966. Evidence from Alaska Glaciers of Major Climatic Change. In *International Symposium on World Climate: World Climate from 8,000 to 0 B. C.*, ed. J. S. Sawyer, pp. 40–53. London: Royal Meteorological Society.

Heusser, C. J. 1960. Late-Pleistocene Environments of North Pacific North America: An Elaboration of Late-Glacial and Postglacial Climatic, Physiographic, and Biotic Changes. Special Publication No. 35. New York: American Geographical Society.

Holloway, R. G. 1989. Analysis of Botanical Materials. In *The Hidden Falls Site, Baranof Island, Alaska*, ed. S. D. Davis, pp. 61–92. Aurora, Vol. 5. Anchorage: Alaska Anthropological Association.

Holmes, C. E., R. J. Dale, and J. D. McMahan. 1989. *Archaeological Mitigation of the Thorne River Site (CRG-177), Prince of Wales Island, Alaska*. State of Alaska, Office of History and Archaeology Report No. 15, Anchorage.

Klein, D. R. 1975. Postglacial Distribution Patterns of Mammals in the Southern Coastal Regions of Alaska. *Journal of the Arctic Institute of North America* 18:7–20.

Lightfoot, R. R. 1989. Cultural Component II. In *The Hidden Falls Site, Baranof Island, Alaska*, ed. S. D. Davis, pp. 199–273. Aurora, Alaska Anthropological Association Monograph Series No. 5. Anchorage.

Mann, D. H. 1983. The Quaternary History of the Lituya Glacial Refugium, Alaska. Ph.D. dissertation, University of Washington.

east Alaska. Paper presented at the 32d annual Northwest Anthropological Conference, Eugene, Oregon.

————. 1986. Wisconsin and Holocene Glaciation of Southeast Alaska. In *Glaciation in Alaska: The Geologic Record*, ed. T. D. Hamilton, K. M. Reed, and R. M. Thorson, pp. 237–65. Anchorage: Alaska Geological Society.

Molina, B. 1986. Glacial History of the Northeastern Gulf of Alaska: A Synthesis. In *Glaciation in Alaska: The Geologic Record*, ed. T. D. Hamilton, K. M. Reed, and R. M. Thorson. Anchorage: Alaska Geological Society.

Moss, M. L. 1989. Analysis of the Vertebrate Assemblage. In *The Hidden Falls Site, Baranof Island, Alaska*, ed. S. D. Davis, pp. 93–129. Aurora, Alaska Anthropological Association Monograph Series No. 5. Anchorage.

Swanston, D. N. 1989. Glacial Stratigraphic Correlations and Late Quaternary Chronology. In *The Hidden Falls Site, Baranof Island, Alaska*, ed. S. D. Davis, pp. 47–60. Aurora, Vol. 5. Anchorage: Alaska Anthropological Association.

GROUND HOG BAY, SITE 2

Robert E. Ackerman

Ground Hog Bay is a locally named, small embayment on the southwestern side of the Chilkat Peninsula about 40 air miles west of Juneau, Alaska. The embayment is protected from the reach of Icy Strait by an offshore reef and headlands to the northwest and southeast. The historic village of Kaxnuwu (Grouse Fort) occupies the southeastern headland while the Ground Hog Bay site 2 (GHB 2) lies upon a marine terrace now set back from the present beach (Fig. 9–4).

PRESENT-DAY VEGETATION

Heavy rainfall supports a forest vegetation dominated by Sitka spruce (*Picea sitchensis*) and western hemlock (*Tsuga heterophylla*) with an understory of Sitka and red alder (*Alnus sinuata, A. rubra*), Alaska blueberry (*Vaccinium alaskaense*), highbush cranberry (*Viburnum edule*), and a floor cover of moss and lichens. Western thimbleberry (*Rubus par-*

viflorus), salmonberry (*Rubus spectabilis*), Pacific red elder (*Samubucus callicarpa*), and devil's club (*Oplopanax horridus*) are found in open areas within the forest cover and along stream banks and the emerging shoreline. In the river valleys, black cottonwood (*Populus trichocarpa*) is a common tree form with willow (*Salix barclayi, S. sitchensis*) or alder as an understory. Poorly drained areas are characterized by muskeg vegetation, e.g., shrubs, sedges, skunk cabbage (*Lysichitum americanum*), and lodgepole pine (*Pinus contorta*).

HISTORY OF RESEARCH

During the last year of an archaeological survey of the Glacier Bay National Park (1963–65) there was sufficient time left in the field season to investigate the cultural history of lands adjacent to the park as a further contribution to the park history. Work was conducted at the historic to late prehistoric Grouse Fort site (site 1) in Ground Hog Bay with the complete excavation of one house pit and the sampling of a late prehistoric midden (Ackerman 1968). Sur-

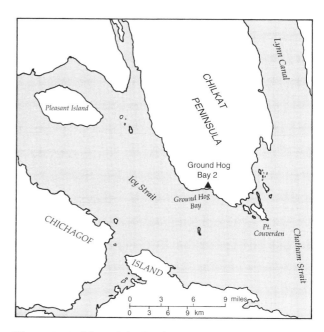

Figure 9–4 Map of the Icy Strait region with the Ground Hog Bay 2 site.

Ground Hog Bay site 2 in middle ground.

vey of the elevated beaches and depositional ter-races north and west of the Grouse Fort site resulted in the recovery of chipped stone material from the eroding face of a terrace along the north shore of Ground Hog Bay. Included among these artifacts was a chert microblade core. A radiocarbon date on charcoal provided a initial date of 10,180 ± 800 BP (Ackerman 1968, 1974; Ackerman et al. 1979). Fur-ther research was conducted at GHB 2 in 1971 and 1973 with funds provided by the National Science Foundation.

SITE CONTEXT

The GHB 2 site is located on a glacio-marine terrace (terrace III, 11–12 m above high tide and 17–18 m above mean sea level) that was formed during a pe-riod of isostatic depression resulting from an ad-vance of a Glacier Bay ice source 11,500 to 11,000 years ago. At this time, till, deposited in a marine environment, covered the bedrock face of the ter-race. Shell incorporated in the till dated to 13,420 ± 130 and 11,630 ± 145 BP (Ackerman et al. 1979). Following local deglaciation the terrace formed the edge of an ancient beach, with the sea depositing layers of sand and gravel over the till surface. Cobbles and boulders from the reworked till surface formed a lag pavement at the back of the beach (Ackerman et al. 1979). Within these beach sands and gravels, a series of cultural occupations dating between 9,200 and 4,200 years ago were un-covered (lower components). Overlying these cul-tural components was a zone of thick forest soil dating between 4,000–1,500 years ago that represented a period when the site was not occupied.

A search was made to find lower terraces that might have been occupied during this time interval. A lower terrace (terrace II) 5–6.5 m above the modern high tide line was found to the west of Ground Hog Bay that dated between 4645 ± 90 BP and 1510 ± 100 BP, the period when terrace III was not occupied. Fire-cracked rock and shell with charcoal were recovered, but no cultural materials were associated. It is likely that the terrace was formed during a period of marine transgression occurring less than 5,000 years ago (Ackerman et al. 1979).

Reoccupation of terrace III at GHB 2 occurred around 900 years ago (component I) and resulted in major alterations to the site integrity. Gravel was brought up from the current beach and spread over the forest floor. Large pits for house posts were dug through the layers of forest humus and beach gravels and on occasion penetrated the underlying till. Considerable mixing of cultural materials occurred in the vicinity of the house post excavations. It is not clear when the site was abandoned in late prehistoric times, perhaps between 400 and 250 years ago when glaciers began expanding in Glacier Bay. The site was reoccupied in historic times as a garden plot, sometime between AD 1848 and 1891 according to local informants.

A list of radiocarbon dates associated with the cultural components and the soil horizons is provided in Figure 9–5.

ARTIFACT ASSEMBLAGE, LOWER COMPONENT

During the field seasons of 1965 and 1971, it appeared that there was a stratigraphic separation between components III and II (Fig. 9–5). Further excavations in 1973 revealed that the separation was less obvious and that it would be better to consider all of the artifacts recovered from the sand/gravel beach deposit dating between 10,000 and 4,200 years ago as a single occupation.

The most diagnostic artifact type in the assemblage is the wedge-shaped, frontally fluted microblade core. Those of obsidian and chert (Fig. 9–6) have a prepared platform created by the removal of platform flakes. No core tablets were recovered, suggesting a conservation of these exotic raw materials during platform preparation. Obsidian appears to have been imported as small nodules from Suemez Island or from Mount Edziza (Fladmark 1985; Erlandson et al. 1992). A frontally fluted microblade core of chert with the base missing has

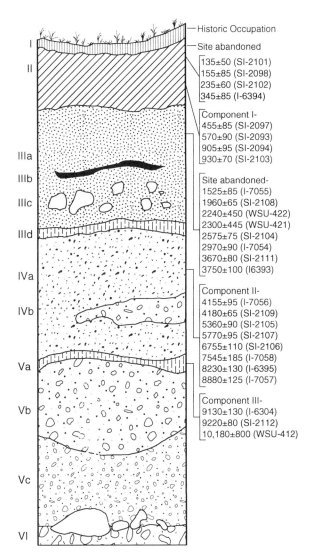

Figure 9–5 Ground Hog Bay, site 2, with radiocarbon dates for terrace III. Stratigraphic profile. Strata descriptions: (I) duff; (II) humus and gravel; (IIIa) dark brown humus; (IIIb) charcoal lens; (IIIc) gravel, pebbles, cobbles in humus; (IIId) brown silt; (IVa) dark brown loam intermixed with beach gravels; (IVb) brown platy gravels; (Va) clay layer; (Vb) reddish brown beach gravels; (Vc) olive-grey sand/gravel; (VI) till. (*Source:* Canadian Journal of Archaeology. *See Acknowledgments.*)

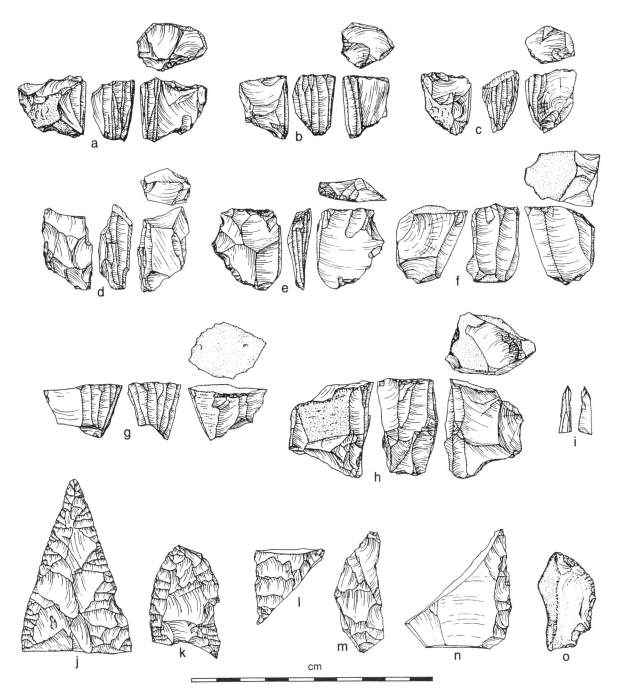

Figure 9–6 Ground Hog Bay, site 2, lower component: wedge-shaped microblade cores (*a–d, f–h*); burin (*e*); microblade with graver-like tip (*i*); biface fragments (*j–m*); side scrapers (*n, o*).

a platform created by the detachment of a single flake. It would appear that the core was created from a split nodule. Argillite microblade cores are also of the frontally fluted type but have blades detached from platforms that retain the original cortex

(Fig. 9–6: *f, h*). The argillite cores appear to have been derived from naturally occurring tabular pieces of raw material which is locally present in large quantity.

Evidence of use retouch on the few microblades

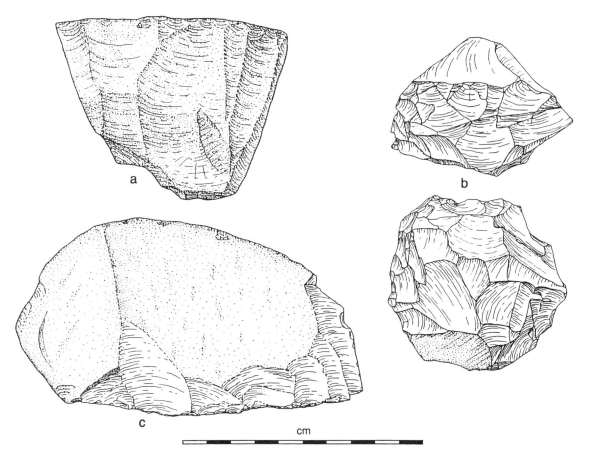

Figure 9–7   Ground Hog Bay, site 2, lower component: argillite blade (*a*) and flake (*b*) cores; andesite cobble chopper (*c*). (*Artifact drawings from originals by Sarah Moore.*)

recovered from the site is rare. A single obsidian microblade (Fig. 9–6: *i*) has been end-retouched to create a graver-like tip.

A burinized flake with a platform prepared by steep transverse flaking on one edge (Fig. 9–6: *e*) is similar to Donnelly burins found in the Denali complex of central Alaska (West 1967, 1981, this volume).

Bifacial fragments (Fig. 9–6: *j–m*) suggest the use of projectile points or knife-like tools in association with the microblade industry. Two obsidian biface fragments shown in Figure 9–6: *l, m* came from the lowest part of the site and indicate that bifacial technology was an early aspect of the complex. The larger triangular biface fragment (Fig. 9–6: *j*) came from the upper part of the component and dated to about 4,200 BP. The bifacial fragment (Fig. 9–6: *k*)

came from the middle of the component where radiocarbon dates suggest an age range of 6,000–5,000 BP.

Side scrapers (Fig. 9–6: *n, o*) as well as notches are part of the lower component assemblage. One of the side scrapers (Fig. 9–6: *o*), from the lowest part of the site, was water-rolled, suggesting an occupation at a time when the beach was still active.

Associated with the previously described tool forms are andesite cobble choppers (Fig. 9–7: *c*), hammerstones of various lithology, and argillite blade and flake cores (Fig. 9–7: *a, b*). Several cores were similar in form to the "horsehoof" core types found in Australia. More than 90 percent of the material recovered from the lower component consists of flakes and shatter derived from argillite cores.

No hearths or evidence of structures were

noted during excavation of the lower component. A random collection of pebbles subjected to study revealed that there might have been some fire alteration of the rock due to heating, but there was no associated charcoal, nor fire-cracked rocks.

CONCLUSIONS

The presence of wedge-shaped, frontally fluted microblade cores with prepared platforms and a Donnelly-type burin at the GHB 2 site suggests an extension southward of tool forms of the Denali complex or Palaeo-Arctic tradition between 10,000 and 9,000 years ago. The discovery of wedge-shaped, frontally fluted microblade cores at the Hidden Falls site on Baranof Island (see regional map) with an estimated age of 9,500 BP (Davis 1989, this volume) provides further support for the early spread of microblade technology into southeastern Alaska.

Additional support for this early age assignment comes from the recovery of microblade core types of a decidedly Northwest Coast type at the Chuck Lake site on Heceta Island with a date of 8,200 BP (Ackerman et al. 1985), at the Thorne River site on Prince of Wales Island dating to 7,500 BP (Dale et al. 1989), and at the Lawn Point and Kasta sites on the Queen Charlotte Islands dating between 5,000 and 8,000 BP (Fladmark 1986). These cores are no longer wedge-shaped as blade removal is carried onto the lateral faces, resulting in a prismatic to conical core form. The absence of these core forms at the GHB 2 and Hidden Falls sites suggests the distribution of a northern microblade core type into southeastern Alaska prior to spread of the Northwest Coast microblade core type from the south some time before 8,000 years ago.

The rapid spread of the northern microblade technology as characterized by the wedge-shaped, frontally fluted microblade core type is indicated by the age cline of this complex from central to southeastern Alaska. The suggested age for the Denali complex in central Alaska has its beginning about 11,000 BP (West 1981, this volume). Moving southward and towards the coast, the wedge-shaped, frontally fluted microblade cores of the Narrows

Phase (9,000–7,600 BP) on the Alaska Peninsula (Henn 1978) may be either equivalent in age or not quite as old as those from the GHB 2 and Hidden Falls sites to the south. As noted earlier (Ackerman 1992), the Anangula blade site complex in the Aleutians (Aigner 1978; McCartney and Veltre, this volume) has been regarded as a cultural continuation of the subsistence patterns established by late Beringian platform populations, but dates somewhat later (8,500 BP) than the early southeastern Alaskan assemblages. The core types and methods of core preparation at the Anangula blade site are technologically similar to those of the Siberian Sumnagin culture complex (Mochanov and Fedoseeva 1984) and hence later than the Dyuktai–Denali cultural assemblages.

Cultural diffusion of the microblade complex from the north thus remains a problem, but the rapid spread of the technological tradition into the Alaskan archipelago suggests the possibility of a coastal transmission. Boats would have been required to reach GHB 2 and Hidden Falls and would have been needed to make use of offshore resources. Within the spruce-hemlock rain forest—established by approximately 10,000 years ago (Holloway 1989)—terrestrial resources would have been inadequate as a resource base even for small groups. Given the maritime adaptations of the entry population, it would seem that a movement of maritime hunters and fishers along the Pacific coast would more likely have been the source of the early southeastern Alaskan culture complexes rather than an expansion of hunter/gatherer peoples down to the coast through the glaciated coastal ranges.

REFERENCES CITED

Ackerman, R. E. 1968. *Archeology of the Glacier Bay Region, Southeast Alaska.* Department of Anthropology Report of Investigations 44. Pullman: Washington State University.

———. 1974. Post-Pleistocene Cultural Adaptations on the Northern Northwest Coast. In *Proceedings, International Conference on the Prehistory and Paleoecology of the Western Arctic and Subarctic,* ed. S. Raymond and P. Schledermann, pp. 1–20. Calgary: University of Calgary Archaeological Association.

———. 1992. Earliest Stone Industries on the North Pacific Coast of North America. *Arctic Anthropology* 29(2):18–27.

Ackerman, R. E., T. D. Hamilton, and R. Stuckenrath. 1979. Early Cultural Complexes on the Northern Northwest Coast. *Canadian Journal of Archaeology* 3:195–209.

Ackerman, R. E., K. C. Reid, J. D. Gallison, and M. E. Roe. 1985. *Archeology of Heceta Island: A Survey of 16 Timber Harvest Units in the Tongass National Forest, Southeastern Alaska.* Project Reports 3, Center for Northwest Anthropology. Pullman: Washington State University.

Aigner, J. 1978. The Lithic Remains from Anangula, an 8,500 Year Old Aleut Coastal Village. *Urgeschichtliche Materialhefte,* Vol. 3. Tübingen: Institut für Urgeschichte der Universität Tübingen.

Dale, R. J., C. E. Holmes, and J. D. McMahan. 1989. *Archaeological Mitigation of the Thorne River Site (CRG-177), Prince of Wales Island, Alaska.* Report Number 15. Anchorage: Office of History and Archeology.

Davis, S. D., ed. 1989. The Hidden Falls Site, Baranof Island, Alaska. Aurora, Vol. 5. Anchorage: Alaska Anthropological Association.

Erlandson, J. M., M. L. Moss, and R. E. Hughes. 1992. Archaeological Distribution and Trace Element Geochemistry of Volcanic Glass from Obsidian Cove, Suemez Island, Southeast Alaska. *Canadian Journal of Archaeology* 16:89–95.

Fladmark, K. R. 1985. *Glass and Ice: The Archaeology of Mt. Edziza.* Department of Archaeology Publication 14. Burnaby: Simon Fraser University.

———. 1986. Lawn Point and Kasta: Microblade Sites on the Queen Charlotte Islands, British Columbia. *Canadian Journal of Archaeology* 10:37–58.

Henn, W. 1978. *Archaeology on the Alaska Peninsula: The Ugashik Drainage, 1973–1975.* Anthropological Papers 14. Eugene: University of Oregon.

Holloway, R. G. 1989. Analysis of Botanical Materials. In *The Hidden Falls Site, Baranof Island, Alaska,* ed. S. D. Davis, pp. 61–92. Aurora, Vol. 5. Anchorage: Alaska Anthropological Association.

Mochanov, Y. A., and S. A. Fedoseeva. 1984. Main Periods in the Ancient History of North-East Asia. In *Beringia in the Cenozoic Era,* ed. V. L. Kontrimavichus, pp. 669–93. New Delhi: Amerind Publishing Co.

West, F. H. 1967. The Donnelly Ridge Site and the Definition of an Early Core and Blade Complex in Central Alaska. *American Antiquity* 32(3): 360–82.

———. 1981. *The Archaeology of Beringia.* New York: Columbia University Press.

ROUND MOUNTAIN MICROBLADE LOCALITY

Douglas R. Reger and Mark E. Pipkin

The Round Mountain microblade locality (49KEN-094) is located on the Kenai Peninsula approximately 90 km south of Anchorage (60°29′22″N, 150°00′47″W). It is situated on a terrace approximately 20 m above the north bank of the Kenai River, just opposite and 0.5 km downstream from the confluence of the Kenai and Russian rivers.

HISTORY OF RESEARCH

Excavation of late prehistoric structures and testing of intervening areas of the Round Mountain locality by the Alaska Division of Geological and Geophysical Surveys (DGGS) during 1984, prior to anticipated road construction, revealed the subsurface lithic scatter which constitutes the microblade locus. Excavations yielded a microblade core, microblades, a transverse burin, burin spalls, core preforms, and numerous flakes (Gibson 1985:110). Further digging at the locality during 1985 resulted in an enlarged collection and a 13-sq.-m total area excavated at that locus.

STRATIGRAPHY AND DATING

The microblade locality is capped by 10 cm of sod overlying 30–40 cm of generally undifferentiated olive brown to yellow brown aeolian silts, which blanket approximately 10 cm of fine colluvial sand intermixed with small pebbles (Fig. 9–8). Immediately below the pebbly sand layer is 10 cm of olive grey sandy silt which caps a reworked glacial till (Pipkin 1989:18).

A single radiocarbon sample which consisted of many small, widely scattered fragments of charcoal produced a date of 1925 ± 145 years before present (WSU-3093), AD 25. Association of that result with the microblade and core collection is highly unlikely based on other archaeological evidence for

The Round Mountain site locality. (*Photo courtesy of Frederick Hadleigh West.*)

the area. The sample probably dates a natural event such as a forest fire (Gibson 1985:111).

The core and blade materials occur just above the contact of the aeolian silts and the lower pebbly sand which is thought to be colluvial. Interpretation of the geologic stratigraphy suggests the aeolian silts were probably deposited after deglaciation (14,500 BP, Rymer and Sims 1982) and before forestation of the stream valley covered the silt source. Boreal forests were established in the area by 7,800 years ago (Ager and Brubaker 1985). That interpretation suggests the occupation should date to early Holocene time.

No faunal or floral remains were recovered in

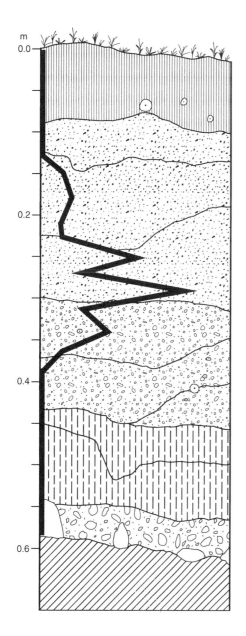

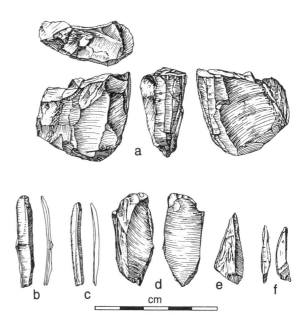

Figure 9–9 Round Mountain site locality: wedge-shaped microblade core (*a*); microblades (*b, c*); burins (*d, e*); burin spall (*f*).

association with the core and blade locus. No features were observed during excavation of the Round Mountain microblade locality. The assemblage consisted of one lithic scatter approximately 4 m in diameter. No spatial patterning within the scatter was discerned.

ARTIFACTS

The assemblage from the locus consisted of 650 items, 80 percent of which are flakes or nondiagnostic chunks of chert or chalcedony. Nine complete microblades, 94 microblade fragments, and a wedge-shaped microblade core were recovered (Fig. 9–9). The core was manufactured on a thick flake, has a bifacially shaped keel extending from the fluted face to the unmodified back of the core, and a single striking platform that shows numerous platform rejuvenation flake scars. One rejuvenation flake fits on the core platform, and two core tablets

	Sod		Olive-grey clayey sandy silt (alluvium?)
	Generally undifferentiated olive-brown to yellow-brown sandy silts (aeolian?)		Reworked glacial till
	Olive-brown to light yellow silty fine sand intermixed with small pebbles (colluvium)		Below limit of excavation
		⊙	Roots
			Relative vertical artifact distribution

Figure 9–8 Round Mountain. Stratigraphic profile. Vertical artifact distributions shown.

from other cores were also found. Two marginally fashioned preforms for microblade cores were also recovered.

A transverse burin with a notch adjacent to the burin facet occurs in the collection along with a burin-faceted rejuvenation flake. Four secondary burin spalls, one possibly retouched on the proximal end, were recovered. Four boulder spalls and a hammerstone of locally available sedimentary rock were also recovered.

Summary

The diagnostic elements of the assemblage show the greatest similarities to assemblages assigned to the Denali complex (West 1967, 1981, this volume). This is further supported by a statistical analysis of the microblade width similar to that outlined by Cook (1968) which shows that this sample has its greatest affinities to interior Denali assemblages (Pipkin 1989). Although not positively dated, the position of the recovered artifacts in the aeolian silts suggests that the assemblage dates to the early Holocene and most likely between 10,000 years BP, when the glacial retreat probably reached an extent where human colonization of the Kenai Peninsula was possible, and ca. 8,000 years BP, when the formation of the boreal forest occurred in the area.

References Cited

Ager, T. A., and L. Brubaker. 1985. Quaternary Palynology and Vegetation History of Alaska. In *Pollen Records of Late Quaternary North America*, ed. V. M. Bryant and R. G. Holloway, pp. 353–84. Dallas: American Association of Stratigraphic Palynologists Foundation.

Cook, J. P. 1968. Some Microblade Cores from the Western Boreal Forest. *Arctic Anthropology* 5(1):121–27.

Gibson, D. 1985. Excavation Results for KEN-092 and KEN-094. In *Progress Report, Project F-021-2(15)/ (A09812) Sterling Highway Archaeological Mitigation: Phase 1 Excavation at Four Sites on the Kenai Peninsula,* ed. C. E. Holmes, pp. 79–140. Public Data File (PDF) 85-04, Alaska Division of Geological and Geophysical Surveys, Fairbanks.

Pipkin, M. E. 1989. The Assemblage from the Round Mountain Microblade Locality 49KEN-094, Kenai Peninsula, Alaska. Master's thesis, University of Oregon.

Rymer, M. J., and J. D. Sims. 1982. Lake Sediment Evidence for the Date of Deglaciation of the Hidden Lake Area, Kenai Peninsula, Alaska. *Geology* 10:314–16.

West, F. H. 1967. The Donnelly Ridge Site and the Definition of an Early Core and Blade Complex in Interior Alaska. *American Antiquity* 32(3):360–82.

———. 1981. *The Archaeology of Beringia.* New York: Co-

BELUGA POINT

Douglas R. Reger

The Beluga Point site (49ANC-054) is located on a low rocky point on the north shore of Turnagain Arm of Cook Inlet, 27.4 km southeast of Anchorage, Alaska (61°00'29"N, 140°41'40"W). The rocky point extends south from the shoreline. The site area has been isolated from the original hillside by rock quarrying and separated into two halves by a bulldozed cut. The cultural deposits in the northern part of the site, which contained the early core and blade component, are being eroded by high tides.

History of Research

The site was discovered and briefly tested during 1975 by the writer. A core and blade component, later designated the Beluga Point North I (or BPN-I) component, was identified. Reger further tested the northern end of the site during 1976 and documented three prehistoric and one historic component. Field school excavations during 1977 by Reger and William B. Workman confirmed the limited areal extent of the core and blade component and extension of the later occupations into the southern half of the site (Reger 1981). The area of the site containing core and blade material has been completely excavated.

Beluga Point site, looking south across Turnagain Arm, Cook Inlet.

STRATIGRAPHY AND DATING

Sediments in the site derived from windblown silts and sands which apparently originated as intertidal sediments to the east of the site. Prevailing westerly winds deposited the sediments on the lee side of a rock knob. The shallow depth to bedrock caused fluctuating groundwater levels to stain the sediments and, combined with forest soil development, mask some detail of the complex site stratigraphy (Fig. 9–10). Later cultural components mixed with the core and blade occupation in several limited areas, indicating probable deflation of site sediments after the early occupation had been buried under aeolian sediments. The core and blade component was buried under an average of about 50 cm of sediment.

No radiocarbon-datable material was collected for the core and blade occupation; dating of the collection depends on typological comparisons with dated collections. Technological traits compare with Denali-related collections in the Tangle Lakes and southwestern Alaska areas. Those comparisons suggest an age for the Beluga Point material of ca. 7,000 to 8,000 years ago.

No faunal remains were recovered with the core and blade component at Beluga Point and no pollen samples were collected. No features were found during excavation and no distribution pattern of artifacts was apparent at the locality.

ARTIFACTS

The northern part of the site yielded three stratigraphically distinct prehistoric cultural levels in ad-

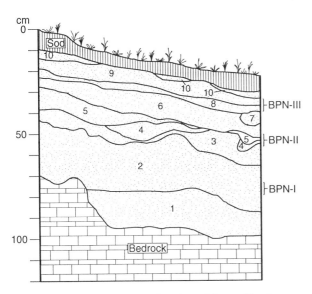

Figure 9–10 Beluga Point. Stratigraphic profile. Strata descriptions: (1) fine sand, olive brown, mottled with greyish brown; (2) very silty fine sand, dark yellowish brown; (3) fine sand, olive brown; (4) fine-to-medium sand, dark greyish brown; (5) fine sand, olive brown; (6) fine-to-medium sand, dark greyish brown; (7) fine sand, dark yellowish brown; (8) silty very fine sand, dark brown; (9) very fine sandy silt, very dark greyish brown; (10) silty very fine sand, strong brown, volcanic ash?

dition to a recent historic layer. The lowermost cultural level, BPN-I, was vertically separated from the other two levels and was horizontally restricted to a very small remnant of an obviously larger initial use area.

BPN-I is the single component at Beluga Point that shows undeniable prepared core and blade stone technology accompanied by bifacial preparation of cutting or thrusting implements (Fig. 9–11). Several grooved, coarse-grained stone flakes indicate sharpening or smoothing of organic tools. Core and blade technology is demonstrated in the BPN-I collection by presence of a platform-and-face fragment of a core, 4 platform rejuvenation flakes, 13 microblades, and 3 blade-like flakes.

The core fragment, reconstructed from frost-rived fragments, was side-struck from the face of a large core at an initial stage of blade or blade-like

flake removal. Too little of the chert piece remains to give a hint of the core shape or any size range. Three of the four platform rejuvenation flakes appear to have been struck from a large core (or cores) of indeterminate shape or total size. Morphology of several rejuvenation flakes suggests relatively narrow cores with blades struck from a narrow end.

The 13 incomplete microblades have a mean width of .74 ± .17 cm. The microblades range from .44 to 1.05 cm wide. The most complete specimen lacks only a small part of the distal end and measures 3.5 cm in length. It has three dorsal facets, is flat in cross section, and shows virtually no longitudinal curvature. This and several other microblades suggest removal from a fairly wide and flat core face. Several specimens have been retouched. Two blade-like flakes were also recovered.

The single bifacial preform in the BPN-I component was fashioned by nonparallel flaking.

SUMMARY

The core and blade component from the Beluga Point site cannot be firmly dated and does not include enough diagnostic artifacts to confidently assign cultural affiliation. Reassembled core fragments illustrate a wide, relatively flattened but still wedge-shaped core face. The striking platforms of the reconstructed cores were apparently formed or

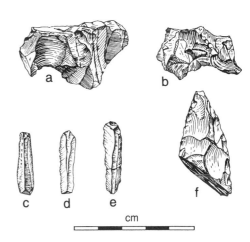

Figure 9–11 Beluga Point site: reconstructed microblade core fragments (*a, b*); microblades (*c–e*); biface tip (*f*).

rejuvenated by a series of side-struck and face-struck blows. Other forms of cores are indicated by core tablets struck from the narrow margin where blades had been removed.

The reconstructed core attributes, microblade width, and biface suggest cultural affiliation with collections from the Long Lake site, ANC-017 (Bacon 1975, 1978; West 1975: Fig. 1; Reger and Bacon, this volume); the Igiugig site, ILI-002 (Dixon and Johnson 1981); the Koggiung phase component at the Graveyard Point site, NAK-018 (Dumond 1981:109); and Tangle Lakes area sites such as Mt. Hayes 122 (West 1981:123, this volume).

REFERENCES CITED

Bacon, G. 1975. Preliminary Testing at the Long Lake Archaeological Site. Paper presented at the 2d annual Alaska Anthropological Association Meeting, Fairbanks.

———. 1978. The Denali Complex as Seen from Long Lake, Southcentral Alaska. Paper presented at the 5th annual Alaska Anthropological Association Meeting, Anchorage.

Dixon, R. G., and W. Johnson. 1981. A Core and Blade Site on the Alaska Peninsula (ILI-002). *Miscellaneous Publications, History and Archaeology Series* 29:144–59. Anchorage: Alaska Division of Parks, Office of History and Archaeology.

Dumond, D. E. 1981. *Archaeology on the Alaska Peninsula: The Naknek Region, 1960–1975.* Anthropological Papers No. 21. Eugene: University of Oregon.

Reger, D. R. 1981. *A Model for Culture History in Upper Cook Inlet, Alaska.* Ph.D. dissertation, Washington State University; University Microfilms, Ann Arbor.

West, F. H. 1975. The Alignment of Late Palaeolithic Chronologies in Beringia. Paper presented at the All-Union Symposium "Correlation of Ancient Cultures of Siberia and Adjoining Territories of the Pacific Coast," Novosibirsk.

———. 1981. *The Archaeology of Beringia.* New York: Columbia University Press.

LONG LAKE

Douglas R. Reger and Glenn H. Bacon

The Long Lake site (49ANC-017) is located at the west end of Long Lake, Mile 86, Glenn Highway, approximately 110 km ENE of Anchorage. The site is a series of loci scattered on knolls and terraces overlooking the lake at an elevation of 460 m. A core and microblade collection was recovered from Locality 1, which is situated on a terrace 250 m west of the lake. The terrace stands 10 m above the nearby lake shore.

HISTORY OF RESEARCH

The locality was discovered by Mr. and Mrs. Gene Chapman of Anchorage. Following their report, F. West and D. Reger made a brief reconnaissance confirming that there were several artifact loci present. Greg Dixon revisited the site with West in 1973 and excavated a 1-sq.-m test pit, hoping to find charcoal. The first published account of the site was by Dixon and Johnson (1973), who recovered a microblade core from a depth of 5 cm. The Dixon–West excavation, meanwhile, resulted in a date of 6606 ± 115 BP (UGa-949) which presumably dates the occupation (West 1975: Fig. 1). Locality 1 (Dixon and Johnson Area E) was investigated by Glenn Bacon during 1974, and seven more pits were excavated (Bacon 1975). An additional 254 flakes and tools were recovered from the locality during 1975. No additional excavation has been accomplished at Locality 1, although others along the lake shore have been investigated.

STRATIGRAPHY AND DATING

The 1975 investigation by Bacon documented a soil profile of aeolian silts overlying gravel and till. The sparsely vegetated locality has 10 cm of yellow loess overlying 20 cm of brown loess. The two loess units blanket a 10-cm-thick gravel layer which overlies the basal till. That profile thins to the eastern part of the excavation. Artifacts from Locality 1 high in the profile were found on edge and at various levels (Bacon 1975:5).

Bacon related the technology and typology of the microblade core with cores assigned to the Denali complex of interior Alaska (Bacon 1978). West has informally questioned the association of the radiocarbon sample with the core and microblade assemblage (personal communication to D. Reger, 1979).

No faunal or floral evidence was recovered from the site nor were there any features. No distributional pattern of the flakes and tools was noted, and flakes on edge indicate displacement from the original positions.

ARTIFACTS

The assemblage from Locality 1 includes a complete blade core, core tablets, core fragments, blades, bifaces, scrapers, retouched flakes, and waste flakes (Fig. 9–12). The microblade core was formed by removal of large flakes to create a roughly wedge-shaped form. The keel of the core has been shaped by secondary removal of flakes along the keel edge. Fragments of the tails of other cores demonstrate bifacial shaping more plainly than does the complete core. The face of the core is broad, with only minimal curvature where the blades have been removed. The platform was formed by removal of multiple flakes from the sides and front. One core tablet struck from the face of a core removed the top of the core from the face all the way to the tail.

A large blade-like flake struck from the front of a core removed much of the face and the front part of the keel. That piece was subsequently used as a scraper, resulting in removal of small flakes around the edge of the tool.

Widths of 206 blades or blade fragments were measured to compare with other microblade collections (Bacon 1975:6). The sample consists of 12 complete blades, 106 medial fragments, 57 proximal fragments, and 31 distal fragments. Complete blades average 11.13 mm wide and the total sample averages 9.63 mm wide. Blade widths vary between

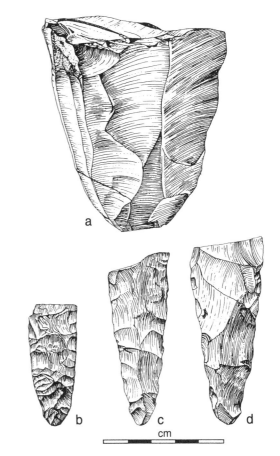

Figure 9–12 Long Lake site: large wedge-shaped microblade core (*a*); bifaces (*b–d*).

3.5 mm and 32.8 mm. A small number of very wide blades in each category skews the average toward the high side of the width curves.

Bifaces in the collection are randomly flaked and have rounded bases. One thick biface appears to be created from a large, blade-like flake.

SUMMARY

The basic technology of blade production and core maintenance in the Locality 1 collection compares favorably with the technology revealed in Denali complex–related collections. The widths of the blades and wide core face suggest temporal placement late in the period during which the Denali complex occurred or perhaps near the limits of typological definition for the complex. The wide blades may indicate closer ties with Anangula or other southwest Alaskan collections.

REFERENCES CITED

Bacon, G. 1975. Preliminary Testing at the Long Lake Archaeological Site. Paper presented at the 2d annual Alaska Anthropological Association meeting, Fairbanks.

———. 1978. The Denali Complex as Seen from Long Lake, Southcentral Alaska. Paper presented at the 5th annual Alaska Anthropological Association meeting, Anchorage.

Dixon, R. G., and W. F. Johnson. 1973. Survey of the Prehistoric and Historic Value of 48 Waysides of the Alaska State Park System. Ms. on file, Alaska Division of Parks, Office of History and Archaeology, Anchorage.

West, F. H. 1975. The Alignment of Late Palaeolithic Chronologies in Beringia. Paper presented at the All-Union Symposium "Correlation of Ancient Cultures of Siberia and Adjoining Territories of the Pacific Coast," Novosibirsk.

RAVINE LAKE LOCALITY

Brian S. Robinson, Frederick H. West, and Douglas R. Reger

The site area lies just west and south of Ravine Lake, a small (ca. 0.6 km) lake with an east-west axis. About 3 km north is the much larger Long Lake. These lakes occupy a high valley, trending east-west, which may represent an ancient course of the Matanuska River. The broad, braided channel of the Matanuska lies less than 2 km south of the Ravine Lake valley but is separated from it by the high ridge that forms the south wall of the Ravine Lake valley. The river lies at an elevation of ca. 273 m; Ravine Lake, at ca. 576 m. The Matanuska River flows from the still-active Matanuska glacier some

15 km to the northeast. The narrow Matanuska valley forms, at this point, a divide between the Chugach Range to its south and the Talkeetna Mountains to the north. Ravine Lake is in the foothills of the Talkeetnas.

The Ravine Lake site locality is unique in the nature of its occurrence. Most interior Alaska sites are on prominences of one kind or another, with good views into some extensive country below. Rightly or wrongly, these are generally taken to represent game lookouts, in addition to whatever other activities are represented. The main site area here is on a small esker-like feature on the south side of this constricted valley. It is separated by a low swale from the land behind (south) which rises fairly steeply to about 636 m. Across the valley—a distance of under 300 m—a nearly vertical wall ascends about 181 m above the valley floor. As will be brought out below, there is little question but that it was the ready accessibility of quarry material that brought people to this place.

The Ravine Lake–Long Lake region is mountainous. Treeline is about 727 m. White spruce, birch, and aspen are common, as are alder and willow. Principal among large mammals are caribou, moose, and fairly abundant grizzly bears.

HISTORY OF RESEARCH

The Ravine Lake locality was discovered in the mid-1960s by Mr. and Mrs. Gene Chapman, then of Anchorage, who generously donated their collections to Alaska Methodist University. In several visits, surface collections were made by them and a series of small excavations carried out. The collecting was somewhat selective, favoring bifaces and more readily recognized forms. Unfortunately, while the bulk of the collection came from one site (49ANC24), some smaller sites in the vicinity were also collected so that the integrity of that designated 49ANC24 is compromised. Lengthy study of the collection and conversation with the collectors led to the conclusion that the greater part was, indeed, from one site. Following descriptions provided by the Chapmans, limited excavations (4 sq. m) were carried out by West in 1968 and again in 1973 by West and Dixon (1 sq. m). This site is designated

View north from site locality.

49ANC25 and it is believed to be within the confines of the main site. Compounding the provenience problems is the fact that the site—situated on a low ridge against the south wall of the valley—had been partially destroyed and some part lost due to the construction of a power line and access road following the ridge line. Clearly, definition of this collection as an assemblage must be viewed with the cautions noted above in mind.

The locality is underlain by the Matanuska Formation, composed of alternating layers of sandstone and shale interbedded with carbonaceous siltstone. Most of the artifacts are made of the siltstone. Gabbro and diabase dikes within the Matanuska Formation may have provided sources for some of the other artifact materials employed (Capps 1961; Ellanna and Buchmeier 1968).

The 1968 and 1973 excavations revealed a podzol consisting of the sequence, A1 (8 cm) black humus; A2 (2 cm) grey ashy layer; B (10–15 cm) yellow-brown zone with grey mottling. Approxi-

mately 7 cm into the B horizon there was a discontinuous, thin tan layer. Artifacts and charcoal were concentrated between the surface of the B horizon and the tan layer. The underlying C horizon was rocky.

In one test unit, 90 percent of the flakes (284/312) came from the B horizon, with the remainder largely displaced to the top of the A horizon by construction activities. Artifact recoveries from the 1968 and 1973 testings include: 2 refitted fragments of a large biface, 2 primary biface fragments, 2 probable large blade core face fragments, 5 large irregular blades, 4 unifacially flaked edge fragments, 3 modified flakes, 1 large heavily battered cobble hammerstone, and 777 flakes. All of the artifacts except for one basalt blank and one jasper flake were made from the locally available siltstone. The excavated materials conform in density and character to those recovered by the original collectors, confirming ANC25 as the probable source for the bulk of their collection.

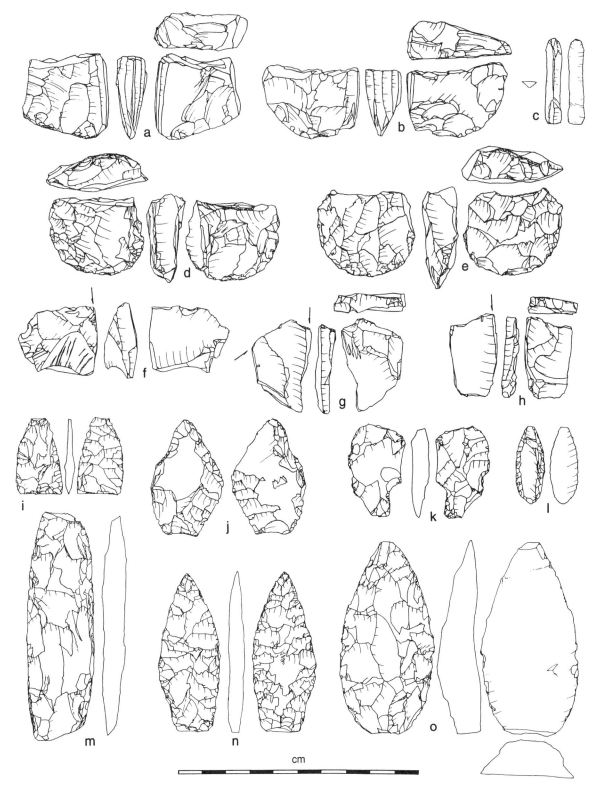

Figure 9–13   Artifacts from the Ravine Lake site in the Chapman collection: wedge-shaped microblade cores (*a, b*) and preforms (*d, e*); microblade (*c*); burins (*f–h*); bifacial points (*i–k, m, n*); limace-like unifaces (*l, o*).

ARTIFACT ASSEMBLAGE

The particular value of this collection lies in the character of the early reduction material rather than in isolated artifact types. The large amount of debitage (over 2,000 flakes) indicates that collecting was comprehensive, but discussions with the Chapmans suggested that certain elements, among them large blocky cores and microblades, were left in the field.

This collection includes 185 flaked tool fragments, 32 blades, and 5 ground stone tools. The flaked tool category includes 131 bifaces and biface fragments, 4 wedge-shaped microblade cores and core preforms, 1 core tablet fragment, 2 blade core parts, 3 burins, 1 possible burin, 4 end scrapers, 4 side scrapers, 4 limace-like lenticular unifaces, 15 unifacially retouched fragments, 9 modified flakes, 2 *tci thos*, and 5 irregular flake cores. Of 32 blades, only 1 may be termed a microblade (Fig. 9–13: *c*). Debitage is predominantly of large size, including large biface edge and overshot flakes measuring up to 18 cm in greatest dimension.

The wedge-shaped cores (Fig. 9–13: *a, b*) and burins (Fig. 9–13: *f–h*) are typical of the Denali complex (West 1967, 1981, this volume). The 2 Denali core preforms (Fig. 9–13: *d, e*) were the first clearcut specimens found in any assemblage. These were recognized by D. R. Reger. Other forms such as lenticular bifaces, burins, large blades, scraper planes or carinate scrapers are also elements that commonly occur in Denali.

In addition to the one microblade, 31 large blades (Fig. 9–14: *f–h*) and blade fragments (average length, 69 mm; width, 25 mm) were isolated from a larger sample of blade-like flakes. The character of the original cores is unknown, however. One core fragment suggests parallel removals from one edge of a tabular core. There is at least 1 large crested blade (Fig. 9–14: *h*) suggesting the presence of a large wedge-shaped core in the size range of the Mt. Hayes 122 specimens (West, this volume).

Of 131 bifaces and fragments, 11 (8%) were classified as "well-thinned." Fifty-seven specimens (44%) were classified as at an intermediate level of reduction and shaping, and 29 (22%) as crude or at an initial stage of testing or reduction. Material types include siltstone (74%), chert (14%), chalced-

ony (1%), sard or jasper (4%), basalt (2%), and felsite (7%). In addition to the siltstone, much of the chert, basalt, and felsite may be of local origin.

A higher proportion of the well-thinned bifaces are of lustrous chert or chalcedony (64%), followed by siltstone (27%) and felsite (18%). Among the thinned bifaces are 2 lenticular specimens with well-ground basal edges (Fig. 9–13: *m, n*), a small triangular specimen (Fig. 9–13: *i*), a small pentagonal form (Fig. 9–13: *j*), and a fragment of a broadly side-notched or stemmed point (Fig. 9–13: *k*). The first four of these are not particularly diagnostic—occurring in both early and late contexts—while the fragmentary side-notched form is later in time. There is also one finely made small biface of chalcedony interpreted to be an inset for a composite tool.

More characteristic of the site is the large number of thick, simple bifaces of rather varied outline. These may represent finished tools, intermediate stages of reduction, or bifacial cores. Of 20 complete specimens, 4 were ovate (Fig. 9–14: *e*), 13 had rounded bases (Fig. 9–14: *a–c*), 2 were bipointed (Fig. 9–14: *d*), and 1 was triangular. They are comparatively large in size, ranging from 54 to 165 mm with an average of 92 mm.

At least 6 complete ovate bifaces may be interpreted as biface-cores (Fig. 9–14: *k–n*), in addition to numerous probable fragments of the same. These are reminiscent of the large biface-cores from the Akmak site (Anderson 1970). These bifaces have steep, unifacial platform preparation for flake removals from the opposite face. On several specimens (Fig. 9–14: *l, n*), opposing edges are unifacially flaked on opposite sides, with broad flakes also removed from both sides. This technology produces broad thin flakes without creating a biconvex cross section, such as would be more typical of bifacial tool reduction.

Four chopper-like bifaces have well-shaped edges and unworked backs. Several additional bifaces exhibit broadly flaked surfaces employing unprepared joint fractures as platforms. The biface category includes a range of functional categories and preform stages.

There are 27 uniface tools and fragments (the latter mostly representing reduction waste): 2 end

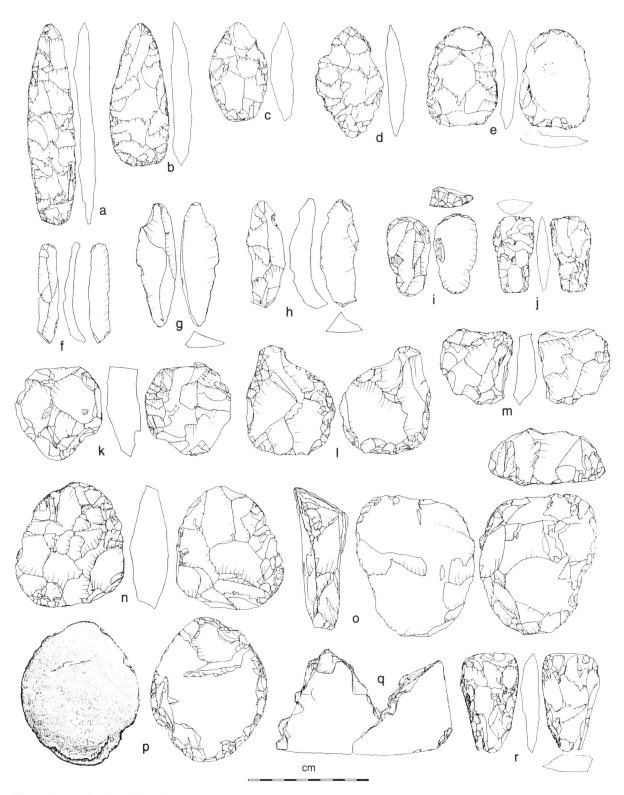

Figure 9–14   Artifacts from the Ravine Lake site in the Chapman collection: intermediate stage bifaces (*a–e*); large blades (*f, g*); crested blade (*h*); end scraper (*i*); biface scraper (*j*); biface cores (*k–n*); large uniface or scraper plane (*o*); *tci tho* (*p*); ground slate knife (*q*); bit-ground adze (*r*).

scrapers (Fig. 9–14: *i*), 2 large end-scraper or scraper-plane-like tools (Fig. 9–14: *o*), 4 side scrapers, 1 stemmed biface-scraper (Fig. 9–14: *j*), and 3 limace-like specimens in a variety of sizes (Fig. 9–13: *l, o*). The latter are well formed on thick flakes, with steep unifacial retouch around most or all edges of the lenticular form.

A small number of ground stone tools suggest a late presence in the area. They include a flaked adze with a ground bit (Fig. 9–14: *r*), a notched pebble, and 2 refitted fragments of a well-made banded slate knife resembling an ulu (Fig. 9–14: *q*). Considering the Denali complex tools, the side-notched point, the chalcedony insert blade, and the ground stone tools, it is apparent that the site region is multicomponent and that characterization is restricted to diagnostic artifacts, the number of which is comparatively small. The potential significance of the collection probably lies mostly in the technological attributes of an initial reduction or quarry site, and the relationship of such sites to other activity area types.

SUMMARY

The Ravine Lake site may be predominantly of the Denali complex, representing a lithic source area at which a good deal of initial reduction was carried out. The artifact assemblage is dominated by early stage reduction (of bifaces and cores) that is generally absent at most Denali sites, but that occurs at sites such as Whitmore Ridge in the Tangle Lakes (West et al., this volume). Although the temporal context of the Ravine Lake collection is not well controlled, the assemblage suggests the contrasts that may stem from activity differences within a technological tradition.

REFERENCES CITED

Anderson, D. D. 1970. Akmak: An Early Archaeological Assemblage from Onion Portage, Northwest Alaska. *Acta Arctica*, Fasc. 16. Copenhagen.

Capps, S. R. 1961. *Geology of Upper Matanuska Valley.* U.S. Geological Survey Bulletin.

Ellanna, L., and D. Buchmeier. 1968. The Ravine Lake Site: A Preliminary Analysis and Classification of

Specimens Submitted. Unpublished report, Alaska Methodist University, Anchorage.

West, F. H. 1967. The Donnelly Ridge Site and the Definition of an Early Core and Blade Complex in Central Alaska. *American Antiquity* 32:3:360–82.

———. 1980. Late Palaeolithic Cultures in Alaska. In *Early Native Americans*, ed. D. L. Browman. World Anthropology. The Hague: Mouton.

———. 1981. *The Archaeology of Beringia.* New York: Columbia University Press.

ANANGULA CORE AND BLADE SITE

Allen P. McCartney and Douglas W. Veltre

The Anangula core and blade site is located on the narrow, southwestern end of Ananiuliak Island, which lies approximately 7 km north-northwest of Nikolski village, southwestern Umnak Island. The large site measures 210 to 300 m long by 75 to 110 m wide, covering an estimated area of between 15,000 and 25,000 sq. m. The site rests on a low plateau whose surface ranges between 12 and 18 m above mean sea level (Black 1976).

PRESENT-DAY LANDSCAPE

This subarctic island has a gently rolling terrain, rocky coastline, and moist tundra vegetation on the lower elevations and alpine tundra on the hilltops. Volcanic ash soils blanket the island, and the ash stratigraphy is over 2 m deep at the core and blade site. Artifacts have been found exposed on the surface of two large and several small blowouts, which are produced by deflation along the northwestern site margin. Additional blowouts are found adjacent to cliff margins at higher elevations. Rabbits, introduced in the 1930s to feed foxes that were trapped by Nikolski Aleuts, burrow into the blowout areas, while seabirds—gulls, kittiwakes, cormorants—nest along the rocky cliffs and puffins burrow into high soil banks above the shore. Foxes

Anangula site locality, looking into Bering Sea.

were removed after World War II to protect the seabird colonies.

SITE LOCALITIES

There are two major sites located on Ananiuliak Island: the ancient core and blade site, and the more recent "village" site, a midden site of some two dozen surface depressions, luxuriant vegetation, and thick midden deposits. The latter site is located approximately .5 km northeast of the core and blade site, on a high (22 m), east-facing bluff. A single radiocarbon date for the village site of 995 BP (Laughlin 1975) is compatible with age estimates for similar eastern Aleutian coastal midden sites, most of which date within the past 1,000–3,000 years.

A smaller third site, the "Neck" or Ikchigh site, is reported by Laughlin (1975) to be located near the northeastern end of Ananiuliak but has not been described.

HISTORY OF RESEARCH

W. S. Laughlin and A. G. May discovered the core and blade site in 1938, while serving on A. Hrdlicka's last Aleutian expedition (Laughlin 1951). In 1952, Laughlin, G. Marsh, P. Spaulding, and A. Ermeloff revisited the site to make a surface collection at the blowouts and to place several test pits into their surfaces (Laughlin and Marsh 1954). While a collection of 1,650 pieces was assembled, the surface artifacts were not tied to *in situ* artifacts in the surrounding stratigraphy. During the 1950s, J. L. Giddings recognized the Mesolithic and Asian character of these core and blade artifacts and, thus, the archaeological importance of the site. This 1952 collection has been studied more recently by M. Pittenger (1986).

In 1962, under the aegis of the University of Wisconsin Aleut–Konyag Prehistory Project, R. F. Black, Laughlin, C. G. Turner II, G. Denniston, and McCartney visited the site to establish the placement of the blowout artifacts in the undisturbed ash

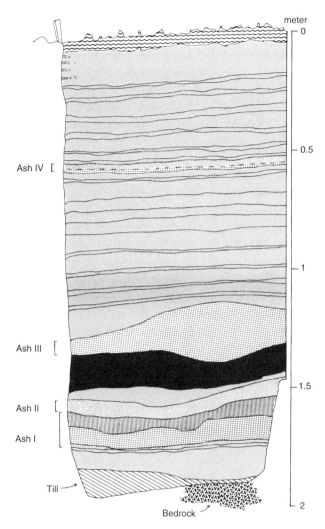

Figure 9–15 Anangula. Stratigraphic profile. Shown are multiple levels of volcanic ash (four of which are widely recognized) and the artifact-bearing stratum (black). (*Source:* Arctic Anthropology. *See Acknowledgments.*)

stratigraphy. A profile was excavated adjacent to one of the blowouts, the cultural horizon was found at the bottom of the profile, and Black and Turner collected charcoal samples which were subsequently dated to 7,660 and 8,425 radiocarbon years ago (Black and Laughlin 1964). In 1963, McCartney, M. Yoshizaki, and R. Nelson returned to expand this profile into a 4 × 12 m trench in the blowout (Fig 9–15; McCartney and Turner 1966a). They also dug five 2 × 2 m test pits outside the blowouts to establish the site's limits. They found layered char-

coal lenses in part of a depression that may represent a house floor.

In 1970, a University of Connecticut crew excavated a 110-sq.-m area at the site as part of a southwestern Umnak project directed by Laughlin, Aigner, and Black. Portions of six depressions or "houses" were found within this trench (Aigner 1974, 1976a). In 1972 and 1973, Laughlin led additional University of Connecticut crews to excavate along the southeastern edge of the site (3 × 8 m and 1 × 6 m trenches) and at the southwestern edge (2 × 5 m trench; S. Laughlin et al. 1975).

In 1974, the most recent excavations at the site and the village site were made by a joint University of Connecticut–Soviet expedition directed by Laughlin and A. P. Okladnikov of the Siberian Branch, U.S.S.R. Academy of Sciences (Laughlin 1975; Okladnikov and Vasilievsky 1976; Okladnikov 1979). Approximately 87 sq. m were excavated at the core and blade site that year.

Site Characterization and Dating

The core and blade site is a single-component but extensive early Holocene locality. Cultural materials are limited to a 10–30-cm-thick horizon that was buried by a thick, coarse volcanic ash layer which likely caused the site's abandonment (McCartney and Turner 1966a). Up to 2 m of ash strata have accumulated over the cultural horizon, protecting it from later disturbance except in the deflated areas referred to above. These thin-to-thick ash strata reflect volcanic eruptions from the surrounding volcanoes on adjacent Umnak Island, such as Okmok volcano, and on the Islands of the Four Mountains, some 75 km due west of Ananiuliak Island (Black 1975).

Black, a geologist with many years of Alaskan experience, began investigating the Anangula site stratigraphy in 1962 as part of a regional study of Pleistocene history, tectonics, and sea level changes (Black 1974a, 1974b, 1975). He established a 12,000-year late Pleistocene–Holocene stratigraphy, including four major ashes which were recognizable at most southwestern Umnak–Ananiuliak Island locales. Ash I dates to ca. 10,000 BP and marks the beginning of permanent deglaciation for the region.

The core and blade horizon falls between Ash II and Ash III, the latter being the coarse, 15–50-cm-thick ash that fell directly upon the active site.

A large suite of 41 dates has been run for the core and blade site, making this one of the best dated early sites in Alaska (Laughlin 1975, 1980; Aigner 1976a). Whereas part of the date range (7,200–8,700) may reflect different periods of site occupation, more likely the range reflects different ages of driftwood used at the site and/or mixing of different aged carbon from the seawater reservoir. Length of site occupation has been estimated to be from several centuries to 1,500 years. Laughlin (1974–75, 1980; S. Laughlin et al. 1975) favors an occupation that might have taken place in the period between 7,200 and 8,700 years ago, while Aigner (Aigner 1976a; Aigner and Del Bene 1982) suggests an occupation of several hundred years between 8,250 and 8,750 radiocarbon years ago.

ASSEMBLAGE

The characteristic assemblage reflects a core and blade technology that ultimately came from the Siberian Palaeolithic tradition (Laughlin and Aigner 1966; Laughlin 1967; Aigner 1970). The core and blade materials are placed under the Anangula technological tradition in order to distinguish them from flakes struck from irregular cores and bifacially flaked tools that characterize the later Aleutian tradition (McCartney 1984).

The slightly acidic pH of the ash soils, in contrast to the slightly alkaline matrix found at Aleutian midden sites, is responsible for the near lack of faunal remains or organic artifacts. While no contemporary bone or ivory tools have been found, the 1970 finds included "rather large quantities of burned bone fragments," made up mostly of whale bones in addition to some large bird bones (Aigner 1976a:42).

The total number of collected or excavated artifacts from this site probably exceeds 50,000 pieces, making this one of the largest early Holocene assemblages found in Alaska. Approximately 30,000 artifacts were found during the 1970 field season (Aigner 1976a; Aigner and Del Bene 1982; Del Bene 1982, 1992). Another 10,000 were recovered in 1974

(Laughlin 1975), and some 6,550 were excavated during 1963 (Aigner 1978). The remainder were found during other years (see above).

Years of excavation, coupled with a multiplicity of investigators and of published reports concerning the site, make it difficult to characterize the artifact collection in precise, quantitative terms. The following outlines essential aspects of the Anangula materials. The predominant artifacts are polyhedral cores, blades, core rejuvenation tablets, and flakes that result from a blade production technology.

There exists no single Anangula "type" of prepared blade core, as cores are quite variable in size and shape (Fig. 9–16: *a, b*). Cores exhibit platform angles from 82° to 92°, blade removal from 20 to 100 percent of the core faces, multiple rejuvenation by means of both new platform flaking and/or core rotation (the latter producing new platforms and new directions for blade removal), and utilization of cores to produce flakes rather than blades, once their usefulness as blade sources had been expended. Although small blades occur, Anangula is not a microblade site, per se, since the distributions of core heights and blade lengths are unimodal, primarily ranging between 13 and 125 mm with an average of 65 mm (Aigner 1978). Blades measuring up to 200 mm long are known from the site.

Primary stone types represented are basalt, felsite, obsidian, chert, cherty shale, and silicified argillite (Laughlin and Marsh 1954; Mason and Aigner 1987). Obsidian was acquired from volcanic sources such as Okmok Caldera, located on northern Umnak Island.

Anangula lithic technology is entirely unifacial, with retouching almost always confined to within some 3–5 mm of the tool edge. Retouched tools rarely alter the original shape of the blade or flake on which they are made, thus allowing unambiguous delineation of the stages of lithic reduction, from core to blade/flake to tool. Blades are clearly the favored tool blank at Anangula. While approximately 20 percent of the total lithic materials are blades, at least 60 percent of tools are made on blades (Aigner 1974).

Some blades were employed as blanks from which segments were snapped and used as tools. These include short obsidian tools with steeply retouched edges, presumably employed as scrapers

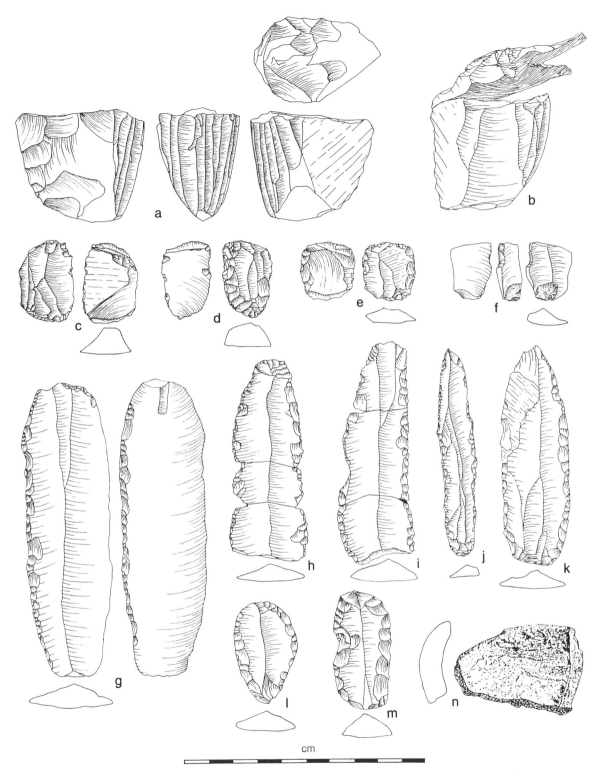

Figure 9–16   Artifacts from Anangula: blade cores (*a, b*), one with refitted rejuvenation tablets (*b*); transverse burins (*c–e*); angle burin (*f*); blades with marginal retouch (*g–k*), one that was notched prior to snapping (*h*); end or side scrapers made from blade segments (*l, m*); stone bowl fragment (*n*). *From:* (*a*) McCartney and Turner 1966b; (*b*) Laughlin and Aigner 1966; (*c–e, h–j*) Aigner and Del Bene 1982; (*g, k, l*) Aigner 1976b. *See Acknowledgments.*

(Fig. 9–16: *l, m*). Other scraping tools were manufactured on blade segments by striking a transverse burin blow across the blade end, creating what are referred to as "Anangula transverse burins" (Fig. 9–16: *c–e*). Spalls removed in this process were not utilized, but they indicate that many of the scraping tools were rejuvenated multiple times. A 2:1 ratio in the direction of spall removal (from left to right across the ventral blade surface when viewing it vertically) has been used as evidence of dominant righthandedness of the toolmakers (Aigner 1970). Some angle burins are also found (Fig. 9–16: *f*).

In addition to the various pointed tools, cutting tools, and scraping tools made on blades (Fig. 9–16: *g–k*) and flakes, a smaller group of other artifacts is also represented at the site. These include fishing line weights, stone bowls (Fig. 9–16: *n*), grinding pallets, faceted ochre grinders, carved stone lamps, grooved and ungrooved scoria abraders, hammerstones, and small incised stones.

Several studies have focused on specialized analyses of the physical and spatial attributes of the Anangula materials. Aigner and Bieber (1976) employed factor analysis on retouched and used tools and tool blanks from a portion of the 1970 excavation area, in an attempt to define tool kits and activity areas. Veltre and Aigner (1976) used factor analysis on a random sample of 500 whole blades, focusing heavily on attributes dealing with use wear. They found several groupings or types of blades based on these attributes. Aigner and Fullem (1976) analyzed 160 blade cores from the 1970 collection. Their results included evidence of manufacturing activity areas, with few cores included on house floors, and possible evidence (based on the rotation of the cores) of handedness among the ancient occupants of the site. Del Bene (1982, 1992) conducted a lengthy study of the Anangula technological system, including aspects of both manufacture and use of lithic materials, while Feder (1982) used factor analysis to establish production, use, and discard areas within the excavated portions of the site. While these studies were preliminary in nature, they offered insights into the Anangula collections at the same time that they explored methodologies for interpreting ancient technological and behavioral patterning.

AUXILIARY EVIDENCE

Although in 1963 a portion of a single depression was excavated (McCartney and Turner 1966b), it was not until the large-scale 1970 excavations that sufficient site area was uncovered to permit a better understanding of possible house features and spatial patterning. In 1970, portions of at least six nonoverlapping depressions were delineated (Aigner 1974, 1976a; Aigner and Fullem 1976; Aigner and Del Bene 1982). These depressions, interpreted by the excavators as house remains, are generally oval in plan, measuring approximately 3 × 5.5 m, and were at least somewhat semi-subterranean in construction. They lack evidence of ancillary rooms, side doors or entrance ways, or superstructures. Presumably roof entries were used as was the case among the historic Aleuts. Interior floor features appear to include three hearths, small sub-floor storage holes, and one stone-lined hole or pit (Aigner 1976a; Aigner and Del Bene 1982). Extrapolations from house size and number yielded a population estimate of some 75 to 125 people for the ancient settlement at Anangula (Aigner and Del Bene 1982), although the number may have been much less.

OTHER PHYSICAL EVIDENCE

Anangula is, geologically, one of the best studied early Holocene sites in Alaska. Black (1974a, 1974b, 1975, 1976) studied the site and Ananiuliak Island in the context of Umnak Island geology and also in the context of the entire Aleutian chain. He particularly studied glacial geology, volcanism, and sea level changes. Relevant to the origins of the core and blade site occupants, Black offered a reconstructed Holocene sea level of 2–3 m above present between 8250 and 3000 radiocarbon years ago. Based on this reconstruction, he suggested that the Anangula people used boats to reach the site from nearby Umnak Island and to cross the several major inter-island passes between the Alaska Peninsula and northeastern Umnak. Further, a maritime economy would have been the only option available for the Anangula peoples.

Black (1975) also supported McCartney and Turner's (1966a) proposition that volcanic activity probably forced the occupants to abandon Anangula, postulating that it was the catastrophic, caldera-forming eruption of Okmok volcano on northeastern Umnak Island that was the source of the overlying coarse Ash III. Although correlations between the Anangula ashes and surrounding volcanoes have not been made, this remains a likely avenue of future research to provide a regional chronology of both major and minor eruptions and periods of human occupation.

OTHER PHYSICAL OBSERVATIONS

While the Anangula core and blade site stood alone for several decades as the only very old and Palaeolithic-related site in the Aleutian Islands, several blade localities are now known in the eastern Aleutians and on the Alaska Peninsula. Excavations near the Anangula village site in 1974 (see above; Laughlin 1975) revealed a mixture of blades and bifacially flaked projectile points that were considered to be intermediate and transitional between the core and blade site and the recent village or midden site. Six radiocarbon dates from the 1974 village site excavation range between 4500 and 5900 BP. Contemporary dates of 4300–5600 BP are also found at the Sandy Beach Bay site on southwestern Umnak (Aigner et al. 1976), some 12 km south of Anangula.

Several core and blade localities are now known from the Unalaska Bay region, some 200 km northeast of Anangula. Reported first in the 1970s and early 1980s, sites containing blades, at least one transverse burin, and one polyhedral microcore were first systematically investigated in 1984 and 1985 (Veltre et al. 1985). In all, six sites appear to contain core and blade artifacts. Most are surface sites, situated several m higher than nearby (and presumably more recent) midden sites, and are not associated with bifacially flaked tools that characterize midden sites of the past 3,000 years. Like Anangula, all lack associated faunal materials. No firm dating is currently available for these sites, but their marginally and unifacially retouched blades, bu-

rins, and cores are clearly in the Anangula technological tradition.

REFERENCES CITED

Aigner, J. S. 1970. The Unifacial Core and Blade Site on Anangula Island, Aleutians. *Arctic Anthropology* 7(2):59–88.

———. 1974. Studies in the Early Prehistory of Nikolski Bay: 1937–1971. *Anthropological Papers of the University of Alaska* 16(1):9–25.

———. 1976a. Dating the Early Holocene Maritime Village of Anangula. *Anthropological Papers of the University of Alaska* 18(1):51–62.

———. 1976b. Early Holocene Evidence for the Aleut Maritime Adaptation. *Arctic Anthropology* 13(2): 32–45.

———. 1978. The Lithic Remains from Anangula, an 8500 Year Old Aleut Coastal Village. *Urgeschichtliche Materialhefte* 3. Tübingen: Institut für Urgeschichte, Universität Tübingen.

Aigner, J. S., and A. M. Bieber, Jr. 1976. Preliminary Analysis of Stone Tool Distributions and Activity Zonation at Anangula, an 8500 B.P. Coastal Village in the Aleutian Islands, Alaska. *Arctic Anthropology* 13(2): 46–59.

Aigner, J. S., and T. Del Bene. 1982. Early Holocene Maritime Adaptations in the Aleutian Islands. In *Peopling of the New World*, ed. J. E. Ericson, R. E. Taylor, and R. Berger, pp. 35–67. Los Altos, Calif.: Ballena Press.

Aigner, J. S., and B. Fullem. 1976. Cultural Implications of Core Distribution and Use Patterns at Anangula, 8500–8000 B.P. *Arctic Anthropology* 13(2):71–82.

Aigner, J. S., B. Fullem, D. Veltre, and M. Veltre. 1976. Preliminary Reports on Remains from Sandy Beach Bay, a 4300–5600 B.P. Aleut Village. *Arctic Anthropology* 13(2):83–90.

Black, R. F. 1974a. Geology and Ancient Aleuts, Amchitka and Umnak Islands, Aleutians. *Arctic Anthropology* 11(2):126–40.

———. 1974b. Late-Quaternary Sea Level Changes, Umnak Island, Aleutians: Their Effects on Ancient Aleuts and Their Causes. *Quaternary Research* 4:264–81.

———. 1975. Late-Quaternary Geomorphic Processes: Effects on the Ancient Aleuts of Umnak Island in the Aleutians. *Arctic* 28(3):159–69.

———. 1976. Geology of Umnak Island, Eastern Aleutian Islands as Related to the Aleuts. *Arctic and Alpine Research* 8(1):7–35.

Black, R. F., and W. S. Laughlin. 1964. Anangula: A Geologic Interpretation of the Oldest Archeologic Site in the Aleutians. *Science* 143(3612):1321–22.

Del Bene, T. A. 1982. The Anangula Lithic Technological System: An Appraisal of Eastern Aleutian Technology Circa 8250–8750 B. P. Ph.D. dissertation, University of Connecticut; Ann Arbor: University Microfilms.

———. 1992. Chipped Stone Technology of the Anangula Core and Blade Site, Eastern Aleutian Islands. *Anthropological Papers of the University of Alaska* 24(1–2):51–72.

Feder, K. L. 1982. The Spatial Dynamics of Activity at Anangula, Aleutians. Ph.D. dissertation, University of Connecticut; Ann Arbor: University Microfilms.

Laughlin, S. B., W. S. Laughlin, and M. E. McDowell. 1975. Anangula Blade Site Excavations, 1972 and 1973. *Anthropological Papers of the University of Alaska* 17(2):39–48.

Laughlin, W. S. 1951. Notes on an Aleutian Core and Blade Industry. *American Antiquity* 17(1):52–55.

———. 1967. Human Migration and Permanent Occupation in the Bering Sea Area. In *The Bering Land Bridge,* ed. D. M. Hopkins, pp. 409–50. Stanford: Stanford University Press.

———. 1974–75. Holocene History of Nikolski Bay, Alaska, and Aleut Evolution. *Folk* 16–17:95–115.

———. 1975. Aleuts: Ecosystem, Holocene History, and Siberian Origin. *Science* 189(4204):507–515.

———. 1980. *Aleuts: Survivors of the Bering Land Bridge.* New York: Holt, Rinehart and Winston.

Laughlin, W. S., and J. S. Aigner. 1966. Preliminary Analysis of the Anangula Unifacial Core and Blade Industry. *Arctic Anthropology* 3(2):41–56.

Laughlin, W. S., and G. H. Marsh. 1954. The Lamellar

Flake Manufacturing Site on Anangula Island in the Aleutians. *American Antiquity* 20(1):27–39.

McCartney, A. P. 1984. Prehistory of the Aleutian Region. In *Handbook of North American Indians.* Vol. 5, *Arctic,* ed. D. Damas, pp. 119–35. Washington, D.C.: Smithsonian Institution.

McCartney, A. P., and C. G. Turner II. 1966a. Stratigraphy of the Anangula Unifacial Core and Blade Site. *Arctic Anthropology* 3(2):28–40.

———. 1966b. Memorandum on the Anangula Core and Blade Complex. Unpublished. ms., Department of Anthropology, University of Arkansas.

Mason, O. K., and J. S. Aigner. 1987. Petrographic Analysis of Basalt Artifacts from Three Aleutian Sites. *American Antiquity* 52(3):595–607.

Okladnikov, A. P. 1979. The Ancient Bridge. *The Alaska Journal* 9(4):42–45.

Okladnikov, A. P., and R. S. Vasilievsky. 1976. *Around Alaska and the Aleutian Islands.* Novosibirsk: Nauka. (In Russian)

Pittenger, M. D. 1986. A Technological Analysis of the 1952 Lithic Collection from the Anangula Core and Blade Site, Eastern Aleutian Islands. M.A. thesis, Washington State University.

Veltre, D. W., and J. S. Aigner. 1976. A Preliminary Study of Anangula Blade Tool Typology and Spatial Clusterings Using a Factor Analytic Approach. *Arctic Anthropology* 13(2):60–70.

Veltre, D. W., A. P. McCartney, M. J. Veltre, and J. S. Aigner. 1984. An Archaeological Site Survey of Amaknak and Unalaska Islands, Alaska. Report submitted to the Alaska Division of Parks and Outdoor Recreation, Anchorage.

Chapter 10

KUSKOKWIM DRAINAGE, SOUTHWESTERN ALASKA

INTRODUCTION

The upland region of southwestern Alaska, which includes part of the Alaska Range and the Kuskokwim Mountains and outliers, was subject to intensive glaciation during the Illinoian and Wisconsinan glaciations (Cady et al. 1955; Coulter et al. 1965; Bundtzen 1980; Kline 1983; Lea 1984; Kline and Bundtzen 1986). In the Farewell area, on the north slope of the Alaska Range, several late Pleistocene glacial morainal features were noted that date to about 25,000 BP (Kline and Bundtzen 1986:136).

As a consequence of glaciation, vast amounts of aeolian sand and silt from outwash rivers were deposited in periglacial zones during the late Pleistocene. Such deposits occur as sand-sheet and sand-loess intergrades and as sand-dune deposits and loess in the lower elevations or as covering deposits over slopes up to 450 m above the river bottoms (Cady et al. 1955; Péwé 1975; Lea 1990; Lea and Waythomas 1990).

By at least 9,500 BP, glaciers had retreated from the upper mountain valleys of the north-central Alaska Range (Kline and Bundtzen 1986:136). General deglaciation had begun somewhat earlier, about 14,000 years ago, as evidenced by the rise in

Betula pollen (Ager 1982, 1983; Ager and Brubaker 1985), by dates on basal peats (Hamilton and Thorson 1983:46), and by increased salinities and temperatures of Bering Sea water (Hopkins 1982:15).

Palynological research by Ager (1982, 1983; Ager and Brubaker 1985) in the Yukon Delta region indicates that full-glacial vegetation was represented by an herbaceous tundra composed of Gramineae, Cyperaceae, *Salix, Artemisia,* and herbs. A rise in *Betula* (*Betula* zone) was noted about 14,000 BP, indicating that "the landscape of much of Beringia underwent a shift from predominantly herbaceous tundra vegetation to a mesic shrub tundra in which Ericaceae, *Betula nana* and *B. glandulosa* (in the interior), *Salix* spp., *Eriophorum* spp. and mosses became major elements" (Ager 1982:87). A *Populus* zone composed of *Populus* and *Salix* along with *Betula* and other species characteristic of the *Betula* zone appeared between 11,000 and 10,000 years ago, to be followed shortly afterward by a *Betula*–Ericaceae zone from 10,000 to 7,500 BP (Ager and Brubaker 1985:369). Spruce (*Picea*) pollen first appears about 5,500 BP, indicating a general westward spread from interior Alaska in mid-Holocene times (Ager 1982, 1983; Ager and Brubaker 1985).

The study of insect fauna recovered from palynological samples from the Nushagak and Holitna lowlands provides further corroboration of the transition from a cold, dry, late-glacial environment to a warmer and wetter climatic cycle with revege-

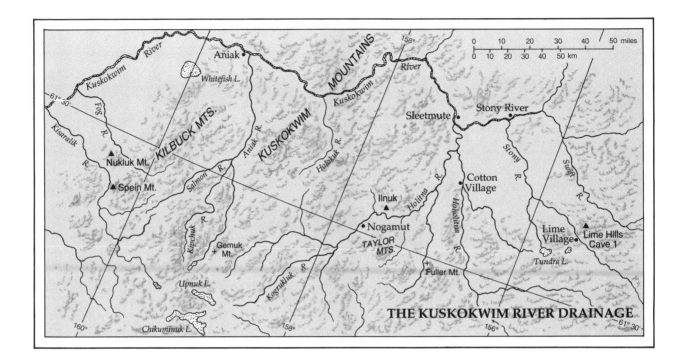

THE KUSKOKWIM RIVER DRAINAGE

tation of the landscape (Short et al. 1992:390). Little change in the insect assemblages was detected between the birch and alder zones, suggesting that the changes in vegetation were due to succession rather than climatic change (Short et al. 1992:391).

SPEIN MOUNTAIN

The Spein Mountain site is located in the western foothills of the Kilbuck Mountains, approximately 60 km east of the town of Bethel (60°35′N, 160°2′W). The site is on a basaltic ridge 210 m above the present course of the Kisaralik River (see regional map). Vegetation in the site area is that of an alpine tundra with a cover of lichen, mosses, fireweed (*Epilobium angustifolium*), mountain avens (*Dryas octopetala*), cut leaf anemone (*Anemone multifica*), cinquefoil (*Potentilla uniflora*), vetch (*Astragulus alpinus*), cotton grass (*Eriophorum* spp.), grasses, and low shrubs such as birch (*Betula nana*) and willow (*Salix* spp.) on the more exposed portion of the ridge. Surrounding the ridgetop are copses of bush alder (*Alnus crispa, A. sinuata*). Farther down the slopes of the ridge stands of alder, aspen (*Populus tremuloides*), poplar (*Populus balsamifera*), and conifers

such as black spruce (*Picea mariana*) are encountered. The river valley bottom is covered by a heavy stand of black spruce, aspen, alder, and willow. Salmon, Arctic char, and grayling can be taken in the river today. Moose, caribou, black bear, and grizzly bear are the prime subsistence animals.

The Spein Mountain site lies beyond the glaciated zone (Hoare and Coonrad 1959; Coulter et al. 1965). The surficial deposits of the low hills and ridgetops of the western margins of the Kilbuk Mountains consist of rocky soil and rubble produced by the weathering of the bedrock with an overlying deposit of silt carried up to 200–300 m above the present stream courses by wind action. The rubble in the higher elevations forms stone nets on the summits and stone streams on the slopes as a result of frost action (Cady et al. 1955). Outwash depositions of silts, sands, and gravels cover the valley bottoms (Hoare and Coonrad 1959).

NUKLUK MOUNTAIN

The Nukluk Mountain sites, like the Spein Mountain site, are located in the western foothills of the Kilbuck Mountains, approximately 60 km east of

In the Kuskokwim drainage, the area between Nenevok and Kagati lakes.

Bethel, Alaska (60°42′N, 160°26′W) (see regional map). The site BTH 069 is on the first bench on the north and west flank of Nukluk Mountain at an elevation of approximately 167 m above the present course of the Kisaralik River (3.7 km to the north). BTH 070 is on a ridge to the south of BTH 069. The Spein Mountain site is 13.5 km to the south and east. Nukluk Mountain is the first significant elevation one encounters coming from the Kuskokwim River lowlands, and it provided hunters with a good location from which to search the surrounding landscape for herds of migratory animals.

Vegetation in the site area is sparse and consists of mountain avens (*Dryas octopetala*), dwarf birch, fireweed (*Epilobium latifolium*), lichens, moss, sedges (*Eriophorum* spp.), willow, alder, and a few black spruce outliers. The nearby river valley is covered with a thicket of alder, aspen, and black spruce. Salmon, arctic char, and grayling occur in the Kisaralik River. The large mammals are moose, caribou, black bear, and grizzly bear.

The Nukluk Mountain area is well beyond the zone of glaciation and is at a transition point between the uplands, characterized by frost-riven, rubble-covered hills, and the massive silt deposits of the Kuskokwim River lowlands. The bedrock forming Nukluk Mountain is andesite with interbedded sedimentary rocks such as graywacke, siltstone, limestone, and pebble conglomerates (Hoare and Coonrad 1959). Sorting of clasts by frost action was evident in the rock stripes seen on the slopes. A thin covering of loess overlies the shattered regolith and, where stabilized by vegetation, is 10–15 cm thick. Exposure of the vegetation to drying winter winds has resulted in the loss of much of the surface vegetation cover in the site area. In the low-lying areas, there is tussock tundra-type vegetation. Trees tend to cluster along stream courses or on the well-drained slopes.

ILNUK SITE

The Ilnuk site is located on the Holitna River, 70 km southwest of the village of Sleetmute (61°6′N, 157°28′W) (see regional map). The site is on a do-lomitic limestone ridge 92 m above the river valley. The site is now covered by a stand of black and white spruce (*Picea mariana, P. glauca*), aspen, poplar, and western paper birch (*Betula papyrifera*). Ground cover consists of blueberry (*Vaccinium ovalifolium*), highbush cranberry (*Viburnum edule*), Labrador tea (*Ledum decumbens*), dwarf dogwood/bunchberry (*Cornus canadensis*), fireweed (*Epilobium angustifolium*), club moss (*Lycopodium* spp.), sphagnum moss, and lichens. The tree cover is somewhat discontinuous with open heath meadows on the well-drained ridgetops or slopes. Valley bottoms are covered with a thicket of willow, alder, and black spruce interspersed with sedge meadows.

The ridge on which the Ilnuk site is situated is part of a group of limestones about 20 miles wide that cuts across the middle course of the Holitna River. The ridge extends eastward after crossing the Hoholitna River, becomes the Door Mountain group, and terminates in the Lime Hills east of the Stony River (Cady et al. 1955). The site locality is on the downslope of the ridge as it angles towards the Holitna River. The site in the past would have provided a view of the Holitna River lying eastward of its present course and of the higher river terraces now covered by tundra vegetation. The southwestward flank of the ridge breaks away sharply, forming the contact zone of a thrust fault (Cady et al. 1955: plate 10). To the rear of the site area, the ridge forms a saddle between the site area and the next higher elevation. The Chuilnuk River valley to the north and east is carpeted by a dense stand of black spruce and deciduous trees.

The Ilnuk, Spein, and Nukluk Mountain sites all lie beyond the extent of glaciation (Cady et al. 1955; Hoare and Coonrad 1959; Coulter et al. 1965).

LIME HILLS KARST REGION*

The Lime Hills, or Door Mountains, are an elongate group of low rolling hills that trend northeastward for about 40 km. Elevations here range from 500 to 2,410 feet (152–735 m). The Stony River, a major tributary of the Kuskokwim, flows northwestward through the study area and creates a canyon that dissects the Door Mountains about 8 km southwest of the Lime Hills caves in which archaeological investigations were conducted in 1993. Lime Village lies on the south bank of Stony River about 12 km due south (see regional map).

The Door Mountains west of the Stony River are composed of a 280-m-thick section of limestone and clastic rocks of Silurian and Cretaceous age. The Silurian limestone was originally described as part of the Holitna Group (Cady et al. 1955) and is now assigned to the Nixon Fork tectono-stratigraphic terrane (Patton et al. 1980; Bundtzen and Gilbert 1983). The limestone units exposed in the Lime Hills karst cavern area are similar in age and depositional environments to the lower Whirlwind Creek formation in the northern Kuskokwim Mountains (Dutro and Patton 1982), unnamed Silurian units near White Mountain (40 km north of the study area [Gilbert 1981]), and the Heceta limestone of Prince of Wales Island, southeastern Alaska (Eberlein and Churkin 1970).

The Lime Hills have been overridden by glacial ice, as evidenced by very large glacial erratics that were dropped on the highest hills, and by mappable units of silty diamicton and drift found throughout the lowlands. The glacial drift and erratics are correlated with the Farewell glaciation to the north, regarded as early Wisconsinan (65,000 to 122,000 yrs BP) in age by Kline and Bundtzen (1986).

During brief geologic reconnaissance work in 1992 and 1993, at least 50 sinkholes, solution caverns, and horizontal or phreatic caves were observed throughout the Lime Hills, but particularly in the uplands north of Lime Village. The age of the cave and karst development is not specifically known. However, Dunaway (personal communication, 1993) noted glacial drift at the bottom of a prominent sinkhole about 2 km southwest of Cave 1, which suggests that sinkhole development predates the glaciation. As the glacial deposits of the study area are correlated with the Farewell I glaciation, we infer that cave and karst development has

*The information in this section was provided by Thomas K. Buntzen.

to be at least 70,000 years old. Ford (1973), Ford et al. (1984), and Lauriol et al. (1989) have estimated the age of cave formation in northern Yukon and Northwest Territories using U-Th isotopic and geomorphologic methods. They conclude that cave formation in the South Nahanni and Bear Cave areas probably took place during warmer interglacial times. Perhaps solution cavern development in the Lime Hills took place during the Sangamon interglacial or between the Farewell I (early Wisconsinan) and Selatna (Illinoian) glaciations 122,000 to 132,000 yrs BP.

Pollen Data

No pollen records for the late Pleistocene Duvanny Yar interval (Hopkins 1982) are available in the Lime Hills region. Herb zone pollen data from the Yukon Delta (Ager 1982) and the Nushagak lowland (Short et al. 1992) indicate that the late Pleistocene was a cold, dry environment characterized by an herbaceous tundra with Gramineae, Cyperaceae, *Salix*, *Artemisia*, and forbs as dominants. The transition from a cold, dry herbaceous tundra to a mesic birch–shrub tundra is recorded in the peaty sediments of the lower Holitna lowland (to the north and west of Lime Hills) and the Kuskokwim–Big River lowland (to the north of Lime Hills) (Short et al. 1992). The birch zone, beginning about 12,000 and ending about 8,500 BP, is characterized by high percentages of *Betula*, Gramineae, Cyperaceae, and Filicales, and low percentages of *Alnus*, *Artemisia*, and Sphagnum. The alder zone (8,500–5,500 BP) is dominated by *Alnus*, *Betula*, Filicales, and Sphag-

num. Grasses and sedges decline from their previous highs. *Artemisia* increases from a low of less than 5 percent in the birch zone to more than 10 percent in the alder zone. *Picea* increases towards the end of the alder zone, invading the region about 5,500 years ago. Short et al. (1992) note that the alder zone pollen spectrum "suggests a mesic environment with alder shrubs and a rich understory of moss, ferns and fern allies and compares well to modern polsters from the same region."

Vegetation

Short et al. (1992:382) have divided the middle zone of the Kuskokwim River region into three zones: (1) low-elevation spruce forest dominated by white spruce and paper birch intermixed with muskeg and bog characterized by tall shrubs—alder (*Alnus crispa*), willow (*Salix alaxensis* and *S. glauca*), and birch (*Betula glandulosa*); low shrubs—willow, dwarf birch and heath (Ericaceae), grass (Gramineae) and sedge (Cyperaceae) meadows with associated herbs and ferns (Filicales); (2) middle-elevation birch–spruce forest to treeline (at 300–360 m); and (3) high-elevation shrub tundra with lichens, mosses, grasses, low shrubs, and heath which includes bog blueberry (*Vaccinium uliginosum*), decumbent birch (*B. nana*), crowberry (*Empetrum nigrum*), Labrador tea (*Ledum decumbens*), alpine bearberry (*Arctostaphylos alpina*), moss campion (*Silene acaulis*), and mountain avens (*Dryas* spp.).

R. E. A.

—⬧✦⬧✕⬧✦⬧—

SPEIN MOUNTAIN

Robert E. Ackerman

Several sites were located during a survey of the Kisaralik River valley conducted in 1979. Most were small lithic scatters left by hunters who intercepted caribou herds in their fall migration from the Kuskokwim–Yukon delta region to their winter quarters in the Nushagak–Mulchatna river uplands. These upriver and midriver assemblages were judged to belong to a cultural period intermediate between the Beringian and Northern Archaic traditions, known locally as the Kagati Lake complex (Ackerman 1985, 1987). At the point where the descent of the river begins to lessen, prior to entering the alluvial plains of the Kuskokwim River valley, the writer and two field assistants, James Gallison and Lance Rennie, surveyed ridgetop locations. These locations on both sides of the valley provided panoramic views of the countryside. Two major site complexes, Spein Mountain and Nukluk Mountain, were discovered as surface exposures on the south side of the Kisaralik River valley, while only small lithic scatters were noted on the north side.

The Spein Mountain site consists of four separate areas (BTH 062–065) distributed along a long ridge where lithic artifacts were noted in deflated areas (Fig. 10–1). The major concentration of artifacts was found at 063, apparently a fall hunting base camp, somewhat out of the wind and out of sight of the caribou moving from the Kuskokwim River lowlands on their fall migration. The other areas—such as 064, which overlooks the Kisaralik River, and 062 and 065, which are oriented towards the tundra uplands between the ridge and Spein Mountain—are likely lookout stations associated with the base camp.

In 1979 artifacts were collected from surface exposures at the four areas. Area 063 contained the greatest number of artifact types. To determine if there were undisturbed portions of the site, three shovel test pits were dug in the vegetated portion of area 063. There flakes were recovered from the upper 15 cm of the loess deposit, indicating that the vegetated zone would provide insights into the site stratigraphy.

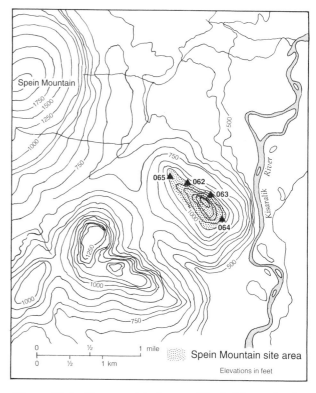

Figure 10–1 Topographical map of Spein Mountain site.

In 1992, the author and three field assistants, Neil Endacott and Elizabeth and Austin Wilmerding, returned to the site to test the vegetated portion of the site in hopes of recovering organic materials suitable for dating, to ascertain the site stratigraphy, and to determine if testing in other areas of the site would reveal different artifact types than had been recovered in 1979. Forty-one 1-×-1-m squares were excavated to bedrock in the area 063 in 1992. Soil depth proved to be thin or nonexistent in areas 062, 064, and 065 of the site. Here collections were made from surface exposures. A total of 4,299 artifacts were recovered from Spein Mountain during the 1979 and 1992 field seasons (Table 10–1).

Site Characteristics

The Spein Mountain site is located on a ridge of highly weathered and shattered regolith of basalt over which is a 15-to-25-cm deposit of dark brown (7.5YR3/2) loess (Cady et al. 1955) which puddles like jelly when wet (thixotropic). Slope features such as shattered rock alignments and rills in surface exposures indicate downslope movement. They may be a result of that and cryogenic processes.

The site stratigraphy indicated an undifferentiated layer of silty loam (loess) that rests directly on a weathered basaltic regolith. Due to soil-formation processes, texture and color changes could be detected, resulting in two to three soil horizons. The artifacts were scattered throughout the loess with most found at 15 cm below the surface. No technological changes in the debitage nor in artifact types were found, which would suggest that the site contained no more than a single component.

Charcoal flecks were found in the surface vegetation mat indicating an historic burn, but not within the mineral soil which would have provided a date for the cultural occupation.

A test off the ridge line on a south-facing slope revealed a pit feature with large angular rocks at its bottom. Artifacts were numerous in and around the pit feature extending down to 41 cm below the ground surface: biface fragments, a polished pebble, a utilized flake, a hammerstone, a raw material chunk, and flakes. Flaking debris were found outside as well as within the pit.

Table 10–1   Artifact Inventory, Spein Mountain Site

Stone Artifacts	*1979*	*1992*
Projectile Points		
Lanceolate		
complete	2	10
base	26	17
tip	5	19
mid	6	11
Pentagonal	1	—
Willow-Leaf-Shaped (Bipoints)	2	—
Bifacial Adze Blades	3	1
Large Bifaces		
half sections	5	5
Bifacial Preforms		
complete	3	1
fragments	6	4
Gravers on Flakes	6	3
Notch on Flake	—	1
Utilized Flakes	13	—
End and Side Scrapers	3	3
Cobble Spall Scrapers	2	—
Chopper (skreblo)	1	—
Hammerstones	2	3
Split Cobbles	3	22
Abrasive Stone	1	—
Debitage	355	3,754
Subtotal	455	3,854
Total		4,299

FAUNAL AND FLORAL REMAINS

A lens of darker sediment was noted in the cross-sectional view of the pit feature, over the rock scatter. A bulk soil sample collected and tested proved to be high in humates containing about 15–20 percent organic carbon (by weight loss on ignition). A liquefied sample put through a very fine mesh screen in search of plant macrofossils yielded small bits of charcoal and calcined bone. A charcoal sample provided an AMS date of 10,050 ± 90 radiocarbon years ago (Beta 64471 [CAMS-8281]). Pollen analysis by Dr. Peter Mehringer revealed that grass was the dominant pollen species (see Table 10–2).

Other species in the pollen spectrum suggest an alpine tundra with shrubs but no tree forms. Today, both black and white spruce are present in the Kisaralik River valley. The pollen data indicate that the site was occupied towards the end of the Birch zone (Ager 1982; Short et al. 1992), but prior to the expansion of alder and spruce into the area. This is supported by the radiocarbon date of 10,050 BP.

Dr. Mehringer noted that the abundance of grass pollen was quite unlike that from natural vegetation, even lush grasslands, and that a grass-lined pit was a reasonable explanation. The grass was apparently laid in the pit after a fire had occurred as none of it was burned.

The fauna would have been similar to that of today, although there is evidence that bison was also hunted by late Pleistocene/early Holocene hunters in interior Alaska 11,000 to 10,000 years ago (Powers et al. 1983).

ARTIFACT ASSEMBLAGE

The assemblage is composed of 190 tools and 4,109 pieces of debitage and consists of lanceolate, pen-

Table 10–2 Pollen Analysis of Soil Sample from Square S9E20, Spein Mountain Site (BTH 063)*

Pollen Count		Spores	
Juniperus	3	Trilete	
Betula	4	*Lycopodium*	2
Salix	2	(41 *Lycopodium* traces recovered from 139,110 tracers added to 4.6 cm³ of sediment)	
Alnus	4	Monolete	
		Athyrium-type	8
		Polypodium	2
Artemisia	8		
Poaceae	465 (85.8% grass pollen)		
Cyperaceae	28		
Ranunculus	3		
Acontium-type	2		
Caryophyllaceae	2		
Gilia-type	2		
Oenothera	1		
Umbelliferae	1		
Other	17		
Total	542		

*Reported by Dr. Peter J. Mehringer

tagonal, and leaf-shaped projectile points, bifacial adze blades, ovate bifaces, bifacial preforms, gravers on flakes, flake knives, scrapers, whetstone/abrader, a notched flake, cobble spall scrapers, choppers, hammerstones, split cobbles, and flakes from bifacial reduction (Fig. 10–2).

The majority of the projectile points are lanceolate in form with square to convex bases, straight to slightly curved sides (Fig. 10–2: *b–e*). Bases are thinned and the edges of the points collaterally flaked. One of the largest points is a leaf-shaped point from area 064 of the site (Fig. 10–2: *a*). The small pentagonal point of chert (Fig. 10–2: *f*) is unique, as only one was recovered and this in a surface context in area 063. It may be either a later addition to the site or modified into its present form by resharpening. Basal edge grinding is rare, but it does occur. In addition, many of the projectile point fragments are heavily weathered so that it is difficult to determine if an edge was once ground. Many of the projectile points were snapped off above the

haft area, indicating an in-camp replacement of broken points. Some points with tips snapped off were resharpened in the haft and later discarded. The bifacial adze blades (Fig. 10–2: *g*) were similar to bifacial, discoidal scrapers, but had the ends beveled in opposite directions, suggesting that this tool form was inserted in an antler or bone sleeve.

Biface preforms, cobble spall scrapers, split cobbles, and a range of bifacial thinning flakes down to the smaller pressure flakes were made of local materials. The primary raw material utilized was a metamorphosed siltstone locally available as river cobbles or as tabular blocks from the Kuskokwim Group formation (Hoare and Coonrad 1959) which rapidly weathered when exposed on the surface of the site. The high frequency of basal point fragments of siltstone clearly demonstrates that this was the material of choice, dictated no doubt by its local availability. Other materials such as chert and chalcedony were recovered as small pressure retouch flakes, but not as larger flakes from initial and

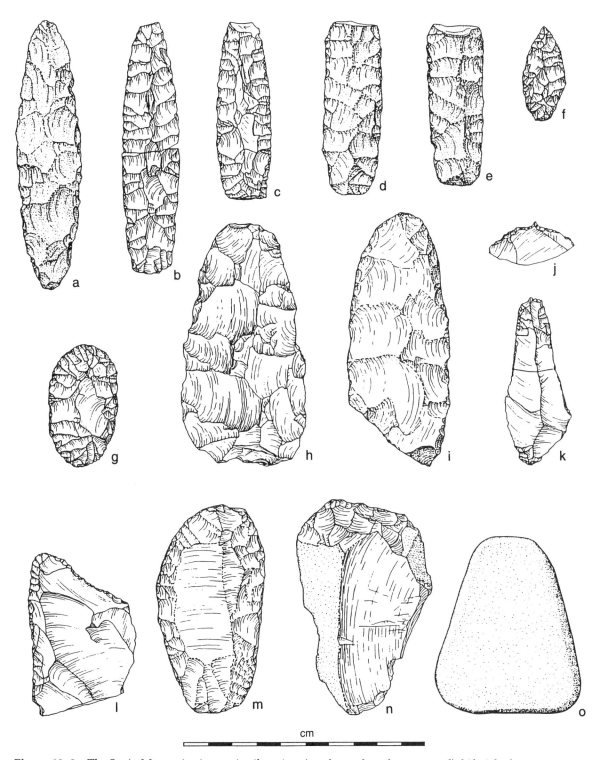

Figure 10–2   The Spein Mountain site: projectile points (*a–e,* lanceolate; *f,* pentagonal); bifacial adze blade insert (*g*); bifacial preform (*h*); bifacial knife or projectile point preform (*i*); gravers on flakes (*j, k*); flake knife (*l*); ovate scraper (*m*); end scraper (*n*); whetstone/abrader (*o*).

secondary stages of biface reduction. Tool forms of chert or chalcedony were rare, indicating that such artifacts were produced elsewhere.

The 1992 excavations did not produce any new artifact types, but materially increased the artifact count in existing categories.

Clearly demonstrated in both the 1979 and 1992 samplings of the site was the complete absence of blade or microblade production. Three tool forms— a pentagonal point, a small bipoint, and a possible side blade—demonstrate that the site was visited by later hunting groups.

Summary

The Spein Mountain complex is markedly different from those recovered from upriver sites on the Kisaralik River or from sites on the Kwetluk, Eek, and Kanetok rivers, as well as from the Kagati Lake region. All of these sites are chiefly characterized by microblade and blade industries with occasional ovate bifaces. Many of the sites are workshops where vast amounts of debitage were recovered. Downriver from the Spein Mountain complex, the Nukluk Mountain assemblage clearly belongs to the Denali complex of the Beringian tradition (West 1967, 1981, this volume).

The Spein Mountain complex, with the emphasis on lanceolate points and the exclusion of a microblade technology, is similar in many ways to the Mesa site cultural complex in north-central Alaska that dates between 11,660 ± 80 and 9730 ± 80 BP (Kunz 1982; Kunz and Reanier 1994, this volume). Projectile points of the Bedwell complex of the Putu site in the Brooks Range (Alexander 1974, 1987; Reanier, this volume) are very similar to those recovered from the Spein Mountain site. A single date of 11,470 ± 500 years BP was obtained from a hearth said to be spatially associated with the base of a non-fluted projectile point (Kunz and Reanier 1994), suggesting that the Bedwell complex may be older than the Putu fluted point complex. The Nenana complex in central Alaska is also within the 11,800 to 10,000 BP date range but the point types are quite different, with the emphasis on a triangular or teardrop shape (Powers and Hoffecker 1989; Hoffecker

et al. 1993). The available data thus suggest a projectile point horizon without microblades that preceded the microblade industry of the Palaeo-Arctic tradition of western and central Alaska (West 1967, 1981; Anderson 1970). Recent dates in excess of 11,000 BP for sites with microblades (C. Holmes, personal communication 1993, this volume), indicate, however, that the early lanceolate to ovate projectile point complex without microblades and the microblade industries of the Denali complex are likely contemporary cultural complexes that derive from rather different cultural traditions.

The Spein Mountain site complex located in the foothills of the Kuskokwim Mountains and the Mesa and Putu site complexes in similar contexts in the Brooks Range are characterized by lanceolate bifacial points, scrapers, flake knives, gravers, and notches, a complex that is very reminiscent of the Palaeoindian site complexes found to the south. Projectile points similar to the Spein Mountain lanceolate type, gravers on flakes, and scrapers have also been recovered from surface sites of the Driftwood Creek complex in the Utukok River region of the western Brooks Range (Humphrey 1970). Like the Putu complex, the Utukok sites also contained a core and blade industry (as well as other artifact types not found at the Spein Mountain site).

The Nenana complex with shorter ovate to teardrop-shaped points seems to be a co-tradition that held sway in the Nenana Valley of central Alaska. The origin of these Palaeoindian-like complexes is as yet unknown, for the contemporary culture complex that dominated Siberia at this time was the Dyuktai culture complex (Mochanov 1977, 1984; Mochanov and Fedoseeva 1984, this volume), which gave rise to the Palaeo-Arctic tradition of western Alaska and the Denali complex of central Alaska. In southwestern Alaska, this tradition is well documented at the Nukluk and Ilnuk sites.

NUKLUK MOUNTAIN

Robert E. Ackerman

In 1979, the writer and two field assistants, James Gallison and Lance Rennie, spent only a few days exploring the Nukluk Mountain area. Two site locations were discovered: site 069, on the north side of the mountain, and site 070 to the south (see regional map). Surface collections were made.

In 1992, the writer with three field assistants, Neil Endacott and Elizabeth and Austin Wilmerding, revisited the site. The goal was to make an extensive survey of the Nukluk Mountain area to determine if there were other occupations and to gain a larger artifact sample from site 069. The survey of the slopes and the flat top of Nukluk Mountain proved to be unrewarding. A single flake was found on a bench above the 069 site.

BTH 069

SITE CHARACTERISTICS

Vegetative cover is sparse, consisting of lichens, mosses, and a few prostrate shrubs. Shallow soil cover is composed largely of windblown silt and sand and overlies shattered regolith. Excavations conducted at site 069 revealed a fairly extensive scatter of lithic artifacts recovered from the southward-facing slope of a small knoll in the middle of a low ridge on the north slope of Nukluk Mountain.

The highly weathered stone tools and debitage recovered from erosion channels that coursed down the knoll and cut down to shattered regolith indicate that the site had been deflated for a considerable period of time. No artifacts were found in the thin loess deposit and remaining vegetative cover.

ARTIFACT ASSEMBLAGE

The 1979 surface collection was composed of 860 lithic items (51 tool forms and 809 pieces of debitage). Excavation efforts in 1992 resulted in 638 lith-

ics (15 tool forms and 623 pieces of debitage). (See Table 10–3.) The principal raw material was argillite, similar to that found at the Spein Mountain site, with chert and quartz as rare tool stones.

The assemblage consists of 5 fragmentary oblanceolate projectile points, with sides expanding towards the tip, and with rounded bases (Fig. 10–3: *d*); 3 frontal, wedge-shaped microblade cores with platforms created by removal of a platform tablet from the flute face (Fig. 10–3: *a*); an asymmetric microblade core (Fig. 10–3: *b*); 3 microblade core preforms (some heavily weathered); 13 microblades; 2 Donnelly-type burins (Fig. 10–3: *c*); 5 macro/microblade core preforms; end and side scrapers (Fig. 10–3: *h, i*); biface preforms (Fig. 10–3: *e, f*); 8 blade-like flakes; abrasive stones with use-wear areas; 1 cobble core (Fig. 10–3: *j*); 1 hammerstone; broken cobbles; a piece of red ochre; and debitage

Table 10–3   Artifact Inventory, Nukluk Mountain Site (BTH 069)

Stone Artifacts	1979	1992
Projectile Points		
Oblanceolate		
base and mid	1	—
base	3	—
tip	1	—
Microblade Cores		
wedge shape, frontal	2	1
asymmetric	1	—
core preform, wedge shape	1	2
Microblades	11	2
Macro/Microblade Core Preforms	4	1
Blade-like Flakes	6	2
Donnelly Burin	1	1
Scrapers		
end	2	—
side (fragments)	4	—
Bifacial Preforms	5	—
Abrasive Stones	7	—
Split Cobbles	1	5
Red Ochre	1	—
Hammerstone	—	1
Debitage	809	623
Subtotal	860	638
Total		1,498

Nukluk Mountain ridge with lithic scatter.

(Table 10–3). The extensive amount of weathering has made artifact illustration difficult.

Most of the debitage appears to be the end result of the reduction of argillite river cobbles into rough cores, bifaces, or choppers. At this stage of reduction it is difficult to determine from the debitage the final intended form. No platform tablets resulting from microblade core platform rejuvenation were recovered. Three "gull-wing" shaped flakes which could have been from platform rejuvenation were noted in the assemblage, but neither of the finished microblade cores showed any evidence of platform edge retouch. Of the few microblades recovered, most are either ridge flakes or thick rejects. The lack of a significant microblade production at the site would suggest that the production of microblades was not an important aspect of the site technology.

In comparison to the Spein Mountain site, few broken projectile points were recovered. The projectile point fragments were also too fragmentary for comparative purposes. One base fragment has edge grinding. Another refitted base-and-mid-section fragment (Fig. 10–3: *d*) is similar to Figure 10–2: *a* of the Spein Mountain site and lacks edge grinding.

The frontally fluted, wedge-shaped microblade cores, the Donnelly burin, and the associated oblanceolate-to-bipointed projectile points and scrapers are sufficiently diagnostic to classify the artifact complex of site 069 as a western expression of the interior Alaskan Denali complex (West 1967, 1981, this volume) which has been dated to approximately 10,600–8,600 BP (Powers and Hoffecker 1989). Since this time range overlaps with that of the Spein Mountain site, it is possible that the Nukluk Mountain and Spein sites could have been occupied at the same time, but, if so, the question arises as to why these two sites, only eight miles apart, should have such different tool kits when both were essentially task-specific sites for intercepting caribou on their fall migration from the Kuskokwim River lowlands to the interior uplands. A difference in age would allow one cultural complex to precede the other and thus be viewed as time-specific complexes. For now, temporal priority must be given to the assemblage from the Spein Mountain site.

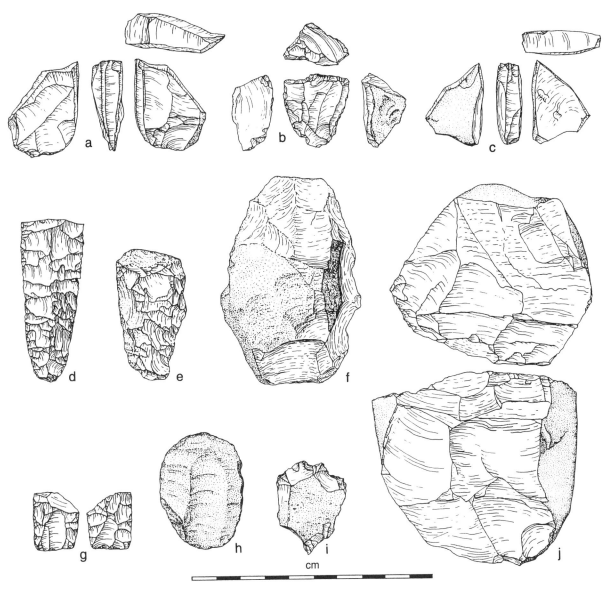

Figure 10–3   Nukluk Mountain sites BTH 069 (*a–f, h–j*) and BTH 070 (*g*): wedge-shaped microblade core (*a*); asymmetric microblade core (*b*); burin (*c*); bifacial projectile point bases (*d, g*); bifaces (*e, f*); scrapers (*h, i*); cobble core (*j*).

BTH 070

Site BTH 070, to the south of BTH 069, consisted of a single basal fragment of a lanceolate point with basal thinning and edge grinding (Fig. 10–3: *g*), recovered from a deflated area on a ridge. No other artifacts were recovered from this location. Flakes were detached perpendicularly from the base on one face and obliquely on the other. The edges were ground from the base to the point of breakage. Since basal thinning by transverse flaking from the base is a technique utilized in a number of cultural phases, the point fragment by itself does not provide any age placement. It is a point type not associated with the Denali complex and probably represents a later use of the area.

In hopes of finding another site with better preservation, the entire mountain area was surveyed. Despite a careful search of many deflation areas in what could be considered choice lookout points, no further sites were discovered. From the position of BTH 069, one can look out over the gently sloping terrain to the Kuskokwim River flats. Hunters on the site would have been in an exposed position, suggesting that the site was a lookout rather than a campsite.

Summary

No direct age assignment has been determined for the Nukluk Mountain sites due to the absence of stratigraphy and datable materials. The site assemblage and the nature of the deposit at BTH 069 suggest a single component. As a Denali complex expression, the Nukluk Mountain site occupation would have preceded that of the Kagati Lake complex (circa 8,000–6,000 BP). Since the eastward migrating caribou would have reached the downriver location of the Nukluk Mountain site before appearing at the Spein Mountain site, it is doubtful that the two sites would have been occupied contemporaneously by different cultural groups. Since better than 50 percent of the debitage at the Nukluk Mountain site is considerably more weathered than the remainder, it is possible that there were two periods of occupation. Individuals from the Spein Mountain group may have used the Nukluk Mountain site as an advance lookout for the main party at Spein Mountain and another group of Denali complex people may have been later users of the site. While this is an attractive hypothesis, it is problematic. Most of the highly weathered artifacts are debitage resulting from the reduction of river cobbles and hence undiagnostic to time period, and of those tool forms that can be associated with the Denali complex, some are as deeply weathered as those found at Spein Mountain.

ILNUK SITE

Robert E. Ackerman

In 1981 the Holitna River valley was surveyed by the writer and field assistants Gail Ackerman, James Gallison, Peter Lea, and Lance Rennie. Several late prehistoric sites were found on open ridgetops which approached the river in the upper part of the drainage, while historic villages were located along the riverbanks. The eroding limestone ridge on which the Ilnuk site is located, while heavily forested, was topographically similar to those upriver on which we had found sites (Fig. 10–4). On the downslope side of USGS triangulation station ILNUK, the survey team put in a series of test pits. One of these in a small clearing produced microblades, a core platform tablet, and a scraper from a buried cultural deposit. A grid was established with datum control provided by the USGS triangulation station as well as the site name. Ten 1-×-1-m squares were laid out in the clearing and excavated during the 1981 season. Tests were also conducted to the east of the site following the ridge line as well as to the north on the next high elevation. Two small flakes of chert were found in a test pit to the east of the site. All other test pits were without cultural materials.

The 1992 survey returned to further test Ilnuk as well as to conduct further tests along the ridge. Twenty 1-×-1-m squares were excavated with the recovery of some new artifact types to add to the site inventory. The distribution of artifacts was linear, following the slope.

Site Characteristics

The site proved to be fairly shallow, with approximately 30 cm of sediment over a weathered limestone regolith. The silty-sand sediments were aeolian in origin as windblown silt deposits extend over the region, up to about 250 m (Cady et al. 1955) and were derived from late Pleistocene glacial outwash deposits (Lea 1990; Lea and Waythomas 1990). The present soil structure is that of a spodosol (Soil

Survey Staff 1975). Since the position of the cultural deposit relative to the soil horizons is critical in terms of dating events, some attention will be given to a description of the soil zones.

The upper horizon is an organic zone containing roots, decomposing organics, humus, and charcoal (O horizon) underlain by what we thought was an A2 horizon—pinkish grey, sandy silt with flecks of charcoal. This turned out to be volcanic ash. Underlying it was a dark organic stained mineral horizon containing rootlets. This may be an A1 or a palaeosol. Underlying that horizon is a 20–30 cm deposit of reddish brown, silty sand (loess) that is high in organics and which had been illuviated to form a series of B horizons—B2HIR, B2IR, B3, B3(1). The lower part of the loess deposit—B3(1)—is very dark grey and may derive its color and texture from the weathered surface of the limestone substrate. Within the basal part of the loess were numerous flakes, microblades, and, occasionally, burned bone. Below the loess deposit was the fractured and weathered regolith (C horizon). Frost-wedge features were noted in the regolith and were filled with silt. Often these features contained an abundance of artifacts.

This stratigraphic sequence occurred throughout the site. There was no evidence of tree-throw disturbance as the buried ash horizon was largely unbroken throughout the site. The crown type of supporting root system of the present trees is aligned parallel to the ash layer and does not penetrate it. Any tree-fall disturbance would be largely restricted to the organic horizons. Within the site, nevertheless, there is clear evidence of the upward movement of artifacts. Although the majority of cores, microblades, and flakes lay below the tephra layer, artifacts were found throughout the loess deposit, through the ash zone and into the bottom of the organic mat as a result of upward movement generated by cryogenic processes. It was evident that the artifacts from the upper part of the site strata were the same as those found in the lowest levels of the site and in terms of assemblage analysis represent only a single occupation event.

In spite of the cryoturbation there seems to have been little horizontal movement, for the plot maps reveal tight concentrations of workshop debris. The

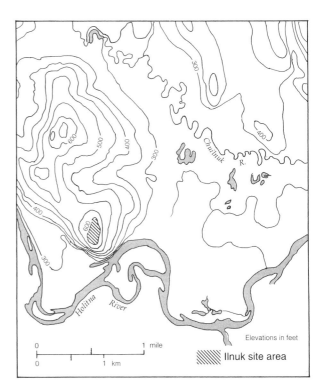

Figure 10–4   Topographical map of the Ilnuk site.

site area was relatively small, measuring approximately 8 by 15 m.

The volcanic ash deposit in the Ilnuk site has been identified as a tephra derived from the Aniakchak crater on the Alaska Peninsula, with an estimated age of eruption between 3,400 and 4,000 BP (Riehle et al. 1987). The Aniakchak crater tephra had a fairly wide distribution as it has been recovered from sites in the Alaska Peninsula, Holitna River (Ilnuk), and Norton Sound areas (Riehle et al. 1987). From the stratigraphy at the Ilnuk site, it can be demonstrated that the deposit of tephra lies above the cultural layer.

Dispersed bits of charcoal were recovered from the organic horizon and the upper part of the tephra layer. Radiocarbon dates are consistent in age: 195 ± 80 (WSU-4475), 1630 ± 80 (WSU-4476), 1830 ± 90 (WSU-4477), and 3684 ± 98 BP (WSU-2759) (Table 10–4). The latter sample of charcoal was taken from the surface of the tephra layer and undoubtedly represents a local wildfire which burned

Table 10–4   Radiocarbon Age Estimates on Charcoal Samples, Ilnuk (SLT 067) Site

Labratory Number	Square	Location within Square	Sample Wt.	¹⁴C Age (BP) Half Life Age 5570 ± 30
4475 WSU	N0W1	N 0.70–0.85, W 1.0–1.44, 10–15 cm below secondary datum. Duff zone.	4.75 g charcoal	195 ± 80
4476 WSU	N0W1	N 0.55–0.75, W 1.2–1.42, 16–17 cm below secondary datum. Charcoal over and in contact with volcanic ash layer	2.6 g charcoal	1630 ± 80
4477 WSU	N0W1	Dispersed charcoal in volcanic ash zone from over surface of square. 14–16 cm below secondary datum.	6.25 g charcoal	1830 ± 90
2759 WSU	N1W0	Sample collected from A2 horizon. Overlying O and A1 horizons also contain charcoal. Probable that charcoal from wild fire.	2.13 g charcoal	3684 ± 98
AA-1721	Collected from several squares	Pyrolysed organic material. No collagen due to burnt bone.	calcined bone	4390 ± 100
Beta 64472 [CAMS 8282]	N0W1	Bottom of B horizon, over limestone regolith, dispersed charcoal at depth of 28–29 cm below secondary datum.	0.62 g charcoal	2220 ± 70

over the site region some time after the volcanic ash fall. It is not clear if the wildfire was an aftermath of the ash fall. An example from a more recent episode of volcanic activity may help illustrate the problem. The tephra from the 1980 eruption of Mount St. Helens in the state of Washington was accompanied by large numbers of charcoal fragments derived from the blasted and burned forest on the sides of the mountain (Mehringer 1985:184).

In other parts of southwestern Alaska, Ager (Ager 1982, 1983) detected volcanic ash layers in the upper part (*Alnus* zone) of his pollen cores from Tungak Lake on the Yukon Delta and from Puyuk and Zagoskin lakes on St. Michael's Island. The ash

from these locations has been identified as tephra from the Aniakchak crater (Riehle et al. 1987). Radiocarbon dates from below and above the ash horizon in two of the pollen cores provided dates of 5645 ± 250 and 3400 ± 750 BP (Tungak Lake) and 6430 ± 90 and 4839 ± 80 BP (Zagoskin Lake) (Ager 1982, 1983). The upper date from Tungak Lake is roughly contemporary with the date from the Ilnuk site given its large standard deviation. An attempt to derive a date from the calcined bone at Ilnuk proved to be unreliable, as only the inorganic carbon fraction remained. The sample yielded a date of 4390 ± 100 BP (AA-1721). (See Table 10–4).

Samples 4476 and 4477 appear to have come

from relatively the same horizon within the site. The date ranges of 1630±80 BP (1550–1710) and 1830±90 BP (1740–1920) do not overlap at one sigma but do at two sigma (1470–1790) and (1650–2010). The date (3684±98 BP) of sample 2759, which is reportedly from charcoal in the A2 (tephra deposit), is older by a factor of two and is puzzling as to the discrepancy. The date for the calcined bone is regarded as unreliable as there was no collagen in the sample and the source of the carbon is questionable. Hopes for an older date from charcoal apparently below the tephra level were dashed when the AMS date of 2220±70 BP (Beta 64472-CAMS 8282) was received.

The tephra is derived from an eruption of the Aniakchak caldera with an estimated age of 4,000 BP. It is apparent that all of the charcoal samples that we have recovered are from wildfires that post-date the tephra event and the earlier archaeological component (Denali complex).

ARTIFACT ASSEMBLAGE

The primary activity at the site was the production of microblades from wedge-shaped frontal microblade cores for insertion into bone or antler arrowheads. The lithic inventory from the Ilnuk site consists of microblade cores, microblades, platform tablets, Donnelly-type burins, burin spalls, a biface midsection, scrapers, gravers, cobble choppers, utilized flakes, trimming flakes, and debitage (Table 10–5).

Cores and Core Parts   There are 12 frontal, wedge-shaped microblade core types of the Denali complex of Palaeo-Arctic tradition–type (Fig.10–5: *a–c*) and 1 conical microblade core (Fig. 10–5: *d*) that represents a variant on the frontal form. The cores had been rejuvenated at the site, for we recovered core tablets (Fig. 10–5: *e–g*) that could be fitted onto microblade cores as well as sections of broken cores that did fit together. Conjoining pieces were found in adjacent squares or in the next one over. The large number of platform-shaping flakes (gull-wing shape) indicated that rejuvenation of the platform by detachment of a core platform tablet often required readjustment to level out the platform.

Table 10–5   Artifact inventory, Ilnuk Site (SLT 067)

Stone Artifacts	1981	1992
Biface		
mid	1	—
Microblade Cores		
wedge shaped, frontal	6	6
conical	1	—
core platform tablets	19	9
platform flakes	122	13*
ridge flakes	8	20
Microblades (total)	577	1,507
Donnelly Burins	5	3
burin spalls	8	1
Scrapers		
end/side	5	1
Utilized Flakes	8	3
Gravers	—	3
Cobble Choppers	—	2
Pebbles (manuports)	—	47
Debitage (flakes)	720	1,510
Bone Artifact		
Bone Arrowhead		
mid-section, slotted for		
insertion of microblades	1	—
Bone Fragments		
burned/calcined	128	16
Subtotal	1,609	3,141
Total	4,750	

*Possibly not accurate count as many may have been included in debitage category

Microblades   The relatively large number of microblades (2,084—whole and fragmentary) recovered indicates an area of concentrated flaking activity. Some of the microblades were snapped and show fine edge retouch, perhaps from use.

Burins and Burin Spalls   The Donnelly burins (Fig. 10–5: *i–k*) exhibit use retouch at the conjunction of the facial edges rather than at the tip, indicating the use of these tools as a type of steep-angled scraper rather than a grooving tool. Burin spalls have side-edge use retouch as well as proximal end-retouch and after detachment from the parent piece were probably used to cut grooves in bone tools.

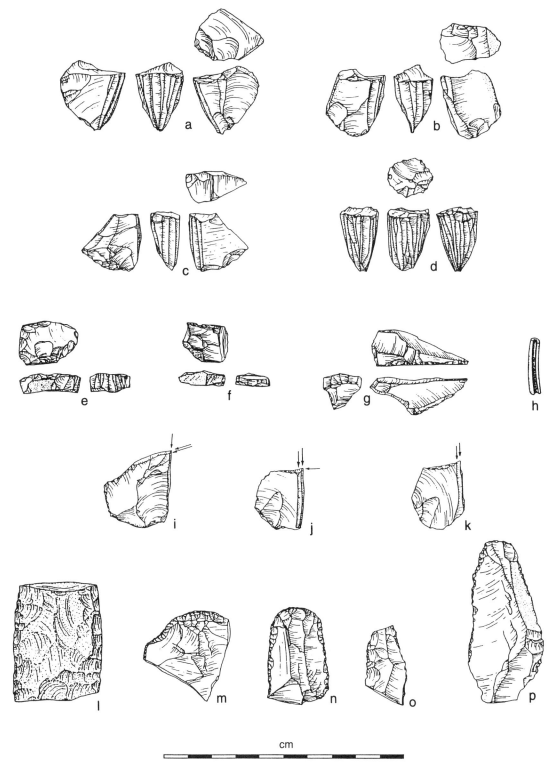

cm

Figure 10–5   The Ilnuk site: wedge-shaped microblade cores, frontal (*a–c*); conical microblade core (*d*); microblade core platform tablets (*e–g*); section of bone projectile point with side slot for microblades (*h*); Donnelly burins (*i–k*); biface mid-section (*l*); end scraper (*m*); end and side scraper (*n*); utilized flakes (*o, p*).

Bone Artifact   A midsection fragment of a bone arrowhead slotted to receive inset microblades (Fig. 10–5: *h*) was recovered with other pieces of broken burned bone from the central area of our excavations. The slot in the arrowhead is 2.0 mm wide, 2.0 mm deep, and rounded at the bottom of the slot. The rounded proximal ends of the burin spalls conform fairly closely to the shape of the bottom of the slot in the bone arrowhead and suggest that these indeed were the tools utilized.

Bifacial Tools   Only one medial section of a wide, relatively thin biface was recovered in two seasons of excavation.

Cobble Tools   One of the 2 cobble choppers found in 1992 was found vertically aligned and surrounded by several hundred microblades, microblade cores, platform tablets, a Donnelly burin, and a graver/notch. The chopper may have served as an anvil stone on which to rest the keels of microblade cores.

Scrapers   End scrapers and utilized flakes are present but not in large numbers.

Gravers   Gravers or notches on flakes are one of the new tool forms recovered during the 1992 excavation.

FAUNAL REMAINS

The several dispersed calcined bone fragments appear to have come from a cervid, but no closer identification was possible (Dr. Carl Gustafson, personal communication 1982). Tooth enamel fragments may have come from either *Bison* or *Cervus* (Wapiti) (Dr. R. Dale Guthrie, personal communication 1982). The remains of large herbivores support the use of the site as a camp associated with the hunting of large game.

SUMMARY

The Ilnuk site, now obscured by forest vegetation, would have had at the time of occupation a cover of low shrub vegetation and afforded a clear view of the eastward slopes of the Kuskokwim Mountains and the Holitna River valley. Alder expanded into southwestern Alaska between 7,500 and 7,000 BP and spruce about 5,500 BP (Ager 1982, 1983). Prior to the alder invasion, a birch shrub, tundra-type vegetation with willow, grasses, sedges, and herbs would have been characteristic.

Close correspondences in tool types and assemblage composition are to be found with the Denali complex sites in central Alaska such as Donnelly Ridge and Dry Creek (West 1967, 1981, this volume; Powers et al. 1983, Hoffecker et al., this volume) as well as with Palaeo-Arctic tradition sites in western Alaska, e.g., Akmak–Kobuk complexes at Onion Portage (Anderson 1988) and the Ugashik Narrows phase on the Alaska Peninsula (Henn 1978). Through typological comparison, the site can be tentatively dated between the late Pleistocene and the early Holocene.

The large number of microblades recovered suggests that the site was a major base camp where the manufacture of microblades for insertion into bone or antler points was an important aspect of the seasonal activity. From the small amount of larger flake debris, it can be hypothesized that the site occupants brought finished microblade cores with them to the site. The bulk of the lithic assemblage resulted from the detachment of microblades, the rejuvenation of microblade cores by the detachment of platform tablets and flakes from around the core platform. Numerous cores were discarded once their mass had been sufficiently reduced in either platform length or flute face height. The use of burin spalls for grooving was noted earlier by Giddings (1964) and indicates, as does the slotted bone fragment, another important task accomplished at the site, the production of bone tools. The few bits of burned bone from cervids and possibly bison indicate that processing of animal bones also occurred at the site.

From their general locations on ridgetops, it is natural to assume that the Ilnuk, Spein Mountain, and Nukluk Mountain sites were hunting camps associated with the procurement of migratory game. The artifact inventory suggests that in addition to game procurement and butchery, there was a large expenditure of time devoted to artifact manufacture. From the amount and range of artifacts recovered from these sites, particularly Spein Mountain and Ilnuk, it would appear that these are camps that were part of the seasonal round and were repeatedly occupied. Such sites could have been long-term summer habitation sites, but, as they would have been exposed to the icy blasts of winter, were not likely to have been winter camps.

CAVE 1, LIME HILLS

Robert E. Ackerman

Archaeological investigations were initiated by Dr. Thomas Bundtzen of the Alaska State Division of Geological and Geophysical Surveys (DGGS). In the course of his 1992 geological reconnaissance of the Lime Hills, he noted that one cave especially among the solution features observed appeared promising as an archaeological site. At the instance of Douglas Reger, Alaska State Office of History and Archaeology, Dr. Bundtzen invited the writer to accompany his survey group and make test excavations in the cave. Fieldwork was conducted June 15 to June 25, 1993 (Ackerman 1993).

The cave is located at the eastern extremity of the Lime Hills at an estimated elevation of 527 m (altimeter reading) above sea level. The present entrance is 6.4 m wide and 2.0–2.5 m high. The cave extends back from the entrance as a narrow, curving corridor for a distance of 17.7 m.

Surface inspection of the cave floor revealed porcupines to be present residents. There were also isolated fragments of chewed bone scattered about and, in addition, a considerable amount of partially burned wood, in small pieces as well as log fragments. The age and source of this material are unknown. Rockfall continues to contribute to the formation of cave sediment.

EXCAVATION

The southern portion of the cave entrance was selected for initial testing. Excavation began with two 1-×-1-m squares laid out on a north-south grid (squares N0E0 and N0E1). The only artifacts recovered during the excavation of squares N0E0 and N0E1 were a stone adze head, a worked piece of antler, and a caribou humerus with possible cutmarks. Faunal material was, however, abundant. Notable was the recovery of a bison astragalus and a left radius, humerus, and metapodial of caribou.

Two additional squares (N3E0 and N4E2) were then excavated in the main corridor of the cave. The upper part of N3E0 contained a few small fragments of bone and a basal fragment of a broad bone point or knife recovered at a depth of 58 cm. Preservation was better in the deeper part of the square with larger bone fragments recovered from a depth of 77 to 115 cm. The lowest item, at a depth of 114–115 cm, was a broken caribou metapodial that may have been used as a fleshing tool.

Square N4E2 contained several bird bones in the upper part of the deposit. At a depth of 36–40 cm, there was recovered a broken bone or antler arrowhead with a beveled tang and narrow grooves on opposing faces for the insertion of microblades. Below the bone arrowhead, fragments of caribou metapodials and a complete caribou pelvis were recovered.

The profile of the east wall of square N3E0 (Fig. 10–6) illustrates the cave stratigraphy. The stratigraphy appears to be undisturbed, with faunal material recovered throughout, although the better preserved is from the bottom levels. A description of the soil horizons accompanies the profile.

Charcoal samples were recovered from squares N3E0 and N4E2. During the cleaning of the charcoal sample from N4E2, a medial fragment of a microblade was recovered. No other microblades were noted during the course of the excavation although small fragments may have been missed since the sediment tended to roll up into clay clumps on the screens.

ARTIFACT ASSEMBLAGE

Base of an Antler Point or Knife   This triangular-shaped specimen (recovered from square N3E0) measures 1.89 cm wide at the break, 5.56 cm in

Area of Lime Hills showing the cave, indicated by arrows.

length, and is rather flat in cross section with rounded sides. Thickness varies from 0.14 cm at the flattened tip to 0.81 cm at the break (Fig. 10–7: *a*).

This appears to be the base of a bone point or knife (rather than the tip) as there is heavy cross-hatched scoring of the faces and along its edges presumably to aid in the hafting. The scoring shows both light and heavy incisions. Polish is also evident along the edges and for about 2 cm from the end of the artifact on the faces, as if it had been held in a haft. Traces of red ochre were noted within the scored areas as well as on one of the polished areas of the tool.

Side-Slotted Antler or Bone Arrowhead This basal fragment was recovered from square N4E2 (Fig. 10–7: *b*). It is oval in cross section and tapers from the midsection, where it is broken, to the tang. Length: 10.7 cm, width: 0.49 to 0.71 cm; thickness: 0.35 to 0.59 cm. The arrowhead was cut from a piece of antler or bone and then ground and polished to its final form. The basal tang (4.9 cm long) was created by beveling one face. Side-blade slots (3.49 and 3.23 cm long) were cut with a sharp-edged tool—such as a flake, rather than a burin—as cut-marks

are visible on the bottoms and sides of the somewhat irregular grooves. The slots are not directly opposite each other, but instead are offset relative to each other (Fig. 10–7: *b*). In spite of this precaution, the arrowhead still broke across the structurally weakened area. The arrowhead is without decoration or owner's marks.

Stone Adze Head A roughly shaped object, possibly an adze head, 19.6 cm long, 4.4 cm wide, and 2.0 cm thick, was made from a banded sedimentary rock. It may be that the piece was once thicker in cross section. Since the object is quite eroded, any traces of possible grinding marks left from its manufacture are gone. The two planes of the bit section meet at a fairly acute angle (42°), and the leading edge of the bit is rounded, as could be expected from use. At the middle part of the upper edge, there is a shallow groove.

Microblade Fragment As noted above, this is a medial section of a microblade, 0.56 cm wide and 0.7 cm long, recovered from a charcoal sample. It has a single arris and is made of an indurated sedimentary stone such as silicified slate or argillite.

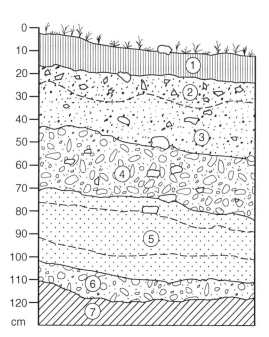

Figure 10–6   Lime Hills. Stratigraphic profile of square N3E0: (1) reddish brown sod intermixed with rockfall, clasts from pebble to cobble size; (2) dark reddish brown sandy to silty loam with angular clasts, probably part of an A horizon; (3) similar to stratum 2 but with fewer clasts; (4) yellow-brown sediment derived from weathering limestone, with numerous clasts, similar to C horizon; (5) reddish brown sandy silty loam with clasts from pebble to cobble size, apparent mineral or organic staining at top and bottom, darker at bottom possibly from accumulation of organics; (6) finely broken-up rockfall with abundant angular clasts; (7) unexcavated.

Flesher or Beamer   This specimen, a caribou metatarsal, is 21.8 cm long and was broken lengthwise, with two impact scars noted along the broken edge. The proximal end was partially intact while the distal articulations had been broken away. Along the length of the split metatarsal and on the distal end there is wear polish. The bone was unusual as it had not been broken into shorter segments like most of the bones in the cave and it did not have the heavy chatter along the edge from carnivore gnawing. Numerous short striations, perpendicular to the broken edges of the metapodial, were noted. At first

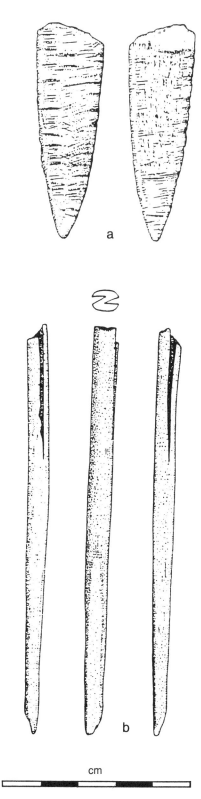

Figure 10–7   Lime Hills, Cave 1: base of bone point (*a*); basal section of slotted bone point (*b*).

Table 10–6   Radiocarbon Dates, Cave 1, Lime Hills

Sample No.	Material	Provenience	Date
Beta 67668 [CAMS 9897]	Partially burned wood and wood chips	N4E2 N4.50, E2.92, depth 64 cm	8150 ± 80 BP
WSU-4504	Charcoal	Sq. N3E0 N3.87, E0.80, depth 55 cm	8480 ± 260 BP
WSU-4505	Charcoal	Sq. N3E0 N3.18, E0.56, depth 60 cm	8480 ± 190 BP
Beta 67667 [CAMS 9896]	Charcoal associated with microblade fragment	Sq. N4E2 N4.32–4.30, E3.0, depth 48 cm	9530 ± 60 BP
Beta 67671	Collagen from broken caribou metapodial used as defleshing tool	N3E0 N3.40, E0.27 and N3.21, E0.14, depth 114–115 cm	13,130 ± 180 BP
Beta 67669	Collagen from caribou humerus with butchery cut-marks	N0E0 N0.51, E0.43 and N0.42, E0.24, depth 91–96 cm	15,690 ± 140 BP
Beta 67670	Collagen from bison astragalus	N0E1 N0.35, E1.38, depth 70–77 cm	27,950 ± 560 BP

these were taken to be use-wear marks that could have come from abrasion against a surface, such as would be found on a scraping tool used to remove the fat and bits of flesh adhering to the inner surface of hides. The amount of polish along the broken edges and on the tip of the metapodial also suggested use-wear polish. Such tools have been identified as fleshers in the archaeological literature (Morlan and Cinq-Mars 1982: Fig. 9). Binford (1981) observed that tethered dogs who were given whole metapodials often spent long hours rolling the metapodials around in their mouths, chewing, licking, and sucking on the ends of the shafts. Such gnawed bones had polish on the ends, and the broken edges were rounded and smooth. While his examples (Binford 1981: Fig. 3.52) are not duplicates of the caribou metapodial from square N3E0, his study does suggest that alternatives to human agencies need to be given careful consideration.

The bone was photographed and then submitted as a source of collagen for radiocarbon dating.

Cut or Worked Bone   The basal section of a caribou antler recovered from the upper 10 cm of square N0E1 had the pedicle cut away just above the burr and before the brow tine. This area was then slightly rounded and smoothed. The brow tine is missing, and there are gnaw marks on the remaining stub. The first beam tine has also been heavily chewed back toward the shaft. Although the shaft beyond the first beam tine is missing, there are heavy gnaw marks on the remaining portion. A section of the obverse face of the shaft may have been removed through a cut-and-splinter technique, as one edge is quite straight and smoothed while the opposite edge has been gnawed. The utilized piece of antler was probably discarded and only later gnawed by carnivores or rodents.

A caribou humerus with several cut-marks that appear to have been produced by stone tools was recovered at a depth of 91–96 cm in square N0E0. The cut marks are straight or curving, at an angle rather than transverse to the shaft, with sharp side walls, and are very narrow across the cut. Such cuts could have been made in defleshing the carcass for drying (Binford 1978:97). Carnivore and rodent tooth marks, in contrast, are broad, generally transverse to the shaft, and generally ragged (skip marks show where the teeth bounced against the bone). The medial and lateral condyles of the distal portion of the shaft had been heavily gnawed, exposing the cancellous portion of the bone.

The bone was photographed and submitted as a source of collagen for radiocarbon dating.

PALAEOENVIRONMENTAL DATA

Fauna   Dr. Carl Gustafson, in a preliminary examination of the faunal materials recovered from the excavations, identified these forms: caribou (*Rangifer tarandus*), hare (*Lepus americanus* or *L. othus*), sheep (*Ovis dalli*), porcupine (*Erethizon dorsatum*), possible grizzly bear (*Ursus arctos*), fox (*Alopex lagopus* or *Vulpes fulva*), possible moose (*Alces alces*), bird, ground squirrel (*Citellus parryii*), and bones of at least three different species of microtine rodent as well as some fish were present. The single bison bone, an astragalus, was stained a darker color than other bones at the same level and proved to be much older than the rest of the bone assemblage.

All of the species noted thus far in our excavation were part of the faunal assemblage of the late Pleistocene and, except for bison, can be found in the region today. The microtine rodent bones may have been brought into the cave mouth by raptors and may by their diversity provide further insights into the nature of the environment.

Flora   Dr. Peter Mehringer analyzed sediments from squares N0E1 and N4E2 for pollen. Pollen grains were present in quantity only from the lowest soil horizon of square N0E1, and this spectrum consisted of *Salix*, sedge, grass, *Gilia*, and *Aster*-type. It is notable that pollen of *Alnus* and *Picea*

were not present. Since these two tree types exist in front of the cave at present and are great pollen producers, it is evident that the deposits were well sealed and that there has been no mixing of the cave sediments. The lack of tree pollen also suggests that the surrounding vegetation was a shrub-tundra type during the earlier part of the cave occupation.

AGE AND CULTURAL DETERMINATION

Charcoal and bone samples from the cave were submitted for radiocarbon age determination. The results are presented in Table 10–6. Two periods of occupation are indicated by these radiocarbon dates—an upper component containing artifacts of bone, antler, and stone, and a lower component which contains a large amount of faunal material, some of which may have been modified by humans.

The earliest dated artifact from the upper component is a microblade fragment recovered with a charcoal sample (Beta 67667/CAMS 9896) from square N4E2 and thus associated with the date of 9530 ± 60 BP. In the same square, a bone or antler arrowhead with side-blade slots (Fig. 10–7b) was recovered. Similar arrowheads dating to 9070 ± 150 BP (Larsen 1968:54–56, Figure 36, Plate III 1–7) have been found in layer III of Cave 2 at the Trail Creek site on Seward Peninsula. The Lime Hills specimen from Cave 1 may be similar in age, as it was found above the 9500 BP charcoal sample. Later dates of 8480 ± 260 and 8480 ± 190 BP immediately above and below the scarfed base or tip of an antler tool from square N3E0 indicate a continued use of the cave into the early part of the Holocene. The date of 8150 ± 80 BP (Beta 67668) raises some questions—given the tool's depth in square N4E2—which need to be addressed by further radiocarbon dating. Carbon samples from the upper 20–30 cm of the site are yet to be dated.

The 9,500 BP date and the presence of a microblade and a side-slotted arrowhead in the lower part of the upper component indicate that the cave occupancy was contemporary with the Denali complex of central Alaska (West 1967, 1975, 1981, this volume; Powers et al. 1983; Powers and Hoffecker 1989; Hoffecker et al. 1993). Until the recovery of a side-slotted bone arrowhead fragment associated with the Denali complex assemblage at the Ilnuk site on the Holitna River, it could not be demonstrated that such arrowheads belonged to the complex. Earlier, Larsen (1968:71–74) had noted that the microblades and side-slotted antler arrowheads from Cave 2 at Trail Creek probably belonged to complexes with frontal, wedge-shaped microblade cores, such as the Akmak and Kobuk assemblages of Onion Portage (Anderson 1970, 1988). The period 8,500–8,000 BP and later is poorly defined thus far in the site. The antler base or tip fragment of a tool, the stone adze-like implement, and the cut antler do not provide any insights into the cultural identity of the later cave occupants.

Three bone samples were selected for dating to determine the age of the lower component. The bison astragalus was submitted for dating to determine when bison would have been present. Two caribou bones from the bottom of the excavations were submitted to determine the lower limit of the cave occupation thus far uncovered by our excavations. Both of these bones were well preserved and provided satisfactory amounts of collagen for dating. The caribou metapodial (a possible fleshing tool) provided a date of $13,130 \pm 180$ BP (Beta 67671), and the caribou humerus with probable cut-marks provided a date of $15,690 \pm 140$ BP (Beta 67669). This indicated that the cave was occupied during the late Pleistocene. The age of the bison bone ($27,950 \pm 560$ BP [Beta 67670]) was quite unexpected and immediately suggested that it had been brought into the cave from a fossil locality.

At the current level of investigation there is some suggestion of a human presence during the time of the lower component, but until further data are obtained an equally valid case could be made that the cave was occupied only by carnivores and rodents. Since no stone artifacts have thus far been recovered from the lower component, the question of whether humans or animals were responsible for the faunal remains cannot be positively resolved. It should be noted, however, that two of the artifacts from the upper component were also made of bone or antler. Studies by Binford (1978, 1981), Bonnichsen (1979), Morlan (1980), Morlan and Cinq-Mars (1982), Dixon (1984), Guthrie (1984), Blumenschine

and Marean (1993), and Enloe (1993) do suggest multiple causes for bone alteration involving humans, animals, or a combination of both.

As noted previously in other studies, such as that of Cinq-Mars (1979) and Morlan and Cinq-Mars (1982) at Bluefish Caves and Dixon (1984) at the Porcupine River caves, a human presence is difficult to determine in the absence of stone tools. It should be noted that only two stone tools were recovered from the upper component of Cave 1, where there is no question of a human presence. These were a microblade fragment and the possible adze head. Microblade fragments from the lower levels of the site could easily have been missed during the test excavations due to the lack of facilities for wet screening. Only through further testing of Cave 1 and other adjacent limestone caves can it be determined if there was a human presence in the Lime Hills region 13,000 to 15,000 years ago.

The assistance provided by Tom Bundtzen, Ellen Harris, Greg Laird, and Curvin Metzer during the excavation of Cave 1 is gratefully acknowledged, as is the mapping of the cave by Curvin Metzer and Sam Dunaway and the original artifact drawings by Sarah Moore.

REFERENCES CITED

Ackerman, R. E. 1985. Southwestern Alaska Archeological Survey. *National Geographic Society Research Reports* 19:67–94.

———. 1987. Mid-Holocene Occupation of Interior Southwestern Alaska. In *Man and the Mid-Holocene Climatic Optimum*, ed. N. A. McKinnon and G. S. L. Stuart, pp. 181–92. Calgary: University of Calgary Archaeological Association.

———. 1993. Investigation of Cave 1, Lime Hills Regions, Southwestern Alaska. Report to Alaska State Office of History and Archaeology, Division of Parks and Outdoor Recreation and Division of Geological and Geophysical Surveys, Anchorage.

Ager, T. A. 1982. Vegetational History of Western Alaska during the Wisconsin Glacial Interval and the Holocene. In *Paleoecology of Beringia*, ed. D. M. Hopkins, J. V. Matthews, Jr., C. E. Schweger, and S. B. Young, pp. 75–94. New York: Academic Press.

———. 1983. Holocene Vegetational History of Alaska. In

Late-Quaternary Environments of the United States, ed. H. E. Wright, Jr. Vol. 2 *The Holocene*, ed. H. E. Wright, Jr., pp. 128–41. Minneapolis: University of Minnesota Press.

Ager, T. A., and L. Brubaker. 1985. Quaternary Palynology and Vegetational History of Alaska. In *Pollen Records of Late Quaternary North American Sediments*, ed. V. M. Bryant and R. G. Holloway, pp. 353–84. Dallas: American Association of Stratigraphic Palynologists Foundation.

Alexander, H. L. 1974. The Association of Aurignacoid Elements with Fluted Point Complexes in North America. In *International Conference on the Prehistory and Paleoecology of Western North American Arctic and Subarctic*, ed. S. Raymond and P. Schledermann, pp. 21–32. Calgary: University of Calgary.

———. 1987. *Putu: A Fluted Point Site in Alaska*. Publication 17. Burnaby, B. C.: Simon Fraser University, Department of Archaeology.

Anderson, D. D. 1970. Akmak: An Early Archeological Assemblage from Onion Portage, Northwest Alaska. *Acta Arctica*, Fasc. 16. Copenhagen.

———. 1988. Onion Portage: The Archaeology of a Stratified Site from the Kobuk River, Northwestern Alaska. *Anthropological Papers of the University of Alaska* 22(1–2).

Binford, L. R. 1978. *Nunamiut Ethnoarchaeology*. New York: Academic Press.

———. 1981. *Bones: Ancient Men and Modern Myths*. New York: Academic Press.

Blumenschine, R. J., and C. W. Marean. 1993. A Carnivore's View of Archaeological Bone Assemblages. In *From Bones to Behavior: Ethnoarchaeological and Experimental Contributions to the Interpretation of Faunal Remains*, ed. J. Hudson, pp. 273–300. Carbondale: Southern Illinois University, Center for Archaeological Investigations.

Bonnichsen, R. 1979. *Pleistocene Bone Technology in the Beringian Refugium*. Archaeological Survey of Canada, Mercury Series, Paper 89. Ottawa: National Museum of Man.

Bundtzen, T. K. 1980. Multiple Glaciation in the Beaver Mountains, Western Interior Alaska. *Geologic Report* 63:11–18. Fairbanks: Alaska Division of Geologic and Geophysical Surveys.

Bundtzen, T. K., and W. G. Gilbert. 1983. Outline of the Geology and Mineral Resources of the Upper Kuskokwim Region, Alaska. *Journal of the Alaska Geological Society* 3:101–18.

Cady, W. M., R. E. Wallace, J. M. Hoare, and E. J. Webber. 1955. *The Central Kuskokwim Region, Alaska*. U.S. Geological Survey Professional Paper 268.

Cinq-Mars, J. 1979. Bluefish Cave: A Late Pleistocene Eastern Beringian Cave Deposit in the Northern Yukon. *Canadian Journal of Archaeology* 3:1–32.

Coulter, H. W., D. M. Hopkins, T. N. V. Karlstrom, T. L.

Péwé, C. Wahrhaftig, and J. R. Williams. 1965. *Map Showing Extent of Glaciations in Alaska*. Miscellaneous Map Investigations, Map I-415. Washington, D.C.: U.S. Geological Survey.

Dixon, E. J. 1984. Context and Environment in Taphonomic Analysis: Examples from Alaska's Porcupine River Caves. *Quaternary Research* 22:201–15.

Dutro, J. T., and W. P. Patton, Jr. 1982. *New Paleozoic Formations in the Northwest Kuskokwim Mountains, West-Central Alaska*. U.S. Geological Survey Bulletin 1529H:H13–H22.

Eberlein, G. E., and M. Churkin, Jr. 1970. *Paleozoic Stratigraphy in the Northwest Coast Area of Prince of Wales Island, Southeastern Alaska*. U.S. Geological Survey Bulletin 1284.

Enloe, J. G. 1993. Ethnoarchaeology of Marrow Cracking: Implications for the Recognition of Prehistoric Subsistence Organization. In *From Bones to Behavior: Ethnoarchaeological and Experimental Contributions to the Interpretation of Faunal Remains*, ed. J. Hudson, pp. 82–97. Carbondale: Center for Archaeological Investigations, Southern Illinois University.

Ford, D. C. 1973. Development of the Canyons of the South Nahanni River, N. W. T. *Canadian Journal of Earth Sciences* 10(3):366–78.

Ford, D. C., J. C. Andrews, T. E. Day, S. A. Harris, J. B. MacPherson, S. Occhietta, W. F. Rannie, and H. O. Slaymaker. 1984. Symposium Canada: How Many Glaciations? *Canadian Geographer* 28(3):205–25.

Giddings, J. L. 1964. *The Archeology of Cape Denbigh*. Providence: Brown University Press.

Gilbert, W. G. 1981. *Preliminary Geologic Map and Geochemical Data of the Cheeneetnuk River Area, Alaska*. Open File Report 153, Alaska Division of Geologic and Geophysical Surveys, Anchorage.

Guthrie, R. D. 1984. The Evidence for Middle-Wisconsin Peopling of Beringia: An Evaluation. *Quaternary Research* 22(2):231–41.

Hamilton, T. D., and R. M. Thorson. 1983. The Cordilleran Ice Sheet in Alaska. In *Late-Quaternary Environments of the United States*, ed. H. E. Wright, Jr. Vol. 1, *The Late Pleistocene*, ed. S. C. Porter, pp. 38–52. Minneapolis: University of Minnesota Press.

Henn, W. 1978. *Archaeology on the Alaska Peninsula: The Ugashik Drainage, 1973–1975*. Anthropological Papers 14. Eugene: University of Oregon.

Hoare, J. M., and W. L. Coonrad. 1959. *Geology of the Bethel Quadrangle, Alaska*. Miscellaneous Geologic Investigations Map I-285. Washington, D.C.: U.S. Geological Survey.

Hoffecker, J. F., W. R. Powers, and T. Goebel. 1993. The Colonization of Beringia and the Peopling of the New World. *Science* 259:46–53.

Hopkins, D. M. 1982. Aspects of the Paleogeography of Beringia during the Late Pleistocene. In *Paleoecology of Beringia*, ed. D. M. Hopkins, J. V. Matthews, Jr.,

C. E. Schweger, and S. B. Young, pp. 3–28. New York: Academic Press.

Humphrey, R. L., Jr. 1970. The Prehistory of the Arctic Slope of Alaska: Pleistocene Cultural Relationships between Eurasia and North America. Ph.D. dissertation, Department of Anthropology, University of New Mexico.

Kline, J. T. 1983. Preliminary Quaternary Glacial Chronology for the Farewell Area, McGrath Quadrangle, Alaska. In *Glaciation in Alaska: Extended Abstracts from a Workshop,* ed. R. M. Thorson and T. D. Hamilton, pp. 57–61. Fairbanks: Alaska Quaternary Center, University of Alaska Museum.

Kline, J. T., and T. K. Bundtzen. 1986. Two Glacial Records from West-Central Alaska. In *Glaciation in Alaska: The Geologic Record,* ed. T. D. Hamilton, K. M. Reed, and R. M. Thorson, pp. 123–50. Anchorage: Alaska Geological Society.

Kunz, M. L. 1982. The Mesa Site: An Early Holocene Hunting Stand in the Iteriak Valley, Northern Alaska. *Anthropological Papers of the University of Alaska* 20(1–2):113–22.

———. 1992. The Mesa Site. *Alaska Anthropological Association Newsletter* 17(2):4–5.

Kunz, M. L., and R. E. Reanier. 1994. Paleoindians in Beringia: Evidence from Arctic Alaska. *Science* 263:660–62.

Larsen, H. 1968. Trail Creek: Final Report on the Excavation of Two Caves on Seward Peninsula, Alaska. *Acta Arctica,* Fasc. 15. Copenhagen.

Lauriol, B., D. C. Ford, and J. Cinq-Mars. 1989. Landscape Development from Caves and Speleothem Data: Preliminary Evidence from Northern Yukon. *Department of Geography Newsletter of the University of Ottawa.*

Lea, P. D. 1984. Paleoclimatic Implications of Late Pleistocene Glacial Asymmetry, Ahklun Mountains, Southwestern Alaska. *Geological Society of America Abstracts with Programs* 18:669.

———. 1990. Pleistocene Periglacial Eolian Deposits in Southwestern Alaska: Sedimentary Facies and Depositional Processes. *Journal of Sedimentary Petrology* 60(4):583–91.

Lea, P. D., and C. F. Waythomas. 1990. Late-Pleistocene Eolian Sand Sheets in Alaska. *Quaternary Research* 34:269–81.

Mehringer, P. J., Jr. 1985. Late-Quaternary Pollen Records from the Interior Pacific Northwest and Northern Great Basin of the United States. In *Pollen Records of Late-Quaternary North American Sediments,* ed. V. M. Bryant, Jr., and R. G. Holloway, pp. 167–89. Austin, Texas: American Association of Stratigraphic Palynologists Foundation.

Mochanov, Y. A. 1977. *Drevneishie Etapy Zaseleniya Chelovekom Severo-Vostochnoi Azii. (The Most Ancient Stages in the Settlement by Man of Northeast Asia.)* Novosibirsk: Nauka.

———. 1984. Paleolithic Finds in Siberia (Resume of Studies). In *Beringia in the Cenozoic Era,* ed. V. L. Kontrimavichus, pp. 694–724. New Delhi: Amerind Publishing Co.

Mochanov, Y. A., and S. A. Fedoseeva. 1984. Main Periods in the Ancient History of North-East Asia. In *Beringia in the Cenozoic Era,* ed. V. L. Kontrimavichus, pp. 669–93. New Delhi: Amerind Publishing Co.

Morlan, R. E. 1980. *Taphonomy and Archaeology in the Upper Pleistocene of the Northern Yukon Territory: A Glimpse of the Peopling of the New World.* Archaeological Survey of Canada, Mercury Series, Paper 94. Ottawa: National Museum of Man.

Morlan, R. E., and J. Cinq-Mars. 1982. Ancient Beringians: Human Occupation in the Late Pleistocene of Alaska and the Yukon Territory. In *Paleoecology of Beringia,* ed. D. M. Hopkins, J. V. Matthews, Jr., C. E. Schweger, and S. B. Young, pp. 329–52. New York: Academic Press.

Patton, W. W., Jr., E. J. Moll, J. T. Dutro, M. L. Silberman, and R. M. Chapman. 1980. *Preliminary Geologic Map of the Medfra Quadrangle, Alaska.* U.S. Geological Survey Open-File Report 80-811A.

Péwé, T. L. 1975. *Quaternary Geology of Alaska.* U.S. Geological Survey Professional Paper 835.

Powers, W. R., R. D. Guthrie, and J. F. Hoffecker. 1983. Dry Creek: Archeology and Paleoecology of a Late Pleistocene Alaskan Hunting Camp. Report submitted to the National Park Service, Washington, D.C.

Powers, W. R., and J. F. Hoffecker. 1989. Late Pleistocene Settlement in the Nenana Valley, Central Alaska. *American Antiquity* 54(2):263–87.

Riehle, J. R., C. E. Meyer, T. A. Ager, D. S. Kaufman, and R. E. Ackerman. 1987. The Aniakchak Tephra Deposit, a Late Holocene Marker Horizon in Western Alaska. In *Geologic Studies in Alaska by the U.S. Geological Survey during 1986.* U.S. Geological Survey Circular 998:19–22.

Short, S. K., S. A. Elias, C. F. Waythomas, and N. E. Williams. 1992. Fossil Pollen and Insect Evidence for Postglacial Environmental Conditions, Nushagak and Holitna Lowland Regions, Southwest Alaska. *Arctic* 45(4):381–92.

Soil Survey Staff. 1975. *Soil Taxonomy: A Basic System of Soil Classification for Making and Interpreting Soil Surveys.* Agriculture Handbook No. 436. Washington, D.C.: Soil Conservation Service, U.S. Department of Agriculture.

West, F. H. 1967. The Donnelly Ridge Site and the Definition of an Early Core and Blade Complex in Central Alaska. *American Antiquity* 32(3):360–82.

———. 1975. Dating the Denali Complex. *Arctic Anthropology* 12(1):76–81.

———. 1981. *The Archaeology of Beringia.* New York: Columbia University Press.

Chapter 11

SEWARD PENINSULA AND BROOKS RANGE, NORTHERN ALASKA

INTRODUCTION

Northern Alaska is characterized by two major climatic regimes: arctic in the north and continental, or subarctic, to the south. Corresponding to, and defining, this dichotomy is the distribution of major vegetation forms: generally, herbaceous tundra in the arctic and forest in the subarctic. These distributions are generalized, of course, and do not take into account the gross differences that accompany differences in altitude. The climate in the south of this region is that characteristic of all of the interior—a mean annual temperature below freezing and an extremely wide annual range of temperatures. A temperature of −62° C, from Prospect Creek on the south slope of the Brooks Range, is the lowest recorded in Alaska. In virtually the same area, an occasional summer high of 37° C is not unknown. In contrast, while arctic coldest temperatures are not as extreme, those of summer are much depressed in comparison with those of the interior.

Precipitation rates are lowest in the arctic (averaging about 17 cm per annum) but are also quite low in the south, averaging about 38 cm per annum. In both instances, however, evaporation rates are quite low so that the latter figure is fully adequate

for forest growth, while in the arctic the effectiveness of the much lower values is augmented by the universal occurrence of permafrost, which greatly enhances the retention of moisture, resulting in the common presence of relatively luxuriant herbaceous tundra.

Wet tundra occurs along the coast of northernmost Alaska and along the northern coast of the Seward Peninsula where permafrost is close to the surface. These areas are saturated, with great expanses of marsh, shallow ponds, and lakes. The presence of underground ice is evidenced by polygonal ground and ridges of peat. Sedges and cottongrass are common. In areas of better drainage, usually indicating a lower permafrost table, moist tundra is the rule. This tundra prevails over most of the arctic. Perhaps the most widely occurring plant in these assemblages is cottongrass (*Eriophorum*), found here in tussocks over extensive areas. Other plants—sedges and dwarf shrubs—occur sparingly.

Permafrost is found throughout the interior but whereas it is continuous in the arctic, south of that region it exists in discontinuous distribution or as isolated masses. True forest growth occurs in the south valleys of the Brooks Range—at low elevations and often ecotonally with shrub tundra. From there southward in the interior, spruce forest, the taiga, is the characteristic forest cover except where that is precluded by altitude.

The Brooks Range dominates the map of north-

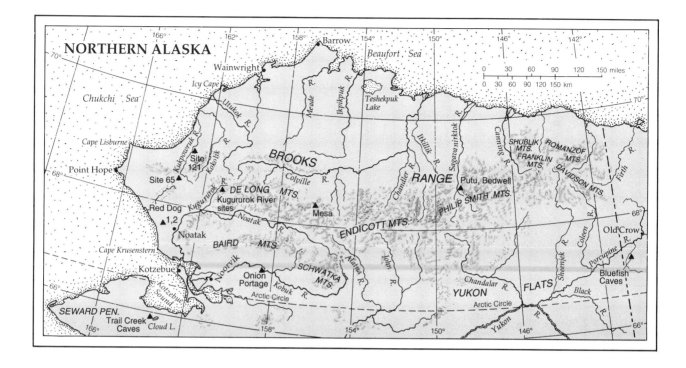

ern Alaska. It is made up of a series of distinct mountain groups. Forming the spine of the range are, in the east, the Davidson, Romanzof, and Philip Smith mountains. In the central sector are the Endicott and the Schwatka mountains, and in the west the Baird and De Long groups. In the east just north of the spine are the Franklin Mountains and north of them the Shubliks and the Sadlerochits. The highest elevations are in the east where, in the Franklins, Mount Chamberlin rises to over 2,700 meters. In the nearby Romanzof group, mounts Michelson and Isto are almost as high. The mountains of the east are rugged and precipitous while those in the west are considerably lower and have more rounded contours. At higher elevations, alpine tundra, fell fields, and permanent snow and ice are found, the latter condition especially in the east.

Flowing from the Brooks Range north to the Arctic Ocean are several relatively short rivers, principal among them the Canning, the Sagavanirktok, and the Itkillik. The latter is a tributary of the Colville, which, rising far in the interior of the eastern Brooks Range, is the largest of all the north-flowing rivers. The Noatak River, the only river whose course is wholly in the Range, flows westward, emptying into Kotzebue Sound. Paralleling it to the

south is the Kobuk. In the east, the Brooks Range is drained by several small streams which flow into the Yukon River after crossing a portion of the vast Yukon Flats. These include the Porcupine and the Chandalar. To the west, the Koyukuk River flows southwestward from the central Brooks Range and also enters the Yukon. Except to landward, this entire region is bounded by the Arctic Ocean.

Faunal differences in this great region are reflective of those in the environment. On the coast proper, on the sea ice, and in coastal waters are mammals—as well as other forms—upon which are built the traditional economies of the Alaskan Eskimos of this region. These include whales, seals of various species, walrus, and polar bear. Of land mammals, traditionally the most important is the caribou, large herds of which, moving seasonally, are found in the mountains and in adjacent lower areas. Caribou have constituted the single most important factor in the economies of the interior Athapaskans and Eskimos of the Brooks Range. Dall sheep are utilized, but play a minor economic role. The brown bear is relatively common in the north but has no economic importance. Wolves are likewise abundant as are various fur-bearers and have various degrees of economic interest.

Foothills of the western Brooks Range. (*Photo courtesy of Michael L. Kunz.*)

The entire coastal area, from the Seward Peninsula to the north coast, is inhabited by Inupiat Eskimos. That tongue is spoken across northernmost North America to Greenland. While these economies were primarily directed to the sea, inland resources were also significant, especially caribou. The Athapaskan-speakers of the Brooks Range are the most northerly of American Indians. Their linguistic family, Na-Dene, includes Tlingit, Eyak, and such distant groups as Navaho and Apache. One of the principal methods used to take caribou was by driving them into elongate "fences" or corrals.

Archaeological sites pertaining to Eskimo prehistory are fairly abundant on the coasts, but those relating to other folk and earlier occupations are much less frequent. Nevertheless, Trail Creek is an early such discovery, as are Kukpowruk and Onion Portage.

F. H. W.

BIBLIOGRAPHY

Gryne, G. 1958. Brooks Range. In *Landscapes of Alaska, Their Geologic Evolution*, ed. H. Williams, pp. 111–18. Berkeley and Los Angeles: University of California Press.

———. 1958. Arctic Slope. In *Landscapes of Alaska, Their Geologic Evolution*, ed. H. Williams, pp. 119–27. Berkeley and Los Angeles: University of California Press.

Hartman, C. W., and P. R. Johnson. 1978. *Environmental Atlas of Alaska*. Fairbanks: University of Alaska.

Hopkins, J. P., and D. M. Hopkins. 1958. Seward Peninsula. In *Landscapes of Alaska, Their Geologic Evolution*, ed. H. Williams, pp. 104–10. Berkeley and Los Angeles: University of California Press.

Manville, R. H., and S. P. Young. 1965. *Distribution of Alaskan Mammals*. Circular 211, Bureau of Sport Fisheries and Wildlife, Fish and Wildlife Service. Washington, D.C.: Department of the Interior.

Rieger, S., D. R. Schoephorster, and C. F. Furbush. 1979. *Exploratory Soil Survey of Alaska*. Washington, D.C.: Soil Conservation Service, U.S. Department of Agriculture.

Viereck, L. A., and E. L. Little, Jr. 1972. *Alaska Trees and Shrubs*. Agricultural Handbook 410. Washington, D.C.: Forest Service, U.S. Department of Agriculture.

———. 1975. *Atlas of United States Trees*. Vol. 2: *Alaska Trees and Common Shrubs*. Misc. Pub. 1293. Washington, D.C.: Forest Service, U.S. Department of Agriculture.

TRAIL CREEK CAVES, SEWARD PENINSULA*

Constance F. West, Editor

The caves are located on Trail Creek one mile above its junction with Cottonwood Creek, near Cloud Lake in the Imuruk region, Seward Peninsula, Alaska. It is an area of low hills rising to approximately 1,500 feet, covered for the most part with herbaceous tundra ranging into low shrub tundra which is dominated by dwarf birch and willow. Limestone cliffs rise sharply from the west bank of Trail Creek to a height of about 300 feet. Numerous natural cavities of various sizes are found in this formation, of which 12 are large enough to permit human entry.

History of Research

The site was originally discovered in 1928 by Taylor Moto and Alfred Karmun, Eskimos from Deering, a village about 30 miles to the northwest. In 1947, David M. Hopkins identified the caves, shown to him from the air, and, in 1948, he was the first to test them. The major excavations were carried out by Helge Larsen, of the Danish National Museum, under the sponsorship of the University of Alaska, the University of Pennsylvania Museum, and his own institution. He directed the investigations of Cave 2 and Cave 9 in 1949 and 1950.

Characterization of the Site

All 12 caves were tested, but only two, Cave 2 and Cave 9, revealed signs of any but the most recent habitation. In Cave 2 only were there found any artifacts older than mid-Holocene age. Located downstream from Cave 9, Cave 2 is situated about 27 m above the creek level. The cave is a narrow tunnel, 21.4 m long, rising slightly at the rear. The height before excavation ranged from 60 cm to 140 cm.

The deposits in the two excavated caves are best characterized as "cave breccia," a poorly sorted gravel-sand-silt mixture derived from weathering of cave walls and roof. In Cave 2, four stratigraphic levels were distinguished, primarily on the basis of color. These were traced, with some difficulty, from the front about 7 m into the back of the cave, and profiles were taken at each meter. From this point back to the 12-m profile, only three layers were distinguished, with layers I and II merging. From the 12-m point to the rear of the cave, only the lowest layer was discernible.

In the rather sparse upper levels were found remains attributed to Denbigh, Choris, and later Eskimo horizons. Only in Cave 2 were there found evidences of early occupation.

Cave 2 Artifact Assemblage

Two hundred and fifty-three lithic and osseous artifacts were recovered from the two caves. In addition, there were found 8 small irregular flakes, the only indication of flaking in the cave. Of this total, only 4 microblades and 7 antler points were considered to pertain to the oldest occupation. These were found in Cave 2, layer III, in the first 5 m from the cave mouth where stratigraphy was most clearly seen. A microblade and an antler point were found in close proximity to each other at a depth of 90 cm. Two antler points found in layer II were probably displaced by ground squirrel activity. A fragment of a chalcedony point was found in the lowest layer (below layer III) outside Cave 2 at 110 cm below surface.

Microblades (4) Four microblade fragments were found in Cave 2 (Fig. 11–1: *a–d*). All are of grey chert. They range in length from 44.5 mm to 17 mm; in width, from 8 to 5 mm. Two are 2 mm thick and the other two 1.5 mm. With one exception they have a straight longitudinal axis, are more or less rectangular in outline, and have two or three full-length

*Abridged, with permission, from Trail Creek: Final Report on the Excavation of Two Caves on Seward Peninsula, Alaska, by Helge Larsen, 1968. *Acta Arctica*, Fasc. 15. Published by the Danish Arctic Institute, Copenhagen.

Cave 2, Trail Creek. (*Photo by Helge Larsen. Courtesy of Danish Arctic Institute.*)

dorsal facets. Three are medial segments; one is proximal. There is an appearance of edge wear on three specimens, seen only on the ventral surface. It is suggested they were struck from either a cylindrical or wedge-shaped core, the latter being the most probable.

Antler Points (7) Of the 7 specimens, one is almost complete and measures 12.1 cm long by 0.8 cm wide and 0.6 cm thick. There are three basal segments, two point segments, and one medial fragment. They are all subrectangular to oval in cross section and are deeply grooved in the narrow side of the oval. The nearly complete specimen (Fig. 11–1: *f*) carries two opposing grooves from near the tip into the beveled base. The grooves vary from 1.5 mm to 2.0 mm in width and are scored to an aver-

age depth of 3 mm (see Fig. 11–1: *f, g*). The dimensions of these slots correspond exactly to those of the microblades.

This is the only occurrence in eastern Beringia of slotted bone points and microblades.

ASSOCIATED ORGANIC REMAINS AND DATING

Both Cave 2 and Cave 9 yielded a large number of animal bones throughout the stratigraphy. The faunal complex was made up of predominantly modern forms, with caribou most frequent at all levels. Caribou bones collected from layer III in Cave 2 were radiocarbon-dated at 9070 ± 250 (K-980). Nonmodern forms found were: *Cervus*, identified as elk, in the lower layer towards the rear of Cave 2; and

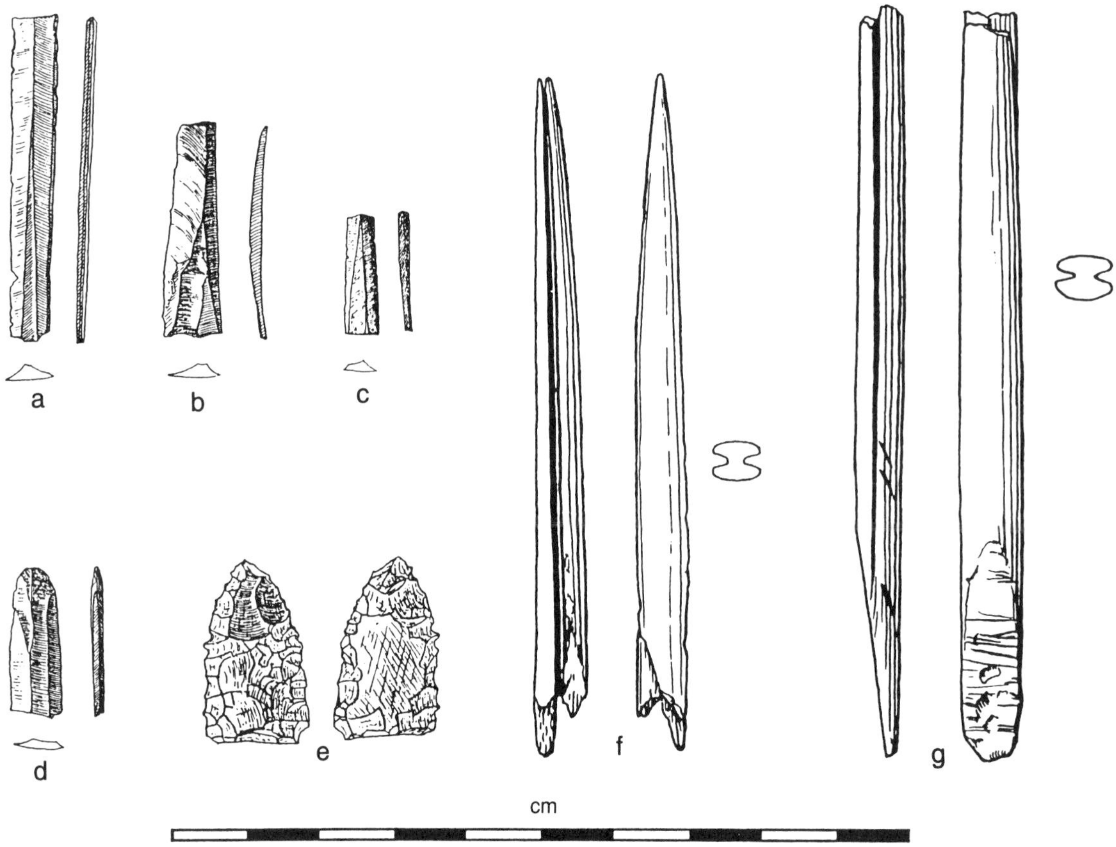

Figure 11–1   Artifacts from layer III (*a–d, f, g*) and below (*e*) at Trail Creek Cave 2: microblades (*a–d*); chalcedony biface from below layer III (*e*); slotted antler points (*f, g*).

bison and horse in the lower (clay) layer outside the south entrance to Cave 9. The organic-collagen fraction of these latter two were radiocarbon-dated at 13,070 ± 280 BP (K-1327) and 15,750 ± 350 BP (K-1210), respectively. There were no artifacts found in association with these bones.

The chronology presented for Trail Creek Cave relied more on typology than on the stratigraphic position of the artifacts. In several instances articulating pieces of broken artifacts were found in widely separated stratigraphic layers. It was concluded that the stratigraphy had been disturbed by frost action, solifluction, and animal burrowing. Of note, however, the artifacts thought to be the oldest—the microblades and associated antler points—were found in the front section of Cave 2. Here the deposits were deepest, and horizontal layering

showed the least disturbance and was most easily distinguished.

Summary

The Trail Creek Cave continues to be somewhat enigmatic. Its importance resides in the association of early Holocene microblade industries with slotted antler points. Until Ackerman's recent discovery at Cave 1, Lime Hills (Ackerman, this volume), this was the only known occurrence in eastern Beringia. The paucity of the artifact assemblage and the difficulties presented by the stratigraphy have placed it almost in a position of ancillary evidence. But, however small the collection, it adds invaluable additional light on the overall picture of the peopling of the New World that is now emerging.

Aerial view of Onion Portage site. (*Source: Anthropological Papers of the University of Alaska. See Acknowledgments.*)

ONION PORTAGE, KOBUK RIVER: AKMAK AND KOBUK COMPONENTS

Compiled by the Editor

The deeply stratified site of Onion Portage is situated in the middle course of the Kobuk River (67°06'N, 158°15'W) about 72 km downstream from Shungnak. Important in the past, the portage allowed the traveler to eliminate almost 16 km of river travel. The Baird Mountains to the north are an extension of the Brooks Range. Generally, the vegetation resembles that of interior Alaska with dense stands of white spruce and birch. Large mammals are those expectable in this setting: moose, black bear, and caribou. The salmon run is quite large, in part attributable to the clear waters of the Kobuk.

HISTORY OF RESEARCH

In 1941, the year following his one-man preliminary survey, J. L. Giddings carried out excavations on four house pits discovered on the portage. In the course of these excavations three microblade cores and one microblade turned up. Giddings (1952) noted their similarity to specimens from the Campus site, which prompted his return in 1961 to conduct further excavations. This led quickly to the recognition of the extraordinary nature of the site; it was deeply stratified, in a part of the world where such occurrences are practically unknown. Intensive work at the site began in 1964 and continued until 1967. Giddings's untimely accidental death in 1964 prevented his learning of the Akmak site. Fortunately, the Kobuk level (band 8), proving the provenance of the wedge-shaped microblade cores,

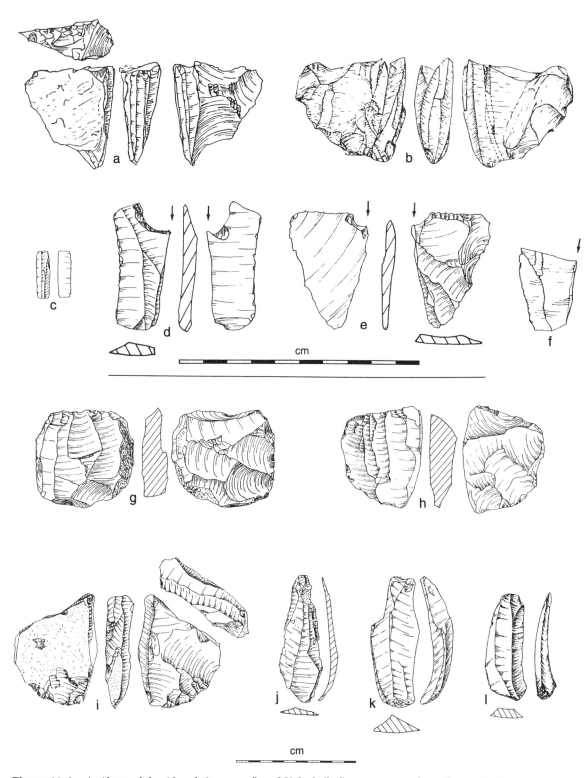

Figure 11–2 Artifacts of the Akmak (*a, c–e, g–l*) and Kobuk (*b, f*) components from Onion Portage: wedge-shaped microblade cores (*a, b*); microblade (*c*); burins (*d–f*); face-faceted blade cores (*g, h*); edge-faceted blade core (*i*); large blades (*j–l*). (*Source: Acta Arctica. See Acknowledgments.*)

was discovered the summer before his death. Akmak was discovered in 1965 after D. D. Anderson had taken over the Onion Portage project. The first find of Akmak artifacts was made by the geologist Sten Florin. The geological sequence was worked through by Thomas D. Hamilton, who was able to determine that Akmak precedes Kobuk and is thus the earliest component at the locality.

CHARACTERIZATION OF THE SITE

In Hamilton's words, "the main occupation area at Onion Portage . . . is marked by a vertical succession of buried ground surfaces, occupied for intervals varying from about 100 to 1,000 years, and separated from each other by fans of sandy alluvium washed down from the adjacent slope along gully systems that periodically became active" (Hamilton 1970:72).

A "band" signifies a more or less discrete series of organic stringers, varying in number from three in band 8 (Kobuk) to as many as 13 in others (e.g., band 6). Each of these levels represents a period of stability during which that surface may have been occupied. Overlying Kobuk is a series of Northern Archaic levels (bands 5, 6, 7). Above this are three levels of Denbigh Flint Complex (bands 3/4–band 5, level 1), over which are Choris levels, Norton–Ipiutak, Itkillik, and finally, in band 1, Arctic Woodland. No other site in the American north has yielded such a stratigraphic array.

The concern here is with the two earliest components, Kobuk and Akmak. Akmak actually lies outside the stratified sequence. It is a separate site, lying on a height of land just north of the stratified Onion Portage site proper. Its relationship to the latter sequence was demonstrated by Hamilton, who showed Akmak to occupy a "stratigraphic position below the third level of band 8" (Hamilton 1970:78)—that occupied by Kobuk. The two assemblages are closely related and are treated together here. Six radiocarbon dates from Band 8 range from 7180 ± 90 BP (P-1111) to 8195 ± 280 BP (GX 1508) with the estimated age of Kobuk considered to be toward the older date. A single date of 9570 ± 150 BP (K 1583, on caribou bone) provides the best approx-

imation of the age of Akmak (Anderson 1988:48, 55).

KOBUK ARTIFACT ASSEMBLAGE (111 ARTIFACTS)

Kobuk is a core-and-burin collection of very small size (111 pieces). It is, however, well-characterized and occupies a clear stratigraphic position at the base of the Onion Portage sequence (band 8, levels 1, 2, and 3).

Microblade Core (1)   This is a wedge-shaped specimen recovered in three pieces. It is entirely conformable with the smaller wedge-shaped cores from Akmak as well as—generally—with the Denali–Gobi forms (Fig. 11–2: *b*).

Microblades   Although reported at 67, it seems a more conservative width delimitation would reduce that number by at least three-quarters.

Larger Blades and Blade-like Flakes (10)   As Anderson noted (1988:70), the presence of larger cores is attested by these specimens. Sizes range up to 9 cm in length and 3 cm in width.

Burins (1)   This is a blade-like flake exhibiting one burin facet (Fig. 11–2: *f*). Four burin spalls are also reported.

Side Scrapers and Unifaces (4)   The two side scrapers appear to have been made on rather heavy blade-like flakes. One specimen shows almost complete unifacial marginal retouch. Both are slightly over 10 cm in length. Of the other unifaces, one resembles a notch or denticulate.

AKMAK ARTIFACT ASSEMBLAGE (>500 ARTIFACTS)

Anderson (1970) recognized eight classes of artifacts—these based on form—in this collection of >500 artifacts and fragments. A notable feature of this blade-and-burin assemblage is the large size of these artifacts when compared to other Alaskan assemblages of similar age and technology. It is possi-

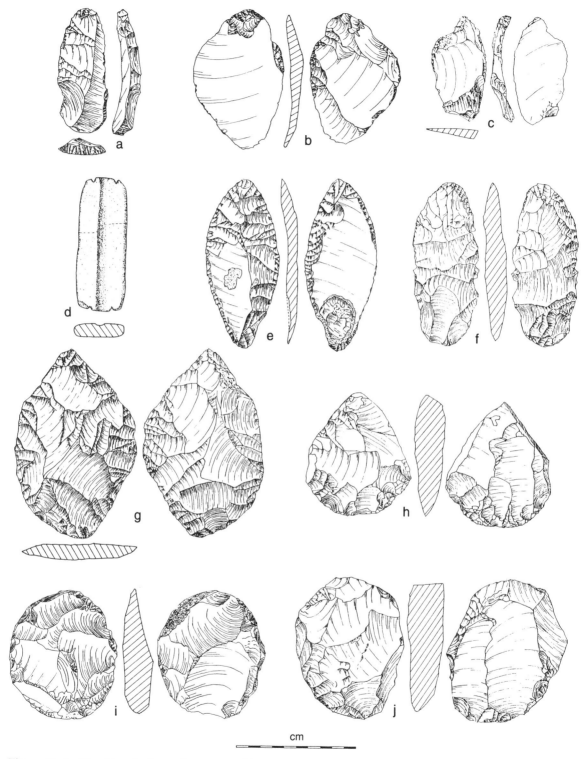

Figure 11–3   The Akmak site: end scraper on blade (*a*); large unifaces (*b, c*); grooved stone (*d*); knife with bifacial edge-retouch (*e*); bifaces (*f, g*); core bifaces (*h–j*). (*Source:* Acta Arctica. *See Acknowledgments.*)

ble that proximity to a source of chert may account for the massive size of these specimens.

Face-Faceted Blade Cores (4) These forms vary somewhat in outline but tend towards rectangularity. They range up to 8 cm in height. Anderson (1970:27) estimates that the original dimensions of some may have been as great as >12 cm. Blades from these specimens would have been wide and somewhat irregular, the billet-struck blades leaving facets of up to 2 cm in width. Two of these are shown as Figure 11–2: *g, h.*

Edge-Faceted Blade Cores (6) As with the preceding, these are quite large. The edge facets are irregular. The platforms were created either by unifacial retouch or by burin blows, in which case the long facets produced a specimen that resembles an inordinately large convergent burin (Fig. 11–2: *i*). Certain of the core bifaces could also be placed in this category.

Microblade Cores (6) These are Denali- or Gobi-style cores showing all the details of manufacture and use common to this group. One of these specimens is shown as Figure 11–2: *a.* Curiously, no core tablets were found here.

Blades, Microblades, and Blade-like Flakes (108) Sixty-one blades and fragments were classed as "large, wide blades." In length this class ranges from 5.5 to 11.9 cm; in width, 3.0–5.4 cm. Percussion flaking on some pieces is suggested by the occurrence of prominent bulbs of percussion. Some of these were used in the manufacture of end scrapers, side scrapers, and "beaked tools." Small, narrow blades number 20 and are considered a subset of this large class. They are derived from the same core form as the larger blades.

Microblades number 32. These are described as being quite uniform in width and thickness. Width measurements range between 4 and 6 mm. One microblade fragment is shown as Figure 11–2: *c.* Large blades and implements made from blade-like flakes are shown in Figures 11–2: *j, k,* and *l* and 11–3: *a.*

Burins (17) One or two of these pieces approximate the notched form found in many Denali col-

lections. The others appear to be more casually made on blade-like flakes and blades, and flakes. Again, unlike burins in good Denali context, each of these specimens bears only a single burin facet. Only five burin spalls were recovered. Two of these burins are shown as Figure 11–2: *d, e.*

Core Bifaces (29) Considered one of the most striking elements of this collection, it is perhaps their massive size that most readily characterizes this grouping. They are taken to have been used for chopping, cutting, or scraping. But, as implied by their designation, perhaps, they also functioned as blanks or proto-tools. As with Ravine Lake, far to the south, it is tempting to read here some continuity of Levallois technique applied to reduction of nodular and tabular chert (Fig. 11–3: *h–j*).

Bifaces (20) These specimens are made on flakes, in contradistinction to the core bifaces. As noted, this is a variable category containing objects of differing morphologies and presumed functions. Examples are shown as Figure 11–3: *f, g.*

large unifaces (7) As with the bifaces, forms are variable, but each specimen will exhibit one or more retouched edges. In greatest dimension the unifaces range from >8 to 10.8 cm (Fig. 11–3: *b, c, e*).

Grooved Stones (3) These are tabular objects (the largest, 11 × 4 cm) made of a coarse sandstone. They presumably functioned as shaft smoothers (Fig. 11–3: *d*).

SUMMARY

The Akmak assemblage is perplexing in a number of ways. The assemblage is unusual in its composition and in the sheer size of the individual objects recovered. Tied to this, in some manner, is the relative paucity of flaking debris. The massive size of the Akmak manufactures could be related to the proximity to source materials; Anderson notes that both tabular chert—probably quarried nearby—and water-rolled cobbles served as raw material. Nonetheless, and this especially applies to the core bifaces, there is engendered an intuitive feeling that these objects were destined for further reduction.

A still more puzzling fact is that the only recognizable evidence of projectile weapon heads consists in the fragmentary form of a relatively few microblades which are interpreted as having served as insets on slotted antler points. (Of these latter, expectably, no evidence survives.) There is assuredly nothing to quarrel with in this interpretation, but in light of their postulated importance the very small number of microblades is even more difficult to comprehend. Perhaps (inscrutably) related to this last observation is the virtual absence of end scrapers.

The bulk of the Akmak artifacts were made from a grey-to-black glassy chert which the Kobuk Eskimos call *akmak*. As Anderson observed, the almost exclusive use of this rock stands in contrast to the band 8 Kobuk assemblage as well as the still later Onion Portage levels.

That Akmak and Kobuk are related seems clear—a view much aided by the small size of the Kobuk assemblage. Akmak is interpreted as a habitation site. Perhaps its size requires that interpretation, but the mechanics by which that habitation was accomplished are less than clear.

References Cited

Anderson, D. D. 1970. Akmak: An Early Archaeological Assemblage from Onion Portage, Northwest Alaska. *Acta Arctica,* Fasc. 16. Copenhagen.

———. 1988. Onion Portage: The Archaeology of a Stratified Site from the Kobuk River, Northwest Alaska. *Anthropological Papers of the University of Alaska* 22(1–2).

Giddings, J. L., Jr. 1952. *The Arctic Woodland Culture of the Kobuk River.* University Monographs. Philadelphia: University Museum, University of Pennsylvania.

Hamilton, T. D. 1970. Appendix: Geologic Relations of the Akmak Assemblage, Onion Portage Area. *In* An Early Archaeological Assemblage from Onion Portage, Northwest Alaska, by D. D. Anderson. *Acta Arctica,* Fasc. 16. Copenhagen.

TWO SITES ON RED DOG CREEK, DE LONG MOUNTAINS

S. Craig Gerlach and Edwin S. Hall, Jr.

The Red Dog Site

The first site, the Red Dog site, is catalogued as DEL-166; the second site, on a terrace about three km from the first, is known only by its catalog designation, DEL-168. Red Dog is located on a prominent knoll oriented roughly southeast to northwest, overlooking the valley of the Red Dog Creek (Fig. 11–4). The knoll is 3 to 5 m above the surrounding tussock tundra but drops sharply to the creek valley on the north and east.

The Red Dog site was discovered in 1982 and systematically tested by the authors in 1985. Approximately 7,942 artifacts were recovered from 67 one-meter tests and 619 m of controlled surface collection.

The topography of the site surface is rough and undulating, with discontinuous sand deposits and thin patches of deflation-lag gravel. Much of the knoll is unvegetated. Vegetation, where it does occur, comprised of lichens, mosses, dryas, sedges, and grass, is sparse and limited by poorly developed soil, wind, and erosion. Six loci (A–F) are scattered across the knoll (Fig. 11–5). Locus A comprises 165 sq. m; locus B , 52 sq. m; locus C, 190 sq. m; locus D, 115 sq. m; locus E, 103 sq. m; and locus F, 26 sq. m. Lithic material was abundant in the unvegetated areas of the knoll. Subsurface materials, however, were recovered only from loci A and B where vegetation was relatively thick.

Artifact Assemblage (7,050 debitage; 892 tools)

Locus A consists of a primary concentration of lithic debris and a thin scatter of lithics around the excavated area. Of the artifacts recovered, 29 were worked tools consisting of: 1 microblade core, 7 microblades, 5 flake burins, 1 burin spall, 2 flake cores, 8 bifaces, 1 projectile point base fragment, and 4 retouched flakes.

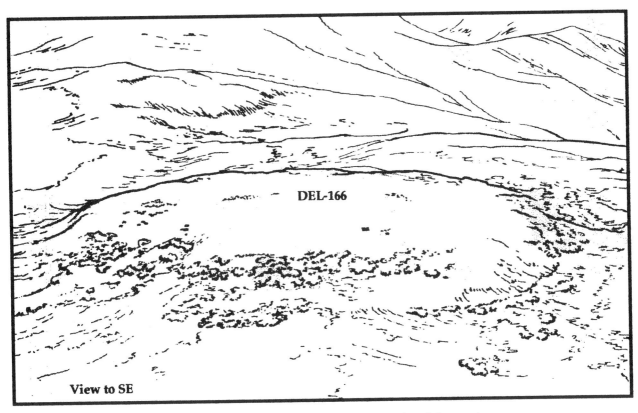

Figure 11–4   Oblique view of DEL-166. (*Source: Cominco Alaska, Inc. See Acknowledgments.*)

The locus A microblade core is narrow and wedge-shaped with large irregular facets on both ends (Fig. 11–6: *a*). It is similar to cores found in possibly late American Palaeo-Arctic sites (Gerlach 1981, 1982; Gal 1982). The flake burins resemble the ubiquitous flake burins and burin-faceted flakes described for coastal or inland Choris, provisionally dating between 2,800 and 2,200 uncalibrated BP (Giddings 1957; Giddings and Anderson 1986; Anderson 1988). The 8 bifaces represent several stages of lithic reduction. One small black chert discoid fragment has use wear on the distal end and is a type not usually found in Brooks Range American Palaeo-Arctic tradition sites. One square, slightly ground projectile point base (Fig. 11–6: *n*) was an arrow point or end blade base. It is similar to those found in the latter phases of the Arctic Small Tool tradition. Locus A appears to contain mixed components.

Locus B yielded 215 worked tools consisting of 8 microblade cores, 4 core tablets, 2 possible rejuvenation flakes, 154 microblades, 7 flake burins, 2 possible blade cores, 1 possible flake core, 2 end scrapers, 22 bifaces, and 13 retouched flakes.

Of the 8 microblade cores, 4 are narrow wedge-shaped cores (Fig. 11–6: *b*) similar to those from various sites in the interior. The remaining 4 represent a range of forms, including a wide-oval form (Fig. 11–6: *c*), an aborted core, and a rotated core (see Bowers 1982; Gal 1982). The core tablets exhibit similar variability. Two are removals from narrow wedge-shaped cores; one is a wide-oval core removal (Fig. 11–6: *i*) and can be compared to some of the core tablets described for the Tunalik site (Gal 1982), to the so-called "aberrant" core from Punyik Point (Irving 1964), and to some of the cores from sites on the Kukpowruk and Killik rivers (Hall 1975). Of the 7 flake burins, 4 are multi-faceted. The burin illustrated (Fig. 11–6: *j*) shows considerable wear on each of the facets as well as on the juncture

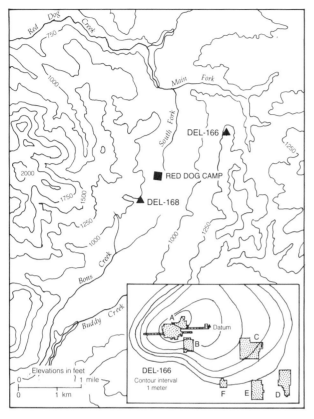

Figure 11–5   Site map, including loci A–F. (*Source: Cominco Alaska, Inc. See Acknowledgments.*)

between two of the faceted edges. The two end scrapers were made on flakes and show continuous overlapping retouch on the length of their margins and distal ends (Fig. 11–6: *m*).

Eight of the bifaces are similar to the large core-bifaces described by Anderson (1970b:18–24) for the Akmak assemblage at Onion Portage. Outlines range from ovate to sub-triangular, semilunar, and amorphous. Although all bifaces exhibit battering and damage on their edges, only one specimen has a segment of relatively fine retouch on an edge (Fig. 11–6: *p*). Some of these specimens may have been used as tools, and some may have functioned as cores for the production of flakes and blades. Six bifaces are classified as projectile points and/or cutting implements (Fig. 11–6: *o*). One of these is a small end blade. Anderson (1970a) records similar bifaces in Brooks Range sites postdating the American Palaeo-Arctic tradition (see also Schoenberg

1995:51–61). The remaining 8 bifaces are very small, irregular, and amorphous fragments and are probably the unusable debris from biface reduction.

Locus B is preliminarily postulated to pertain to a post-American Palaeo-Arctic tradition.

Locus C is primarily a single concentration of lithic debris with a small quantity of lithic material distributed downslope to the south and southeast of the main concentration. Four hundred and twenty-five worked tools were recovered: 11 microblade cores, 9 core tablets, 316 microblades, 2 flake cores, 10 flake burins, 36 burin spalls, 2 irregularly shaped bifaces, and 39 retouched flakes.

All 11 of the microblade cores found in this locus are narrow wedge-shaped microblade cores. Within this category there is some variation in the treatment of the keel, the form of the edge opposite the faceted end—"flat-backed"—following Gal (1982:69) and one which is classified as a core-burin and exhibits use wear on one facet and the ventral surface of the flake (Fig. 11–6: *d, e*).

Of the 10 burins, two show use wear on the convex edges. One burin spall also shows use wear. These specimens are similar to the flake burins and burin-faceted flakes found in loci A and B. Locus C appears to pertain to an early phase of the American Palaeo-Arctic tradition.

A very small assemblage including 30 identifiable implements was recovered from locus D. These are: 2 microblade cores, 2 core tablets, 1 rejuvenation flake, 23 microblades, 1 burin spall, and 1 retouched flake. The microblade cores are similar to the narrow wedge-shaped cores found in locus C. Locus D appears most closely to resemble the classically defined American Palaeo-Arctic tradition.

Locus E is a small concentration of lithic debris located on a low bench at the southern end of the knoll. Downslope movement may have affected the integrity of the original distribution. The assemblage contains 2 microblade cores, 74 microblades, 6 flake burins, 1 burin-faceted blade, 7 burin spalls, 1 biface, 10 retouched flakes, and 2 angular raw material fragments.

Of the 2 cores recovered, one is clearly of the Campus type. The other (Fig. 11–6: *f*) is cylindrical or pencil-shaped. Its form may, in fact, indicate that the core is expended. One of the burins has a pre-

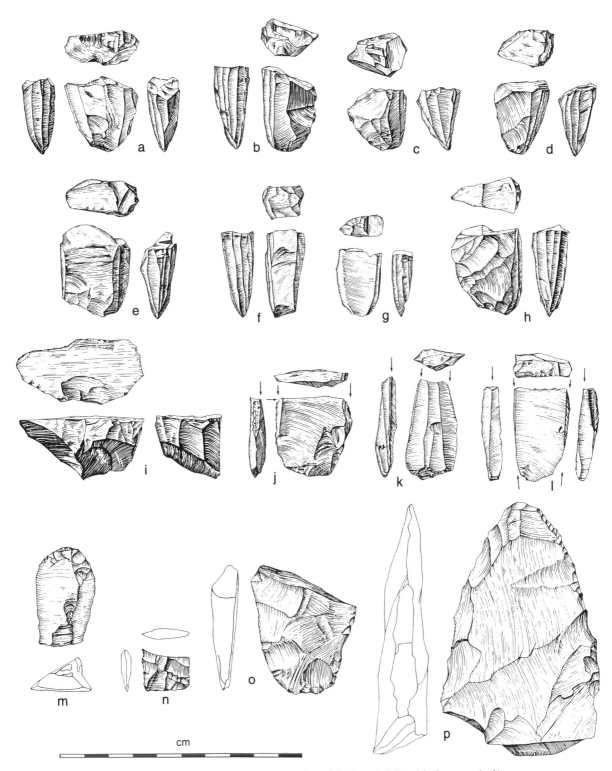

Figure 11–6   Artifacts from DEL-166 (the Red Dog site) and DEL-168. Microblade cores (*a–h*); microblade core tablet (*i*); burins (*j–l*); end scraper (*m*); bifaces (*n–p*). DEL-166, locus A (*a, n*); locus B (*b, c, i, j, m, o, p*); locus C (*d, e*); locus E (*f*); locus F (*g, k*). DEL-168 (*h, l*).

pared notch platform from which the burin blow was struck, a diagnostic trait in both the earlier and later phases of the American Palaeo-Arctic tradition (Anderson 1970a, 1970b; Gal 1982). The locus E assemblage appears to be related to the American Palaeo-Arctic tradition and most likely represents a single occupation with only a limited number of activities.

Locus F is a small clustered association of lithic scatters located on a level area of a lower southern bench. Total identifiable artifacts number 90 and include 2 microblade cores, 73 microblades, 5 flake burins, 2 burin spalls, 3 bifaces, and 5 retouched flakes. Of the cores, one is a narrow, wedge-shaped form with a foreshortened back, indicating that it is probably expended (Fig. 11–6: *g*). The other is a core-burin on a small blade (Fig. 11–6: *k*). The burins are made on intentionally snapped flakes. One of the bifaces is a small discoid fragment which has fine collateral bifacial flaking as well as tiny pressure retouch scars on the convex margin. This is similar to the small discoids present in the later phases of the Arctic Small Tool tradition sites (Norton and late Ipiutak) on the coast and in the interior of the Brooks Range foothills. Although the assemblage from locus F is small, it appears that it may be a mixed assemblage.

SUMMARY

To analyze the lithics from the six loci at the Red Dog site (DEL-166) one may try to establish cultural and temporal relationships on the basis of "type" artifacts in the assemblages. The assemblages from some of the loci appear to represent the early American Palaeo-Arctic tradition, while others may represent the later American Palaeo-Arctic tradition and the Arctic Small Tool tradition. The six loci may represent a number of different short-term occupations of small groups of mobile hunter-gatherers who may have been related culturally.

Twenty-six microblade cores, representing several different forms, were recovered from the six tested loci. Some of the microblade core forms occur at all six loci, but the associations among them are not necessarily replicated at all loci. Narrow,

wedge-shaped cores with foreshortened backs are present only in loci A, B, and C. Although the core from locus C is similar to the "flat-back" cores described by Gal (1982) for the hypothesized post-American Palaeo-Arctic tradition, it is associated in this locus with a core-burin, a form that is more typical for the American Palaeo-Arctic tradition. The narrow, wedge-shaped microblade cores from loci E and F are within the range of variability for the Denali complex microblade cores in interior Alaska (West 1981, this volume; Powers 1983; Powers and Hoffecker 1989; Mobley 1991).

Of the total of 36 bifaces in the site, 22 were recovered from locus B. From analysis of the unmodified flake debris it seems doubtful that any of these bifaces were produced on site. When these specimens are compared to the core-bifaces from the Akmak assemblage at Onion Portage on the Kobuk River (Anderson 1968, 1970b), there are some typological similarities; they are also similar to specimens recovered from several lithic concentrations at the Lisburne site (Bowers 1982), and to WAI-109, a small lithic site located between Icy Cape and Wainwright on the Arctic coastal plain (Gerlach 1981).

The flake burins from locus B may be more similar to the burin-faceted flakes from Choris or other late Arctic Small Tool tradition expressions than they are to the formal burins described for the earlier phases of the American Palaeo-Arctic tradition. The end blade from this locus is not easily placed in any known American Palaeo-Arctic tradition assemblage, but might on typological grounds alone be expected to occur in a transitional assemblage (Larsen 1968).

The end scrapers from locus B are similar to those from area 25 at the Lisburne site (Bowers 1982), the Palisades complex at Onion Portage (Anderson 1968), and from several of the small shoreline bluff sites (Gerlach 1981, 1982). Steep-angled end scrapers on flakes also have been identified in a late Ipiutak context at the Croxton site located at Tukuto Lake in the western Brooks Range (Gerlach 1982, 1989; Gerlach and Hall 1988), from the Feniak Lake Ipiutak site (Gerlach and Hall 1988), and from other late prehistoric sites in the Brooks Range.

Among the lanceolate bifaces, biface fragments,

Surface of DEL-168
View to E

Figure 11–7   View to east, DEL-168. (*Source: Cominco Alaska, Inc. See Acknowledgments.*)

end scrapers, and retouched flakes are many elements that are reminiscent of the Denali complex of
the interior. A question remains as to whether they
are associated with the microblade core technology
at locus B or whether they represent an earlier or
possibly later component.

The real problem of interpretation of the locus
B assemblage turns on whether this assemblage
represents one of the earlier phases of the American
Palaeo-Arctic tradition or the postulated transitional post-American Palaeo-Arctic tradition. On
the basis of associations discovered at other Brooks
Range sites, particularly Onion Portage (Anderson
1968, 1970a, 1970b), the Noatak River sites (Anderson 1972), the Tunalik site (Gal 1982), the Lisburne
site (Bowers 1982), and some of the shoreline bluff
sites (Gerlach 1981, 1982), the co-occurrence of the
microblade cores, core tablets, flake burins, blade
cores, and the core-bifaces may typologically represent a single-component assemblage. As the spatial
association of the artifact types is tightly clustered,
and as the specific physical setting within the site
does not seem to invite multiple occupations, it
would appear a single-component locus is not inconsistent with the typological association.

Comparison of the assemblage from locus C to
other loci at DEL-166, and to other sites in the
Brooks Range, suggests that the closest typological
similarities are probably in one of the earlier phases
of the American Palaeo-Arctic tradition. This is

based on the large number of narrow wedge-
shaped microblade cores and the absence of the
wide-oval core forms. Added to this, the presence
of generalized flake burins and the relatively large
number of microblade core tablets indicate the
lithic assemblage from locus C may represent a regional variant of one of these earlier phases or an
assemblage that is typologically and temporally intermediate between the earlier American Palaeo-
Arctic tradition assemblages represented at the
Noatak River sites (Anderson 1972), the Tunalik site
(Gal 1982), and possibly some of the shoreline bluff
sites on the Arctic coastal plain (Gerlach 1981,
1982).

Of the six loci discovered at the Red Dog site,
loci A and F appear to have mixed components.
Loci B, C, D, and E are tentatively identified as unmixed components, with, respectively, affinities to
a post-American Palaeo-Arctic tradition, an early
American Palaeo-Arctic tradition, and a classic
American Palaeo-Arctic tradition.

DEL-168

DEL-168 is located about 2.8 km southwest of the
Red Dog site approximately 400 m from the Red
Dog Creek on a well-defined terrace (Fig. 11–7). The
terrace slopes gently from west to east and provides
an unobstructed view of the valley. The site consists
of a large concentration of lithic material scattered

across the gently sloping eastern side.

The site was discovered in 1982 and systematically surface-collected and tested in the summer of 1985 by the authors. A grid system was laid out and a total of 29 m were surface-collected and 33 1-m units were excavated. Materials were recovered down to a depth of 40 cm.

Vegetation is sparse with scattered stands of dwarf birch, grass, lichens, and moss predominating. Wind-polished inclusions of large and small gravel and weathered natural chert nodules comprise the unvegetated terrace surface. Vertical stratigraphy is absent, but sediments consist of packed light brown to tan windblown sands with abundant gravel and chert inclusions. Wind erosion resulted in terrace deflation, indicated by the presence of a 4–5-cm-thick deflated lag deposit in several grid squares. No organic remains or radiocarbon dates were obtained from the site.

ARTIFACT ASSEMBLAGE

Of the 1,709 artifacts recovered from the site, 147 exhibited working. Among these are 8 microblade cores, 2 core tablets, 1 rejuvenation flake, 38 microblades, 7 flake burins, 6 bifaces, 5 flake core fragments, 1 tabular split flake core, and 79 retouched flakes.

The 8 microblade cores are made on thick bifacially worked flakes and have bifacially flaked keels (Fig. 11–6: *h*). They fairly uniformly conform to the narrow, wedge-shaped type, and are most similar to the American Palaeo-Arctic tradition cores described by Anderson (1970a, 1970b, 1988) from the Akmak and Kobuk assemblages.

Two narrow wedge-shaped core tablets were recovered from the site. One large tablet (Fig. 11–6: *l*) with large regular facets from previous blade removals is also a flake burin. It is similar to the burins on thick tabular flakes described for the Tunalik site (Gal 1982). The other core tablet shows retouch flaking along an edge, suggesting that it was used as a tool in its own right after being struck from the core.

The 7 burins include a dihedral and a lateral form, both typical of the American Palaeo-Arctic

tradition. The others are burin-faceted flakes whose irregular morphologies suggest that the burin technology was one of expedience and utility. Many of the waste flakes are similarly snapped, and have spurs, or *bec*-like projections, even though definite burin facets are not present.

One of the 6 bifaces and fragments is similar to the Type 1-A core biface in the Akmak assemblage (Anderson 1970b:18–21). The 5 remaining are small, crude, irregular fragments, the debris from biface reduction.

SUMMARY

There is no indication that more than a single occupation is represented at DEL-168. This interpretation is supported by the uniform utilization of raw material at the site, consistency in the microblade core forms, and the concentration and spatial integrity of the lithic distribution. By typology, the site can be assigned to some phase of the American Palaeo-Arctic tradition.

CONCLUSION

These two sites, DEL-166, the Red Dog site, and DEL-168, were the largest discovered in the course of surveys conducted by the authors between 1982 and 1986. Together, they contain roughly half of all microblade cores described from the western Brooks Range, Alaska. In spite of the difficulty of interpretation caused by the lack of stratigraphy and organic remains for radiocarbon dating, these sites can cast light on the culture history of the region if analyzed by typological means. The tentative interpretation is that the sites cover a period from the early American Palaeo-Arctic tradition to a postulated post–American Palaeo-Arctic tradition.

REFERENCES

Anderson, D. D. 1968. A Stone Age Campsite at the Gateway to America. *Scientific American* 218(6):24–33.

————. 1970a. Microblade Traditions in Northwestern Alaska. *Arctic Anthropology* 7(2):2–15.

————. 1970b. Akmak: An Early Archaeological Assemblage from Onion Portage, Northwest Alaska. *Acta Arctica*, Fasc. 16.

————. 1972. An Archaeological Survey of the Noatak Drainage. *Arctic Anthropology* 9(1):66–117.

————. 1988. Onion Portage: The Archaeology of a Stratified Site from the Kobuk River, Northwestern Alaska. *Anthropological Papers of the University of Alaska* 27(1–2).

Bowers, P. M. 1982. The Lisburne Site: Analysis and Culture History of a Multi-Component Lithic Workshop in the Iteriak Valley, Arctic Foothills, Northern Alaska. *Anthropological Papers of the University of Alaska* 20(1–2):70–112.

Gal, R. 1982. Excavation of the Tunalik Site, Northwestern National Petroleum Reserve in Alaska. *Anthropological Papers of the University of Alaska* 20(1–2):61–78.

Gerlach, S. C. 1981. An Archaeological Survey of an Ancient Beach Ridge, Wainwright Quadrangle, Alaska: A Description of the Lithic Industries from Seventeen Prehistoric Sites. Ms. on file, Department of Anthropology, University of Alaska, Fairbanks.

————. 1982. Small Site Archaeology in Northern Alaska: The Shoreline Bluff Survey. *Anthropological Papers of the University of Alaska* 20(1–2):15–49.

————. 1989. Models of Caribou Exploitation, Butchery, and Processing at the Croxton Site, Tukuto Lake, Alaska. Ph.D. dissertation, Brown University.

Gerlach, S. C., and E. S. Hall, Jr. 1988. The Later Prehistory of Northern Alaska: The View from Tukuto Lake. In *The Late Prehistoric Development of Alaska's Native People*, ed. R. D. Shaw, R. K. Harritt, and D. E. Dumond, pp. 107–35. Aurora, No. 4. Alaska Anthropological Association. Anchorage.

Giddings, J. L., Jr., 1957. Round Houses in the Western Arctic. *American Antiquity* 23(2):121–35.

Giddings, J. L., Jr. and D. D. Anderson. 1986. *Beach Ridge Archeology of Cape Krusenstern: Eskimo and Pre-Eskimo Settlements around Kotzebue Sound, Alaska*. Publications in Archeology 20, National Park Service. Washington, D.C.: U.S. Department of the Interior.

Hall, E. S., Jr. 1975. An Archaeological Survey of Interior Northwest Alaska. *Anthropological Papers of the University of Alaska* 17(2):13–30.

Irving, W. 1964. Punyik Point and the Arctic Small Tool Tradition. Ph.D. dissertation, University of Wisconsin.

Larsen, H. 1968. Trail Creek: Final Report on the Excavation of Two Caves on Seward Peninsula, Alaska. *Acta Arctica* 15:7–79.

Mobley, C. M. 1991. *The Campus Site*. Fairbanks: University of Alaska Press.

Powers, W. R. 1983. The 1977 Survey for Pleistocene Age Archeological Sites. In Dry Creek: Archeology and Paleoecology of a Late Pleistocene Alaskan Hunting Camp. Unpublished report to the National Park Service.

Powers, W. R., and J. F. Hoffecker. 1989. Late Pleistocene Settlement in the Nenana Valley, Central Alaska. *American Antiquity* 54(2):263–87.

Schoenberg, K. M. 1995. The Post-Paleoarctic Interval in the Central Brooks Range. *Arctic Anthropology* 32(1):51–61.

West, F. H. 1981. *The Archaeology of Beringia*. New York: Columbia University Press.

THE MESA SITE, ITERIAK CREEK

Michael L. Kunz and Richard E. Reanier

The Mesa site was discovered in 1978 during a cultural resources inventory designed to ameliorate the impacts of oil and gas exploration activities in the National Petroleum Reserve–Alaska. The site is located at the northern flank of the Brooks Range in arctic Alaska on the east side of the Iteriak Creek valley 570 km northwest of Fairbanks and 240 km north of the Arctic Circle.

PRESENT-DAY LANDSCAPE

The Mesa site lies outside the limits of the mid-Wisconsinan Itkillik glaciation and also beyond the visible limits of the mid-Pleistocene Anaktuvuk glaciation (Detterman et al. 1958; Chapman et al. 1964; Hamilton and Porter 1975; Hamilton 1979, 1980a, 1980b, 1981; Hamilton et al. 1980). These data indicate the Mesa site was not glaciated during the late Pleistocene and the Iteriak valley may have been ice-free for at least the last 500,000 years.

As is typical of most arctic sites, the soil at the Mesa site is shallow, varying from 5 to 35 cm deep. In general the soils can be classified as arctic brown shallow phase soils because of their well-drained

The Mesa site, Iteriak Creek. *(Photo courtesy of Richard E. Reanier.)*

nature, degree of profile development, and geomorphic situation (Tedro 1977; Reanier 1982). Surface cover on the mesa consists of lichen, moss, grasses, herbs, and ground-hugging woody plants and shrubs (Kunz 1982).

The region experiences a harsh, dry, arctic climate with a mean annual temperature of −9.6°C and an annual precipitation average of 208 mm. February is the coldest month, averaging −38.2°C, while July is the warmest, averaging 7.7°C. Precipitation reaches a monthly peak in August with an average of 49.5 mm (Reanier 1982).

Subsistence animals in the area today include grizzly bear, caribou, Dall sheep, and moose. Common furbearers are wolf, fox, wolverine, mink, weasel, arctic fox, marmot, and ground squirrel. Waterfowl and ptarmigan are present but not numerous. During the late Pleistocene, mammoth, bison, and antelope were known to inhabit the area and musk-ox may have been present until the recent past (Guthrie 1982).

Site Locality

The site lies atop a mesa-like ridge which rises 60 meters above the surrounding treeless tundra, providing a 360° unobstructed view of the surrounding countryside. The "mesa" is an erosional remnant of a medium-grained gabbro and basalt sliver that was intruded into thin-bedded chert and siliceous shale as part of the mountain-building event that formed the Brooks Range about 125 million years ago (C. G. Mull, personal communication 1994). More erosion resistant and much darker in color than the sedimentary country rock, the mesa physically and visually dominates the valley.

The Mesa site has been divided into four localities which generally mirror the mesa's natural physiographic areas. Because excavation is tied to the original engineering survey, the site has been excavated using the English measurement system.

Locality A encompasses the most elevated portion of the site and is located on the sparsely vegetated south side of the mesa's median ridge crest. This is the largest locality at the site and encompasses more than 1,400 square meters. To date, 28

4' × 4' units have been excavated in this locality.

Locality B lies in the northeastern quadrant of the site, 6 meters lower and slightly more than 30 meters northeast of locality A. The central portion of this locality is perched upon a slight rise from which the ground slopes gently away in all directions. The total area encompassed by this locality approaches 930 square meters. The vegetation cover in locality B is robust and the site's only growth of dwarf birch occurs there. The soil depth varies dramatically across this locality, ranging from 5 to 20 cm deep. In most cases, soil depth appears to reflect subsurface bedrock topography. To date, 50 4' × 4' units have been excavated in locality B.

The Saddle locality lies just to the southwest of locality B and encompasses an area of slightly more than 740 square meters. Most of this locality lies in a shallow basin and displays the site's thickest soil profile, up to 35 cm, and most robust vegetation mat. To date, 8 4' × 4' units have been excavated in this locality.

The East Ridge locality is separated from locality B by a 4-meter-high rock outcrop and encompasses the narrow eastern one-quarter of the mesa. Approximately 60 square meters of relatively level surface area lie along this constricted, steep-sided portion of the site. Although the surface is reasonably well vegetated, the soil is shallow and rocky, comparable to that in locality A. To date, 4 4' × 4' units have been excavated in the East Ridge locality.

When compared with typical non-arctic sites, the soil at the mesa is shallow, yet by arctic standards most of the site can be considered well buried. Excavation indicates there is no identifiable cultural stratigraphy and lithic cultural material ranges from surface through the soil profile to the underlying bedrock rubble. With the exception of a small, distinctly delineated microblade scatter, apparently the result of a single stop-over episode of incidental site use, the artifacts and debitage remain stylistically and technologically homogeneous across the site, indicating regular use of the site by people of the Mesa culture. To date, excavation at the mesa has demonstrated that only Mesa complex artifacts are uniformly present throughout the site and that other regional archaeological cultures, which are present at nearby sites in the Iteriak valley, are, with the previously noted exception, not represented at the mesa.

ARTIFACT ASSEMBLAGE

The artifact assemblage is dominated by bifacial tools, primarily finished projectile points, of which there are more than 80 complete, reworked, or fragmentary examples and over 70 fragmentary examples of large bifaces in various stages of manufacture. The bifacial tools account for about 80 percent of the formal artifact inventory, while the remaining 20 percent is comprised of unifacial tools, mostly single- or multi-spurred gravers made on flakes, although at least one tool, a spurred end scraper, was made on a blade or blade-like flake. Two large flake cores, more than 50 retouched flakes, over 20,000 unretouched waste flakes, 10 stream-cobble hammer and anvil stones, several possible pieces of hematite, and a few quartz crystals complete the assemblage.

Classic Mesa projectile points are lanceolate in shape with sides that gently taper outward from a concave base to the point's greatest breadth, usually at about two-thirds of its length. A well-controlled percussion flaking technique was employed to produce a high-quality preform or blank. Completing the manufacturing sequence, a robust pressure flaking technique was utilized to detach broad flakes that are purposefully set up to terminate along the projectile point's median line. This type of flaking creates an intentionally robust projectile point with a subdued diamond-shaped cross section. A finer pressure flaking technique was employed to resharpen impact-fractured projectile points while still mounted in the haft; however, fine secondary lateral edge retouch is rare on any of the projectile points. Extremely heavy edge grinding is common along one-half to two-thirds of the projectile point's length, from the base forward, suggesting the point was set deeply in the haft. Grinding of the basal concavity also occurs but varies in its degree of intensity. A few of the specimens have flat or convex bases which are also ground. The size range of the complete specimens is very tight, with length ranging from 60 to 69 mm, width from 18 to 23 mm, and thickness from 8 to 10 mm. Measurements from fragmentary projectile points indicate that, except for length (some projectile points may have been in excess of 100 mm in length), the thickness and width range remains constant throughout the assemblage (Kunz and Reanier 1994, 1995).

Large fragmentary bifaces, which appear to be lance/spear points, knives, or other butchering implements, are common in the assemblage. These artifacts are manufactured entirely by a well-controlled percussion technique and are remarkably thin in relation to their length and breadth. A typical finished biface reconstructed from fragments measures $140 \times 35 \times 10$ mm.

The lithic debitage recovered at the site demonstrates that biface reduction was the primary manufacturing activity occurring there. Few cortex flakes are present in the collection, suggesting that the initial stage of the reduction sequence was being carried out at the lithic source locales. The few cortex flakes that have been recovered indicate that the raw material, local chert for the most part, was being obtained directly from area outcrops, or more commonly, as cobbles from nearby Iteriak Creek.

Four complete unaltered projectile points have been recovered from the site, and several additional examples have been reconstructed from fragments. Some specimens which appear complete at first glance are actually points which were broken during use and resharpened in the haft. The resharpened points appear to be completely serviceable but were discarded by the hunters who made and used them, possibly because they were considered lacking in some functional or aesthetic sense. The number of resharpened projectile points recovered from the site (n = 8) suggests that resharpening was a field expedient only and, at the first opportunity, hunters replaced the resharpened points with new ones. Most of the projectile point assemblage is comprised of bases of finished points that were broken during use and were too short to permit in-haft resharpening. A single complete specimen (Fig. 11–8: *g*) exhibits bipolar fluting on both faces. Although projectile point technology was utilized to manufacture the tool, it is unclear what its function may have been (Kunz and Reanier 1995).

The excavation of 15 hearths at the site has pro-

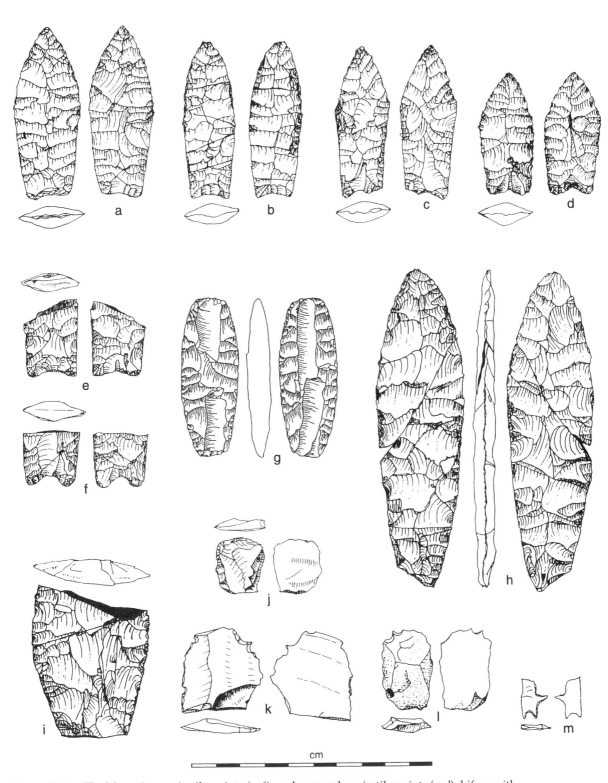

Figure 11–8   The Mesa site: projectile points (*a–f*); resharpened projectile points (*c, d*); biface with bipolar fluting (*g*); large bifaces (*h, i*); end scraper (j); spurred graver/perforators (*k–m*).

Table 11–1 Mesa Site AMS Radiocarbon Dates

Laboratory Sample No.	Date	Locality Provenience	Material Dated Comment
Beta-36805 ETH-6570	9730 ± 80	Saddle N117-121/E96-100 8–13 cm below surface	Soil charcoal; flakes directly associated with charcoal; collected 1989, dated 1990
Beta-50428 ETH-9086	$10,090 \pm 85$	A S1-5/E16-20 10–20 cm below surface	Hearth charcoal: projectile point, tools, and flakes directly associated with hearth; collected 1991, dated 1992
Beta-50430 ETH-9087	9945 ± 75	Saddle N103-107/E94-98 8–15 cm below surface	Hearth charcoal: projectile point and flakes associated with hearth; collected 1991, dated 1992
Beta-52606 CAMS-2688	$10,060 \pm 70$	B N179-183/E146-150 6–14 cm below surface	Hearth charcoal: recovered from archived charcoal/ soil bulk sample which provided original site date (DIC-1589); projectile points, tools, and flakes in direct association with hearth; collected 1979, dated 1992
Beta-55282 CAMS-3568	9990 ± 80	Saddle N109-111/E88-90 8–15 cm below surface	Hearth charcoal: flakes directly associated with hearth; collected 1992, dated 1992
Beta-55283 CAMS-3569	$10,240 \pm 80$	B N209-211/E184-186 6–13 cm below surface	Hearth charcoal: flakes and tools directly associated with hearth; collected 1992, dated 1992
Beta-55284 CAMS-3570	9930 ± 80	B N213-215/E180-182 6–15 cm below surface	Hearth charcoal: projectile point and flakes directly associated with hearth; collected 1992, dated 1992
Beta-55285 CAMS-3571	$10,000 \pm 80$	B N217-219/E176-178 5–12 cm below surface	Hearth charcoal: projectile point, tools, and flakes directly associated with hearth; collected 1992, dated 1992
Beta-55286 CAMS-3572	$11,660 \pm 80$	B N217-219/E180-182 5–23 cm below surface	Hearth charcoal: projectile point, tools, and flakes directly associated with hearth, split sample same source as Beta-57430; collected 1992, dated 1992
Beta-57429 CAMS-4146	9900 ± 70	B N209-211/E176-178 12–23 cm below surface	Hearth charcoal: flakes directly associated with hearth; collected 1992, dated 1992
Beta-57430 CAMS-4147	$11,190 \pm 70$	B N217-219/E180-182 5–23 cm below surface	Hearth charcoal: projectile point, tools, and flakes directly associated with hearth, split sample same source as Beta-55286; collected 1992, dated 1993
Beta-69898 CAMS-11035	$10,070 \pm 60$	Saddle N111-115/E98-102 10–20 cm below surface	Hearth charcoal: flakes directly associated with hearth; collected 1993, dated 1994
Beta-69899 CAMS-11036	9900 ± 80	B N211-215/E174-178 14–20 cm below surface	Hearth charcoal: flakes and tools directly associated with hearth; collected 1993, dated 1994
Beta-69900 CAMS-11037	$10,050 \pm 90$	B N211-215/E182-186 9–19 cm below surface	Hearth charcoal: preform and flakes directly associated with hearth; collected 1993, dated 1994

duced a suite of 14 accelerator mass spectrometry ^{14}C dates ranging from 9700 to 11,700 years BP (Kunz and Reanier 1994). Distinctive lanceolate projectile points have been found in direct association with many of the hearths. Pot-lid fractures occur on four projectile point bases recovered from hearths, indicating these impact-fractured points were discarded into the hearths while the fires were still burning. In addition, pot-lid fractures are often found on lithic waste flakes recovered from the charcoal/soil matrix of many of the hearths. These examples graphically demonstrate the direct association between the artifacts and the charcoal used to date site occupation (see Table 11–1).

SUMMARY

Based upon a suite of factors—including the site's morphology, location, preponderance of use-fractured projectile points, and amount and type of lithic debitage—we believe the site was used primarily as a hunting lookout and retooling station. Hunters, probably camped nearby in Iteriak Creek's riparian zone, would climb to the top of the mesa to take advantage of its unobstructed 360° view to scan the area for game animals traversing the treeless countryside. While scanning the surrounding 100 square kilometers from this vantage point, the hunters manufactured and repaired the weapons needed to hunt and butcher big game.

Generally, pre-ceramic archaeological cultures are identified by their most diagnostic artifact, usually a particular projectile point style. This is especially true of the Palaeoindian complexes (Irwin and Wormington 1970; Frison 1978). The Palaeoindian tradition is typified by lanceolate projectile points which exhibit heavy edge grinding along the lower one-half of the blade, masterful workmanship, and consistency of style (Wormington 1957; Irwin and Wormington 1970; Judge 1973; Frison 1988; Lynch 1991). Single- and multi-spurred gravers and spurred end scrapers are also diagnostic of the Palaeoindian cultures to the extent that these tools have sometimes been used as criteria to identify surface Palaeoindian sites in the absence of projectile points (Judge 1973; Boast 1983). The practice

of in-haft resharpening of projectile points is also a trait common to Palaeoindian cultures (Irwin and Wormington 1970; Judge 1973; Frison and Stanford 1982; Frison 1988). Given the range of dates from the site, projectile point morphology, manufacturing technique, and associated artifacts, the Mesa complex fits comfortably within the technological and temporal parameters of the Palaeoindian tradition.

Technologically the Mesa assemblage appears closely related to the Agate Basin complex of the North American high plains. The strongest similarities can be seen between the projectile points of these two Palaeoindian cultures although the associated artifacts such as spurred gravers and end scrapers are no less similar. Agate Basin points commonly display straight or convex bases while Mesa projectile point bases are generally concave; otherwise, the flake removal pattern (horizontal-opposed-parallel), overall shape, and extent of edge grinding are remarkably similar among the projectile points of both groups. Agate Basin points exhibit a considerably greater range of length than Mesa points, a circumstance which is probably due to raw material size. However, the range of width and thickness is the same for Mesa and Agate Basin projectile points, with the averages being no more than 1 mm apart (Kunz and Reanier 1994; Kunz and Shelley 1994).

Mesa complex radiocarbon dates from the Mesa as well as other sites (Reanier 1995) encompass the temporal span of 10,000 to 10,500 BP assigned to the Agate Basin culture (Frison and Stanford 1982). However, the oldest radiocarbon dates from the Mesa site predate the Agate Basin dates by at least 1,000 years. Based on these ^{14}C data as well as the technological and morphological similarities of the two complexes, it is possible that the Mesa culture could be ancestral to Agate Basin.

The data strongly suggest the Mesa complex can be considered one of the most ancient of the Palaeoindian groups. As is the case with the other Palaeoindian cultures of the North American high plains and desert southwest, the Mesa complex remains unique unto itself within the fabric of the Palaeoindian tradition. More specifically, the Mesa complex is the first well-documented component of, and the keystone for, the establishment of the Northern Palaeoindian tradition.

The Mesa site is an arctic Palaeoindian site which has undergone extensive ^{14}C documentation and is not contaminated by the remains of more recent cultures. Because of this, the Mesa complex enjoys excellent temporal and technological integrity and its distinctive projectile point style can be used as a horizon marker, identifying Palaeoindian period occupations at other arctic sites (Reanier 1995). As is often the case in archaeology, a new discovery can unexpectedly alter the complexity of a segment of a region's culture history. The Mesa site has done this by demonstrating an eleventh-millennium arctic presence for the oldest well-documented cultural tradition in North America.

REFERENCES CITED

Boast, R. B. 1983. The Folsom Gravers: A Functional Determination through Microwear Analysis. M.A. thesis, University of Colorado.

Chapman, R. M., R. L. Detterman, and M. D. Mangus. 1964. *Geology of the Killik–Etivluk Rivers Region, Alaska*. U.S. Geological Survey Professional Paper 303-F.

Detterman, R. L., A. L. Bowher, and J. T. Dutro, Jr. 1958. Glaciation on the Arctic Slope of the Brooks Range, Northern Alaska. *Arctic* 11:43–61.

Frison, G. C. 1978. *Prehistoric Hunters of the High Plains*. New York: Academic Press.

———. 1988. Paleoindian Subsistence and Settlement during Post-Clovis Times on the Northwestern Plains, the Adjacent Mountain Ranges, and Intermontane Basins. In *Americans before Columbus: Ice Age Origins*, ed. R. C. Carlisle. Ethnology Monographs No. 12. Pittsburgh: Department of Anthropology, University of Pittsburgh.

Frison, G. C., and D. J. Stanford. 1982. *The Agate Basin Site: A Record of the Paleoindian Occupation of the Northwestern High Plains*. New York: Academic Press.

Guthrie, R. D. 1982. Mammals of the Mammoth Steppe as Paleoenvironmental Indicators. In *Paleoecology of Beringia*, ed. D. M. Hopkins, J. V. Matthews, Jr., C. E. Schweger, and S. B. Young, pp. 307–26. New York: Academic Press.

Hamilton, T. D. 1979. Late-Cenozoic Glaciations and Erosion Intervals, North-Central Brooks Range. In *The United States Geological Survey in Alaska— Accomplishments during 1978*, ed. K. M. Johnson and J. R. Williams, pp. B27–B29. U.S. Geological Survey Circular 804B.

———. 1980a. *Surficial Geologic Map of the Killik River Quadrangle, Alaska*. Geological Survey Miscellaneous Field Studies Map 1234. Washington, D.C.: U.S. Geological Survey.

———. 1980b. *Quaternary Stratigraphic Sections with Radiocarbon Dates, Wiseman Quadrangle, Alaska*. Geological Survey Open-File Report 80-791. Washington, D.C.: U.S. Geological Survey.

———. 1981. Multiple Moisture Sources and the Brooks Range Glacial Record. *Abstracts, Tenth Annual Arctic Workshop, March 12, 13, and 14, 1981*, pp. 16–18. Boulder: Institute of Arctic and Alpine Research, University of Colorado.

Hamilton, T. D., and S. C. Porter. 1975. Itkillik Glaciation in the Brooks Range, Northern Alaska. *Quaternary Research* 5:471–79.

Hamilton, T. D., R. Stuckenrath, and M. Stuiver. 1980. Itkillik Glaciation in the Central Brooks Range: Radiocarbon Dates and Stratigraphic Record. *Geological Society of America, Abstracts with Programs* 12:109.

Irwin, H. T., and H. M. Wormington. 1970. Paleo-Indian Tool Types in the Great Plains. *American Antiquity* 35(1):24–34.

Judge, W. J. 1973. *Paleoindian Occupation of the Central Rio Grande in New Mexico*. Albuquerque: University of New Mexico Press.

Kunz, M. L. 1982. The Mesa Site: An Early Holocene Hunting Stand in the Iteriak Valley, Northern Alaska. *Anthropological Papers of the University of Alaska* 20(1–2):113–22.

Kunz, M. L., and R. E. Reanier. 1994. Paleoindians in Beringia: Evidence from Arctic Alaska. *Science* 263:660–62.

———. 1995. The Mesa Site: A Paleoindian Hunting Lookout in Arctic Alaska. *Arctic Anthropology* 32(1):5–30.

Kunz, M. L., and P. H. Shelley. 1994. The Mesa Complex: A Possible Beringian Precursor to the Agate Basin Culture. Paper presented at the 52d Annual Plains Anthropological Conference, Lubbock, Texas.

Lynch, T. F. 1991. Paleoindians in South America: A Discrete and Identifiable Cultural Stage? In *Clovis Origins and Adaptations*, ed. R. Bonnichsen and K. L. Turnmire, pp. 255–60. Corvallis: Oregon State University, Center for the Study of the First Americans.

Reanier, R. E. 1982. An Application of Pedological and Palynological Techniques at the Mesa Site, Northern Brooks Range, Alaska. *Anthropological Papers of the University of Alaska* 20(1–2):123–39.

———. 1995. The Antiquity of Paleoindian Materials in Northern Alaska. *Arctic Anthropology* 32(1):31–50.

Tedro, J. C. F. 1977. *Soils of the Polar Landscapes*. New Brunswick, N.J.: Rutgers University Press.

Wormington, H. M. 1957. *Ancient Man in North America*. Popular Series No. 4. Denver: Denver Museum of Natural History.

Sagavanirktok River from the Putu and Bedwell site locality.

PUTU AND BEDWELL

Richard E. Reanier

The Putu and Bedwell sites are situated on the north side of the Brooks Range, atop a hill above the Sagavanirktok River, some 8 km downstream from its confluence with the Atigun River. The hill affords a commanding, unobstructed view up and down the Sagavanirktok Valley. To the north the Gallagher Flint Station is visible 15 km in the distance. The sites were discovered in 1970 by Herbert L. Alexander during surveys by the Alyeska Archeology Project (Cook 1970:112). The Bedwell and Putu sites were first considered separate localities of a single site, but were subsequently given separate names by Alexander (1974, 1987). In the 1970 field notes the Bedwell site is listed as site S-111, and the Putu site is S-111, lower terrace. The Aly-

eska Archeology Project referred to Bedwell as S-13A and to Putu as S-13B (Cook 1970:112). Later the sites were given a single state AHRS number, PSM-027. Informally, the Bedwell site is referred to as PSM-027A and the Putu site as PSM-027B. Topographically, the Bedwell site sits on top of the hill, and the Putu site is situated on a bench about 30 m lower and 100 m south of the crest of the hill (Alexander 1987). The sites were briefly reported upon by Alexander in 1973 (Alexander 1974), and a short monograph on the Putu site was published in 1987 (Alexander 1987). Reanier (1994, 1995) has reported revisions to the radiocarbon chronologies of the sites.

PUTU

The Putu site was the first discovery in Alaska to produce radiocarbon dates said to be in association with fluted points (Alexander 1974, 1987). Because

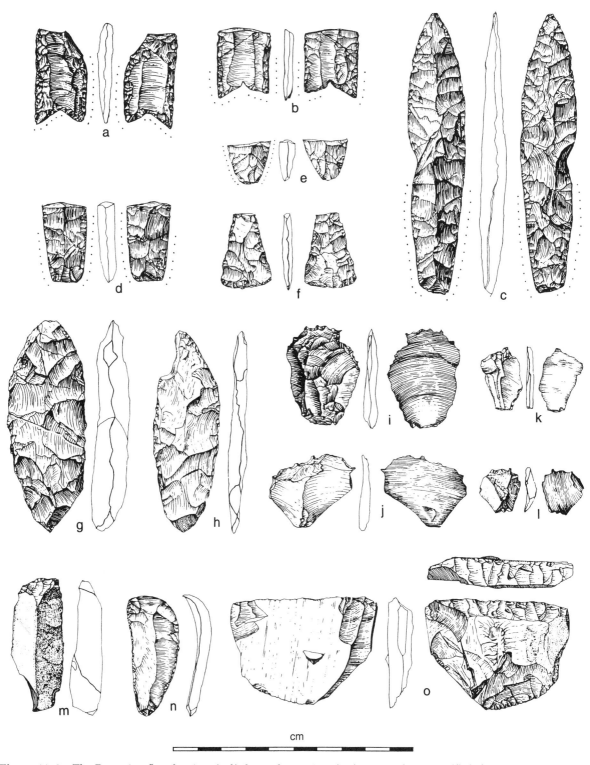

cm

Figure 11–9   The Putu site: fluted points (*a, b*); lanceolate points (*c–e*); triangular point (*f*); bifaces (*g, h*); spurred gravers (*i–l*); end scrapers (*m, n*); side scraper with blade removals (*o*).

of this, the site has had an important impact on discussions of regional chronology (Morlan 1977; Dumond 1980; West 1981; Haynes 1982; Morlan and Cinq-Mars 1982; Clark and Clark 1983; Clark 1984, 1991; Hoffecker et al. 1993; Kunz and Reanier 1994). Excavated by Alexander in 1970 and 1973, the Putu site and the adjacent Bedwell site have produced a number of lanceolate projectile points, including at least two that are fluted (Alexander 1974, 1987). Alexander considered a date of 11,470 ± 500 (SI-2382) to be associated with the fluted points (Alexander 1987), and for some this became evidence for northern origins of North American fluted points (e.g., Morlan 1977). Alexander reported evidence for two components at the Putu site, the ancient one discussed here and a recent one consisting of cracked bone, saw-cut antler, and fresh, unweathered charcoal recovered from the sod layer (Alexander 1987:9).

Artifacts

The original fluted point find at the Putu locality was a thin base of grey-tan chert found on the surface (Fig. 11–9: *b*). Each face bears three channel flake scars, and the basal edges and basal concavity have been ground. An obsidian projectile point base was discovered about three inches below the surface during the 1970 excavation (Fig. 11–9: *a*). It also has three channel flake scars on each face and grinding on the basal concavity and one of the edges. The other, longer, edge displays damage suggestive of re-use of the point as a knife or scraper. Alexander (1987:14) also reported fragments of two additional fluted points. Alexander recovered a series of lanceolate projectile points from the site which he termed Putu points (Alexander 1987:15). These points are nonfluted, have flat (Fig. 11–9: *c, d*) to convex (Fig. 11–9: *e*) bases and lenticular to diamond-shaped cross sections, and display heavy grinding on the basal edges. A complete example (Fig. 11–9: *c*), 11.5 cm long, has a base very similar to another projectile point base (Fig. 11–9: *d*) that was associated with a hearth. These lanceolate forms are technologically similar to those from the Mesa site (KIR-102) and bear technological

similarities to Agate Basin and Hell Gap points from the midcontinent.

In addition to these projectile points from Putu, other bifacial tools include a thin triangular point (Fig. 11–9: *f*) and a large series of crude bifaces, nearly all of which are broken (Fig. 11–9: *g, h*). Unifacial tools include multiple- and single-spurred gravers (Fig. 11–9: *i–l*) similar to those from the Mesa site and other Palaeoindian assemblages. Two end scrapers on blades or blade-like flakes are reported by Alexander (Fig. 11–9: *m, n*), and one side scraper has blade facets on its ventral surface (Fig. 11–9: *o*). A number of tools are said to display burin blows (Fig. 11–9: *h*). Cores and blades are also reported from Putu, many on local tabular chert or silicified mudstone. One partially reconstructed example (Fig. 11–10: *a*) reveals a number of refit blade-like forms, some of which seem to have fractured through natural fracture planes in the mudstone. Blade widths reported by Alexander range from 5 mm to 29 mm (Alexander 1987:35). The Putu assemblage does contain a number of flake cores (Fig. 11–10: *b*), but lacks any of the formal blade and microblade core types familiar from other sites in the region.

Stratigraphy and Dating

Alexander identified three stratigraphic units at the Putu site (Fig. 11–11). Zone I consists of the root mat and dark, organic-rich surface soil horizons ranging from 3 to 6 inches in thickness. Zone II is a tan-colored unit 9 to 20 inches thick from which most of the cultural material was recovered. Zone III is a grey, gravelly unit reported to be culturally sterile. Features reported from the excavations included both flake concentrations and hearths. Five hearths were reported from Zone I and two from Zone II.

Recently the radiocarbon chronology from the Putu site has been re-evaluated (Reanier 1994, 1995). The first radiocarbon date derived for Putu was from a sample recovered in 1970. This date, 8450 ± 130 (WSU-1318), was reported by the excavator to have come from approximately 16 inches below the surface near the bottom of Zone II and

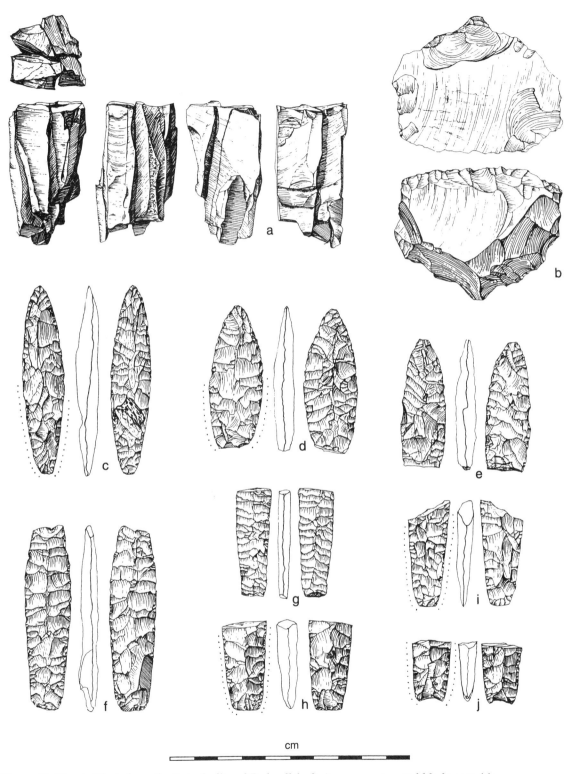

Figure 11–10   Artifacts from the Putu (*a, b*) and Bedwell (*c–j*) sites: reconstructed blade core (*a*); flake core (*b*); lanceolate projectile points (*c–j*).

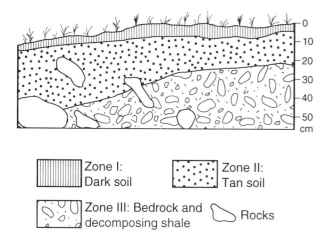

Zone I: Dark soil   Zone II: Tan soil

Zone III: Bedrock and decomposing shale   Rocks

Figure 11–11 Putu. Stratigraphic profile. (*Source: Simon Fraser University. See Acknowledgments.*)

was associated with several chert flakes (Reanier 1995). The 1973 excavation season produced a number of radiocarbon samples, many of which were submitted for dating. A series of samples was submitted to the Gakushuin University laboratory in Tokyo in 1973. A sample from Feature 7, a shallow hearth near the surface, produced a date of 650 ± 100 BP, and is consistent with the finding of saw-cut antler in the surface levels of the site (Alexander 1987:36). A second date, 5700 ± 190 (GaK-4941), was run on charcoal combined by Alexander from seven different samples that had come from various locations and depths within Zone II. A third date of 6090 ± 430 (GaK-4939) came

from three soil samples that had come from different locations within Zone II (Table 11–2).

The last radiocarbon sample submitted by Alexander from the Putu site was processed by the Smithsonian Radiation Biology Laboratory in late 1975. It is this date upon which arguments for the antiquity of Putu have been based (Morlan 1977; Dumond 1980; Haynes 1982; Morlan and Cinq-Mars 1982; Alexander 1987). The date, 11,470 ± 500 BP (SI-2382), was reported by Alexander to have come from Feature 9, a hearth about 12 inches below the surface in excavation unit 25-30S, 15-20W (Alexander 1987:36; Seymour n.d.). However, records in the archives of the Radiation Biology Laboratory indicate the date was run on a sample from Feature 3, described by the excavator as coming from within the gravel (Zone III) at a depth of about 18 to 19 inches below the surface, at or below the bottom of the culture-bearing stratum, in a different excavation unit (35-40S, 0-5W) (Reanier 1995; Wilson n.d.:6). Test excavations at Putu in 1993 revealed organic-rich smears within the upper part of the Zone III gravel that may be surface organics buried by deposition of the culture-bearing stratum (Zone II). Feature 3 could have been a pocket of such material.

Feature 9, the hearth formerly thought to date to 11,470 ± 500 BP, had yielded three charcoal samples in 1973, one of which had been archived. In 1994 this sample was submitted for AMS dating, producing a date of 8810 ± 60 BP (Beta-69901, CAMS-11038)—closely in line with the original date from the site of 8450 ± 130 BP. Associated with this hearth is the base of an edge-ground lanceolate pro-

Table 11–2 Radiocarbon Dates from the Putu Site*

Lab Number	Date BP	Material	Locality
GaK-4940	650 ± 100	Charcoal	Feature 7, hearth, Zone I
GaK-4941	5700 ± 190	Charcoal	7 combined samples, Zone II
GaK-4939	6090 ± 430	Soil	3 combined samples, Zone II
WSU-1318	8450 ± 130	Charcoal	lower Zone II
Beta-69901 (CAMS-11038)	8810 ± 60	Charcoal	Feature 9, hearth, Zone II
SI-2382	11,470 ± 500	Charcoal	Feature 3, Zone II/III

*Alexander 1987:35; Reanier 1994, 1995

jectile point (Fig. 11–9: *d*). The point was recovered from within 5 inches of the mapped boundary of Feature 9 and at the same depth (Seymour n.d.:52). The complete edge-ground lanceolate projectile point (Fig. 11–9: *c*) was found 3' 5" southwest of the point base at a depth below surface of about 8 inches and only ¹/₂ to 1 inch above the sterile grey gravel (Seymour n.d.). In contrast, the fluted points from Putu seem to lack any clear association with the two reliable old radiocarbon dates from the site. The chert fluted point (Fig. 11–9: *b*) was found on the surface, probably in what later became excavation unit 40-45S, 0-5W. The obsidian fluted point (Fig. 11–9: *a*) was recovered from only about 3 inches below the surface in square 30-35S, 10-15W, approximately 5 feet from the mapped edge of Feature 9. Another possible fragment of a fluted point was found in square 30-35S, 5-10W, about 10 feet southeast of Feature 9 and 6 to 7 inches below the surface (Alexander 1987: Fig. 8*c*).

BEDWELL

The Bedwell site is discussed briefly in Alexander's 1974 paper, but since no radiocarbon dates were forthcoming and it yielded no fluted points, it has not played much of a role in the discussions of northern Palaeoindian materials. The Bedwell site has yielded a series of edge-ground lanceolate projectile points from shallow excavations along the hilltop. Most of the points have flat to convex bases (Fig. 11–10: *c–h*), but one is concave, like most from the Mesa site (Fig. 11–10: *j*). The general style of flaking is like that of the Mesa site points, with large collateral flakes forming lenticular to diamond-shaped cross sections. One of the specimens is not edge-ground and contains a flaw near its base, suggesting it is unfinished (Fig. 11–10: *f*). Another point is unlike the others in that it is not edge-ground and has fine parallel collateral flaking.

Charcoal samples were recovered in the 1970 excavations but were not submitted for dating by Alexander. A review of the field notes indicated that some charcoal samples came from excavation units that also yielded projectile points. One of the lanceolate projectile points (Fig. 11–10: *d*) was found

in excavation unit 10-15N, 0-5E, situated at the top of the hill. A soil and charcoal sample (70-84-253) was recovered from the same unit. Although the horizontal provenience within the excavation unit was not recorded for either the projectile point or the sample, both were reported to have come from the same "yellowish soil" (similar to Putu Zone II) above the sterile gravelly substrate (McMurdo n.d.). In 1994, small pieces of charcoal from the sample were submitted for AMS dating and produced a date of 10,490 ± 70 BP. (Beta-69895, CAMS-11032). Although it is likely that this radiocarbon date is associated with the projectile point, additional testing will be required at the Bedwell site to recover charcoal samples unequivocally associated with artifacts. At the very least, the date documents a late Pleistocene occupation at the Bedwell site and suggests that a similar occupation also may have occurred at the adjacent Putu site, an occupation not presently indicated by dates from Putu itself.

The revised chronology from Putu and Bedwell casts a new light on the place of these sites in northern prehistory. Since the 11,470 ± 500 BP date probably relates to the time of deposition of Zone II, and likely is not a cultural date at all, Putu is no longer a candidate for a late Pleistocene occupation of the Brooks Range. Feature 9, together with its associated nonfluted lanceolate projectile point base, has now been shown to be younger than formerly thought on the basis of the newly run AMS date of 8810 ± 60 BP from the feature. The two mid-Holocene dates from the site, produced from combined samples, should be ignored since they cannot be said to date individual cultural events. Most importantly for northern culture history, the fluted points from Putu, which for so long had been held out as the sole example of dated northern forms, can no longer be considered so. Not only are they not 11,470 years old, but because of the lack of demonstrable spatial association between the fluted points and any of the reliable radiocarbon dates, they must be considered essentially undated by radiocarbon. Taken together, the reliable dates from Putu and Bedwell suggest substantial longevity for edge-ground, nonfluted lanceolate projectile point forms in the Brooks Range.

REFERENCES CITED

Alexander, H. L. 1974. The Association of Aurignacoid Elements with Fluted Point Complexes in North America. In *International Conference on the Prehistory and Paleoecology of Western North American Arctic and Subarctic*, ed. S. Raymond and P. Schlederman, pp. 21–31. Calgary: University of Calgary Archaeological Association.

———. 1987. *Putu: A Fluted Point Site in Alaska*. Publication No. 17, Department of Archaeology. Burnaby, B.C.: Simon Fraser University.

Clark, D. W. 1984. Northern Fluted Points: Paleo-Eskimo, Paleo-Arctic, or Paleo-Indian. *Canadian Journal of Anthropology* 4:65–81.

———. 1991. The Northern (Alaska–Yukon) Fluted Points. In *Clovis: Origins and Adaptations*, ed. R. Bonnichsen and K. Turnmire, pp. 35–47. Corvallis: Center for the Study of the First Americans, Oregon State University.

Clark, D. W., and A. M. Clark. 1983. Paleo-Indians and Fluted Points: Subarctic Alternatives. *Plains Anthropologist* 28–102 (Pt. 1):283–92.

Cook, J. P., ed. 1970. Report of Archeological Survey and Excavations along the Alyeska Pipeline Service Company Haulroad and Pipeline Alignments. Unpublished report submitted to Alyeska Pipeline Service Company, Anchorage.

Dumond, D. E. 1980. The Archeology of Alaska and the Peopling of America. *Science* 209:984–91.

Haynes, C. V. 1982. Were Clovis Progenitors in Beringia? In *Paleoecology of Beringia*, ed. D. M. Hopkins, J. V. Matthews, Jr., C. E. Schweger, and S. B. Young, pp. 383–98. New York: Academic Press.

Hoffecker, J. F., W. R. Powers, and T. Goebel. 1993. The Colonization of Beringia and the Peopling of the New World. *Science* 259:46–53.

Kunz, M. L. and R. E. Reanier. 1994. Paleoindians in Beringia: Evidence from Arctic Alaska. *Science* 263:660–62.

McMurdo, A. n.d. Unpublished 1970 Field Notes. On file, Bureau of Land Management, Fairbanks.

Morlan, R. E. 1977. Fluted Point Makers and the Extinction of the Arctic-Steppe Biome in Eastern Beringia. *Canadian Journal of Archaeology* 1:95–108.

Morlan, R. E., and J. Cinq-Mars. 1982. Ancient Beringians: Human Occupation in the Late Pleistocene of Alaska and the Yukon Territory. In *Paleoecology of Beringia*, ed. D. M. Hopkins, J. V. Matthews, Jr., C. E. Schweger, and S. B. Young, pp. 383–98. New York: Academic Press.

Reanier, R. E. 1994. The Putu Site: Pleistocene or Holocene? *Current Research in the Pleistocene* 11:148–50.

———. 1995. The Antiquity of Paleoindian Materials in Northern Alaska. *Arctic Anthropology* 32(1):31–50.

Seymour, B. n.d. Unpublished 1973 Field Notes. On file, Bureau of Land Management, Fairbanks.

West, F. H. 1981. *The Archaeology of Beringia*. New York: Columbia University Press.

Wilson, P. n.d. Unpublished 1973 Field Notes. On file, Bureau of Land Management, Fairbanks.

BLUEFISH CAVES

Edited by Robert E. Ackerman

Bluefish Caves (Caves 1–3) (67°09′N, 140°45′W) are located 54 km south of Old Crow village along the middle course of the Bluefish River, a tributary of the Porcupine in the northwestern Yukon. The caves were found at the base of either end of a Devonian limestone ridge in the Keele Mountains south of the Bluefish glacial lake basin. The ridge contains numerous karst features which have been enlarged by frost fracturing (Cinq-Mars and Lauriol 1985). Bluefish Caves 1–3 are small cavities within the limestone ridge with volumes between 10 and 30 cubic m.

Local vegetation consists of stands of white and black spruce (*Picea glauca* and *P. mariana*) which extend over the valley bottoms and lower slopes. Alpine tundra vegetation is found in higher elevations along the crests of ridges, extending up to 750 m. Ritchie et al. (1982) note that the predominance of spruce is typical for unglaciated areas of calcareous bedrock and contrasts strongly with typical forest types in the northern Yukon and Alaska where mixed forests of spruce, poplar, and birch are more common.

The Bluefish Caves were first tested in 1978, with more extensive excavation conducted in 1979 and 1981 (Morlan and Cinq-Mars 1982). Excavations were conducted inside the caves and downslope below the driplines (Cinq-Mars 1979, 1982, 1990).

STRATIGRAPHY

The depositional sequence of the cave floors is as follows, proceeding from the base of the deposit to the top (Cinq-Mars 1979, 1982, 1990):

Unit A   Frost-spalled and lag-covered bedrock.

Unit B   Sometime after 25,000 years ago, unit A began to be covered by a following series of three different aeolian silt or loess layers. The source of the aeolian silt is thought to be the fluctuating margins of the glacial lakes that filled the Old Crow and Bluefish basins. Rockfall is incorporated into these layers.

Within unit B are preserved bones of mammoth (*Mammuthus* sp.), bison (*Bison* cf. *B. priscus*), horse (*Equus lambei*), sheep (*Ovis dalli*), caribou (*Rangifer tarandus*), moose (cf. *Alces*), wapiti (*Cervus elaphus*), saiga (*Saiga tatarica*), musk-ox (*Ovibos moschatus*), lion (*Panthera leo atrox*), cougar (*Felis concolor*), bear

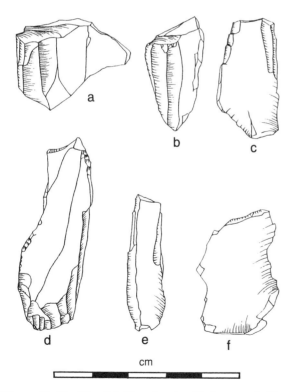

Figure 11–12   Bluefish Cave 2: microblade cores (*a, b*); angle burins (*c–f*). (*Source:* Revista de Arqueología Americana. *See Acknowledgments.*)

(*Ursus arctos*), wolf (*Canis lupus*), and numerous small mammals, birds, and fish (Cinq-Mars 1990:Table 1).

Unit B formed over a period of approximately 15,000 years. The pollen record indicates a shift from a dry, herbaceous steppe-tundra to a moist shrub tundra (Cinq-Mars 1979, 1990).

Unit B in Caves 1 and 2 contained lithic artifacts and altered bone (reflecting butchering episodes, bone tool production, and bone reduction by flaking).

Unit C   Thick, humus-rich sediment with rock fall inclusions suggesting wetter boreal conditions (invasion of boreal forest). Unit C was formed after 10,000 years ago and contains only Holocene fauna typical of the northwestern Cordilleran interior.

Unit D   Modern humus, litter, and vegetation.

No cultural material was reported from units C and D. All of the cultural materials recovered are from unit B.

CULTURAL MATERIALS

Cinq-Mars and Morlan (n.d.) note that there are three classes of lithic materials. The first group consists of microblade cores, microblades, core tablets, burins, burin spalls, and debitage. All are made of high quality chert and all but two of these artifacts came from unit B of Caves 1 and 2 (Fig. 11–12). The microblade cores appear to have been made on flakes with the platform created by the removal of a platform tablet flake and the keel formed by bifacial edge retouch. Angle burins were made on flakes or blades. One has an edge-retouched platform. The complex is similar to that recovered from Denali complexes in central Alaska (West 1967, 1981, this volume). No directly associated datable material was recovered to provide dates for these artifacts. The only correlation is stratigraphic.

The second class of lithic materials consists of microflakes measuring 1–3 mm in length or width. They were recovered throughout unit B with the greatest concentration at the point where the palynological information records the shift from herbaceous to shrub tundra, dated elsewhere to about 13,500 BP (Ritchie et al. 1982; Ritchie 1984).

The third class is made up of small cobbles and pebbles which could have been brought into the caves by palaeokarst stream transport and may not be cultural materials.

Split long bones that show traces of "whittling or shaving" and with areas of high polish have been advanced as possible tool forms. One, a split caribou tibia that could have been a broken fleshing tool, has an AMS date of 24,800 BP.

A mammoth bone core and reattached flake from the bottom of unit B in Cave 2 is regarded as evidence of a well-developed bone technology. Separate dates (AMS) for bone core and flake provide an average age of 23,500 BP. Cinq-Mars and Morlan (n.d.) conclude that distinctions can be made between bone that was culturally modified and that modified by natural agencies.

Brief Conclusions

While there is considerable emphasis on the bone technology and its implied antiquity based upon AMS and standard ^{14}C dating, the appearance of a Denali lithic complex in the Bluefish Caves is of considerable importance. Cinq-Mars (1990) believes that due to the stratigraphic position of the artifacts the complex dates to 12,000 BP, and possibly somewhat before. The distribution of microflakes through unit B has also prompted Cinq-Mars (1990) to suggest that the caves were perhaps occupied as early as 25,000 years ago.

References Cited

Cinq-Mars, J. 1979. Bluefish Cave I: A Late Pleistocene Eastern Beringian Cave Site in the Northern Yukon. *Canadian Journal of Archaeology* 3:1–32.

———. 1982. *Les Grottes du Poisson-Bleu. GEOS* 11(1): 19–21.

———. 1990. *La Place des Grottes du Poisson-Bleu dans la Prehistoire Beringienne. Revista de Arquelog'a Americana* 1:9–32.

Cinq-Mars, J., and B. Lauriol. 1985. *Le Karst de Tsi-It-Toh-Choh: Notes Preliminaires sur Quelques Phenomenes Karstiques du Yukon Septentrional, Canada. Annales de la Société de Belgique* 108:185–95.

Cinq-Mars, J., and R. E. Morlan. n. d. Bluefish Caves and Old Crow Basin: A New Rapport. Manuscript in possession of author.

Morlan, R. E., and J. Cinq-Mars. 1982. Ancient Beringians: Human Occupation in the Late Pleistocene of Alaska and the Yukon Territory. In *Paleoecology of Beringia*, edited by D. M. Hopkins, J. V. Matthews, Jr., C. E. Schweger, and S. B. Young, pp. 353–81. New York: Academic Press.

Ritchie, J. C. 1984. *Past and Present Vegetation of the Far Northwest of Canada.* Toronto: University of Toronto Press.

Ritchie, J. C., J. Cinq-Mars, and L. C. Cwynar. 1982. *L'Environement Tardiglaciaire du Yukon Septentrional, Canada. Géographie physique et Quaternaire* 36(1–1):241–50.

West, F. H. 1967. The Donnelly Ridge Site and the Definition of an Early Core and Blade Complex in Central Alaska. *American Antiquity* 32(3):360–82.

———. 1981. *The Archaeology of Beringia.* New York: Columbia University Press.

PRISMATIC CORE SITES ON THE KUKPOWRUK AND KUGURUROK RIVERS

Ralph S. Solecki

In the summer of 1949 the writer joined a party of geologists surveying the Kukpowruk and Kokolik rivers from their headwaters down to the Arctic Ocean (Solecki 1950, 1951). One of the geologists, Edward Sable, had earlier found the famous Utukok River fluted point reported by Raymond Thompson (1948). This was the immediate reason for an archaeologist being asked to join the party.

In addition to a report of this survey, information is included here on finds made by Milton C. Lachenbruch on the Kugururok River the following year. The lithic collection from the Kukpowruk–Kokolik survey was deposited with the Smithsonian Institution (with the exception of a small sample collection left with the University of Alaska [accession no. 44]). Artifacts recovered from the Kugururok survey were, presumably, also deposited with the Smithsonian.

View down Kukpowruk River in the foothills of the Brooks Range.

Of the three physiographic provinces of northern Alaska, the arctic foothills province contributed the major share of archaeological findings, particularly lithics. This province is a hilly region between the mountainous Brooks Range province and the nearly flat coastal plain province (Fig. 11–13). The foothills province is subdivided into southern and northern sections, which differ somewhat in both topography and geology. The rocks are composed of alternating layers of sandstone and shales which have been folded into anticlines and synclines. As a result, there are marked differences in their resistance to erosion, and these differences are indicated by the general east-west alignment of the ridges. Near the mountains, in the southern section, the folding has been more intense and in places the beds stand nearly vertical. Farther away from the mountains, in the north, where the folding is more gentle, these beds are nearly horizontal. The harder layers form steps on the hillsides or cap the hills (Smith and Mertie 1930; Chapman and Sable 1960). This land north of the Brooks Mountain Range is unglaciated and marked by an arid climate with freezing temperatures nearly all year. This condition has helped to preserve artifacts of historic age and contributed to the integrity of the ridgetop sites.

The very prominent eastward-trending sandstone ridges of nearly uniform height (about 180 meters in the northern foothills section) are barren of vegetation and obstacles, making for very fine foot-travel routes. Summertime travel in the arctic coastal plain and in the intervening swampy valleys between the hills in the foothills province is to be avoided. Of course, once the short summer season ends and the terrain is frozen, travel in any direction is easily accomplished.

Raw material for the production of chipped stone tools is plentiful in the area. In the Kukpowruk formation, Chapman and Sable (1960) report finding conglomerate beds containing chert pebbles and cobbles. These lenses occur within and at the base of thick sandstone beds. They are most abundant in the southern part of the region, and are rare or absent in the northern part. Along the upper Kukpowruk River, raw chert pebbles are very easily dislodged from their silty and sandy matrix. Chert cobbles are also readily available in the river gravels. Chapman and Sable (1960) report that black chert is the most common material, followed by grey, greyish green, and greyish red colored chert, in that order. White quartz, argillite, and quartzite ranging in color from black to reddish grey are also to be found. However, despite this range of color and type of lithic raw material, greyish green chert appears to be the most predominant material in the artifacts recovered from the Kukpowruk and Kokolik basins. Whether this is due to the size of the raw material cobbles, their lithic characteristics, or other unknown preferences is not fully understood.

Kukpowruk River Site 65

The site is located at ca. 68°45′N, 163°10′W on a ridgetop bordering the right bank of the Kukpowruk River, at an elevation of 235 m. It commands an excellent view of the surrounding terrain and, like other ridgetop sites, was a strategic location where hunters could refurbish their stone armament while waiting for game to appear in the valley below. (U.S. Geological Survey triangulation station no. 18 is located on the site.) The top of the ridge and its adjacent slopes were littered with flint chips. No other occupational evidence was noted. Camps were undoubtedly made near the river, below the ridgetop. No signs of early occupation sites were found in the entire survey of both river valleys. The changing river course would have assuredly wiped out any such traces over the years.

Artifacts

Recovered from the site were the following specimens: 2 prismatic cores of grey-green chert; 1 core

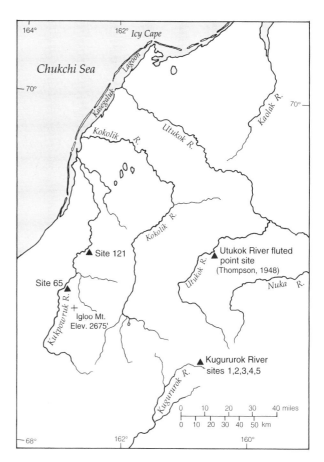

Figure 11–13 Arctic foothills province showing sites discussed in text.

blank or possible reject of grey-green chert; 4 blades—1 use-retouched of grey-green chert, 1 of light green chert, 2 of black chert; 1 well-chipped triangular projectile point of black chert; 3 bifaces or knives—2 of grey-green chert, 1 crudely chipped, 1 broken, 1 of black chert, a large half section; and more than 180 flakes, 154 of grey-green chert, 25 of black chert, and several of varicolored chert.

The two prismatic cores (Core No. 1 [Fig. 11–14: *a*] and Core No. 2 [Fig. 11–14: *c*]) were found partially buried in the soil down to a depth of about 5 cm. They lay point end to point end, as though purposefully placed in position. Both are patinated. Core No. 1, illustrated in Solecki (1950), measures about 7.5 cm long and 6.0 cm wide across its greatest diameter at the striking platform. The largest blade detachment scar measures 8.5 cm long and 1 cm in greatest width. The blade scars or flutes ex-

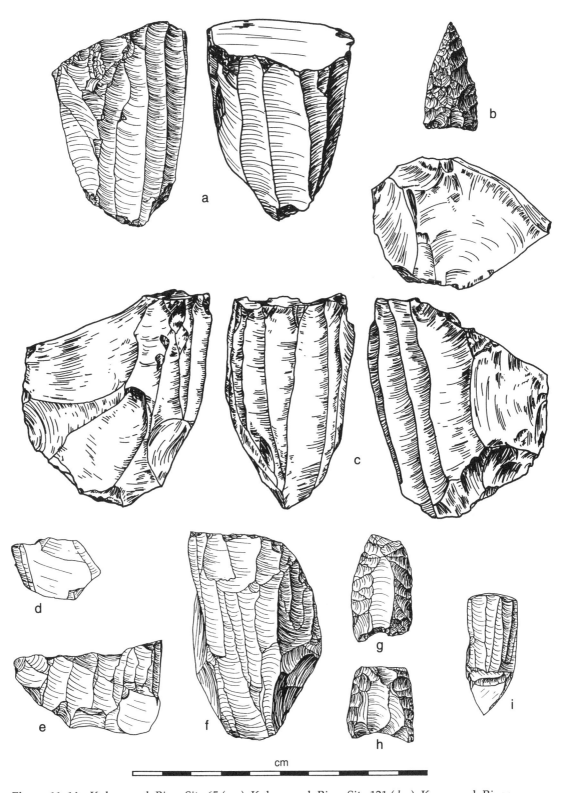

Figure 11–14   Kukpowruk River Site 65 (*a–c*), Kukpowruk River Site 121 (*d, e*), Kugururok River Site 1 (*f–h*), Kugururok River Site 2 (*i*): large prismatic cores (*a, c, f*); triangular bifacial projectile point (*b*); boat-shaped bladelet cores (*d, e*); fluted projectile points (*g, h*); prismatic bladelet core (*i*).

tend halfway around the outer face of the specimen. The back face of the core does not bear blade detachment scars, beyond the rough preparation or trimming. The striking plane surface is shallowly concave, evidently the result of the removal of a thick flake or tablet. There is some roughening or burring of the striking platform edge, presumably done purposefully in order to make a suitable purchase for the blade removal blow. The blade scars run more or less true down the forward sides of the specimen, curving inward to nearly meet at the apex or bottom of the core. This blade core is very reminiscent of Eurasiatic Upper Palaeolithic cores. It is quite possible that at least some of the collected blades and flakes could be refitted to this core.

The second blade core from Site 65 (Fig. 11–14: *c*), which was deposited with a sample collection from the Kukpowruk–Kokolik survey at the University of Alaska, was studied by John V. Matthews, Jr. (1964). It is of grey chert, according to Matthews (which others might call grey-green like the first specimen), and measures 7.3 cm long by 6.1 × 4.0 cm. It is roughly ovate in section. There is a small bit of cortex on one side. The core bears blade scars on the frontal end. The back of the specimen shows rough preparation, and was not utilized for blade removal. As with the first core, the forward edge of the striking platform exhibits burring or blunting with a hammer, presumably providing purchase for blade removal. Matthews was able to find a blade in the lot which fitted on to the core face and also records what he calls a "blade surface renewal spall" (Matthews 1964). It is of grey-green chert, measures 6.1 × 2.3 × 0.7 cm, and features remnants of the blade scars on the short dimension.

Since raw material was readily available in the river gravels, and some of the flakes and blades exhibit traces of cortex, we may assume that the early hunters fabricated their stone tools on the spot. The black chert blades were struck off from a prismatic core which was not recovered by the survey.

The bifacial triangular projectile point (Fig. 11–14: *b*) found at Site 65 was made of black flint or chert, probably local. It is the only specimen of this type found during the survey of the two rivers. The point is well made, measures 3.5 cm long, and is moderately tapered to a point from a base 1.8 cm wide. The base is shallowly concave. The chipping

is not distinctive, although it appears to have been made by the pressure technique.

Since this was a surface site, with no stratigraphy evident, it can be assumed to be a multicomponent site because a variety of artifacts and technologies are present among the finds.

Kukpowruk Site 121

This site, located about 69°14′N and 162°45′W, was found on the east side of a sandstone ridge running north-south to a tributary of the Kukpowruk River. On the right bank of the river, the site occupies a prominent observation point on a rock outcrop covered with a sparse growth of grass. Its elevation is about 230 m.

ARTIFACTS

One hundred and twenty-five flakes were found in three small concentrations, which measured 1 m or less in diameter and as much as 7 cm in depth. Several stray flakes were found scattered in the nearby vicinity, within 20–30 m of the chipping stations. In one of the three chipping concentrations was found what looked like Alaska Campus material as described by Nelson (1937:270–72), mixed with nondescript chips and flakes. With the exception of two cores (Fig. 11–14: *d, e*) and one large flake, all the material was grey-green chert. The smaller core is made of translucent light brown chalcedony, which appears to have been an import. The larger specimen is made of a light tan chert, probably of local origin. The large flake is made of a light tan chert as well. These semi-prismatic cores probably were not derived from original nodules, but secondarily from other artifacts. It appears that a large broken chert knife or other implement was selected to take advantage of the break which could now function as a striking platform. The blades were punched off one end of the specimen. Nelson (1937) comments that nondescript spalls were used for the manufacture of the Campus cores, and that the flaking exhibits three stages of manipulation: the striking platform, the preparatory transverse flaking, and the lateral fluting.

The smaller light brown chalcedony core has a boat shape, measuring 4.0 cm long by 2.5 cm wide and 1.4 cm thick (Fig. 11–14: *d*). The forward edge has six bladelet detachment scars. The rear parts of the specimen exhibit flake scars. Judging by the scars, the longest bladelet detached was about 2.0 cm long. Several small scars or breaks are present at the forward part of the core at the blade detachment end, which was the presumable result of a scoring or burring of the core edge in order to facilitate purchase for blade detachment. The striking platform had been rejuvenated, forming a step. The surface of the specimen at the striking platform has three plateaus, possibly indicating a rejuvenation of the striking platform three times.

The larger specimen, made of light brown chert (Fig. 11–14: *e*) has a thick light tan patina. It has a long narrow boat-shaped profile and measures 6.8 cm long by 3.8 cm wide by 1.5 cm thick. There are four blade scars at the broader forward edge, which was the only area worked on the core. The rear end of the core exhibits flake scars. The striking platform at the forward blade detachment end is hollowed. There is a step about two-thirds down the face from the striking platform, the fault that may have occasioned the core's discard. There are a number of minute chip scars at the forward edge of the striking platform at its junction with the blade scars, which presumably represent preparatory burring or abrasion to facilitate blade detachment.

Inspection of the cores from Sites 65 and 121 shows real differences, suggesting two different cultural traditions. The Site 65 cores were obviously made from native local cobbles specially selected to produce large nodules for the purpose of extracting large blades. These cores are larger, thicker, and closer to round in plan. Blades were struck off from up to half of the circumference.

The cores from Site 121 appear to be reutilized large flakes or broken implements which have been adapted as bladelet cores. These cores are boat-shaped in profile and elongate in plan. Only the extreme forward edge of these cores is utilized for the extraction of small blades or bladelets. These bladelets, more delicate in nature than the blades from the larger Site 65 cores, are half as long—even shorter than the blades from the latter site.

We postulate that the Kukpowruk River finds represent two different and noncontemporaneous techniques of extracting blades and bladelets from cores. The less refined antler punch and antler hammer technique was probably used with the larger cores. Examination of the ends of the blades reveals that soft hammer and punch were used. Smaller prismatic cores, because of their size, could not be effectively manipulated using the more massive punch and hammer technique. Detachment of bladelets from the cores had to be done with precision. The crutch method and vise-held core satisfied this requirement. As to which method came first in the Arctic, it is probably safe to say that the technique evidenced in the larger cores was the earlier of the two.

Kugururok River Site 1

Highly important data were recovered during the summer of 1950 by Milton C. Lachenbruch, a geologist with the U.S. Geological Survey in the Kugururok River valley, just bordering the northern side of the Brooks Range mountain province. The area of his finds was in the pass route through the De Long Mountains (Fig. 11–13). The area of the finds is about 100 km east-southeast of Kukpowruk Site 65. The Kugururok River is one of the upper branches of the Noatak River. The headwaters of the Kugururok River are about 8 km from the headwaters of the Utukok River, on which the fluted point described by Thompson (1948) was found. The connecting pass between these two respectively southward- and northward-flowing drainages is at an elevation of about 900 m. There are towering mountains up to 300 to 600 m higher on either side. On prominent knolls within the glacial debris-strewn valley, Lachenbruch discovered five sites yielding several hundred flint chippings and a few tools. These were all surface finds. Among the tools were two fluted projectile points and two prismatic cores.

The site was found on the east bank of the Kugururok River between two tributaries on a rubble-covered knoll about 8 m high. This lies just north of a former ice field where the Kugururok River narrows to enter the mountains. The valley walls narrow at this point and the knoll provides a good view of this pass. The site measured about 30 by 90 m in area.

ARTIFACTS

Lachenbruch recovered two fluted points, 1 large prismatic fluted core (Fig. 11–14: *f*), 1 large end scraper, 3 side scrapers, three-quarters of a knife blade or biface, 27 use-retouched flakes, 5 broken blanks, and some 204 flakes and chips from this hilltop site. Nearly all of the specimens were patinated to varying degrees. The raw material included a variety of chert, ranging from light tan color through reds and brown, and green to black. A sedimentary cherty stone, an unusual material, was also utilized.

Fluted Points   One fluted projectile point (Fig. 11–14: *g*) of black chert measured 3.5 cm long but as it was broken at the tip, was probably about 4.2 cm long originally. It is 2.1 cm wide. Flutings measuring 2.5 cm long and 0.8 cm wide were present on both sides. One side has a single flute, while the opposite side is double-channeled with flutes side by side. The flutes were struck from the basal end, which evidences additional chipping over the channel at the base. The flutes were struck off after the sides had been prepared. The flakes had been removed at right angles to the axis of the specimen but cannot be called true "ripple flakes." The midsection when viewed in longitudinal cross section is thicker (0.6 cm) than the rest of the specimen, which is tapered at both ends. The projectile point viewed in outline is expanded forward of the center, tapering toward the tip end, which had been shattered. The opposing basal sides and the concave base had been abraded or dulled moderately. The "ears" are short, and one of them is broken. There is basal chipping at the incurvate base of the specimen. Frank H. H. Roberts, Jr., of the Bureau of American Ethnology, considered this point typologically admissible as a Folsom point.

The other fluted point was identified as the Plainview type (Fig. 11–14: *h*). It is made of a light green chert with translucent edges. Like the first fluted point, it is broken at the tip end. It measures 2.6 cm long but was originally about 4 cm long. The

basal end is 2.3 cm wide. It is 0.5 cm thick. One face bears three attempts at channeling, while the other bears two. The channels or flutings are short. The longest measures 1.8 cm long and 0.8 cm wide on one face, and on the obverse face there is a channel 1.4 cm long and 0.8 cm wide. The sides are curvate toward the point end. The base is shallowly concave with short ears. The maximum breadth is at the base. The ripple flakes on the surface are marred and obscured by the flutings. The concave base and sides had been rubbed or dulled smooth to a moderate degree.

Other Artifacts   Other tools include a beveled end scraper which shows considerable use wear. Of the three side scrapers, one has a parrot-like beak on one end. This feature, however, may be merely an accident of flaking, since no marked use wear could be seen on it.

Kugururok Site 2

Another site was found at the fork of the Kugururok River and a tributary. The site was found on the west bank of the river, on the north end of a 150 × 60 m rubble-covered knoll. The northern end is 15 m higher than the southern end. Most of the surrounding valley is low, hence this point offered good observation of game. Below was another knoll, both knolls showing signs of glaciation. On the adjacent lower knoll were found two chipping stations and several small depressions, each of which measures about 30 cm deep and almost 1 m in diameter. One chert flake was found in one of the depressions.

ARTIFACTS

From this site were recovered one prismatic core (Fig. 11–14: *i*), 12 use-retouched flakes, 1 piece of smooth elongate sandstone, 2 pieces of hematite, 5 scraps of unidentifiable bleached bones, and 56 chert flakes. Some of the flakes are of a peculiar reddish silty chert which is native to the locale.

The core is much smaller and more refined-looking than the larger core from Kugururok Site 1. It measures about 4.0 cm long and 1.6 cm wide.

From its size, only bladelets could be pressed off from its forward face. The core is made of a light grey-green chert, which is probably derived from the local river cobbles. This specimen has a pronounced cupped striking platform, which appears to have been intentionally produced by the detachment of several flakes. The flutings are found on one half of the core circumference. The remainder is only roughly blocked out. This area of the core was presumably prepared by the percussion technique whereas the bladelets on the face were probably pressed off with an intermediate punch and hammer. The usefulness of the core seems to have been impaired by a fault in the form of an impeding bulge on the lower part of the core face. There is evidence of some abrasion around the lip edge of the core face. Typologically, this core looks like the Alaskan Campus type, whereas the large core from Site 1 looks like the Kukpowruk Site 65 type.

Kugururok Site 3

A chipping station was found on a knoll on the right, or east, bank of the river from Site 1. The area of the knoll measures about 30 by 90 m, is about 8 m high with a flat top, and is prominently situated with steeply sloping sides. The site is covered with rubble, and shows the existence of a former ice field. Lachenbruch recovered from this site two use-retouched flakes and 16 ordinary chert flakes. These artifacts were found northeast of the knoll's center in a wide, flat area.

Kugururok Site 4

On another knoll about 360 m from Site 1 were recovered one broken blank, 5 use-retouched chert flakes, and 11 ordinary flakes, predominantly of green chert.

Adjacent to this site was another knoll, from which were recovered two fragments of chert blades, 5 use-retouched flakes, and 31 waste flakes. The chipping debris from this site was predominantly black in color.

SUMMARY

The majority of the artifacts found in the Kukpowruk–Kokolik and the Kugururok surveys appear to be pre-Eskimo in type and age. Early finds in Siberia from the Chukchi Peninsula as reported by A. P. Okladnikov and I. A. Nekrasov (1959), particularly the elongate bifaces, look quite similar to the Kukpowruk–Kokolik bifaces. Prismatic cores were also recovered from the Lake El'gytkhyn site on the central Anadyr plateau.

The smaller, finer-looking prismatic cores like those from the Campus site appear to be part of a technological tradition separate from that which produced the larger Kukpowruk and Kugururok cores. Because all of the finds were surface collected, it cannot be determined which artifacts were really associated with the fluted points from the Kugururok River.

Edwin S. Hall, Jr. (1975) illustrates lithic finds made by Elmer Schell of a U.S. Geological Survey party of geologists "relatively near where the Kukpowruk River breaks out from the De Long Mountains into the coastal plain." This description of locale is ambiguous, because the De Long Mountains are many kilometers from the Arctic coastal plain. Nevertheless, the Kukpowruk River cores illustrated by Hall (1975) closely resemble the Site 65 blade (*not* microblade) cores. Hall's illustrated Plate 6 No. 2 core may be classified as a microblade core of the Campus type. Hall does not give measurements of these specimens, although some idea of their proportions may be obtained from the scaled photograph. He does mention that the two large Kukpowruk cores recovered by Schell have teardrop cross sections rather than the almost circular sections of the cores from Site 65 (Hall 1975:18–19).

West (1981:90–91) has commented that "the typing of the blade core technology will eventually prove to be one of the most useful diagnostic devices in sorting through these Beringian assemblages." He presents a formal classification of Beringian blade cores and discusses the complexity of the arguments surrounding fluted points in the north (West 1981:184–86). In any event, we are on firm ground in one respect: Lachenbruch states unequivocally that the fluted point finds were made on glaciated features and hence must postdate glaciation in the Kugururok valley.

REFERENCES CITED

Chapman, R. M., and E. G. Sable. 1960. *Geology of the Utukok–Corwin Region, Northwestern Alaska. Exploration of Naval Petroleum Reserve No. 4 and Adjacent Areas, Northern Alaska 1944–1953.* U.S. Geological Survey Professional Paper 303-C.

Hall, E. S., Jr. 1975. An Archaeological survey of Interior Northwest Alaska. *Anthropological Papers of the University of Alaska* 17(2):13–30.

Matthews, J. V., Jr. 1964. Description of a Portion of the Collection of Artifacts from the Kukpowruk and Kokolik Rivers, Northwestern Alaska. Class report, University of Alaska, Fairbanks.

Nelson, N. C. 1937. Notes on Cultural Relations between Asia and America. *American Antiquity* 2(4):267–76.

Okladnikov, A. P., and I. A. Nekrasov. 1959. New Traces of an Inland Neolithic Culture in the Chukotsk (Chukchi) Peninsula. *American Antiquity* 25(2): 247–56.

Smith, P. S., and J. B. Mertie, Jr. 1930. *Geology and Mineral Resources of Northwestern Alaska.* U.S. Geological Survey Bulletin No. 815.

Solecki, R. S. 1950. A Preliminary Report of an Archaeological Reconnaissance of the Kukpowruk and Kokolik Rivers in Northwest Alaska. *American Antiquity* 16(1):66–69.

———. 1951. Notes on Two Archaeological Discoveries in Northern Alaska, 1950. *American Antiquity* 17(1): 55–57.

Thompson, R. M. 1948. Notes on the Archaeology of the Utukok River. *American Antiquity* 14(1):62–65.

West, F. H. 1981. *The Archaeology of Beringia.* New York: Columbia University Press.

Part Three

———————

CONCLUSIONS

Chapter 12

BERINGIA AND NEW WORLD ORIGINS
I.

THE LINGUISTIC EVIDENCE

Joseph H. Greenberg

There are other approaches to the question of the first peopling of the Americas. It becomes a matter of interest for linguists, growing naturally out of their classifications of native languages. For the archaeologist, the interest in linguistic results derives from the linguists' evidence being completely independent of archaeological or anthropological thinking. The following article by one of the most important linguists of this age is striking in its parallelism to the archaeological evidence presented in this volume.

F. H. W.

What can the languages of Native Americans tell us about how the New World was settled? In seeking an answer to this question, I will as far as possible rely on linguistic evidence only and, not being an archaeologist or geneticist by profession, and therefore not capable of independent evaluation in these other fields, I will refrain as far as possible from using hypotheses which derive from them. Conclusions which derive jointly from a number of scholarly disciplines are reliable and convincing only

insofar as they are based on independent evidence from each field.

Two preliminary observations are in order. One is that in considering linguistic evidence, there are first the languages themselves, that is, their vocabulary, sound systems, morphological and syntactic structures and their similarities and differences, which lead to hypotheses of common origin and of language contacts over time. But one major apparently external factor does inevitably play a role: the present and presumed former geographical distribution of languages and their speakers.

Moreover, the geography plays a similar role as a formally external, yet basic, intrinsic factor in considering archaeological and genetic evidence also. It forms, as it were, the universal background in all of these fields, yet what is presumed to have happened within these spatial configurations is based on different and independent evidence. If the results converge to produce reasonable and harmonious conclusions, we are naturally gratified. The other preliminary observation is one that presumably just about every linguist and nonlinguist at all acquainted with historical linguistics takes for granted, yet it deserves some consideration. This is the role of the so-called genetic classification of languages. In orthodox historical linguistics this is constituted by a taxonomy of inclusion basically similar to that of biological classification. Just as in

525

biology, this basic family-tree approach has its problems. For example, at the lowest taxonomic level, difficulty in distinguishing separate species from varieties of the same species has its parallel in distinguishing dialect from language. This is because we are dealing with a dynamic process, speciation and language formation. Ultimately we have what are indubitably separate languages and separate species. Moreover, as has long been seen, it is the assumption that this dynamic process of the development of ever more internal variation and ultimate separation takes place in a similar fashion in the present and past that leads to an evolutionary interpretation in biology on the one hand, and in genetic linguistics on the other.

Not all historical deductions from language are based directly on the genetic hierarchy of languages, e.g., conclusions regarding language contact and those regarding the cultural implications of proto-vocabularies, but they all require a genetic classification as a prerequisite. These may be illustrated from examples. When a group of languages presently or formerly adjacent have had intrinsic contacts with each other we talk of areal factors. One well-known example is that of the Balkans. Not only borrowed words, but typological characteristics such as the existence of a suffixed definite article and a future tense formed from a verb meaning 'to wish' are among the Balkan linguistic characteristics (Sandfeld 1968). These latter items are often called loan translations or calques.

In regard to the suffixed article, Bulgarian, Romanian, and Albanian all share the structural similarity to which we have alluded. But in each language the article itself is based on inherited material, different in each case. It is only on the basis of previous genetic classification that we can identify its sources, and only by comparison with other languages of the same genetic level that we can identify it as a change in inherited forms. Thus, the Romanian suffixed article -*l*, as in *calu-l* 'the horse' has the same origin as the French article *le/la/les* which occurs before the noun, and they have the same immediate source in the Latin *ille* 'that.' Similarly, standard Bulgarian has a suffixed article -*t* which has a common Slavic origin with non-Balkan Slavic languages and can be seen to be a convergent structural feature like that of the Romanian post-

posed article. Without a background, then, in genetic classification, such conclusions could not be reached.

So also for cultural vocabulary. Here comparison and reconstruction within a genetically defined group is of the essence. Without Proto-Indo-European, we have no right to posit a former linguistic community. Without the existence of reconstructed forms for numerals at least as high as 100 and of a reconstructible vocabulary which includes 'horse' and a number of other domesticated pastoral species, we would not be able to make statements about the economic subsistence type of Indo-European speakers. Genetic classification is, then, the indispensable background, directly or indirectly, for all historical inferences drawn from languages in the absence of direct, written, historical records.

The chief types of inference with which we are concerned in regard to the settlement of the Americas is the number of such settlements, the relative—and, if possible, the absolute—chronology of their arrival, and the area of first settlement and that of subsequent groups derived by successive splits from the original group.

We may note that at the very beginning of the paper the question raised, naturally enough, was what the languages of Native Americans can tell us about the settlement of the Americas.

From this it might be thought that were it not for the abundance of direct historical evidence regarding European settlement, we would not be able to determine that the European languages—which for the sake of simplicity will be confined here to Spanish, Portuguese, French, English, and Dutch—were recent in the New World, nor would we be able to determine, on purely linguistic grounds, that the corresponding languages in Europe and elsewhere in the world originated in Europe only a few centuries ago.

If we consider any one of the European languages earlier mentioned in relation to its present differentiation and the geographical location of these variant forms, we always arrive at the same conclusions. For example, the still-surviving local dialects of French that are spoken in France show far more drastic and fundamental differences from each other than from the French spoken in Quebec,

Louisiana, Martinique, and, for that matter, the French spoken in other parts of the world, such as, for example, Tahiti. In fact, when a language spreads it is normally only one variant, frequently some version of the "standard" language, that is involved.

This leads us naturally to the conclusion that French spoken in the Americas is the result of recent migration from France. Similar results will follow from an examination of Spanish, Portuguese, English, and Dutch.

There are additional linguistic facts which greatly strengthen this already well-established conclusion. If we proceed to compare all of the five languages mentioned, we find a second-stage application of the same principle. It is almost immediately obvious that these five languages fall into two groupings: Spanish, Portuguese, and French, which are Romance languages, and English and Dutch, which are Germanic languages.

If we look at the geographical distribution of the remaining Romance languages, e.g., Catalan, Italian, Romansch, and Romanian, we see here a solid block of languages, all spoken in Europe and all showing considerable internal differentiation into local dialect forms. On the other hand, English and Dutch go together as members of what is usually called the Germanic linguistic stock. Once more we find a solid block of languages, all spoken in Western Europe and each with considerable antiquity and internal differentiations.

In fact, we can divide Germanic languages genetically into North Germanic (Scandinavian, Icelandic, Faroese) and West Germanic (English, Frisian, Dutch, Low German, and High German). Dutch was also carried into South Africa where in its changed form it is called Afrikaans. The same kind of reasoning described earlier will lead to the conclusion that Dutch/Afrikaans originated in the Dutch/Flemish area of Europe and not in Africa. Within West Germanic, Anglo-Frisian forms a separate subgroup, while Dutch goes with Low German (Plattdeutsch). The general distribution of Germanic to the east of England on the continent, and the location of the Frisian Islands off the coast of Holland, Germany, and Denmark, suggests once more a continental European origin for English from the coasts and islands closest to the coast.

We can carry this kind of reasoning one or, probably, two steps further. Both the Germanic and Romance languages are members of the Indo-European family of languages which stretches, with interruptions, through the presence of Turkish and Semitic languages as far as India. Hence Germanic and Romance are among the western outliers of a vast distribution that centers further east. We can probably extend this analysis yet one more step in that Indo-European appears to be the westernmost branch of a vast family, Eurasiatic. Closest to Indo-European on the east are the Uralic languages which apparently have a special relationship to Yukaghir, still further east in the central and eastern parts of northern Siberia, and after that to the Altaic family with its three branches: Turkic, Altaic, and Tungus-Manchu. A further discussion of this vast stock which I have called Eurasiatic is postponed until later for its bearing on the settling of the Americas.

It will be noted that each successive higher genetic node in linguistic classification strengthens the case for the European rather than American origin of the Western European languages which now are spoken by the vast majority of the population of the New World. We have seen how, even in the absence of written attestation, it would have been possible to deduce the extreme recency of these languages in the Americas, which would lead us back to the numerous languages of the Native Americans in our search for linguistic evidence regarding the peopling of the New World.

Throughout this reasoning, a number of principles have been tacitly employed. One of these is that the existence of a hierarchic taxonomy in genetic linguistic classification leads us to evaluate each level separately in terms of the distribution of the constituent languages as language groups of the next lower levels. The second is that we make no assumption regarding any inherent tendency of peoples to move in one direction or the other. This leads to the general notion that the most central area of the distribution of the component members is the most likely area of origin. This is based on "the principle of least moves" (Dyen 1956).

Of course, external nonlinguistic reasons, including geographic conformations in terms of land and water and considerations of climate, as well as linguistic factors regarding earlier locations and evidence from contact shown through loan words, play a role in modifying or specifying more exactly the original area and subsequent spread of a linguistic grouping.

For example, the existence of a stratum of words in Proto-Finno-Ugric, which are undoubtedly Indo-European and which point specifically to the Indo-Iranian branch of Indo-European, show that these peoples must once have been in contact, and our historical scenario must somehow account for this. Finno-Ugric is one of the two branches of Uralic, Samoyed being the other. There are no Indo-European loan words in Samoyed, and there are a sufficient number of words common to Finno-Ugric and Samoyed, as well as other, often basic, terms and grammatical markers in either Finno-Ugric or Samoyed which agree with those of Indo-European to indicate that Indo-European and Uralic are related and are probably both members of a larger grouping, in fact the one that I have called Eurasiatic.

What might be called the "center-of-gravity method," which has just been sketched, is most effective and convincing when one stock of low genetic rank is peripheral in distribution and its fellow members are all located in some other area. This method was effectively used by Sapir in his well-known paper regarding the northern origins of the Navaho (Sapir 1936). Another example is that of the Bantu, who occupy almost all of the southern third of Africa and whose origin is to the northwest in the border area of Nigeria and the Cameroons (Greenberg 1963).

One further principle should, in fact, not even need mention from the scientific point of view, but in practice is often a major barrier. This principle might be called linguistic democracy. The number of speakers and the geographical expanse of a grouping are irrelevant. What counts are sheer linguistic differences and similarities. Thus, at the same genetic level, English, a world language with hundreds of millions of speakers, and Frisian, a language confined to small islands with a small and dwindling number of speakers, are of equal weight in determining the Anglo-Frisian homeland.

Guthrie, a well-known Bantuist, was outraged at the notion that Bantu should have any external connections or that it should be ranked at a very low level genetically within Niger-Kordofanian and have, as its closest relations at its own level, minor languages of Nigeria with small numbers of speakers.

One major factor that has been considered up to now in the exposition only in an incidental way is chronology. We wish to know not only where and whence, but also when. It is clear that a valid, genetically based taxonomy of languages contains an inherent relative chronology. By definition, Proto-Germanic must be earlier than the present Germanic languages. This correlates with the degree of genetically relevant similarity, of course, precisely because such degrees of similarity are themselves the basis of the hierarchies incorporated in the genetic classification.

For absolute dating, the only purely linguistic method we have is glottochronology, which has many weaknesses, especially for long dates. Further discussion is here deferred to the point at which it becomes applicable to the analysis of the Native American linguistic situation, to which we now turn.

In 1960 the present writer outlined in brief form a linguistic classification of the native languages of the Americas, and this was published in much fuller detail in the book *Language in the Americas* (1987). According to this analysis there are three basic linguistic stocks of very unequal size: Eskimo-Aleut in the extreme north; Na-Dene, spoken exclusively in North America, for the most part in the northwestern part of that continent; and the large Amerind stock, including all of South and Central America and most of North America.

Before proceeding to the evidence from each of these three stocks regarding the settling of the Americas, I shall use a set of linguistic assumptions at slight variance with those adduced earlier to indicate that, with an extremely high probability, the Americas must have been settled from Asia. This is contrary to the view that extralinguistic data are required for such a statement.

Geographically, there are only three possibilities. The Americas were settled from Asia, from islands in the Pacific, or from Africa or Europe. It has already been shown that languages from Europe represent recent branches of Indo-European. The same is true for languages from Africa. Interestingly enough, earlier enumerations of indigenous language stocks of South America included a language called Arda, spoken in Colombia. Rivet (1925), alerted by the resemblance of Arda to Ardra, the language spoken in the slave-trading port of Dahomey, discovered that Arda, evidently spoken by a community of runaway slaves, was virtually identical with Ewe, a West African Niger-Congo language spoken in and around that area. Note that, even in the absence of other evidence, the same kind of reasoning illustrated repeatedly above would show the recent African provenance of Arda, a member of the Niger-Congo subgroup of Niger-Kordofanian, the most widespread of the four major African language families.

In the Pacific islands, there are three major linguistic groups: Indo-Pacific, Australian, and Austronesian. Of these three, the Austronesian that is spoken on the islands closest to the Americas, e.g., Easter Island, the Hawaiian Islands, etc., has the least internal depth. Moreover, all of the islands that are closest to the Americas have languages of the Polynesian subgroup. There is much difference of opinion regarding the subgrouping of Austronesian, but there is unanimity that the Polynesian languages, which are very similar to each other, are at the lowest genetic level among the major subgroupings of Austronesian. According to Pawley and Green (1985) Polynesian would be at an intermediate genetic level, coordinate with Fiji and Rotuma; the center of genetic diversity within Austronesian as a whole is in the area comprising Taiwan and the Philippines. The next step in the analysis concerning Austronesian brings us directly to the mainland of Southeast Asia since, as shown by Benedict (1942), its closest relative is the Thai-Kadai group of languages of which Thai, the standard language of Thailand, is the only one with a large group of speakers.

The Indo-Pacific family shows far deeper internal differences than Austronesian and is doubtless much earlier in the Pacific. Its major groupings are almost all found on New Guinea and neighboring islands. The farthest eastern extension is in Mel-

anesia, in the Solomon Islands, a vast distance indeed from the mainlands of North and South America. Australian is confined to the continent of Australia.

This leaves Asia as the only plausible source for the pre-European settling of the Americas. In first discussing this topic I stated it in terms of very high probability, not certainty. For example, it is not *impossible* that groups that left no linguistic relatives in their homeland emigrated at a very early date from Europe, Africa, or the Pacific, and these died out in the Americas. However, there is no positive evidence in the form of actual languages. The same holds for Asia, so that the three-migration theory enunciated in the remainder of this paper is really a "3 + n" theory.

Assuming the theory of three linguistic groups among the speakers of Native American languages, does this ensure that all three of these came from Asia separately and at different times? Once again it is possible to give an answer based on language alone. For there to have been only two or even one population movement with subsequent differentiation within the New World, some two or all three of these linguistic stocks must be shown not only to be related but to form a complete stock—that is, to have formed a valid linguistic entity without other members—thus presupposing a single population which then differentiated into two or three branches. To answer the question, then, we must consider the external relations of these three families.

It is clear that Eskaleut, Na-Dene, and Amerind are not branches of the same stock, much less the *only* branches of such a stock. If this is true, then there cannot have been just one migration followed by the subsequent differentiation of this family into three branches.

It will be convenient to consider Na-Dene first. Genetically, Na-Dene is the most divergent of the three stocks. Sapir believed that Na-Dene was related to Sino-Tibetan. This idea has been taken up in a wider context by several Russian linguists, notable Starostin (1984) and Nikolaev (1989). Much of their evidence is as yet unpublished, but they have already made a plausible case for a widespread family of "leftovers" which they call Sino-Caucasian. The language families they have connected are Ket, an isolated language in northeastern Siberia; Sino-Tibetan; North Caucasian (i.e., the non-Khartvelian Caucasian languages); and Na-Dene. If we assume the validity of this family, three of its four branches are in the Eastern Hemisphere, and the center-of-gravity type of reasoning already discussed suggests an Old World origin, perhaps in northern China or Manchuria.

With regard to Eskaleut and Amerind, an ultimate connection is highly probable, but not a direct genetic one in the sense that there is a single family which has Eskaleut and Amerind as branches.

In Greenberg (1987) it was hypothesized that Eskimo-Aleut was the easternmost branch of a Eurasiatic family for which the evidence will be presented in Greenberg (in press). Its membership consists of (1) Indo-European, (2) Uralic-Yukaghir, (3) Altaic, (4) Ainu-Korean-Japanese, (5) Gilyak, (6) Chukotian, (7) Eskimo-Aleut. Since Eskimo-Aleut is at the eastern end of a vast extension that centers in Asia, we are once more led to postulate an origin in Asia, presumably in central or western Siberia and to the north of the ancestral area of Na-Dene.

Amerind as a whole shows clear indication of a closer relation to Eskimo-Aleut than to Na-Dene. However, Amerind is, for a number of linguistic reasons not discussed here, to be viewed as coordinate with Eurasiatic as a whole, not as one of its branches. Once more Asia is indicated as the ultimate source, but at a greater linguistic time depth than Eskaleut, which was presumably not yet a distinct linguistic entity at the time the Amerindian population entered the New World.

The external linguistic evidence thus indicates an Asian provenance for all three stocks, and a more recent one for Eskimo-Aleut than for Amerind. This is reinforced by the internal linguistic evidence. The Amerind stock has the greatest internal diversity of the three, followed by Na-Dene and Eskimo-Aleut.

For the moment, however, we can see that the geographical position of the three stocks reinforces the hypothesis that is beginning to emerge from our consideration of the data up to this moment: namely, that the Western Hemisphere was successively occupied by the Amerind, the Na-Dene, and the Eskaleut branch of Eurasiatic. This might be

called the "sock argument." If we take out Christmas presents from a sock that hangs over the mantelpiece on Christmas morning, we will deduce that the present on the bottom was first put in, and so on. The geographical location of the three stocks conforms quite closely to a southern-northern progression: Amerind, Na-Dene, Eskaleut.

We may now consider the internal divisions of each of these stocks. Eskaleut divides into two branches, Eskimo and Aleut. While the difference between Eskimo and Aleut is not trivial, Rask recognized this grouping in the early nineteenth century and it has not been called into serious doubt since. Eskimo, in turn, divides into two groups based on the word for 'people', which became a kind of shibboleth in this matter. On this basis we have the Inuit and Yuit. There is a sharp division at Unalakleet on the central coast of western Alaska. Everything to the north, occupying not only northwestern Alaska and Canada, but also the west and east coasts of Greenland, is hardly more than one language. Its spread must therefore be very recent, within the last 1,500–2,000 years. It has been identified with the archaeological Thule culture.

The Yuit include not only population in southwestern Alaska but also a few communities in Siberia, not far from the Bering Strait, which are considered to be a recent reflux. The position of Aleut and the small internal diversity of Inuit suggest southwestern Alaska as the area in which Proto-Eskaleut began to divide into Eskimo and Aleut.

The Athabaskans are by far the largest population speaking a Na-Dene language. They dominate the northwestern interior of Canada, with outliers in California and the American Southwest (Navaho-Apache). The greatest number of distinct branches of Athabaskan are in the interior of Alaska, and the supposition of their northwestern origin is strengthened by the fact that the language most closely related to Athabaskan is that of the virtually extinct small group of the Eyak in southeastern Alaska. At the next genetic remove is Tlingit, which ranks with Athabaskan-Eyak. Finally, most distant of all—in fact, not recognized by some as Na-Dene at all—is Haida, spoken on Queen Charlotte Island. The distribution of Na-Dene therefore suggests a central or insular origin in the southeastern extension of Alaska adjacent to Canada.

It should be interjected at this point that the earlier arrival of the Na-Dene, as compared to Eskaleut, is disputed by some who see Eskaleut settlement as somewhat earlier. The linguistic evidence suggests, however, the explanation described above: basically, the greater internal linguistic differentiation of Na-Dene than of Eskaleut and a more southerly geographical center of gravity.

The internal genetic divisions of Amerind and the historical inferences to be drawn from them present some complex problems. In Greenberg (1987) the etymologies confined to single subgroups are present in terms of 11 subgroups. Following this, etymologies found in two or more of these 11 are presented. However, of these 11, two in South America, Macro-Ge (1) and Macro-Panoan (2), are especially close, and then at a further remove Macro-Carib (3), thus forming a Ge-Pano-Carib group. Since there were no previous internal comparisons within these groups and they had never before been defined fully in the literature, individual etymologies for each group were provided. In addition to the lexicon, grammatical evidence points to a special relationship among these three groups. The most important are the pronominal possessive prefixes: *i-* first person singular palatalizing; *a-* second person singular; *i-* third person nonpalatalizing and recurring before consonants, alternating with *t-* before vowels. Parts of the pattern are pan-Amerindian in origin, but the particular configuration is Ge-Pano-Carib. No special Ge-Pano-Carib section is contained in the book.

Two further South American groups, Equatorial (4) and Macro Tucanoan (5), are suggested as having a special relationship. In addition, there is an Andean group in South America (6), and Chibchan-Paezan (7) which itself falls into two parts, Chibchan and Paezan, extending into Central America with one Paezan outlier, Timucua, found even in Florida.

Further, there is a Central Group (8) (Oto-Mangue, Uto-Aztecan, and Kiowa-Tanoan), extending from Central America to the American Southwest. Finally there are three groups, Penutian (9), Hokan (10), and Almosan-Keresiouan (11), found

chiefly in North America with a few outliers further south, collectively called Northern Amerind. Once more, in Greenberg (1987) the intermediate groupings Equatorial-Tucanoan and Northern Amerind do not have separate etymological sections. Groupings of this level seemed evident to me in the course of working through the whole classification.

In the book, however, there is a matrix in Appendix C which shows the distribution of each etymology in relation to the 11 groups mentioned above. In Ruhlen (1991), this matrix is analyzed mathematically. It supports strongly all the intermediate groups mentioned earlier: (Ge-Pano-) Carib, Equatorial-Tucanoan, and Northern Amerind. It also, though somewhat less clearly, suggests two further conclusions. The first is that the four most southern groups, found mainly, or in some instances exclusively, in South America, form a southern division of Amerind consisting of (1) Ge-Pano-Carib, (2) Equatorial-Tucanoan, (3) Andean, (4) Chibchan-Paezan. The second main conclusion is that Central Amerind (Oto-Mangue, Uto-Aztecan, Tanoan-Kiowa) stands apart from the rest and therefore probably presents the first cleavage within Amerind. Both these conclusions seem plausible.

If, as we assume, the Amerind stock came from Asia, we would expect its deepest internal diversity to be in the north. While this is not so, since the unity of Northern Amerind is strongly supported, the probable existence of a single southern group is gratifying as it shows that the deepest divisions are not within southern Amerind languages.

It would seem that the spread of Amerind must have been fairly rapid, i.e., within one or two thousand years, since numerous separate northern subbranches of highest genetic rank within Amerind did not develop. The first split between Central and the rest seems to indicate an initial movement of Central Amerind into the American Southwest while the main body moved separately, leaving behind a Northern Amerind branch while the southern groups split up within South America. This topic is not pursued here further. For a more detailed reconstruction of the presumed movements of Amerind peoples, the reader is referred to Ruhlen (1991).

The general picture of three genetic groups

moving separately from Asia into their present locations is supported by the existence of an almost
identical three-fold division based on fossil teeth
and population genetics (Greenberg, Turner, and
Zegura 1986). The correlation of language with
population genetics is greatly strengthened by the
monumental work of Cavalli-Sforza and his associates on the world distribution of genes in populations whose languages are spoken world-wide.
Insofar as these results concern the Americas, they
are strongly confirmatory. Recent work on mitochondrial DNA does not always give a clear picture,
some of it agreeing and some disagreeing, but the
technique is new, and I believe that we shall have to
wait some time before this evidence can be reliably
assessed.

What is the alternative to the view presented
here in relation to the linguistic evidence? It would
be to accept the view of what is probably the majority of linguists working on American Indian languages, according to which there are somewhere
between 100 and 200 pre-Columbian linguistic
stocks, among which no affinity can be traced. An
approximate figure can only be given. There is no
precise classification accepted by everyone. To take
but one example, some Americanists accept part or
all of the wider Hokan grouping, while some break
it down into a dozen or more families. Something
like an "official" list is found in Campbell and Mithun (1979) which, in spite of its title, *The Languages
of Native America*, does not include South America
or indeed parts of Central America. Here we find,
with some hedging—for example regarding the relationship of Tlingit to Athabaskan-Eyak, said to be
"perhaps distantly related" (ibid.:39), but included
within Na-Dene in the listing—62 independent
stocks compared to Powell's 1891 listing of 58 north
of Mexico. Regarding South America, the closest
thing to an "official" listing I can find is Voegelin
and Voegelin (1965:146–50), with 93 independent
stocks. In addition, one would have to add Central
American languages not found in Campbell and
Mithun, of which 12 are listed in Voegelin and
Voegelin. Six of these are now universally accepted
as forming a single Otomanguean family.

Assuming that this represents an approximately true picture—namely, about 150 separate

families with no known connection to each other—
what are the historical implications for the settlement of the Americas?

A recent discussion, based on the assumption
that my classification is wrong and the "official"
doctrine is true, is that of Nichols (1990). She refers
at the very outset of her paper to Austerlitz (1980) as
strikingly original, and since her paper is basically a
continuation of the same approach, though with
some important differences which will be noted, it
will be discussed here first.

Austerlitz's paper appeared, of course, before
Greenberg (1987) and as a matter of course assumes
a very large number of families in the Americas. Actually Austerlitz only compares Eurasia and North
America, omitting Africa, Oceania, Australia, and
South America. He believes that "there is something like an ideal density of language families
. . . differences among continents are not likely to
be dramatic." He proposes a measure called the
GUDR (genetic unit density ratio): the number of
linguistic genetic units in a continent divided by its
area. Even with his highly conservative views, there
are only 37 linguistic genetic units in Eurasia. This
is then compared to 71 in North America, which has
roughly half the area of Eurasia. It is assumed that
genetic classification is based on the comparative
method and that linguists apply it in the same manner in each area. It turns out then that there are
about four times as many linguistic stocks per unit
area in North America as in Eurasia. How to account for this "dramatic discrepancy"? The reason
he offers as the most plausible is that it would appear that the Old World, admittedly a much older
inhabited area, was drained of more than half of its
original stocks by migration to the New World.
These groups migrated in toto, not leaving related
languages behind. The reason for this potent *Drang
nach Osten* is not given. Nichols in fact rejects this
explanation (1990:487) on the grounds that "available evidence suggests that it is most typical for migrations into new territory to produce distributions
where part of the group moves and part stays behind." Regarding South America, Austerlitz notes
that it is enigmatic and that as many as 70 genetic
units have been proposed, which would produce an
even higher genetic density than North America

and make the discrepancy between Eurasia and the Western Hemisphere "still more dramatic."

Nichols's paper continues the basic approach of Austerlitz but is more inclusive geographically. In her appendix she gives a linguistic survey of the entire world in which languages are classified on two levels: stocks and families. Thus, Indo-European is a stock, but Germanic and Celtic are among its constituent families. Many stocks have no major internal branching, e.g., Basque, and are therefore isolates. Stocks (ibid.:477) are the oldest groupings reachable by the comparative method, are mostly in the vicinity of about 6,000 years old, and display regular phonological correspondences. The Niger-Congo stock is one of the major families which I was the first to distinguish as part of my African work (Greenberg 1963). In my classification of the languages of native America, Amerind, and even "Hokan," a subgroup of Amerind, were grouped above the stock level. These higher groups, which Nichols obviously views as speculative and uncertain, are not reckoned in her calculations, which, along the lines laid down by Austerlitz, consist of ratios of linguistic stocks and isolates to land areas.

Accepting a large number of different stocks in the New World, she therefore confronts the same problem as Austerlitz: namely, how to account for the great discrepancy between the far greater density of "lineages" in the Eastern as opposed to the Western Hemisphere while rejecting the emptying-of-the-Old-World hypothesis of Austerlitz as unrealistic? Her answer can best be given by a citation from the abstract at the beginning of her paper:

> The unmistakable testimony of the linguistic evidence is that the New World has been inhabited nearly as long as Australia or New Guinea, perhaps some 35,000 years. Genetic unity for 'Amerind' is incompatible with the chronology demanded by the linguistic facts.

The denial of the existence of an Amerind family is thus asserted by a proof *per impossibile*. Historical linguistics can only attain the stock level of about 6,000 years. Since such a vast number of separate stocks are found in America, the conventional chronology of the archaeologists is wrong. " 'Amerind' presents us with a chronological paradox. . . . If 'Amerind' is a single genetic lineage, it is at least

50,000 years old." On the other hand, the colonization of the New World by numerous independent lineages, which she of course favors, would also go well beyond the usually accepted chronology as we have seen. Hence linguistic conservatism becomes allied with archaeological radicalism in chronology!

The reader may have noted by now a major contradiction in Nichols's exposition. A stock is about 6,000 years old and attainable by the comparative method. Niger-Congo and even the larger Niger-Kordofanian (of which Niger-Congo is a constituent) are listed as stocks together with the other three families that I distinguished in Africa: Nilo-Saharan, Afroasiatic, and Khoisan. However, Afroasiatic, for which we have written attestation for Semitic and Egyptian, is surely well beyond the 6,000-year limit.

According to footnote 1 on page 477 of Nichols's article, regarding stocks, "Greenberg (1987) makes clear that he believes that such groupings [that is, those higher than stocks] cannot be reached by the standard comparative method; a wholly different method, mass comparison, is required."

But I reached my African classification by mass comparison, and I employed an identical method of classification in the Americas (and elsewhere). If she accepts my conclusions for the Americas, the pseudo-paradox she is seeking to explain simply disappears. No doubt more than 6,000 years are required, but from what reasoning does this limit of 6,000 years come? It has never been explained or justified.

If one looks at the table of common words in the languages of Europe in Greenberg (1987:12), one sees that by the time one gets to the third word, the division among Indo-European, Finno-Ugric, and Basque is clear and confirmed again and again by further lexical items. If a correct hypothesis can be generated by such a small part of the evidence, it can clearly go further and generate chronologically deeper classifications. Indeed, if one were to continue with such highly stable items as the first- and second-person pronouns, the interrogative pronoun, and the word for 'water,' it would even be clear that Indo-European and Finno-Ugric group against Basque. Indo-European and Uralic (which includes Finno-Ugric) belong to a large group of languages, Eurasiatic, which includes Eskaleut, as noted earlier.

Finally, the question of methodology raised by Nichols's statement quoted above should be discussed since it involves widely shared but false assumptions about the relation between mass (or multilateral, as I now call it) comparison and the comparative method. To either my credit or discredit, I am reputed to have abandoned the comparative method in making language classifications. Although Nichols states that I "make clear that such groupings cannot be reached by the comparative method," she cites no specific statement of mine. The whole topic is treated at some length in Chapter 1 of Greenberg (1987). It is, I believe, important to discuss it here, if only briefly, because linguists and nonlinguists alike have misunderstood the whole question.

It is pardonable, therefore, for an archaeologist like Renfrew, repeating what he has heard from linguists, to make inaccurate statements. Thus in a recent paper (1991) Renfrew makes the following statement about multilateral comparison, after outlining my African classification:

> The methodology employed proved controversial among linguists depending on what Greenberg terms 'multilateral comparison,' that is to say, on lexical similarities studied in a number of languages at the same time. Previously, most linguistic comparisons had been made two at a time, but Greenberg claims to reach greater time depths with his multiple comparisons.

The first misunderstanding has to do with the notion that only lexical (and therefore not grammatical) items are compared. As early as my first essay on African linguistic classification (1949), in which I was seeking to exclude the irrelevant typological criteria which had confused earlier African classification, I sought to distinguish typological from genetically relevant criteria. Genetic criteria involve resemblances in form and meaning simultaneously, whether lexical, e.g., German *Nase*, English *nose*, or grammatical, as the German and English adjective comparatives in the suffix *-er*. These contrast with typological resemblances involving sound only, e.g., having glottalized stops, and meaning only, e.g., having a future tense. In fact,

the resemblance in noun class systems, a grammatical set of criteria, played a major role in my first work on Niger-Congo, and the relation to Kordofanian was based more on these resemblances than on lexical items, few of which were available at the time. In my volume *Language in the Americas* (1987) there is an entire chapter on grammatical resemblances, and they play an important role in the book.

Secondly, and of at least equal significance, comparing only two languages at a time can never lead to a taxonomy of languages. It is no doubt widely practiced by American Indianists but does not represent orthodox historical linguistics. The work which is universally recognized as inaugurating historical linguistics is Bopp (1816), which compares Germanic, Latin, Greek, Sanskrit, and Persian. It was by this very fact that it was novel and epoch-making. All general treatments of Indo-European compare all the branches simultaneously. When Buck (1933) wrote a comparative grammar of Latin and Greek, he explained in his preface that while treating these two languages together was linguistically unjustified, the cultural relations between their speakers and the fact that they were the common concern of a specific body of scholars, the Classicists, made such a treatment convenient. However, all the forms in the two languages are explained by reference to Proto-Indo-European, which is, of course, reconstructed with the aid of many other languages and thus involves tacit reference to multilateral comparison. Grammars of Balto-Slavic are different, and no apologia for them is given since they are generally presumed to stem from a common intermediate unity between Baltic and Slavic on the one hand and Proto-Indo-European on the other.

Finally, what is the relationship between multilateral comparison and the comparative method? There is no contradiction between the two. By the comparative method is meant the comparison of lexical and grammatical forms between members of the same genetic group of languages in order to reconstruct as far as possible the ancestral forms and the changes these forms underwent in becoming the forms of the later languages. Such a theory may be deemed explanatory in the historic sense. Later forms are explained from the earlier forms and the manner in which they gave rise to them. The most important fact about languages that makes this possible is the fact that most sound change is regular.

But to do this one must first determine which languages belong to the same genetic units at various levels. Since the number of ways of classifying *n* objects rapidly becomes astronomical, one needs a classification to begin with, and, as in any empirical science, one must first observe the objects to be classified. It is in this preliminary stage that multilateral comparison is clearly the only viable method.

In general, forms which descend from a more immediate common ancestor will be more alike both in sound and meaning than those from languages with a more remote ancestry. Phonetic and semantic change is, on the whole, from similar to similar. There are literally hundreds, perhaps thousands, of well-attested examples of "similar" changes from *p*, a labial sound, to *f*, in contrast to strange or exotic changes: from 'nose' to 'nostril,' in contrast to the change from 'bead' to 'prayer.' Hence the observation of "surface resemblances" leads to the correct classification.

In fact, it makes sense to see in classification based on multilateral comparison the first step in the comparative method itself. Sound correspondences do not spring like Athena from the head of Zeus. They are based at the beginning on the resemblances found in the initial stage of classification. At this stage, if we compare English *four* with German *vier*, the meaning is identical, and *f* corresponds to German *v* (pronounced *f*), *r* to *r*, and the vowels to each other.

At a later stage we posit the ancestral form and the regular recurrent correspondences. But whether we do this or not, the classification is valid, and the detailed application of the comparative method in the usual sense only becomes possible on its basis.

This point has now become clear at least to some comparative linguists. In a recent publication, Watkins, a distinguished Indo-Europeanist, cites with approval the remarks of Newman, an Africanist linguist whom I had previously cited on this matter (Watkins 1990). Newman asserts that the comparative method is only applied to languages already presumed to be related. But how do we presume them to be related except by looking at a broad array of them and noting how they group genetically? This, precisely, is multilateral comparison.

Note that, because in biology nothing corresponds to sound laws (Greenberg 1987:34; Dyen 1987:101–108), reconstruction of such entities as the proto-feline or proto-mammal is not feasible. But who will claim that biology is less taxonomically advanced than linguistics?

REFERENCES CITED

Austerlitz, R. 1980. Language Family Density in Asia and North Eurasia. *Ural-Altaische Jahrbuecher* 52:1–10.

Benedict, P. 1942. Thai, Kadai and Indonesian: A New Alignment in Southwestern Asia. *American Anthropologist* 44:576–601.

Bopp, F. 1816. *Ueber das Conjugationssytem der Sanskritsprache in Vergleichung mit dem der griechischen, lateinischen, persischen und germanischen Sprache.* Frankfurt: Andreaeisch Buchhandlung.

Buck, C. D. 1933. *Comparative Grammar of Latin and Greek.* Chicago: Chicago University Press.

Campbell, L., and M. Mithun, eds. 1979. *The Languages of Native America: Historical and Comparative Assessment.* Austin: University of Texas Press.

Dyen, I. 1956. Language Distribution and Migration Theory. *Language* 32:11–26.

———. 1987. Genetic Classification in Linguistics and Biology, In *Festschrift for Henry Hoenigswald*, ed. George Cardona and Norman H. Zide, pp. 101–108. Tübingen: Gunter Narr.

Greenberg, J. H. 1949. Studies in African Linguistic Classification: I. The Niger-Congo Family. *Southwestern Journal of Anthropology* 5:79–100.

———. 1963. *The Languages of Africa.* Bloomington: University of Indiana Press.

———. 1987. *Language in the Americas.* Stanford: Stanford University Press.

———. In press. *Indo-European and Its Closest Relatives: The Eurasiatic Family.* Stanford: Stanford University Press.

Greenberg, J. H., C. G. Turner II, and S. Zegura. 1986. The Settlement of the Americas: A Comparison of the Linguistic, Dental and Genetic Evidence. *Current Anthropology* 25:477–97.

Nichols, J. 1990. Linguistic Diversity and the First Settlement of the New World. *Language* 66:475–521.

Nikolaev, S. 1989. Eyak-Athabaskan—North Caucasian Sound Correspondences. In *Reconstructing Languages and Cultures*, ed. V. Shevoroshkin, pp. 63–65. Bochum: Brockmeyer.

Pawley, A., and R. C. Green. 1985. The Proto-Oceanic Language Community. In *Out of Asia*, ed. R. Kirk and E. Szathmary, pp. 147–60. Canberra: Journal of Pacific History.

Renfrew, C. 1991. Before Babel: Speculations on the Origins of Linguistic Diversity. *Cambridge Archaeological Journal* 1(1):3–23.

Rivet, P. 1925. *La langue arda ou une plaisante méprise.* Proceedings of the 21st International Congress of Americanists 2:388–90. Lichtenstein: Nendeln.

Ruhlen, M. 1991. The Amerind Phylum and the Prehistory of the New World. In *Sprung from Some Common Source*, ed. S. M. Lamb and E. D. Mitchell, pp. 328–50. Stanford: Stanford University Press.

Sandfeld, K. 1968. *Linguistique balkanique.* Paris: Klincksieck.

Sapir, E. 1936. Internal Linguistic Evidence Suggestive of the Northern Origin of the Navaho. *American Anthropologist.* 38:224–35.

Starostin, S. A. 1984. *Gipoteza o genetiskechix svjazax sinotebetskix jazykov s enisejskimi i severnokavkazkimi jazikami.* Konferencija: Nostratichiskije jazyki i nostraticheskoje jazykoznanije. 4:74–94. Moscow.

Voegelin, C. F., and F. Voegelin. 1965. Native America. *Anthropological Linguistics* 7.7, fascicle 2.

Watkins, C. 1990. Etymologies, Equations and Comparanda: Types and Values and Criteria for Judgment. In *Linguistic Change and Reconstruction*, ed. P. Baldi, pp. 289–304. Berlin: Mouton de Gruyter.

BERINGIA AND
NEW WORLD ORIGINS
II.

THE ARCHAEOLOGICAL EVIDENCE

Frederick H. West

INTRODUCTION: PRELIMINARY MATTERS AND DIGRESSIONS

To unravel the prehistory of the Americas is an enormous task to be accomplished only through the slow, methodical processes that characterize most of science. In archaeological practice this translates to the search for sites in areas where perhaps little or nothing is known, in order to gain some notion of their character. After the analysis of the often puzzling bits and pieces of evidence so found, excavations may subsequently be undertaken at those sites that the preliminary survey has indicated are likely to produce the kind of information sought. To the non–field scientist this may sound a somewhat cumbrous set of procedures and to an extent that is true. Archaeologists are so frequently asked how they know where to look for sites that it becomes difficult sometimes to avoid a certain smugness—discreetly concealed, of course—that is, until experience reminds that archaeologist that this very question, phrased in appropriately scientific terminology, had been niggling at that very archaeologist

as he prepared to launch himself into the unknown.

In the far north, into the recent past, it was possible to study a map covering many hundreds of square kilometers and find, to one's amazement, that *nothing whatsoever* was known of its archaeology; and, obviously, *nothing* was known of its prehistory. And, of course, the archaeology does come first. Prehistory—i.e., the knowledge of prehistory—is the product of archaeology. The knowledge of prehistory is the objective, the goal.

Archaeology is extraordinarily time-consuming. The requirements of field research and laboratory analysis, coupled with the often concurrent demand of further field research to confirm or refute conclusions tentatively reached, result in a very considerable lag between the initiation of field research and the attainment of coherent results. Fortunately, in that long interim, knowledge of finds made and tentative conclusions about them are shared among those working at common problems. It is in this way that advances are made, perceptions are sharpened, and some comprehension of an unchronicled past is gained. Such has been the course of Beringian archaeology, as, it is hoped, the pages of this volume reveal.

We have attempted in this work to present as much concrete evidence as could be marshalled on the matter of the first settlement of the New World. The work began with a bias which was enunciated earlier, that the only way archaeologically to approach this question was to look in the north where the roots of those origins had to lie. Thanks to the remarkable cooperation of a great many researchers it has been possible to achieve this compilation. If it be said the whole is somewhat short on theory, the response is that this was the design. Large pictures painted on, effectively, diminutive canvases are readily found. Theories come and theories go. The objective here has been not so much to present a theory, or theories, as such, as to provide a body of evidence which might serve as a reference compendium and which might equally, by its objectivity, be employed to establish different ways of interpretation. But it must be admitted that some involved in this work feel strongly that the evidence presented here really does speak for itself and, with the interpretive aid of Joseph Greenberg, should present a truer picture of American beginnings.

The settlement of the New World has been described as one of the greatest human adventures. That these were the last continents to be entered cannot be surprising. All hinged on the advances made in the Upper Palaeolithic. It seems quite clear that it was those advances alone that allowed the successful adaptation to the far north—environments previously totally unknown to the species. It seems clear, as well, that while the Siberian arctic was inhabited early, population there did not burgeon. It certainly expanded, but, compared with other regions being occupied by Upper Palaeolithic people, population density in the far north remained low. The movement into America was a result of population pressure, but a pressure that might never have been exerted but for the environmental changes that began some 5,000 years prior to the Holocene. It cannot be known whether eastern Beringia would have been occupied at all had that environmental impetus not been present. Had there been no corridor southward or had that corridor lain to the east of the continental ice sheet, it seems certain that the discovery of America would have had to await the coming of navigators from over the seas.

Special Nature of the Record

It is not possible to speak of the first peopling of Beringia—American Beringia—without simultane-

ously confronting the question of the first peopling of the hemisphere. Influences from other quarters apparently did, much later, reach the Americas, but those probings had to await advances in technology that, in the later Pleistocene, were in an unimaginable future.

Not for a very long time has there been voiced any serious scholarly opposition to the idea that the first peopling of the Americas took place by way of Bering Strait. And yet, as observed earlier, the preponderance of discussion about that first peopling most often takes place with small reference to the presence or absence of corroborative evidence in this, the signally crucial region. This curious inattention may be largely ascribable to the relative inaccessibility of information on the archaeology of far northwestern North America. In turn, that may be partly the result of there not being present here any sort of overall scheme or sequence of cultural development comparable to those worked out for other regions. In addition to their primary function of portraying the prehistory of a region, such schemes have the virtue of providing ready assignment of archaeological assemblages as they turn up in the course of field research.

The construction of a developmental framework, of course, is dependent upon the possibility of discerning change through time. Marked, rapid, cultural change may produce elaborate, well-characterized sequences. The changes that, in temperate regions, followed upon the advent of agriculture were of just this sort—rapid, progressive, *and* cumulative in nature. The environmental components operative in that watershed event are evident: the agricultural Neolithic developed in, and was restricted to, subtropical to temperate regions. Even now, with the enormous advances in knowledge and technology accreted over several thousands of years, food production remains effectively nonexistent in the regions considered here. Lacking comparable changes, the record in the far north is much more subtle and difficult to read.

The lifeways recorded among native peoples of the Alaskan interior revealed a precise articulation with their stressful, demanding environments. This was seen in the size and flexibility of their social groupings, in their intimate knowledge of all aspects of their world, in their specially adapted skin clothing, their habitations, their transport, and particularly in their economies, which were able to exploit fully the restricted resources available and to survive, usually, the periods of unavailability. Where the growing season for native plants was constrained within three to four months and the range of species rather meager, plant foods could not figure importantly in the diet. In regions where fishing was possible, the number of species and their seasonal availability were likewise restricted, and these resources must have been equally of small importance. This last assertion should be read in a *relative* sense: seasonally, gathering, fishing, and the taking of small game played a relatively significant, augmentive role, but those activities could never, of themselves, have sustained human existence. In short, these recent people were hunters following a way of life whose foundations were laid down in the Upper Pleistocene of Beringia. Their ancestors had made a certainly difficult transition from the relative abundance of the late Pleistocene to the relative scarcity of the Holocene, but the passage of that event is barely discernible in the archaeology.

Granting readily the insufficiencies of the archaeological record, it is nevertheless suggested that there is, then, an additional reason for the seeming inaccessibility of information on interior Alaskan archaeology. This lies in there having been, effectively, no change in basic economic pursuits over the entire period of the Holocene. Populations were always small, lacked permanent settlements, and made use of vast hunting ranges. In fact, there are quite interesting changes in material culture recorded over this span of 8,000 to 10,000 years, including some rather more subtle ones, but these were all very much in the nature of variations on a theme—that of inland hunters pursuing large game animals. Well-characterized, progressive archaeological sequences with marked differences one from the other are difficult to discern, and, thus, schemes depicting cultural sequences have tended to be local rather than broadly applicable and, partly in consequence, have suffered some degree of ephemerality. It is these characteristics of the northern record that have conferred upon it a certain opacity, rendering it difficult even for the specialist to read. Yet, while acknowledging these difficulties, still it must be

said that to begin an understanding of the settlement of the Americas, it is this record that must be consulted.

The Case for Very Early Settlement

Fundamental to all that follows is the conviction that there is *no* convincing evidence in sub-Laurentide America for *any* human presence prior to the appearance of Clovis, which is to say 11,500–11,000 years ago. But there are a number of serious claimants to earlier status, and therefore it is necessary to indicate the bases upon which the authenticity of purported pre-Clovis evidence is dismissed.

At present there are appear to be three important sites that, by their researchers and advocates, are interpreted to provide clear evidence of human settlement and to well pre-date Clovis. Two of these sites are in South America: Pedra Furada in eastern Brazil and Monte Verde in southern Chile; in North America there is Meadowcroft Rockshelter in western Pennsylvania.

These are chosen because among those who espouse the pre-Clovis position these appear to be accorded the greatest attention and, perhaps, the greatest confidence. The character of each site is unique, and they are thus non-comparable except, perhaps, in their proposed ages.

The validity of the upper levels of the Pedra Furada rockshelter (Guidon 1987, 1989) appear to be firmly established. They date well into the Holocene and are not controversial. The lower levels, from which dates ranging from 32,000 to 17,000 BP are recorded, suffer from the fact that the quartz and quartzite cobbles, identified as artifacts, appear, rather, to have been naturally fractured (Lynch 1990; Meltzer et al. 1994).

The site of Monte Verde (Dillehay 1989) presents a series of rather more complex problems. Two levels are construed as archaeological. The lower level (Monte Verde I), radiocarbon dated at about 33,000 BP, contains charred plant remains and some fractured gravels. It is considered problematical but not impossible even though the 20,000-year hiatus between it and Monte Verde II is seen as difficult to explain. Monte Verde II is dated to 13,000 BP. Remains here are largely organic: logs, other plant fragments, mastodon bones (large ribs), eggshells, insect remains, feathers, and bits of hide. Certain of the log arrangements are interpreted as indicative of rectilinear house floors with small, irregular basins considered as hearths or "braziers." The clearest of several indubitably human-manufactured stone specimens appears to lack provenience. In sum, the Monte Verde II assemblage is extraordinarily heterogeneous, consisting largely of elements not clearly identifiable as artifactual. The relevance of the dating is therefore unclear (cf. Lynch 1990:26–27; West 1993).

There is no question as to the validity and coherence of Meadowcroft Rockshelter as an archaeological site (Adovasio et al. 1978, 1985). The problem here is dating. The lowermost cultural stratum contains one small, non-diagnostic simple biface, several small irregular blades, and a few flakes. The several radiocarbon dates from this stratum now accepted are in the range of 13,000–14,000 years ago. The abundant plant and animal remains are Holocene to modern in character, and while the biface is mostly unrevealing, it—and, like it, the rest of the assemblage—is, of itself, not amenable to chronological assignment. These observations have led to the conclusion in some quarters that there has been a systematic contamination of the soil column from which these assays were obtained (Mead 1980; Haynes 1980; Tankersley et al. 1987). The excavations and analyses of Meadowcroft have been exemplary. The investigators have vigorously defended their age assignments (Adovasio et al., 1980, 1990) but they remain quite dubious: the Holocene–modern biota of itself should provide entirely sufficient refutation.

There are, of course, other claimants to pre-Clovis status. Those above were chosen to exemplify the general problem because they are among those most often cited as possible examples of pre-Clovis sites. But they also illustrate nicely a fundamental problem that underlies the entire proposition: each of these sites is unlike either of the other two. That condition will be found to characterize the entire corpus of pre-Clovis evidence. No two sites are alike; it is a body of purported evidence of extraordinary and inexplicable heterogeneity defying all that has been learned of the regularity of cultural evolution.

The universe of archaeological data is, to be sure, not altogether orderly, *but* anywhere that sizeable bodies of evidence are recorded patterning is discerned. Middle Palaeolithic does not resemble Upper Palaeolithic. Mesolithic and Iron Age are easily recognizable and readily characterized. These are, after all, historical growths and eventually the logic of their development and composition becomes apparent.

There are, among these, bona fide sites that present genuine problems demanding serious attention. Meadowcroft is a case in point. Many more, however, are of quite a different nature providing, at best, diversion. (This is by no means a veiled reference to the other two sites considered above.) The most controversial sites seem almost to gain currency in direct proportion to their improbability. A review of recent scientific news in the popular press, as well—sadly—as in some journals devoted generally to science, will provide testimony. The basic difficulty of the pre-Clovis argument is made more acute when one dubious site is called upon to provide support for another. In any event, nothing said or reported here is apt to have any effect upon those truly committed advocates of ancient, pre-Clovis settlement.

Time, Space, Typology, and the Domino Effect

Having set aside, as it were, possible obfuscating pre–Late Pleistocene corporalities, it becomes possible to concentrate on a consideration of the archaeological record from both sides of Bering Strait.

The earliest settlement of the Americas must be seen as purely and simply an aspect of the earliest settlement of Beringia. From this it follows that earliest American settlement can only be understood in that context. The ramifications of this position make it necessary that the central components of this argument be set forth unambiguously.

Before it was possible to make a case of the sort presented here, it was necessary that a number of independent variables be rectified and aligned. These variables may be thought of as an array of dominoes. They are the dimensions of time, typology, and place. Although the discovery and description of the elements comprised by each term proceeded more or less erratically (which is to say that the discovery of archaeological sites is unpredictable), there was a logical order that needed to be found in order to establish properly the required relationships. That order would be *place*, *typology*, and *time*. Thus, to establish an historically appropriate relationship between western and eastern Beringia it was necessary to find, in those two places, site assemblages in which the comparative typologies would reveal a close formal resemblance such as would suggest a genetic relationship. This was accomplished and the crucial question then became that of time. Was the "time slope," in Giddings's phrase, appropriate? Definitive proof on this point was to await the radiometric dating of sites in Siberia and Alaska and the filling in of the evidence from these sites. This research is continuing and will continue well into the future.

Sufficient evidence is now available to show that, indeed, the resemblances are there, in geographical proximity, and the age differential between the Siberian sites and those of Alaska and the Yukon is exactly appropriate. There are no bizarre, inexplicable reversals. Those dominoes have fallen. The greater part of this volume has been concerned with presenting these evidences and making known their character.

The resemblances between the various core and blade assemblages of Alaska and the Yukon and their counterparts in the central and coastal regions of Siberia are detailed and specific. These are Upper Palaeolithic core and blade complexes. It is no longer useful to hedge on that designation out of deference to their distance from the rest of the Upper Palaeolithic world or out of an apprehension that the resemblances to accepted Upper Palaeolithic may be spurious—the result of accidental convergence or inadequate comparative analysis. The dating of the Siberian evidence is exactly appropriate to the level of typological resemblance and leaves no doubt of the relationship to the more familiar Eurasian Upper Palaeolithic complexes. Moreover, the relative ages of the western and eastern Beringian groupings leave no doubt as to the derivation of the latter, the American. These are two parts of a whole and it is for this reason that they are collectively known as Beringian Tradition.

Rather more difficult has been the matter of applying these same criteria to the resemblances between Beringian Tradition and Clovis. The reasons for this have partly to do with *time* and *typology*. Perhaps more, however, the difficulty has to do with the fact that Clovis has been on record for a long time and, quite naturally, there has accreted about it a substantial shell of thought and speculation, virtually none of which looks to Beringian Tradition as having anything to do with its origins. Some of these speculations are built around the acceptance of certain of the pre-Clovis evidences which have been dismissed here. Even by those who do not consider it earliest, Clovis is acknowledged as having taken the pivotal position in the populating of sub-Laurentide America. As will be brought out, it is maintained here that the criteria of time and typology can be effectively demonstrated here as well and that the relationship to Beringian Tradition is there to be seen, and with it the fall of the last and most difficult domino.

THE SETTLEMENT OF BERINGIA

Beringia is conceived as having constituted a biotic province. By definition, the name refers to the completed province. The central sector consists of the "Bering Land Bridge" or "Bering Sea Platform" which formed the connection between east and west. It came into existence at any time in the Pleistocene when, as a result of world-wide glaciation, sea levels lowered sufficiently for the shallow sea floor to form a dry land connection. For the flora and fauna that came to occupy it, there would have been no discernible difference between this land and those now forming parts of Siberia and Alaska. This point, while obvious, requires emphasis. It appears that at no time in the Würm was the central sector transgressed by the sea. In that last episode, Beringia was in existence for some 70,000 years. Except as determined by normal environmental controls, such as topography, central Beringia is unlikely to have differed floristically or faunistically from any of the adjacent lands.

That it is treated as a biotic province testifies to Beringia having had a character that differentiated it from adjoining regions. Some of the features conferring that differentiation are referred to in the pages of this volume. But even though there is this consensus on this point among Beringian researchers, it does not follow that a boundary line could be easily drawn around the province. Proof of this may be found in the varying statements as to what should be its extent. Perhaps this is characteristic of the biotic province that is not insular; it is identified by a series of attributes having distributions that are not necessarily coterminous but rather coexistent in the heart of the province.

The Spread of the Upper Palaeolithic

At the present time there is only one site in western Beringia that runs counter to the pattern seen in the evidence presented in this volume. The site of Diring in Sakha has produced pebble tools in great quantity and is interpreted by Mochanov and Fedoseeva as Lower Palaeolithic and representing the work of *Homo erectus*. The site is unique and controversial. In no case, at any rate, would it have any bearing on the subject of this volume. As such, it is simply set aside.

The archaeological sites described here that define Siberian Beringia were not selected from some larger and more heterogeneous body of data. Rather, they are a true and accurate representation of the whole of northeast Asian archaeology and prehistory. Assuredly a great deal more will be learned of Siberian prehistory than is now known, but it is unlikely that the general depiction, as presented by these Russian researchers, will be substantially changed. The earliest sites here are all distinctly Upper Palaeolithic in character.

With the provisional exception of Diring, no credible evidence of human presence earlier than the Upper Palaeolithic has ever been discovered in Siberian Beringia. It is suggested that this is an accurate reflection of the prehistory and that the reasons are inherent in the systems: one is that of the Upper Palaeolithic itself, the other that of the Beringian environment.

The transformation from Middle Palaeolithic to Upper Palaeolithic was one of the most remarkable and perplexing events in all of prehistory. The sequence recording this transformation is repeated broadly through the upper mid-latitudes of Eurasia but is best known, of course, in Europe, hence the

convention of employing that terminology. From the standpoint of the archaeology alone, the transition is that from the Mousterian, with its characteristic flake and simple biface industries, to the Châtelperronian–early Aurignacian, with their varied, advanced, and highly differentiated manufactures. In the former, some Mousterian elements are distinguished along with some of those that came to characterize the typically Upper Palaeolithic Aurignacian. These include retouched blades and bladelets struck from prepared cores, burins, thick (carinate) scrapers, strangulated scrapers, end scrapers on blades, *pièces esquilleés*, and, perhaps most notably, antler points interpreted as having been used on spears. This enhanced complexity of material culture seen in bone and stone industries is certainly a reflection of a greatly elaborated intellectual, social, and adaptive culture. In turn, these new capabilities would translate to more effective means of adaptation, impelling the extraordinarily rapid geographic spread of the Upper Palaeolithic.

Despite some evidence of biological and cultural overlap in a handful of western European sites, the appearance in general is that of an abrupt and radical transition that seems to correlate precisely with the appearance of *Homo sapiens sapiens*, i.e., completely modern man. (It is, fortunately, appropriate here to ignore the current great debate on the earliest emergence of the modern form as this involves areas far afield from those of concern here.) Using the term broadly, wherever Aurignacian is found with the skeletal remains of its bearers, those remains are Cro-Magnon—completely modern; wherever Mousterian is found so accompanied, the remains are of Neanderthalers.

Although their roots may have been present earlier, it is at the onset of the Würm glaciation that Neanderthalers appear in Europe. There they held forth as the only hominids from roughly 70,000 years ago until replaced about 37,000 years ago. It has long been suggested that their primary means of cold adaptation was anatomical, exhibited in certain body proportions and in the form of the face. That of the Upper Palaeolithic folk is considered to have been essentially cultural.

Recent research in south-central Siberia has brought to light important information on this whole question. Sites in the Altai region have demonstrated conclusively this transition and provided dates corresponding exactly to those in Europe (Derevianko 1990; Mochanov 1992; Goebel et al. 1993; Goebel 1995). Moreover, if the dating reported for two sites in the upper Yenisei drainage is correct, it would appear that the Upper Palaeolithic made its appearance there 6,000 years earlier than in Europe (Goebel and Aksenov 1995). Finally, amongst fragmentary human bone in two of the Altai caves (Okladnikov and Denisova), teeth found have been preliminarily identified as Neanderthal and proximate morphologically to the European forms (Turner 1990:66).

The Peopling of Western Beringia: The Way North

The most ancient sites in Siberian Beringia are known from the middle and lower Aldan River of Sakha (Ust-Mil 2 and the Ikhine sites). These date to the base of the Upper Palaeolithic, i.e., ca. 35,000 years ago. To the south and east, Derevianko reports that the bottom strata of his Selemdga River series are older than 23,000 to 25,000 years; this by reference to overlying radiocarbon dates (Derevianko 1990:58). There is a unity among the Siberian sites in their virtually universal possession of characteristic Upper Palaeolithic artifact assemblages. Where there appear to be exceptions, e.g., the much more recent high arctic site of Berelekh, the causes are presumably due to the small size of the *in situ* collection. Otherwise the only exception to this general pattern is the Layer VII assemblage from the Ushki locality in Kamchatka, where the stemmed bifaces of this lowest component are quite at variance. Because of its location, well down the great Kamchatka Peninsula, Ushki VII could be dismissed except that the overlying stratum *is* in accordance with the pattern and, like the Selemdga sites, is said by its investigator to fit into the Dyuktai category.

The dating of certain of the earliest Aldan sites has been questioned and there has been some interest in assigning later ages to certain elements of these assemblages, but it would appear that Mochanov's counter-arguments, and his marshalling of supporting evidence, carries the day. Certainly, the increasingly persuasive evidence from neighboring areas in Siberia, as previously described, provides significant support for the dating. In sum, it does appear that western Beringia was occupied by Upper Palaeolithic folk by 35,000 years ago and that their equipage included essentially the same artifact forms that would characterize the entire 25,000-year span of this Dyuktai variant of the Beringian Tradition.

If there can be said to be a special character to this western Beringian Upper Palaeolithic pattern, it is, in part, the low incidence of implements made on blades and the virtual ubiquity of microblade technology. This is not to say that microblades are necessarily numerous in particular assemblages, but rather that the technology, as evidenced by the cores as well as microblades, is an element in virtually all and that its occasional absence is unremarkable.

The Mammoth Fauna of Western Beringia

Where faunal remains are found at these western Beringian sites, they consistently point to an emphasis on that large, diverse, ungulate fauna termed collectively the "mammoth fauna." In addition to the more frequently encountered remains of mammoth, bison, and horse, this fauna included the woolly rhinoceros, yak, musk-ox, saiga antelope, and the cave lion. It is suggested that even if western Beringia were seen as simply an appendix to the great Eurasian "mammoth steppe," this representation would be found to have been unusually rich. There can be no doubt that the economic prompting it provided was decisive in bringing early Upper Palaeolithic people into the subarctic of northeastern Siberia.

However, even an extraordinary richness of quarry could not alone account for hunters moving into such a region, nor could population pressure. What is indicated in this particular Upper Palaeolithic record is a completely new adaptation. To be sure, its fundamentals must have been worked out elsewhere in regions of cold, but it must be emphasized that the cold of northeastern Siberia is totally unlike that of regions conventionally thought of as "northern." Periods of prolonged, intense cold are the rule in northeastern Siberia. It is not uncommon to find temperatures remaining in the range of −40° to −45°C for protracted periods, with occasional descents to −60°C. And, into this equation there must be factored the considerably colder conditions of Würmian time. Even for people who live in these regions today and know the rigorous demands imposed, it is difficult to imagine how people might have sustained themselves under the much more severe conditions of 20,000 years ago. Clearly, what is shown is an extraordinarily high degree of adaptive sophistication together with a need not just to survive, but to succeed under the most arduous conditions ever faced by humans. The basis for that adaptation lay, once more, in the fundamental nature of the Upper Palaeolithic. Whether it was entirely cultural or a combination of that with the anatomical changes considered to have appeared with modern man, the Upper Palaeolithic spread quickly, explosively in some cases, into new areas.

The evidence does indicate that Siberian Beringia was occupied quite early in the Upper Palaeolithic period. Yet the move into this vast territory seems to have been somewhat tentative and the population level at any one time relatively low. If these impressions are valid, it could help account for an apparent conservatism evidenced by the broad spatial and temporal distribution of the Dyuktai.

No region of the world could have provided anything approaching the rigorous testing of these new adaptive capacities that Beringia did. That the descendants of these settlers accomplished, ultimately, the conquest of an entire hemisphere is perhaps, then, not surprising. Another way of viewing this matter is expressed thus: "Beringia, as a cultural province, was really the farthest eastern and last remaining remnant of the greater Upper Palaeolithic culture sphere of Eurasia. It is only an accident of fate that this area has an interest beyond that of terminal Upper Palaeolithic, that it contains traces of that culture that was to become ancestral to an entire hemisphere" (C. F. West, personal communication, 1995).

The cultural changes seen in the final Würm–early Holocene were much less pronounced than those in many parts of the world. Dyuktai is replaced, but by the Sumnagin (termed "Final Upper Palaeolithic") in which the classes of artifacts remain essentially as before while the forms are changed. The contrast between Dyuktai and Sumnagin is reminiscent of that between late Aurignacian and Magdalenian, showing important stylistic changes, but within the same context. (An astonishing example of this Holocene Palaeolithic—or Mesolithic—phase is the recently described site on Zhokov Island in the DeLong Archipelago in the Siberian high arctic [approximately 76°N, 153°E]. The assemblage appears to represent inland hunters—in this instance pursuing polar bear and reindeer—dates to 8,000 BP, and is identified as Sumnagin [Pitulko 1993]).

The Peopling of Eastern Beringia: The Way East

The earliest occupation dates for American Beringia are in the millennium 11,000 to 12,000 before the present. In the light of what is now known of the settlement of Siberian Beringia, the question might be raised, why was this region settled so late in that sequence? In order to address this question, a terminological matter must first be considered, thus: to term *migrations* the movements that brought people to Siberia and then to Alaska is, at best, a convenience. These were surely expansions of population into available adjacent niches. A consideration of the number of generations involved in the more than 20,000-year previous history of western Beringia accomplishes two things: it immediately conveys the certainty that, in the most conservative population theory, there were seemingly more than enough people available to account for an expansion into adjacent lands to the east; but the second matter is that puzzling question just posed but now informed: with so many people theoretically available, why, indeed, was the accomplishment of this occupation so delayed?

Herewith some thoughts on this seeming paradox. Although the settlement of western Beringia was early, and the fauna relatively rich and diverse, it may be posited that its carrying capacity was limited. As a result, human breeding patterns would have been maintained at compensatorily low rates,

distances between populations would have remained relatively stable, and the competition for
hunting territory, which would have encouraged
expansion, while surely present, would have remained relatively inconsequential. The combination
of these interrelated factors would have resulted in
relatively slow expansion in northeastern Siberia
throughout the late Pleistocene. If the mammoth
fauna were sporadic in occurrence and numbers, a
powerful additional control on human population
growth would be exerted. Such a sporadic distribution pattern would be compatible with the presence
of a vegetation mosaic of which, in some associations, an important component would have been
that of *Artemisia*–grass. There is some evidence that
this sporadic condition became more prevalent
with the winding down of the Pleistocene. If correct, the diminished resources might have impelled
the eastward expansion.

A very significant complication to the foregoing
comes from Tomirdiaro's interpretation of the central Beringian region (the land bridge) as having
been subject to wide climatic extremes. This might
suggest—contrary to what was said above—that the
central region was not a favorable environment,
that movement into it was late, and that relatively
small numbers of people ever reached the American side.

As discussed previously by Vereshchagin and
Baryshnikov (1982), habitat changes well before the
beginning of the Holocene led to the die-off of
major elements of the mammoth fauna. Lozhkin de-
scribes, for the period radiocarbon-dated at 17,000
to ca. 13,000 BP, direct evidence of a dominance of
grass, *Artemisia,* and pine at the Berelekh River site,
on the Kolyma River, and elsewhere in the northeast. He finds these deposits overlain by ones
containing ''larch forest–tundra and light coniferous–larch forest [occurring] ca. 12,500–12,000 BP''
(Lozhkin 1993:432). There is indicated ''a substantial warming that played a decisive role in the extinction of mammoth fauna'' (Lozhkin op. cit.).
Vereshchagin and Baryshnikov (1982:278) speak of
a general pattern of extinction at the end of the Sartan, bracketing the occurrence at an estimated
11,000–10,000 years ago. Ukraintseva attributes this
pattern of extinction to the loss of habitat, especially

the diminution of proper fodder (Ukraintseva et al.,
this volume). There is some suggestion that the
woolly rhinoceros became extinct early in the final
Pleistocene oscillations, which may partly explain
why that form failed to reach Alaska.

In a recent note on the phenomenon of extinction by Lister and Sher (1995), they emphasize the
singularity of the ecosystem. They suggest that the
environmental changes that were occurring in the
period just before 11,000 years ago were of such
magnitude that the range of the previously pan-
Eurasian mammoth became vastly reduced such
that after that time, until their final demise, they
were to be found uniquely in arctic Siberia and
Alaska. According to those authors, the warming
event of ca. 11,500 BP signalled the final dissolution
of the tundra-steppe biome. The difference in terms
of effect between the Holocene warming and earlier
interglacials (which these animals survived) is ascribed to there having been, during the interglacials, quite different vegetation from that of the
Holocene. There seems to be evidence in support of
this. Lister and Sher's brief discussion seems especially germane:

> It has been argued, on circumstantial evidence, that
> the higher biotic diversity of the Pleistocene might
> in some way reflect greater climatic variability than
> in the Holocene. Recent discoveries seem to sup-
> port this model. The Greenland ice cores indicate
> almost constant and rapid climatic fluctuations
> over the 100 kyr before the Holocene warming and,
> in strong contrast, a stable climatic signal for the
> Holocene itself. Conceivably, the constant fluctua-
> tions of climate in the Late Pleistocene were partly
> responsible for maintaining a mosaic of plant com-
> munities, by a constant ''stirring,'' favouring a pio-
> neering character of vegetation which supported
> the grazing megafauna. In contrast, the unique sta-
> bility of Holocene climate, indicated by the ice
> cores, may have contributed to the development of
> today's strong zonation of climax vegetation types,
> such as tundra and coniferous forest, unsuitable for
> large grazers like the woolly mammoth.*

(The recent discovery of dwarf mammoths on
Wrangel Island [73°N] dating to ca. 4,000 years ago

*Reprinted with permission from A. M. Lister and A. V. Sher and
from *Nature,* Vol. 378, p. 23. Copyright 1995 Macmillan Maga-
zines Limited.

is fascinating of itself as a narrative of survival but appears to have no immediate bearing on this discussion [see Vartanyan et al. 1993 and Martin and Stuart 1995].)

In the light of all this it is possible to suggest that the peopling of America was, indeed, an accident—an environmental accident. The progressive deterioration of the western Beringian ecosystem, and certainly the central as well, caused these small hunting groups to shift their accustomed ranges to the east, eventually into American Beringia. It goes without saying that if the western portion of Beringia was undergoing such alteration then so certainly was the eastern. As may be seen in some of the archaeological discussions in early chapters, in but one or two American instances is there even an equivocal association with mammoth proper. The palaeoecological record for earlier stages of the Wisconsin is clear: the mammoth fauna flourished in Alaska–Yukon, but where the record attenuates (just before the transition to the Holocene) it is clear that conditions had changed. Of the late Pleistocene forms—the old mammoth fauna—in final Wisconsin, it would seem that only a few survived to become parts of the Holocene–modern fauna. Bison was still to be found in Alaska into the recent past. The same is true of musk-ox and elk (wapiti). Caribou and moose, of course, made the transition successfully, although it would be several thousand years before their recent numbers were attained.

The question that has become integral to all discussions of Pleistocene extinctions must be addressed here. Is there evidence of man's role in the extirpation of the American Beringia Pleistocene fauna? The answer to that must be negative. The evidence is quite absent, and perhaps this comes back to what was suggested previously. Even as early (in the Alaskan time scale) as 11,500 BP, eastern Beringia was no longer a significant refuge for those forms at risk. The breakdown of the old, opulent, Beringian ecosystem had affected all portions of the province.

The ultimate destruction of the province of Beringia came some one to two thousand years later with the disintegration and inundation of the central land bridge. There must be archaeological sites on the floor of the Bering and Chukchi seas; the

likelihood is that they are few. It is not possible even to surmise what the effect of the gradual flooding might have been for humans. In a previous work, the writer spoke of this as a "catastrophe" (West 1981) and that is surely an accurate description even if the effect were solely physiographic, not affecting any life forms. For animals capable of migrating, the effect was probably imperceptible to nil, for by the time full submergence came, their ancestors would have long since removed to more productive regions so the catastrophe consisted primarily in the loss of habitat.

Sites and Complexes of Eastern Beringia, 12,000–9,000 BP

Given the number of researchers who have contributed to this work, it is exceedingly unlikely that all would subscribe to one and the same interpretation of this body of data. It is even more likely that many hold no particular position with respect to the interpretation and that is, in this writer's view, a great virtue. It may be taken as a measure of those individuals' involvement in their own research. It is that attention to these restricted tasks that gives each contribution the stamp of genuine authority.

But while it is obvious that there could be no consensus on the meaning of it all, it does not follow that, for this volume, it would suffice to simply forego any attempt at synthesis. Thus the thoughts that follow. The editor takes full responsibility for them, but at the same time it may be said that, in the main, they represent a synthesis of ideas—perhaps imperfectly absorbed—that have been discussed for years by northern researchers.

As is seen in earlier chapters, there are some differences among certain of the sites that fall into this earliest period. To be sure, there is some diversity of interpretive opinion as well. While it is perhaps a bit beyond the mark to speak of these as "schools of thought," they will be treated as such in the cursory remarks to follow.

The longest series of sites with sure classificatory status are those assigned to the Denali complex (West 1967). This was the first serious proposal of a genuine archaeological culture ("complex" being a cloak to conceal the proposer's true intent). From its humble beginnings in 1967 (consisting of only four sites), it has increased in numbers and distribution. Its dating was at first based almost entirely on comparative typology—to the dismay of some. Its suggested age (12,000–8,000) has been borne out fairly closely for the most part by radiocarbon dating, except that the more recent end of the proposed scale is clearly too recent. This should probably more nearly be 9,000–9,500 BP. The earlier end is still being worked out: Denali is certainly present by at least 10,600 BP but may be earlier by a thousand years. Its relationship to Dyuktai is clear: that of direct derivative. Put another way, if one of the Tangle Lakes sites or the Component II assemblage from Dry Creek had been found in northeastern Siberia by Yuri Mochanov or Svetlana Fedoseeva there is no question but that it would be classified as Dyuktai. Equally, an errant (geographically) Dyuktai site discovered in interior Alaska would assuredly be identified as Denali.

If the dating of Bluefish Caves is allowed, then this site would represent earliest Denali. It remains—at this time—to be seen how the Swan Point lowest level will be classified. To some degree, here, as in other sciences concerned with morphology, this decision will reflect the proclivities of the classifier: some split, some lump. Swan Point and Broken Mammoth are two of the most important discoveries made in Alaska and will add immeasurably to an understanding of the earliest period.

The Akmak site still stands essentially alone as a very distinctive but puzzling assemblage. It is possible that the large sizes seen in Akmak artifacts—that may, in part, account for its distinctive appearance—simply reflect the fact of the site's proximity to a lithic source—in this instance, river cobbles. The one radiocarbon date for Akmak seems too recent. Anderson has classified Akmak as a constituent of the American Palaeoarctic tradition.

The Nenana complex has been described as "pre-microblade"; "non-microblade" might be a better appellation in view of recent discoveries. Nenana is described as "Palaeoindian" and is taken to derive from sources in East Asia other than those which gave rise to Denali. Berelekh is considered possibly one of these sources, as is, perhaps, the Layer VII assemblage from Ushki. The Walker Road

site of this complex is one of the earliest dated sites in Alaska.

The recently reevaluated Mesa site is also termed Palaeoindian and is interpreted as being ancestral to the Agate Basin complex of the Great Plains. A long series of radiocarbon dates bracket a considerable period of intermittent occupation. While bifaces resembling the Mesa form have been found elsewhere in Alaska, their preponderance and numbers here are striking.

Healy Lake in some respects also stands alone, with an assemblage from its lowest level (especially the teardrop-form "Chindadn" point, as well as microblades) that is unusual. With these occurrences there are now credible early dates (an earlier bone apatite date having been deleted from the record). A relationship with Nenana has been suggested by some; with Denali by others.

The small collections thus far made in the Kuskokwim drainage are proving to be quite important. Some of the Spein Mountain bifaces, for example, resemble those from Mesa. The Ilnuk assemblage is classified as Denali, and at Lime Hills the find of a slotted bone projectile now joins the only other such find in Alaska, that at Trail Creek. Its provenience and associations suggest it is Denali as well.

The Kukpowruk collection, while lacking stratigraphic context, is nevertheless of great interest and importance precisely because of its typology. It represents, as well, a collection made early in the brief history of Alaskan archaeology.

The Putu site has been difficult to interpret for quite a long time, so a genuine service has been rendered with its sorting out, particularly with regard to the status of the fluted bifaces.

Anangula, on record for many years, is another assemblage that is virtually unique. The collection is very large, and its setting is unusual. This is one of several sites whose classification is rendered difficult principally because of its location.

Of all these sites, Beluga Point, Ground Hog Bay, and Hidden Falls are perhaps the most difficult to comprehend. The assemblages appear to be almost at one with interior core and blade sites (e.g., Denali), the dating of Ground Hog and Hidden Falls is in accordance, but the coastal setting with

THE SETTLEMENT OF AMERICAN BERINGIA

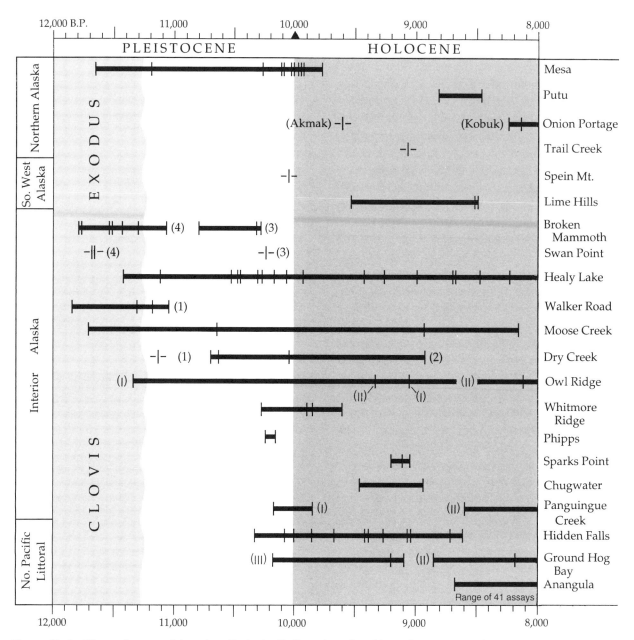

Figure 12–1   The settlement of American Beringia. Radiocarbon-dated late Pleistocene–early Holocene assemblages. Vertical lines represent central point of assays; standard deviations excluded. Dates having standard deviations exceeding 500 years are not reproduced, nor are those rejected by investigators. *(Compiled by Brian S. Robinson.)*

its implication of a completely different sort of economy requires serious reflection. Perhaps, with Anangula, there is seen here a first transition by Beringian descendants to a littoral economy. The small Beluga Point assemblage, despite its setting, could represent an outpost of caribou hunters.

With only the noted littoral exceptions, all these were the sites of inland hunters—completely compatible with their western Beringian ancestry. In most of these sites, organic preservation is poor so that direct evidence of hunting practices is wanting. Still, the artifacts themselves allow inferences with a reasonable level of confidence. The Phipps site in the Tangle Lakes is such an example.

The interpretation of these people as hunters is occasionally qualified by some, seemingly because logic suggests that other means of securing food must have been available. But this is not supported by the ethnographic analogy. As noted previously, on that time level other options, on examination, are shown as inadequate in themselves. Other food sources might have provided seasonal variety, but it was the hunting of caribou and moose primarily that kept these small bands intact. The essential validity of these assertions may be confirmed by consulting the ethnographic record of interior Alaska. This is a reasonable model for reading these archaeological records—*except* that it is likely that conditions of the terminal Pleistocene–early Holocene were even more difficult. Figure 12–1 presents Alaskan sites of this period for which radiocarbon dates are available.

The Beringian Tradition

The writer first proposed the term Beringian Tradition in 1981. It still seems the uniquely appropriate designation. It encompasses all sites in all of Beringia which are clearly within the Upper Palaeolithic fold. In this writer's opinion, that, in fact, translates to all sites, west and east, that date in the period 35,000 to 9,500 years ago. An examination of the archaeological record as presented in this volume will, moreover, show that there is in general a high degree of coherence in this large grouping. While it would not be appropriate to infer a single

great complex from this record, nevertheless it is clear that Dyuktai (considered broadly, including the Selemdga sites and certain elements of Ushki) and its descendant Denali tend to dominate the record. Terminology is important. It is useful to be able to distinguish between Early Beringian Tradition (35,000–9,500 BP) and a Late Beringian Tradition (which is largely beyond the scope of this volume). The alignments of their chronologies differ, obviously, between Siberia and Alaska–Yukon. In the former the span is 35,000 to 10,500 BP for Early Beringian; 10,500 to 7,000 BP for Late Beringian Tradition (in which the dominant entity is Sumnagin). In America these are, respectively 12,000 to ~9,500 BP for Early Beringian Tradition and ~8,500 to 7,500 BP for the Holocene Late Beringian Tradition.

On the subject of terms, perhaps it would be appropriate to indicate some difficulties this writer sees in certain alternative nomenclature. The term "Palaeoarctic" is unusual in that it appears to derive from "Palaeoindian," but where the latter refers to ancient people, the former is not parallel, referring to an ancient area. It was surely intended as an interim usage. In a somewhat similar vein, the writer would question the utility of applying "Palaeoindian" to sites that seem to have in common with that conventional application in sub-Laurentide America only the matter of *time* while excluding others that share that attribute. In the north its implications are, at best, unclear.

Within the Alaskan Early Beringian constituent there is a certain degree of assemblage variability in terms of artifact occurrences and frequencies. In the comparison of one site assemblage with another, it is often difficult to assess the nature of these differences. To what extent does the collection reflect the totality of the archaeological culture? How controlling has been the role of chance: to what extent is there reflected the randomness of deposition; to what extent the variability of preservation? It is taken as certain by northern researchers that bone, antler, and ivory were extensively used, yet the discovery of direct corroboration is extraordinarily rare. Can there be discerned, in a given site, the evidence of highly specific activities which would explain a restricted range of artifact forms present? A few suggestions of this sort have been made, as example those sites at which microblade and burin production seem to dominate. But the scarcity of sites amenable to such interpretation allows, at best, only tenuous statistical support. How may length of occupation be, even approximately, determined? Can a short-term residence be differentiated from a longer on the basis of the artifacts recovered? Can paucity of recovery alone be read as reflecting the former condition? Season of occupation ought to be more readily recognizable here than anywhere else in the world. To date, little has been said on this subject.

In various degrees, of course, the operation of these variables is common to archaeology everywhere, but in the north in late Pleistocene times the landscape was sparsely populated—even if recent ethnographic analogies are applied—by very small bands of nomadic hunters possessing a meager material culture. If these sites tell us anything at all of how these societies functioned, it is this and almost uniquely this. However, that should not be read as a counsel of despair; a very great deal of information can be extracted from this parsimonious record. Artifact forms tend to be quite distinctive and lend themselves well to formal analysis. Thus, while it is true that most individual site assemblages are small, the addition of more and more sites to the inventory, the great majority of which are quite resemblant one to the other, has had the effect of creating a very coherent and secure data base from which to draw conclusions of the kind that are suggested here.

The clear elimination of a pre-microblade stage should go far towards clarifying the record for eastern Beringia. As example, since Components I and II of the Dry Creek site are temporally proximate, and since the former must now be termed *non-microblade*, there is surely urged a comparison of the lower component with the several non-microblade loci of the overlying—Denali—level. In the early phases of the delineation of Denali there were some who maintained that in addition to the Denali of late Wisconsin–early Holocene, there was another Denali, perhaps separated from the earlier by an hiatus of thousands of years such that this late Denali actually occurred well into the Christian era. The

actually few such claims have subsequently been shown to have been the result of too uncritical a reliance on radiocarbon samples collected under conditions recognized at the time as unlikely to produce reliable dates, coupled with an unwillingness to abandon the results even when they flew in the face of other evidence. The writer obtained samples of this kind at Donnelly Ridge later to find that the apparently charred wood was a piece of root that could be traced up to the surface. Expectably, the dates bore no relation to the age of the site. Because of its historical significance, it is especially unfortunate that the Campus site appears to suffer from this same problem. Radiocarbon dates obtained in the shallow brown forest soil of the interior all too often record one condition: the ubiquity of natural burning in this great region. These instances, all involving very shallow, usually unstratified deposits, charitably may be taken to reflect a grasping at chronological straws impelled by frustration. When given the stamp of scientific reliability their effects can be pernicious in the extreme. Problems of these kinds will be solved as further research brings with it a larger corpus of evidence. In like manner, the problem of the incompleteness of any given collection is, in part, resolved by the capability of considering the greater range of artifact forms across the spectrum of many sites of a particular complex—obviously a device to be used with discretion. In any case, there is a moral here, which is that typology cannot be ignored.

It is not possible at this point for this writer to suggest how many separate complexes may ultimately be determined to characterize the Early Beringian Tradition of Alaska. Presently, that most easily identified is the widely distributed Denali. It appears unlikely that such other complexes as may come to be defined will differ significantly from the form clearly seen in the Selemdga-Dyuktai-Denali continuum.

Although its consideration is outside the scope of the present volume, reference must be made to the increasingly clear evidence of a later, related tradition in Alaska—a Late Beringian Tradition entity equivalent to Sumnagin. This entity is characterized by conoidal microblade-to-blade cores, burins (present but infrequent), and well-made bifaces often showing diagonal and collateral flaking. It appears that a variant of the wedge-shaped microblade or small blade core somewhat irregular in form also occurs in some sites of this proposed complex—either representing a continuity from Denali or a trait that came in with these other elements. Its dating, noted above, is at present somewhat tenuous. Ackerman (1987) suggested the name "Kagati Lake" for this complex and included in it the Kagati Lake site in the Kuskokwim drainage and several sites discussed above, the datings of which are somewhat equivocal. These include Anangula, Long Lake, and Mount Hayes 122. Ackerman includes as well what is described here as the second component of Whitmore Ridge (Mount Hayes 72) and Teklanika West, both of which, undated in 1981, were considered to be at one end or the other of the Denali sequence (West 1981:Table 3.3) in recognition of their departure from the Denali norm. The second component of Panguingue Creek probably would fit here, leaving the earlier to the Denali category, as was originally suggested by its investigators. There is little question but that the source for this late Beringian grouping is the Sumnagin culture or tradition which succeeded Dyuktai in Siberia about 10,500 years ago.

When earliest Swan Point and Walker Road were occupied, final inundation of central Beringia was still some 1,500 years in the future. When the early components at Dry Creek were occupied, that event was some 700 to 1,000 years off. These simple computations can be made for other sites as well. The fundamental point, however, is that for a long time after the first settlement of Alaska, the province of Beringia remained physically intact. While the evidence would indicate a continuing degradation of its ecosystem, the Beringians were certainly pursuing those species that remained of the old mammoth fauna. There is some suggestion that later sites of the Early Beringian should be found in high alpine settings where the vegetational characteristics provided refuges for the diminishing game species. The Tangle Lakes likely served as one such area.

Some 2,000 to 3,000 years after the arrival of the Early Beringian hunters, Sumnagin influences made their appearance. Whether this represented a movement of people or the diffusion of ideas through preexisting populations cannot be known at this time. This inability is unfortunate but it is at least certain that, however those influences made their way to Alaska, it was at a time when the land connection still existed, though presumably in deteriorated state.

Now a curious fact emerges: the American Early Beringian, typified by Denali, seems to simply disappear from the Alaskan record. On being drawn to this fact while preparing *The Archaeology of Beringia* (West 1981), the writer concluded that nothing less than a population crash was recorded. This was attributed to the vastly changed environment of the later Holocene, specifically the onset, at about 7,000 years ago, of the hypsithermal warming that was taken to have eradicated the last vestiges of the old Beringian environment. The problem with that theory was one of insufficient knowledge: the possibility of a later phase of the Beringian Tradition simply had not yet been fully realized. In consequence, it was grouped with the older, numerically dominant, Beringian sequence typified by Denali. One result of that, as may be seen, is that the chronology erred seriously on the recent end.

To return now to the premise just enunciated, but with prudent modifications: the Denali complex (this standing for the American Early Beringian) does, indeed, seem to drop out of the record. To say the causes are elusive is to understate the case. Whereas it clearly prevailed throughout the late Pleistocene and earliest Holocene, at least by ca. 9,000 BP it is gone. Whether it can be seen to have been transformed into Kagati or whether it disappears completely before any such contact was made is the great question. Obviously, if overlap between the traditions can be shown, it will suggest cultural replacement—perhaps a "swamping"—by the now better-adapted Sumnagin. Cultural change by contact is assuredly more common than cultural change by demise. If, on the other hand, the gap that now on thin evidence seems to separate the two cannot be convincingly closed, then an altogether different interpretation must be constructed. Either reading of the record presents very large questions. Moreover, it appears that the later tradition followed a trajectory similar to the first. At least in the interior of Alaska it, too, disappears from the record at about 7,000 years ago. There is, furthermore, a hiatus of about 3,000 years before new occupations appear and they—Northern Archaic, deriving from points east and south—are totally unlike either Beringian Tradition.

But the two cases are not identical. The early tradition may have died out completely *in the north* (see text following). In contrast, complexes probably deriving from, or related to, Kagati continue far to the east in Yukon Territory, appearing first in the Little Arm phase and continuing through subsequent phases until ca. 4,000 years ago (MacNeish 1964; Workman 1978).

The two traditions differed in other important ways. The earlier was the end point of a highly developed Upper Palaeolithic lifeway of very long duration, seemingly stable and in equilibrium with its peculiar faunal environment. The short duration of the second might suggest an inability to achieve the kind of stability that characterized its predecessor, a stability dependent upon constant fluctuations of climate (Lister and Sher 1995:23). Perhaps in the environmental flux of the early Holocene, prior to the establishment of subarctic zonation, effective, stable, cultural responses were precluded. There simply were not here the kinds of options to be found far to the south at this critical time—options that allowed, and encouraged, a shift into foraging economies.

Everywhere in the world the Holocene imposed changes that required immediate and effective response. In lower latitudes, in regions of more genial climate, the transitions appear to have been made smoothly, aided, probably, by the fact that most of these were economies already based on hunting and foraging. In the far north, economies were, of necessity, more narrowly focused. It is useful to speculate on how these events may have played out in the light of what is known.

Foraging was never a significant option in the north; life was dependent upon the hunting of those animals that composed the mammoth fauna, and even as the Pleistocene wore down and game became more difficult to secure, it must have still been possible to extend hunting ranges to compensate, in part, for the diminished harvest. But the Holocene brought in extensive vegetational changes that

withdrew vast areas of the remaining steppe tundra and, with it, the range on which the remnants of the mammoth fauna depended. There is some possibility that two representatives of the old fauna—bison and elk—survived in small numbers in restricted areas virtually into modern times, but their numbers seem always to have been quite low and it is difficult to assess their possible economic importance. There is evidence of their mid-Holocene presence in the southwest Yukon but virtually no such evidence for the interior of Alaska. Moose and caribou, which had been very minor constituents of the Pleistocene fauna, must have been present as well in the early Holocene, but seemingly in low numbers. These would become dominant by the late Holocene and into modern times, but that was far in the future.

Given what may well have amounted to famine conditions for some small groups, there would have come into play certain inevitable concomitants affecting, eventually, all elements of the population: increased susceptibility to disease; excessively high mortality rates from disease and other causes; low birth rates conceivably aided by cultural sanctions such as infanticide, but certainly induced by the involuntary suppression of fertility known to accompany extreme conditions of malnutrition. With population base numbers that would have been now far lower than formerly, and with what would have been a negative replacement rate, it does not seem extravagant to suggest that some of these small bands likely died out.

These speculations arise from what this writer sees as objective facts of the archaeological record. They represent an attempt to make comprehensible that record. Whether the result of those conditions that attended the transition into the Holocene was a general population crash or, instead, a major decline, cannot be determined on the basis of present evidence. But to return to the point at which this began, the withdrawal of major segments of the old economy, by extinctions and sweeping reductions of surviving game herds, had to have induced reductions in the populations of their Beringian hunters.

One final note here. There will probably be those who will continue to seek earlier, and perhaps more exotic, origins for the beginnings of America, but the array of evidence presented in this volume would seem to render that an uncommonly fruitless task.

Amerind and the Early Beringian Tradition: A Footnote on Greenberg's Linguistic Evidence

Greenberg sees three "migrations" that account for the historic distribution of all native American languages. Using biological criteria, Turner and Zegura likewise arrived at a theory of three migrations. This resulted in their joint article with Greenberg (1986) which was reviewed by this writer in the following year (West 1987). The correspondence between Greenberg's reconstruction and the present interpretation of earliest American prehistory is too striking to be passed over without comment. The reader is requested to enter the requisite qualifiers to the following remarks.

Early Beringian Tradition is proposed as the source for Amerind. Amerind does not exist in Alaska or the Yukon. Whether Early Beringian/Ancestral Amerind died out literally in these regions or was linguistically swamped by the Sumnagin cannot be known at present. Whatever the case, it left no descendants, culturally or linguistically, in the north, disappearing finally there in the earliest Holocene. But, well prior to that event (and well prior to the advent of Sumnagin), i.e., before 11,500 BP, one group seems to have made its way south where, as Clovis, it spread explosively and widely. The rapidity with which the Clovis descendants spread through the New World provided the fullest possible time—almost 11,000 years or nearly 450 generations—for the Amerind linguistic differentiations to take place.

The recent studies in biogenetic differentiation via mitochondrial DNA, et al., will certainly contribute importantly to the questions of relationships addressed by Greenberg. As with the latter, however, the translation of the measured divergences to something like absolute time will require reference to the findings of archaeology.

(Although it is perhaps too early to broach this problem, and this is not the proper temporal venue, nevertheless, this writer feels strongly that when all evidence is in and assessed, with its predominantly far northwestern distribution, Na-Dene will prove to descend from the New World Sumnagin.)

The Antecedent Clovisians

Clovis is taken to be the basal, the founding, population for the Americas. With the aid of recently acquired radiocarbon dates in the north, it is possible now to make a far clearer case for the Beringian derivation of Clovis. The respective assemblage typologies can be shown as closely similar, which is, of course, the first requirement.

The clarity of the demonstration of similarity depends in part upon the question of whether Clovis is one entity or several. It is possible to suggest that the prevailing thought has always favored Clovis as a single entity in which there is seen but little variation from one part of its distributional range to the other. That, of course, is completely in keeping with the evidence of its extraordinarily rapid spread. However that may be, Tankersley's recent demonstration of Clovis, the single entity, as a "colonizing" population is certainly convincing (Tankersley 1994).

Viewed from the northern perspective, Clovis has a distinctly Upper Palaeolithic appearance, more specifically a *Beringian* Upper Palaeolithic appearance. The latter part of this assertion of similarity should not occasion surprise. If the Upper Palaeolithic cast is granted, then it would assuredly be more efficient to derive it from Alaska by way of the Mackenzie corridor than from some other—perhaps unstated—source. Figure 12–2 is intended to suggest these assemblage similarities.

The Mackenzie Corridor: The Way South

The probability, indeed the necessity, of an Ice Age passageway southward from northwesternmost North America was early recognized by palaeontologists, archaeologists, and others concerned with biogeography. The presence of Eurasian elements in the North American mammalian fauna, for example, constituted prima facie evidence that a passage existed. Where, exactly, this passageway might have been was likewise answered by the observation that the margins of the two great North Ameri-

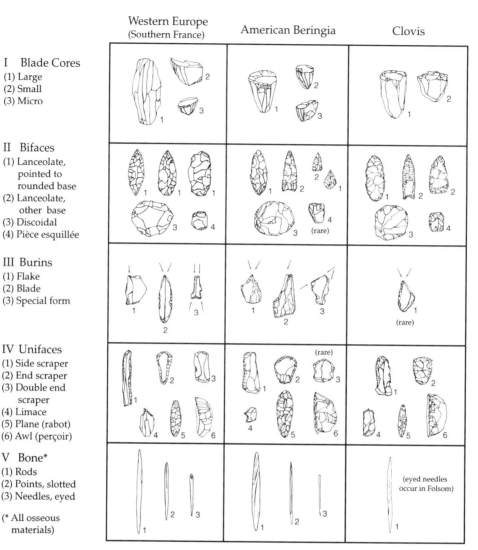

	Western Europe (Southern France)	American Beringia	Clovis
I Blade Cores (1) Large (2) Small (3) Micro			
II Bifaces (1) Lanceolate, pointed to rounded base (2) Lanceolate, other base (3) Discoidal (4) Pièce esquillée			
III Burins (1) Flake (2) Blade (3) Special form			
IV Unifaces (1) Side scraper (2) End scraper (3) Double end scraper (4) Limace (5) Plane (rabot) (6) Awl (perçoir)			
V Bone* (1) Rods (2) Points, slotted (3) Needles, eyed (* All osseous materials)			

Figure 12–2 A comparison of selected Upper Palaeolithic artifact classes from western Europe (variously from Aurignacian, Gravettian, and Solutrean levels of southern France) with those from American Beringia and Clovis, illustrating the continuity of forms and technology. (Col. 1 from de Sonneville-Bordes; col. 3 from Gramly. *See Acknowledgments*.)

can ice sheets, the vast continental Laurentide and the much smaller, far western Cordilleran, moving from their centers of accumulation, joined roughly along the axis of the Canadian Rocky Mountains. With this knowledge and without any reasonable alternative, the Mackenzie corridor came to be accepted as the route by which faunal interchanges via Beringia were effected—and this was long before any direct supporting evidence was even sought. The acceptance of this route by archaeolo-

gists has, perhaps, not been quite so ready. This may be in part because it was held by some that a coastal route would constitute a reasonable alternative, but perhaps more because of the nature of archaeological evidence and the subtleties of its proofs.

The corridor itself was essentially a theoretical construct and as such it became one of those quicksilver variables entangled in the discussions of the earliest Americans. That there had been almost no field research directly on the problem was due to the fact that, until World War II and the construction of the Alaska Highway, much of this great swath of western Canada was some of the most inaccessible wilderness in North America. Recent research in parts of this region is now casting important light on the ice-free corridor.

Eleven thousand five hundred years ago, ice still covered much of northernmost North America. The front of the Laurentide ice sheet extended from a point near the mouth of the Mackenzie River in Yukon Territory south and east through Northwest Territories, into northeastern British Columbia to about the middle of Alberta Province. From there it trended eastward across south central Saskatchewan, dropping a bit farther south in Manitoba, then continued trending to the south across western Ontario. From there it descended to its southernmost points in the western Great Lakes of Wisconsin and Michigan before turning east and north to parallel the course of the St. Lawrence River.

The corridor was formed, of course, between the two bodies of ice, the Cordilleran augmented locally by montane glaciers. Major advance by either or both might effect coalescence or closure. Its availability as a passage depended firstly upon the opposite condition obtaining, but equally upon the presence within it of biota sufficient to support those traversing it. There appears to be general agreement that the primary determinant of coalescence was the Cordilleran ice and that its pulsations were not synchronous with those of the Laurentide (Catto and Mandryk 1990:3 et seq.; Bobrowsky et al. 1990:117). Both these sets of authors, while granting that unknowns still exist, nevertheless agree that the sum of present knowledge allows the conclusion that the " 'Ice-Free Corridor' may have existed along the eastern flank of the Cordillera throughout late Wisconsin time" (Bobrowsky et al. 1990:92). Catto and Mandryk go somewhat further in suggesting the corridor "remained geographically available . . . throughout most of Wisconsin time" (1990:85). The slightly different interpretation by Bobrowsky and Rutter (1990) of the availability of the corridor finds it open until about 15,000 years ago, then open again from 13,500 BP, following a possible coalescence.

A useful summary of late Pleistocene environ-

mental conditions in the southern portions of the corridor is provided by Ives, Beaudoin, and Magne (1989). These records derive from western Alberta and east central British Columbia. The general picture is one of early *Artemisia* and grass dominance, this dating often >11,000 BP. Slightly later these tend to be augmented in some sequences by willow, poplar, or birch, or a combination of these. Mandryk's Mitchell Lake pollen record from western Alberta reveals a similar sequence, this one terminating ca. 11,400 BP (Mandryk 1990). Where these records continue into the early Holocene, there is seen an influx of spruce or pine, the latter often succeeding the spruce.

As reported by these authors, data on fauna here tend to be somewhat sparse and to pertain to the mid-Wisconsin. They do, however, refer to findings from two areas of western Alberta of the terminal Pleistocene (11,500–10,000 BP). Here have been found mammoth, horse, bison (*priscus*), caribou, and small forms. The latter condition was found also in the faunal remains of final Pleistocene age from the Charlie Lake Cave archaeological site (Fladmark et al. 1988). In all these instances the records are read to indicate more open, perhaps prairie-like, conditions.

The lower, fluted-point, level of Charlie Lake Cave is acknowledged to much postdate the Clovis migration. Its estimated age is ca. 10,500–10,000 BP. The authors' evaluation of the environment at that time would seem to hold for most of these records: "The later stages of the ice-free corridor clearly were not inhospitable" (Fladmark et al. 1988:382). Equally clearly, it is no longer necessary for prehistorians to tread this trail quite so gingerly.

The Antecedent Clovis Migration

It is not possible simply to present the route by which the nascent Clovisians traveled southward without engaging in further speculation. The appearance in this instance is that of a *true migration* (or, perhaps, a *trek*) in the sense of a determined move from one point to another. It has, moreover, the appearance of a migration by one small group, rather than a population spread, as previously suggested by this writer (West 1983:376) and others. An expansion of population would have left the obvi-

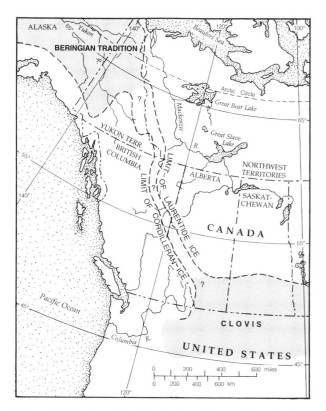

Figure 12–3 American Beginnings. The settlement of North America ca. 11,000 BP.

ous signs—a series of sites of appropriate age along the corridor. However, despite some relatively intensive archaeological survey along some portions of the corridor, no evidence whatsoever has been found that would substantiate any sort of movement at 11,500 or perhaps 11,300 years ago. Confronted, then, with the fact that such a move *must* have been made and the fact that the first evidence of a successful passage is found beyond the southern end of the corridor with nothing intervening, the conclusion of a rapid, determined passage seems inevitable. Carrying this one giant speculative step further, it is difficult to avoid the feeling that there must have been a strong element of desperation in this passage. (That may equally have been the case for the presumed mammoth fauna elements being hunted.) The movement need not have taken more than a few years; it may have been accomplished in much shorter time. It may have

been made easier by the warming (Allerød) that preceded the intense cold of the Younger Dryas event. If this reconstruction has any validity it may be speculated further—and regretfully—that direct evidence may *never* be found for the reason that virtually nothing would have been left behind by a small group that spent little time at any one encampment.

What impelled the move south? From where in present-day Alaska or Yukon might such a group have originated? Was this a small band that, pressed to seek new hunting grounds, had wandered far to the east and found itself in a virtual cul-de-sac—a restricted, linear region with severely constrained possibilities? In such a situation perhaps the only means of extending their hunting territory would have been either to retreat in the direction from which they came or to follow such game as was available southward into perhaps more promising, unoccupied country. While the lateral boundaries could not be described as pressing in on this hypothetical small band, nevertheless those barriers— the Laurentide and the Cordilleran ice masses— were there. Even though it is likely that both sides of the corridor were vegetated, the appearance would have been that of being constrained by continuous mountain walls, sometimes near, sometimes far away. At any rate, in this vacuum of direct evidence, one thing is probably safely asserted: Beringian perseverance won through.

FURTHER THOUGHTS

Parenthetically, this notion may be put forward. Firstly, it is obvious that this writer sees Clovis as Upper Palaeolithic. The correspondences are there in the assemblage, the economy, and the dating. But beyond this it may be suggested that the entire Palaeoindian tradition is the American equivalent of later Upper Palaeolithic. The timing and the economies provide some support for such an argument, although the typical Upper Palaeolithic artifacts seem to have mostly dropped out. As observed by a number of writers, the transition to Archaic mirrors perfectly the Old World transition to Mesolithic, in both continents a response to the changed conditions of the Holocene. In the north, however, the change is constrained and not readily comparable.

There may be some hesitation in accepting the argument presented here based upon seemingly important differences between Beringian and Clovis artifact assemblages. But then, how to explain the differences between Clovis and Folsom or between either of those and their later Palaeoindian descendants? It is difficult to see gradual, progressive change but then perhaps it is not there to be seen; perhaps that is not the way those changes proceeded. Possibly, instead, this kind of cultural change was not gradual, not *incremental*, but rather *mutational*. The evidence cited seems to point that way.

Where did the distinctive Clovis projectile point originate? This author and others (e.g., Morlan 1987; Carlson 1991) have suggested that its fabrication entails techniques essentially like that of blade production and thus could arise readily among such people as the Beringians. Why it arose, and why microblades (which is to say bone or antler points armed with microblade edges) dropped out, simply cannot be known at this time. Clearly, though, one preferred form of projectile weapon gave way to another.

At this point Beringians became Clovisians and the narrative of their careers falls within the purview of other scholars. Northern researchers may figuratively bid them farewell, knowing full well that their future is assured.

———◆◆◆———

There is no Rosetta Stone with which to unlock the secrets of the most ancient Americas. Conceptually, instead, there is an enormous mosaic in the form of the two continents. The pieces of which it is composed are those bits of its prehistory as they have been painstakingly brought forth. The mosaic is far from complete and quite surely will never be so. But even in its incompleteness it does reveal an unparalleled richness of indigenous cultures.

It is curious to reflect on how that all began. Twelve thousand years is not very long ago as the history of the species is written. A small band of resilient and successful arctic hunters entirely accidentally found themselves in an unimaginably rich,

and unoccupied, New World. The Clovisians and their immediate descendants swept the hemisphere in a few hundred years—a move parallel to, but much more far-reaching in its effects than, that of earlier Upper Palaeolithic colonists in other lands. Their genes and their language spread through the Americas and their descendants included the civilizations of the Aztec, the Maya, and the Inca. That resilience that brought those first Americans through one of the most arduous passages in the human record must surely have played some part in those developments.

BIBLIOGRAPHY

Ackerman, R. E. 1987. Mid-Holocene Occupation of Interior Southwestern Alaska. In *Man and the Mid-Holocene Climatic Optimum*, ed. N. A. McKinnon and G. S. L. Stuart, pp. 181–92. Calgary: Archaeological Association of the University of Calgary.

Adovasio, J. M., J. Donahue, and R. Stuckenrath. 1990. The Meadowcroft Rockshelter Radiocarbon Chronology 1975–1990. *American Antiquity* 55(2):348–54.

Adovasio, J. M., J. D. Gunn, J. Donahue, and R. Stuckenrath. 1978. Meadowcroft Rockshelter, 1977: An Overview. *American Antiquity* 43:632–51.

Adovasio, J. M., J. D. Gunn, J. Donahue, R. Stuckenrath, J. E. Guilday, and K. Volman. 1980. Yes Virginia, It Really Is That Old: A Reply to Haynes and Mead. *American Antiquity* 45:588–95.

Adovasio, J. M., R. C. Carlisle, K. A. Cushman, J. Donahue, J. E. Guilday, W. C. Johnson, K. Lord, P. W. Parmalee, R. Stuckenrath, and P. W. Wiegman. 1985. Paleoenvironmental Reconstruction at Meadowcroft Rockshelter, Washington County, Pennsylvania. In *Environments and Extinctions: Man in Late Glacial North America*, ed. J. I. Mead and D. J. Meltzer, pp. 73–110. Orono: Center for the Study of Early Man, University of Maine.

Bobrowsky, P. T., and N. W. Rutter. 1990. Geologic Evidence for an Ice-Free Corridor in Northeastern British Columbia, Canada. *Current Research in the Pleistocene* 7:133–35.

Bobrowsky, P. T., N. R. Catto, J. W. Brink, B. E. Spurling, T. H. Gibson, and N. W. Rutter. 1990. Archaeological Geology of Sites in Western and Northwestern Canada. In *Archaeological Geology of North America*, ed. N. P. Lasca and J. Donahue, pp. 87–122. Boulder: Geological Society of America.

Carlson, R. L. 1991. Clovis from the Perspective of the Ice-Free Corridor. In *Clovis Origins and Adaptations*, ed. R. Bonnichsen and K. Turnmire, pp. 81–90. Corvallis:

Center for the Study of the First Americans, Oregon State University.

Catto, N. R., and C. A. Mandryk. 1990. Geology of the Postulated Ice-Free Corridor. In *Megafauna and Man: Discovery of America's Heartland*, ed. L. D. Agenbroad, J. I. Mead, and L. W. Nelson, pp. 80–85. Scientific Papers, Vol. 1. Hot Springs: The Mammoth Site of Hot Springs, South Dakota, Inc.

Cavalli-Sforza, L. L., P. Menozzi, and A. Piazza. 1994. *The History and Geography of Human Genes*. Princeton, N.J.: Princeton University Press.

Derevianko, A. P. 1990. *Palaeolithic of North Asia and the Problem of Ancient Migrations*. Novosibirsk: Institute of History, Philology, and Philosophy, Siberian Branch, Academy of Sciences of the USSR.

Dillehay, T. D. 1989. *Monte Verde: A Late Pleistocene Settlement in Chile*. Washington, D.C.: Smithsonian Institution Press.

Fladmark, K. R., J. C. Driver, and D. Alexander. 1988. The Paleoindian Component at Charlie Lake Cave (HbR 39), British Columbia. *American Antiquity* 53(2):371–84.

Goebel, T. 1995. The Record of Human Occupation of the Russian Subarctic and Arctic. *BPRC Miscellaneous Series M-335*, pp. 133–35. Columbus: Byrd Polar Research Center, Ohio State University.

Goebel, T., and M. Aksenov. 1995. Accelerator Radiocarbon Dating of the Initial Upper Palaeolithic in Southeast Siberia. *Antiquity* 69(263):349–57.

Goebel, T., A. P. Derevianko, and V. T. Petrin. 1993. Dating the Middle-to-Upper Palaeolithic Transition at Kara-Bom. *Current Anthropology* 34:452–58.

Gramly, R. M. 1992. *Guide to the Palaeo-Indian Artifacts of North America*, Buffalo, New York: Persimmon Press.

Greenberg, J. H., C. G. Turner II, and S. L. Zegura. 1986. The Settlement of the Americas: A Comparison of the Linguistic, Dental, and Genetic Evidence. *Current Anthropology* 27(5):477–97.

Guidon, N. 1987. Cliff Notes. *Natural History* 96(8):6–13.

———. 1989. On Stratigraphy and Chronology at Pedra Furada. *Current Anthropology* 30:641–42.

Haynes, C. V. 1980. Paleoindian Charcoal from Meadowcroft Rockshelter: Is Contamination a Problem? *American Antiquity* 45(3):582–87.

Ives, J. W., A. B. Beaudoin, and M. P. R. Magne. 1989. Evaluating the Role of a Western Corridor in the Peopling of the Americas. Paper presented at the Circum-Pacific Prehistory Conference, Seattle.

Lister, A. M., and A. V. Sher. 1995. Ice Cores and Mammoth Extinction. *Nature* 378 (2 November):23–24.

Lozhkin, A. V. 1993. Geochronology of Late Quaternary Events in Northeastern Russia. *Radiocarbon* 35(3):429–33.

Lynch, T. F. 1990. Glacial-Age Man in South America? A Critical Review. *American Antiquity* 55(1):12–36.

MacNeish, R. S. 1964. Investigations in Southwest Yukon: Archaeological Excavations, Comparisons and Speculations. *Papers of the R. S. Peabody Foundation for Archaeology* 6(2):201–488.

Mandryk, C. A. 1990. Could Humans Survive the Ice Free Corridor? Late-Glacial Vegetation and Climate in West Central Alberta. In *Megafauna and Man: Discovery of America's Heartland*, ed. L. D. Agenbroad, J. I. Mead, and L. W. Nelson, pp. 67–79. Scientific Papers, Vol. 1. Hot Springs: The Mammoth Site of Hot Springs, South Dakota, Inc.

Martin, P. S., and A. J. Stuart. 1995. Mammoth Extinction: Two Continents and Wrangel Island. *Radiocarbon* 37(1):7–10.

Mead, J. I. 1980. Is It Really That Old? A Comment about the Meadowcroft Rockshelter "Overview." *American Antiquity* 45(3):579–82.

Meltzer, D. J., J. M. Adovasio, and T. D. Dillehay. 1994. On a Pleistocene Human Occupation at Pedra Furada, Brazil. *Antiquity* 68:695–714.

Mochanov, Y. A. 1992. The Late Palaeolithic of North Asia (35,000–10,500 Years Ago). Paper presented at the 45th Annual Northwest Anthropological Conference, Simon Fraser University, Burnaby, British Columbia.

Morlan, R. E. 1987. The Pleistocene Archaeology of Beringia. In *The Evolution of Human Hunting*, ed. M. H. Nitecki and D. V. Nitecki, pp. 267–307. New York and London: Plenum Press.

Pitulko, V. V. 1993. An Early Holocene Site in the Siberian High Arctic. *Arctic Anthropology* 30(1):13–21.

Sonneville-Bordes, D. de. 1960. *Le Paléolithique supérieur en Périgord*. Tome 1. Published in conjunction with Centre National de la Recherche Scientifique, Paris. Bordeaux: Delmas.

Tankersley, K. B. 1994. Was Clovis a Colonizing Population in Eastern North America? In *The First Discovery of America*, ed. W. S. Dancey, pp. 95–116. Columbus: Ohio Archaeological Council, Inc.

Tankersley, K. B., C. A. Munson, and D. Smith. 1987. Recognition of Bituminous Coal Contaminants in Radiocarbon Samples. *American Antiquity* 52(2):318–30.

Turner, C. G. II. 1990. Palaeolithic Siberian Dentition from Denisova and Okladnikov Caves, Altayskiy Kray, USSR. *Current Research in the Pleistocene* 7:65–66.

Vartanyan, S. L., V. E. Garutt, and A. V. Sher. 1993. Holocene Dwarf Mammoths from Wrangel Island in the Siberian Arctic. *Nature* 362 (25 March): 337–40.

Vereshchagin, N. K., and G. F. Baryshnikov. 1982. Paleoecology of the Mammoth Fauna in the Eurasian Arctic. In *Paleoecology of Beringia*, ed. D. M. Hopkins, J. V. Matthews, Jr., C. E. Schweger, and S. B. Young, pp. 267–79. New York: Academic Press.

West, F. H. 1967. The Donnelly Ridge Site and the Definition of an Early Core and Blade Complex in Central Alaska. *American Antiquity* 32(3):360–82.

———. 1981. *The Archaeology of Beringia*. New York: Columbia University Press.

———. 1983. The Antiquity of Man in America. In *Late-Quaternary Environments of the United States*, ed. H. E. Wright, Jr.; Vol. 1, *The Late Pleistocene*, ed. S. C. Porter, pp.364–382. Minneapolis: University of Minnesota Press.

———. 1987. Migrationism and New World Origins. (Re-view of Greenberg, Turner, and Zegura.) *The Quarterly Review of Archaeology* 8(1):11–14.

———. 1993. Review of *Palaeoenvironment and Site Context. Monte Verde: A Late Pleistocene Settlement in Chile, Vol. 1* by Tom D. Dillehay. *American Antiquity* 58(1):166–67.

Workman, W. B. 1978. *Prehistory of the Aishihik–Kluane Area, Southwest Yukon Territory*. Archaeological Survey of Canada Paper No. 74, Mercury Series. Ottawa: National Museums of Canada.

NAME INDEX

SUBJECT INDEX